Xian-Da Zhang
Modern Signal Processing
De Gruyter STEM

Also of interest

5G
An Introduction to the 5th Generation Mobile Networks
Ulrich Trick, 2021
ISBN 978-3-11-072437-0, e-ISBN (PDF) 978-3-11-072450-9,
e-ISBN (EPUB) 978-3-11-072462-2

Digital Electronic Circuits
Principles and Practices
Shuqin Lou, Chunling Yang, 2019
ISBN 978-3-11-061466-4, e-ISBN (PDF) 978-3-11-061491-6,
e-ISBN (EPUB) 978-3-11-061493-0

Metrology of Automated Tests
Static and Dynamic Characteristics
Viacheslav Karmalita, 2020
ISBN 978-3-11-066664-9, e-ISBN (PDF) 978-3-11-066667-0,
e-ISBN (EPUB) 978-3-11-066669-4

Communication, Signal Processing & Information Technology
Series: Advances in Systems, Signals and Devices, 12
Edited by Faouzi Derbel, Nabil Derbel, Olfa Kanoun, 2020
ISBN 978-3-11-059120-0, e-ISBN (PDF) 978-3-11-059400-3,
e-ISBN (EPUB) 978-3-11-059206-1

Signal Processing and Data Analysis
Tianshuang Qiu, Ying Guo, 2018
ISBN 978-3-11-046158-9, e-ISBN (PDF) 978-3-11-046508-2,
e-ISBN (EPUB) 978-3-11-046513-6

Xian-Da Zhang

Modern Signal Processing

DE GRUYTER

清華大学出版社
TSINGHUA UNIVERSITY PRESS

Author
Prof. Xian-Da Zhang
Tsinghua University
Dept. of Automation
Haidian District
1 Tsinghua Park
100084 Beijing
People's Republic of China

Translation by

Dong-Xia Chang
Beijing Jiaotong University
Ling Zhang
Ocean University of China
Dao-Ming Zhang
LEIHUA Electronic Technology Institute

ISBN 978-3-11-047555-5
e-ISBN (PDF) 978-3-11-047556-2
e-ISBN (EPUB) 978-3-11-047566-1

Library of Congress Control Number: 2022931618

Bibliographic information published by the Deutsche Nationalbibliothek
The Deutsche Nationalbibliothek lists this publication in the Deutsche Nationalbibliografie; detailed bibliographic data are available on the Internet at http://dnb.dnb.de.

Cover image: bestdesigns / iStock / Getty Images Plus
Printing and binding: CPI books GmbH, Leck

www.degruyter.com

Acknowledgements

We would like to thank Mr. Yan-Da Li, Academician of Chinese Academy of Sciences, IEEE Fellow, and Professor of Department of Automation, Tsinghua University, for his encouragement and strong support to the translation of this book. We are very grateful to our editor Yi-Ling Wang for her patience, understanding, and help in the course of our translating this book. We are grateful to Dr. Xi-Yuan Wang and Dr. Fang-Ming Han for their help. We would also like to thank Ji-Min Zheng, Ming-Nuan Qin, and Hai-Zhou Wu for their help.

https://doi.org/10.1515/9783110475562-202

Contents

1 Random Signals

Signal carries information. The information a signal conveys can be the parameter, impulse response, and power spectral of the system either physical or biological, or in the form of features for identifying artificial targets such as airplanes and vessels, weather or hydrological forecast, and abnormality in electrocardiogram. A signal whose values or observations are random variables is called a random signal. The term "random" refers to the fact that the samples of the signal are distributed according to certain probability law, which could be fully known, partially known, or completely unknown. Stochastic process, random function, and random sequence are other names for random signal. This chapter will focus on the representation of stationary random signals in two domains: the time domain and the frequency domain, which are complementary and of equal importance in characterizing random signals.

1.1 Signal Classifications

Mathematically, a signal is expressed by a series of variables. Let $\{s(t)\}$ be an array of real or complex numbers. Then, the sequence $\{s(t)\}$ is a signal. When time t is defined on interval of continuous variable, i.e., $t \in (-\infty, \infty)$ or $t \in [0, \infty]$, $\{s(t)\}$ is a continuous-time signal. Many artificial and natural signals such as those arising from radar, sonar, radio, telecommunications, control systems, and biomedical engineering are examples of the continuous-time signals. However, when being processed in a digital computer, a continuous-time signal must be first converted into a discrete-time signal. If the time variable t of the signal takes value from the integer set, i.e., $t = 0, \pm 1, \ldots,$ or $t = 0, 1, \ldots,$ the sequence of variables $\{s(t)\}$ is a discrete-time signal.

The sequence $\{s(t)\}$ is called a deterministic signal if its value at any time is not random but can be specified by a certain deterministic function. Followings are several deterministic signals commonly used.

- Step signal

$$U(t) = \begin{cases} 1, & t \geq 0, \\ 0, & t < 0. \end{cases} \tag{1.1.1}$$

- Sign signal

$$\text{sgn}(t) = \begin{cases} 1, & t \geq 0, \\ -1, & t < 0. \end{cases} \tag{1.1.2}$$

- Rectangular impulse

$$P_a(t) = \begin{cases} 1, & |t| \leq a, \\ 0, & |t| > a. \end{cases} \tag{1.1.3}$$

https://doi.org/10.1515/9783110475562-001

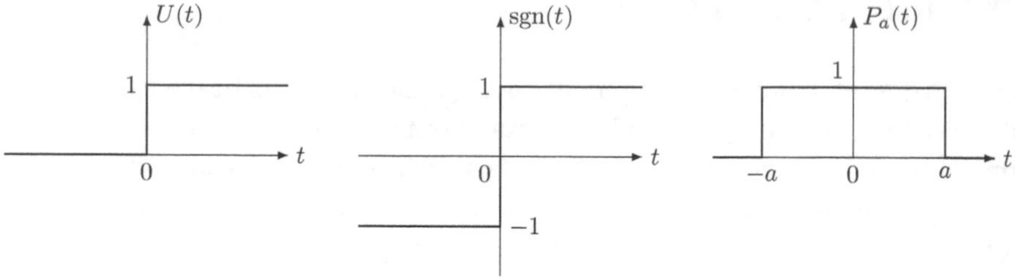

Fig. 1.1.1: Step signal, sign signal, and rectangular impulse.

– Sinusoidal signal (or harmonic signal)

$$s(t) = A\sin(\omega_c t + \theta_0), \tag{1.1.4}$$

where θ_0 is the given initial phase.

As an illustration, Fig. 1.1.1 depicts the waveforms of step signal, sign signal, and rectangular impulse.

Sequence $\{s(t)\}$ is called a random signal, if different from deterministic signal, the value of the sequence $\{s(t)\}$ at any given time is a random variable. For instance, the sinusoidal signals with random phase

$$s(t) = A\cos(\omega_c t + \theta), \quad \text{(real harmonic signal)} \tag{1.1.5}$$

$$s(t) = A\exp(\omega_c t + \theta). \quad \text{(complex harmonic signal)} \tag{1.1.6}$$

are random signals. Here, θ is a random variable distributed uniformly within interval $[-\pi, \pi]$, which has probability density function (PDF)

$$f(\theta) = \begin{cases} \frac{1}{2\pi}, & -\pi \le \theta \le \pi, \\ 0, & \text{otherwise}. \end{cases} \tag{1.1.7}$$

Random signal, also known as stochastic process, has the following properties.

(1) The value of the random signal at any time is random and cannot be determined in advance.

(2) Although its exact value at any time cannot be determined in advance, the random signal is governed by statistical laws. In other words, a random signal or stochastic process can be characterized statistically by probability distribution, which is referred to as the statistical property of the signal.

Let $x(t)$ be a complex stochastic process in continuous time. The stochastic process $x(t)$ at each time instant t is a random variable $X = x(t)$, which has the mean $\mu(t)$ expressed by

$$\mu(t) = \mathrm{E}\left\{x(t)\right\} \stackrel{\text{def}}{=} \int_{-\infty}^{\infty} x f(x,t)\,\mathrm{d}x, \tag{1.1.8}$$

where $f(x,t)$ is the PDF of random variable $X = x(t)$ at time t. The autocorrelation function $R_x(t_1,t_2)$ of complex random signal $x(t)$ is defined as the correlation between observations of $x(t)$ at time t_1 and t_2. Namely,

$$\begin{aligned} R_x(t_1,t_2) &\stackrel{\text{def}}{=} \mathrm{E}\left\{x(t_1)x^*(t_2)\right\} \\ &= \int_{-\infty}^{\infty}\int_{-\infty}^{\infty} x_1 x_2^* f(x_1,x_2;t_1,t_2)\,\mathrm{d}x_1\,\mathrm{d}x_2 \\ &= R_x^*(t_2,t_1), \end{aligned} \tag{1.1.9}$$

where the superscript $\star$ is the complex conjugate, and $f(x_1,x_2;t_1,t_2)$ is the joint PDF of random variables $X_1 = x(t_1)$ and $X_2 = x(t_2)$. In general, autocorrelation function depends upon both t_1 and t_2. Given arbitrary set of complex numbers α_k $(k = 1,\ldots,n)$, define

$$Y = \sum_{k=1}^{n} \alpha_k x(t_k). \tag{1.1.10}$$

Obviously, Y is a random variable and $\mathrm{E}\left\{|Y|^2\right\} \geq 0$. Thus, we have

$$\begin{aligned} \mathrm{E}\left\{|Y|^2\right\} &= \sum_{i=1}^{n}\sum_{k=1}^{n} \alpha_i\alpha_k^* \mathrm{E}\left\{x(t_i)x^*(t_k)\right\} = \sum_{i=1}^{n}\sum_{k=1}^{n} \alpha_i\alpha_k^* R_x(t_i,t_k) \\ &= [\alpha_1,\alpha_2,\cdots,\alpha_n] \begin{bmatrix} R_x(t_1,t_1) & R_x(t_1,t_2) & \cdots & R_x(t_1,t_n) \\ R_x(t_2,t_1) & R_x(t_2,t_2) & \cdots & R_x(t_2,t_n) \\ \vdots & \vdots & \vdots & \cdots \\ R_x(t_n,t_1) & R_x(t_n,t_2) & \cdots & R_x(t_n,t_n) \end{bmatrix} \begin{bmatrix} \alpha_1^* \\ \alpha_2^* \\ \vdots \\ \alpha_n^* \end{bmatrix} \\ &\geq 0, \end{aligned}$$

or

$$\begin{bmatrix} R_x(t_1,t_1) & R_x(t_1,t_2) & \cdots & R_x(t_1,t_n) \\ R_x(t_2,t_1) & R_x(t_2,t_2) & \cdots & R_x(t_2,t_n) \\ \vdots & \vdots & \vdots & \cdots \\ R_x(t_n,t_1) & R_x(t_n,t_2) & \cdots & R_x(t_n,t_n) \end{bmatrix} \succeq 0,$$

where $\mathbf{R} \succeq 0$ means that $\mathbf{R}$ is a positive semidefinite matrix. That is, all the eigenvalues of $\mathbf{R}$ are non-negative. Note that the term "non-negative definite" is a synonym for "positive semidefinite". With Eq. (1.1.9), the above equation is reduced to

$$\begin{bmatrix} R_x(t_1,t_1) & R_x(t_1,t_2) & \cdots & R_x(t_1,t_n) \\ R_x^*(t_1,t_2) & R_x(t_2,t_2) & \cdots & R_x(t_2,t_n) \\ \vdots & \vdots & \vdots & \cdots \\ R_x^*(t_1,t_n) & R_x^*(t_2,t_n) & \cdots & R_x(t_n,t_n) \end{bmatrix} \succeq 0, \tag{1.1.11}$$

where the matrix on the left hand side (LHS) of the inequality is a conjugate symmetric matrix or a Hermitian matrix.

In particular, when $n = 2$, Eq. (1.1.11) yields

$$\begin{bmatrix} R_x(t_1,t_1) & R_x(t_1,t_2) \\ R_x^\star(t_1,t_2) & R_x(t_2,t_2) \end{bmatrix} \succeq 0,$$

or

$$|R_x(t_1,t_2)|^2 \le R_x(t_1,t_1)R_x(t_2,t_2), \tag{1.1.12}$$

which is called Schwartz inequality.

The mean and autocorrelation function $R_x(t_1,t_2)$ are respectively the first and second order moments of random signal $x(t)$. Analogously, we can define the kth order moment of the random signal $x(t)$ as

$$\mu(t_1,\cdots,t_k) \stackrel{\text{def}}{=} \mathrm{E}\left\{x(t_1)\cdots x(t_k)\right\}. \tag{1.1.13}$$

By the dependence of kth order moment on time, random signals can be further categorized into two classes: stationary and non-stationary signals.

Definition 1.1.1 (nth order stationarity). *Random signal $\{x(t)\}$ is nth order stationary if for all integers $1 \le k \le n$, $t_1,\ldots,t_k$ and τ, $\{x(t)\}$ has finite kth moment which satisfies*

$$\mu(t_1,\cdots,t_k) = \mu(t_1+\tau,\cdots,t_k+\tau). \tag{1.1.14}$$

Specifically, a random signal is wide-sense stationary if it is second order stationary.

Definition 1.1.2 (wide-sense stationarity). *Complex random signal $\{x(t)\}$ is wide-sense stationary, if*

(1) *its mean is constant, i.e.,* $\mathrm{E}\left\{x(t)\right\} = \mu_x$;
(2) *its second order moment is bounded, i.e.,* $\mathrm{E}\left\{x(t)x^*(t)\right\} = \mathrm{E}\left\{|x(t)|^2\right\} < \infty$;
(3) *its correlation function is time independent, i.e.,*

$$C_{xx}(\tau) = \mathrm{E}\left\{[x(t)-\mu_x][x(t-\tau)-\mu_x]^*\right\}.$$

Wide-sense stationarity is also known as covariance stationarity and weak stationarity. A wide-sense stationary signal is called a stationary signal for brevity.

Definition 1.1.3 (strict stationarity). *Random signal $\{x(t)\}$ is strictly stationary if the sets of random variables $\{x(t_1+\tau),\cdots,x(t_k+\tau)\}$ and $\{x(t_1),\cdots,x(t_k)\}$ have identical joint probability distribution for any $\tau > 0$ and $t_1,t_2,\cdots,t_k$, where $k = 1,2,\ldots$.*

In plain words, the signal $x(t)$ whose joint probability distribution is invariant with time is strictly stationary.

The relation of nth order stationarity, wide-sense stationarity, strict stationarity, and non-stationarity is summarized below.

(1) Wide-sensing stationarity is nth order stationarity for $n = 2$.

(2) Strict stationarity must be wide-sensing stationarity. On the contrary, wide-sensing stationarity does not necessarily imply strict stationarity.
(3) As a stochastic process that is not wide-sense stationary cannot be either nth ($n > 2$) order stationary or strictly stationary, random signal not being wide-sense stationary is called non-stationary signal.

The stationary signal is usually termed time-invariant signal, underlining the fact that its statistics do not change with time. The non-stationary signal is similarly called time-variant signal, since at least one of its statistics such as mean and autocorrelation function is a function of time. Note that the concepts of the time-invariant and time-variant signals are not related to whether or not the sample or waveform of a signal is constant in time.

In wireless communications, the transmit signals are generally stationary, and there are two types of wireless channel: Gaussian channel and Rayleigh fading channel. Gaussian channel is non-fading and time-invariant and thus is stationary. Consequently, passing through Gaussian channel, the communication signals at the receiver is still stationary. Different from the Gaussian channel, the Rayleigh fading channel is non-stationary and time-variant. So, the signal transmitted through the Rayleigh fading channel becomes non-stationary at the receiver.

Ergodicity is another important property of random signals. It is connected to the crucial question whether the statistics of a random signal such as autocorrelation function and power spectral can be estimated from its single realization. A thorough discussion of ergodicity requires advanced probability theory. So we only present the concept of ergodicity in its most used form, i.e., the mean-square ergodicity.

Let $\{x(t)\}$ be a stationary signal whose moments of nth and lower orders are independent of time. The signal is nth order mean-square ergodic if for all integers $k = 1, \cdots, n$ and arbitrary $t_1, \cdots, t_k$, the following equation of mean-square limit holds.

$$\lim_{N\to\infty} \mathrm{E}\left\{\left|\frac{1}{2N+1}\sum_{t=-N}^{N} x(t+t_1)x(t+t_2)\cdots x(t+t_k) - \mu(t_1,\cdots,t_k)\right|^2\right\} = 0, \tag{1.1.15}$$

from which comes the term "mean-square ergodicity".

The statistical average of the nth and all the lower orders of a signal which is stationary and nth order mean-square ergodic is identical to the corresponding time average. In other words, the statistics can be estimated based on one realization of the signal. In this book, we assume that all random signals under study are mean-square ergodic.

For a mean-square ergodic and stationary signal $x(t)$ with N samples $x(1), \cdots, x(N)$, its mean μ_x can be estimated from the time average

$$\mu_x = \frac{1}{N}\sum_{n=1}^{N} x(n). \tag{1.1.16}$$

1.2 Correlation Function, Covariance Function, and Power Spectral Density

As stated in the previous section, a random signal can be characterized by its statistical properties, which can be further classified as statistical properties of the first order, second order, and high order (third order and higher). The aforementioned mean value belongs to the first order statistics and is the mathematical expectation of the signal. The second and higher-order statistics of a signal, which are more useful than the first order statistics, can be obtained by taking the expectation of second and higher-order (third order and higher) products of signal samples. The first few chapters of the book will extensively utilize second order statistics as the mathematical tool for the analysis and treatment of stationary random signals. The higher-order statistics of stationary random signals will be the subject of Chapter 6.

Correlation function, covariance function, and power spectral density are among the most used second order statistics. In this section, we will give the definitions of the autocorrelation function, autocovariance function, and power spectral density of a single stationary random signal along with the cross correlation function, cross covariance function, and cross power spectral density between two stationary random signals.

1.2.1 Autocorrelation Function, Autocovariance Function, and Power Spectral Density

Let $x(t)$ be a wide-sense stationary random signal of complex value with time variable $t \in (-\infty, \infty)$ or $t \in [0, \infty]$. The random signal $x(t)$ has a constant mean independent of time t, which is

$$\mu_x = \mathrm{E}\left\{x(t)\right\}, \tag{1.2.1}$$

and has autocorrelation and autocovariance functions dependent solely on the difference of time $\tau = t_1 - t_2$, which are defined by

$$R_{xx}(\tau) \stackrel{\text{def}}{=} \mathrm{E}\left\{x(t)x^*(t-\tau)\right\}, \tag{1.2.2}$$

$$\begin{aligned} C_{xx}(\tau) &\stackrel{\text{def}}{=} \mathrm{E}\left\{[x(t)-\mu_x][x(t-\tau)-\mu_x]^*\right\} \\ &= R_{xx}(\tau) - \mu_x\mu_x^* = R_{xx}(\tau) - |\mu_x|^2. \end{aligned} \tag{1.2.3}$$

Autocorrelation and autocovariance functions are also called correlation and covariance functions for brevity. The quantity $\tau = t_1 - t_2$, which is the time difference between two signal samples, is called lag.

The correlation and covariance functions of complex signals are complex in general and have the following properties.

$$R_{xx}^{*}(\tau) = R_{xx}(-\tau), \tag{1.2.4}$$

$$C_{xx}^{*}(\tau) = C_{xx}(-\tau), \tag{1.2.5}$$

$$|C_{xx}(\tau)| \le C_{xx}(0), \quad \forall\tau, \tag{1.2.6}$$

where the notation $\forall\tau$ means that the equations hold for any τ.

We give the proof for the first equation, and leave the rest as exercises. According to the definition, we have

$$R_{xx}^{*}(\tau) = \mathrm{E}\left\{x(t)x^{*}(t-\tau)\right\}.$$

After the variable substitution $u = t - \tau$, the above equation turns into

$$R_{xx}^{*}(\tau) = \mathrm{E}\left\{x(u+\tau)x^{*}(u)\right\} = R_{xx}(-\tau).$$

In particular, for real signal $x(t)$, we have

$$R_{xx}(\tau) = R_{xx}(-\tau), \tag{1.2.7}$$

$$C_{xx}(\tau) = C_{xx}(-\tau), \tag{1.2.8}$$

$$R_{xx}(\tau) \le R_{xx}(0). \tag{1.2.9}$$

The relation between the autocorrelation function and the autocovariance function is summarized below.

(1) For random signal $x(t)$ with zero mean, the autocorrelation function and autocovariance function are identical:

$$R_{xx}(\tau) = C_{xx}(\tau). \tag{1.2.10}$$

(2) When $\tau = 0$, the autocorrelation function of signal $x(t)$ is reduced to the second order moment, i.e.,

$$R_{xx}(0) = \mathrm{E}\left\{x(t)x^{*}(t)\right\} = \mathrm{E}\left\{|x(t)|^2\right\}. \tag{1.2.11}$$

(3) When $\tau = 0$, the autocovariance function of signal $x(t)$ is reduced to the variance of $x(t)$[1], i.e.,

$$\begin{aligned} C_{xx}(0) &= \mathrm{var}\left[x(t)\right] = \mathrm{E}\left\{[x(t)-\mu_x][x(t)-\mu_x]^{*}\right\} \\ &= \mathrm{E}\left\{|x(t)-\mu_x|^2\right\} = \mathrm{E}\left\{|x(t)|^2\right\} - |\mu_x|^2 \\ &= R_{xx}(0) - |\mu_x|^2. \end{aligned} \tag{1.2.12}$$

1 The variance of complex random variable is denoted as $\mathrm{var}[\xi]$ which is defined by $\mathrm{var}[\xi] = \mathrm{E}\{[\xi - \mathrm{E}\{\xi\}][\xi - \mathrm{E}\{\xi\}]^{*}\} = \mathrm{E}\{|\xi - \mathrm{E}\{\xi\}|^2\}$, with $\sigma = \sqrt{\mathrm{var}[\xi]}$ as the standard deviation.

Consider the stochastic process $x(t)$ on finite interval $-T < t < T$. The Fourier transform of $x(t)$ is

$$X_T(f) = \int_{-T}^{T} (x(t) - \mu_x)\mathrm{e}^{-\mathrm{j}2\pi ft}\,\mathrm{d}t,$$

and the power spectral on the interval is

$$\frac{|X_T(f)|^2}{2T} \geq 0.$$

Taking the ensemble average of the power spectral function yields

$$\begin{aligned} P_T(f) &= \mathrm{E}\left\{\frac{|X_T(f)|^2}{2T}\right\} \\ &= \frac{1}{2T}\int_{-T}^{T}\int_{-T}^{T} \mathrm{E}\left\{[x(t_1) - \mu_x][x(t_2) - \mu_x]^*\right\}\mathrm{e}^{-\mathrm{j}2\pi f(t_1-t_2)}\,\mathrm{d}t_1\,\mathrm{d}t_2 \\ &= \int_{-2T}^{2T} C_{xx}(\tau)\left(1 - \frac{|\tau|}{2T}\right)\mathrm{e}^{-\mathrm{j}2\pi f\tau}\,\mathrm{d}\tau \\ &\geq 0. \end{aligned}$$

Evidently, $P_T(f)$ represents how the average power of stochastic process $x(t)$ on interval $(-T, T)$ is distributed with frequency f. As $T \to \infty$, the distribution yields the power spectral density as

$$P_{xx}(f) = \lim_{T\to\infty} P_T(f) = \int_{-2T}^{2T} C_{xx}(\tau)\mathrm{e}^{-\mathrm{j}2\pi f\tau}\,\mathrm{d}\tau. \tag{1.2.13}$$

The above equation is essential in that it is the definition of power spectral density and at the same time shows that the power spectral density must be non-negative. More properties of power spectral density are listed below.

(1) Power spectral density $P_{xx}(f)$ is real.

(2) Power spectral density is non-negative, i.e., $P_{xx}(f) \geq 0$.

(3) The autocovariance function is the inverse Fourier transform of power spectral density. That is,

$$C_{xx}(\tau) = \int_{-\infty}^{\infty} P_{xx}(f)\mathrm{e}^{\mathrm{j}2\pi f\tau}\,\mathrm{d}f. \tag{1.2.14}$$

(4) Integration of power spectral density over frequency gives the variance of signal $\{x(t)\}$. That is,

$$\int_{-\infty}^{\infty} P_{xx}(f)\,\mathrm{d}f = \mathrm{var}\left[x(t)\right] = \mathrm{E}\left\{|x(t) - \mu_x|^2\right\}. \tag{1.2.15}$$

(5) If $\{x(t)\}$ is stochastic process with zero mean, the covariance function and the correlation function are identical, i.e., $C_{xx}(\tau) = R_{xx}(\tau)$. Then, Eqs. (1.2.13) and (1.2.14) are respectively equivalent to

$$P_{xx}(f) = \int_{-\infty}^{\infty} R_{xx}(\tau)\mathrm{e}^{-\mathrm{j}2\pi f\tau}\,\mathrm{d}\tau, \tag{1.2.16}$$

$$R_{xx}(\tau) = \int_{-\infty}^{\infty} P_{xx}(f)\mathrm{e}^{\mathrm{j}2\pi f\tau}\,\mathrm{d}f. \tag{1.2.17}$$

The relation determined by Eqs. (1.2.16) and (1.2.17) is referred to as Wiener-Khinchine theorem, which states that for a wide-sense stationary stochastic process with zero mean, the power spectral density $P_{xx}(f)$ and the autocorrelation function $R_{xx}(\tau)$ constitute a Fourier transform pair.

(6) For a stochastic process $\{x(t)\}$ with zero mean, integration of its power spectral density equals the value of correlation function at zero lag ($\tau = 0$). Namely,

$$\int_{-\infty}^{\infty} P_{xx}(f)\,\mathrm{d}f = \mathrm{E}\left\{|x(t)|^2\right\} = R_{xx}(0). \tag{1.2.18}$$

We now prove Property (1) of power spectral density. According to the definition of power spectral density of complex stochastic process $x(t)$, we immediately have

$$P_{xx}^{*}(f) = \int_{-\infty}^{\infty} C_{xx}^{*}(\tau)\mathrm{e}^{\mathrm{j}2\pi f\tau}\,\mathrm{d}\tau = \int_{-\infty}^{\infty} C_{xx}(-\tau)\mathrm{e}^{\mathrm{j}2\pi f\tau}\,\mathrm{d}\tau,$$

which after variable change $\tau' = -\tau$ gives

$$P_{xx}^{*}(f) = -\int_{\infty}^{-\infty} C_{xx}(\tau')\mathrm{e}^{-\mathrm{j}2\pi f\tau'}\,\mathrm{d}\tau' = \int_{-\infty}^{\infty} C_{xx}(\tau')\mathrm{e}^{-\mathrm{j}2\pi f\tau'}\,\mathrm{d}\tau' = P_{xx}(f).$$

That is, power spectral density $P_{xx}(f)$ must be a real function in frequency f.

The proof for Property (2) of power spectral density is more involved and will be given later.

Further, if $x(t)$ is real, its power spectral density $P_{xx}(f)$ must be a real even function.

If the power spectral density is constant, i.e., $P_{xx}(f) = N_0$, the stochastic process $\{x(t)\}$ is called white noise since its power (or energy) is independent of frequency and is similar to the energy distribution of white light. In contrast, noise whose power spectral density is not constant is called colored noise.

Example 1.2.1 Let $\{x(t)\}$ be a real sequence whose samples at any time are independent and have zero mean and variance σ^2. Then, $\{x(t)\}$ is a sequence of white noise.

Solution. Let vector $\mathbf{x} = [x(1), \cdots, x(N)]$. Based on given facts, for each element $x(i)$ in the vector, we have $\mathrm{E}\{x(i)\} = 0$ and

$$\mathrm{E}\{x(i)x(i-\tau)\} = R_{xx}(\tau) = \begin{cases} \sigma^2, & \tau = 0, \\ 0, & \tau \neq 0. \end{cases}$$

Since $\{x(t)\}$ has zero mean, its covariance and correlation functions are the same, which gives

$$C_{xx}(\tau) = R_{xx}(\tau) = \sigma^2\delta(\tau), \tau = 0, \pm 1, \pm 2, \cdots .$$

Thus, $\{x(t)\}$ has power spectral density as

$$P_{xx}(f) = \int_{-\infty}^{\infty} C_{xx}(\tau)\mathrm{e}^{-\mathrm{j}2\pi f\tau}\,\mathrm{d}\tau = \int_{-\infty}^{\infty} \sigma^2\delta(\tau)\mathrm{e}^{-\mathrm{j}2\pi f\tau}\,\mathrm{d}\tau = \sigma^2,$$

which shows that $\{x(t)\}$ is indeed a white noise sequence. □

A function is positive definite if its Fourier transform is positive everywhere. A function is non-negative definite or positive semidefinite if its Fourier transform is non-negative. As the power spectral density is non-negative, the covariance function is positive semidefinite.

Consider a discrete-time stationary random signal $x(n)$ $(n = 1, \cdots, N)$. Denote $\mathbf{x}(n) = [x(1), \cdots, x(N)]^T$ as the observation vector of random signal $\{x(n)\}$. Then, the correlation function matrix is defined by

$$\mathbf{R} \stackrel{\text{def}}{=} \mathrm{E}\left\{\mathbf{x}(n)\mathbf{x}^H(n)\right\} = \begin{bmatrix} R_{xx}(0) & R_{xx}(-1) & \cdots & R_{xx}(-N+1) \\ R_{xx}(1) & R_{xx}(0) & \cdots & R_{xx}(-N+2) \\ \vdots & \vdots & \vdots & \vdots \\ R_{xx}(N-1) & R_{xx}(N-2) & \cdots & R_{xx}(0) \end{bmatrix}, \tag{1.2.19}$$

where $\mathbf{x}^H(n)$ is the conjugate transpose of $\mathbf{x}(n)$.

The structure of the correlation function matrix in Eq. (1.2.19) is rather special in that not only the principal diagonal of the matrix has the same elements, but each diagonal other than the principal one has a constant element. The matrix with such property is called the Toeplitz matrix.

1.2.2 Cross Correlation Function, Cross Covariance Function, and Cross Power Spectral Density

Next we discuss the statistics involving two stationary complex random signals $x(t)$ and $y(t)$. Let

$$\mu_x = \mathrm{E}\{x(t)\} \quad \text{and} \quad \mu_y = \mathrm{E}\{y(t)\}, \tag{1.2.20}$$

which are both constant.

The cross correlation and cross corvarance functions between two complex random signals $x(t)$ and $y(t)$ are respectively defined by

$$R_{xy}(t_1, t_2) \stackrel{\text{def}}{=} \mathrm{E}\left\{x(t_1)y^*(t_2)\right\}, \tag{1.2.21}$$

$$C_{xy}(t_1, t_2) \stackrel{\text{def}}{=} \mathrm{E}\left\{[x(t_1) - \mu_x][y(t_2) - \mu_y]^*\right\} = R_{xy}(t_1, t_2) - \mu_x\mu_y^*. \tag{1.2.22}$$

If both $R_{xy}(t_1, t_2) = R_{xy}(t_1 - t_2)$ and $R_{yx}(t_1, t_2) = R_{yx}(t_1 - t_2)$ depend solely on the lag $t_1 - t_2$, $x(t)$ and $y(t)$ are jointly stationary.

The cross correlation and cross covariance functions of two complex signals which are jointly stationary are complex in general and have the properties below.

$$R_{xy}^*(\tau) = R_{yx}^*(-\tau), \tag{1.2.23}$$

$$|R_{xy}(\tau)|^2 \leq |R_{xx}(0)| \cdot |R_{yy}^*(0)|, \quad \forall\tau, \tag{1.2.24}$$

$$C_{xy}^*(\tau) = C_{yx}^*(-\tau). \tag{1.2.25}$$

When $\mu_x = 0$ and $\mu_y = 0$, cross covariance and cross correlation functions are identical:

$$C_{xy}(\tau) = R_{xy}(\tau). \tag{1.2.26}$$

With the cross covariance function, the cross correlation coefficient can be defined as

$$\rho_{xy}(\tau) = \frac{C_{xy}(\tau)}{\sqrt{C_{xx}(0)C_{yy}(0)}}. \tag{1.2.27}$$

It can be shown that

$$|\rho_{xy}| \leq 1, \quad \forall\tau. \tag{1.2.28}$$

We now give a physical interpretation of the cross correlation coefficient. Notice that the cross covariance function involves the product of two different signals $x(t)$ and $y(t)$. In general, the two signals which have their means removed could have common parts, which correspond to the deterministic components, and non-common parts, which are the stochastic components. The sample values from the product of the common parts always have consistent signs which have the common parts strengthened and preserved, whereas the non-common parts of the two signals are random and thus result in product values with both positive and negative signs, which after taking expectation to get smoothed out. This indicates that the cross covariance function can extract the common parts of two signals and suppress the non-common parts. Hence, the cross covariance function can be utilized to characterize the similarity between two signals. Nevertheless, the similarity measure defined by the cross covariance function could be awkward in application as it is in the form of absolute value. In contrast, the cross correlation coefficient derived from normalizing the cross covariance function is more effective in measuring the similarity between two signals. To be more specific, the two signals are more alike as the cross correlation coefficient comes close to one. Conversely, the difference between the two signals becomes more evident when the cross correlation coefficient goes to zero.

Example 1.2.2 Consider two complex harmonic signals $x(t) = A\mathrm{e}^{\mathrm{j}\omega_1 t}$ and $y(t) = B\mathrm{e}^{\mathrm{j}\omega_2 t}$ where A and B are Gaussian random variables with probability densities

$$f_A(a) = \frac{1}{\sqrt{2\pi}\sigma_1}\mathrm{e}^{-a^2/(2\sigma_1^2)}, \quad f_B(b) = \frac{1}{\sqrt{2\pi}\sigma_2}\mathrm{e}^{-b^2/(2\sigma_2^2)}.$$

Suppose that A and B are independent random variables. Derive the autocovariance function $R_{xx}(\tau)$ and the cross covariance function $C_{xy}(\tau)$.

Solution. As $\mathrm{e}^{\mathrm{j}\omega_i t}$ $(i = 1, 2)$ are deterministic functions, and the amplitudes A and B are random variables with $\mathrm{E}\{A\} = \mathrm{E}\{B\} = 0$, $\mathrm{E}\{A^2\} = \sigma_1^2$, and $\mathrm{E}\{B^2\} = \sigma_2^2$, the means of $x(t)$ and $y(t)$ are also those of random variables A and B, respectively. That is,

$$\mu_x = \mathrm{E}\left\{A\mathrm{e}^{\mathrm{j}\omega_1 t}\right\} = \mathrm{E}\{A\}\,\mathrm{e}^{\mathrm{j}\omega_1 t} = 0,$$
$$\mu_y = \mathrm{E}\left\{B\mathrm{e}^{\mathrm{j}\omega_2 t}\right\} = \mathrm{E}\{B\}\,\mathrm{e}^{\mathrm{j}\omega_2 t} = 0.$$

Therefore, $C_{xx}(\tau) = R_{xx}(\tau)$ and $C_{xy}(\tau) = R_{xy}(\tau)$.

Direct derivation gives

$$R_{xx}(\tau) = \mathrm{E}\left\{A\mathrm{e}^{\mathrm{j}\omega_1 t}[A\mathrm{e}^{\mathrm{j}(\omega_1 t-\omega_1\tau)}]^*\right\} = \mathrm{E}\left\{A^2\mathrm{e}^{\mathrm{j}\omega_1\tau}\right\} = \mathrm{E}\left\{A^2\right\}\mathrm{e}^{\mathrm{j}\omega_1\tau} = \sigma_1^2\mathrm{e}^{\mathrm{j}\omega_1\tau}.$$

Noticing that A and B are independent, we have

$$\begin{aligned} C_{xy}(\tau) &= \mathrm{E}\left\{A\mathrm{e}^{\mathrm{j}\omega_1 t}[B\mathrm{e}^{\mathrm{j}(\omega_2 t-\omega_2\tau)}]^*\right\} = \mathrm{E}\left\{AB\mathrm{e}^{\mathrm{j}(\omega_1-\omega_2)t}\mathrm{e}^{\mathrm{j}\omega_2\tau}\right\} \\ &= \mathrm{E}\{A\}\,\mathrm{E}\{B\}\,\mathrm{e}^{\mathrm{j}(\omega_1-\omega_2)t}\mathrm{e}^{\mathrm{j}\omega_2\tau} = 0\cdot 0\cdot \mathrm{e}^{\mathrm{j}(\omega_1-\omega_2)t}\mathrm{e}^{\mathrm{j}\omega_2\tau} = 0, \end{aligned}$$

which shows that $x(t)$ and $y(t)$ are uncorrelated. □

The cross power spectral density of complex signals $x(t)$ and $y(t)$ is defined as the Fourier transform of the cross covariance function. That is,

$$P_{xy}(f) = \int_{-\infty}^{\infty} C_{xy}(\tau)\mathrm{e}^{-\mathrm{j}2\pi f\tau}\,\mathrm{d}\tau. \tag{1.2.29}$$

Unlike power spectral density $P_{xx}(f)$ which is a real function of frequency f, cross power spectral density is a complex function, whose real part is called cospectrum and imaginary part is called quadrature spectrum.

Denote

$$P_{xy}(f) = |P_{xy}(f)|\exp[\mathrm{j}\phi_{xy}(f)], \tag{1.2.30}$$
$$\dot{\phi}_{xy}(f) = \frac{\mathrm{d}}{\mathrm{d}f}\phi_{xy}(f), \tag{1.2.31}$$

where $|P_{xy}(f)|$ and $\phi_{xy}(f)$ are respectively the amplitude and phase of cross power spectral density, and $\dot{\phi}_{xy}(f)$ is the group delay.

Based on cross power spectral density, coherence function can be defined as

$$C(f) \stackrel{\text{def}}{=} \frac{|P_{xy}(f)|}{\sqrt{P_{xx}(f)P_{yy}(f)}}, \tag{1.2.32}$$

which is real and satisfies

$$|C(f)| \leq 1. \tag{1.2.33}$$

Correlation function, covariance function, and correlation coefficient are statistics of signal in the time domain, belonging to the time-domain feature of signal, whereas power spectral density and coherence function are statistics of signal in the frequency domain, and are classified as the frequency-domain feature of the signal. Since covariance function and power spectral density can be converted into each other through Fourier transform, the signal features in both time and frequency domains are of equal importance in applications.

The random signal with samples distributed according to the normal distribution is called Gaussian random signal, while the random signal following non-normal distribution is called a non-Gaussian random signal. For non-Gaussian signals, sufficient characterization of statistics cannot be achieved by correlation function and power spectral density alone and calls for the employment of statistics of third order or higher, which are collectively called high-order statistics. High-order statistics consist of high-order moments, high-order cumulants, and high-order spectrum, which will be primarily covered in Chapter 6.

Before processing a stationary signal, the mean of the signal should be estimated and removed from samples. This procedure is called the centralization of the stationary signal. From now on, unless otherwise stated, we assume that signals or additive noise all have zero mean. Due to the obligatory centralization as the preprocessing of signals, the correlation and covariance functions are generally interchangeable in literature, since both are equivalent for signals of zero mean.

1.3 Comparison and Discrimination between Two Random Signals

In the previous section, we have discussed the second order statistics of two signals. In many applications, we are more interested in the comparison of statistics of two signals, which covers issues such as whether two signals are statistically independent, uncorrelated, or orthogonal. Some applications may require discrimination between two signals, to determine whether two signals are strongly correlated or coherent, for instance.

1.3.1 Independence, Uncorrelatedness, and Orthogonality

1. Independence

Consider two random variables y_1 and y_2. The two random variables are independent if the information of y_1 contains no information about y_2 and vice versa.

The independence of two random variables y_1 and y_2 can also be defined by their PDFs. Let $p(y_1, y_2)$ be the joint PDF of y_1 and y_2, $p(y_1)$ be the marginal PDF of y_1, which is

$$p(y_1) = \int p(y_1, y_2)\,\mathrm{d}y_2, \tag{1.3.1}$$

and similarly $p(y_2)$ be the marginal PDF of y_2 given by

$$p(y_2) = \int p(y_1, y_2)\,\mathrm{d}y_1. \tag{1.3.2}$$

From the equations above, it can be seen that the two random variables y_1 and y_2 are independent if and only if

$$p(y_1, y_2) = p(y_1)p(y_2). \tag{1.3.3}$$

To generalize, the components of random vector $\mathbf{y} = [y_1, \cdots, y_m]^T$ are independent if and only if

$$p(\mathbf{y}) = p(y_1, \cdots, y_m) = p(y_1)\cdots p(y_m). \tag{1.3.4}$$

If random variables y_1 and y_2 are independent, and $h_1(y_1)$ and $h_2(y_2)$ are functions in y_1 and y_2 respectively, we have

$$\mathrm{E}\left\{h_1(y_1)h_2(y_2)\right\} = \mathrm{E}\left\{h_1(y_1)\right\}\mathrm{E}\left\{h_2(y_2)\right\}, \tag{1.3.5}$$

whose proof is given below[109].

$$\begin{aligned}
\mathrm{E}\left\{h_1(y_1)h_2(y_2)\right\} &= \int\int h_1(y_1)h_2(y_2)p(y_1, y_2)\,\mathrm{d}y_1\,\mathrm{d}y_2 \\
&= \int\int h_1(y_1)h_2(y_2)p(y_1)p(y_2)\,\mathrm{d}y_1\,\mathrm{d}y_2 \\
&= \int h_1(y_1)p(y_1)\,\mathrm{d}y_1 \int h_2(y_2)p(y_2)\,\mathrm{d}y_2 \\
&= \mathrm{E}\left\{h_1(y_1)\right\}\mathrm{E}\left\{h_2(y_2)\right\}.
\end{aligned}$$

Extending the concept of independence of random variables to the case of two stochastic processes $x(t)$ and $y(t)$, we say signals $x(t)$ and $y(t)$ are statistically independent if the joint PDF $f_{X,Y}(x, y)$ equals the product of marginal PDFs $f_X(x)$ for $x(t)$ and $f_Y(y)$ for $y(t)$:

$$f_{X,Y}(x, y) = f_X(x)f_Y(y). \tag{1.3.6}$$

2. Uncorrelatedness

Two random signals $y_1(t)$ and $y_2(t)$ are uncorrelated if

$$\mathrm{E}\left\{y_1(t)y_2(t)\right\} = \mathrm{E}\left\{y_1(t)\right\}\mathrm{E}\left\{y_2(t)\right\}. \tag{1.3.7}$$

3. Orthogonality

Two random variables y_1 and y_2 are orthogonal if y_1 contains no component of y_2 and vice versa. The orthogonality of two random variables y_1 and y_2 are denoted as $y_1 \perp y_2$, which mathematically is defined by

$$\mathrm{E}\{y_1 y_2\} = 0. \tag{1.3.8}$$

By $y_1 = x(t)$ and $y_2 = y(t-\tau)$, the orthogonality of two random variables can be directly extended to the orthogonality of two random processes or signals. That is, two random processes or signals $x(t)$ and $y(t)$ are orthogonal if for any lag τ, the correlation function of $x(t)$ and $y(t)$ is identical to zero:

$$R_{xy}(\tau) = \mathrm{E}\left\{x(t)y^*(t-\tau)\right\} = 0, \quad \forall\tau. \tag{1.3.9}$$

For simplicity, the orthogonal signals are denoted as $x(t) \perp y(t)$. The notion of orthogonality can also be expressed in the form of inner product as

$$\langle x(t), y(t-\tau)\rangle = \int_{-\infty}^{\infty} x(t)y^*(t-\tau)\,\mathrm{d}t = 0, \quad \forall\tau. \tag{1.3.10}$$

Next, we summarize the relations among statistical independence, uncorrelatedness, and orthogonal.

(1) Statistical independence necessarily implies uncorrelatedness, but the converse does not hold in general. The only exception is the case of the Gaussian stochastic process, where the statistical independence and uncorrelatedness are equivalent for arbitrary two Gaussian stochastic processes.
(2) If the means of $x(t)$ and $y(t)$ are both zero, uncorrelatedness and orthogonality are equivalent.

Therefore, for two Gaussian signals with zero mean, the three concepts of statistical independence, uncorrelatedness, and orthogonality are equivalent.

It can be seen from the definition of correlation coefficient that as $C_{xy}(\tau) = 0, \quad \forall\tau$, the correlation coefficient $\rho_{xy}(\tau) = 0, \forall\tau$. Hence $\rho_{xy}(\tau) = 0\ (\forall\tau)$ implies that signals $x(t)$ and $y(t)$ are uncorrelated.

Now consider the correlation coefficient of two special signals $x(t)$ and $y(t) = c \cdot x(t-\tau_0)$, where c is a complex constant and τ_0 is a real constant. Consequently, $x(t)$ and $y(t)$ have the following properties.

(1) $y(t)$ and $x(t)$ differ by a complex amplitude c. For $c = |c|e^{j\phi_c}$, $y(t)$ differs from $x(t)$ by amplification or attenuation of $|c|$ times and phase shift of ϕ_c.
(2) $y(t)$ and $x(t)$ differ by a time delay τ_0.

The signals with both properties above are called coherent signals. Naturally, $y(t)$ can be called the coherent signal of $x(t)$ and $x(t)$ the coherent signal of $y(t)$. To underline the fact that coherent signals are a replica of each other, coherent signal is called replica signal occasionally. Since

$$\mu_y = \mathrm{E}\left\{cx(t)\right\} = c\mathrm{E}\left\{x(t-\tau)\right\} = c\mu_x, \tag{1.3.11}$$

$$\begin{aligned} C_{yy}(0) &= \mathrm{E}\left\{[y(t)-\mu_y][y(t)-\mu_y]^*\right\} \\ &= \mathrm{E}\left\{[cx(t)-c\mu_x][c^*x^*(t)-c^*\mu_x^*]\right\} \\ &= |c|^2[R_{xx}(0)-|\mu_x|^2] \\ &= |c|^2 C_{xx}(0), \end{aligned} \tag{1.3.12}$$

and

$$\begin{aligned} C_{xy}(\tau) &= \mathrm{E}\left\{[x(t)-\mu_x][y(t-\tau)-\mu_y]^*\right\} \\ &= \mathrm{E}\left\{[x(t)-\mu_x][c^*x^*(t-\tau_0-\tau)-c^*\mu_x^*]\right\} \\ &= c^*\mathrm{E}\left\{[x(t)-\mu_x][x(t-\tau_0-\tau)-\mu_x]^*\right\} \\ &= c^*C_{xx}(\tau+\tau_0), \end{aligned} \tag{1.3.13}$$

the correlation coefficient can be expressed as

$$\rho_{xy}(\tau) = \frac{C_{xy}(\tau)}{\sqrt{C_{xx}(0)C_{yy}(0)}} = \frac{c^*C_{xx}(\tau+\tau_0)}{\sqrt{C_{xx}(0)|c|^2C_{xx}(0)}} = \frac{c^*}{|c|}\frac{C_{xx}(\tau+\tau_0)}{C_{xx}(0)}.$$

Obviously, if $\tau = -\tau_0$, the correlation coefficient of two coherent signals has unit modulus, i.e.,

$$|\rho_{xy}(-\tau_0)| = 1, \tag{1.3.14}$$

which suggest that if the cross correlation coefficient of signals $x(t)$ and $y(t)$ is unit for some $\tau = -\tau_0$, $y(t)$ must be the coherent signal of $x(t)$ and is time-delayed $x(t)$ by τ_0. That is to say, apart from being a tool for detecting signal coherence, the cross correlation coefficient also provides means for estimating the time delay of two signals. Coherent signal detection and delay estimation are crucial in many engineering practices of radar, wireless communications, geophysics, etc. Take radar and wireless communications for example, where the transmit signal reaches the receiver after multipath transmission. The multipath signals are largely coherent but with reduced power after attenuation. If the coherent signals are collected to achieve the so-called "coherent combination", the signal receiving can be beneficial from the combination, which can effectively increase the energy of the received signal and boost the signal-to-noise ratio,

When signals $x(t)$ and $y(t)$ are coherent, it is easy to verify by Eqs. (1.3.12) and (1.3.13) that

$$P_{yy}(f) = |c|^2 P_{xx}(f) \quad \text{and} \quad P_{yy}(f) = c^* P_{xx}(f).$$

So by definition, the coherence function of coherent signals satisfies

$$C(f) = \frac{|c^*P_{xy}(f)|}{\sqrt{P_{xx}(f)|c|^2P_{xx}(f)}} = 1, \quad \forall f, \tag{1.3.15}$$

hence the name.

We summarize the discussion above as follows.

(1) If the cross correlation coefficient $\rho_{xy}(\tau)$ is identical to zero for any τ, signals $x(t)$ and $y(t)$ are incoherent.

(2) If the cross correlation coefficient $\rho_{xy}(\tau)$ has unit modulus for some τ, signals $x(t)$ and $y(t)$ are coherent.

(3) The coherence function of coherent signals is identical to one for any frequency f.

To sum up, independence, uncorrelatedness, orthogonality, and coherence are of great importance in characterizing statistical relation between two random signals.

In many applications, the task of signal processing may involve more than one signal but many signals make up a signal set. For instance, in multiple access of wireless communications, each user is assigned a unique spreading sequence as its "transmitting id" which is called the characteristic signal of the user. Then, under what condition can each signal in the set be used as a characteristic signal? Usually, the following two conditions are employed.

(1) It should be easy to distinguish each signal $x(t)$ in the set from its time delayed version $x(t-\tau)$.

(2) It should be easy to distinguish each signal $x(t)$ in the set from any other signal with or without delay.

As a matter of fact, apart from $x(t)$ and $y(t-\tau)$, significant difference is also required by the discrimination of $x(t)$ and $-y(t-\tau)$. For example, $y(t)$ and $-y(t)$ must be both treated when binary symbols is carried by $y(t)$ or $y(t)$ is modulated by some carrier. Understandably, the larger the variance of signal difference is, the easier the two signals can be discriminated or identified. Hence, the variance is a proper measure of distinction between two signals, and the quantity

$$\begin{aligned} r(\tau) &= \mathrm{E}\left\{|x(t) \pm y(t-\tau)|^2\right\} \\ &= \mathrm{E}\left\{[x(t) \pm y(t-\tau)][x(t) \pm y(t-\tau)]^*\right\} \\ &= \mathrm{E}\left\{|x(t)|^2\right\} + \mathrm{E}\left\{|y(t)|^2\right\} \pm \mathrm{E}\left\{x(t)y^*(t-\tau)\right\} \pm \mathrm{E}\left\{y(t-\tau)x^*(t)\right\} \\ &= E_x + E_y \pm R_{xy}(\tau) \pm R_{yx}(-\tau) \\ &= E_x + E_y \pm R_{xy}(\tau) \pm R_{xy}^*(\tau) \end{aligned} \tag{1.3.16}$$

is defined as the identifiability of two signals.

As can be seen from Eq. (1.3.17), since the real and imaginary parts of cross correlation function $R_{xy}(\tau)$ can be either positive or negative at different τ, the amplitude of cross correlation function $R_{xy}(\tau)$ must be uniformly small for any τ so that $r(\tau)$ can be maximized at all values of τ. Accordingly, the distinction between signals $x(t)$ and $y(t)$ together with its delayed version becomes more evident with a smaller value of correlation. Under the ideal condition of $R_{xy}(\tau) = 0$ $(\forall\tau)$, $x(t)$ and $y(t)$ can be perfectly discriminated. In other words, $x(t)$ and $y(t)$ can be perfectly discriminated when they are orthogonal.

Eq. (1.3.17) also applies to the special case of $y(t) = x(t)$, where to fulfill Condition (1) the signal must have autocorrelation function $R_{xx}(\tau) = 0$ ($\forall \tau \neq 0$) since $R_{xx}(0) = \mathrm{E}\left\{|x(t)|^2\right\}$ is non-zero.

Applying the results above to code division multiple access (CDMA) system in wireless communications, we know that the characteristic signal allocated to each user should be close to white noise and the characteristic signals of different users should be orthogonal to each other.

1.3.2 Gram-Schmidt Orthogonalization Process of Polynomial Sequence

Let functions $f_i(x)$ ($i = 1, \cdots, n$) be polynomials of x. $\{f_i(x)\}$ is called a polynomial sequence. If each element $f_i(x)$ in the sequence cannot be expressed by a linear combination of the others, the polynomials are linearly independent. $\{f_1(x), f_2(x), f_3(x)\}$ with $f_1(x) = 1$, $f_2(x) = x$, and $f_3(x) = x^2$ is an example of linearly independent polynomial sequence.

Let x take values in interval $[a, b]$ and define the inner product between $f_i(x)$ and $f_k(x)$ as

$$\left\langle f_i(x), f_k(x)\right\rangle \stackrel{\text{def}}{=} \int_a^b f_i(x) f_k^{\star}(x)\,\mathrm{d}x. \tag{1.3.17}$$

So, if linearly independent polynomial sequence $\{f_i(x)\}$ satisfies

$$\left\langle f_i(x), f_k(x)\right\rangle = 0, \quad \forall i \neq k, \tag{1.3.18}$$

$\{f_i(x)\}$ is an orthogonal polynomial sequence. Further, if it also holds along with Eq. (1.3.18) that

$$\left\langle f_i(x), f_i(x)\right\rangle = 1, \quad \forall i = 1, \cdots, n, \tag{1.3.19}$$

$\{f_i(x)\}$ is called an orthonormal polynomial sequence.

The orthonormal polynomial sequence $\{\phi_i(x)\}$ can be generated from a linearly independent polynomial sequence $\{f_i(x)\}$ by Gram-Schmidt orthogonalization process.

Let

$$\left\|f(x)\right\| = \left\langle f(x), f(x)\right\rangle^{1/2} = \left[\int_a^b |f(x)|^2\,\mathrm{d}x\right]^{1/2} \tag{1.3.20}$$

be the norm of the function $f(x)$.

Gram-Schmidt orthogonalization process then operates as follows.

$$\phi_1(x) = \frac{f_1(x)}{\|f_1(x)\|}, \tag{1.3.21}$$

$$\phi_2(x) = \frac{f_2(x) - \langle f_2(x), \phi_1(x)\rangle \phi_1(x)}{\|f_2(x) - \langle f_2(x), \phi_1(x)\rangle \phi_1(x)\|}, \tag{1.3.22}$$

$$\vdots$$

$$\phi_k(x) = \frac{f_k(x) - \sum_{i=1}^{k-1} \langle f_k(x), \phi_i(x)\rangle \phi_i(x)}{\left\|f_k(x) - \sum_{i=1}^{k-1} \langle f_k(x), \phi_i(x)\rangle \phi_i(x)\right\|}, \quad k = 2, \cdots, n. \tag{1.3.23}$$

1.4 Linear System with Random Input

The previous discussion is related to the statistics of two random signals. But in a large number of signal processing applications, one may be interested in the statistics of the input and output of a linear system, especially the power spectral density of the system output.

1.4.1 The Power Spectral Density of System Output

Suppose that a linear system is time-invariant with input random signal $\{x(t)\}$. Then the system output $y(t)$, which is also random, is expressed by the convolution between the input and the impulse response of the system $h(t)$, i.e.,

$$y(t) = x(t) * h(t) = \int_{-\infty}^{\infty} h(u)x^*(t-u)\,\mathrm{d}u \tag{1.4.1}$$

$$= h(t) * x(t) = \int_{-\infty}^{\infty} x(u)h^*(t-u)\,\mathrm{d}u. \tag{1.4.2}$$

Obviously, if the input is an impulse signal $x(t) = \delta(t)$, the output or response is given by

$$y(t) = \int_{-\infty}^{\infty} h(t-\tau)\delta(\tau)\,\mathrm{d}\tau = h(t), \tag{1.4.3}$$

which is exactly the reason why $h(t)$ is named impulse response.

A linear time-invariant system is called causal system if

$$h(t) = 0, \quad \tau < 0, \tag{1.4.4}$$

which suggests that before there is any input (cause), there is no system output (result). If bounded input yields bounded output, the system is stable, which requires impulse response to satisfy

$$\int_{-\infty}^{\infty} |h(t)|\,dt < \infty. \tag{1.4.5}$$

The condition requires that the impulse response is absolute integrable, whose proof is left as an exercise.

The Fourier transform of the impulse response

$$H(f) = \int_{-\infty}^{\infty} h(t)e^{-j2\pi ft}\,dt \tag{1.4.6}$$

is called the transfer function of a linear system. Next, we derive the first order (mean) and second order statistics (covariance function and power spectral density) of system output.

Taking expectation of both sides of Eq. (1.4.1) yields[2]

$$\mathrm{E}\left\{y(t)\right\} = \int_{-\infty}^{\infty} \mathrm{E}\left\{x(t-u)\right\} h(u)\,du = \mathrm{E}\left\{x(t)\right\} * h(t). \tag{1.4.7}$$

Noting that the impulse response $h(t)$ of linear time-invariant system is non-stochastic, we have from Eq. (1.4.7) that the mean of output $y(t)$ is the convolution of the mean $\mathrm{E}\left\{x(t)\right\}$ of system input $x(t)$ and system impulse response.

If $\mathrm{E}\left\{x(t)\right\} = \mu_x$ is constant, Eq. (1.4.7) gives

$$\mathrm{E}\left\{y(t)\right\} = \mu_x \int_{-\infty}^{\infty} h(t)\,dt = \mu_x H(0) = \text{constant}, \tag{1.4.8}$$

where $H(0) = \int_{-\infty}^{\infty} h(t)\,dt$ is the value of transfer function $H(f)$ at zero frequency. In the special case of $x(t)$ being zero-mean random signal, the output $y(t)$ is a random signal with zero-mean as well.

Now consider the autocovariance function of system output when the input $x(t)$ is a wide-sense stationary stochastic process. By the convolutions in Eqs. (1.4.1) and (1.4.7), it is straightforward to derive that

$$\begin{aligned}&[y(t)-\mu_y][y(t-\tau)-\mu_y]^* \\ &= \int_{-\infty}^{\infty}\int_{-\infty}^{\infty} [x(t-u_1)-\mu_x][x^*(t-\tau-u_2)-\mu_x^*]h(u_1)h^*(u_2)\,du_1\,du_2.\end{aligned}$$

2 When the function $f(x)$ is absolute integrable, i.e., $\int_{-\infty}^{\infty} f(x)\,dx < \infty$, expectation and integration can be exchanged. In other words, $\mathrm{E}\left\{\int_{-\infty}^{\infty} f(x)\,dx\right\} = \int_{-\infty}^{\infty} \mathrm{E}\left\{f(x)\right\}dx$.

Hence, the autocovariance function of system output is given by

$$\begin{aligned} C_{yy}(\tau) &= \mathrm{E}\left\{[y(t)-\mu_y][y(t-\tau)-\mu_y]^*\right\} \\ &= \int_{-\infty}^{\infty}\int_{-\infty}^{\infty} \mathrm{E}\left\{[x(t-u_1)-\mu_x][x^*(t-\tau-u_2)-\mu_x^*]\right\} h(u_1)h^*(u_2)\,\mathrm{d}u_1\,\mathrm{d}u_2 \\ &= \int_{-\infty}^{\infty}\int_{-\infty}^{\infty} C_{xx}(\tau-u_1+u_2)h(u_1)h^*(u_2)\,\mathrm{d}u_1\,\mathrm{d}u_2. \end{aligned} \tag{1.4.9}$$

Applying Fourier transform with respect to lag τ to both sides of the equality in Eq. (1.4.9) gives the power spectral density of the system output:

$$P_{yy}(f) = \int_{-\infty}^{\infty}\left[\int_{-\infty}^{\infty}\int_{-\infty}^{\infty} C_{xx}(\tau-u_1+u_2)h(u_1)h^*(u_2)\,\mathrm{d}u_1\,\mathrm{d}u_2\right]\mathrm{e}^{-\mathrm{j}2\pi f\tau}\,\mathrm{d}\tau,$$

which with variable change $\tau' = \tau - u_1 + u_2$ is reduced to

$$\begin{aligned} P_{yy}(f) &= \int_{-\infty}^{\infty} C_{xx}(\tau')\mathrm{e}^{-\mathrm{j}2\pi f\tau'}\,\mathrm{d}\tau' \int_{-\infty}^{\infty} h(u_1)\mathrm{e}^{-\mathrm{j}2\pi fu_1}\,\mathrm{d}u_2 \int_{-\infty}^{\infty} h^*(u_2)\mathrm{e}^{\mathrm{j}2\pi fu_2}\,\mathrm{d}u_2 \\ &= P_{xx}(f)H(f)H^*(f), \end{aligned}$$

or equivalently

$$P_{yy}(f) = P_{xx}(f)|H(f)|^2. \tag{1.4.10}$$

As is revealed by Eq. (1.4.10), for the linear system $H(f)$ excited by random input $x(t)$, the power spectral density $P_{yy}(f)$ of system output is the product of $P_{xx}(f)$, the power spectral density of system input, and $|H(f)|^2$, the squared modulus of system transfer function. The goal of power spectral analysis is to extract the output power spectral density from N observations of system input. We will present a detailed discussion of power spectral analysis and estimation in Chapter 4.

In particular, when system input $x(t)$ with zero mean produces system output $y(t)$ with the same mean of zero, the statistics

$$\mathrm{E}\left\{|y(t)|^2\right\} = R_{yy}(0) = \int_{-\infty}^{\infty} P_{xx}(f)|H(f)|^2\,\mathrm{d}f \tag{1.4.11}$$

gives the average power of the output signal.

Example 1.4.1 Langevin equation and Brown motion. If $y(0) = 0$ and $y(t)$ obeys the following differential equation, known as the Langevin equation,

$$y'(t) + \alpha y(t) = n(t), \quad t \geq 0,$$

$y(t)$ is called Brown motion. $y(t)$ can be viewed as an output of a linear system whose input is $x(t) = n(t)U(t)$ and impulse response is $h(t) = \mathrm{e}^{-\alpha t}U(t)$ with $U(t)$ as the unit step function. $n(t)$ is stationary white noise with zero mean and covariance function $C_{nn}(\tau) = \sigma_n^2\delta(\tau)$. Find the covariance function $C_{yy}(\tau)$ and the average power of $y(t)$.

Solution. As $C_{nn}(\tau) = \sigma_n^2\delta(\tau)$, the power spectral density of the linear system input is $P_{nn}(f) = \sigma_n^2$. The system transfer function system can be derived from the impulse response as

$$H(f) = \int_{-\infty}^{\infty} \mathrm{e}^{-\alpha t}U(t)\mathrm{e}^{-\mathrm{j}2\pi ft}\,\mathrm{d}t = \int_{0}^{\infty} \mathrm{e}^{-(\alpha+\mathrm{j}2\pi f)t}\,\mathrm{d}t = \frac{1}{\alpha + \mathrm{j}2\pi f}.$$

Thus, the power spectral density of output is

$$P_{yy}(f) = P_{nn}|H(f)|^2 = \sigma_n^2\left|\frac{1}{\alpha + \mathrm{j}2\pi f}\right|^2 = \frac{\sigma_n^2}{\alpha^2 + 4\pi^2 f^2},$$

which by Fourier transform yields the covariance function of system output

$$C_{yy}(\tau) = \int_{-\infty}^{\infty} \frac{\sigma_n^2}{\alpha^2 + 4\pi^2 f^2}\mathrm{e}^{\mathrm{j}2\pi ft}\,\mathrm{d}f = \frac{\sigma_n^2}{2\alpha}\mathrm{e}^{-\alpha|\tau|}.$$

As the input is a stochastic process with zero mean, the output of the linear system is also of zero mean and has average energy as

$$\mathrm{E}\left\{|y(t)|^2\right\} = R_{yy}(0) = C_{yy}(0) = \frac{\sigma_n^2}{2\alpha}.$$

□

1.4.2 Narrow Band Bandpass Filter

Consider a narrow band bandpass filter, which has an ideal transfer function

$$H(f) = \begin{cases} 1, & a \le f \le b, \\ 0, & \text{otherwise}, \end{cases} \tag{1.4.12}$$

where $b - a$ takes a small value.

From Eq. (1.4.10), the output power spectral density of the signal $x(t)$ passing through the narrow band bandpass filter can be expressed as

$$P_{yy}(f) = \begin{cases} P_{xx}(f), & a \le f \le b, \\ 0 & \text{otherwise}. \end{cases} \tag{1.4.13}$$

For $x(t)$ with zero mean, the output $y(t)$ is of zero mean and has an average power

$$\mathrm{E}\left\{|y(t)|^2\right\} = \int_{-\infty}^{\infty} P_{yy}(f)\,\mathrm{d}y = \int_{a}^{b} P_{xx}(f)\,\mathrm{d}f. \tag{1.4.14}$$

It can be seen that the signal power is restricted to a narrow band $[a, b]$. The concentration of power is referred to as power localization.

Invoking the result of power localization in Eq. (1.4.14), we can effortlessly prove that the power spectral density of any stationary stochastic process $x(t)$ is non-negative at all frequency f, i.e., $P_{xx}(f) \geq 0$. Applying the narrow band bandpass filter to signal $x(t)$, we have from Eq. (1.4.14) that the average power $\mathrm{E}\left\{|y(t)|^2\right\}$ is always non-negative. So,

$$\mathrm{E}\left\{|x(t)|^2\right\} = \int_{a}^{b} P_{xx}(f)\,\mathrm{d}f \geq 0$$

holds for any a and b, which leads to $P_{xx}(f) \geq 0$.

The white noise is of wide band since it has uniform power throughout the frequency range. But in signal processing, one may encounter noise of a different character, which is of narrow band and is thus called narrow band noise. Clearly, narrow band noise can be viewed as the output of a narrow band bandpass filter with wide band noise as the input.

Formally, random signal $\{x(t)\}$ is called a narrow band noise process if its power spectral density is non-zero within an extremely narrow range of frequency whose bandwidth $\Delta f \leq f_c$ with f_c as the center of the frequency range (or the center frequency for short). That is, the power spectral density has an expression as

$$P_{xx}(f) \begin{cases} \neq 0, & \text{if } f \in \left(\pm f_c - \frac{\Delta f}{2}, \pm f_c + \frac{\Delta f}{2}\right), \\ = 0, & \text{otherwise.} \end{cases} \tag{1.4.15}$$

An alternative expression for the narrow band noise process $x(t)$ is

$$x(t) = x_{\mathrm{I}}(t)\cos(2\pi f_c t) + x_{\mathrm{Q}}(t)\cos(2\pi f_c t), \tag{1.4.16}$$

where $x_{\mathrm{I}}(t)$ and $x_{\mathrm{Q}}(t)$ are orthogonal stationary processes with zero mean. That is,

$$\mathrm{E}\left\{x_{\mathrm{I}}(t)\right\} = 0, \tag{1.4.17}$$
$$\mathrm{E}\left\{x_{\mathrm{Q}}(t)\right\} = 0, \tag{1.4.18}$$
$$\mathrm{E}\left\{x_{\mathrm{I}}(t)x_{\mathrm{Q}}(t-\tau)\right\} = 0, \quad \forall \tau, \tag{1.4.19}$$
$$\mathrm{E}\left\{x_{\mathrm{Q}}(t)x_{\mathrm{I}}(t-\tau)\right\} = 0, \quad \forall \tau. \tag{1.4.20}$$

Due to orthogonality, $x_{\mathrm{I}}(t)$ and $x_{\mathrm{Q}}(t)$ respectively are known as the inphase and quadrature components of narrow band noise $x(t)$. Note that as both $x_{\mathrm{I}}(t)$ and $x_{\mathrm{Q}}(t)$ have zero mean, the orthogonal components of $x_{\mathrm{I}}(t)$ and $x_{\mathrm{Q}}(t)$ are uncorrelated as well.

Define

$$R_{x_I,x_I}(\tau) \stackrel{\text{def}}{=} \mathrm{E}\left\{x_I(t)x_I(t-\tau)\right\},$$
$$R_{x_Q,x_Q}(\tau) \stackrel{\text{def}}{=} \mathrm{E}\left\{x_Q(t)x_Q(t-\tau)\right\}.$$
$$R_{x_I,x_Q}(\tau) \stackrel{\text{def}}{=} \mathrm{E}\left\{x_I(t)x_Q(t-\tau)\right\},$$
$$R_{x_Q,x_I}(\tau) \stackrel{\text{def}}{=} \mathrm{E}\left\{x_Q(t)x_I(t-\tau)\right\}.$$

As both $\cos(2\pi f_c t)$ and $\sin(2\pi f_c t)$ are deterministic, we have

$$\begin{aligned}\mathrm{E}\left\{x(t)\right\} &= \mathrm{E}\left\{x_I(t)\cos(2\pi f_c t)\right\} + \mathrm{E}\left\{x_Q(t)\sin(2\pi f_c t)\right\}\\ &= \mathrm{E}\left\{x_I(t)\right\}\cos(2\pi f_c t) + \mathrm{E}\left\{x_Q(t)\right\}\sin(2\pi f_c t)\\ &= 0+0\\ &= 0,\end{aligned} \tag{1.4.21}$$

and

$$R_{x_I,x_Q}(\tau) = 0, \quad \forall\tau, \tag{1.4.22}$$
$$R_{x_Q,x_I}(\tau) = 0, \quad \forall\tau. \tag{1.4.23}$$

For the narrow band noise process, the correlation function and covariance function are equivalent as a result of zero mean. Then, from the definition of correlation function and based on Eqs. (1.4.19) and (1.4.20), the covariance function of narrow band noise process can be derived as

$$\begin{aligned}C_{xx}(\tau) =& R_{xx}(\tau)\\ =& \mathrm{E}\left\{[x_I(t)\cos(2\pi f_c t) + x_Q(t)\sin(2\pi f_c t)]\right.\\ &\left.\times[x_I(t-\tau)\cos(2\pi f_c(t-\tau)) + x_Q(t-\tau)\sin(2\pi f_c(t-\tau))]\right\}\\ \stackrel{(a)}{=}& \cos(2\pi f_c t)\cos(2\pi f_c(t-\tau))\mathrm{E}\left\{x_I(t)x_I(t-\tau)\right\}\\ &+ \sin(2\pi f_c t)\sin(2\pi f_c(t-\tau))\mathrm{E}\left\{x_Q(t)x_Q(t-\tau)\right\}\\ =& \cos(2\pi f_c t)\cos(2\pi f_c(t-\tau))R_{x_I,x_I}(\tau)\\ &+ \sin(2\pi f_c t)\sin(2\pi f_c(t-\tau))R_{x_Q,x_Q}(\tau),\end{aligned} \tag{1.4.24}$$

where the equality (a) follows from the orthogonality of x_I and x_Q.

It is worth mentioning that in this section we confine the discussion to the second order statistics of the input and output of linear system excited by Gaussian signal and the high order statistics have not been touched upon. When a linear system is excited by non-Gaussian signals, compared with second order statistics, the role of higher-order statistics of the system output will be more prominent. Chapter 6 will be devoted to related topics.

Summary

In this chapter, we first reviewed the basic concept of the random signal along with the definitions and properties of covariance function and power spectral density. From the perspective of the four fundamental statistical relations of independence, uncorrelatedness, orthogonality, and coherence, the comparison and discrimination of two random signals were further discussed. Subsequently, the Gram-Schmidt orthogonalization procedure for the polynomial sequence was introduced. Finally, focusing on the linear system excited by random signals, the statistics of system input and output were analyzed, which shed more insight into the relationship between two random signals.

The basic statistics of random signals described in this chapter will provide a theoretical foundation for topics of random signal processing in succeeding chapters.

Exercises

1.1 A discrete-time random signal is the superposition of two sinusoidal signals:

$$x(t) = A\sin(\omega_1 t) + B\sin(\omega_2 t), \quad \omega_i = 2\pi f_i,\, i = 1, 2,$$

where the amplitudes A and B are independent Gaussian random variables with PDFs

$$f_A(a) = \frac{1}{\sqrt{2\pi}\sigma_1}\mathrm{e}^{-a^2/(2\sigma_1^2)},$$
$$f_B(b) = \frac{1}{\sqrt{2\pi}\sigma_2}\mathrm{e}^{-b^2/(2\sigma_2^2)}.$$

Give the condition for $x(t)$ being strictly stationary.

1.2 The stochastic process expressed by

$$x(t) = \begin{cases} 1\cdot q(t), & \text{with probability } p, \\ -1\cdot q(t), & \text{with probability } (1-p), \end{cases}$$

is called the Bernoulli process. In the expression, $q(t) = [u(t-(n-1)T) - u(t-nT)]$ with integer n and parameter T, and $u(t)$ is the step function, i.e.,

$$u(t) = \begin{cases} 1, & t \geq 0, \\ 0, & t < 0. \end{cases}$$

Find the PDF of $x(t)$.

1.3 Consider the signal

$$x(t) = A\cos(\omega_c t + \pi/2)$$

where $0 \le t \le T$, $f_c = \frac{1}{2T}$, and $\omega_c = 2\pi f_c$ is the carrier frequency. The amplitude of the signal is random, i.e.,

$$A = \begin{cases} 1 \cdot q(t), & \text{with probability } p, \\ -1 \cdot q(t), & \text{with probability } (1-p), \end{cases}$$

where $q(t) = [u(t-(n-1)T) - u(t-nT)]$ with integer n and parameter T, and $u(t)$ is the step function. The signal is termed as amplitude shift keying signal. Find the joint PDF of $x(t)$.

1.4 Let random signal $x(t) = \sin(\alpha t)$, where random variable α has finite fourth order moment, i.e., $\mathrm{E}\{|\alpha|^4\} < \infty$. Find the mean of random variable $m_x = \frac{\mathrm{d}x(t)}{\mathrm{d}t}$.

1.5 The time average of a real stochastic process $\{x(t)\}$ is given by

$$\bar{x} = \frac{1}{T}\int_0^T x(t)\,\mathrm{d}t.$$

Furthermore, we have

$$m(t) = \mathrm{E}\{x(t)\} = \nu\mu, \quad \forall t$$
$$C(t,s) = \mathrm{E}\{[x(t)-m(t)][x(s)-m(s)]\} = \nu\mu \mathrm{e}^{-(t-s)/\mu}, \quad \forall t, s \ge 0.$$

Find $\mathrm{E}\{\bar{x}\}$ and $\mathrm{var}[\bar{x}]$.

1.6 Let harmonic (sinusoidal) signal $x(t) = A\cos(\omega_0 t - \phi)$, where the frequency ω_0 is real and fixed, phase ϕ is a random variable uniformly distributed on $[0, 2\pi]$. Consider the two cases below.

(1) Amplitude A is real and fixed;

(2) Amplitude A is a random variable of Rayleigh distribution, which is independent of ϕ and has PDF

$$f_A(a) = \frac{a}{\sigma^2}\mathrm{e}^{=a^2/(2\sigma^2)}, \quad a \ge 0.$$

Question: is the harmonic signal wide sense stationary in the two cases?

1.7 Prove that the covariance function of the wide sense stationary stochastic process $x(t)$ has the following properties.

$$C_{xx}^{\star}(\tau) = C_{xx}(-\tau),$$
$$|C_{xx}(\tau)| \le C_{xx}(0).$$

1.8 Prove the following properties of the cross correlation and covariance functions of two wide sense stationary stochastic processes:

$$C_{xy}^{\star}(\tau) = C_{yx}(-\tau),$$
$$R_{xy}^{\star}(\tau) = R_{yx}(-\tau),$$
$$|R_{xy}(\tau)| \le R_{xx}(0)R_{yy}(0).$$

1.9 Consider two harmonic signals $x(t)$ and $y(t)$:

$$x(t) = A\cos(\omega_C t + \phi),$$
$$y(t) = B\cos(\omega_C t),$$

where A and ω_C are positive constants, ϕ is a random variable from uniform distribution and has PDF

$$f(\phi) = \begin{cases} \frac{1}{2\pi}, & 0 \le \phi \le 2\pi, \\ 0, & \text{otherwise,} \end{cases}$$

and B is a normal random variable with zero mean and unit variance and has PDF

$$f_B(b) = \frac{1}{\sqrt{2\pi}} e^{-b^2/2}, \quad -\infty < b < \infty.$$

(1) Find the mean $\mu_X(t)$, variance $\sigma_X^2(t)$, autocorrelation function $R_{XX}(\tau)$, and autocovariance function $C_{XX}(\tau)$ of $x(t)$.
(2) If ϕ and B are independent random variables, find the cross correlation function $R_{Xy}(\tau)$ and autocovariance function $C_{Xy}(\tau)$ of $x(t)$ and $y(t)$.

1.10 Let random signal $z(t)$ be the sum of two random signals $x(t)$ and $y(t)$, i.e., $z(t) = x(t) + y(t)$. Suppose that both $x(t)$ and $y(t)$ have zero mean. Find the covariance function $C_{zz}(\tau)$ of random signal $z(t)$.

1.11 Let

$$x(t) = A\cos(2\pi f_c t + \phi) + n(t)$$

where ϕ is a random variable uniformly distributed on $[-\pi, \pi]$ and has PDF

$$f(\phi) = \begin{cases} \frac{1}{2\pi}, & -\pi \le \phi < \pi, \\ 0 & \text{otherwise,} \end{cases}$$

and $n(t)$ is stationary Gaussian noise with zero mean and has power spectral density

$$P_n(f) = \begin{cases} \frac{N_0}{2}, & |f - f_c| \le B/2, \\ 0, & \text{otherwise.} \end{cases}$$

ϕ and $n(t)$ are independent. Have $x(t)$ as the input of a "square-law circuit" and $y(t) = x^2(t)$ as the output. Find the mean and autocorrelation function of output signal $y(t)$.

Hint: the third order moment of zero-mean Gaussian random process is identical to zero, i.e., $\mathrm{E}\left\{n(t)n^2(t-\tau)\right\} = \mathrm{E}\left\{n^2(t)n(t-\tau)\right\} = 0, \forall\tau$.

1.12 Random signal $x(t)$ has zero mean and power spectral density

$$P_X(f) = \begin{cases} \frac{\sigma^2}{B}, & -\frac{B}{2} \le f \le \frac{B}{2}, \\ 0, & \text{otherwise,} \end{cases}$$

where $\sigma^2 > 0$. Find the autocorrelation function and power of the signal.

1.13 Given the joint PDF of random variables x and y

$$f(x,y) = \alpha \exp\left\{-\frac{1}{2(1-r^2)}\left(\frac{(x-\mu_x)^2}{\sigma_x^2} - 2r\frac{(x-\mu_x)(y-\mu_y)}{\sigma_x\sigma_y} + \frac{(y-\mu_y)^2}{\sigma_y^2}\right)\right\},$$

where r is the correlation coefficient of x and y, i.e.,

$$r = \frac{\mathrm{E}\left\{(x-\mu_x)(y-\mu_y)\right\}}{\sigma_x\sigma_y},$$

prove the following results.

(1) The marginal PDF of x and y are

$$f(x) = \frac{1}{\sqrt{2\pi}\sigma_x}\exp\left(-\frac{(x-\mu_x)^2}{2\sigma_x^2}\right),$$

$$f(y) = \frac{1}{\sqrt{2\pi}\sigma_y}\exp\left(-\frac{(y-\mu_y)^2}{2\sigma_y^2}\right),$$

which shows that both x and y are normal random variables.

(2) If random variables x and y are uncorrelated, they are also independent.

1.14 Let

$$y(t) = \int_{-\infty}^{\infty} h(u)x(t-u)\,\mathrm{d}u.$$

Prove that for the system to be stable, the impulse response should be absolute integrable, i.e.,

$$\int_{-\infty}^{\infty} |h(t)|\,\mathrm{d}t < \infty.$$

1.15 The linearly independent polynomial sequence $\{f_i(x)\}$ is composed of polynomials $f_1(x) = 1, f_2(x) = x$, and $f_3(x) = x^2$ with x taking value on interval $[-1, 1]$.

(1) Use Gram-Schmidt orthogonalization algorithm

$$\phi_k = \frac{f_x - \sum_{i=1}^{k-1}\langle f_k, \phi_i\rangle \phi_i}{\left\| f_x - \sum_{i=1}^{k-1}\langle f_k, \phi_i\rangle \phi_i \right\|}, \quad k = 1, \cdots, n,$$

to transform $\{f_i(x)\}$ into orthogonal sequence $\{\phi_i(x)\}$;

(2) Use Gram-Schmidt orthogonalization algorithm in the form of matrix norm

$$d_k = \left\| \begin{matrix} \langle f_1, f_1\rangle & \langle f_1, f_2\rangle & \cdots & \langle f_1, f_k\rangle \\ \langle f_2, f_1\rangle & \langle f_2, f_2\rangle & \cdots & \langle f_2, f_k\rangle \\ \vdots & \vdots & \vdots & \vdots \\ \langle f_{k-1}, f_1\rangle & \langle f{-1k}, f_2\rangle & \cdots & \langle f_{k-1}, f_k\rangle \\ f_1 & f_2 & \cdots & f_k \end{matrix} \right\|,$$

$$\phi_k = \frac{d_k}{\langle d_k, d_k\rangle^{1/2}}, \quad k = 1, \cdots, n,$$

to transform $\{f_i(x)\}$ into sequence $\{\phi_i(x)\}$, and verify that the sequence $\{\phi_i(x)\}$ is orthonormal.

1.16 Use finite series expansion

$$\hat{x}(t) = \sum_{i=1}^{m} c_i \phi_i(t)$$

to approximate signal $x(t)$. Suppose that the continuous-time basis functions $\phi_1(t)$, $\cdots$, $\phi_m(t)$ are known. Find the coefficients $c_1, \cdots, c_m$.

1.17 Use finite series expansion

$$\hat{x}(t) = \sum_{i=1}^{m} c_i \phi_i(t)$$

to approximate signal $x(t)$. Suppose that the values of basis functions $\phi_1(t_k)$. $\cdots$, $\phi_m(t_k)$ at discrete time $k = 1, \cdots, N$ are known. Find the coefficients $c_1, \cdots, c_m$.

2 Parameter Estimation Theory

The fundamental task of signal processing is to make a statistical decision related to the characteristics of signals and (or) systems based on observational data. The statistical decision theory is mainly concerned with two large categories of problems: hypothesis testing and estimation. Signal detection and radar maneuvering target detection are typical problems of hypothesis testing. Estimation theory has a scope much wider covering both parametric and non-parametric methods. The parametric method assumes that the data are generated from some probabilistic model which has a known structure but unknown parameters. Close related to system identification, the parametric method has optimization theory as its basis, including the criteria under which the optimality of the estimated parameter can be established and the approaches by which the optimal parameter estimation can be reached. Contrary to the parametric method, the non-parametric method does not hold the assumption that the data are from a certain probabilistic model. Power spectral density estimation and high order spectral estimation based on discrete Fourier transform are examples of the non-parametric method.

Regarding signal processing, we have classic and modern signal processing. Known as non-parametric signal processing, classic signal processing relies on Fourier transform with no reference to the system generating signals. Modern signal processing, also called parametric signal processing, treats signals as the output of excited systems and largely employs methods that estimate the model parameters of systems and signals. Therefore, before the introduction of theories, methods, applications of modern signal processing, it is necessary to first present the general theory of parameter estimation as a unified basis and framework.

2.1 Performance of Estimators

The problem parameter estimation theory concerns how to identify the exact probability distribution of random variable x whose presumed cumulative distribution function is in some given family of distributions. Now, suppose we perform an experiment that keeps records of realizations, also known as samples or observations, of random variables, and expect to guess from N samples $x_1, \cdots, x_n$ the parameter θ which determines the probability distribution of x. For example, let $x_1, \cdots, x_N$ be N samples drawn from normal distribution $N(\theta, \sigma^2)$ where the mean parameter θ is to be estimated from the samples. Undoubtedly there exist various functions of the data which can be adopted to estimate θ. Among the functions, using the first sample x_1 as the estimate of θ is the most straightforward. Although x_1 has a mean equal to θ, it is evident that the estimate of θ from averaging more samples would be much better than the one using x_1 alone. We may further conjecture that the sample mean $\bar{x}_N = \frac{1}{N}\sum_{i=1}^{N} x_i$

https://doi.org/10.1515/9783110475562-002

could be the optimal estimate of θ. In parameter estimation theory, the estimate or the estimated value of actual parameter θ is normally called the estimator of θ. An estimator is a statistic and in a sense is the best approximation of the parameter. Then, how to evaluate or measure the proximity between an estimator and the true parameter? Further, how to estimate the proximity? The study of these problems constitutes the two key subjects of parameter estimation theory:

(1) Give quantitative definition of the proximity between the estimator and true value.
(2) Study different estimation methods and compare their performance.

2.1.1 Unbiased and Asymptotic Unbiased Estimation

Above all, we give the definition of estimator.

Definition 2.1.1. *The estimator of true parameters $\theta_1, \cdots, \theta_p$ from N samples is a function T that maps the N-dim sample space $\mathcal{X}^N$ to the p-dim parameter space Θ, i.e., $T : \mathcal{X}^N \to \Theta$.*

For convenience, we only treat the case of $p = 1$. The estimator $T(x_1, \cdots, x_N)$ of θ is usually denoted as $\hat{\theta}$ for simplicity. As an approximate to the parameter θ, the estimator $\hat{\theta}$ is thus expected to have proper proximity. The simplest measure of proximity is the error $\hat{\theta} - \theta$ of the estimator $\hat{\theta}$. Since the N samples observed under varied circumstances are random variables and the estimate of θ is random as well, the estimation error is a random variable. Obviously, using random variable for evaluation can be rather difficult. Therefore, the estimation error should be made into non-random quantity.

Definition 2.1.2. *The bias of estimator $\hat{\theta}$ for parameter θ is defined as the expectation of estimation error, i.e.,*

$$b(\hat{\theta}) \stackrel{\text{def}}{=} \mathrm{E}\left\{\hat{\theta} - \theta\right\} = \mathrm{E}\left\{\hat{\theta}\right\} - \theta. \tag{2.1.1}$$

The estimator $\hat{\theta}$ is said to be unbiased if the bias $b(\hat{\theta})$ is zero or $\mathrm{E}\left\{\hat{\theta}\right\} = \theta$, that is, the expectation of the estimator equals the true parameter.

Example 2.1.1 Let $x(1), \cdots, x(N)$ be N independent samples of random signal $x(n)$,

$$\bar{x} = \frac{1}{N}\sum_{n=1}^{N} x(n) \tag{2.1.2}$$

be the estimate of mean of $x(n)$ obtained from samples. Taking the expectation of $\bar{x}$, we have

$$\mathrm{E}\{\bar{x}\} = \mathrm{E}\left\{\frac{1}{N}\sum_{n=1}^{N} x(n)\right\} = \frac{1}{N}\sum_{n=1}^{N}\mathrm{E}\{x(n)\} = \frac{1}{N}\sum_{n=1}^{N} m_x = m_x,$$

where $m_x = \mathrm{E}\{x(n)\}$ is the mean of $x(n)$. Hence, the estimate in Eq. (2.1.2) is an unbiased estimate of the mean of $x(n)$. With the mean estimate $\bar{x}$, we have

$$\operatorname{var}[x] = \frac{1}{N}\sum_{n=1}^{N}[x(n) - \bar{x}]^2,$$

which is the estimate of the variance of $x(n)$.

The estimation that is not unbiased is called biased estimation. Being unbiased is an important feature that we expect an estimator to have. Nevertheless, it does not mean that the biased estimate is worthless. As a matter of fact, biased estimate that is asymptotically unbiased can still be "good" and even be better than unbiased ones.

Definition 2.1.3. *Estimator $\hat{\theta}$ of true parameter θ is asymptotically unbiased if the bias $b(\hat{\theta}) \to 0$ as the sample size $N \to \infty$. That is,*

$$\lim_{N\to\infty} \mathrm{E}\left\{\hat{\theta}_N\right\} = \theta, \tag{2.1.3}$$

where $\hat{\theta}_N$ is the estimator obtained from N samples.

Notice that an unbiased estimator must be asymptotically unbiased but an asymptotically unbiased estimator is not necessarily biased.

Example 2.1.2 As a typical example, consider the following two estimators of the autocorrelation function of real random signal $x(n)$

$$\hat{R}_1(\tau) = \frac{1}{N-\tau}\sum_{n=1}^{N-\tau} x(n)x(n+\tau), \tag{2.1.4}$$

$$\hat{R}_2(\tau) == \frac{1}{N}\sum_{n=1}^{N-\tau} x(n)x(n+\tau). \tag{2.1.5}$$

Suppose the samples of $x(n)$ are independent. It is easy to verify that

$$\mathrm{E}\left\{\hat{R}_1(\tau)\right\} = \mathrm{E}\left\{\frac{1}{N-\tau}\sum_{n=1}^{N-\tau} x(n)x(n+\tau)\right\} = \frac{1}{N-\tau}\sum_{n=1}^{N-\tau}\mathrm{E}\{x(n)x(n+\tau)\} = R_x(\tau), \tag{2.1.6}$$

$$\mathrm{E}\left\{\hat{R}_2(\tau)\right\} = \mathrm{E}\left\{\frac{1}{N}\sum_{n=1}^{N-\tau} x(n)x(n+\tau)\right\} = \frac{1}{N}\sum_{n=1}^{N-\tau}\mathrm{E}\{x(n)x(n+\tau)\} = \left(1-\frac{\tau}{N}\right)R_x(\tau), \tag{2.1.7}$$

where $R_x(\tau) = \mathrm{E}\{x(n)x(n+\tau)\}$ is the true correlation function of random signal $x(n)$. As is demonstrated by Eqs. (2.1.6) and (2.1.7), the estimator $\hat{R}_1(\tau)$ is unbiased whereas $\hat{R}_2(\tau)$ is biased. However, $\hat{R}_2(\tau)$ is asymptotically unbiased, according to Eq. (2.1.7) which gives

$$\lim_{N\to\infty}\mathrm{E}\left\{\hat{R}_2(\tau)\right\} = R_x(\tau).$$

The asymptotically unbiased estimator $\hat{R}_2(\tau)$ is positive semidefinite while the unbiased estimator $\hat{R}_1(\tau)$ is indefinite. As it is desirable to have the feature of positive semidefiniteness in many signal processing applications, the biased but asymptotically unbiased estimator $\hat{R}_2(\tau)$ is much preferred by researchers in comparison with the scarce usage of the unbiased $\hat{R}_1(\tau)$.

Bias is the expected value of error. But a zero bias does not guarantee that a small error of estimator occurs with high probability. The notion that reflects the probability of small estimator error is consistency.

Definition 2.1.4. *The estimator $\hat{\theta}$ of parameter θ is said to be consistent with θ in probability if the estimator converges to the true parameter θ with probability one as $N \to \infty$. That is,*

$$\hat{\theta} \xrightarrow{p} \theta, \quad \text{when } N \to \infty, \tag{2.1.8}$$

where $\xrightarrow{p}$ stands for convergence in probability.

2.1.2 Effectiveness of Estimators

Unbiasedness, asymptotical unbiasedness, and consistency are statistical properties an estimator is expected to have. The latter two are related to the behavior of an estimator when sample size goes to infinity and are known as large sample properties which involve complicated theoretical analysis. When it comes to small data set with N samples which is more frequent in practice, how to assess the performance of estimators is an issue to be addressed.

1. Comparison of Two Unbiased Estimators.

If $\hat{\theta}_1$ and $\hat{\theta}_2$ are two unbiased estimators obtained from N samples, we prefer the one with a smaller variance. For example, suppose $\hat{\theta}_1$ has variance larger than $\hat{\theta}_2$, i.e., $\text{var}[\hat{\theta}_1] > \text{var}[\hat{\theta}_2]$. This means that values of $\hat{\theta}_2$ are more concentrated than those of $\hat{\theta}_1$ around the true parameter θ. In other words, the probability of $\hat{\theta}_2$ being in the region $(\theta - \epsilon, \theta + \epsilon)$ is higher than that of $\hat{\theta}_1$. Thus, $\hat{\theta}_2$ is said to be more effective than $\hat{\theta}_1$. As the widely-used measure of the effectiveness of two estimators, the relative effectiveness of $\hat{\theta}_2$ with respect to $\hat{\theta}_1$ is defined as

$$\text{RE} = \left[\frac{\text{var}[\hat{\theta}_1]}{\text{var}[\hat{\theta}_2]} \times 100 \right] \%. \tag{2.1.9}$$

For instance, if $\text{var}[\hat{\theta}_1] = 1.25, \text{var}[\hat{\theta}_2]$, RE = 125%.

2. Comparison of Unbiased and Asymptotically Unbiased Estimators.

Roughly speaking, any estimator that is not asymptotically unbiased (note that an unbiased estimator must be asymptotically unbiased) is not a "good" estimator. This means the missing of asymptotic unbiasedness for any estimator is regarded as a serious flaw. Suppose that one of $\hat{\theta}_1$ and $\hat{\theta}_2$ is unbiased and another is asymptotically unbiased, or both estimators are unbiased. Under such circumstances, variance is not the elusive measure that is fit for evaluating the effectiveness of estimators. Consider the case where $\hat{\theta}_1$ has larger bias but smaller variance compared to $\hat{\theta}_2$. In such case, how should one choose between $\hat{\theta}_1$ and $\hat{\theta}_2$? A reasonable solution is to take both bias and variance into account and introduce the mean square error of estimator.

Definition 2.1.5. *The mean square error $M^2(\hat{\theta})$ of estimator $\hat{\theta}$ for the parameter θ is defined as the expectation of the squared error between the estimator and true parameter. That is,*

$$M^2(\hat{\theta}) = \mathrm{E}\left\{(\hat{\theta} - \theta)^2\right\}. \tag{2.1.10}$$

It can be derived from the definition that

$$\begin{aligned} M^2(\hat{\theta}) &= \mathrm{E}\left\{\left[\hat{\theta} - \mathrm{E}\left\{\hat{\theta}\right\} + \mathrm{E}\left\{\hat{\theta}\right\} - \theta\right]^2\right\} \\ &= \mathrm{E}\left\{\left[\hat{\theta} - \mathrm{E}\left\{\hat{\theta}\right\}\right]^2\right\} + \mathrm{E}\left\{\left[\mathrm{E}\left\{\hat{\theta}\right\} - \theta\right]^2\right\} + 2\mathrm{E}\left\{\left[\hat{\theta} - \mathrm{E}\left\{\hat{\theta}\right\}\right]\left[\mathrm{E}\left\{\hat{\theta}\right\} - \theta\right]\right\}, \end{aligned} \tag{2.1.11}$$

where $\mathrm{var}[\hat{\theta}] = \mathrm{E}\{[\hat{\theta}-\mathrm{E}\{\hat{\theta}\}]^2\}$ is the variance of estimator $\hat{\theta}$. Note that $\mathrm{E}\{\hat{\theta}-\theta\} = \mathrm{E}\{\hat{\theta}\}-\theta$ is constant, it can be derived that $\mathrm{E}\{[\mathrm{E}\{\hat{\theta}\} - \theta]^2\} = [\mathrm{E}\{\hat{\theta}\} - \theta]^2$ is the square of bias $\mathrm{E}\{\hat{\theta}\} - \theta$ and $\mathrm{E}\{[\hat{\theta} - \mathrm{E}\{\hat{\theta}\}][\mathrm{E}\{\hat{\theta} - \theta\}]\} = [\mathrm{E}\{\hat{\theta}\} - \theta]\mathrm{E}\{\hat{\theta} - \mathrm{E}\{\hat{\theta}\}\}$. Substitute the results into Eq. (2.1.11) to get

$$M^2(\hat{\theta}) = \mathrm{var}[\hat{\theta}] + b^2(\hat{\theta}) + 2\left[\mathrm{E}\left\{\hat{\theta}\right\} - \theta\right]\mathrm{E}\left\{\hat{\theta} - \mathrm{E}\left\{\hat{\theta}\right\}\right\}.$$

Due to $\mathrm{E}\{\hat{\theta} - \mathrm{E}\{\hat{\theta}\}\} = \mathrm{E}\{\hat{\theta}\} - \mathrm{E}\{\hat{\theta}\} = 0$, the above equation is reduced to

$$M^2(\hat{\theta}) = \mathrm{var}[\hat{\theta}] + b^2(\hat{\theta}), \tag{2.1.12}$$

which reveals that the mean square error $\mathrm{E}\{(\hat{\theta} - \theta)^2\}$ of estimator $\hat{\theta}$ is the sum of variance $\mathrm{E}\{[\hat{\theta} - \mathrm{E}\{\hat{\theta}\}]^2\}$ and square of bias $\mathrm{E}\{\hat{\theta} - \theta\}$. Notably, when both estimators are unbiased, the mean square errors of the estimators are reduced to their respective variances owing to zero bias.

In summary, as the loss function (or cost function) for the error of estimator, the mean square error is more appropriate than either variance or bias. According to their mean square errors, different estimators of θ can be evaluated and their performances can be compared.

Definition 2.1.6. *Estimator $\hat{\theta}_1$ is said to be better than estimator $\hat{\theta}_2$ if the inequality*

$$\mathrm{E}\left\{(\hat{\theta}_1 - \theta)^2\right\} \le \mathrm{E}\left\{(\hat{\theta}_2 - \theta)^2\right\} \tag{2.1.13}$$

holds for any θ.

The concept of effectiveness is mostly useful when comparing the performance of two estimators, but gives no definite answer to the question that whether an estimator is the best among all possible estimators. To resolve the question, it is necessary to consider whether an unbiased estimator of parameter θ has the minimum variance, which will be the focus of the next section.

2.2 Fisher Information and Cramér-Rao Inequality

Suppose that the parameter θ is hidden behind the random signal $x(t)$. One realization x of the signal gives an estimator of θ. The natural question then arises: is this estimator optimal? In fact, an equivalent question is: given actual parameter θ, which criterion should be used to determine the optimal estimator obtained from the observation x.

2.2.1 Fisher Information

To answer the above question, one might as well treat x as a random variable and assess the quality of conditional PDF $f(x|\theta)$. The measure for such assessment is called score function of random variable x.

Definition 2.2.1. *For given actual parameter θ, the score function $V(x)$ of random variable x is defined as the partial derivative of the log conditional PDF $\log f(x|\theta)$ with respect to the parameter θ. That is,*

$$V(x) = \frac{\partial}{\partial\theta} \log f(x|\theta) = \frac{\frac{\partial}{\partial\theta} f(x|\theta)}{f(x|\theta)}. \tag{2.2.1}$$

Definition 2.2.2. *The variance of the score function, denoted by $J(\theta)$, is called Fisher information and is defined by*

$$J(\theta) = \mathrm{E}\left\{\left[\frac{\partial}{\partial\theta} \log f(x|\theta)\right]^2\right\} = -\mathrm{E}\left\{\frac{\partial^2}{\partial\theta\partial\theta} \log f(x|\theta)\right\}. \tag{2.2.2}$$

According to probability theory, the expectation of any function $g(x)$ can be expressed in the conditional PDF of x as

$$\mathrm{E}\{g(x)\} = \int_{-\infty}^{\infty} g(x) f(x|\theta)\,\mathrm{d}x. \tag{2.2.3}$$

Substitute Definition 2.2.1 into Eq. (2.2.3) to have the expectation of score function as

$$\mathrm{E}\{V(x)\} = \int_{-\infty}^{\infty} \frac{\frac{\partial}{\partial\theta} f(x|\theta)}{f(x|\theta)} f(x|\theta)\,\mathrm{d}x = \frac{\partial}{\partial\theta} \int_{-\infty}^{\infty} f(x|\theta)\,\mathrm{d}x = 0,$$

where we have used the familiar result in probability theory

$$\int_{-\infty}^{\infty} f(x|\theta)\,\mathrm{d}x = 1.$$

Due to zero expectation, the second moment of the score function is the variance, i.e., $\text{var}\left[V(x)\right] = \mathrm{E}\left\{V^2(x)\right\}$, which plays a central role in evaluating unbiased estimators.

Now consider the case of N samples $x_1, \cdots, x_N$. Let sample vector $\mathbf{x} = [x_1, \cdots, x_N]$ and the conditional PDF can be expressed as

$$f(\mathbf{x}|\theta) = f(x_1, \cdots, x_N|\theta).$$

Therefore, the Fisher information for N random samples $x_1, \cdots, x_N$ should be defined by

$$J(\theta) = \mathrm{E}\left\{\left[\frac{\partial}{\partial\theta}\log f(\mathbf{x}|\theta)\right]^2\right\} = -\mathrm{E}\left\{\frac{\partial^2}{\partial\theta\partial\theta}\log f(\mathbf{x}|\theta)\right\}. \tag{2.2.4}$$

2.2.2 Cramér-Rao Lower Bound

The significance of Fisher information is revealed by the following theorem.

Theorem 2.2.1 (Cramér-Rao Inequality). *Let $x = [x_1, \cdots, x_N]$ be the sample vector. If the estimate $\hat{\theta}$ of actual parameter θ is unbiased, and both $\frac{\partial f(\mathbf{x}|\theta)}{\partial\theta}$ and $\frac{\partial^2 f(\mathbf{x}|\theta)}{\partial\theta^2}$ exist, the lower bound, aka Crameér-Rao bound, of mean square error of $\hat{\theta}$ is the inverse of Fisher information. That is,*

$$\text{var}\left[\hat{\theta}\right] = \mathrm{E}\left\{(\hat{\theta} - \theta)^2\right\} \ge \frac{1}{J(\theta)}, \tag{2.2.5}$$

where Fisher information $J(\theta)$ is defined in Eq. (2.2.2). *In Eq.* (2.2.5), *the equality holds if and only if*

$$\frac{\partial}{\partial\theta}\log f(x|\theta) = K(\theta)(\hat{\theta} - \theta), \tag{2.2.6}$$

where $K(\theta)$ is a positive function of θ and does not depend on samples $x_1, \cdots, x_N$.

Proof. Based on the assumption that $\mathrm{E}\{\hat{\theta}\} = \theta$ or $\mathrm{E}\{\hat{\theta} - \theta\} = 0$, we have

$$\mathrm{E}\left\{\hat{\theta} - \theta\right\} = \int_{-\infty}^{\infty} \cdots \int_{-\infty}^{\infty} (\hat{\theta} - \theta) f(\mathbf{x}|\theta)\,\mathrm{d}x_1 \cdots \mathrm{d}x_N = 0.$$

Taking partial derivative of both sides of the above equation with respect to θ gives

$$\frac{\partial}{\partial\theta}\mathrm{E}\left\{\hat{\theta} - \theta\right\} = \frac{\partial}{\partial\theta}\int_{-\infty}^{\infty} (\hat{\theta} - \theta) f(\mathbf{x}|\theta)\,\mathrm{d}\mathbf{x} = \int_{-\infty}^{\infty} \frac{\partial}{\partial\theta}\left[(\hat{\theta} - \theta) f(\mathbf{x}|\theta)\right]\mathrm{d}\mathbf{x} = 0,$$

which further yields

$$-\int_{-\infty}^{\infty} f(\mathbf{x}|\theta)\,\mathrm{d}\mathbf{x} + (\hat{\theta} - \theta)\int_{-\infty}^{\infty} \frac{\partial}{\partial\theta} f(\mathbf{x}|\theta)\,\mathrm{d}\mathbf{x} = 0. \tag{2.2.7}$$

On the other hand, from the derivative of compound function, we have

$$\frac{\partial}{\partial\theta}f(\mathbf{x}|\theta) = \left[\frac{\partial}{\partial\theta}\log f(\mathbf{x}|\theta)\right] f(\mathbf{x}|\theta). \tag{2.2.8}$$

As $f(\mathbf{x}|\theta)$ is conditional PDF of $\mathbf{x}$, we get

$$\int_{-\infty}^{\infty} f(\mathbf{x}|\theta)\,\mathrm{d}\mathbf{x} = 1. \tag{2.2.9}$$

Substitute Eqs. (2.2.8) and (2.2.9) into Eq. (2.2.7) to get

$$\int_{-\infty}^{\infty}\left[\frac{\partial}{\partial\theta}\log f(\mathbf{x}|\theta)\right] f(\mathbf{x}|\theta)(\hat{\theta}-\theta)\,\mathrm{d}\mathbf{x} = 1,$$

or equivalently

$$\int_{-\infty}^{\infty}\left[\frac{\partial}{\partial\theta}\log f(\mathbf{x}|\theta)\sqrt{f(\mathbf{x}|\theta)}\right]\left[\sqrt{f(\mathbf{x}|\theta)}(\hat{\theta}-\theta)\right]\mathrm{d}\mathbf{x} = 1. \tag{2.2.10}$$

From Cauchy-Schwartz inequality, for any two complex functions $f(x)$ and $g(x)$, the inequality

$$\left|\int_{-\infty}^{\infty} f(x)g(x)\,\mathrm{d}x\right|^2 \le \int_{-\infty}^{\infty}|f(x)|^2\,\mathrm{d}x\int_{-\infty}^{\infty}|g(x)|^2\,\mathrm{d}x \tag{2.2.11}$$

always holds, and the equality holds if and only if $f(x) = cg^*(x)$. Apply Cauchy-Schwartz inequality to Eq. (2.2.10) to get

$$\int_{-\infty}^{\infty}\left[\frac{\partial}{\partial\theta}\log f(\mathbf{x}|\theta)\right]^2 f(\mathbf{x}|\theta)\,\mathrm{d}\mathbf{x}\int_{-\infty}^{\infty}(\hat{\theta}-\theta)^2 f(\mathbf{x}|\theta)\,\mathrm{d}\mathbf{x} \ge 1,$$

or equivalently

$$\int_{-\infty}^{\infty}(\hat{\theta}-\theta)^2 f(\mathbf{x}|\theta)\,\mathrm{d}\mathbf{x} \ge \frac{1}{\int_{-\infty}^{\infty}\left[\frac{\partial}{\partial\theta}\log f(\mathbf{x}|\theta)\right]^2 f(\mathbf{x}|\theta)\,\mathrm{d}\mathbf{x}}. \tag{2.2.12}$$

In addition, from the condition for equality in Cauchy-Schwartz inequality, the equality in Eq. (2.2.12) holds if and only if $\frac{\partial}{\partial\theta}\log f(\mathbf{x}|\theta)\sqrt{f(\mathbf{x}|\theta)} = K(\theta)(\hat{\theta}-\theta)\sqrt{f(\mathbf{x}|\theta)}$. In other words, only when Eq. (2.2.6) is satisfied, the equality in Eq. (2.2.12) is achieved.

Notice that $\mathrm{E}\{\hat{\theta}\} = \theta$ which gives

$$\mathrm{var}\left[\hat{\theta}\right] = \mathrm{E}\left\{(\hat{\theta}-\theta)^2\right\} = \int_{-\infty}^{\infty}(\hat{\theta}-\theta)^2 f(\mathbf{x}|\theta)\,\mathrm{d}\mathbf{x}. \tag{2.2.13}$$

Also by Eq. (2.2.3), we have

$$\mathrm{E}\left\{\left[\frac{\partial}{\partial\theta}\log f(\mathbf{x}|\theta)\right]^2\right\} = \int_{-\infty}^{\infty}\left[\frac{\partial}{\partial\theta}\log f(\mathbf{x}|\theta)\right]^2 f(\mathbf{x}|\theta)\,\mathrm{d}\mathbf{x}. \tag{2.2.14}$$

The inequality in Eq. (2.2.5) directly follows from substituting Eqs. (2.2.13) and (2.2.14) into Eq. (2.2.12), and Eq. (2.2.6) is both necessary and sufficient to achieve the equality in Eq. (2.2.5). □

Cramér-Rao lower bound provides the minimum variance that any unbiased estimator can achieve and can be used to determine the most efficient estimator.

Definition 2.2.3. *Unbiased estimator $\hat{\theta}$ is called most efficient if its variance attains Cramér-Rao lower bound, i.e., var$[\hat{\theta}] = 1/J(\theta)$.*

For biased estimator $\hat{\theta}$, Cramér-Rao inequality is

$$\mathrm{E}\left\{(\hat{\theta}-\theta)^2\right\} \geq \frac{\left(1+\frac{\mathrm{d}b(\theta)}{\mathrm{d}\theta}\right)^2}{\mathrm{E}\left\{\left[\frac{\partial}{\partial\theta}\log f(\mathbf{x}|\theta)\right]^2\right\}}, \tag{2.2.15}$$

where $b(\theta)$ is the bias of estimator $\hat{\theta}$ satisfying $\mathrm{E}\{\hat{\theta}\} = \theta + b(\theta)$, and is assumed to be differentiable in θ.

Fisher information measures the amount of information about θ that can be retrieved from observations, and for estimating parameter θ from observed data, it also determines the lower bound of the variance of an estimator. However, it is noteworthy that the estimator achieving the lower bound may not exist.

In the case of multiple parameters $\theta_1, \cdots, \theta_p$, let $\boldsymbol{\theta} = [\theta_1, \cdots, \theta_p]^{\mathrm{T}}$. Fisher information then becomes Fisher information matrix $\mathbf{J}(\boldsymbol{\theta})$ whose element $J_{ij}(\boldsymbol{\theta})$ is given by

$$J_{ij}(\boldsymbol{\theta}) = -\int f(x|\boldsymbol{\theta})\frac{\partial^2 \log f(x|\boldsymbol{\theta})}{\partial\theta_i\partial\theta_j}\,\mathrm{d}x = -\mathrm{E}\left\{\frac{\partial^2 \log f(x|\boldsymbol{\theta})}{\partial\theta_i\partial\theta_j}\right\}, \tag{2.2.16}$$

and Cramér-Rao inequality becomes the matrix inequality

$$\boldsymbol{\Sigma} \geq \mathbf{J}^{-1}(\boldsymbol{\theta}), \tag{2.2.17}$$

where $\boldsymbol{\Sigma}$ is the covariance matrix of p unbiased estimators $\hat{\theta}_1, \cdots, \hat{\theta}_p$, $\mathbf{J}^{-1}(\boldsymbol{\theta})$ is the inverse of Fisher information matrix $\mathbf{J}(\boldsymbol{\theta})$, and the matrix inequality $\boldsymbol{\Sigma} \geq \mathbf{J}^{-1}(\boldsymbol{\theta})$ means that the matrix $\boldsymbol{\Sigma} - \mathbf{J}^{-1}(\boldsymbol{\theta})$ is positive semidefinite.

2.3 Bayes Estimation

The quality of an estimator is determined by what criterion and method are adopted for parameter estimation. As for the parameter estimation method, we have two categories:

one merely fits specific problems, and another one, in contrast, is applicable to a large body of problems. Since the estimation techniques of the former category are too narrow to be of any generic guidance, we only concentrate upon methods from the latter category in this book. Actually, it turns out that the parameter estimation methods for general problems are few and from this section we will begin to introduce them one by one. Note that the well-known moment method is of the latter category, but we will not give it much attention in the book and would like to refer the readers to [176] for the description of the method.

2.3.1 Definition of Risk Function

The estimation $\hat{\theta}$ of parameter θ typically has non-zero estimation error $\theta - \hat{\theta}$. Therefore, the extent to which the estimation error approaches zero directly affects the quality of estimation $\hat{\theta}$, and apart from measures such as bias, variance, and mean square error, which were mentioned in previous sections, other measures of estimation error reflecting the error range can also be employed. Such measures are collectively termed as cost functions or loss functions, which are denoted by $C(\hat{\theta}, \theta)$.

Definition 2.3.1. *Let θ be the parameter in the parameter space Θ, and $\hat{\theta}$ be the estimation taking values in the decision space A. $C(\hat{\theta}, \theta)$ is called loss function or cost function if $C(\hat{\theta}, \theta)$ is a real function in $\hat{\theta}$ and θ and satisfies the following two conditions:*
(1) *$C(\hat{\theta}, \theta) \geq 0$ for any $\hat{\theta} \in A$ and $\theta \in \Theta$;*
(2) *there exists at least one $\hat{\theta}$ in the decision space A for each $\theta \in \Theta$ such that $C(\hat{\theta}, \theta) = 0$.*

Followings are three common loss functions, where $C(\hat{\theta}, \theta)$ is the loss function for scalar parameter estimation $\hat{\theta}$, and $C(\hat{\boldsymbol{\theta}}, \boldsymbol{\theta})$ is the loss function for vector parameter estimation $\hat{\boldsymbol{\theta}}$.
(1) Absolute loss function

$$C(\hat{\theta}, \theta) = |\hat{\theta} - \theta|, \quad \text{(scalar parameter)} \tag{2.3.1}$$

$$C(\hat{\boldsymbol{\theta}}, \boldsymbol{\theta}) = \|\hat{\boldsymbol{\theta}} - \boldsymbol{\theta}\|, \quad \text{(vector parameter)} \tag{2.3.2}$$

where $\|\hat{\boldsymbol{\theta}} - \boldsymbol{\theta}\|$ represents the norm of estimation error vector $\hat{\boldsymbol{\theta}} - \boldsymbol{\theta}$.
(2) Quadratic loss function

$$C(\hat{\theta}, \theta) = |\hat{\theta} - \theta|^2, \quad \text{(scalar parameter)} \tag{2.3.3}$$

$$C(\hat{\boldsymbol{\theta}}, \boldsymbol{\theta}) = \|\hat{\boldsymbol{\theta}} - \boldsymbol{\theta}\|^2. \quad \text{(vector parameter)} \tag{2.3.4}$$

(3) Uniform loss

$$C(\hat{\theta}, \theta) = \begin{cases} 0, & |\hat{\theta} - \theta| < \Delta, \\ 1, & |\hat{\theta} - \theta| \geq \Delta. \end{cases} \quad \text{(scalar parameter)} \tag{2.3.5}$$

$$C(\hat{\boldsymbol{\theta}}, \boldsymbol{\theta}) = \begin{cases} 0, & \|\hat{\boldsymbol{\theta}} - \boldsymbol{\theta}\| < \Delta, \\ 1, & \|\hat{\boldsymbol{\theta}} - \boldsymbol{\theta}\| \geq \Delta. \end{cases} \quad \text{(vector parameter)} \tag{2.3.6}$$

It is noteworthy that as a function of the random observed data x, the loss function is also random. As random function is inconvenient for evaluating parameter estimators, it is necessary to convert the random loss function into a deterministic function. To this end, the mathematical expectation of loss function

$$R(\hat{\theta}, \theta) = \mathrm{E}\left\{C(\hat{\theta}, \theta)\right\} \tag{2.3.7}$$

is taken as the performance measure of a parameter estimator, and is named risk function. The parameter estimation that minimizes risk function $R(\hat{\theta}, \theta)$ is called Bayes estimation.

2.3.2 Bayes Estimation

Next, we discuss the Bayes estimation under quadratic and uniform risk functions.

1. Quadratic Risk Function

The quadratic risk function is defined by

$$R_{\mathrm{MMSE}} \stackrel{\mathrm{def}}{=} \mathrm{E}\left\{(\hat{\theta} - \theta)^2\right\} = \int_{-\infty}^{\infty} \cdots \int_{-\infty}^{\infty} (\hat{\theta} - \theta)^2 f(x_1, \cdots, x_N, \theta)\, \mathrm{d}x_1 \cdots \mathrm{d}x_N\, \mathrm{d}\theta, \tag{2.3.8}$$

which in fact is the mean square error between parameter estimation $\hat{\theta}$ and actual parameter θ. Accordingly, the estimation that minimizes the quadratic risk function is called the minimum mean square error (MMSE) estimation.

To obtain MMSE estimation, first recall the well-established equality in probability theory

$$f(x_1, \cdots, x_N, \theta) = f(\theta|x_1, \cdots, x_N) f(x_1, \cdots, x_N) = f(x_1, \cdots, x_N) f(\theta), \tag{2.3.9}$$

where $f(\theta|x_1, \cdots, x_N)$ is the posterior PDF of θ given N observed data $x_1, \cdots, x_N$. Consequently, Eq. (2.3.8) can be rewritten as

$$R_{\mathrm{MMSE}} = \int_{-\infty}^{\infty} \cdots \int_{-\infty}^{\infty} \left[\int_{-\infty}^{\infty} (\hat{\theta} - \theta)^2 f(\theta|x_1, \cdots, x_N)\, \mathrm{d}\theta\right] f(x_1, \cdots, x_N)\, \mathrm{d}x_1 \cdots \mathrm{d}x_N,$$

where both the integral and the PDF $f(x_1, \cdots, x_N)$ are non-negative. To minimize the risk R_{MMSE}, take the derivative of R_{MMSE} with respect to $\hat{\theta}$ and let the derivative equal zero to get

$$\begin{aligned}\frac{\partial R_{\text{MMSE}}}{\partial\hat{\theta}} &= \int_{-\infty}^{\infty}\cdots\int_{-\infty}^{\infty}\left[2\int_{-\infty}^{\infty}(\hat{\theta}-\theta)f(\theta|x_1,\cdots,x_N)\,\mathrm{d}\theta\right]f(x_1,\cdots,x_N)\,\mathrm{d}x_1\cdots\mathrm{d}x_N\\ &= \int_{-\infty}^{\infty}\cdots\int_{-\infty}^{\infty}\left[2\int_{-\infty}^{\infty}\hat{\theta}f(\theta|x_1,\cdots,x_N)\,\mathrm{d}\theta-2\int_{-\infty}^{\infty}\theta f(\theta|x_1,\cdots,x_N)\,\mathrm{d}\theta\right]\times\\ &\quad f(x_1,\cdots,x_N)\,\mathrm{d}x_1\cdots\mathrm{d}x_N\\ &= 0.\end{aligned}$$

Rearranging the terms in the equation above yields

$$\hat{\theta}_{\text{MMSE}}\cdot\int_{-\infty}^{\infty}f(\theta|x_1,\cdots,x_N)\,\mathrm{d}\theta=\int_{-\infty}^{\infty}\theta f(\theta|x_1,\cdots,x_N)\,\mathrm{d}\theta.$$

From $\int_{-\infty}^{\infty}f(\theta|x_1,\cdots,x_N)\,\mathrm{d}\theta=1$, the MMSE estimation can then be solved as

$$\hat{\theta}_{\text{MMSE}}=\int_{-\infty}^{\infty}\theta f(\theta|x_1,\cdots,x_N)\,\mathrm{d}\theta=\mathrm{E}\{\theta|x_1,\cdots,x_N\}. \tag{2.3.10}$$

In other words, for quadratic risk function, i.e., mean square error, the Bayesian estimation of unknown parameter θ, or the MMSE estimation $\hat{\theta}_{\text{MMSE}}$, is the posterior mean of θ given samples $x_1,\cdots,x_N$.

2. Uniform Risk Function

Denote the uniform loss function as $C_{\text{unif}}(\hat{\theta},\theta)$. From the definition in (2.3.5), we have

$$\begin{aligned}\int_{-\infty}^{\infty}C_{\text{unif}}(\hat{\theta},\theta)f(\theta|x_1,\cdots,x_N)\,\mathrm{d}\theta &= \int_{\theta\notin[\hat{\theta}-\Delta,\hat{\theta}+\Delta]}f(\theta|x_1,\cdots,x_N)\,\mathrm{d}\theta\\ &= \left[\int_{-\infty}^{\infty}-\int_{\hat{\theta}-\Delta}^{\hat{\theta}+\Delta}\right]f(\theta|x_1,\cdots,x_N)\,\mathrm{d}\theta\\ &= 1-\int_{\hat{\theta}-\Delta}^{\hat{\theta}+\Delta}f(\theta|x_1,\cdots,x_N)\,\mathrm{d}\theta.\end{aligned} \tag{2.3.11}$$

Then, the uniform risk function can be written as

$$\begin{aligned}
R_{\text{unif}} &= \mathrm{E}\left\{C_{\text{unif}}(\hat{\theta},\theta)\right\} \\
&= \int_{-\infty}^{\infty}\cdots\int_{-\infty}^{\infty} C_{\text{unif}}(\hat{\theta},\theta) f(x_1,\cdots,x_N,\theta)\,\mathrm{d}x_1\cdots\mathrm{d}x_N\,\mathrm{d}\theta \\
&= \int_{-\infty}^{\infty}\cdots\int_{-\infty}^{\infty} C_{\text{unif}}(\hat{\theta},\theta) f(\theta|x_1,\cdots,x_N) f(x_1,\cdots,x_N)\,\mathrm{d}x_1\cdots\mathrm{d}x_N\,\mathrm{d}\theta \\
&= \int_{-\infty}^{\infty}\cdots\int_{-\infty}^{\infty}\left[\int_{-\infty}^{\infty} C_{\text{unif}}(\hat{\theta},\theta) f(\theta|x_1,\cdots,x_N)\,\mathrm{d}\theta\right] f(x_1,\cdots,x_N)\,\mathrm{d}x_1\cdots\mathrm{d}x_N.
\end{aligned}$$

Substituting Eq. (2.3.11) into the above equation gives

$$R_{\text{unif}} = \int_{-\infty}^{\infty}\cdots\int_{-\infty}^{\infty} f(x_1,\cdots,x_N)\left[1-\int_{\hat{\theta}-\Delta}^{\hat{\theta}+\Delta} f(\theta|x_1,\cdots,x_N)\,\mathrm{d}\theta\right]\mathrm{d}x_1\cdots\mathrm{d}x_N. \tag{2.3.12}$$

The necessary condition for R_{unif} to attain the minimum is $\frac{\partial R_{\text{unif}}}{\partial \hat{\theta}} = 0$, which by (2.3.12) is equivalent to

$$f(\hat{\theta}+\Delta|x_1,\cdots,x_N) - f(\hat{\theta}-\Delta|x_1,\cdots,x_N) = 0. \tag{2.3.13}$$

By Taylor series expansion at $\hat{\theta}$, Eq. (2.3.13) becomes

$$2\frac{\partial f(\hat{\theta}|x_1,\cdots,x_N)}{\partial \theta}\Delta + \mathcal{O}(\Delta) = 0. \tag{2.3.14}$$

As $\Delta \to 0$, from the equation above, one may expect the optimum $\hat{\theta}$ that minimizes the risk should satisfy the stationary equation

$$\frac{\partial}{\partial \theta} f(\hat{\theta}|x_1,\cdots,x_N) = 0. \tag{2.3.15}$$

The solution of (2.3.15) $\hat{\theta}$ corresponds to the maximum of posterior PDF $f(\theta|x_1,\cdots,x_N)$ in the limit case of $\Delta \to 0$. The resulting estimation, denoted as $\hat{\theta}_{\text{MAP}}$, is called maximum a posterior (MAP) estimation.

Now we have shown that MAP estimation is the limit of Bayesian estimation with uniform loss. However, the derivation above is intuitive and does not constitute a formal proof. In fact, certain conditions must be imposed upon the posterior PDF to guarantee the validity of the claim and exceptions do exist if the conditions are missing.

More common form of Eq. (2.3.15) is

$$\frac{\partial}{\partial \theta}\log f(\theta|x_1,\cdots,x_N) = 0. \tag{2.3.16}$$

Substitute Eq. (2.3.9) into Eq. (2.3.16) to get

$$\frac{\partial}{\partial\theta}\left[\log f(x_1,\cdots,x_N|\theta)+\log f(\theta)-\log f(x_1,\cdots,x_N)\right]=0. \tag{2.3.17}$$

For θ with uniform distribution, its PDF $f(\theta)$ satisfies

$$\frac{\partial\log f(\theta)}{\partial\theta}=0.$$

From the above equation and the fact that $f(x_1,\cdots,x_N)$ does not contain unknown parameter θ, Eq. (2.3.17) can be reduced to

$$\frac{\partial}{\partial\theta}\log f(x_1,\cdots,x_N|\theta)=0. \tag{2.3.18}$$

Since $\log f(x_1,\cdots,x_N|\theta)$ is the (logarithm) likelihood of samples $x_1,\cdots,x_N$, the estimation obtained from Eq. (2.3.18) is called maximum likelihood estimation, which is denoted as $\hat{\theta}_{ML}$. To conclude, for unknown parameter θ with uniform distribution, the Bayesian estimation under uniform loss coincides with the maximum likelihood estimation, i.e.,

$$\hat{\theta}_{MAP}=\hat{\theta}_{ML}. \tag{2.3.19}$$

In the following section, we will discuss the maximum likelihood estimation for θ with general PDF $f(\theta)$.

2.4 Maximum Likelihood Estimation

Maximum likelihood estimation (MLE) is one of the most popular and effective estimation methods. The idea of MLE is to estimate the parameter from available samples without any prior information about the unknown quantity or parameter. Thus, in the MLE method, the parameter to be estimated is assumed to be constant but unknown, whereas the observed data are supposed to be random.

Let $x_1,\cdots,x_N$ be the N samples of random variable x and $\{f(x_1,\cdots,x_N|\theta),\theta\in\Theta\}$ be the joint conditional PDF of samples $\{x_1,\cdots,x_N\}$ given parameter θ. Assuming that the joint conditional PDF exists and is bounded, we now consider the estimation problem for the unknown and deterministic parameter θ. When treated as a function of true parameter θ, the joint conditional PDF is called likelihood function. Simply put, the likelihood function is a function incorporating information on the possibility of parameter θ values, from which the term “likelihood” comes.

Strictly speaking, $f(x_1,\cdots,x_N|\theta)$ multiplied by arbitrary function of samples $x_1,\cdots,x_N$ gives a likelihood function. But in this book, we only call the joint conditional PDF $f(x_1,\cdots,x_N|\theta)$ likelihood function. Evidently, different realizations $x_1,\cdots,x_N$ of random variable x yield different joint conditional PDF $f(x_1,\cdots,x_N|\theta)$. So the global maximum of likelihood is determined by sample values $x_1,\cdots,x_N$.

MLE is the estimate $\hat{\theta}$ that maximizes the likelihood $f(x_1, \cdots, x_N|\theta)$, which is denoted by

$$\hat{\theta}_{\mathrm{ML}} = \arg\max_{\theta \in \Theta} f(x_1, \cdots, x_N|\theta), \tag{2.4.1}$$

and can be viewed as the global maximum of the joint conditional PDF $f(x_1, \cdots, x_N|\theta)$.

Due to monotonicity of logarithmic function, $\log f(x_1, \cdots, x_N|\theta)$ and $f(x_1, \cdots, x_N|\theta)$ have the same maximum. So, the logarithm of likelihood $\log f(x_1, \cdots, x_N|\theta)$, called log-likelihood, is usually used as a replacement of likelihood $f(x_1, \cdots, x_N|\theta)$. In signal processing literature, $\log f(x_1, \cdots, x_N|\theta)$ is also called likelihood for brevity.

For convenience, denote

$$L(\theta) = \log f(x_1, \cdots, x_N|\theta). \tag{2.4.2}$$

The optimality condition for MLE of θ is then

$$\frac{\partial L(\theta)}{\partial \theta} = 0. \tag{2.4.3}$$

In the case of vector parameter $\boldsymbol{\theta}$, i.e., $\boldsymbol{\theta} = [\theta_1, \cdots, \theta_p]^{\mathrm{T}}$, the MLEs $\hat{\theta}_{i,\mathrm{ML}}$ $(i = 1, \cdots, p)$ of p unknown parameters are determined by equations

$$\frac{\partial L(\boldsymbol{\theta})}{\partial \theta_i} = 0, \quad i = 1, \cdots, p. \tag{2.4.4}$$

If $x_1, \cdots, x_N$ are samples drawn independently, the likelihood can be expressed as

$$L(\theta) = \log f(x_1, \cdots, x_N|\boldsymbol{\theta}) = \log\left(f(x_1|\boldsymbol{\theta}) \cdots f(x_N|\boldsymbol{\theta})\right) = \sum_{i=1}^{N} \log f(x_i|\boldsymbol{\theta}). \tag{2.4.5}$$

Then, $\hat{\theta}_{i\mathrm{ML}}$ $(i = 1, \cdots, p)$ can be solved from equations

$$\left.\begin{aligned} \frac{\partial L(\boldsymbol{\theta})}{\partial \theta_1} &= \frac{\partial}{\partial \theta_1} \sum_{i=1}^{N} \log f(x_i|\boldsymbol{\theta}) = 0 \\ &\vdots \\ \frac{\partial L(\boldsymbol{\theta})}{\partial \theta_p} &= \frac{\partial}{\partial \theta_p} \sum_{i=1}^{N} \log f(x_i|\boldsymbol{\theta}) = 0 \end{aligned}\right\}. \tag{2.4.6}$$

MLE has properties as follows.

(1) MLE is not unbiased in general, but its bias can be corrected by multiplying the estimate by a proper constant.
(2) MLE is consistent.
(3) If exists, the most efficient estimation is given by MLE.
(4) For large N, MLE $\hat{\theta}_{\mathrm{ML}}$ asymptotically follows Gaussian distribution, which has mean θ and variance

$$\frac{1}{N}\left[\mathrm{E}\left\{\frac{\partial}{\partial \theta}[f(x_1, \cdots, x_N|\theta)]^2\right\}\right]^{-1}.$$

Example 2.4.1 Let $x_1, \cdots, x_N$ be random samples drawn from a normal distribution with PDF

$$f(x|\mu, \sigma^2) = \frac{1}{\sqrt{2\pi}\sigma} e^{-(x-\mu)^2/(2\sigma^2)}.$$

Find the MLEs of mean μ and variance σ^2.

Solution. As the likelihood is the function of mean μ and variance σ^2, we have

$$\begin{aligned} f(x_1, \cdots, x_N|\mu, \sigma^2) &= \prod_{i=1}^{N} \frac{1}{\sqrt{2\pi}\sigma} e^{-(x_i-\mu)^2/(2\sigma^2)} \\ &= (2\pi\sigma^2)^{-N/2} \exp\left(-\frac{1}{2\sigma^2} \sum_{i=1}^{N} (x_i - \mu)^2\right). \end{aligned}$$

The log-likelihood is then

$$L(\mu, \sigma^2) = \log f(x_1, \cdots, x_N|\mu, \sigma^2) = -\frac{N}{2}\log(2\pi) - \frac{N}{2}\log(\sigma^2) - \frac{1}{2\sigma^2}\sum_{i=1}^{N}(x_i - \mu)^2.$$

Take the derivatives of $L(\mu, \sigma^2)$ with respective to μ and σ^2 and let the derivatives equal to zero to obtain

$$\begin{aligned} \frac{\partial L}{\partial \mu} &= \frac{2}{2\sigma^2}\sum_{i=1}^{N}(x_i - \mu) = 0, \\ \frac{\partial L}{\partial \sigma^2} &= -\frac{N}{2\sigma^2} + \frac{1}{2\sigma^4}\sum_{i=1}^{N}(x_i - \mu)^2 = 0. \end{aligned}$$

Solve $\frac{\partial L}{\partial \mu} = 0$ to get

$$\hat{\mu}_{\mathrm{ML}} = \frac{1}{N}\sum_{i=1}^{N} x_i = \bar{x}.$$

Substitute the solution of μ into $\frac{\partial L}{\partial \sigma^2}$ to get

$$\hat{\sigma}^2_{\mathrm{ML}} = \frac{1}{N}\sum_{i=1}^{N}(x_i - \bar{x})^2.$$

Note that the sample mean

$$\bar{x} = \frac{1}{N}\sum_{i=1}^{N} x_i, \tag{2.4.7}$$

and the sample variance

$$s^2 = \frac{1}{N-1}\sum_{i=1}^{N}(x_i - \bar{x})^2 \tag{2.4.8}$$

are both unbiased. As a result, the MLE of mean $\hat{\mu}_{\text{ML}}$ is unbiased but the MLE of variance $\hat{\sigma}^2_{\text{ML}}$ is biased. Nevertheless, the modified estimate obtained from multiplying $\hat{\sigma}^2_{\text{ML}}$ by constant $\frac{N}{N-1}$ is unbiased and the bias in the original estimate $\hat{\sigma}^2_{\text{ML}}$ can then be eliminated. □

Example 2.4.2 The received signal is given by

$$y_i = s + w_i, \quad i = 1, \cdots, N.$$

Suppose $w_i \sim N(0, \sigma^2)$ is Gaussian white noise, i.e.,

$$\mathrm{E}\{w_i\} = 0, \quad i = 1, \cdots, N,$$

$$\mathrm{E}\{w_i w_j\} = \begin{cases} \sigma^2, & i = j, \\ 0, & i \neq j. \end{cases}$$

Find the MLE $\hat{s}_{\text{ML}}$ of s and the Cramér-Rao bound for the variance of estimate $\hat{s}_{\text{ML}}$, and determine if $\hat{s}_{\text{ML}}$ is most efficient.

Solution. First, note that the received signal y_i and additive noise w_i are both Gaussian and only have a difference in mean s. Thus,

$$\begin{aligned} f(y_1, \cdots, y_N | s, \sigma^2) &= \prod_{i=1}^{N} f(w_i) = \frac{1}{(2\pi\sigma^2)^{N/2}} \exp\left(-\sum_{i=1}^{N} \frac{w_i^2}{2\sigma^2}\right) \\ &= \frac{1}{(2\pi\sigma^2)^{N/2}} \exp\left(-\sum_{i=1}^{N} \frac{(y_i - s)^2}{2\sigma^2}\right), \end{aligned}$$

by which the log-likelihood can be expressed as

$$L(s, \sigma^2) = \log f(y_1, \cdots, y_N | s, \sigma^2) = -\frac{N}{2}\log(2\pi\sigma^2) - \sum_{i=1}^{N} \frac{(y_i - s)^2}{2\sigma^2}. \tag{2.4.9}$$

Consequently, the MLE of signal s can be solved by

$$\frac{\partial L}{\partial s} = 2\sum_{i=1}^{N} \frac{(y_i - s)}{2\sigma^2} = 0,$$

which gives

$$\hat{s}_{\text{ML}} = \frac{1}{N}\sum_{i=1}^{N} y_i = \bar{y}. \tag{2.4.10}$$

The expectation of $\hat{s}_{\text{ML}}$ is

$$\mathrm{E}\{\hat{s}_{\text{ML}}\} = \mathrm{E}\{\bar{y}\} = \mathrm{E}\left\{\frac{1}{N}\sum_{i=1}^{N} y_i\right\} = \mathrm{E}\left\{\frac{1}{N}\sum_{i=1}^{N}(s + w_i)\right\} = s + \frac{1}{N}\sum_{i=1}^{N}\mathrm{E}\{w_i\} = s.$$

Therefore, $\hat{s}_{\mathrm{ML}}$ is unbiased.

Take the second-order derivative of log-likelihood $\log f(y_1, \cdots, y_N|s, \sigma^2)$ with respect to s to get

$$\frac{\partial^2}{\partial s^2} \log f(y_1, \cdots, y_N|s, \sigma^2) = -\sum_{i=1}^{N} \frac{1}{\sigma^2} = -\frac{N}{\sigma^2}.$$

From Eq. (2.2.4), the Fisher information is

$$J(s) = -\mathrm{E}\left\{\frac{\partial^2}{\partial s^2} \log f(y_1, \cdots, y_N|s, \sigma^2)\right\} = -\sum_{i=1}^{N} \frac{1}{\sigma^2} = \frac{N}{\sigma^2},$$

From Theorem 2.2.1 the Cramér-Rao inequality is

$$\mathrm{var}\left[\hat{s}\right] = \mathrm{E}\left\{(\hat{s} - s)^2\right\} \geq \frac{1}{\frac{N}{\sigma^2}} = \frac{\sigma^2}{N}, \tag{2.4.11}$$

and the equality holds if and only if

$$\frac{\partial}{\partial s} \log f(y_1, \cdots, y_N|s, \sigma^2) = K(s)(\hat{s} - s). \tag{2.4.12}$$

Further from

$$\frac{\partial}{\partial s} \log f(y_1, \cdots, y_N|s, \sigma^2) = \frac{\sum_{i=1}^{N} y_i - Ns}{\sigma^2} = \frac{N\bar{y} - Ns}{\sigma^2} = \frac{N}{\sigma^2}(\bar{y} - s) = \frac{N}{\sigma^2}(\hat{s}_{\mathrm{ML}} - s),$$

it can be deduced that Eq. (2.4.12) can be satisfied by taking $K(s) = N/\sigma^2$. That is,

$$\mathrm{var}\left[\hat{s}_{\mathrm{ML}}\right] = \mathrm{E}\left\{(\hat{s}_{\mathrm{ML}} - s)^2\right\} = \frac{\sigma^2}{N},$$

which shows that MLE $\hat{s}_{\mathrm{ML}}$ is most efficient. □

2.5 Linear Mean Squares Estimation

Bayesian estimation requires posterior PDF $f(\theta|x_1, \cdots, x_N)$, and MLE likelihood function $f(x_1, \cdots, x_N|\theta)$. However, under many circumstances, both functions could be unavailable. Besides, MLE can sometimes result in a nonlinear estimation problem, which is difficult to solve. Therefore, a linear estimation method that does not require prior knowledge and is easy to implement is much more appealing. Linear mean squares (LMS) estimation and least squares estimation are both of such categories. We will discuss LMS estimation in this section.

In LMS estimation, the estimator for the unknown parameter is expressed as the linearly weighted sum of observed data. That is,

$$\hat{\theta}_{\mathrm{LMS}} = \sum_{i=1}^{N} w_i x_i, \tag{2.5.1}$$

where $w_1, \cdots, w_N$ are weights to be determined. The target of LMS estimation is to minimize the mean square error $E\{(\hat{\theta} - \theta)^2\}$. Namely, the weights w_i are solved from

$$\min E\left\{(\hat{\theta} - \theta)^2\right\} = \min E\left\{\left(\sum_{i=1}^{N} w_i x_i - \theta\right)^2\right\} = \min E\left\{e^2\right\}, \tag{2.5.2}$$

where $e = \hat{\theta} - \theta$ is the estimation error.

Setting the partial derivative of the objective in Eq. (2.5.2) with respect to w_k to zero gives

$$\frac{\partial E\{e^2\}}{\partial w_k} = E\left\{\frac{\partial e^2}{\partial w_k}\right\} = 2E\left\{e\frac{\partial e}{\partial w_k}\right\} = 2E\{ex_k\} = 0,$$

or

$$E\{ex_i\} = 0, \quad i = 1, \cdots, N. \tag{2.5.3}$$

This result is called the orthogonality principle. In plain words, the orthogonality principle states that the mean square error is minimized if and only if the error e is orthogonal to each observation x_i $(i = 1, \cdots, N)$.

To derive the optimal weights, rewrite Eq. (2.5.3) as

$$E\left\{\left(\sum_{k=1}^{N} w_k x_k - \theta\right) x_i\right\} = 0, \quad i = 1, \cdots, N. \tag{2.5.4}$$

Notice that samples x_i $(i = 1, \cdots, N)$ and θ are correlated such that $E\{\theta x_i\}$ does not equal $\theta E\{x_i\}$.

let

$$g_i = E\{\theta x_i\} \quad \text{and} \quad R_{ik} = E\{x_i x_k\}.$$

Eq. (2.5.4) can be expressed as

$$\sum_{k=1}^{N} R_{ik} w_k = g_i, \quad i = 1, \cdots, N, \tag{2.5.5}$$

which is called a normal equation. Denoting

$$\mathbf{R} = [R_{ij}]_{i,j=1}^{N,N},$$
$$\mathbf{w} = [w_1, \cdots, w_N]^{\mathrm{T}},$$
$$\mathbf{g} = [g_1, \cdots, g_N]^{\mathrm{T}},$$

we can change Eq. (2.5.5) into a more concise form as $\mathbf{Rw} = \mathbf{g}$, which has solution

$$\mathbf{w} = \mathbf{R}^{-1}\mathbf{g}. \tag{2.5.6}$$

The independence of samples $x_1, \cdots, x_N$ is one of the conditions that guarantee a non-singular $\mathbf{R}$.

Being a class of important estimation method, LMS estimation cannot be applied to the situation in which the correlation function $g_i = \mathrm{E}\{\theta x_i\}$ is not readily accessible. Interestingly, in the applications of filtering, one can encounter problems analogous to Eq. (2.5.1). Specifically, one may expect to design a set of filter coefficients $w_1, \cdots, w_M$ such that the linear combination of the coefficients and a random signal $x(n)$ with its delays $x(n-i)$

$$\hat{d}(n) = \sum_{i=1}^{M} w_i x(n-i) \tag{2.5.7}$$

is an approximation to a target signal $d(n)$. In such case, the LMS estimation is applicable since $g_i = \mathrm{E}\left\{d(n)x(n-i)\right\}$ can be estimated. We will give a detailed discussion on the LMS filter design problem in Chapter 5.

Due to the usage of the MMSE criterion, LMS estimation in essence gives an MMSE estimator. Both orthogonality principle and LMS estimation are widely adopted in signal processing applications and will be frequently revisited in this book.

2.6 Least Squares Estimation

In addition to LMS estimation, least squares (LS) estimation is another estimation method that is prior-free.

2.6.1 Least Squares Estimation and Its Performance

In many applications, the unknown parameter vector $\boldsymbol{\theta} = [\theta_1, \cdots, \theta_p]^{\mathrm{T}}$ is modeled as a matrix equation

$$\mathbf{A}\boldsymbol{\theta} = \mathbf{b}, \tag{2.6.1}$$

where $\mathbf{A}$ and $\mathbf{b}$ are coefficient matrix and vector associated with observed data and are both known. The data model incorporates the following three cases.

(1) The number of the unknown parameters is equal to that of equations, and the matrix $\mathbf{A}$ is non-singular. In this case, matrix equation (2.6.1) is called a well-determined equation and has unique solution $\boldsymbol{\theta} = \mathbf{A}^{-1}\mathbf{b}$.

(2) The matrix $\mathbf{A}$ is “tall” and has more rows than columns. That is, the number of equations is larger than that of unknown parameters. In this case, matrix equation (2.6.1) is called a overdetermined equation.

(3) The matrix $\mathbf{A}$ is “fat” and has more columns than rows. That is, the number of equations is less than that of unknown parameters. In this case, matrix equation (2.6.1) is called a underdetermined equation.

Overdetermined equation is frequently used in spectral estimation and system identification and is the focus of our book.

To find out the estimation of the parameter vector $\hat{\theta}$, we consider the criterion under which the sum of square errors

$$\sum_{i=1}^{N} e_i^2 = \mathbf{e}^T\mathbf{e} = (\mathbf{A}\hat{\boldsymbol{\theta}} - \mathbf{b})^T(\mathbf{A}\hat{\boldsymbol{\theta}} - \mathbf{b}) \tag{2.6.2}$$

is minimized. The resulting estimation is called LS estimation, denoted as $\hat{\theta}_{\text{LS}}$.

Expanding the loss or cost function $J = \mathbf{e}^T\mathbf{e}$ gives

$$J = \hat{\boldsymbol{\theta}}^T\mathbf{A}^T\mathbf{A}\hat{\boldsymbol{\theta}} + \mathbf{b}^T\mathbf{b} - \hat{\boldsymbol{\theta}}^T\mathbf{A}^T\mathbf{b} - \mathbf{b}^T\mathbf{A}\hat{\boldsymbol{\theta}}.$$

Take the derivative of J with respect to $\hat{\boldsymbol{\theta}}$ and set the derivative to zero to get

$$\frac{\partial J}{\partial \hat{\boldsymbol{\theta}}} = 2\mathbf{A}^T\mathbf{A}\hat{\boldsymbol{\theta}} - 2\mathbf{A}^T\mathbf{b} = 0,$$

Accordingly, the LS estimation must satisfy the equation

$$\mathbf{A}^T\mathbf{A}\hat{\boldsymbol{\theta}} = \mathbf{A}^T\mathbf{b}, \tag{2.6.3}$$

which has two cases regarding its solution:

(1) **A** is full column rank. As $\mathbf{A}^T\mathbf{A}$ is non-singular, the LS estimation is uniquely determined by

$$\hat{\boldsymbol{\theta}}_{\text{LS}} = (\mathbf{A}^T\mathbf{A})^{-1}\mathbf{A}^T\mathbf{b}, \tag{2.6.4}$$

and the parameter vector $\boldsymbol{\theta}$ is said to be uniquely identifiable.

(2) **A** is rank-deficient and different values of $\boldsymbol{\theta}$ can yield identical value of $\mathbf{A}\boldsymbol{\theta}$. As a result, although the vector **b** contains information about $\mathbf{A}\boldsymbol{\theta}$, we cannot distinguish different values of $\boldsymbol{\theta}$ for the fixed value of $\mathbf{A}\boldsymbol{\theta}$. In this regard, we say that the parameter vector $\boldsymbol{\theta}$ is unidentifiable. To generalize, if different values of parameter give identical distribution in sample space, the parameter is unidentifiable[68].

The following theorem states that if the error vector has uncorrelated components of equal variance, LS estimation of the parameter $\boldsymbol{\theta}$ in linear model (2.6.1) is optimal in the sense of minimum variance.

Theorem 2.6.1 (Gauss-Markov Theorem). *Let* **b** *be the random vector expressed by* $\mathbf{b} = \mathbf{A}\boldsymbol{\theta} + \mathbf{e}$, *where* $N \times p$ $(N > p)$ *matrix* **A** *has rank* p, *and the error vector* **e** *has mean* $\mathrm{E}\{\mathbf{e}\} = \mathbf{0}$ *and variance (covariance) matrix* $\mathrm{var}[\mathbf{e}] = \sigma^2\mathbf{I}$ *with unknown variance* σ^2. *Then, given the linear function of parameter* $\boldsymbol{\beta} = \mathbf{c}^T\boldsymbol{\theta}$, *for any unbiased estimator* $\tilde{\boldsymbol{\beta}}$ *of the function, it always holds that* $\mathrm{E}\{\hat{\boldsymbol{\theta}}_{LS}\} = \boldsymbol{\theta}$ *and* $\mathrm{var}[\mathbf{c}^T\hat{\boldsymbol{\theta}}_{LS}] \le \mathrm{var}[\hat{\boldsymbol{\beta}}]$.

Proof. Because $\mathrm{E}\{\mathbf{e}\} = \mathbf{0}$ and $\mathrm{var}[\mathbf{e}] = \sigma^2\mathbf{I}$, it can be derived that

$$\mathrm{E}\{\mathbf{b}\} = \mathrm{E}\{\mathbf{A}\boldsymbol{\theta}\} + \mathrm{E}\{\mathbf{e}\} = \mathbf{A}\boldsymbol{\theta}$$

and

$$\mathrm{var}[\mathbf{b}] = \mathrm{var}[\mathbf{A}\boldsymbol{\theta} + \mathbf{e}] = \mathrm{var}[\mathbf{A}\boldsymbol{\theta}] + \mathrm{var}[\mathbf{e}] = \mathrm{var}[\mathbf{e}] = \sigma^2\mathbf{I}.$$

Therefore,

$$\mathrm{E}\left\{\hat{\boldsymbol{\theta}}_{\mathrm{LS}}\right\} = \mathrm{E}\left\{(\mathbf{A}^{\mathrm{T}}\mathbf{A})^{-1}\mathbf{A}^{\mathrm{T}}\mathbf{b}\right\} = (\mathbf{A}^{\mathrm{T}}\mathbf{A})^{-1}\mathbf{A}^{\mathrm{T}}\mathrm{E}\left\{\mathbf{b}\right\} = (\mathbf{A}^{T}\mathbf{A})^{-1}\mathbf{A}^{\mathrm{T}}\mathbf{A}\boldsymbol{\theta} = \boldsymbol{\theta},$$

from which we have

$$\mathrm{E}\left\{\mathbf{c}^{\mathrm{T}}\hat{\boldsymbol{\theta}}_{\mathrm{LS}}\right\} = \mathbf{c}^{\mathrm{T}}\mathrm{E}\left\{\hat{\boldsymbol{\theta}}_{\mathrm{LS}}\right\} = \mathbf{c}^{\mathrm{T}}\boldsymbol{\theta} = \boldsymbol{\beta}.$$

Therefore, $\mathbf{c}^{\mathrm{T}}\hat{\boldsymbol{\theta}}_{\mathrm{LS}}$ is unbiased. As $\hat{\boldsymbol{\beta}}$ is a linear estimator, it can be expressed by $\hat{\boldsymbol{\beta}} = \mathbf{w}^{\mathrm{T}}\mathbf{b}$, where $\mathbf{w}$ is a constant vector. Further, from the fact that $\hat{\boldsymbol{\beta}}$ is an unbiased estimator of $\boldsymbol{\beta}$, for any $\boldsymbol{\theta}$ we have

$$\mathbf{w}^{\mathrm{T}}\mathbf{A}\boldsymbol{\theta} = \mathbf{w}^{\mathrm{T}}\mathrm{E}\left\{\mathbf{b}\right\} = \mathrm{E}\left\{\mathbf{w}^{\mathrm{T}}\mathbf{b}\right\} = \mathrm{E}\left\{\hat{\boldsymbol{\beta}}\right\} = \boldsymbol{\beta} = \mathbf{c}^{\mathrm{T}}\boldsymbol{\theta}.$$

It then can be deduced that $\mathbf{w}^{\mathrm{T}}\mathbf{A} = \mathbf{c}^{\mathrm{T}}$.

From the variances

$$\begin{aligned} \mathrm{var}\left[\tilde{\boldsymbol{\beta}}\right] &= \mathrm{var}\left[\mathbf{w}^{\mathrm{T}}\mathbf{b}\right] = \mathbf{w}^{\mathrm{T}}\mathrm{var}\left[\mathbf{b}\right]\mathbf{w} = \sigma^{2}\mathbf{w}^{\mathrm{T}}\mathbf{w}, \\ \mathrm{var}\left[\mathbf{c}^{\mathrm{T}}\hat{\boldsymbol{\theta}}_{\mathrm{LS}}\right] &= \sigma^{2}\mathbf{c}^{\mathrm{T}}(\mathbf{A}^{\mathrm{T}}\mathbf{A})^{-1}\mathbf{c} = \sigma^{2}\mathbf{w}^{\mathrm{T}}\mathbf{A}(\mathbf{A}^{\mathrm{T}}\mathbf{A})^{-1}\mathbf{A}^{\mathrm{T}}\mathbf{w}, \end{aligned}$$

in order to prove $\mathrm{var}[\mathbf{c}^{\mathrm{T}}\hat{\boldsymbol{\theta}}_{\mathrm{LS}}] \le \mathrm{var}[\tilde{\boldsymbol{\beta}}]$, we only need to prove

$$\mathbf{w}^{\mathrm{T}}\mathbf{A}(\mathbf{A}^{\mathrm{T}}\mathbf{A})^{-1}\mathbf{A}^{\mathrm{T}}\mathbf{w} \le \mathbf{w}^{\mathrm{T}}\mathbf{w},$$

or equivalently $\mathbf{F} = \mathbf{I} - \mathbf{A}(\mathbf{A}^{\mathrm{T}}\mathbf{A})^{-1}\mathbf{A}^{\mathrm{T}}$ is positive semidefinite[107]. It is straightforward to verify that $\mathbf{F}^{2} = \mathbf{F}\mathbf{F} = \mathbf{F}$ so that F is idempotent. Since any idempotent matrix is positive semidefinite, the theorem is proved. □

2.6.2 Weighted Least Squares Estimation

Theorem 2.6.1 reveals that when the components of error vector $\mathbf{c}$ are both of identical variance and uncorrelated, the LS estimator $\hat{\boldsymbol{\theta}}_{\mathrm{LS}} = (\mathbf{A}^{\mathrm{T}}\mathbf{A})^{-1}\mathbf{A}^{\mathrm{T}}\mathbf{b}$ has the minimum variance among all linear estimators and is thus optimal. However, when the components of the error vector have unequal variances or are correlated, the LS estimation could fail to attain minimum variance and its optimality no longer holds. Now the question is: in such circumstance how to obtain the estimator of minimum variance?

To remedy the defect of LS estimation, we modify the original loss function of the sum of square errors and adopt the "weighted sum of square errors" as the new loss function:

$$Q(\hat{\boldsymbol{\theta}}) = \mathbf{e}^{\mathrm{T}}\mathbf{W}\mathbf{e}, \tag{2.6.5}$$

where the weight matrix $\mathbf{W}$ is symmetric. Eq. (2.6.5) is named weighted error function for short. Instead of minimizing $J = \mathbf{e}^{T}\mathbf{e}$ as in LS method, minimizing weighted error function $Q(\hat{\boldsymbol{\theta}})$ is now the criterion. The resulting estimator from the criterion is called weighted LS (WLS) estimator and is denoted by $\hat{\boldsymbol{\theta}}_{\text{WLS}}$. To obtain the WLS estimate, expand $Q(\hat{\boldsymbol{\theta}})$ as

$$Q(\hat{\boldsymbol{\theta}}) = (\mathbf{b} - \mathbf{A}\hat{\boldsymbol{\theta}})^{\mathrm{T}}\mathbf{W}(\mathbf{b} - \mathbf{A}\hat{\boldsymbol{\theta}}) = \mathbf{b}^{\mathrm{T}}\mathbf{W}\mathbf{b} - \hat{\boldsymbol{\theta}}^{\mathrm{T}}\mathbf{A}^{\mathrm{T}}\mathbf{W}\mathbf{b} - \mathbf{b}^{T}\mathbf{W}\mathbf{A}\hat{\boldsymbol{\theta}} + \hat{\boldsymbol{\theta}}^{T}\mathbf{A}^{T}\mathbf{W}\mathbf{A}\hat{\boldsymbol{\theta}}.$$

Take derivative of $Q(\hat{\boldsymbol{\theta}})$ with respect to $\hat{\boldsymbol{\theta}}$ and set the derivative to zero to get

$$\frac{\partial Q(\hat{\boldsymbol{\theta}})}{\partial \hat{\boldsymbol{\theta}}} = -2\mathbf{A}^{\mathrm{T}}\mathbf{W}\mathbf{b} + 2\mathbf{A}^{\mathrm{T}}\mathbf{W}\mathbf{A}\hat{\boldsymbol{\theta}} = \mathbf{0},$$

which gives the equation for the WLS estimator as

$$\mathbf{A}^{\mathrm{T}}\mathbf{W}\mathbf{A}\hat{\boldsymbol{\theta}}_{\text{WLS}} = \mathbf{A}^{\mathrm{T}}\mathbf{W}\mathbf{b}.$$

Suppose $\mathbf{A}^{\mathrm{T}}\mathbf{W}\mathbf{A}$ is non-singular and $\hat{\boldsymbol{\theta}}_{\text{WLS}}$ is given by

$$\hat{\boldsymbol{\theta}}_{\text{WLS}} = (\mathbf{A}^{\mathrm{T}}\mathbf{W}\mathbf{A})^{-1}\mathbf{A}^{\mathrm{T}}\mathbf{W}\mathbf{b}. \tag{2.6.6}$$

Next, we consider how to choose $\mathbf{W}$.

Suppose the variance var $[\mathbf{e}]$ of error vector has a generic form of $\sigma^2\mathbf{V}$ where $\mathbf{V}$ is known and positive definite. $\mathbf{V}$ can be expressed as $\mathbf{V} = \mathbf{P}\mathbf{P}^{T}$ with non-singular matrix $\mathbf{P}$ since it is positive definite. Let $\boldsymbol{\epsilon} = \mathbf{P}^{-1}\mathbf{e}$ and $\mathbf{x} = \mathbf{P}^{-1}\mathbf{b}$. Multiplying both sides of the observation equation $\mathbf{b} = \mathbf{A}\boldsymbol{\theta} + \mathbf{e}$ by $\mathbf{P}^{-1}$ gives

$$\mathbf{x} = \mathbf{P}^{-1}\mathbf{A}\boldsymbol{\theta} + \boldsymbol{\epsilon} = \mathbf{B}\boldsymbol{\theta} + \boldsymbol{\epsilon}, \tag{2.6.7}$$

where $\mathbf{B} = \mathbf{P}^{-1}\mathbf{A}$. Noticeably, the error vector in the new observation model (2.6.7) has variance matrix

$$\text{var}\,[\boldsymbol{\epsilon}] = \text{var}\left[\mathbf{P}^{-1}\mathbf{e}\right] = \mathbf{P}^{-1}\text{var}\,[\mathbf{e}]\,\mathbf{P}^{-\mathrm{T}} = \mathbf{P}^{-1}\sigma^2\mathbf{P}\mathbf{P}^{\mathrm{T}}\mathbf{P}^{-\mathrm{T}} = \sigma^2\mathbf{I}, \tag{2.6.8}$$

where $\mathbf{P}^{-\mathrm{T}} = (\mathbf{P}^{-1})^{\mathrm{T}}$. Eq. (2.6.8) shows that the new error vector $\boldsymbol{\epsilon} = \mathbf{P}^{-1}\mathbf{e}$ has uncorrelated components with equal variance. Consequently, with the defined $\mathbf{x}$, $\mathbf{B}$ and $\boldsymbol{\epsilon}$, the observation model $\mathbf{x} = \mathbf{B}\boldsymbol{\theta} + \boldsymbol{\epsilon}$ becomes exactly the same model in Theorem 2.6.1 and satisfies the conditions in theorem. So, Theorem 2.6.1 applies to the new model, and the LS estimator

$$\hat{\boldsymbol{\theta}}_{\text{LS}} = (\mathbf{B}^{\mathrm{T}}\mathbf{B})^{-1}\mathbf{B}^{\mathrm{T}}\mathbf{x} = (\mathbf{A}^{\mathrm{T}}\mathbf{V}^{-1}\mathbf{A})^{-1}\mathbf{A}^{\mathrm{T}}\mathbf{V}^{-1}\mathbf{b} \tag{2.6.9}$$

must have the minimum variance and is optimal. Comparing Eqs. (2.6.9) and (2.6.6), it follows that to obtain the optimal WLS estimate of $\boldsymbol{\theta}$, one may set the weight matrix $\mathbf{W}$ to be

$$\mathbf{W} = \mathbf{V}^{-1}, \tag{2.6.10}$$

which is the inverse matrix of $\mathbf{V}$. The matrix $\mathbf{V}$ is given by the variance matrix var $[\mathbf{e}] = \sigma^2\mathbf{V}$ of the error vector.

Apart from LS and WLS methods, there are two variations of LS estimation: generalized LS and total LS methods, which will be given full coverage in Chapter 4.

Summary

Many problems in signal processing can be attributed to parameter estimation. In this chapter, several basic properties regarding the performance of parameter estimator were discussed first, including unbiasedness, asymptotic unbiasedness, and effectiveness. Next, from the criterion of optimal estimator, we introduced the variance of score function–Fisher information, and Cramér-Rao inequality indicating the lower bound of variance. In the succeeding sections, Bayesian estimation, maximum likelihood estimation, linear mean squares estimation, and least squares estimation, which are among the most commonly used parameter estimation methods, are introduced in sequence.

Like Chapter 1, this chapter involves theoretical fundamentals of modern signal processing. The basic knowledge in this chapter will pave way for a better understanding of theories, methods, and applications of modern signal processing in the subsequent chapters.

Exercises

2.1 x is a normal or Gaussian random variable with PDF

$$f(x) = \frac{1}{\sqrt{2\pi}\sigma}\mathrm{e}^{-\frac{(x-\mu)^2}{2\sigma^2}}.$$

Prove that μ and σ^2 are the mean and variance of x, respectively.

2.2 A normal random variable x of zero mean has PDF

$$f(x) = \frac{1}{\sqrt{2\pi}\sigma}\mathrm{e}^{-\frac{x^2}{2\sigma^2}}.$$

Prove that the nth order moment of x is given by

$$\mathrm{E}\left\{x^n\right\} = \begin{cases} 0, & n = 2k+1, \\ 1\cdot 3\cdots(n-1)\sigma^2, & n = 2k. \end{cases}$$

2.3 A random signal $x(t)$ has observations $x(1), x(2), \cdots$. Let $\bar{x}_k$ and s_k^2 be the sample mean $\bar{x}_k$ and variance $s^2 = \frac{1}{k}\sum_{i=1} k[x(i) - \bar{x}_k]^2$, respectively, both obtained from k samples $x(1), \cdots, x(k)$. With a new sample $x(k+1)$ available, try to calculate $\bar{x}_{k+1}$ and s_{k+1}^2 using $x(k+1)$, $\bar{x}_k$, and s_k^2. Find the updating formula for the calculation of sample mean $\bar{x}_{k+1}$ and variance s_{k+1}^2.

2.4 Let $\{x(n)\}$ be a stationary process with mean $\mu = \mathrm{E}\left\{x(n)\right\}$. Given N independent samples $x(1), \cdots, x(N)$, prove that

(1) sample mean $\bar{x} = \frac{1}{N}\sum_{n=1}^{N} x(n)$ is unbiased estimation of mean μ;

(2) sample variance $s^2 = \frac{1}{N-1}\sum_{n=1}^{N}[x(n) - \bar{x}]^2$ is unbiased estimation of the true variance $\sigma^2 = \mathrm{E}\left\{|x(n) - \mu|^2\right\}$.

2.5 A random process is defined by $x(n) = A + v(n)$, where A is an unknown constant, $v(n)$ is Gaussian white noise with zero mean and variance σ^2. If $\hat{A}$ is the estimate of A from samples $x(1), \cdots, x(N)$, find the Cramér-Rao lower bound for the variance of the estimate.

2.6 A random process is described by $x(n) = A + Bn + v(n)$, where $v(n)$ is Gaussian white noise with zero mean and variance σ^2, and both A and B are unknown and need to be estimated. Find the Cramér-Rao lower bounds for the variances of the estimates $\hat{A}$ and $\hat{B}$.

2.7 The received signal is $y_i = s + w_i$ $(i = 1, \cdots, N)$, where $w_i \sim N(0, \sigma^2)$ is Gaussian white noise with zero mean and variance σ^2. Suppose both signal s and noise variance σ^2 are unknown. Find the MLEs of s and σ^2, and derive the Cramér-Rao lower bound for the variance $\mathrm{var}[\hat{\sigma}^2]$ of $\hat{\sigma}^2$.

2.8 The observed data are given by $y_i = s + w_i$ $(i = 1, \cdots, N)$, where w_i is Gaussian white noise of zero mean and unit variance. The signal s has PDF

$$f(s) = \frac{1}{\sqrt{2\pi}} \mathrm{e}^{-\frac{s^2}{2}}.$$

Find the MMSE estimation $\hat{s}_{\mathrm{MMSE}}$ and MAP estimation $\hat{s}_{\mathrm{MAP}}$ respectively.

2.9 The received signal is given by

$$y_i = A\cos(\omega_c i + \theta) + w_i, \quad i = 1, \cdots, N,$$

where $w_i \sim N(0, 1)$ is Gaussian noise of zero mean and unit variance, ω_c is the angular frequency of the carrier, and θ is the unknown phase. Suppose $w_1, \cdots, w_N$ are independent. Give the MLE $\hat{\theta}_{\mathrm{ML}}$ of the unknown phase.

2.10 Consider a real random vector $\mathbf{x}$. Find its LMS estimator $\hat{\mathbf{x}}$.

2.11 The unknown random variable follows uniform distribution and has PDF

$$f(x) = \begin{cases} 1, & 0 \le x \le 1, \\ 0, & \text{otherwise}. \end{cases}$$

Without any prior information, we can use a constant as the LMS estimation of the random variable x. Find the constant.

2.12 Given n independent random variables $x_1, \cdots, x_n$ with identical mean μ and different variances $\sigma_1^2, \cdots, \sigma_n^2$, consider the sum of the random variables weighted by n constants $a_1, \cdots, a_n$:

$$z = \sum_{i=1}^{n} a_i x_i.$$

If $\mathrm{E}\{z\} = \mu$ and σ_z^2 is minimized, give the values of the weights $a_1, \cdots, a_n$.

2.13 An aircraft moves along a straight line during a period of time with initial location α and constant velocity β. The observed location of the aircraft is given by $y_i = \alpha + \beta i + w_i$ $(i = 1, \cdots, N)$, where w_i is random and has zero mean. There are ten samples of observed locations $y_1 = 1, y_2 = 2, y_3 = 2, y_4 = 4, y_5 = 4, y_6 = 8, y_7 = 9, y_8 = 10, y_9 = 12, y_{10} = 13$. Find the LS estimates of the initial location α and velocity β of the aircraft.

2.14 Consider the discrimination of signals of multi-class targets[77]. Suppose there are c targets, and $N = N_1 + \cdots + N_c$ feature vectors $\mathbf{s}_{11}, \cdots, \mathbf{s}_{1N_1}, \cdots, \mathbf{s}_{c1}, \cdots, \mathbf{s}_{cN_c}$ are retrieved during training phase, where $Q \times 1$ vector $\mathbf{s}_{ij}$ represents the jth feature vector of ith target. Define the with-in class scatter matrix

$$\mathbf{S}_W = \sum_{i=1}^{c} \mathbf{S}_i,$$

where

$$\mathbf{S}_i = \sum_{j=1}^{N_i} (\mathbf{s}_{ij} - \mathbf{m}_i)(\mathbf{s}_{ij} - \mathbf{m}_i)^{\mathrm{T}},$$

and $\mathbf{m}_i = \frac{1}{N_i}\sum_{j=1}^{N_i} \mathbf{s}_{ij}$ is the mean or center of feature vectors from ith target. Similarly, define the between-class scatter matrix

$$\mathbf{S}_b = \sum_{i=1}^{c} N_i(\mathbf{m}_i - \mathbf{m})(\mathbf{m}_i - \mathbf{m})^{\mathrm{T}},$$

where

$$\mathbf{m} = \frac{1}{N}\sum_{i=1}^{c}\sum_{j=1}^{N_i} \mathbf{s}_{ij} = \frac{1}{N}\sum_{i=1}^{c} N_i \mathbf{m}_i$$

is the mean vector of $N = N_1 + \cdots + N_c$ feature vectors. Let $\mathbf{U}$ be the $Q \times Q$ discriminant matrix. Denote the vector containing the diagonal elements of a matrix $\mathbf{X}$ as diag $\{\mathbf{X}\}$. The function

$$J(\mathbf{U}) = \frac{\prod \mathrm{diag}\left\{\mathbf{U}^T\mathbf{S}_b\mathbf{U}\right\}}{\prod \mathrm{diag}\left\{\mathbf{U}^T\mathbf{S}_w\mathbf{U}\right\}}$$

is an effective measure of the separation of classes with N given feature vectors. Find the optimal discriminant matrix that maximizes the function $J(\mathbf{U})$.

2.15 Let $\mathbf{y}$ be the observed vector which follows the observation equation below

$$\mathbf{y} = \mathbf{H}\mathbf{x} + \mathbf{v},$$

where $\mathbf{H}$ is the observation matrix, $\mathbf{x}$ is the unknown state vector, and $\mathbf{v}$ is the additive observation noise vector. Suppose the observation noise follows the Gaussian distribution

$$f(\mathbf{v}) = \frac{1}{\sqrt{(2\pi)^p|\mathbf{R}|}}\exp\left(-\frac{1}{2}\mathbf{v}^T\mathbf{R}\mathbf{v}\right),$$

where $\mathbf{R}$ is the covariance matrix of observation noise and $|\mathbf{R}|$ is the determinant of $\mathbf{R}$. Give the MLE $\hat{\mathbf{x}}$ of the unknown state vector $\mathbf{x}$ and the covariance matrix $\mathbf{P}_e$ of the estimation error $\mathbf{e} = \mathbf{x} - \hat{\mathbf{x}}$. The problem is known as the maximum likelihood estimation for linear Gaussian model[136].

2.16 The following figure depicts a circuit that simulates MLE under Gaussian noise [136]. In the figure, V_i is the voltage drop between one end of the resistance R_i and the ground, and the other ends of each resistance are connected to a common point. If $R_{\text{in}} = \infty$, find the output voltage V.

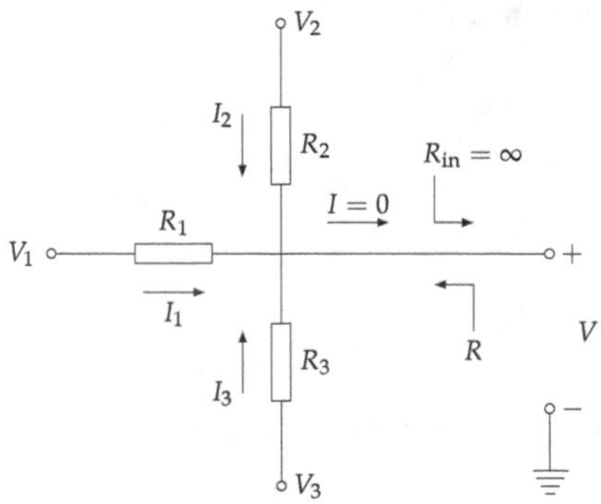

Fig-Exer. The circuit that simulates MLE

2.17 Suppose the observed data vector is expressed by $\mathbf{y} = \mathbf{A}\boldsymbol{\theta} + \mathbf{e}$, where the rank of $N \times p$ matrix $\mathbf{A}$ is p and $\mathrm{E}\{\mathbf{e}\} = \mathbf{0}$ and $\mathrm{var}[\mathbf{e}] = \sigma^2\boldsymbol{\Sigma}$ with positive definite matrix $\boldsymbol{\Sigma}$. Let

$$\hat{\boldsymbol{\theta}}_{\text{WLS}} = (\mathbf{A}^T\boldsymbol{\Sigma}^{-1}\mathbf{A})^{-1}\mathbf{A}^T\boldsymbol{\Sigma}^{-1}\mathbf{y}$$

be the WLS estimate of $\boldsymbol{\theta}$. Prove that

$$\hat{\sigma}^2 = \frac{1}{N-p}(\mathbf{y} - \mathbf{A}\hat{\boldsymbol{\theta}}_{\text{WLS}})^T\boldsymbol{\Sigma}^{-1}(\mathbf{y} - \mathbf{A}\hat{\boldsymbol{\theta}}_{\text{WLS}})$$

is an unbiased estimate of σ^2.

3 Signal Detection

Signal detection is referred to as the problem of inferring the existence of a particular signal by using observed data, and in essence belongs to statistical hypothesis testing. The so-called statistical hypothesis is simply a declaration about some unknown character of a population under investigation. The fundamental task of testing a statistical hypothesis is to decide whether the declaration about the unknown character can be supported by the samples from a random test. Generally speaking, such declaration is associated with some unknown parameters or functions of sample distribution, and the question of whether the declaration is statistically supported by samples is settled based on probability. In a word, with the evidence of observed data, the declaration is accepted if its correctness is highly probable, and is rejected otherwise.

From the perspective of statistical hypothesis testing, signal detection in various circumstances can be analyzed and studied in a unified mathematical framework. It is from this aspect that we start our discussion on the theory, method, and applications of signal detection in this chapter.

3.1 Statistical Hypothesis Testing

In principle, signal detection theory (SDT) is a theory of statistical hypothesis testing. By analyzing data from experiment, SDT can make decisions on uncertain stimulus (or response) as to whether the stimulus is a signal generated by some known process, or merely noise. SDT has broad applications of great diversity, including psychology (psychophysics, perception, memory), medical diagnosis (whether the symptoms match the diagnosis or not), wireless communications (whether the transmitted binary signal is zero or one), and radar (whether the blip on a radar display is an aircraft, a missile, or clutter interference), etc.

Psychologists are among the earliest to apply SDT to the study of cognition for the purpose of differentiating signal (stimulus present) from noise (stimulus absent) with memory recognition (to identify an object as being seen before or not) and lie test (to identify the answers as being truth or lies) as examples.

3.1.1 Basic Concepts of Signal Detection

Fig. 3.1.1 depicts the basic procedure of signal detection.

In transmission and sensing, the physical (signal) source is first transmitted and converted into observable data and then is observed and collected by sensors.

The observed physical process or phenomenon is called stimulus, and the output of sensing is called observed data. Single sample from data is inadequate for making

https://doi.org/10.1515/9783110475562-003

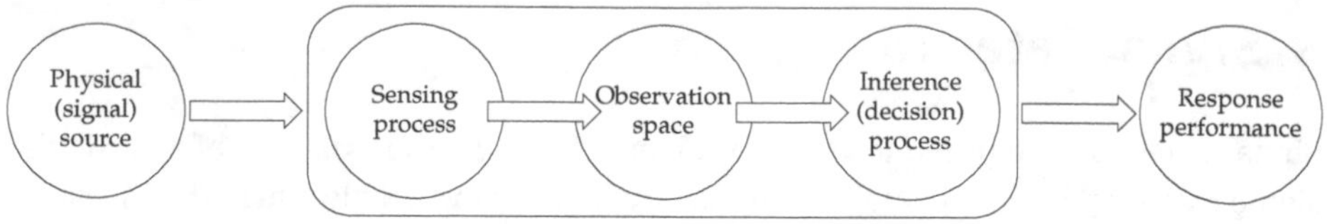

Fig. 3.1.1: Prcedure of Signal Detection

correct statistical decision. Thus, it is necessary to use a collection of observed data $y_1, \cdots, y_N$ for detector to make an inference or decision. The result of inference or decision is called response, which is further shown by display equipment.

From the perspective of computation, SDT is a computational framework describing how to extract signal from noise and explicating any deviation and factors that could affect the extraction. Recently, SDT has been successfully applied in explaining how human brain perceives environment and detects signal by suppressing noise through inherent functions of brain[92].

It should be mentioned that the concepts of signal and noise do not necessarily refer to actual signal and noise. Rather, they are used metaphorically sometimes. In general, common event is conventionally regarded as signal, and sporadic event is treated as noise. For example, in the experiment of memory recognition, the participant is required to decide whether the current stimulus has been presented before or not[4]. In such case, signal is the familiar stimulus in memory, and noise is the perception of a novel stimulus. As another example, in tumor diagnosis (by ultrasound), signal of clear echo usually suggests benign lump. In contrast, indistinct echo or irregular response could be a sign of a malignant tumor (cancer). In the following, we use the terms of signal and noise to label different objects in detection.

In signal detection, there are two scenarios: the presence and absence of a signal, where the absence of a signal also means the presence of noise. The output of decision or detection is binary with either "yes" (affirmation) or "no" (rejection). Thus, the result is one of the following four categories.

(1) The signal is present and the decision is "yes." The decision is a success and is called a "hit".
(2) The signal is present but the decision is "no" which rejects the presence of signal. The erroneous decision is called a missed detection or a "miss." In radar or other military applications, the result is also called "missing alarm".
(3) The signal is absent but the decision is "yes" which affirms the presence of signal. The incorrect decision is called a "false alert" or a "false alarm" (FA).
(4) The signal is absent and the decision is "no" and rejects the presence of signal. The decision is called a correct denial or a "correct rejection" (CR).

Tab. 3.1.1 lists all possible results above in SDT[4].

Tab. 3.1.1: Four Possible Results in SDT

Signal	Decision	
	Yes (Affirmation)	No (Rejection)
Presence	hit	miss
Absence	false alert, false alarm	correct denial, correct rejection

Note that the consequences of incorrect decisions in different scenarios can differ greatly. In wireless communications, bit errors occur in binary communications for deciding 1 when 0 is transmitted or decide 0 when 1 is transmitted. Both detection errors have almost identical consequences. However, in military applications such as radar and sonar, the incorrect decision (missing alarm) when the signal (unexpected aircraft or missile) is present could incur more severe consequence than the incorrect decision (false alarm) when the signal is absent. Similarly, in the diagnosis of tumor, deciding a benign tumor to be cancerous could cause much psychological stress to patient. In contrast, diagnosing malignant tumor as being benign can delay the treatment of the patient and may have fatal consequence.

The proportion of a stimulus or response in total number of experiment is called relative frequency or probability. In SDT, the relative frequencies of four results are related. For instance, when the signal is present, the sum of rates of hit and miss is one (third row of Tab. 3.1.1). If the signal is absent, the sum of rates of FA and CR is also one (fourth row of Tab. 3.1.1).

The observer that declares "no" on all occasions can achieve a CR rate of 100% but never hit, with always missing alarms (third column of Tab. 3.1.1). On the contrary, the observer that keeps declaring "yes" can make correct guess when signal is present but never have CR, i.e., it has FAs all the time (second column in Tab. 3.1.1).

Following are three primary causes of error in signal and noise discrimination.

(1) The stimulus is beyond the standard detection window. For instance, the physical or physiological detector might has a narrower frequency range than the detected signal when detecting ultraviolet signal using visible light image sensors with color vision deficiency.
(2) The stimulus is "masked" by external noise, which reduces SNR. In such case, there are two cases as below.
① Noise increase. The external noise can severely affect detection.
② Signal decrease. The external noise could excite suppression causing lower energy of signal.
(3) Change in converter or sensor (including the effect of gain control induced by experiment or neurons and failure of normalization).

The so-called "statistical inference" is to form judgment on the hypothesis related to the statistical properties (e.g., PDF) of random variable. The hypothesis could arise from practical observation or theoretical analysis of the stochastic phenomena. Basically, the statistical hypothesis makes statements on the character of the studied population and is denoted by H. Several simple examples are given below.

Example 3.1.1 On detecting the presence of a target, we have two hypotheses below.

$$\begin{cases} H_0: & \text{The target is not present.} \\ H_1: & \text{The target is present.} \end{cases}$$

Example 3.1.2 For an athlete doping test, one may have two declarations below.

$$\begin{cases} H_0: & \text{The test result is negative.} \\ H_1: & \text{The test result is positive.} \end{cases}$$

Example 3.1.3 For the probability distribution of a given random variable ξ, one may have the following declarations or assumptions.

$$\begin{cases} H_0: & \xi \text{ follows normal distribution.} \\ H_1: & \xi \text{ follows exponential distribution.} \\ H_2: & \xi \text{ follows } \chi^2 \text{ distribution.} \end{cases}$$

In the above examples, the sets of target signals, doping tests, and probability distributions are the populations we are interested in, and the presence of signal, positiveness of test, and specific distribution are respectively the characters of the populations.

Binary hypothesis testing is the testing problem that has two statistical hypotheses. Examples 3.1.1 and 3.1.2 are instances of binary hypothesis testing. In most cases, the two statistical hypotheses cannot be true at the same time: either H_0 is true and H_1 is false or H_1 is true and H_0 is false. Namely, the two hypotheses are mutually exclusive as one is true and the other must be false. The random event that catches our attention is mostly accidental and the declaration of its not occurring is thus taken as the default hypothesis. For example, in radar, the presence of an aircraft target is accidental. In doping test, the positive samples are also rare. As probability model, Gaussian (normal) distribution is more common. It is the convention to take H_0 as the default hypothesis of the random event not occurring and H_1 as the hypothesis of the occurrence of the random event. The default hypothesis H_0 is called null or original hypothesis and the opposite hypothesis H_1 is called alternative hypothesis.

Hypothesis testing can have more than two hypotheses, e.g., Example 3.1.3. Most part of this chapter will be devoted to binary hypothesis testing but the last two sections will introduce the subjects of testing multiple hypotheses and multiple hypothesis testing respectively.

3.1.2 Signal Detection Measures

The SDT measure can be expressed by rate or probability.

1. Measure of Rate

Hit rate and false-alarm rate are both rate measures.

(1) Hit rate

$$R_{\mathrm{H}} = \frac{\text{Number of hits}}{\text{Number of signal events}}. \tag{3.1.1}$$

(2) False-alarm rate

$$R_{\mathrm{FA}} = \frac{\text{Number of false alarms}}{\text{Number of noise events}}. \tag{3.1.2}$$

2. Measure of Probability

Probability measure is usually determined by certain measure function. ϕ function and inverse ϕ function are two commonly used measure functions.

(1) ϕ function maps z score into probability. The value of ϕ function determine the area under the PDF of normal distribution that lies on the left side of vertical line set by z score. The larger z score is, the higher the probability gets. For example, $\phi(-1.64) = 0.05$ means that the probability that corresponds to z score -1.64 is 0.05.
On the other hand, z test can yield the area of normal distribution on the right side of z score. As z score gets larger, z test gives smaller probability.

(2) Inverse ϕ function is the inversion of ϕ function denoted by ϕ^{-1}. Inverse ϕ function maps probability into z score. For example, $\phi^{-1}(0.05) = -1.64$ means that the left side probability of 0.05 has a z score of -1.64.

In binary hypothesis testing, the statistical evidence Y of one experiment is usually first calculated from N observations $y_1, \cdots, y_N$. Then, the test decision is made by comparing Y against threshold λ, which is called decision parameter. As the statistical evidence Y is a random variable rather than a constant, it is natural to consider odds or probability as the measure that quantifies the consequence of the decision based on certain statistical evidence.

The decision of accepting hypothesis H_i under hypothesis H_j is denoted by $(H_i|H_j)$ and the corresponding probability measure is given by conditional probability $P(H_i|H_j)$, which reads "the probability of accepting H_i when H_j is true". In the case of binary hypothesis testing, with such notation, there are four types of probabilities:

(1) Probability of false alarm $P(H_1|H_0) = P_{\mathrm{F}}$: the probability of deciding H_1 when H_0 actually occurs, which can be expressed as

$$P(H_1|H_0) = P_{\mathrm{F}} = P(Y > \lambda|H_0) = \int_{\lambda}^{\infty} p(y|H_0)\,\mathrm{d}y. \tag{3.1.3}$$

In the equations, we define $p(y|H_0)$ and $p(y|H_1)$ as the PDFs of observation y conditioned on H_0 and H_1, respectively.

(2) Probability of correct rejection: the probability of deciding H_0 when H_0 is true, expressed as

$$P(H_0|H_0) = P(Y < \lambda|H_0) = \int_{-\infty}^{\lambda} p(y|H_0)\,dy. \tag{3.1.4}$$

(3) Probability of hit: the probability of deciding H_1 when H_1 is true, which has expression as

$$P(H_1|H_1) = P_H = P(Y > \lambda|H_1) = \int_{\lambda}^{\infty} p(y|H_1)\,dy. \tag{3.1.5}$$

(4) Probability of miss detection or missing alarm: the probability of choosing H_0 when H_1 is true, which is expressed by

$$P(H_0|H_1) = P(Y < \lambda|H_1) = \int_{-\infty}^{\lambda} p(y|H_1)\,dy. \tag{3.1.6}$$

The probabilities of false alarm and correct rejection sum to one, and so do the probabilities of hit and missing alarm. That is,

$$P(H_1|H_0) + P(H_0|H_0) = 1 \quad \text{and} \quad P(H_1|H_1) + P(H_0|H_1) = 1. \tag{3.1.7}$$

As is evident from the definitions above, all four types of probabilities depend upon

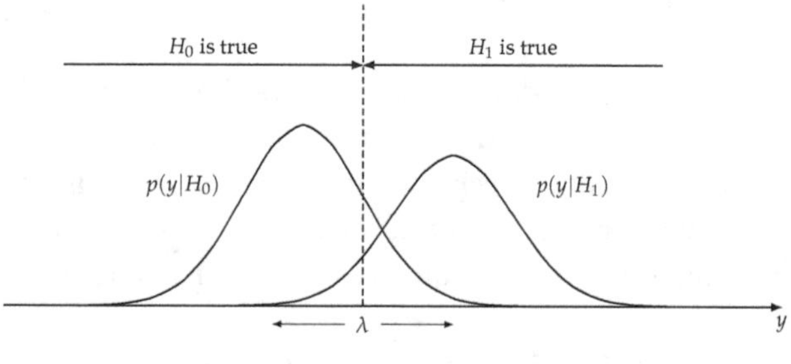

Fig. 3.1.2: Influence of parameter λ on four probabilities

the decision parameter λ in the signal detection model (see Fig. 3.1.2). Fig. 3.1.2 gives an illustration of the conditional PDFs $p(y|H_0)$ and $p(y|H_1)$ and the decision regions determined by λ. From Fig. 3.1.2, several observations can be summarized.

(1) The effect of overlap area of PDFs. Smaller overlap area of conditional PDFs $p(y|H_0)$ and $p(y|H_1)$ results in a smaller hit probability P_H and a higher false alarm probability P_F. On the contrary, larger overlap area of the two PDFs will increase the hit probability and decrease the false alarm probability.

(2) The effect of decision parameter λ. As λ moves left towards $p(y|H_0)$, i.e., λ is decreased, both P_H and P_F will grow. On the other hand, as λ moves right towards $p(y|H_1)$, i.e., λ is increased, both P_H and P_F will decrease accordingly.

As the null hypothesis H_0 is the default hypothesis, the rejection of H_0 is the focus of signal detection. Regarding the decision about H_0, one may either reject H_0 or fail to reject H_0. Depending on what actually happens, there are two results for each decision:

$$\text{reject } H_0 \begin{cases} \text{when } H_0 \text{ is true,} \\ \text{when } H_0 \text{ is false.} \end{cases} \qquad \text{fail to reject } H_0 \begin{cases} \text{when } H_0 \text{ is true,} \\ \text{when } H_0 \text{ is false.} \end{cases}$$

Note that failing to reject H_0 should not be identified with accepting H_0. For instance, in the early-warning radar network, as a simple rule of data fusion, the rejection of H_0 cannot be made until the decision is settled by the vote of multiple radars. It does not means that the whole radar network should reach the decision of absence of target when only a minority of radars fails to reject H_0. Similarly, in Example 3.1.2, being indecisive to reject H_0 in one doping test of an athlete gives at most a "suspected negative" case, which requires more tests to completely rule out the possibility of a positive doping test.

Furthermore, it should be emphasized that a decision in statistical hypothesis testing is in essence a procedure of inference based on statistics of the observation. Owing to the presence of observation noise or error and the limitation in observation length, the estimation errors in statistics are inevitable. Consequently, our statistical inference is fallible. In other words, deciding the rejection of H_0 does not necessarily imply H_0 being false. In fact, H_0 could still be true and it is our inference that can accidentally be incorrect. Likewise, failing to reject H_0 cannot be taken as the truth of H_0 being definitely confirmed since our decision can have error.

If our decision is to reject H_0, we are either rejecting the true hypothesis (erroneous decision) or rejecting the false hypothesis (correct decision). Likewise, if we are unable to reject H_0, the decision either fails to reject the true hypothesis (correct decision), or fails to reject the false hypothesis (erroneous decision). Therefore, no matter the decision is to reject or not to reject H_0, the final inference could be a deviation from the reality, which is called inference error.

The error of rejecting a true H_0 hypothesis is called type I error or error of the first kind, and the error for failing to reject a false H_0 hypothesis is called type II error or error of the second kind. Noticeably, type I error occurs only when H_0 is true, and type II error only when H_0 is untrue. As H_0 cannot not be both true and untrue, the inference error must belong to either type of error, and cannot be both.

To quantify the inference error, a numeric parameter is required to measure the odds of the two types of error, and the error probability is exactly the relevant metric to use.

Definition 3.1.1. *The probability of rejecting H_0 when H_0 is true is called the probability of type I error, which is denoted as α.*

Definition 3.1.2. *The probability of failing to reject H_0 when H_0 is false is called the probability of type II error, which is denoted as β.*

Obviously, the two types of error probabilities both lie between 0 and 1, i.e., $0 \le \alpha \le 1$ and $0 \le \beta \le 1$.

From the definitions above, the probabilities of type I and II errors can be expressed as

$$P(\text{reject } H_0 | H_0 \text{ is true}) = \alpha, \tag{3.1.8}$$

$$P(\text{fail to reject } H_0 | H_0 \text{ is false}) = \beta. \tag{3.1.9}$$

From Eqs. (3.1.8) and (3.1.9), both error probabilities are conditional probabilities. In general, due to the limitation in the knowledge of real world, we cannot get the precise probabilities of type I and II errors, and instead the estimates of the probabilities α and β of type I and II errors can be obtained in practice.

The probability of type I error α is also called level of statistical significance, which indicates that with adequate sample, the evidence of rejecting H_0 is sufficient (or significant) and the maximum error probability of rejection is at most α.

The fundamental problem in statistical hypothesis testing is how to make decision about the truth of hypothesis H_0. To this end, a decision rule must be designed to form judgment from experiment data on whether rejecting H_0 or not. Such decision rule constitutes the procedure of statistical hypothesis testing.

Definition 3.1.3. *The statistical hypothesis testing based on the statistic of a population is a decision rule, following which the decision on whether to reject H_0 can be made once samples from random experiment are obtained.*

Decision rule in statistical hypothesis testing is mostly in the form of a function. The decision procedure can be reduced to the comparison between the statistic of observed data and threshold. That is, the decision relies upon some proper statistic, which is called test statistic.

Which statistic can be employed as test statistic is completely determined by the model used in hypothesis testing problem and the decision rule. The sample average $\bar{y} = \frac{1}{N}\sum_{n=1}^{N} y_n$ of N observations $y_1, \cdots, y_N$ is one of the most simple and frequently used test statistics.

Now consider the simple null hypothesis about parameter θ

$$H_0 \; : \; \theta = \theta_0,$$

where θ_0 is the parameter under investigation.

If the alternative hypothesis is of the form

$$H_1 \; : \; \theta > \theta_0 \quad \text{or} \quad H_1 \; : \; \theta < \theta_0,$$

H_1 is called one-sided alternative hypothesis, since the value of θ under H_1 is only on one side of the parameter θ_0.

Different from one-sided alternative hypothesis, the alternative hypothesis

$$H_1 \ : \ \theta \neq \theta_0$$

is called two-sided alternative hypothesis, since the value of θ under H_1 can lie on both sides of the parameter θ_0.

When H_1 is a composite alternative hypothesis, which is in a more complex form, θ that rejects H_0 takes value in a parameter set. As H_0 is the null hypothesis, deciding whether to reject H_0 is the fundamental task of hypothesis testing.

Definition 3.1.4. *If the decision of rejecting H_0 is made by the result of comparison $g > Th$, where g is test statistic and Th is preset threshold, the range (Th, ∞) is called the critical region of the test statistic.*

The critical region is denoted by $R_c = (\text{Th}, \infty)$. Evidently, as the parameter θ traverses the critical region R_c, the probability of type II error $\beta(\theta)$ also varies to yield a function $\beta(\theta)$, which is called operating characteristic function. The curve showing the variation of β with θ is called operating characteristic curve.

$\beta(\theta)$ is the probability of θ falling outside critical region and failing to reject H_0 in the case of false H_0 (incorrect decision) , and accordingly $1 - \beta(\theta)$ is the probability of test statistic in the critical region and rejecting H_0 for false H_0 (correct decision).

Definition 3.1.5. *The function $P(\theta) = 1 - \beta(\theta)$ is defined as the power function of statistical hypothesis testing, which is the probability of rejecting H_0 for a false H_0 (correct decision).*

Notice that in binary hypothesis testing, false H_0 does not necessarily imply H_1 being true. Rejecting H_0 or accepting H_1 is the basic decision in binary hypothesis testing. Thus, $1 - \beta(\theta)$ represents the effectiveness (power) of the basic decision of correctly rejecting H_0 or accepting H_1, which is the reason for the name of power function.

The power function $1-\beta(\theta)$ and the characteristic function $\beta(\theta)$ are complementary in that their sum is one.

3.1.3 Decision Space

Now we turn to the detection problem involving two signals S_1 and S_0, which belongs to the binary hypothesis testing. In varied scenarios of application, what the two signals actually refer to could be quite different. In target detection of radar or sonar, S_1 and S_0 are the presence and absence of the target, respectively. However, in signal detection of digital communications, $S_1 = g(t)$ and $S_0 = -g(t)$ are two transmitted signals with opposite polarity.

The collection of all signals is called signal or parameter space, and is denoted by $\mathcal{S}$. In binary hypothesis testing, the signal space is composed of two signals S_0 and S_1, that is,

$$\mathcal{S} = \{S_0, S_1\}.$$

Transmitted through channel, the signal is mixed with additive noise w_n and the observed data are:

$$\begin{cases} H_0 \ : \ y_n = S_0 + w_n, \\ H_1 \ : \ y_n = S_1 + w_n, \end{cases} \qquad n = 1, 2, \cdots , \tag{3.1.10}$$

where H_0 is the null hypothesis and H_1 is the alternative hypothesis.

The additive noise w_n samples are assumed to be independent and stationary. In common applications, the additive noise w_n is also assumed to be white Gaussian with mean $\mathrm{E}\{w_n\} = \mu_0$ and variance σ^2.

Due to the existence of additive noise w_n, the observed data y_n are random. As a consequence, correct decision about signal detection cannot be achieved with single random sample y_n, and the judgment on the presence of signal is made based on N observed samples $y_1, \cdots , y_N$ in the signal detection problem. The judgment is called decision in the detection problem.

For convenience, denote

$$\mathbf{y} = (y_1, \cdots , y_N) \tag{3.1.11}$$

as the array of observed samples $y_1, \cdots , y_N$. Note that $\mathbf{y}$ is not vector notation here.

The set of all possible observations constitute the space of observed samples, and is denoted as Ω. The observation array $\mathbf{y}$ is an element of Ω, and belongs to the space of observed samples, which is denoted as $\mathbf{y} \in \Omega$.

Still, $\mathbf{y}$ is an array of random variables, which is tedious to be used as a test statistic for the detection problem. As such, the observation array is needed to be reduced into a well-defined statistic, which is called decision statistic for the detection problem and is expressed by

$$t = g(\mathbf{y}) = g(y_1, \cdots , y_N). \tag{3.1.12}$$

The average of N samples is one of the most simple decision statistic

$$t = g(\mathbf{y}) = \bar{y} = \frac{1}{N}\sum_{n=1}^{N} y_n. \tag{3.1.13}$$

Suppose that the distribution is characterized by some unknown parameter θ. N samples $y_1, \cdots , y_N$ are drawn from the distribution (to estimate θ). Let $t = g(y_1, \cdots , y_N)$ be the statistic (or random variable) obtained from observations $y_1, \cdots , y_N$, and P_θ be the distribution family parameterized by θ.

Definition 3.1.6. [256] *Statistic $t = g(y_1, \cdots , y_N)$ is called the sufficient statistic of distribution family P_θ if and only if any one of the following equivalent conditions holds:*

(1) For arbitrary event A, conditional probability $P(A|t)$ is independent from θ.
(2) For any random variable α whose expectation $\mathrm{E}\{\alpha\}$ exists, the conditional expectation $\mathrm{E}\{\alpha|t\}$ is independent from θ.
(3) Given t, the conditional PDF $p(\alpha|t)$ of any random variable α, which always exists, does not depend on θ.

In general, conditional probability $P(A|t)$, conditional expectation $\mathrm{E}\{\alpha|t\}$, and conditional distribution $p(\alpha|t)$ are all connected with θ. But for sufficient statistic t fixed as T, the conditional distribution and expectation are independent from θ, which suggests that letting $t = T$ also has θ fixed. To put it in simple words, the statistic t plays the same role as θ, as what the term "sufficient statistic" indicates. Comprehensibly, one may expect test statistic t is a sufficient statistic in the hypothesis testing.

In some textbooks, decision statistic is also named decision function. Nevertheless, in this book, we refer to the term of decision function as the function defined by decision rule and denote the function by $L(\mathbf{y}) = L(y_1, \cdots, y_N)$. One of the most popular decision function is the ratio of two conditional PDFs: $p(\mathbf{y}|H_1)/p(\mathbf{y}|H_0)$, which is usually named as likelihood ratio. Generally, the derivation of decision function from observed data is not straightforward. In contrast, decision statistic can be estimated directly from the data.

Let Th be the threshold used in the decision. If the decision statistic is greater than the threshold, H_0 is rejected and H_1 is accepted to decide that signal S_1 is present. Otherwise, claim that H_0 is true and decide that signal S_0 is present. The decision rule can written as

$$\begin{cases} H_0 & : \text{ if } g(\mathbf{y}) \le \mathrm{Th}, \text{ decide signal } S_0 \text{ is present}, \\ H_1 & : \text{ if } g(\mathbf{y}) > \mathrm{Th}, \text{ decide signal } S_1 \text{ is present}, \end{cases} \tag{3.1.14}$$

or in a combined form,

$$g(\mathbf{y}) \underset{H_1}{\overset{H_0}{\lesseqgtr}} \mathrm{Th}. \tag{3.1.15}$$

Decision space is the set of all possible values of decision statistic, and is denoted as D. With threshold Th as the boundary, the decision space $D = (-\infty, \infty)$ is divided into two subspaces: $D = D_0 \cup D_1$, where

$$D_0 = (-\infty, \mathrm{Th}) \quad \text{and} \quad D_1 = (\mathrm{Th}, \infty).$$

All the decision results constitute an action space or outcome space. The outcome space of binary hypothesis testing is composed of two elements A_0 and A_1, i.e.,

$$\mathcal{A} = \{A_0, A_1\},$$

where A_0 and A_1 are the outcomes of decision under hypotheses H_0 and H_1, respectively.

A decision can be either hard or soft. A hard decision is simply a claim about the presence of a signal or signal characteristic (e.g., signal polarity). While a soft decision gives numeric value, i.e., probability, that measures the uncertainty in the presence of a signal or signal characteristic. In binary hypothesis testing, hard decision is more common.

If the signal space $\mathcal{S} = \{S_1, S_2, \cdots, S_M\}$ contains more than two elements, the corresponding detection problem requires testing multiple hypothesis and is called M-ary testing.

To sum up, the four spaces in decision theory are

$$\begin{aligned} \mathcal{S} &= \text{signal or parameter space,} \\ \Omega &= \text{sample space,} \\ D &= \text{decision space,} \\ \mathcal{A} &= \text{outcome space.} \end{aligned}$$

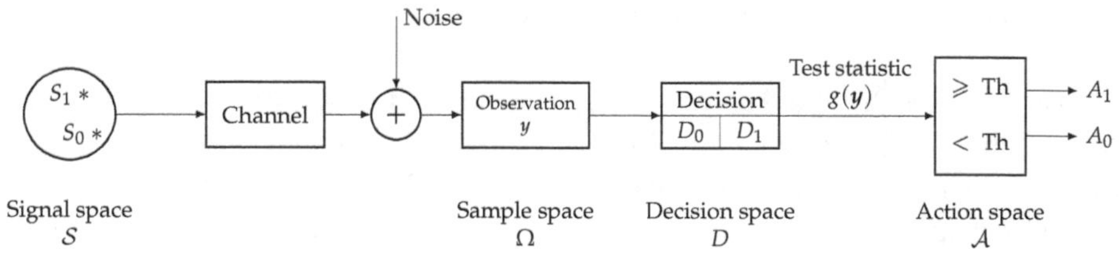

Fig. 3.1.3: Decison space of binary hypothesis testing

Fig 3.1.3 depicts the relationship among decision theory spaces of binary hypothesis testing as the example.

The random variable following normal distribution $\mathcal{N}(\mu, \sigma^2)$ with mean μ and variance σ^2 plays a central role in binary hypothesis testing. In fact, random variables in many practical problems of science and engineering are normally distributed or approximately normally distributed.

In the next section, we will concentrate on the derivation of the decision rule for the detection of a signal.

3.2 Probability Density Function and Error Function

In signal detection theory and method, one may repeatedly encounter calculations involving PDF and error function. So, before the discussion of signal detection, it is necessary to review the basics of PDF and error function which are closely related to signal detection.

3.2.1 Probability Density Function

Let x be a random variable. The probability distribution function of random variable x is defined as

$$F(x) = \int_{-\infty}^{x} p(u)\,\mathrm{d}u, \tag{3.2.1}$$

and $p(x)$ is the distribution density function of x, or distribution density for short.

Distribution density function has the following properties:

$$p(x) \geq 0 \quad \text{(nonnegative)}, \tag{3.2.2}$$

$$\int_{-\infty}^{\infty} p(x)\,\mathrm{d}x = 1 \quad \text{(normalized)}. \tag{3.2.3}$$

In this chapter, we assume the occurrence of discrete events H_0 and H_1 is random and observations $y_1, \cdots, y_N$ are continuous random variables.

In the detection problem for the presence of signal,

$$\begin{cases} H_0 \ : \ y_n = w_n, \\ H_1 \ : \ y_n = s + w_n, \end{cases} \qquad n = 1, \cdots, N, \tag{3.2.4}$$

the additive noise w_n is assumed to be Gaussian white with zero mean and variance σ^2. Thus, conditioned on either hypothesis of H_0 and H_1, the observed data y_n is normally distributed. The means of the normal distributions of y_n are 0 and s, respectively, and the variances have identical value of σ^2. That is, the conditional distributions are

$$p(y_n|H_0) = \frac{1}{\sqrt{2\pi}\sigma} \mathrm{e}^{-y_n^2/(2\sigma^2)}, \tag{3.2.5}$$

$$p(y_n|H_1) = \frac{1}{\sqrt{2\pi}\sigma} \mathrm{e}^{-(y_n-s)^2/(2\sigma^2)}, \tag{3.2.6}$$

which can be written in a compact form as

$$p(y_n|H_i) = \frac{1}{\sqrt{2\pi}\sigma} \mathrm{e}^{-(y_n-\mu_i)^2/(2\sigma^2)}, \quad i = 0, 1. \tag{3.2.7}$$

In (3.2.7), $\mu_0 = 0$ and $\mu_1 = s$.

For an array of samples $\mathbf{y} = (y_{1,}, \cdots, y_N)$, the conditional PDF $p(y_n|H_i)$ is in fact a marginal PDF conditioned on H_i.

In binary hypothesis testing, the decision is based on all N observed samples rather than single sample. As the samples are independent, the joint conditional PDF $p(\mathbf{y}|H_i)$ of the sample array $\mathbf{y} = (y_1, \cdots, y_N)$ is the product of the marginal PDFs $p(y_n|H_i)$ of each sample y_n:

$$\begin{aligned} p(\mathbf{y}|H_i) &= p(y_1,\cdots,y_N|H_i) = \prod_{n=1}^{N} p(y_n|H_i) = \prod_{n=1}^{N} \frac{1}{\sqrt{2\pi}\sigma} \mathrm{e}^{-(y_n-\mu_i)^2/(2\sigma^2)} \\ &= \frac{1}{(2\pi\sigma^2)^{N/2}} \exp\left(-\sum_{n=1}^{N} \frac{(y_n-\mu_i)^2}{2\sigma^2}\right) \quad i=1,2. \end{aligned} \tag{3.2.8}$$

When discrete random event H_i occurs, the conditional PDF $p(y_1,\cdots,y_N|H_i)$ of continuous samples $\mathbf{y} = (y_1,\cdots,y_N)$ and the conditional probability $P(H_i|y_1,\cdots,y_N)$ have different meanings.

Conditional PDF. The conditional PDF $p(y_1,\cdots,y_N|H_i)$ given that random event H_i i has occurred is called the likelihood function of samples $\mathbf{y}$, since it represents the plausibility, or the likelihood in terms of probability theory, of the random event H_i generating the samples $\mathbf{y}$, after the occurrence of event H_i. Accordingly, the ratio of two likelihood functions $p(y_1,\cdots,y_N|H_1)$ and $p(y_1,\cdots,y_N|H_0)$ is called likelihood ratio.

Conditional Probability. The conditional probability $P(H_i|y_1,\cdots,y_N)$ is the plausibility or probability of H_i to happen given samples $\mathbf{y}$, and thus represents the decision maker's confidence in the occurrence of random event H_i after collecting samples $y_1,\cdots,y_N$.

The joint PDF of the sample and the random event H $p(y_1,\cdots,y_N,H)$ is defined as

$$p(y_1,\cdots,y_N,H) = p(y_1,\cdots,y_N|H)P(H) = P(H|y_1,\cdots,y_N)p(y_1,\cdots,y_N), \tag{3.2.9}$$

from which we have

$$P(H|y_1,\cdots,y_N) = \frac{p(y_1,\cdots,y_N|H)P(H)}{p(y_1,\cdots,y_N)}. \tag{3.2.10}$$

The integral of PDF over interval $(-\infty,\infty)$ is identical to one. That is,

$$\int_{-\infty}^{\infty}\cdots\int_{-\infty}^{\infty} p(y_1,\cdots,y_N)\,\mathrm{d}y_1\cdots\mathrm{d}y_N = 1 \quad \text{(joint PDF)}, \tag{3.2.11}$$

$$\int_{-\infty}^{\infty}\cdots\int_{-\infty}^{\infty} p(y_1,\cdots,y_N|H)\,\mathrm{d}y_1\cdots\mathrm{d}y_N = 1 \quad \text{(likelihood function)}. \tag{3.2.12}$$

The mean or expectation of function $g(y_1,\cdots,y_N)$ is defined as the N-fold integral of the product of this function and the joint PDF $p(y_1,\cdots,y_N)$:

$$\mathrm{E}\left\{g(y_1,\cdots,y_N)\right\} \stackrel{\text{def}}{=} \int_{-\infty}^{\infty}\cdots\int_{-\infty}^{\infty} g(y_1,\cdots,y_N)p(y_1,\cdots,y_N)\,\mathrm{d}y_1\cdots\mathrm{d}y_N. \tag{3.2.13}$$

Particularly, when $H = \theta$ with θ as some random parameter, the integral involving posterior PDF

$$\mathrm{E}\left\{\theta|y_1,\cdots,y_N\right\} \stackrel{\text{def}}{=} \int_{-\infty}^{\infty} \theta p(\theta|y_1,\cdots,y_N)\,\mathrm{d}\theta \tag{3.2.14}$$

is the conditional mean or expectation of random parameter θ given N observed samples $y_1, \cdots, y_N$. Clearly, conditional expectation $\mathrm{E}\{\theta|y_1, \cdots, y_N\}$ is the function of observation array $\mathbf{y} = (y_1, \cdots, y_N)$.

Next we analyze the relation between the expectation $\mathrm{E}\{\theta\}$ and the conditional expectation $\mathrm{E}\{\theta|y_1, \cdots, y_N\}$.

Let $\mathrm{E}_{y_1,\cdots,y_N}$ and E_θ be the notations for the expectations over random array $(y_1, \cdots, y_N)$ and random parameter θ, respectively. We have

$$\mathrm{E}\{\theta\} = \mathrm{E}_{y_1,\cdots,y_N}\{\mathrm{E}_\theta\{\theta|y_1, \cdots, y_N\}\}, \tag{3.2.15}$$

which is known as the law of conditional expectation. From

$$\mathrm{E}_{y_1,\cdots,y_N}\{g(y_1, \cdots, y_N)\} = \int_{-\infty}^{\infty} g(y_1, \cdots, y_N)p(y_1, \cdots, y_N)\,\mathrm{d}y_1 \cdots \mathrm{d}y_N, \tag{3.2.16}$$

the expectation of θ is obtained by substituting (3.2.14) and (3.2.16) into (3.2.15) as

$$\begin{aligned}
\mathrm{E}\{\theta\} &= \mathrm{E}_{y_1,\cdots,y_N}\{\mathrm{E}_\theta\{\theta|y_1, \cdots, y_N\}\} = \mathrm{E}\left\{\int_{-\infty}^{\infty} \theta p(\theta|y_1, \cdots, y_N)\,\mathrm{d}\theta\right\} \\
&= \int_{-\infty}^{\infty} \cdots \int_{-\infty}^{\infty}\left[\int_{-\infty}^{\infty} \theta p(\theta|y_1, \cdots, y_N)\,\mathrm{d}\theta\right] p(y_1, \cdots, y_N)\,\mathrm{d}y_1 \cdots \mathrm{d}y_N \\
&= \int_{-\infty}^{\infty} \cdots \int_{-\infty}^{\infty}\int_{-\infty}^{\infty} \theta p(\theta|y_1, \cdots, y_N)p(y_1, \cdots, y_N)\,\mathrm{d}\theta\,\mathrm{d}y_1 \cdots \mathrm{d}y_N,
\end{aligned} \tag{3.2.17}$$

which is an $N + 1$-fold integral.

3.2.2 Error Function and Complementary Error Function

In signal detection, it is often needed to calculate the detection probabilities and error, which has a close relationship with error and complementary error functions of Gaussian random variable.

For Gaussian random variable x, its PDF is

$$p(x) = \frac{1}{\sqrt{2\pi}\sigma_x}\mathrm{e}^{-(x-m_x)^2/(2\sigma_x^2)}, \tag{3.2.18}$$

where m_x and σ_x^2 are respectively the mean and variance of Gaussian random variable x. The cumulative distribution function (CDF) is defined by

$$F(x) = \int_x^{\infty} p(u)\,\mathrm{d}u = \frac{1}{\sqrt{2\pi}\sigma_x}\int_x^{\infty}\mathrm{e}^{-(u-m_x)^2/(2\sigma_x^2)}\,\mathrm{d}u = \frac{1}{2}\frac{2}{\sqrt{\pi}}\int_{(x-m_x)/(\sqrt{2}\sigma_x)}^{\infty}\mathrm{e}^{-t^2}\,\mathrm{d}t. \tag{3.2.19}$$

The error function is given by

$$\mathrm{erf}(z) = \frac{2}{\sqrt{\pi}} \int_0^z \mathrm{e}^{-t^2}\,\mathrm{d}t, \tag{3.2.20}$$

and the complementary error function is given by

$$\mathrm{erfc}(z) = \frac{2}{\sqrt{\pi}} \int_z^\infty \mathrm{e}^{-t^2}\,\mathrm{d}t = 1 - \mathrm{erf}(z). \tag{3.2.21}$$

The CDF of Gaussian random variable x in (3.2.19) can then be expressed by error and complementary error functions as

$$F(x) = \frac{1}{2}\mathrm{erfc}\left(\frac{x - m_x}{\sqrt{2}\sigma_x}\right) = \frac{1}{2}\left[1 - \mathrm{erf}\left(\frac{x - m_x}{\sqrt{2}\sigma_x}\right)\right]. \tag{3.2.22}$$

So, the error function is also called probability integral.

Error function has the following symmetry properties

$$\mathrm{erf}(-z) = -\mathrm{erf}(z) \quad \text{and} \quad \mathrm{erf}(z^*) = (\mathrm{erf}(z))^*. \tag{3.2.23}$$

In particular, for $z \to \infty$, it holds that

$$\lim_{z\to\infty} \mathrm{erf}(z) = 1 \quad \text{and} \quad \lim_{z\to-\infty} \mathrm{erf}(z) = -1, \quad |\arg(z)| < \frac{\pi}{4}. \tag{3.2.24}$$

Thus, we have

$$\lim_{z\to\infty} \mathrm{erfc}(z) = 0 \quad \text{and} \quad \lim_{z\to-\infty} \mathrm{erfc}(z) = -2, \quad |\arg(z)| < \frac{\pi}{4}, \tag{3.2.25}$$

which is an important result and will be used in the following sections.

The area under the right tail of Gaussian PDF is denoted as $Q(x)$, defined as

$$Q(x) = \frac{1}{\sqrt{2\pi}} \int_x^\infty \mathrm{e}^{-t^2}\,\mathrm{d}t, \tag{3.2.26}$$

which is called Q-function.

Comparing (3.2.20) and (3.2.26) yields the relation of Q-function, error and complementary error functions:

$$Q(x) = \frac{1}{2}\mathrm{erf}\left(\frac{x}{\sqrt{2}}\right) = 1 - \frac{1}{2}\mathrm{erfc}\left(\frac{x}{\sqrt{2}}\right). \tag{3.2.27}$$

Appendix 3A gives a table of error function. Both error function value for given x and variable value x for given $\mathrm{erf}(x)$ can be obtained by looking up the table. Besides, values of inverse complementary error function and Q-function can be read from the table.

Example 3.2.1 Find the value of inverse complementary error function $\mathrm{erfc}^{-1}(0.02)$.

Solution. Let $x = \mathrm{erfc}^{-1}(0.02)$. The complementary error function has a value of $\mathrm{erfc}(x) = 0.02$. Then, the error function

$$\mathrm{erf}(x) = 1 - \mathrm{erfc}(x) = 1 - 0.02 = 0.98,$$

for which we get $x = 1.64$ on the table. So, the value of inverse complementary error function $\mathrm{erfc}^{-1}(0.02) = 1.64$. □

Example 3.2.2 Find the value of inverse complementary error function $\mathrm{erfc}^{-1}(1.8)$.

Solution. Let $x = \mathrm{erfc}^{-1}(1.8)$. The complementary error function has a value of $\mathrm{erfc}(x) = 1.8$, from which the error function is

$$\mathrm{erf}(x) = 1 - \mathrm{erfc}(x) = 1 - 1.8 = -0.8.$$

Look up the table to get $\mathrm{erf}(0.91) = 0.8$. Using the symmetry property of error function $\mathrm{erf}(-x) = -\mathrm{erf}(x)$ immediately gives $x = -0.91$ for $\mathrm{erf}(x) = -0.8$. So, the value of inverse complementary error function $\mathrm{erfc}^{-1}(1.8) = -0.91$. □

Example 3.2.3 Find the values of Q-function $Q(0.1)$ and $Q(0.3)$.

Solution. For $x = 0.1$, $x/\sqrt{2} = 0.0707$ and for $x = 0.3$, $x/\sqrt{2} = 0.2121$. From the table, we have

$$\mathrm{erf}\left(\frac{x}{\sqrt{2}}\right) = \mathrm{erf}\left(\frac{0.1}{\sqrt{2}}\right) = \mathrm{erf}(0.0707) \approx 0.07885,$$
$$\mathrm{erf}\left(\frac{x}{\sqrt{2}}\right) = \mathrm{erf}\left(\frac{0.3}{\sqrt{2}}\right) = \mathrm{erf}(0.2121) \approx 0.2336.$$

Therefore,

$$Q(0.1) = Q(x) = 1 - \frac{1}{2}\mathrm{erf}\left(\frac{x}{\sqrt{2}}\right) = 1 - \frac{1}{2} \times 0.07885 = 0.9606,$$
$$Q(0.3) = Q(x) = 1 - \frac{1}{2}\mathrm{erf}\left(\frac{x}{\sqrt{2}}\right) = 1 - \frac{1}{2} \times 0.2336 = 0.8832.$$

□

As will be evident in the following section, error and complementary error functions are essential in expressing the probabilities of detection and false alarm.

3.3 Probabilities of Detection and Error

In Section 3.1, various spaces involved in decision theory are discussed, and in Section 3.2, probability theory in signal detection is briefly reviewed. In this section, the interaction between outcome space and signal space will be discussed in more depth.

In binary hypothesis testing, if the result of the decision is either $A_1 = S_1$ or $A_0 = S_0$, that is, the outcome space $\mathcal{A} = \{A_0, A_1\}$ coincides with the signal space $\mathcal{S} = \{S_0, S_1\}$, the signal is detected correctly. However, owing to the effect of observational noise (or error) or finite size of data, the error in the estimation of decision statistic is inevitable, and will give rise to the inconsistency of outcome space and signal space, which could cause either $A_1 = S_0$ or $A_0 = S_1$ and a detection error. Next we present the analytic theory on correct and erroneous detection.

3.3.1 Definitions of Detection and Error Probabilities

The correct decision about the presence of S_1 or S_0 is correct detection in binary hypothesis testing. The probability of correct decision is called probability of (signal) detection, which has the following two types.

1. Detection Probability of S_1

The probability of correctly detecting the presence of signal S_1 is the detection probability of S_1, and is determined by the conditional probability $P(g > \mathrm{Th}|H_1)$ of decision statistic $g = g(\mathbf{y})$ exceeding the threshold Th given hypothesis H_1, or equivalently by the conditional probability $P(g \in D_1|H_1)$ of decision statistic g in the decision space $D_1 = (\mathrm{Th}, \infty)$ given hypothesis H_1. That is,

$$P_{\mathrm{D}_1} = P(g > \mathrm{Th}|H_1) = P(g \in D_1|H_1) = \int_{\mathrm{Th}}^{\infty} p(g|H_1)\,\mathrm{d}g. \tag{3.3.1}$$

2. Detection Probability of S_0

The probability of correctly detecting signal S_0 is the detection probability of S_0, and is the conditional probability $P(g > \mathrm{Th}|H_0)$ of decision statistic $g = g(\mathbf{y})$ below the threshold Th given hypothesis H_0, or equivalently by the conditional probability $P(g \in D_0|H_0)$ of decision statistic g in the decision space $D_0 = (-\infty, \mathrm{Th})$ given hypothesis H_0. That is,

$$P_{\mathrm{D}_0} = P(g < \mathrm{Th}|H_0) = P(g \in D_0|H_0) = \int_{-\infty}^{\mathrm{Th}} p(g|H_0)\,\mathrm{d}g. \tag{3.3.2}$$

Opposite to the above two types of correct decisions, both mistaking S_1 for S_0 and mistaking S_0 for S_1 are incorrect decisions. The probability of making wrong decisions is called error probability in signal detection, which can be classified as error probabilities of type I and type II.

3. Error Probability of Type I

Rejecting true H_0 (i.e., the presence of signal S_0) and deciding the presence of signal S_1 is type I error. The probability of type I error is denoted as α. In early-warning radar system, type I error is also known as false alarm, since it is an incorrect alert to decide the presence of a target when actually no target exists. The probability of false alarm is denoted by P_F. However, in detecting wireless communication signals of binary pulse amplitude modulation (PAM), $S_1 = p(t)$ and $S_0 = -p(t)$, where $p(t)$ is an impulse with positive amplitude, are two transmitted signals of opposite polarity. In this case, the type I error is to mistake the signal of negative polarity $S_0 = -p(t)$ for the signal $S_1 = p(t)$ of positive polarity.

Being the error probability of rejecting H_0, the error probability of type I is given by the conditional probability $P(g > \mathrm{Th}|H_0)$ of decision statistic g above threshold Th under hypothesis H_0, or equivalently, the conditional probability $P(g \in D_1|H_0)$ of decision statistic g in the decision space $D_1 = (\mathrm{Th}, \infty)$ given hypothesis H_0. That is,

$$P_F = \alpha = P(g > \mathrm{Th}|H_0) = P(g \in D_1|H_0) = \int_{\mathrm{Th}}^{\infty} p(g|H_0)\,\mathrm{d}g. \tag{3.3.3}$$

4. Error Probability of Type II

For the false H_0 (i.e., the absence of signal S_0), to decide the presence of signal S_0 is type II error of failing to reject the false H_0. Similar to type I error, type II error has different interpretation depending the application.

Recall that in radar detection S_1 represents the presence of a target and S_0 represents the absence of a target. So, in terms of radar early warning, falsely taking target for no target is called probability of miss, which is denoted as P_M or β. Differently, in detecting wireless communication signals of binary PAM, the type II error is to mistake the signal of positive polarity $S_1 = p(t)$ for the signal $S_0 = p(t)$ of negative polarity.

Being the error probability of failing to reject H_0, the error probability of type II is given by the conditional probability $P(g < \mathrm{Th}|H_1)$ of decision statistic g below threshold Th under true hypothesis H_1, or equivalently, the conditional probability $P(g \in D_0|H_1)$ of decision statistic g in the decision space $D_0 = (-\infty, \mathrm{Th})$ under hypothesis H_1. That is,

$$P_M = \beta = P(g < \mathrm{Th}|H_1) = P(g \in D_0|H_1) = \int_{-\infty}^{\mathrm{Th}} p(g|H_1)\,\mathrm{d}g. \tag{3.3.4}$$

Now we discuss the relation between detection and error probabilities. First notice that

$$\int_{g \in D} p(g|H_i)\,\mathrm{d}g = \int_{-\infty}^{\infty} p(g|H_i)\,\mathrm{d}g = 1, \quad i = 1, 0.$$

Thus, the detection and error probabilities have the important properties

$$P_{D_0} = \int_{-\infty}^{\mathrm{Th}} p(g|H_0)\,\mathrm{d}g = \int_{-\infty}^{\infty} p(g|H_0)\,\mathrm{d}g - \int_{\mathrm{Th}}^{\infty} p(g|H_0)\,\mathrm{d}g = 1 - \alpha, \tag{3.3.5}$$

$$P_{D_1} = \int_{\mathrm{Th}}^{\infty} p(g|H_1)\,\mathrm{d}g = \int_{-\infty}^{\infty} p(g|H_1)\,\mathrm{d}g - \int_{-\infty}^{\mathrm{Th}} p(g|H_1)\,\mathrm{d}g = 1 - \beta. \tag{3.3.6}$$

The physical interpretation of the above two equations is as follows. Since the probabilities of detecting and missing a signal sums to one, the probability P_{D_0} of signal S_0 being correctly detected is one minus the error probability α. Likewise, the probability P_{D_1} of signal S_1 being correctly detected is one minus the error probability β.

The complete detection probability P_D in binary hypothesis testing incorporates both detection probabilities of hypotheses H_0 and H_1. Denote p_i $(i = 1, 0)$ be the prior probability of signal S_i being present, with $p_0 + p_1 = 1$. The detection probability in binary hypothesis testing is

$$P_D = p_0 P_{D_0} + p_1 P_{D_1} = p_0 \int_{-\infty}^{\mathrm{Th}} p(g|H_0)\,\mathrm{d}g + p_1 \int_{\mathrm{Th}}^{\infty} p(g|H_1)\,\mathrm{d}g. \tag{3.3.7}$$

Similarly, the error probability in binary hypothesis testing incorporates both probability of false alarm α and probability of miss β:

$$\begin{aligned} P_E &= p_0 P_F + p_1 P_M = p_0\alpha + p_1\beta \\ &= p_0 \int_{\mathrm{Th}}^{\infty} p(g|H_0)\,\mathrm{d}g + p_1 \int_{-\infty}^{\mathrm{Th}} p(g|H_1)\,\mathrm{d}g. \end{aligned} \tag{3.3.8}$$

Detection probability P_D and error probability P_E satisfy

$$P_D = 1 - P_E. \tag{3.3.9}$$

It is noteworthy that in radar signal detection, since S_0 represents no target, the probability of detecting signal is usually referred to as the probability of detecting S_1, and the detection probability of S_0 is less cared about. But when it comes to error probability, both error probability for S_0 (probability of false alarm α) and error probability for S_1 (probability of miss β) are received equal attention. In such case, the detection probability is defined as the probability of detecting S_1

$$P_D = \int_{\mathrm{Th}}^{\infty} p(g|H_1)\,\mathrm{d}g = 1 - \beta = \gamma, \tag{3.3.10}$$

where γ is called power of test and depends upon the alternative hypothesis H_1.

The physical interpretation of (3.3.10) is as follows. As H_0 and H_1 are respectively the hypotheses corresponding to the presence and absence of a target, $\gamma = 1 - \beta$ is the detection probability of a true target and thus represents the power of binary hypothesis testing.

With radar signal detection as the example, Fig 3.3.1 depicts the conditional PDFs $p(g|H_1)$ and $p(g|H_0)$ of the decision statistic g and the relation among detection probability $P_D = \gamma$, probability of miss α, probability of false alarm β, and decision space $D = D_0 + D_1$.

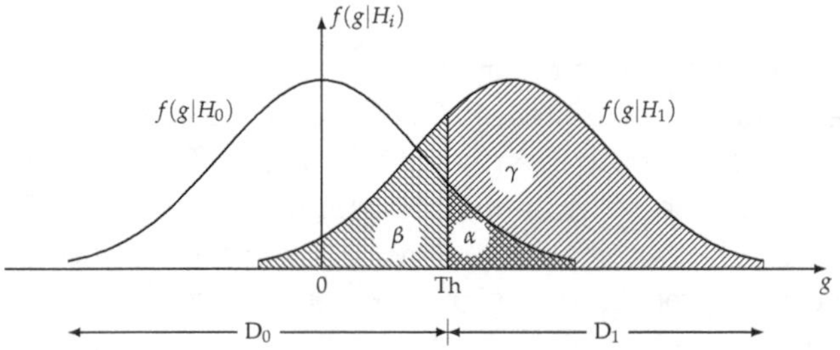

False alarm probability: $P_F = \alpha =$ area of grid line region
Missing alarm probability: $P_M = \beta =$ area of backslash region excluding grid line region
Detection probability: $P_D = \gamma =$ area of forwardslash region including grid line region

Fig. 3.3.1: Relation among P_F ,P_M and P_D

3.3.2 Power Function

Eq. (3.3.10) defines the power of the binary hypothesis testing

$$\begin{cases} H_0 \ : \ y_n = w_n, \\ H_1 \ : \ y_n = S_1 + w_n, \end{cases}$$

which is to decide the presence of a target signal. Now consider the more general problem of binary hypothesis testing

$$\begin{cases} H_0 \ : \ y_n = S_0 + w_n, \\ H_1 \ : \ y_n = S_1 + w_n, \end{cases} \tag{3.3.11}$$

where w_n is a Gaussian random variable with mean μ and variance σ^2.

From (3.3.4), the probability β of type II error is a function of threshold Th, which is further related to parameter S_1 in hypothesis H_1. As a result, β is in essence a function of parameter S_1 in H_1, and can be written as $\beta(S_1)$.

Definition 3.3.1. *For the binary hypothesis testing based on decision function $L(\mathbf{y})$ or decision statistic $g(\mathbf{y})$, $P(\mathbf{y}) = 1 - \beta(S_1)$ is called the power function of the hypothesis testing, which is the probability of rejecting H_0 with the decision function or statistic in the case of a false H_0.*

From the above definition, power function is the detection probability of signal S_1 with $\beta(S_1)$ as the probability of missing S_1. As such, it is desirable to have larger value of power function $P(\mathbf{y}) = 1-\beta(S_1)$ in theory. The decision statistic $g(\mathbf{y})$ is the most powerful if its power function has greatest value. In other words, the most powerful decision statistic is optimal in the sense of power function. However, in different applications, the criterion of optimal power may differ. So, the decision criterion should be adapted for detecting different target of interest.

1. Neyman-Pearson Criterion

In applications such as early warning radar, H_0 is the hypothesis of no target, i.e., $S_0 = 0$, and H_1 is the hypothesis of existence of a target. As the damage caused by type II error (missing a target) is much severe than type I error (falsely warning a target), the probability of miss should be minimized with the probability of false alarm (probability of type I error) kept below certain level. Or equivalently, the power function, which is the detection probability of signal S_1, is maximized. Such is the renowned Neyman-Pearson criterion.

2. Uniformly Most Power Criterion

In wireless communication, when the transmitted signals are reverse polarity signal $S_1 = s(t)$ and $S_0 = -s(t)$, or when the transmitted characters are binary characters 1 and -1, type I and type II errors are both error codes, which causes basically the same consequences with no difference in severity. Since the power function represents the probability of detecting signal S_1, it is a function of the parameter S_1 in H_1. So in such application problems, if the power function can be made independent of S_1 parameter, it is possible to achieve the same maximum power function for all possible values of parameter S_1. Such a maximum power function is called a uniformly (or consistent) most power function, and the corresponding decision criterion is called a uniformly most power criterion.

3. Bayesian Criterion

Any decision rule is accompanied by risk, and the decision rule with optimal power can give rise to higher risk. Therefore, in some applications the optimality of power is less emphasized, and the decision that has minimum risk is much favored. The criterion that minimizes risk function is called Bayesian criterion.

In the following sections, our discussion will revolve around the above three criteria and their performance analysis.

3.4 Neyman-Pearson Criterion

In the applications of signal detection, many practical problems can be attributed to binary hypothesis testing where the null hypothesis represents that the observed data contains noise only and the alternative hypothesis represents that the data contains signal of interest. Typical example of such problem is the signal detection in radar and sonar.

3.4.1 Probabilities of False Alarm and Miss alarm in Radar Signal Detection

In radar signal detection, the radar echo signal can be described by the following binary hypothesis testing model:

$$\begin{cases} H_0 \;:\; y_n = w_n, & \text{The target is absent,} \\ H_1 \;:\; y_n = S + w_n, & \text{The target is present,} \end{cases} \tag{3.4.1}$$

where S denotes the amplitude of the radar echo pulse, which is a deterministic signal. The additive noise w_n is white Gaussian with zero mean and variance σ^2, i.e.,

$$\mathrm{E}\{w_n\} = 0, \quad \mathrm{var}[w_n] = \sigma^2, \quad \mathrm{E}\{w_n w_k\} = 0, \; (\text{if } n \neq k).$$

For the above model, the sample mean of N samples $y_1, \cdots, y_N$

$$\bar{y} = \frac{1}{N}\sum_{n=1}^{N} y_n \tag{3.4.2}$$

is used as the decision function. Thus, the decision rule is

$$\begin{cases} H_0 \;:\; \text{decide target is absent,} & \text{if } \bar{y} \le \mathrm{Th}, \\ H_1 \;:\; \text{decide target is present,} & \text{if } \bar{y} > \mathrm{Th}. \end{cases} \tag{3.4.3}$$

Therefore, it is crucial to choose a proper threshold Th.

Note that S is deterministic and w_n is white Gaussian noise with zero mean and variance σ^2. So it follows that like w_n, the sample mean $\bar{y}$ is Gaussian as well.

Conditioned on the hypotheses, the expectations of sample mean $\bar{y}$ are

$$\mathrm{E}\{\bar{y}|H_0\} = \mathrm{E}\{w_n\} = 0, \quad \mathrm{E}\{\bar{y}|H_1\} = \mathrm{E}\{S\} = S,$$

and the variances are

$$\mathrm{var}[\bar{y}|H_0] = \mathrm{var}[\bar{y}|H_1] = \frac{\sigma^2}{N}.$$

In short, sample mean $\bar{y}$ is normally distributed with zero mean and variance σ^2/N under H_0 and with mean S and variance σ^2 under H_1. Then, the conditional PDFs of the Gaussian random variable $\bar{y}$ are

$$\begin{aligned} p(\bar{y}|H_0) &= \frac{1}{\sqrt{2\pi \mathrm{var}[\bar{y}|H_0]}} \exp\left[-\frac{(\bar{y} - \mathrm{E}\{\bar{y}|H_0\})^2}{2\mathrm{var}[\bar{y}|H_0]}\right] \\ &= \frac{1}{\sqrt{2\pi/N}\sigma} \exp\left(-\frac{\bar{y}^2}{2\sigma^2/N}\right), \end{aligned} \tag{3.4.4}$$

$$\begin{aligned} p(\bar{y}|H_1) &= \frac{1}{\sqrt{2\pi \mathrm{var}[\bar{y}|H_1]}} \exp\left[-\frac{(\bar{y} - \mathrm{E}\{\bar{y}|H_1\})^2}{2\mathrm{var}[\bar{y}|H_1]}\right] \\ &= \frac{1}{\sqrt{2\pi/N}\sigma} \exp\left(-\frac{(\bar{y} - S)^2}{2\sigma^2/N}\right). \end{aligned} \tag{3.4.5}$$

So, the probability of false alarm is

$$\alpha = \int_{\mathrm{Th}}^{\infty} p(\bar{y}|H_0)\,\mathrm{d}\bar{y} = \int_{\mathrm{Th}}^{\infty} \frac{1}{\sqrt{2\pi/N}\sigma} \exp\left(-\frac{\bar{y}^2}{2\sigma^2/N}\right) \mathrm{d}\bar{y} = \frac{1}{2}\frac{2}{\sqrt{\pi}} \int_{\frac{\mathrm{Th}}{\sqrt{2/N}}\sigma}^{\infty} \mathrm{e}^{-u^2}\,\mathrm{d}u,$$

which can be expressed by the complementary error function in (3.2.21) as

$$\alpha = \frac{1}{2}\mathrm{erfc}\left(\frac{\mathrm{Th}}{\sqrt{2/N}\sigma}\right). \tag{3.4.6}$$

Therefore, the threshold can be derived from the probability of false alarm as

$$\mathrm{Th} = \frac{\sqrt{2}\sigma}{\sqrt{N}}\mathrm{erfc}^{-1}(2\alpha). \tag{3.4.7}$$

Similarly, noticing $\int_{-\infty}^{\infty} p(\bar{y}|H_1)\,\mathrm{d}\bar{y}$, the probability of miss alarm is

$$\begin{aligned}\beta &= \int_{-\infty}^{\mathrm{Th}} p(\bar{y}|H_1)\,\mathrm{d}\bar{y} = 1 - \int_{\mathrm{Th}}^{\infty} p(\bar{y}|H_1)\,\mathrm{d}\bar{y} \\ &= 1 - \int_{\mathrm{Th}}^{\infty} \frac{1}{\sqrt{2\pi/N}\sigma} \exp\left(-\frac{(\bar{y}-S)^2}{2\sigma^2/N}\right) \mathrm{d}\bar{y} \\ &= 1 - \frac{1}{2}\frac{2}{\sqrt{\pi}} \int_{\frac{\mathrm{Th}-S}{\sqrt{2/N}\sigma}} \mathrm{e}^{-u^2}\,\mathrm{d}u,\end{aligned}$$

which can be expressed by complementary error function as

$$\beta = 1 - \frac{1}{2}\mathrm{erfc}\left(\frac{\mathrm{Th}-S}{\sqrt{2/N}\sigma}\right). \tag{3.4.8}$$

Finally, the detection probability is

$$P_{\mathrm{D}} = \int_{\mathrm{Th}}^{\infty} p(\bar{y}|H_1)\,\mathrm{d}\bar{y} = \int_{\mathrm{Th}}^{\infty} \frac{1}{\sqrt{2\pi/N}\sigma} \exp\left(-\frac{(\bar{y}-S)^2}{2\sigma^2/N}\right) \mathrm{d}\bar{y} = \frac{1}{2}\frac{2}{\sqrt{\pi}} \int_{\frac{\mathrm{Th}-S}{\sqrt{2/N}\sigma}}^{\infty} \mathrm{e}^{-u^2}\,\mathrm{d}u,$$

which gives

$$P_{\mathrm{D}} = \frac{1}{2}\mathrm{erfc}\left(\frac{\mathrm{Th}-S}{\sqrt{2/N}\sigma}\right) = 1-\beta \tag{3.4.9}$$

and is consistent with (3.3.10).

In radar or sonar signal detection, the value of threshold Th and sample size N are to be specified such that the requirement from application on the probability α of false alarm and/or probability β of miss alarm is met.

Example 3.4.1 The received signal of the radar is

$$y_n = \begin{cases} H_0 : w_n, & \text{target is absent,} \\ H_1 : 1 + w_n, & \text{target is present,} \end{cases}$$

where the additive noise w_n is Gaussian white with zero mean and unit variance. For the probability of false alarm fixed as $\alpha = 0.01$, find the threshold Th, probability of miss β, and detection probability P_D at sample sizes $N = 20$ and $N = 25$, respectively.

Solution. From (3.4.7) and $\sigma = 1$, we have

$$\text{Th} = \frac{\sqrt{2}\sigma}{\sqrt{N}} \text{erfc}^{-1}(2\alpha) = \frac{\sqrt{2}}{\sqrt{N}} \text{erfc}^{-1}(2\alpha).$$

Let $\text{erfc}^{-1}(2\alpha) = x$,

$$\text{erf}(x) = 1 - \text{erfc}(x) = 1 - 2\alpha = 1 - 2 \times 0.01 = 0.98.$$

From the table of error function (Appendix 3A), we get $x = 1.64$. Thus, the thresholds are given by

$$\text{Th} = \sqrt{2/N} \text{erfc}^{-1}(2\alpha) = \sqrt{2/N} \times 1.64 = \begin{cases} 0.5186, & N = 20, \\ 0.4639, & N = 25. \end{cases}$$

From (3.4.9) with $S = 1$ and the above threshold values, the detection probabilities are

$$\begin{aligned} P_D &= \frac{1}{2} \text{erfc}\left(\frac{\text{Th} - S}{\sqrt{2/N}\sigma} \right) \\ &= \begin{cases} 0.5\text{erfc}(-1.5223) = 0.5[1 + \text{erf}(1.5223)] \approx 0.984, & N = 20, \\ 0.5\text{erfc}(-1.8954) = 0.5[1 + \text{erf}(1.8954)] \approx 0.996, & N = 25, \end{cases} \end{aligned}$$

where $\text{erf}(1.5223) \approx 0.9864$ and $\text{erf}(1.8954) \approx 0.9926$ are from the table of error function. The probabilities of miss alarm is

$$\beta = 1 - P_D = \begin{cases} 0.016, & N = 20, \\ 0.004, & N = 25. \end{cases}$$

In conclusion, when the sample size $N = 20$, threshold $\text{Th} = 0.5186$, probability of miss $\beta = 0.016$, and detection probability $P_D = 0.984$. When the sample size $N = 25$, threshold $\text{Th} = 0.4639$, probability of miss $\beta = 0.004$, and detection probability $P_D = 0.996$. □

Example 3.4.2 The model for the radar data is the same as in Example 3.4.1. For the probability of false alarm $\alpha = P_F = 0.01$ and the probability of miss $\beta = P_M = 0.05$, find the value of threshold Th and sample size N satisfying the requirement of error probabilities.

Solution. From Example 3.4.1, the probability of false alarm is

$$\alpha = \frac{1}{2}\operatorname{erfc}\left(\frac{\mathrm{Th}}{\sqrt{2/N}}\right)$$

and the probability of miss is

$$\beta = 1 - \frac{1}{2}\operatorname{erfc}\left(\frac{\mathrm{Th}-1}{\sqrt{2/N}}\right).$$

From the above equations and the given conditions $\alpha = 0.01$ and $\beta = 0.05$, we have

$$\mathrm{Th} = \frac{\sqrt{2}}{\sqrt{N}}\operatorname{erfc}^{-1}(2\alpha) = \frac{\sqrt{2}}{\sqrt{N}}\operatorname{erfc}^{-1}(0.02),$$
$$\mathrm{Th} - 1 = \frac{\sqrt{2}}{\sqrt{N}}\operatorname{erfc}^{-1}(2(1-\beta)) = \frac{\sqrt{2}}{\sqrt{N}}\operatorname{erfc}^{-1}(1.9).$$

The ratio of the two equations is

$$\frac{\mathrm{Th}}{1-\mathrm{Th}} = \frac{\operatorname{erfc}^{-1}(0.02)}{\operatorname{erfc}^{-1}(1.9)} \approx \frac{1.64}{-1.16} = -1.1438,$$

which gives Th = 0.5857. The sample size can be solved by substitution of the threshold

$$\mathrm{Th} = \frac{\sqrt{2}}{\sqrt{N}}\operatorname{erfc}^{-1}(0.02) = \frac{\sqrt{2}}{\sqrt{N}} \times 1.64 = 0.5857$$

to yield $N = 15.6808$ and we take $N = 16$. Therefore, in the detection of radar signal, by setting the threshold Th = 0.5857, the sample size $N = 16$ for the observed array, and using the decision rule based on sample mean $\bar{y}$

$$\begin{cases} H_1 : \text{ decide the target is present, if } \bar{y} > 0.5857, \\ H_0 : \text{ decide the target is absent, if } \bar{y} < 0.5857, \end{cases}$$

the requirements of probability of false alarm $\alpha = 0.01$ and probability of miss $\beta = 0.05$ can be met. □

As the detection probability P_D is a monotonic function of the probability of false alarm P_F, the relation illustrated by the $P_\mathrm{D} \sim P_\mathrm{F}$ curve of the function is called receiver operating characteristic (ROC).

3.4.2 Neyman-Pearson Lemma and Neyman-Pearson Criterion

In military, missing alarm can have more detrimental consequence in comparison with false alarm. Therefore, in applications of radar and sonar signal detection, the criterion of choosing decision statistic or decision function is to minimize the probability of miss β or to maximize the detection probability to improve the power $\gamma = 1 - \beta$, with probability of false alarm kept below a value of α_0. Specifically, if the conditional PDF $p(\mathbf{y}|H_1)$ is chosen as the decision function, the criterion of maximizing power under constraint gives the following optimization problem

$$\gamma = \max \int_{\mathrm{Th}}^{\infty} p(\mathbf{y}|H_1)\,\mathrm{d}\mathbf{y} = \max \int_{\mathbf{y}\in R_c} p(\mathbf{y}|H_1)\,\mathrm{d}\mathbf{y}, \tag{3.4.10}$$

with constraint

$$\int_{\mathbf{y}\in R_c} p(\mathbf{y}|H_0)\,\mathrm{d}\mathbf{y} = \alpha_0. \tag{3.4.11}$$

Strictly speaking, the inequality constraint $\int_{\mathbf{y}\in R_c} p(\mathbf{y}|H_0)\,\mathrm{d}\mathbf{y} \le \alpha_0$ should be used. However, the inequality constraint could result in a more difficult problem than the one with equality constraint. So, the above problem is solved under the equality constraint of the false alarm probability α_0.

The central task in the criterion of maximum power is to solve the optimal critical region, which is depicted in Fig 3.4.1, and requires the following Neyman-Pearson lemma.

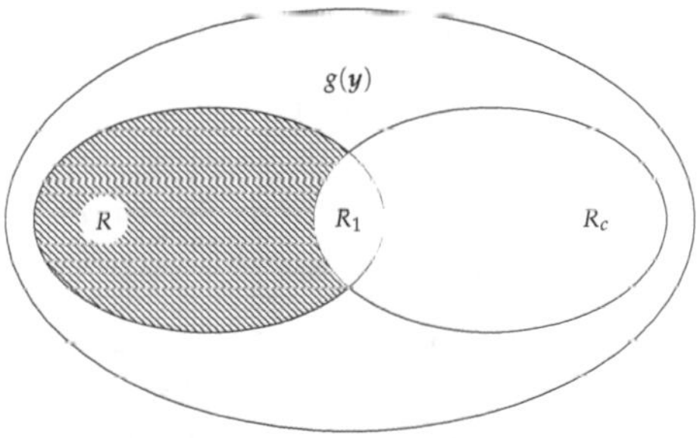

Fig. 3.4.1: Illustration of critical regions

Corollary 3.4.1. *Neyman-Pearson Lemma*

Let

$$R_c = \{\mathbf{y} : p(\mathbf{y}|H_1) > \eta p(\mathbf{y}|H_0)\}, \tag{3.4.12}$$

where η is the constant such that Eq. (3.4.11) *is satisfied. Let R be arbitrary region in the sample space Ω and*

$$\int_{\mathbf{y}\in R} p(\mathbf{y}|H_0)\,\mathrm{d}\mathbf{y} = \alpha_0. \tag{3.4.13}$$

Then

$$\int_{\mathbf{y}\in R_c} p(\mathbf{y}|H_1)\,\mathbf{dy} \geq \int_{\mathbf{y}\in R} p(\mathbf{y}|H_1)\,\mathbf{dy}. \tag{3.4.14}$$

Neyman-Pearson lemma shows that the power of any region R in the sample space Ω cannot exceed the power of the critical region R_c. So, R_c is optimal in the sense of power maximization and the region defined in (3.4.12) is the optimal solution of the constrained problem in (3.4.10).

The decision rule derived from Neyman-Pearson is called the Neyman-Pearson criterion.

Taking the likelihood ratio function from the sample array $\mathbf{y} = (y_1, \cdots, y_N)$

$$L(\mathbf{y}) = \frac{p(\mathbf{y}|H_1)}{p(\mathbf{y}|H_0)} \tag{3.4.15}$$

as the decision function, the Neyman-Pearson criterion can be recast as

$$L(\mathbf{y}) = \frac{p(\mathbf{y}|H_1)}{p(\mathbf{y}|H_0)} \underset{H_1}{\overset{H_0}{\lesseqgtr}} \eta, \tag{3.4.16}$$

where η is the threshold. Accordingly, the solution of critical region in (3.4.12) has the equivalent form

$$R_c = \{\mathbf{y} \;:\; L(\mathbf{y}) > \eta\}. \tag{3.4.17}$$

Again rewrite the likelihood ratio function by the common logarithm transform

$$L_1(\mathbf{y}) = \log[L(\mathbf{y})] = \log\left[\frac{p(\mathbf{y}|H_1)}{p(\mathbf{y}|H_0)}\right] = \log[p(\mathbf{y}|H_1)] - \log[p(\mathbf{y}|H_0)]. \tag{3.4.18}$$

With the form above, the Neyman-Pearson criterion gives the equivalent decision rule as

$$L_1(\mathbf{y}) = \log[L(\mathbf{y})] = \log\left[\frac{p(\mathbf{y}|H_1)}{p(\mathbf{y}|H_0)}\right] \underset{H_1}{\overset{H_0}{\lesseqgtr}} \log\eta, \tag{3.4.19}$$

which decides hypothesis H_1 to be true for the decision function greater than the threshold $\log\eta$ and hypothesis H_0 to be true otherwise.

Note that the threshold η or $\log\eta$ is determined by the preset probability of false alarm α_0, which is also known as level of test and is the performance measure of detection of radar and sonar signal.

Next we give examples demonstrating how to apply Neyman-Pearson criterion in practice.

Example 3.4.3 Let $y_1, \cdots, y_N$ be the N observed samples of a Gaussian random variable with unknown mean μ and known variance σ^2. Find the optimal critical region of the hypothesis testing with $\mu_1 > \mu_0$

$$\begin{cases} H_0 \;:\; \mu = \mu_0, \\ H_1 \;:\; \mu = \mu_1, \end{cases}$$

under level of test α_0.

Solution. From the given conditions, the observed sample y_n is normally distributed with mean μ and variance σ^2 and has conditional PDF as

$$p(y_n|H_i) = \frac{1}{\sqrt{2\pi}\sigma}\exp\left(-\frac{(y_n-\mu_i)^2}{2\sigma^2}\right), \quad i=0,1.$$

Since y_n ($n = 1,\cdots,N$) are independent, the joint conditional PDF of N samples $y_1,\cdots,y_N$ is the product of marginal (conditional) PDFs

$$p(y_1,\cdots,y_N|H_i) = \prod_{n=1}^{N} p(y_n|H_i) = \frac{1}{(\sqrt{2\pi}\sigma)^N}\exp\left(-\sum_{n=1}^{N}\frac{(y_n-\mu_i)^2}{2\sigma^2}\right).$$

Thus, from the Neyman-Pearson criterion, the decision function

$$L(y_1,\cdots,y_N) = \frac{p(y_1,\cdots,y_N|H_1)}{p(y_1,\cdots,y_N|H_0)} = \frac{\exp\left(-\sum_{n=1}^{N}\frac{(y_n-\mu_1)^2}{2\sigma^2}\right)}{\exp\left(-\sum_{n=1}^{N}\frac{(y_n-\mu_0)^2}{2\sigma^2}\right)} > k.$$

Taking logarithm of both sides, the above equation is reduced to

$$-\sum_{n=1}^{N}(y_n-\mu_1)^2 + \sum_{n=1}^{N}(y_n-\mu_0)^2 > 2\sigma^2\log(k), \tag{3.4.20}$$

The difference of the two summations in left hand side of (3.4.20) is

$$-\sum_{n=1}^{N}(y_n-\mu_1)^2 + \sum_{n=1}^{N}(y_n-\mu_0)^2 = -N(\mu_1^2-\mu_0^2) + 2(\mu_1-\mu_0)\sum_{n=1}^{N}y_n.$$

Substituting the above equation into (3.4.20) gives

$$-N(\mu_1^2-\mu_0^2) + 2(\mu_1-\mu_0)\sum_{n=1}^{N}y_n > 2\sigma^2\log(k),$$

or

$$2(\mu_1-\mu_0)\bar{y} > 2\sigma^2\log(k) + N(\mu_1^2-\mu_0^2),$$

where

$$\bar{y} = \frac{1}{N}\sum_{n=1}^{N}y_n$$

is the mean of N random samples. Noticing that $\mu_1 > \mu_0$ and $2(\mu_1-\mu_0)$ is always positive, we have

$$\bar{y} > \frac{N(\mu_1^2-\mu_0^2) + 2\sigma^2\log(k)}{2(\mu_1-\mu_0)},$$

which reveals that the decision statistic is the sample mean, i.e., $g(y_1, \cdots, y_N) = \bar{y}$, and the threshold is

$$\text{Th} = \frac{N(\mu_1^2 - \mu_0^2) + 2\sigma^2 \log(k)}{2(\mu_1 - \mu_0)}.$$

Namely, the optimal critical region for the decision statistic $g(y_1, \cdots, y_N) = \bar{y}$ is

$$R_c = \left(\frac{N(\mu_1^2 - \mu_0^2) + 2\sigma^2 \log(k)}{2(\mu_1 - \mu_0)}, \infty \right),$$

where $\mu_1 > \mu_0$. □

As a summary, in the above example, the decision function is the likelihood ratio $L(y_1, \cdots, y_N) = p(y_1, \cdots, y_N|H_1)/p(y_1, \cdots, y_N|H_0)$ and the decision statistic $g(y_1, \cdots, y_N)$ is the sample mean $\bar{y}$.

3.5 Uniformly Most Power Criterion

In Section 3.4, taking the radar signal detection as an example, the problem of binary hypothesis test with different results between missing alarm and false alarm is discussed. Taking the binary pulse amplitude modulation (PAM) communication system as an example, this section studies another type of binary hypothesis testing when the consequences caused by two wrong decisions are almost or exactly the same.

3.5.1 Communication Signal Detection Problem

Considering the binary PAM communication system, the transmitted signal waveforms are $S_1(t) = p(t)$ and $S_0(t) = -p(t)$, where $p(t)$ is an arbitrary pulse, which is zero in the symbol interval $0 \leq t \leq T$, but is zero at other times. Since $S_1(t) = -S_0(t)$, these two signals are called antipodal signals.

The binary hypothesis testing problem of binary PAM communication can be described as

$$\begin{cases} H_0 : r_0(t) = S_0(t) + w(t) \\ H_1 : r_1(t) = S_1(t) + w(t) \end{cases} \quad 0 \leqslant t \leqslant T. \tag{3.5.1}$$

Different from radar signal detection, the communication transmission signal $S_1(t) = p(t)$ being wrongly judged as $S_0(t) = -p(t)$ by the decision criteria, or $S_0(t) = -p(t)$ being wrongly judged as $S_1(t) = p(t)$, the impact of these two wrong decisions is the same, and there are no serious consequences such as false alarm and missing alarm. In the problem of communication signal detection, all detection errors are collectively referred to as bit error, and there are no missing alarm and false alarm. Therefore, in the wireless communication system, there are only two indexes: detection probability P_D and error probability P_E. Bit error probability is abbreviated as bit error rate. Due to

$P_D + P_E = 1$, only the bit error rate needs to be analyzed. In practical applications, the bit error rate is often used

$$\text{BER} = \frac{\text{Number of bit error}}{\text{Total number of bits transmitted}} \times 100\%. \quad (3.5.2)$$

The detection problem shown in Eq.(3.5.1) is not easy to solve. Therefore, we consider the results of the correlation demodulation of the observed signal using the known pulse signal $p(t)$

$$r_i = \int_0^T r_i(t)p(t)\mathrm{d}t = \int_0^T S_i(t)p(t)\mathrm{d}t, \ i = 0, 1. \quad (3.5.3)$$

In this way, the detection problem Eq.(3.5.1) of the binary PAM communication signal becomes

$$\begin{cases} H_0 : r_0 = S_0 + n = -E_p + n, \\ H_1 : r_1 = S_1 + n = E_p + n, \end{cases} \quad (3.5.4)$$

where

$$S_0 = \int_0^T S_0(t)p(t)\mathrm{d}t = -\int_0^T p^2(t)\mathrm{d}t = -E_\mathrm{p}, \quad (3.5.5)$$

$$S_1 = \int_0^T S_1(t)p(t)\mathrm{d}t = \int_0^T p^2(t)\mathrm{d}t = E_\mathrm{p}, \quad (3.5.6)$$

$$n = \int_0^T w(t)p(t)\mathrm{d}t, \quad (3.5.7)$$

and E_p represents the energy of the pulse signal $p(t)$

$$E_\mathrm{p} = \int_0^\infty p^2(t)\mathrm{d}t = \int_0^T p^2(t)\mathrm{d}t. \quad (3.5.8)$$

Because the pulse signal $p(t)$ is a deterministic signal, it is statistically uncorrelated with any random signal including Gaussian white noise, i.e., $\mathrm{E}\{w(t)p(t)\} = 0$, so it has

$$\mathrm{E}\{n\} = \int_0^T \mathrm{E}\{w(t)p(t)\}\mathrm{d}t = 0. \quad (3.5.9)$$

Note that n is a Gaussian random variable with zero mean.

Based on the above discussion, the decision rule of binary PAM communication signal detection can be obtained: set the threshold to zero, if the output signal r of the correlation demodulator is greater than zero, the transmitted signal is judged as $S_1(t) = p(t)$; Otherwise, it is judged as $S_0(t) = -p(t)$.

It should be noted that although the mean of Gaussian random variable n is zero, using zero as the threshold will still cause wrong judgment due to its random change.

3.5.2 Uniformly Most Power Test

Eq.(3.5.4) is extended to a more general binary hypothesis testing model

$$\begin{cases} H_0 : y_n = s_0 + w_n = -\sqrt{E_p} + w_n, \\ H_1 : y_n = s_1 + w_n = \sqrt{E_p} + w_n, \end{cases} \tag{3.5.10}$$

where w_n is a Gaussian white noise with zero mean and variance σ^2.

Assuming that the signal $s_1 = \sqrt{E_p}$ is transmitted, since the threshold of the decision is zero, the error probability $P(e|H_1)$ under H_1 hypothesis testing is the probability of the output $y < 0$ directly, i.e.,

$$\begin{aligned} P(e|H_1) &= \int_{-\infty}^{0} p(y|H_1)\mathrm{d}y = \frac{1}{\sqrt{\pi N_0}} \int_{-\infty}^{0} \exp\left(-\frac{(y-\sqrt{E_p})^2}{N_0}\right) \mathrm{d}y \\ &= \frac{1}{\sqrt{2\pi}} \int_{-\infty}^{-\sqrt{2E_p/N_0}} \mathrm{e}^{-x^2/2}\mathrm{d}x = \frac{1}{\sqrt{2\pi}} \int_{\sqrt{2E_p/N_0}}^{\infty} \mathrm{e}^{-x^2/2}\mathrm{d}x \\ &= Q\left(\sqrt{\frac{2E_p}{N_0}}\right), \end{aligned} \tag{3.5.11}$$

where $N_0 = \frac{1}{2}\sigma^2$ represents the energy of the additive white Gaussian noise $w(n)$, and $Q(x)$ denotes Q function.

Similarly, when the signal $s_0 = -\sqrt{E_p}$ is transmitted, the error probability under H_0 hypothesis testing is the probability of $r > 0$, and $P(e|H_0) = Q(\sqrt{2E_p/N_0})$ also holds. Since the binary signals $s_1 = \sqrt{E_p}$ and $s_0 = -\sqrt{E_p}$ are usually transmitted with equal probability, i.e. $p_1 = p_0 = \frac{1}{2}$, the average error probability (bit error rate) is

$$P_E = p_1 P(e|H_1) + p_0 P(e|H_0) = \frac{1}{2}P(e|H_1) + \frac{1}{2}P(e|H_0) = Q\left(\sqrt{\frac{2E_p}{N_0}}\right). \tag{3.5.12}$$

Two important facts can be observed from the above:
(1) The bit error rate is only related to the ratio $\frac{2E_p}{N_0}$, but has nothing to do with other characteristics of the signal and noise.
(2) The ratio $\frac{2E_p}{N_0}$ represents the output signal-to-noise ratio (SNR) of the receiver. Since the output includes two bit symbols, $\frac{E_p}{N_0}$ is called the SNR of each bit.

The critical region R_c or the threshold Th is generally related to the parameters of the alternative hypothesis H_1.

Definition 3.5.1. *For the binary hypothesis testing described in Eq.(3.5.10), if the critical region R_c or threshold Th is independent of the parameter s_1 of the alternative hypothesis H_1, the hypothesis testing is called uniformly most power (UMP) test.*

Interpreting the above definition, "the critical region R_c or threshold Th is independent of the parameter s_1 of alternative hypothesis H_1" means that the power function of the hypothesis testing is the same or consistent for all different parameters s_1. It always seeks the maximum power of the hypothesis testing, so it is called UMP test.

If the binary hypothesis testing is the UMP test, the test statistic $g(y_1, \cdots, y_N)$ used is called the UMP test statistic.

The question is, how to construct UMP statistics for a binary hypothesis testing? We still consider the hypothesis testing in the case of additive Gaussian white noise, but w_n is not zero mean, but the mean is m and the variance is still σ^2.

Because w_n is white noise and s_1 and s_0 are definite, $y_1, \cdots, y_N$ are independent under the two assumptions H_0 and H_1, and $\{y_n\}$ is a Gaussian random process with mean $\bar{s}_i$ and variance σ^2 under the assumption H_i, i.e.,

$$P(y_n|H_1) = \frac{1}{\sqrt{2\pi\sigma^2}} \exp\left(-\frac{(y_n - \bar{s}_1)^2}{2\sigma^2}\right), \tag{3.5.13}$$

$$P(y_n|H_0) = \frac{1}{\sqrt{2\pi\sigma^2}} \exp\left(-\frac{(y_n - \bar{s}_0)^2}{2\sigma^2}\right), \tag{3.5.14}$$

where $\bar{s}_1 = s_1 + m$ and $\bar{s}_0 = s_0 + m$.

Using the independence of $\{y_i\}$ under hypotheses H_0 and H_1, the conditional distribution density functions of the observation data $\mathbf{y} = (y_1, \cdots, y_N)$ under H_1 and H_0 respectively is

$$P(\mathbf{y}|H_1) = \prod_{n-1}^{N} p(y_n|H_1) = \frac{1}{(2\pi\sigma^2)^{N/2}} \exp\left(-\sum_{n=1}^{N} \frac{(y_n - \bar{s}_1)^2}{2\sigma^2}\right), \tag{3.5.15}$$

$$P(\mathbf{y}|H_0) = \prod_{n=1}^{N} p(y_n|H_0) = \frac{1}{(2\pi\sigma^2)^{N/2}} \exp\left(-\sum_{n=1}^{N} \frac{(y_n - \bar{s}_0)^2}{2\sigma^2}\right). \tag{3.5.16}$$

The likelihood ratio function is obtained

$$\frac{p(y|H_1)}{p(y|H_0)} = \frac{\exp\left(-\sum_{n=1}^{N} \frac{(y_n - \bar{s}_1)^2}{2\sigma^2}\right)}{\exp\left(-\sum_{n=1}^{N} \frac{(y_n - \bar{s}_0)^2}{2\sigma^2}\right)}. \tag{3.5.17}$$

Consider using the log likelihood ratio function as the decision function

$$L_1(\mathbf{y}) = \ln\left[\frac{p(y|H_1)}{p(y|H_0)} = -\sum_{n=1}^{N} \frac{(y_n - \bar{s}_1)^2}{2\sigma^2} + \sum_{n=1}^{N} \frac{(y_n - \bar{s}_0)^2}{2\sigma^2}\right]. \tag{3.5.18}$$

Thus, similar to the Neyman-Pearson criterion, if the decision function $L_1(\mathbf{y}) > k_1$, i.e.,

$$-\sum_{n=1}^{N} \frac{(y_n - \bar{s}_1)^2}{2\sigma^2} + \sum_{n=1}^{N} \frac{(y_n - \bar{s}_0)^2}{2\sigma^2} > k_1,$$

judge that the transmitted signal is s_1. Otherwise, judge that signal s_0 is transmitted. By simplifying the above expression, the following decision criteria can be obtained: if

$$(\bar{s}_1 - \bar{s}_0)\sum_{n=1}^{N} y_n - \frac{N}{2}(\bar{s}_1^2 - \bar{s}_0^2) > \sigma^2 k_1,$$

it is judged that the transmitted signal is s_1; Otherwise, judge that signal s_0 is transmitted. This decision criterion can be described equivalently as: judging that signal s_1 is transmitted, if

$$\frac{1}{N}\sum_{n=1}^{N} y_n > \frac{1}{N}\left[\sigma^2 k_1 + \frac{N}{2}(\bar{s}_1^2 - \bar{s}_0^2)\right]\frac{1}{\bar{s}_1 - \bar{s}_0}. \tag{3.5.19}$$

Let the test statistics be

$$g(\mathbf{y}) = \bar{y} = \frac{1}{N}\sum_{n=1}^{N} y_n, \tag{3.5.20}$$

and the threshold be

$$\mathrm{Th} = \frac{1}{N}\left[\sigma^2 k_1 + \frac{N}{2}(\bar{s}_1^2 - \bar{s}_0^2)\right]\frac{1}{\bar{s}_1 - \bar{s}_0}, \tag{3.5.21}$$

then the decision criterion Eq.(3.5.19) can be described as: if the decision statistic

$$g(\mathbf{y}) = \bar{y} = \frac{1}{N}\sum_{n=1}^{N} y_n > \mathrm{Th}, \tag{3.5.22}$$

judge signal s_1 is transmitted; Otherwise, judge signal s_0 is transmitted.

The properties of the decision statistic $g(\mathbf{y}) = \bar{y}$ are analyzed below. Firstly, because the decision statistic is the sample mean of the observation data, and this mean obeys the same normal distribution as the observation data, when the decision statistic takes a random value g, the conditional distribution density functions of $g(\mathbf{y})$ under hypotheses H_1 and H_0 are normal distribution, respectively

$$p(g|H_1) = \frac{1}{\sqrt{2\pi\sigma^2/N}}\exp\left(-\frac{(g - \bar{s}_1)^2}{2\sigma^2/N}\right), \tag{3.5.23}$$

$$p(g|H_0) = \frac{1}{\sqrt{2\pi\sigma^2/N}}\exp\left(-\frac{(g - \bar{s}_0)^2}{2\sigma^2/N}\right). \tag{3.5.24}$$

Follow the false alarm rate in Section 3.4, define

$$\begin{aligned}\alpha &= \int_{\mathrm{Th}}^{\infty} p(g|H_0)\mathrm{d}g = \int_{\mathrm{Th}}^{\infty} \frac{1}{\sqrt{2\pi\sigma^2/N}}\exp\left[-\frac{(g - \bar{s}_0)^2}{2\sigma^2/N}\right]\mathrm{d}g \\ &= \frac{1}{2}\frac{2}{\sqrt{\pi}}\int_{\frac{\mathrm{Th}-\bar{s}_0}{\sqrt{2}\sigma/\sqrt{N}}}^{\infty} \mathrm{e}^{-t^2}\mathrm{d}t = \frac{1}{2}\mathrm{erfc}\left(\frac{\mathrm{Th} - \bar{s}_0}{\sqrt{2}\sigma/\sqrt{N}}\right).\end{aligned} \tag{3.5.25}$$

That is, the false alarm rate α is half of the value of the complementary error function of the conditional distribution density function $p(g|H_0)$ under $\frac{\text{Th}-\bar{s}_0}{\sqrt{2}\sigma/\sqrt{N}}$, which is independent of parameter s_1 of alternative hypothesis H_1.

From the false alarm rate Eq.(3.5.25), the threshold can be obtained

$$\text{Th} = \frac{\sqrt{2}\sigma}{\sqrt{N}}\text{erfc}^{-1}(2\alpha) + \bar{s}_0, \tag{3.5.26}$$

where $\text{erfc}^{-1}(z)$ is the inverse function of the complementary error function erfc(z). Since the false alarm rate α is independent of parameter s_1, the selection of threshold Th is independent of parameter s_1 of alternative hypothesis H_1. Therefore, the optimal decision statistic $g(\mathbf{y}) = \bar{y}$ given in Eq.(3.5.22) is a UMP statistic. The corresponding decision criterion is also independent of parameter s_1 of alternative hypothesis H_1, which is a UMP criterion.

Conclusion: if the noise w_n in the binary hypothesis testing Eq. (3.5.10) is a Gaussian white noise with mean m and variance σ^2, and the likelihood ratio function or log likelihood ratio function is used as the decision function, the sample mean of N observation data $y_1, \cdots, y_N$ is a UMP test statistic.

3.5.3 Physical Meaning of UMP Criterion

Similar to the derivation of the false alarm rate Eq.(3.5.25), the detection probability is

$$P_\text{D} = \int_{\text{Th}}^{\infty} p(g|H_1)\text{d}g = \frac{1}{2}\frac{2}{\sqrt{\pi}}\int_{\frac{\text{Th}-\bar{s}_1}{\sqrt{2}\sigma/\sqrt{N}}}^{\infty} \text{e}^{-t^2}\text{d}t = \frac{1}{2}\text{erfc}\left(\frac{\text{Th}-\bar{s}_1}{\sqrt{2}\sigma/\sqrt{N}}\right). \tag{3.5.27}$$

Therefore, the probability of missing alarm is

$$P_\text{M} = 1 - P_\text{D} = 1 - \frac{1}{2}\text{erfc}\left(\frac{\text{Th}-\bar{s}_1}{\sqrt{2}\sigma/\sqrt{N}}\right). \tag{3.5.28}$$

Consider the special case of signal $s_0 = 0$ presence or absence detection. Let's analyze the physical meaning of the UMP criterion in this case. Without losing generality, it is assumed that the mean of the additive Gaussian white noise is zero ($m = 0$). Then, in the threshold equation Eq.(3.5.26) $\bar{s}_0 = 0$, so the threshold is

$$\text{Th} = \frac{\sqrt{2}\sigma}{\sqrt{N}}\text{erfc}^{-1}(2\alpha). \tag{3.5.29}$$

Substituting the above equation and $\bar{s}_1 = s_1 + m = s_1$ into Eq.(3.5.27), there is

$$P_\text{D} = \frac{1}{2}\text{erfc}\left(\text{erfc}^{-1}(2\alpha) - \frac{\bar{s}_1}{\sqrt{2}\sigma/\sqrt{N}}\right). \tag{3.5.30}$$

Let

$$B = \frac{\bar{s}_1}{\sqrt{2}\sigma/\sqrt{N}}, \tag{3.5.31}$$

then B^2 can be regarded as SNR. Therefore, Eq.(3.5.30) can be abbreviated as

$$P_D = \frac{1}{2}\text{erfc}[\text{erfc}^{-1}(2\alpha) - B]. \tag{3.5.32}$$

This shows that the detection probability P_D is a function of the false alarm rate α and the SNR B^2. When SNR $B^2 \to 0$ or equivalently $B \to 0$, the detection probability

$$P_D = \frac{1}{2}(2\alpha) = \alpha \quad (\text{If SNR} = 0). \tag{3.5.33}$$

When SNR $B^2 \to \infty$ or equivalently $B \to \infty$ and use Eq.(3.2.25) to get the detection probability

$$P_D = \lim_{B\to\infty} \frac{1}{2}\text{erfc}(-B) = \frac{1}{2} \lim_{x\to-\infty} \text{erfc}(x) = 1 \quad (\text{If SNR} \to \infty). \tag{3.5.34}$$

From the above analysis, it can be concluded that the physical meaning of the UMP criterion in the detection of signal presence and absence is as follows:

(1) When the SNR is zero, the detection probability of signal s_1 is equal to the false alarm probability.
(2) When the SNR is infinite, the detection probability of signal s_1 is 1, and s_1 can be 100% correctly detected.
(3) The higher the SNR, the greater the detection probability of signal s_1. Therefore, in the detection of the signal presence or absence, the UMP criterion is equivalent to maximizing the SNR when the false alarm probability α is limited to a certain level, or equivalent to maximizing the detection probability.

In the detection of the signal presence or absence, UMP criterion is equivalent to Neyman-Pearson criterion.

3.6 Bayes Criterion

If the test statistic is regarded as a parameter, the decision-making process is also a parameter estimation process in essence. Therefore, we can also discuss the decision criteria of binary hypothesis testing from the perspective of parameter estimation. Such a criterion is called Bayes criterion. Different from Neyman-Pearson criterion, which maximizes the detection probability when the false alarm probability is limited to a certain level, and the UMP criterion pursuing that the threshold is independent of the parameters of the alternative hypothesis, Bayes criterion aims to minimize the risk of decision-making.

3.6.1 Bayes Decision Criterion

Judging H_j hypothesis as H_i hypothesis requires cost, which is expressed by a cost factor C_{ij}.

The cost factor C_{ij} has the following basic properties:

(1) The cost factor is always nonnegative, that is, $C_{ij} \geq 0, \forall i, j$.

(2) For the same hypothesis testing H_j, the cost of wrong decision is always greater than that of correct decision, i.e.,

$$C_{ij} > C_{jj}, \quad j \neq i. \tag{3.6.1}$$

For example, $C_{10} > C_{00}$ and $C_{01} > C_{11}$.

The conditional probability $P(H_i|H_j)$ represents the probability that the decision result is H_i hypothesis under the condition that H_j hypothesis is true. The corresponding cost is represented by $C_{ij}P(H_i|H_j)$. If the prior probability $P(H_j)$ of the occurrence of H_j hypothesis is known, the cost of deciding H_i hypothesis under H_j hypothesis is $C_{ij}P(H_j)P(H_i|H_j)$. Therefore, in the binary hypothesis testing, the cost of correct decision under H_j hypothesis is $C_{jj}P(H_j)P(H_j|H_j)$, and the cost of wrong decision is $C_{ij}P(H_j)P(H_i|H_j)$, $i \neq j$. The sum of the cost of correct and wrong decisions is called the total cost under H_j hypothesis, i.e.,

$$C(H_j) = C_{0j}P(H_j)P(H_0|H_1) + C_{1j}P(H_j)P(H_1|H_j), \; j = 0, 1. \tag{3.6.2}$$

The first term on the right of the above equation represents the cost of wrong decision under H_j hypothesis, and the second term is the cost of correct decision under H_j hypothesis.

The total average cost of statistical decision of binary hypothesis testing problem consists of the sum of the total cost of hypothesis H_0 and the total cost of hypothesis H_1:

$$C = C(H_0) + C(H_1) = \sum_{j=0}^{1}\sum_{i=0}^{1} C_{ij}P(H_j)P(H_i|H_j). \tag{3.6.3}$$

The total average cost is referred to as the average cost, also known as average risk.

If N observations $y_1, \cdots, y_N$ constitute the evidence $Y = g(y_1, \cdots, y_N)$ of hypothesis testing. For example, the sampling mean $Y = \frac{1}{N}\sum_{i=1}^{N} y_i$ can be used as the amount of evidence. Let λ be the threshold when making statistical decision by using the amount of evidence. Substitute the definition Eq.(3.1.3)~ Eq.(3.1.6) of the conditional probability $P(H_i|H_j)$ into Eq.(3.6.3) to obtain

$$C = C_{10}p_0 + C_{11}p_1 + \int_{-\infty}^{\lambda} [(C_{01} - C_{11})p_1 p(Y|H_1) - (C_{10} - C_{00})p_0 p(Y|H_0)]\mathrm{d}Y, \tag{3.6.4}$$

where $R_0 = (-\infty, \lambda)$ represents the decision region of H_0 hypothesis corresponding to $Y < \lambda$.

Bayes Criterion: when the prior probability $p_j = P(H_j)$ is known and the cost factor C_{ij} is given, minimize the average cost C.

Due to the non negativity of a priori probability and cost factor, in order to minimize the average cost, the integral function in Eq.(3.6.4) should be negative, i.e.,

$$R_0: \quad (C_{01} - C_{11})p_1 p(Y|H_1) < (C_{10} - C_{00})p_0 p(Y|H_0). \tag{3.6.5}$$

Then, the amount of evidence Y will be located in the decision area $R_0 = (-\infty, \lambda)$, that is, the decision with the H_0 hypothesis should be made. Where, the evidence $Y = p(y_1, \cdots, y_N)$ is a statistic obtained from N observations.

On the contrary, if

$$R_0: \quad (C_{01} - C_{11})p_1 p(Y|H_1) > (C_{10} - C_{00})p_0 p(Y|H_0), \tag{3.6.6}$$

the amount of evidence Y is located in the decision area $R_1 = (\lambda, \infty)$, judge H_1 hypothesis holds.

Therefore, the Bayes criterion of binary hypothesis testing problem can be written as

$$\frac{p(Y|H_1)}{p(Y|H_0)} > \eta \quad \text{judge } H_1 \text{ holds,}$$
$$\frac{p(Y|H_1)}{p(Y|H_0)} \leq \eta \quad \text{judge } H_0 \text{ holds.}$$

Or, combine the above two expressions into one decision expression

$$\frac{p(Y|H_1)}{p(Y|H_0)} \underset{H_1}{\overset{H_0}{\lesseqgtr}} \eta, \tag{3.6.7}$$

where

$$\eta = \frac{(C_{10} - C_{00})p_0}{(C_{01} - C_{11})p_1}, \tag{3.6.8}$$

is the threshold of Bayes decision criterion.

Take logarithm on both sides of Eq.(3.6.1) to get the common form of Bayes criterion

$$L(Y) \underset{H_1}{\overset{H_0}{\lesseqgtr}} \ln \eta. \tag{3.6.9}$$

The Bayes criterion Eq.(3.6.1) is also called likelihood ratio criterion. The logarithm of the ratio of conditional distribution density

$$L(Y) = \ln \frac{p(Y|H_1)}{p(Y|H_0)}, \tag{3.6.10}$$

is called the likelihood function of observation data y.

When the prior probability $p_j = p(H_j), j = 0, 1$ is known and the cost factor $C_{ij}, i = 0, 1$ is given, the threshold η of Bayes criterion can be determined for a binary hypothesis testing problem.

3.6.2 Detection of Binary Signal Waveform

As an application example of Baye criterion, consider binary communication signal

$$\begin{cases} H_0 : y(t) = S_0(t) + n(t), \ 0 \le t \le T, \\ H_1 : y(t) = S_1(t) + n(t), \ 0 \le t \le T, \end{cases} \tag{3.6.11}$$

where $n(t) \sim \mathcal{N}(0, \sigma_n^2)$ is a white Gaussian noise with zero mean and variance σ_n^2; T is bit interval (symbol interval); $S_0(t)$ and $S_1(t)$ are the modulation waveform with energy

$$\begin{cases} E_0 = \int_0^T |S_0(t)|^2 \mathrm{d}t, \\ E_1 = \int_0^T |S_1(t)|^2 \mathrm{d}t, \end{cases} \tag{3.6.12}$$

respectively.

The waveform correlation coefficient between modulated signals $S_0(t)$ and $S_1(t)$ is

$$\rho = \frac{1}{\sqrt{E_0 E_1}} \int_0^T S_0(t) S_1(t) \mathrm{d}t. \tag{3.6.13}$$

To obtain the discrete signal, Karhunen-Loeve (K-L) transform is performed on the analog binary communication signal to get

$$y(t) = \lim_{N \to \infty} \sum_{k=1}^{N} y_k f_k(t). \tag{3.6.14}$$

Taking the sum of the first N terms as approximation, we can get

$$y_N(t) = \sum_{k=1}^{N} y_k f_k(t), \tag{3.6.15}$$

where y_k is the K-L expansion coefficient

$$y_k = \int_0^T y(t) f_k(t) \mathrm{d}t. \tag{3.6.16}$$

$f_k(t)$ is orthogonal basis function, which can be constructed by Gram-Schmidt standard orthogonalization

$$\begin{cases} f_1(t) = \frac{1}{\sqrt{E_1} S_1(t)}, & 0 \le t \le T, \\ f_2(t) = \frac{1}{\sqrt{(1-\rho^2)E_0}} \left[S_0(t) - \rho \sqrt{\frac{E_0}{E_1}} S_1(t) \right], & 0 \le t \le T, \end{cases} \tag{3.6.17}$$

and the other orthogonal basis functions $f_k(t), k = 3, 4, \cdots$ are all orthogoanl to $f_1(t), f_2(t)$.

Using the K-L expansion coefficient r_k, the analog expression Eq.(3.6.11) of the original binary hypothesis testing problem can be written equivalently as

$$\begin{cases} H_0: & y_k = S_0 + n_k, \ k = 1, 2, \cdots, \\ H_1: & y_k = S_1 + n_k, \ k = 1, 2, \cdots. \end{cases} \tag{3.6.18}$$

From Eq.(3.6.18), we have

$$\mathrm{E}\{y_k|H_0\} = \mathrm{E}\{S_{0k} + n_k\} = S_{0k}, \tag{3.6.19}$$

$$\mathrm{var}(y_k|H_0) = \mathrm{E}\{(y_k - s_0)^2\} = \mathrm{E}\{n_k^2\} = \sigma_n^2, \tag{3.6.20}$$

$$\mathrm{E}\{y_k|H_1\} = \mathrm{E}\{S_{1k} + n_k\} = S_{1k}, \tag{3.6.21}$$

$$\mathrm{var}(y_k|H_1) = \mathrm{E}\{(y_k - s_1)^2\} = \mathrm{E}\{n_k^2\} = \sigma_n^2. \tag{3.6.22}$$

Since y_k follows normal distribution like Gaussian white noise n_k, the conditional distribution density function of observation data y_k is

$$p\{y_k|H_0\} = \frac{1}{\sqrt{2\pi}\sigma_n}\exp\left(-\frac{(y_k - S_{0k})^2}{2\sigma_n^2}\right), \tag{3.6.23}$$

$$p\{y_k|H_1\} = \frac{1}{\sqrt{2\pi}\sigma_n}\exp\left(-\frac{(y_k - S_{1k})^2}{2\sigma_n^2}\right). \tag{3.6.24}$$

Taking a function of the observation data $y_1, \cdots, y_n$ as evidence $Y = g(y_1, \cdots, y_n)$, the joint conditional distribution density functions of the evidence Y are

$$\begin{aligned} p\{Y|H_0\} &= \prod_{k=1}^{N} p\{y_k|H_0\} \\ &= \left(\frac{1}{\sqrt{2\pi}\sigma_n}\right)^N \exp\left[-\frac{1}{2\sigma_n^2}\left(\sum_{k=1}^{N} y_k^2 - 2\sum_{k=1}^{N} y_k S_{0k} + \sum_{k=1}^{N} S_{0k}^2\right)\right], \end{aligned} \tag{3.6.25}$$

$$\begin{aligned} p\{Y|H_1\} &= \prod_{k=1}^{N} p\{y_k|H_1\} \\ &= \left(\frac{1}{\sqrt{2\pi}\sigma_n}\right)^N \exp\left[-\frac{1}{2\sigma_n^2}\left(\sum_{k=1}^{N} y_k^2 - 2\sum_{k=1}^{N} y_k S_{1k} + \sum_{k=1}^{N} S_{1k}^2\right)\right], \end{aligned} \tag{3.6.26}$$

respectively. Note that $\sum_{k=1}^{N} S_{0k}^2 = E_0$, $\sum_{k=1}^{N} S_{1k}^2 = E_1$ and

$$R_{yS_0} = \sum_{k=1}^{N} y_k S_{0k} \Leftrightarrow R_{yS_0} = \int_0^T y(t) S_0(t) \mathrm{d}t, \tag{3.6.27}$$

$$R_{yS_1} = \sum_{k=1}^{N} y_k S_{1k} \Leftrightarrow R_{yS_1} = \int_0^T y(t) S_1(t) \mathrm{d}t, \tag{3.6.28}$$

repectively represent the cross-correlation function of observation signal and modulation signal, so the original form of Bayes decision criterion is

$$\frac{p(Y|H_1)}{p(Y|H_0)} = \exp\left[\frac{1}{\sigma_n^2}\left(\int_0^T y(t)S_1(t)\mathrm{d}t - \int_0^T y(t)S_0(t)\mathrm{d}t\right) - \frac{1}{2\sigma_n^2}(E_1 - E_0)\right] \underset{H_1}{\overset{H_0}{\lessgtr}} \eta. \tag{3.6.29}$$

Taking logarithm on both sides of the above equation, the Bayes decision criterion can be written as

$$\int_0^T y(t)S_1(t)\mathrm{d}t - \int_0^T y(t)S_0(t)\mathrm{d}t \underset{H_1}{\overset{H_0}{\lessgtr}} \sigma^2 \ln \eta + \frac{1}{2}(E_1 - E_0), \tag{3.6.30}$$

or abbreviated as

$$Y \underset{H_1}{\overset{H_0}{\lessgtr}} \lambda, \tag{3.6.31}$$

where

$$Y = \sum_{k=1}^{N} y_k S_{1k} - \sum_{k=1}^{N} y_k S_{0k} = \int_0^T y(t)S_1(t)\mathrm{d}t - \int_0^T y(t)S_0(t)\mathrm{d}t, \tag{3.6.32}$$

$$\lambda = \sigma_n^2 \ln \eta + \frac{1}{2}(E_1 - E_0), \tag{3.6.33}$$

are the amount of evidence and threshold of the binary hypothesis testing problem, respectively.

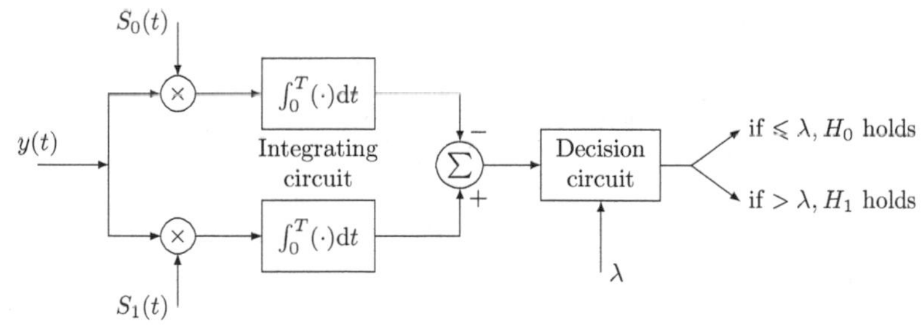

Fig. 3.6.1: Detection system of binary communication signals

Fig 3.6.1 shows the detection system for binary communication signals.

From Eq.(3.6.32) and Eq.(3.6.13), the conditional mean

$$\mathrm{E}\{Y|H_0\} = \mathrm{E}\left\{\sum_{k=1}^{N}(S_{0k}+n_k)S_{1k} - \sum_{k=1}^{N}(S_{0k}+n_k)S_{0k}\right\} = \rho\sqrt{E_0E_1} - E_0 \tag{3.6.34}$$

$$\mathrm{E}\{Y|H_1\} = \mathrm{E}\left\{\sum_{k=1}^{N}(S_{1k}+n_k)S_{1k} - \sum_{k=1}^{N}(S_{1k}+n_k)S_{0k}\right\} = E_1 - \rho\sqrt{E_0E_1} \tag{3.6.35}$$

and conditional variance

$$\mathrm{var}\{Y|H_0\} = \mathrm{var}\left\{\sum_{k=1}^{N}(S_{0k}+n_k)S_{1k} - \sum_{k=1}^{N}(S_{0k}+n_k)S_{0k}\right\} = \rho\sqrt{E_0E_1} - E_0 \tag{3.6.36}$$

$$\mathrm{var}\{Y|H_1\} = \mathrm{var}\left\{\sum_{k=1}^{N}(S_{1k}+n_k)S_{1k} - \sum_{k=1}^{N}(S_{1k}+n_k)S_{0k}\right\} = E_1 - \rho\sqrt{E_0E_1} \tag{3.6.37}$$

of the evidence Y can be obtained.

3.6.3 Detection Probability Analysis

1. Sample Statistics Analysis

Calculate the conditional mean and conditional variance of evidence Y.

(1) The conditional mean: under H_1 assumption, the conditional mean of the amount of evidence Y is

$$\mathrm{E}\{Y|H_1\} = \mathrm{E}\left\{\frac{1}{N}\sum_{k=1}^{N}y_k|H_1\right\} = \mathrm{E}\left\{\frac{1}{N}\sum_{k=1}^{N}(S_1+n_k)\right\} = S_1, \tag{3.6.38}$$

while the conditional mean of evidence Y under H_0 hypothesis is

$$\mathrm{E}\{Y|H_0\} = \mathrm{E}\left\{\frac{1}{N}\sum_{k=1}^{N}y_k|H_0\right\} = \mathrm{E}\left\{\frac{1}{N}\sum_{k=1}^{N}(S_0+n_k)\right\} = S_0. \tag{3.6.39}$$

(2) The conditional variance: under H_1 assumption, the conditional variance of the amount of evidence Y is

$$\mathrm{var}(Y|H_1) = \mathrm{var}\left(\frac{1}{N}\sum_{k=1}^{N}y_k|H_1\right) = \mathrm{var}\left(\frac{1}{N}\sum_{k=1}^{N}(S_1+n_k)\right) = \frac{1}{N}\sigma_n^2, \tag{3.6.40}$$

while the conditional variance of evidence Y under H_0 hypothesis is

$$\mathrm{var}(Y|H_0) = \mathrm{var}\left(\frac{1}{N}\sum_{k=1}^{N}y_k|H_0\right) = \mathrm{var}\left(\frac{1}{N}\sum_{k=1}^{N}(S_0+n_k)\right) = \frac{1}{N}\sigma_n^2. \tag{3.6.41}$$

2. Detection Probability Analysis

From the above conditional mean and conditional variance, it is easy to know that the conditional distribution density of the sample mean evidence respectively is

$$p\{Y|H_1\} = \frac{\sqrt{N}}{\sqrt{2\pi}\sigma_n}\exp\left(-\frac{N(Y-S_1)^2}{2\sigma_n^2}\right), \tag{3.6.42}$$

$$p\{Y|H_0\} = \frac{\sqrt{N}}{\sqrt{2\pi}\sigma_n}\exp\left(-\frac{N(Y-S_0)^2}{2\sigma_n^2}\right). \tag{3.6.43}$$

Once the conditional distribution density of the sample mean evidence is obtained, the detection probability in the sample mean Bayes criterion can be calculated.

(1) False alarm probability

$$p(H_1|H_0) = \int_{\lambda}^{\infty} p(Y|H_0)\mathrm{d}Y = \int_{\frac{\sigma_n^2}{N(S_1-S_0)}\ln\eta+\frac{(S_1+S_0)}{2}}^{\infty} \frac{\sqrt{N}}{\sqrt{2\pi}\sigma_n}\exp\left(-\frac{N(Y-S_0)^2}{2\sigma_n^2}\right)\mathrm{d}Y.$$

Replace the variable $u = \frac{\sqrt{N}(Y-S_0)}{\sqrt{2}\sigma_n}$, and the above equation can be expressed as

$$p(H_1|H_0) = \int_{\frac{\sigma_n}{\sqrt{2N}(S_1-S_0)}\ln\eta+\frac{\sqrt{N}(S_1-S_0)}{2\sqrt{2}\sigma_n}}^{\infty} \frac{1}{\sqrt{\pi}}\exp(-u^2)\mathrm{d}u$$

$$= \frac{1}{2}\mathrm{erfc}\left(\frac{\sigma_n}{\sqrt{2N}(S_1-S_0)}\ln\eta+\frac{\sqrt{N}(S_1-S_0)}{2\sqrt{2}\sigma_n}\right), \tag{3.6.44}$$

where $\mathrm{erfc}(x) = \int_x^{\infty}\frac{2}{\sqrt{\pi}}\exp(-u^2)\mathrm{d}u$ is the complement error function of x.

(2) Hit probability

$$p(H_1|H_1) = \int_{\lambda}^{\infty} p(Y|H_1)\mathrm{d}Y = \int_{\frac{\sigma_n^2}{N(S_1-S_0)}\ln\eta+\frac{S_1+S_0}{2}}^{\infty} \frac{\sqrt{N}}{\sqrt{2\pi}\sigma_n}\exp\left(-\frac{N(Y-S_1)^2}{2\sigma_n^2}\right)\mathrm{d}Y.$$

Replace the variable $u = \frac{\sqrt{N}(Y-S_1)}{\sqrt{2}\sigma_n}$, and we get

$$p(H_1|H_1) = \int_{\frac{\sigma_n}{\sqrt{2N}(S_1-S_0)}\ln\eta+\frac{\sqrt{N}(S_1-S_0)}{2\sqrt{2}\sigma_n}}^{\infty} \frac{1}{\sqrt{\pi}}\exp(-u^2)\mathrm{d}u$$

$$= \frac{1}{2}\mathrm{erfc}\left(\frac{\sigma_n}{\sqrt{2N}(S_1-S_0)}\ln\eta-\frac{\sqrt{N}(S_1-S_0)}{2\sqrt{2}\sigma_n}\right). \tag{3.6.45}$$

(3) The rejection probability $p(H_0|H_0) = 1-p(H_1|H_0)$ and the missed alarm probability $p(H_0|H_1) = 1-p(H_1|H_1)$ respectively are

$$p(H_0|H_0) = 1-\frac{1}{2}\mathrm{erfc}\left(\frac{\sigma_n}{\sqrt{2N}(S_1-S_0)}\ln\eta+\frac{\sqrt{N}(S_1-S_0)}{2\sqrt{2}\sigma_n}\right), \tag{3.6.46}$$

$$p(H_0|H_1) = 1-\frac{1}{2}\mathrm{erfc}\left(\frac{\sigma_n}{\sqrt{2N}(S_1-S_0)}\ln\eta-\frac{\sqrt{N}(S_1-S_0)}{2\sqrt{2}\sigma_n}\right). \tag{3.6.47}$$

Table 3.6.1 compares the signal model, decision criterion, decision function and threshold of Neyman-Pearson criterion, UMP criterion and Bayes criterion. In the table, a_0 is the preset false alarm probability.

The following conclusions can be drawn from the table:

(1) The three decision criteria all use the likelihood ratio function as the decision function, but the choice of the threshold of the decision function is different: in the Neyman-Pearson criterion and UMP criterion, the threshold of the decision function depends on the allowable false alarm probability a_0, while the threshold of the decision function of Bayes criterion is directly determined by the ratio p_0/p_1 of the a priori probability of H_0 and H_1 assumptions.

(2) The three decision criteria all use the mean of the observation samples as detection statistics, and their thresholds are different.

In the detection of binary digital communication signals, the cost factor $C_{10} = C_{01}$ and $C_{11} = C_{00}$ is usually assumed, so from Eq.(3.6.8), the original threshold of Bayes decision criterion is

$$\eta = \frac{(C_{10} - C_{00})p_0}{(C_{01} - C_{11})p_1} = \frac{p_0}{p_1}. \tag{3.6.48}$$

Then, when the sample mean Y of N observation data is used as the evidence, the actual threshold of Bayes decision criterion is

$$\lambda = \frac{\sigma_n^2}{N(S_1 - S_0)}[\ln p_0 - \ln p_1] + \frac{S_1 + S_0}{2}. \tag{3.6.49}$$

3.7 Bayes Derived Criteria

Depending on the choice of the cost factor, Bayes criterion can derive several other decision criteria.

3.7.1 Minimum Error Probability Criterion

In some applications (such as binary digital communication), both correct rejection and hit are correct detection, which has no cost, that is, the cost factor $C_{00} = C_{11} = 0$. On the other hand, although the wrong decision needs to pay a price, the cost of false alarm and missed alarm is the same, that is, the cost factor $C_{10} = C_{01}$. Under these assumptions of cost factor, the average cost C is simplified to the error probability P_E, i.e.,

$$C = p_0P(H_1|H_0) + p_1P(H_0|H_1) = p_0\alpha + p_1\beta = P_E. \tag{3.7.1}$$

Tab. 3.6.1: Comparison of three decision criteria

Criterion	Neyman-Pearson Criterion	UMP Criterion	Bayes Criterion
Signal model	$\begin{cases} H_1 : y_n = S + w_n \\ H_0 : y_n = w_n \end{cases}$	$\begin{cases} H_1 : y_n = S_1 + w_n \\ H_0 : y_n = S_0 + w_n \end{cases}$	$\begin{cases} H_1 : y_n = S_1 + w_n \\ \max \gamma = 1 - S(\beta) \end{cases}$
Decision criterion	$\begin{cases} \alpha \leqslant \alpha_0 \\ H_0 : y_n = S_0 w_n \end{cases}$	Th should be independent of parameter S_1	Minimum risk
Decision function	$L(y) = \ln \frac{p(y\|H_1}{p(y\|H_0} > \ln \eta$	$L(y) = \ln \frac{p(y\|H_1}{p(y\|H_0} > \ln \eta$	$L(y) = \ln \frac{p(y\|H_1}{p(y\|H_0} > \ln \eta$
Threshold	η is related to α_0	η is related to α_0	$\eta = \frac{(C_{10}-C_{00})p_0}{(C_{11}-C_{01})p_1}$
Amount of evidence	$g(\mathbf{y}) = \frac{1}{N}\sum_{n=1}^{N} y_n > \text{Th}$	$g(\mathbf{y}) = \frac{1}{N}\sum_{n=1}^{N} y_n > \text{Th}$	$g(\mathbf{y}) = \frac{1}{N}\sum_{n=1}^{N} y_n > \lambda$
Threshold	$\text{Th} = \frac{\sqrt{2}\sigma_n}{\sqrt{N}} \text{erfc}^{-1}(2\alpha_0)$	$\text{Th} = \frac{\sqrt{2}\sigma_n}{\sqrt{N}} \text{erfc}^{-1}(2\alpha_0) + \bar{S}_0$	$\lambda = \frac{\sigma_n^2}{N(S_1-S_0)} \ln \eta + \frac{S_1+S_0}{2}$
Miss alarm probability	$\beta = 1 - \frac{1}{2}\text{erfc}\left(\frac{\text{Th}-S}{\sqrt{2/N}\sigma_n}\right)$	$\beta = 1 - \frac{1}{2}\text{erfc}\left(\frac{\text{Th}-\bar{S}_1}{\sqrt{2/N}\sigma_n}\right)$	$\beta = 1 - \frac{1}{2}\text{erfc}\left(\frac{\sigma_n}{\sqrt{2N}(S_1-S_0)} \ln \eta - \frac{\sqrt{N}(S_1-S_0)}{2\sqrt{2}\sigma_n}\right)$

Thus, the Bayes criterion for minimizing the average cost becomes the minimum error probability criterion

$$\frac{p(Y|H_1)}{p(Y|H_0)} \underset{H_1}{\overset{H_0}{\lessgtr}} \eta = \frac{p_0}{p_1}. \tag{3.7.2}$$

In other words, the calculation method and steps of minimum error probability criterion and Bayes criterion are exactly the same, but their threshold η is different.

In addition, sometimes the cost factor is normalized, so the assumption of the above cost factor is often expressed as $C_{00} = C_{11} = 0$ and $C_{01} = C_{10} = 1$.

3.7.2 Maximum A Posteriori Probability Criterion

In some applications, the difference between the cost of wrong and correct decisions under different test assumptions H_j is the same, that is, $C_{01} - C_{11} = C_{10} - C_{00}$. Under this assumption, Bayes criterion can be reduced to

$$\frac{p(Y|H_1)}{p(Y|H_0)} \underset{H_1}{\overset{H_0}{\lessgtr}} \frac{p_0}{p_1}(= \eta), \tag{3.7.3}$$

which is the same as the minimum error probability criterion. Note that the above criterion can be written equivalently as

$$p_1 p(Y|H_1) \underset{H_1}{\overset{H_0}{\lessgtr}} p_0 p(Y|H_0). \tag{3.7.4}$$

Using $p(H_i|Y) = p(Y)p(H_i)p(Y|H_i) = p(Y)p_i p(Y|H_i)$, the above equation can be expressed as a posteriori probability $p(H_i|Y)$

$$p(H_1|Y) \underset{H_1}{\overset{H_0}{\lessgtr}} p(H_0|Y). \tag{3.7.5}$$

Eq.(3.7.5) shows that the decision criterion of binary hypothesis testing is: the hypothesis with the largest a posteriori probability is tenable. Specifically, if $p(H_1|Y) = p(H_0|Y)$, judge H_1 hypothesis holds; On the contrary, judget H_0 hypothesis holds. Therefore, the criterion shown in Eq.(3.7.5) is conventionally called the maximum a posteriori probability criterion.

Example 3.7.1 The observation data model is

$$\begin{aligned} y_n &= 1 + w_n \quad (H_1 : \text{ with signal}), \\ y_n &= w_n \quad (H_0 : \text{ without signal}), \end{aligned}$$

where $n = 1, \cdots, 24$, and w_n is a Gaussian white noise with zero mean and variance 1. If a priori probability $p_0 = 1/5$, $p_1 = 4/5$, try to find the error probability P_E and the detection probability P_D.

Solution. The threshold is

$$\eta = \ln \frac{p_0}{p_1} = \ln(0.25) = -1.3863.$$

Since the probability distribution of y_n under H_0 hypothesis is the same as that of additive Gaussian white noise w_n, and y_n is a Gaussian white noise with mean 1 and variance 1 under H_1 hypothesis, the conditional probability density functions of the observation array $\mathbf{y} = (y_1, \cdots, y_N)$ are

$$p(\mathbf{y}|H_0) = \prod_{n=1}^{24} p(y_n|H_0) = \frac{1}{(2\pi)^{24/2}} \exp\left(-\sum_{n=1}^{24} y_n^2/2\right),$$

and

$$p(\mathbf{y}|H_1) = \prod_{n=1}^{24} p(y_n|H_1) = \frac{1}{(2\pi)^{24/2}} \exp\left(-\sum_{n=1}^{24} (y_n - 1)^2/2\right).$$

Calculating the log likelihood ratio function yields

$$L(\mathbf{y}) = \ln\left(\frac{p(y|H_1)}{p(y|H_0)}\right) = -\sum_{n=1}^{24} \frac{(y_n - 1)^2}{2} + \sum_{n=1}^{24} \frac{y_n^2}{2} = \sum_{n=1}^{24} y_n - \frac{24}{2} = 24\bar{y} - 12.$$

Substitute this result directly into Bayes criterion

$$L(\mathbf{y}) = \ln\left(\frac{p(y|H_1)}{p(y|H_0)}\right) > \ln\left(\frac{p_0}{p_1}\right) = -1.3863,$$

so that $24\bar{y} - 12 > -1.3863$, thus

$$\bar{y} > 0.4422$$

is obtained. In other words, when using the mean of the observation data as decision statistics, the threshold is $\lambda = 0.4422$. Therefore, if the sample mean of the observation data is greater than 0.4422, judge the signal present, otherwise judge the signal absent.

On the other hand, the sample mean

$$\bar{y} = \frac{1}{24} \sum_{n=1}^{24} y_n,$$

is also a Gaussian distributed random variable with conditional mean

$$\mathrm{E}\{\bar{y}|H_0\} = 0, \quad \mathrm{E}\{\bar{y}|H_1\} = 1,$$

and variance

$$\mathrm{var}\{\bar{y}|H_0\} = \mathrm{var}\{\bar{y}|H_1\} = \frac{1}{\sigma_w^2} = \frac{1}{25}.$$

Therefore, the conditional distribution density functions of the sample average $\bar{y}$ are

$$p(\bar{y}|H_0) = \frac{1}{(2\pi/24)^{1/2}}\exp\left(-\frac{\bar{y}^2}{2/24}\right),$$
$$p(\bar{y}|H_1) = \frac{1}{(2\pi/24)^{1/2}}\exp\left(-\frac{(\bar{y}-1)^2}{2/24}\right).$$

Therefore, the detection probabilities under H_0 and H_1 hypothesis testing are

$$\begin{aligned} p_{D_0} &= \int_{-\infty}^{\eta} p(\bar{y}|H_0)\mathrm{d}\bar{y} = \int_{-\infty}^{\infty} p(\bar{y}|H_0)\mathrm{d}\bar{y} - \int_{\eta}^{\infty} p(\bar{y}|H_0)\mathrm{d}\bar{y}, \\ &= 1 - \int_{\eta}^{\infty} p(\bar{y}|H_0)\mathrm{d}\bar{y} = 1 - \frac{1}{2}\mathrm{erfc}\left(\frac{0.4422}{\sqrt{2/24}}\right) = 1 - 0.5\mathrm{erfc}(1.53) \\ &= 0.9847 \\ p_{D_1} &= \int_{\eta}^{\infty} p(\bar{y}|H_1)\mathrm{d}\bar{y} = \frac{1}{2}\mathrm{erfc}\left(\frac{0.4422-1}{\sqrt{2/24}}\right) = 0.5\mathrm{erfc}(-1.93) \\ &= 0.9968. \end{aligned}$$

Therefore, the false alarm probability and missed alarm probability are respectively

$$\alpha = 1 - p_{D_0} = 1 - 0.9847 = 0.0153,$$
$$\beta = 1 - p_{D_1} = 1 - 0.9968 = 0.0032.$$

The error probability is

$$P_E = p_0\alpha + p_1\beta = \frac{1}{5}\times 0.0153 + \frac{4}{5}\times 0.0032 = 0.0056.$$

The detection probability is

$$P_D = p_0 p_{D_0} + p_1 p_{D_1} = \frac{1}{5}\times 0.9847 + \frac{4}{5}\times 0.9968 = 0.9944.$$

□

3.7.3 Minimax Criterion

Bayes criterion and its derived minimum total error probability criterion and maximum a posteriori probability criterion all assume that the priori probability p_1 or p_0 is known and fixed. Since $p_1 = 1 - p_0$ and H_1 is the alternative hypothesis of interest, the unknown a priori probability p_1 must be regarded as a variable rather than a fixed value. The problem is: when the prior probability p_1 is unknown or fluctuates greatly, how to select its guessed value, so as to control the error of Bayes criterion within a certain range and avoid large error fluctuation.

To solve this problem, note $\alpha = P_F = P(H_1|H_0)$ and $\beta = P_M = P(H_0|H_1)$. So, there is $P(H_0|H_0) = 1 - \alpha$ and $P(H_1|H_1) = 1 - \beta$. Let $\alpha(p_1)$ and $\beta(p_1)$ represent the false alarm probability and missed alarm probability corresponding to the unknown p_1 respectively, then the average cost or average risk can be rewritten as

$$\begin{aligned} C(p_1) = \; & C_{00} + (C_{10} - C_{00})\alpha(p_1) \\ & + p_1[C_{11} - C_{00} + (C_{01} - C_{11})\beta(p_1) - (C_{10} - C_{00})\alpha(p_1)]. \end{aligned} \tag{3.7.6}$$

It can be proved that when the likelihood ratio $p(Y|H_1)/p(Y|H_0)$ obeys a strictly monotonic probability distribution, the average cost $C(p_1)$ is a strictly convex function of the variable p_1, as shown in Fig 3.7.1. Let $C_{min}(p_1)$ represent the minimum average cost

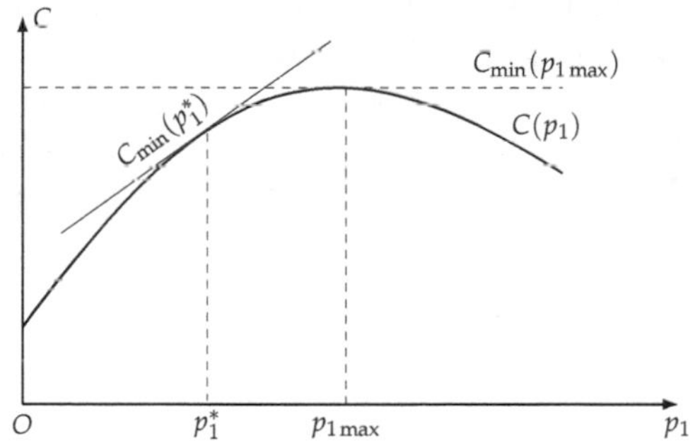

Fig. 3.7.1: Average cost function

function required by Bayes criterion, i.e.,

$$C_{min}(p_1) = \min C(p_1) \quad \Leftrightarrow \quad \frac{\partial C(p_1)}{\partial p_1} = 0, \tag{3.7.7}$$

where

$$\frac{\partial C(p_1)}{\partial p_1} = C_{11} - C_{00} + (C_{01} - C_{11})\beta(p_1) - (C_{10} - C_{00})\alpha(p_1) = 0.$$

Sort out the above equation to get

$$C_{10}\alpha(p_1) + C_{00}[1 - \alpha(p_1)] = C_{01}\beta(p_1) + C_{11}[1 - \beta(p_1)]. \tag{3.7.8}$$

This is the minimax equation of the unknown a priori probability p_1.

Note that using p_0 instead of p_1 in Eq.(3.7.8), we can get the unknown priori probability $p_0 = P(H_0)$. That is, the minimax equation of the unknown priori probability p_0 is

$$C_{10}\alpha(p_0) + C_{00}[1 - \alpha(p_0)] = C_{01}\beta(p_0) + C_{11}[1 - \beta(p_0)]. \tag{3.7.9}$$

From $\alpha = P(H_1|H_0)$ and $\beta = P(H_0|H_1)$, the left side of minimax equation Eq.(3.7.8) represents the cost under H_0 hypothesis and the right side represents the cost under H_1 hypothesis. The solution of minimax equation Eq.(3.7.8) is to balance these two costs. In particular, if the cost function satisfies the minimum error probability criterion, because $C_{11} = C_{00} = 0$ and $C_{10} = C_{01} = 1$, the minimax equation is reduced to $\alpha(p_1) = \beta(p_1)$, i.e., the false alarm and missed alarm probabilities are the same. A typical example of this situation is that in digital communication, the binary 0 is determined to be 1 and 1 is determined to be 0, and the two bit error rates are the same.

Eq.(3.7.7) shows that the minimum average cost function $C_{\min}(p_1)$ is actually the tangent equation of the average cost function $C(p_1)$ at the priori probability p_1 point, which is a straight line.

Since $C(p_1)$ is strictly convex, $C_{\min}(p_1)$ has one and only one maximum $C_{\max\min}(p_1) = \max C_{\min}(p_1)$. This maximum is actually the most unfavorable average cost in Bayes criterion. The corresponding prior probability p_1 is denoted as $p_{1\max}$, which is called the most unfavorable prior probability. Thus, $C_{\min}(p_{1\max})$ is the tangent of the average cost function $C(p_1)$ at the point of the most unfavorable priori probability $p_{1\max}$. This is a horizontal line parallel to the horizontal axis, as shown by the dotted line in Fig 3.7.1.

The most unfavorable priori probability $p_{1\max}$ is actually the maximum value of minimizing the average cost function, i.e.,

$$p_{1\max} = \arg\max\min C(p_1) = \arg\max C_{\min}(p_1) \tag{3.7.10}$$

According to the order of optimization, the optimization process is called max min. It should be noted that there are also references called max min maximum minimization according to the order of symbols.

The optimization criterion of minimax method is to use the most unfavorable a priori probability $p_{1\max}$ as the guess value of unknown a priori probability p_1. This is called minimax criterion.

Contrary to the most unfavorable priori probability, let $C_{\min}(p_1^*)$ be the minimum average cost function corresponding to the unknown real a priori probability p_1^*. It is the tangent of the average cost function $C(p_1)$ at the point of the real priori probability p_1^*, which is an oblique line, as shown in Fig 3.7.1.

By comparing the horizontal tangent $C_{\min}(p_{1\max})$ with the oblique straight line $C_{\min}(p_1^*)$, it is easy to know:

(1) Compared with the most unfavorable a priori probability $p_{1\max}$, the result $C_{\min}(p_{1g})$ of any other guess value p_{1g} of the unknown priori probability p_1 is the same as $C_{\min}(p_{1\max})$, which is neither optimal nor worst. Therefore, "using the most unfavorable a priori probability as a guess" is not good enough, but it is a safe choice.

(2) Because the prior probability p_1 is unknown, its true value p_1^* can not be obtained. If the difference between the guessed priori probability p_{1g} and the real priori probability p_1^* is relatively small, the guessed average cost $C_{\min}(p_{1g})$ is not much different from the ideal average cost $C_{\min}(p_1^*)$, which is naturally a good choice.

Tab. 3.7.1: Comparison of Bayes criterion and its derived criteria

Method	Priori probability p_1, p_0	Known conditions	Decision criterion
Bayesian	known, fixed	$C_{01}, C_{11}, C_{10}, C_{00}$	$\frac{p(\mathbf{y}\|H_1)}{p(\mathbf{y}\|H_0)} \underset{H_1}{\overset{H_0}{\lessgtr}} \frac{p_0(C_{10}-C_{00})}{p_1(C_{01}-C_{11})}$
Minimum error probability	known, fixed	$C_{00} = C_{11} = 0,$	$\frac{p(\mathbf{y}\|H_1)}{p(\mathbf{y}\|H_0)} \underset{H_1}{\overset{H_0}{\lessgtr}} \frac{p_0}{p_1}$
Maximum a posteriori probability	known, fixed	$C_{01} - C_{11} = C_{10} - C_{00}$	$\frac{p(\mathbf{y}\|H_1)}{p(\mathbf{y}\|H_0)} \underset{H_1}{\overset{H_0}{\lessgtr}} \frac{p_0}{p_1}$
Minimax	unknown or fluctuating	$\alpha(p_1), \beta(p_1)$	$C_{10}\alpha(p_1) + C_{00}[1-\alpha(p_1)] = C_{01}\beta(p_1) + C_{01}[1-\beta(p_1)]$

However, if the guess value p_{1g} deviates from the real priori probability p_1^*, the corresponding average cost $C_{\min}(p_{1g})$ may be greatly different from the ideal average cost $C_{\min}(p_1^*)$. Therefore, trying to pursue a better guess than the most unfavorable priori probability is likely to cause a large error fluctuation of Bayes criterion, resulting in great risk.

In a word, minimax criterion is a Bayes decision method to determine the threshold according to the most unfavorable a priori probability.

The characteristic or advantage of minimax criterion is that the average cost is constant and does not change with the fluctuation of a priori probability.

3.8 Multivariate Hypotheses Testing

In some complex hypothesis testing problems, there are often multiple hypotheses. For example, in blind signal separation, it is necessary to judge whether the signal is Gaussian, sub-Gaussian or super-Gaussian distribution:

$$\begin{cases} H_0: & \text{Gaussian distribution,} \\ H_1: & \text{sub-Gaussian distribution,} \\ H_2: & \text{super-Gaussian distribution.} \end{cases} \tag{3.8.1}$$

For another example, the multivariate hypotheses model of tumor diagnosis is

$$\begin{cases} H_0: & \text{Benign tumor,} \\ H_1: & \text{Malignant tumor: early stage,} \\ H_2: & \text{Malignant tumor: metaphase,} \\ H_3: & \text{Malignant tumor: middle-late stage,} \\ H_4: & \text{Malignant tumor: late stage.} \end{cases} \tag{3.8.2}$$

3.8.1 Multivariate Hypotheses Testing Problem

The testing of several hypotheses is called multivariate hypothesis testing.

Consider the M-ary hypothesis testing problem

$$\begin{aligned} H_0: &\quad y_k = S_0 + n_k, \qquad k = 1, \cdots, N, \\ H_1: &\quad y_k = S_1 + n_k, \qquad k = 1, \cdots, N, \\ &\qquad\vdots \\ H_{M-1}: &\quad y_k = S_{M-1} + n_k, \quad k = 1, \cdots, N. \end{aligned} \tag{3.8.3}$$

Let $Y = g(y_1, \cdots, y_N)$ be the amount of evidence obtained from N observations when H_j is assumed to be true. If the hypothesis H_j is true, the amount of evidence Y determines that the hypothesis H_i is true, then the judgment result is marked as $(H_i|H_j)$. Therefore, in the M-ary hypothesis testing, there are M possible decision results $(H_0|H_j), (H_1|H_j), \cdots, (H_{M-1}|H_j)$ under H_j hypothesis. Under the M-ary hypothesis, there are $M \times M = M^2$ possible decision results

$$(H_0|H_j), (H_1|H_j), \cdots, (H_{M-1}|H_j), \quad j = 0, 1, \cdots, M-1.$$

The corresponding decision probability is

$$P(H_0|H_j), P(H_1|H_j), \cdots, P(H_{M-1}|H_j), \quad j = 0, 1, \cdots, M-1,$$

where, there are only M correct decisions, and the decision probability is $P(H_j|H_j), j = 0, 1, \cdots, M-1$. The other $M(M-1)$ decisions are wrong, and the probability of wrong decisions is $P(H_i|H_j), i \neq j (i, j = 0, 1, \cdots, M-1)$.

Let R_i indicate the area where the evidence is located when the judgment is H_i hypothesis. It is generally assumed that these areas are non empty and non crosslinked:

$$R_i \neq \Phi, \quad R_i \cap R_j = \Phi \quad (i \neq j). \tag{3.8.4}$$

If the amount of evidence Y falls in the decision domain $R_i, i = 0, 1, \cdots, M-1$, judge H_i hypothesis to be true.

The union of M decision domains $R_0, R_1, \cdots, R_{M-1}$ forms the decision domain of M-ary hypothesis testing

$$R = \cup_{i=0}^{M-1} R_i. \tag{3.8.5}$$

3.8.2 Bayes Criteria for Multiple Hypotheses Testing

Assuming that the prior probability $p_i = P(H_i), i = 0, 1, \cdots, M-1$ is known, the cost factors $C_{ij}, i, j = 0, 1, \cdots, M-1$ of various decisions have also been determined.

Bayes average cost

$$\begin{aligned} C &= \sum_{j=0}^{M-1}\sum_{i=0}^{M-1} C_{ij}p_jP\left(H_i \mid H_j\right) \\ &= \sum_{j=0}^{M-1}\sum_{i=0}^{M-1} C_{ij}p_j \int_{R_i} p\left(Y \mid H_j\right) \mathrm{d}Y \\ &= \sum_{i=0}^{M-1}\left[C_{ii}p_i \int_{R_i} p\left(Y \mid H_j\right) \mathrm{d}Y + \sum_{j=0,\neq i}^{M-1} C_{ij}p_j \int_{R_i} p\left(Y \mid H_j\right) \mathrm{d}Y\right]. \end{aligned} \tag{3.8.6}$$

From $\int_R p\left(Y \mid H_0\right) \mathrm{d}Y = 1$, we have

$$\int_{R_i} p\left(Y \mid H_i\right) \mathrm{d}Y + \int_{\bigcup_{j=0,\neq i}^{M-1} R_j} p\left(Y \mid H_i\right) \mathrm{d}Y = 1. \tag{3.8.7}$$

Substituting Eq. (3.8.7) into Eq. (3.8.6), it is easy to obtain

$$\begin{aligned} C &= \sum_{i=0}^{M-1} C_{ii}p_i\left[1 - \sum_{j=0,j\neq i}^{M-1} \int_{R_j} p\left(Y \mid H_j\right) \mathrm{d}Y\right] + \sum_{i=0}^{M-1}\sum_{j=0,\neq i}^{M-1} C_{ij}p_j \int_{R_i} p\left(Y \mid H_j\right) \mathrm{d}Y \\ &= \sum_{i=0}^{M-1} C_{ii}p_i - \sum_{i=0}^{M-1}\sum_{j=0,\neq i}^{M-1} C_{ii}p_i \int_{R_i} p\left(Y \mid H_i\right) \mathrm{d}Y + \sum_{i=0}^{M-1}\sum_{j=0,\neq i}^{M-1} C_{ij}p_j \int_{R_i} p\left(Y \mid H_j\right) \mathrm{d}Y \\ &= \sum_{i=0}^{M-1} C_{ii}p_i + \sum_{i=0}^{M-1} \int_{R_i} \sum_{j=0,\neq i}^{M-1} p_j\left(C_{ij} - C_{jj}\right) p\left(Y \mid H_j\right) \mathrm{d}Y. \end{aligned} \tag{3.8.8}$$

In the above equation, the first term on the right is the fixed cost, which is irrelevant with the judgment area R_i. The second term is the cost function, which is related to the decision domain R_i.

Bayes criterion minimizes the average cost C. For this reason, let

$$I_i(Y) = \sum_{j=0,\neq i}^{M-1} p_j\left(C_{ij} - C_{jj}\right) p\left(Y \mid H_j\right). \tag{3.8.9}$$

Then Eq. (3.8.8) can be abbreviated as

$$C = \sum_{i=0}^{M-1} C_{ii}p_i + \sum_{i=0}^{M-1} \int_{R_i} I_i(Y)\mathrm{d}Y. \tag{3.8.10}$$

Since the first term is a constant, there is

$$\min C \Longleftrightarrow \min\left\{I_0(Y), I_1(Y), \cdots, I_{M-1}(Y)\right\}. \tag{3.8.11}$$

Bayes criterion of M-ary hypothesis testing: if

$$I_i(Y) = \min\left\{I_0(Y), I_1(Y), \cdots, I_{M-1}(Y)\right\}, \tag{3.8.12}$$

judge $Y \in R_i$, i.e., $H_i$hypothesis holds.

In particular, if $C_{ii} = 0$ and $C_{ij} = 1, j \neq i$, then the average cost of Bayes criterion for M-ary hypothesis testing becomes the error probability, that is

$$C = \sum_{i=0}^{M-1} \sum_{j=0,\neq/i}^{M-1} p_j P\left(H_i \mid H_j\right) = P_{\mathrm{E}}. \tag{3.8.13}$$

So, there is

$$\min C \Longleftrightarrow \min P_{\mathrm{E}}. \tag{3.8.14}$$

In other words, if $C_{ii} = 0$ and $C_{ij} = 1, j \neq i$, then the Bayes criterion of M-ary hypothesis testing is simplified to the minimum error probability criterion.

On the other hand, when $C_{ii} = 0$ and $C_{ij} = 1, j \neq i$, $I_i(y)$ difined by Eq. (3.8.9) becomes

$$I_i(Y) = \sum_{j=0,\neq/i}^{M-1} p_j p\left(Y \mid H_j\right) = \sum_{j=0,\neq/i}^{M-1} p(Y) p\left(H_j \mid Y\right) = \left[1 - p\left(H_i \mid Y\right)\right] p(Y), \tag{3.8.15}$$

where $\sum_{j=0,\neq/i}^{M-1} p\left(H_j \mid Y\right) = \sum_{j=0}^{M-1} p\left(H_j \mid Y\right) - P\left(H_i \mid Y\right) = 1 - p\left(H_i \mid Y\right)$ is used. So, again

$$\min I_i(Y) \Leftrightarrow \max p\left(H_i \mid Y\right). \tag{3.8.16}$$

That is, if $C_{ii} = 0$ and $C_{ij} = 1, j \neq i$, then the Bayes criterion of M-ary hypothesis testing is simplified to the maximum a posteriori probability criterion.

3.9 Multiple Hypothesis Testing

In multiple hypothesis testing problem, there is only one original hypothesis H_0, but there are $M - 1$ alternative assumptions $H_1, \cdots H_{M-1}$. In other words, all the M assumptions in the multiple hypothesis are independent and do not duplicate each other. However, in some important applications, although there are $m > 2$ hypotheses, they are not independent, but there are a lot of repetitions, with only two different hypotheses.

For example, in biostatistics, there are $m = S + V$ differentially expressed genes, of which S are truly differentially expressed genes, and V are actually non differentially expressed genes, which are false positive. Thus, M genes obey only two kinds of assumptions

$$\begin{cases} H_0\text{: non differentially expressed genes,} \\ H_1\text{: differentially expressed genes.} \end{cases} \tag{3.9.1}$$

The basic task of biostatistics is to select the number of differentially expressed genes through judgment, and hope that the average value of error ratio (error rate) $Q = V/m$ does not exceed a preset value (e.g. 0.05 or 0.01).

Broadly speaking, due to repeated assumptions, there are only two types of assumptions H_0 and H_1 in multiple hypotheses, which is called multiple hypothesis testing problem. In biostatistics, gene expression profile and genome-wide association analysis need tens of thousands or even millions of multiple testing. Therefore, the M-multiple hypothesis testing can be regarded as a special case of M-ary hypothesis testing in the case of repeated hypotheses, with only the original hypothesis H_0 and alternative hypothesis H_1.

Multiple hypothesis testing is a common problem in data analysis, which widely exists in the fields of internet communication, social economics, medicine and health statistics, etc.

3.9.1 Error Rate of Multiple Hypothesis Testing

Table 3.9.1 summarizes possible results of M-ary hypothesis testing[26].

Tab. 3.9.1: Number of correct and wrong decisions of M-ary hypothesis testing

Original hypothesis	Not reject H_0 hypothesis (non significant testing)	Reject H_0 hypothesis (significance testing)	Number
H_0 is true	U (Hit)	V (type I error)	m_0: number of H_0 assumed to be true
H_0 is false	T(type II error)	S (Correct rejection)	$m - m_0$: number of H_0 assumed to be false
Number of Testing results	$m - R$	R	m: assumption number

In the table, m_0 $\boldsymbol{H}_0$ assumptions among the m hypotheses are true, and the other $m - m_0$ $\boldsymbol{H}_0$ are assumed to be false, which corresponds to $m - m_0$ $\boldsymbol{H}_1$ assumptions to be true. In the m-multiple hypothesis testing, the number of the original hypotheses hit are U, and the number of the original hypotheses misjudged by type I are V, where $U + V = m_0$. On the other hand, the number of original assumptions misjudged by type II is T, and the number of the original assumptions correctly rejected is S. Let $R = V + S$ represent the total number of rejected original assumptions, then the number of original assumptions not rejected is $m - R$.

If each original hypothesis takes α as the testing significance level, when the amount of evidence is less than or equal to α (i.e. non significant testing), the judgment result is not to reject H_0 assumption. On the contrary, if the amount of evidence is larger than α (i.e. significant testing), then H_0 assumption will be rejected. Obviously, the number R of H_0 hypotheses rejected is an observable random variable $R(\alpha)$ related to the testing significance level α, while U, V, S, T are unobservable random variables.

When conducting multiple testing on m hypotheses, it is demanded to judge how many H_1 hypotheses hold. At this time, an important question is: how to minimize the error of the final testing.

In the general binary hypothesis testing, there is only one original hypothesis. At this time, the Neyman-Pearson criterion is to minimize the probability of type II error (miss alarm) β, or maximize the equivalent power function, while allowing the probability of type I error (false alarm) to be limited to the level α.

However, if the Neyman-Pearson criterion is directly applied to the multiple hypothesis testing, and the probability α of type I error is still used to measure the overall error of the multiple hypothesis testing, it will lead to invalidity. In other words, in the multiple hypothesis testing problem, a new measure must be used to test the significance level α.

In multiple hypothesis testing, it is necessary to treat the m test as a whole, and take the proportion of type I error or type II error in all errors as the error measurement standard. Here are five common error metrics[140].

1. Per-family Error Rate (PFER)

The PFER is defined as

$$\text{PFER} = \text{E}\{V\}. \tag{3.9.2}$$

Since the number V of assumption of making type I error rate is an unpredictable random variable, its expected value (average) $\text{E}\{V\}$ is used as the error measure.

An obvious disadvantage of this error measure is that the total number of original assumptions m is not considered, and the total number of original assumptions is closely related to the final error control.

2. Per-comparison Error Rate (PCER)

The PCER is defined as

$$\text{PCER} = \text{E}\{V\}/m. \tag{3.9.3}$$

This measure considers the ratio of PFER in the total number of assumptions m, which is an improvement of PFER.

The disadvantage of PCER is that each hypothesis testing in the m-fold hypothesis testing is carried out under α without considering the "totality" of the multiple hypothesis testing problem, which makes the testing standard too “loose”.

3. Family-wise Error Rate (FWER)

The FWER is defined as

$$\text{FWER} = P(V \geqslant 1). \tag{3.9.4}$$

Different from PCER, FWER is a probability value, which represents the probability of making at least one type I error in the m-ary test.

4. False Discovery Rate (FDR)

The FDR is defined as

$$\text{FDR} = \begin{cases} E\{\frac{V}{V+S}\} = E\{\frac{V}{R}\}, R \neq 0, \\ 0, R = 0. \end{cases} \tag{3.9.5}$$

This is proposed by Benjamini and Hochberg in 1995[26].

Let $Q = \frac{V}{V+S}$ indicate the ratio of the number of original hypotheses making type I errors to the total number of rejected original hypotheses, then naturally specify $Q = 0$ if $V + S = 0$. The expected value of ratio Q is

$$Q_E = \text{E}\{Q\} = \text{E}\{V/(V+S)\} = \text{E}\{V/R\} \tag{3.9.6}$$

The error detection rate has the following important properties[26]:

(1) If all the original assumptions are true, i.e. $m_0 = m$, then FDR = FWER. Thus, $S = 0$ and $V = R$. Then, if $V = 0$, then $Q = 0$; And if $V > 0$, then $Q = 1$. This means $P(V \geqslant 1) = \text{E}\{Q\} = Q_E$. Therefore, controlling the FDR means that the FWER can be controlled under weak controlled conditions.

(2) When $m_0 < m$, the FDR is less than or equal to the FWER. At this time, if $V > 0$, then $\frac{V}{R} \leqslant 1$, resulting in the probability distribution $\mathcal{X}_{(V \geqslant 1)} \geqslant Q$. Take the mathematical expectation on both sides and get $P(V \geqslant 1) \geqslant Q_E$. There is a great difference between the two error rates. As a result, any process that controls FWER also controls FDR. If a process can only control FDR, this may be "less stringent". Since most of the original assumptions often encountered in the actual test are not true, the power of this process that only controls FDR will be improved.

5. Positive False Discovery Rate (PFDR)[200]

The PFDR is defined as

$$\text{PFDR} = \text{E}\left\{\frac{V}{R}|R > 0\right\}. \tag{3.9.7}$$

Obviously, PFDR is a special case of FDR when $R \neq 0$. It should be noted that[26], when all m original assumptions are true, $PFDR = 1$ because $V = R$. In this case, significance level α cannot be selected so that PFDR<α. In other words, PFDR fails when all m original assumptions are true. However, in practical problems, there are very few cases where m original hypotheses are true. PFDR is widely used in hypothesis testing problems with untrue original hypotheses.

3.9.2 Error Control Method of Multiple Hypothesis Testing

Here are four error control methods for m-multiple hypothesis testing[140].

1. Classical Bonferroni Multiple Testing[31]

Consider the m-fold hypothesis testing $\{H_1, H_2, \cdots, H_m\}$. If P_i is the prior probability (p value for short) corresponding to each original hypothesis H_i, there is a prior probability set $\{P_1, P_2, \cdots, P_m\}$. Given a significance level α, treat each original hypothesis equally, that is, divide the significance level α by m, based on α/m. For the p-value set $\{P_1, P_2, \cdots, P_m\}$, the classical Bonferroni multiple test method is as follows:

$$\text{if } P_i \leqslant \alpha/m, \text{ reject } H_i \quad (i = 1, \cdots, m). \tag{3.9.8}$$

All rejected hypotheses belong to the alternative hypothesis H_1 of the two types of hypotheses in the multiple hypothesis testing, while all other hypotheses that have not been rejected belong to the original hypothesis H_0 of the multiple hypothesis testing.

The probability can be obtained from the Bonferroni inequality $P_i \leqslant \alpha/m$

$$P\left\{\bigcup_{i=1}^{m}(P_i \leqslant \alpha/m) < \alpha \quad (0 \leqslant \alpha \leqslant 1)\right\}. \tag{3.9.9}$$

The classical Bonferroni multiple test method was proposed by Bonferroni in 1930. Its advantage is simple and intuitive. Since the assumption of density distribution of random variables is not involved, it is easy to apply.

2. Improved Bonferroni Multiple Test

There are three ways to improve the classical Bonferroni multiple test method:

(1) Holm step-down control method[105]: Before the hypothesis testing, the p-value $P_1, P_2, \cdots, P_m$ is rearranged into $P_{(1)}, P_{(2)}, \cdots, P_{(m)}$ from small to large. Then, for all $j = 1, \cdots, i$, judge whether the inequality $P_{(j)} \leqslant \alpha/(m - j + 1)$ holds? If this inequality holds, the hypothesis $H_{(j)}$ is rejected.

(2) Simes control method[195]: On the basis of Holm test, the control process is improved: for all $j = 1, \cdots, m$, judge whether the inequality $P_{(j)} \leqslant j\alpha/m$, is true. If the inequality holds for a certain j, reject all the reordering original assumptions $H_{(1)}, H_{(2)}, \cdots, H_{(j)}$.

(3) Hochberg step-up control method[104]: if

$$k = \max_i \left\{P_{(i)} \leqslant \frac{1}{m - i + 1}\alpha\right\} \quad (0 \leqslant \alpha \leqslant 1), \tag{3.9.10}$$

then reject the original reordered hypotheses $H_{(1)}, H_{(2)}, \cdots, H_{(k)}$.

3. FDR Multiple Test[26]

FDR multiple test method, also known as FDR error control method, is proposed by Benjamini and Hochberg in 1995. This method determines the domain of P value by controlling the FDR. For example, select $R = V + S$ differentially expressed genes, of which S are truly differentially expressed, and the other V are not differentially expressed and are false positive genes. In biostatistics, it is hoped that the expected or average value of error rate $Q = V/R$ cannot exceed a preset threshold (e.g. 0.05 or 0.01). Statistically, this is equivalent to controlling the FDR no more than 5% or 1%.

FDR multiple test is a step-down control method.

Algorithm 3.9.1. *FDR multiple test algorithm*

Step 1 Sort: arrange the p-value $P_1, P_2, \cdots, P_m$ *into* $P_{(1)} < P_{(2)} < \cdots < P_{(m)}$ *from small to large, and record the corresponding original hypotheses as* $H_{(1)}, H_{(2)}, \cdots, H_{(m)}$.

Step 2 let $i = m, m-1, \cdots, 1$ *and test inequality* $P_{(i)} < \frac{i}{m}\alpha$. *If* k *is the maximum* i *value satisfying this inequality, all the original assumptions* $H_{(1)}, H_{(2)}, \cdots, H_{(k)}$ *are rejected.*

Step 3 if no i *satisfies the inequality* $P_{(i)} < \frac{i}{m}\alpha$, *the original hypotheses* $H_{(1)}, H_{(2)}, \cdots, H_{(m)}$ *are not rejected.*

For example, in biostatistics, if H_0 is assumed to be non differentially expressed genes, the selection result of FDR error control method is k differentially expressed genes $H_{(1)}, H_{(2)}, \cdots, H_{(k)}$.

In recent ten years, the applied research results of FDR multiple test method have been reported in Science and Nature, two top academic journals in the world[140]. In 2001, astrophysicists and statisticians jointly published a paper on confirming the big bang theory of the origin of the universe by FDR method in Science[156]; In 2005, in Nature, geneticists cooperated with statisticians to apply FDR method to the study of the effect of interaction between genetic polymorphisms on gene expression[33].

4. Storey multiple test method[199, 200]

The above three error control methods follow a common pattern: on the premise of giving the error control level (i.e. fixing type I error level), based on a single hypothesis testing, the rejection domain of the test is constructed through the error control method, and the test results can be obtained.

In 2002 and 2003, Storey put forward a new idea of hypothesis testing: give a rejection domain empirically, then estimate the error rate. If the error rate can be accepted, the test is considered to be valid; If the error rate is large, the rejection domain needs to be readjusted until the error rate is controlled at a satisfactory level.

The theory and method of multiple hypothesis testing are especially suitable for the statistical analysis of complex data such as gene chip. There are also a large number of similar complex data in the fields of Internet, social economics, medicine and health statistics. Therefore, multiple hypothesis testing is also widely used in Internet communication, econometrics and statistical analysis of medical and health data such as epidemiology and health statistics.

The following introduces two important applications of multiple hypothesis testing: multiple linear regression and the test of equality of multi population mean.

3.9.3 Multiple Linear Regression

Consider the multiple linear regression model

$$y_i = \beta_0 + \beta_1 x_{1i} + \cdots + \beta_m x_{mi},\ i = 1, \cdots, N, \tag{3.9.11}$$

where, y_i and $x_{1i}, \cdots, x_{mi}, i = 1, \cdots, N$ are called explained variable and explanatory variables respectively; $\beta_0, \beta_1, \cdots, \beta_m$ are called linear regression parameters and m is called linear regression order.

Multiple linear regression, also known as multiple linear fitting, has two main purposes:

(1) Determine whether the multiple regression model is linear or nonlinear.
(2) If the multiple linear regression model holds, the order of the multiple linear regression model (i.e., the number of explanatory variables) m needs to be determined.

The above objectives are essentially equivalent to testing whether the regression parameter $\beta_j, j = 1, \cdots, m$ in the multiple regression model is significantly non-zero? Therefore, m-ary linear regression is essentially an m-multiple hypothesis test

$$\begin{cases} H_0 : \forall j \in \{1, 2, \cdots, m,\}\ \beta_j = 0 & \text{(linear regression does not hold)},\\ H_1 : \exists j \in \{1, 2, \cdots, m,\}\ \beta_j \neq 0 & \text{(linear regression holds)}. \end{cases} \tag{3.9.12}$$

Suppose the true value of the explained variable is Y_i, the linear regression or observation value is

$$\hat{y}_i = \hat{\beta}_0 + \hat{\beta}_1 x_{1i} + \cdots + \hat{\beta}_m x_{mi}, \quad i = 1, \cdots, N. \tag{3.9.13}$$

And the mean of the linear regression is

$$\bar{y} = \frac{1}{N}\sum_{i=1}^{N} \hat{y}_i. \tag{3.9.14}$$

There are three measures of regression or fitting quality.

(1) Total sum of squares (TSS) measure: Sum of squares of regression dispersion $v_i = \hat{y}_i - \bar{y}$

$$\text{TSS} = \sum_{i=1}^{N} (\hat{y}_i - \bar{y})^2 = \sum_{i=1}^{N} v_i^2. \tag{3.9.15}$$

(2) Regression sum of squares (RSS) measure: Sum of squares of error between final approximate value and the true value $\delta_i = \bar{y} - y_i$

$$\text{RSS} = \sum_{i=1}^{N} (\bar{y} - y_i)^2 . \tag{3.9.16}$$

(3) Error sum of squares (ESS) measure: Sum of squares of the observation error (error for short) $e_i = y_i - \hat{y}_i$

$$\text{ESS} = \sum_{i=1}^{N} e_i^2, \tag{3.9.17}$$

where the mean of the error $e_i = y_i - \hat{y}_i$ is zero, i.e., $\frac{1}{N}\sum_{i=1}^{N} e_i = 0$, and the optimal regression error e_i is orthogonal to the known variables $x_{ki}, k = 1, \cdots, m$, i.e., $e_i \perp x_{ki}, k = 1, \cdots, m$.

Consider the total sum of squares

$$\begin{aligned}\text{TSS} &= \sum_{i=1}^{N} (\hat{y}_i - \bar{y})^2 = \sum_{i=1}^{N} \left[(\hat{y}_i - y_i) + (y_i - \bar{y})\right]^2 \\ &= \sum_{i=1}^{N} (\hat{y}_i - y_i)^2 + 2\sum_{i=1}^{N} (\hat{y}_i - y_i)(y_i - \bar{y}) + \sum_{i=1}^{N} (y_i - \bar{y})^2 .\end{aligned}$$

According to the nature of regression error, the second summation term of the above equation is equal to zero, so we have

$$\text{TSS} = \sum_{i=1}^{N} (\hat{y}_i - y_i)^2 + \sum_{i=1}^{N} (y_i - \bar{y})^2 = \text{ESS} + \text{RSS}. \tag{3.9.18}$$

That is, TSS is equal to the sum of ESS and RSS.

Consider the relationship between the explanatory variable (independent variable) $x_1, \cdots, x_m$ and the explained variable (dependent variable) y. Let the observed values obtained from N experiments be $(x_{1i}, \cdots, x_{mi}; y_i), i = 1, \cdots, N$. Let

$$\bar{x}_i = \frac{1}{N}\sum_{k=1}^{N} x_{ik}, \quad i = 1, \cdots, m, \tag{3.9.19}$$

$$\bar{y} = \frac{1}{N}\sum_{k=1}^{N} y_k, \tag{3.9.20}$$

$$l_{ij} = \sum_{k=1}^{N} (x_{ik} - \bar{x}_i)(x_{jk} - \bar{x}_j), \quad i, j = 1, \cdots, m, \tag{3.9.21}$$

$$r_i = \sum_{k=1}^{N} (x_{ik} - \bar{x}_i)(y_k - \bar{y}), \quad i = 1, \cdots, m, \tag{3.9.22}$$

then we get the covariance matrix

$$\boldsymbol{L} = \begin{bmatrix} l_{11} & \cdots & l_{1m} \\ \vdots & \ddots & \vdots \\ l_{m1} & \cdots & l_{mm} \end{bmatrix}, \tag{3.9.23}$$

and its inverse matrix

$$\boldsymbol{L}^{-1} = \begin{bmatrix} c_{11} & \cdots & c_{1m} \\ \vdots & \ddots & \vdots \\ c_{m1} & \cdots & c_{mm} \end{bmatrix}. \tag{3.9.24}$$

Sum all $i = 1, \cdots$ and N on both sides of Eq. (3.9.11), and then divide them by N to obtain the average value

$$\bar{y} = \beta_0 + \beta_1 \bar{x}_1 + \cdots + \beta_m \bar{x}_m. \tag{3.9.25}$$

Subtract Eq. (3.9.11) from (3.9.3) to get

$$y_k - \bar{y} = \beta_1 \left(x_{1k} - x_1\right) + \cdots + \beta_m \left(x_{mk} - \bar{x}_m\right), \quad k = 1, \cdots, N. \tag{3.9.26}$$

In equation (3.9.26), multiply both sides by $\left(x_{jk} - \bar{x}_j\right)$, and sum $k = 1, \cdots, N$, then we have

$$r_j = \beta_1 l_{j1} + \cdots + \beta_m l_{jm}, \quad j = 1, \cdots, m. \tag{3.9.27}$$

Or in matrix form

$$\begin{bmatrix} l_{11} & \cdots & l_{1m} \\ \vdots & \ddots & \vdots \\ l_{m1} & \cdots & l_{mm} \end{bmatrix} \begin{bmatrix} \beta_1 \\ \vdots \\ \beta_m \end{bmatrix} = \begin{bmatrix} r_1 \\ \vdots \\ r_m \end{bmatrix}. \tag{3.9.28}$$

Then the solution of the regression parameters can be obtained

$$\begin{bmatrix} \hat{\beta}_1 \\ \vdots \\ \hat{\beta}_m \end{bmatrix} = \begin{bmatrix} l_{11} & \cdots & l_{1m} \\ \vdots & \ddots & \vdots \\ l_{m1} & \cdots & l_{mm} \end{bmatrix}^{-1} \begin{bmatrix} r_1 \\ \vdots \\ r_m \end{bmatrix} = \begin{bmatrix} c_{11} & \cdots & c_{1m} \\ \vdots & \ddots & \vdots \\ c_{m1} & \cdots & c_{mm} \end{bmatrix} \begin{bmatrix} r_1 \\ \vdots \\ r_m \end{bmatrix}. \tag{3.9.29}$$

By substituting these regression parameters into equation (3.9.3), we can get the solution of β_0

$$\hat{\beta}_0 = \bar{y} - \hat{\beta}_1 \bar{x}_1 - \cdots - \hat{\beta}_m \bar{x}_m. \tag{3.9.30}$$

"The overall linear relationship of the multiple linear regression is significant" does not mean that each explanatory variable $x_{1i}, \cdots, x_{mi}, i = 1, \cdots, N$ has a significant effect on the explained variable y_i. Therefore, a significance test must be performed on each explanatory variable to determine whether each explanatory variable should be retained in the multiple linear regression model. This test is achieved by the t test for the explanatory variables.

Definition 3.9.1. *If the random variables $X \sim \mathcal{N}(0, 1)$ and $Y \sim \chi_n$ are independent, then*

$$T = \frac{\sqrt{n} X}{Y} \tag{3.9.31}$$

obeys t distribution with a degree of freedom of n, denoted as $T \sim t_n$.

Define the t statistic for the explanatory variable x_i as

$$t_i = \frac{\hat{\beta}_i - \beta_i}{\sqrt{c_{ii} \frac{\mathbf{e}^T \mathbf{e}}{N-m-1}}}. \tag{3.9.32}$$

Then the t statistic follows the t distribution with a degree of freedom of $N - m - 1$, i.e.,

$$t_i = \frac{\hat{\beta}_i - \beta_i}{\sqrt{c_{ii} \frac{\mathbf{e}^T \mathbf{e}}{N-m-1}}} \sim t(N - m - 1), \tag{3.9.33}$$

where $\boldsymbol{e} = [e_1, \cdots, e_m]^{\mathrm{T}}$ is the error vector.

Thus, the multiple hypothesis testing equation (3.9.3) for multiple linear regression becomes a t test: given the significance level α, the critical value $t_{\alpha/2}(N - m - 1)$ can be found from the t distribution table[97]. Then, the value of the statistic t_i, $i = 0, 1, \cdots, t_m$ can be calculated from samples, and finally by testing

$$|t_i| \underset{H_1}{\overset{H_0}{\lesseqgtr}} t_{\alpha/2}(N - m - 1), \quad i = 0, 1, \cdots, m, \tag{3.9.34}$$

judge whether the corresponding explanatory variables should be included in the multiple linear regression model.

The following is the t test algorithm for multiple hypothesis testing of multiple linear regression[97].

Algorithm 3.9.2. *t-test algorithm of multiple linear regression*
Known: experimental observations $(x_{1i}, \cdots, x_{mi}; y_i)$, $i = 1, \cdots, N$, and given the significance level α.
Step 1 Calculate the regression parameter vector $\hat{\boldsymbol{\beta}} = \left[\hat{\beta}_1, \cdots, \hat{\beta}_m\right]^T$ from equation (3.9.3) or $\hat{\boldsymbol{\beta}} = (\mathbf{L}^T\mathbf{L})^{-1}\mathbf{L}^T\mathbf{r}$, and then calculate the constant term $\hat{\beta}_0$ from equation (3.9.30)
Step 2 Calculate

$$S_{re} = \sum_{i=1}^{N} (\hat{y}_i - \bar{y})^2 . \tag{3.9.35}$$

Step 3 Calculate the residual standard deviation

$$s = \sqrt{\frac{S_{re}}{N - m - 1}}, \tag{3.9.36}$$

and partial regression square sum

$$p_i = \frac{\hat{\beta}_i^2}{c_{ii}}, i = 1, \cdots, m, \tag{3.9.37}$$

where c_{ii} is the diagonal element of the inverse matrix $\boldsymbol{L}^{-1}$.
Step 4 Calculate the t-statistic

$$t_i = \frac{\sqrt{p_i}}{s}, i = 1, \cdots, m. \tag{3.9.38}$$

Step 5 Look up the t distribution table according to the given significance level α to obtain the critical value $t_{\alpha/2}(N - m - 1)$, and then use the t test of equation (3.9.3) to judge whether each explanatory variable should be included in the multiple linear regression model.

For example, in the ternary linear regression, given the significance levels $\alpha = 0.05$ and $N = 23$, the degree of freedom of t distribution is $N - m - 1 = 23 - 3 - 1 = 19$. The corresponding critical value $t_{0.025}(19) = 2.093$ can be found through the t distribution table. If the absolute values $|t_0|$, $|t_1|$, $|t_2|$ and $|t_3|$ of all calculated t values are greater than the critical value $t_{0.025}(19)$, then the four explanatory variables including the constant term β_0, x_1, x_2 and x_3 are significant at the level of 95%, that is, the four explanatory variables have passed the significance test. This shows that the explanatory variables have obvious explanatory power to the explained variable y, i.e., the ternary linear regression model holds.

3.9.4 Multivariate Statistical Analysis

In economic and medical multivariate statistical analysis, it is often necessary to evaluate whether there are differences in the results? For example, different personnel of several appraisal institutions have evaluated the investment environment in China. Let the political environment score of the jth appraiser of the ith institution is $x_{j1}^{(i)}$, the legal environment score is $x_{j2}^{(i)}$, the economic environment score is $x_{j3}^{(i)}$ and the cultural environment score is $x_{j4}^{(i)}$. According to these scores, it is necessary to analyze whether there are differences in the evaluation of China's investment environment by these evaluation institutions? For another example, in order to study a disease, several groups of people of different ages and genders were subjected to biochemical examination, in which the total cholesterol (CHO) of the jth person in group i was $x_{j1}^{(i)}$, triglyceride (TG) was $x_{j2}^{(i)}$, low density lipoprotein cholesterol (LDL) was $x_{j3}^{(i)}$, and high density lipoprotein cholesterol (HDL) was $x_{j4}^{(i)}$. At this time, it is necessary to analyze whether there are significant differences in biochemical indexes among these groups?

The essence of m-ary statistical analysis is to analyze m related variables at the same time: compare whether the mean or covariance matrix of these correlation variables are equal. Note that multivariate statistical analysis is not a multivariate hypothesis testing, but a multiple hypothesis testing, because multivariate statistical analysis has only two hypotheses H_0 and H_1.

Just as univariate normal distribution $\mathcal{N}(\mu, \sigma^2)$ is the basic hypothesis of binary hypothesis testing and statistical analysis, multivariate normal distribution is the basic hypothesis of multivariate and multiple hypothesis testing and multivariate statistical analysis.

Definition 3.9.2. *Let $\boldsymbol{x} = [x_1, \cdots, x_q]^T$, where $x_1, \cdots, x_q$ is an independent univariate normal distribution $\mathcal{N}(0, 1)$. If $\boldsymbol{\mu}$ is a p-dimensional constant vector and $\boldsymbol{A}$ is a $q \times p$ constant matrix, then*

$$\boldsymbol{y} = \boldsymbol{\mu} + \boldsymbol{A}^T \boldsymbol{x} \tag{3.9.39}$$

follows the p-ary normal distribution, denoted as $\boldsymbol{x} \sim \mathcal{N}_p(\boldsymbol{\mu}, \boldsymbol{\Sigma})$, *where* $\boldsymbol{\mu}$ *and* $\boldsymbol{\Sigma} = \boldsymbol{A}^T\boldsymbol{A}$ *are called the mean vector and variance matrix of p-ary normal distribution respectively.*

In fact, the distribution of many practical problems is often multivariate normal distribution or approximate multivariate normal distribution; Or even if it is not multivariate normal distribution, its sample mean is approximately multivariate normal distribution.

The test methods of multivariate statistical analysis are introduced in three cases.

(1) Test for equality of multivariate normal mean vectors

The mathematical description of the test problem of equality of multivariate normal mean vectors is

$$\begin{cases} H_0 : \mu_1 = \cdots = \mu_m \\ H_1 : \text{otherwise} \end{cases} \tag{3.9.40}$$

Consider m variables $X_1, \cdots, X_m$. Let $x_{kj}^{(i)}$ represent the j-th observation data in the i-th (where $i = 1, \cdots, p$) observation data group of X_k, and let the k-th variable X_k of the i-th group have $N_i(i = 1, \cdots, p)$ observation data.

The i-th sample mean of the k-th variable X_k is

$$\bar{X}_k^{(i)} = \frac{1}{N_i}\sum_{j=1}^{N_i} x_{kj}^{(i)}, \quad i = 1, \cdots, p; k = 1, \cdots, m. \tag{3.9.41}$$

The group i sample mean of m variables constitutes the group i multivariate sample mean vector

$$\bar{\boldsymbol{x}}^{(i)} = \left[\bar{X}_1^{(i)}, \cdots, \bar{X}_m^{(i)}\right]^{\mathrm{T}}, \quad i = 1, \cdots, p. \tag{3.9.42}$$

Let $N = N_1 + \cdots + N_m$, then the population sample mean of the k-th variable X_k is

$$\bar{X}_k = \frac{1}{N}\sum_{i=1}^{p}\sum_{j=1}^{N_i} x_{kj}^{(i)}, \quad k = 1, \cdots, m. \tag{3.9.43}$$

They form the total mean vector

$$\bar{\boldsymbol{x}} = \left[\bar{X}_1, \cdots, \bar{X}_m\right]^{\mathrm{T}}. \tag{3.9.44}$$

Define the data matrix for group i

$$\boldsymbol{X}^{(i)} = \begin{bmatrix} x_{11}^{(i)} & \cdots & x_{1N_i}^{(i)} \\ \vdots & \ddots & \vdots \\ x_{m1}^{(i)} & \cdots & x_{mN_i}^{(i)} \end{bmatrix} = \left[\boldsymbol{x}_1^{(i)}, \cdots, \boldsymbol{x}_{N_i}^{(i)}\right], \tag{3.9.45}$$

where

$$\boldsymbol{x}_j^{(i)} = \left[x_{1j}^{(i)}, \cdots, x_{mj}^{(i)}\right]^{\mathrm{T}}, \quad j = 1, \cdots, N_i. \tag{3.9.46}$$

From the above definition, the sample covariance matrix in group i is

$$\begin{aligned}\boldsymbol{E}_i &= \left[\boldsymbol{x}_1 - \bar{\boldsymbol{x}}^{(i)}, \cdots, \boldsymbol{x}_{N_i}^{(i)} - \bar{\boldsymbol{x}}^{(i)}\right]\left[\boldsymbol{x}_1 - \bar{\boldsymbol{x}}^{(i)}, \cdots, \boldsymbol{x}_{N_i}^{(i)} - \bar{\boldsymbol{x}}^{(i)}\right]^{\mathrm{T}} \\ &= \sum_{j=1}^{N_i}\left(\boldsymbol{x}_j^{(i)} - \bar{\boldsymbol{x}}^{(i)}\right)\left(\boldsymbol{x}_j^{(i)} - \bar{\boldsymbol{x}}^{(i)}\right)^{\mathrm{T}}, \quad i = 1, \cdots, p. \end{aligned} \tag{3.9.47}$$

The covariance matrix of total samples within the group is

$$\boldsymbol{E} = \sum_{i=1}^{p} \boldsymbol{E}_i = \sum_{i=1}^{p}\sum_{j=1}^{N_i}\left(\boldsymbol{x}_j^{(i)} - \bar{\boldsymbol{x}}^{(i)}\right)\left(\boldsymbol{x}_j^{(i)} - \bar{\boldsymbol{x}}^{(i)}\right)^{\mathrm{T}}. \tag{3.9.48}$$

The total sample covariance matrix between groups is

$$\boldsymbol{B} = \sum_{i=1}^{p} N_i\left(\bar{\boldsymbol{x}} - \bar{\boldsymbol{x}}^{(i)}\right)\left(\bar{\boldsymbol{x}} - \bar{\boldsymbol{x}}^{(i)}\right)^{\mathrm{T}}. \tag{3.9.49}$$

The sum of total sample covariance matrix within group and total sample covariance matrix between groups is

$$\boldsymbol{A} = \boldsymbol{E} + \boldsymbol{B}, \tag{3.9.50}$$

which is called the total sample covariance matrix.

The statistical meaning[256] of the determinant of the above matrix is as follows.

$|\boldsymbol{B}|$: inter group variance of m sets of sample points divided by the population sample of m variables;

$|\boldsymbol{E}|$: the sum of the inter group variances of each group from m groups;

$|\boldsymbol{A}|$: total variance of m groups of sample points.

If the overall mean of m variables is equal, the sample points of m groups should be very close. The intra group variance is large, but the inter group variance is small, that is, the intra group variance is almost the total variance. At this time, $\lambda^{2/n} = \frac{|\boldsymbol{E}|}{|\boldsymbol{B}+\boldsymbol{E}|}$ should be close to 1.

Definition 3.9.3. [256] *Let the data matrix* $\boldsymbol{X}_{n\times p} \sim \mathcal{N}_{n\times p}(\boldsymbol{0}, \boldsymbol{I}_n \otimes \boldsymbol{\Sigma})$*, then the covariance matrix* $\boldsymbol{W}$ *follows the Wishart distribution and is recorded as* $\boldsymbol{W} \sim W_p(n, \boldsymbol{\Sigma})$.

Definition 3.9.4. [256] *Let* $\boldsymbol{E} \sim W_p(n, \boldsymbol{\Sigma})$ *and* $\boldsymbol{B} \sim W_p(m, \boldsymbol{\Sigma})$ *be independent of each other, and* $m > p$, $n > p$. *Matrix* Σ *is positive definite, then*

$$\Lambda = \frac{|\boldsymbol{E}|}{|\boldsymbol{E}+\boldsymbol{B}|} \text{ or } \lambda_1 = \frac{|\boldsymbol{E}|^{N/2}}{|\boldsymbol{E}+\boldsymbol{B}|^{N/2}}, \tag{3.9.51}$$

obeys Wilks distribution, denoted as $\lambda_1 \sim \Lambda_{p.n,m}$. *The critical point or quantile* α *of Wilks distribution* $\Lambda_{p,n,m}$ *can be obtained by looking up the table in reference [239].*

Theorem 3.9.1. [256] *When the mean of multivariate normal variables* $\mu_1 = \cdots = \mu_m$*, there are*

$$\boldsymbol{A} \sim W_p(N-1, \boldsymbol{\Sigma}), \quad \boldsymbol{E} \sim W_p(N-m, \boldsymbol{\Sigma}), \quad \boldsymbol{B} \sim W_p(m-1, \boldsymbol{\Sigma}).$$

***E** and **B** are independent of each other. Thus λ_1 defined by equation (3.9.51) obeys Wilks distribution, namely $\lambda_1 \sim \Lambda_{p,N-m,m-1}$.*

Algorithm 3.9.3. *Test for equality of multivariate normal mean vectors[256] Given: Significance level α.*

Step 1 Calculate the i-th set of multivariate mean vectors $\bar{\mathbf{x}}^{(i)}$ and the total sample mean vector $\bar{\mathbf{x}}$ using equations (3.9.41) $\sim$ (3.9.44).

*Step 2 Calculate the within-group total sample covariance matrix **E** and the inter-group total sample covariance matrix **B** using equation (3.9.48) and equation (3.9.49), respectively.*

Step 3 Calculate the parameter λ_1 from equation (3.9.51), and then perform the following test decision

$$\lambda \underset{H_1}{\overset{H_0}{\lessgtr}} \Lambda_{p,N-m,m-1}(\alpha). \tag{3.9.52}$$

(2) Test for equality of multivariate normal covariance matrices

The problem formulation of the test for equality of multivariate normal covariance matrices[256] is

$$\begin{cases} H_0: & \boldsymbol{\Sigma}_1 = \cdots = \boldsymbol{\Sigma}_m, \\ H_1: & \boldsymbol{\Sigma}_1, \cdots, \boldsymbol{\Sigma}_m \text{ not all equal.} \end{cases} \tag{3.9.53}$$

Let $x_{ij_k}^{(k)}$ denote the jth observation of the ith data set of the kth variable, where $i = 1, \cdots, p; j_k = 1, \cdots, N_k; k = 1, \cdots, m$. Let the sample vector of the kth variable obey a multivariate normal distribution

$$\boldsymbol{x}_j^{(k)} = \left[x_{1j}^{(k)}, \cdots, x_{pj}^{(k)}\right]^{\mathrm{T}} \sim \mathcal{N}_p\left(\boldsymbol{\mu}_k, \boldsymbol{\Sigma}_k\right), \quad j = 1, \cdots, N_k, \tag{3.9.54}$$

that is

$$\begin{cases} \boldsymbol{x}_1^{(1)}, \cdots, \boldsymbol{x}_{N_1}^{(1)} & \sim \mathcal{N}_p\left(\boldsymbol{\mu}_1, \boldsymbol{\Sigma}_1\right) \\ & \vdots \\ \boldsymbol{x}_1^{(m)}, \cdots, \boldsymbol{x}_{N_m}^{(m)} & \sim \mathcal{N}_p\left(\boldsymbol{\mu}_m, \boldsymbol{\Sigma}_m\right) \end{cases} \tag{3.9.55}$$

Let $N = N_1 + \cdots + N_m$. It can be proven that[256], the likelihood ratio statistic of H_0 to H_1 is

$$\lambda_2 = \frac{\prod_{k=1}^{m} |\boldsymbol{A}_k/N_k|^{N_k/2}}{|\boldsymbol{A}/N|^{N/2}}, \tag{3.9.56}$$

where

$$\boldsymbol{A}_k = \sum_{j=1}^{N_k} \left(\boldsymbol{x}_j^{(k)} - \bar{\boldsymbol{x}}^{(k)}\right)\left(\boldsymbol{x}_j^{(k)} - \bar{\boldsymbol{x}}^{(k)}\right)^{\mathrm{T}}, \quad k = 1, \cdots, m, \tag{3.9.57}$$

$$\bar{\boldsymbol{x}}^{(k)} = \frac{1}{N_k}\sum_{j=1}^{N_k} \boldsymbol{x}_j^{(k)}, \quad k = 1, \cdots, m, \tag{3.9.58}$$

$$\mathbf{A} = \mathbf{A}_1 + \cdots + \mathbf{A}_m. \tag{3.9.59}$$

Algorithm 3.9.4. *Algorithm for testing equality of multivariate normal covariance matrices*
Given the significance level α.
Step 1 Compute the matrices $\boldsymbol{A}_1, \cdots, \boldsymbol{A}_m$ and $\boldsymbol{A}$ using Eq. (3.9.57) ~ (3.9.59).
Step 2 Compute the likelihood ratio statistic λ_2 of H_0 to H_1 using equation (3.9.56).
Step 3 If $\lambda_2 > \Lambda_{p,N-m,m-1}(\alpha)$, then reject H_0 hypothesis; otherwise, judge that H_0 hypothesis holds.

(3) Test for equality of multivariate normal mean vectors and covariance matrices

Consider the test of equality of multivariate normal mean vectors and covariance matrices[256]

$$\begin{cases} H_0: & \boldsymbol{\mu}_1 = \cdots = \boldsymbol{\mu}_m; \boldsymbol{\Sigma}_1 = \cdots = \boldsymbol{\Sigma}_m, \\ \boldsymbol{H}_1: & \boldsymbol{\mu}_1, \cdots, \boldsymbol{\mu}_m \text{ not all equal and } \boldsymbol{\Sigma}_1, \cdots, \boldsymbol{\Sigma}_m \text{ not all equal.} \end{cases} \tag{3.9.60}$$

It can be proved that[256] the likelihood ratio statistic of H_0 to H_1 is

$$\lambda_3 = \frac{N^{pN/2} \prod_{k=1}^{m} |\boldsymbol{A}_k|^{N_k/2}}{|\boldsymbol{T}|^{N/2} \prod_{k=1}^{m} N_k^{pN_k/2}}, \tag{3.9.61}$$

where

$$\boldsymbol{A}_k = \sum_{j=1}^{N_k} \left(\boldsymbol{x}_j^{(k)} - \bar{\boldsymbol{x}}^{(k)}\right)\left(\boldsymbol{x}_j^{(k)} - \bar{\boldsymbol{x}}^{(k)}\right)^{\mathrm{T}}, \quad k = 1, \cdots, m, \tag{3.9.62}$$

$$\boldsymbol{T} = \sum_{k=1}^{m}\sum_{j=1}^{N_k} \left(\boldsymbol{x}_j^{(k)} - \bar{\boldsymbol{x}}\right)\left(\boldsymbol{x}_j^{(k)} - \bar{\boldsymbol{x}}\right)^{\mathrm{T}}, \tag{3.9.63}$$

$$\bar{\boldsymbol{x}} = \frac{1}{N}\sum_{k=1}^{m}\sum_{j=1}^{N_k} \mathbf{x}_j^{(k)}. \tag{3.9.64}$$

Algorithm 3.9.5. *Test for equality of multivariate normal mean vectors and covariance matrices Given: significance level α.*
Step 1 Use equations (3.9.62) ~ (3.9.64) to calculate the matrices $\boldsymbol{A}_1, \cdots, \boldsymbol{A}_m$ and $\boldsymbol{T}$.
Step 2 Use equation (3.9.61) to calculate the likelihood ratio statistics λ_3 of H_0 to H_1.
Step 3 If $\lambda_3 > \Lambda_{p,N-m,m-1}(\alpha)$, then reject H_0 assumption; Otherwise, judge H_0 hypothesis holds.

Summary

This chapter focuses on the basic theory and methods of binary hypothesis testing, first classifying signal detection problems into three different types, such as radar target detection, communication signal detection and signal detection problems emphasizing risk minimization. Then for these three types of signal detection problems, the corresponding decision criteria are highlighted: the Neyman-Pearson criterion, the UMP criterion and the Bayes criterion. Although the starting points and core ideas of the three decision criteria are different, they all ultimately use the likelihood ratio function as the decision function, but the choice of the threshold value is different. The last three sections introduce the theory, methods and applications of the Bayes-derived criterion, multivariate hypothesis testing and multiple hypothesis testing, respectively.

Exercises

3.1 The observed data model is

$$\begin{cases} H_1 : y_n = 4 + w_n & \text{when signal is present,} \\ H_0 : y_n = w_n & \text{when signal is absent,} \end{cases}$$

where $n = 1, \cdots, 16$, and w_n is a Gaussian white noise with mean of 1 and variance of 4. If the false alarm probability $\alpha = 0.05$ is required, try to find the detection probability using the Neyman-Pearson criterion.

3.2 The binary phase-shift keying (BPSK) signal is observed in additive Gaussian white noise $w(t)$:

$$\begin{cases} H_1 : y(t) = A\cos(\omega_c t + \theta) + w(t), \\ H_0 : y(t) = -A\cos(\omega_c t + \theta) + w(t), \end{cases}$$

where $0 \leqslant t \leqslant 2\mu$ s, the mean value of Gaussian white noise $w(t)$ is 0 and the power spectral density is 10^{-12} W/Hz. If the BPSK signal is emitted with equal probability and the amplitude of the carrier $A = 10$mV, find the BER.

3.3 Let $y_1, \cdots, y_N$ is from the Poisson distribution

$$p(y;\lambda) = \begin{cases} \frac{e^{-\lambda}\lambda^y}{y}, y = 0, 1, 2, \cdots; \lambda > 0 \\ 0, \text{ other} \end{cases}$$

of a random sample of observations, where λ is unknown at the end. Try to determine the optimal critical region for the binary hypothesis testing

$$\begin{cases} H_0 : \lambda = \lambda_0 \\ H_1 : \lambda = \lambda_1 \end{cases}$$

at the test level α, where $\lambda_1 > \lambda_0$.

3.4 Observed data by the model

$$\begin{cases} H_1 : y_n = 1 + w_n & \text{when signal } +1 \text{ is transmitted} \\ H_0 : y_n = -1 + w_n & \text{when signal } -1 \text{ is transmitted} \end{cases}$$

where $n = 1, \cdots, 16$, and the mean value of Gaussian white noise w_n is 1 and the variance is 9. If the emission probability $p(1) = 0.75$ for +1 signal and $p(-1) = 0.25$ for −1 signal, try to determine the detection probability and false alarm probability using Bayes criterion.

3.5 BPSK signal is observed in a Gaussian white noise $w(t)$ with zero mean and power spectral density of $\sigma_0/2$:

$$\begin{cases} H_1 : y(t) = A\cos(\omega_c t + \theta) + w(t) \\ H_0 : y(t) = -A\cos(\omega_c t + \theta) + w(t) \end{cases}$$

where θ is an unknown constant. If $p_0 = 0.25, p_1 = 0.75$. Compute the error probability p_E.

3.6 Continuous observation data are

$$\begin{cases} H_1 : y(t) = s(t) + w(t) & (\text{ signal is present }) \\ H_0 : y(t) = w(t) & (\text{ signal is absent }) \end{cases} \qquad t = 1, \cdots, T.$$

where $w(t)$ is a Gaussian white noise with an average value of 0 and a power spectral density of $N_0/2$. Let $E = \int_0^T s^2(t)dt$ represent the energy of signal $s(t)$ in the observation time $[0, T]$.

(1) If the probability of false alarm does not exceed α, prove: when

$$\int_0^T y(t)s(t)dt \geqslant \text{Th},$$

the detector designed by Neyman- Pearson criterion will decide signal $s(t)$ present. In the above equation, Th is a threshold determined by the complementary error function $\text{erfc}(\sqrt{\frac{2}{N_0 E}}\text{Th}) = \alpha$.

(2) Prove: detection probability

$$P_D = \text{erfc}\left(\text{erfc}^{-1}(\alpha) - \sqrt{\frac{2E}{N_0}}\right).$$

3.7 Suppose $y_1, \cdots, y_N$ be samples taken from distribution $p(y)$. Consider the hypothesis testing of distribution $p(y)$

$$\begin{cases} H_0 : p(y) = \frac{1}{\sqrt{2\pi}}\exp(-\frac{1}{2}y^2) \\ H_1 : p(y) = \frac{1}{2}\exp(-|y|) \end{cases}.$$

Use the Neyman-Pearson criterion to determine the test statistics and their decision areas when the significance level of the test is α_0.

3.8 Suppose $T = T(y_1, y_2, \cdots, y_N)$ be a statistic. If the conditional distribution of samples $y_1, y_2, \cdots, y_N$ is independent of hypothesis H when T is given, then T is called a sufficient statistic about hypothesis H. Let the samples $y_1, y_2, \cdots, y_N$ be independently and homogeneously distributed and obey the exponential distribution. Consider the hypothesis testing of distribution

$$\begin{cases} H_0 : p(y) = \frac{1}{\lambda_0}\exp\left(-y/\lambda_0\right) \\ H_1 : p(y) = \frac{1}{\lambda_1}\exp\left(-y/\lambda_1\right) \end{cases}.$$

Prove that the sample mean $\bar{y} = \frac{1}{N}\sum_{i=1}^{N} y_i$ is a sufficient statistic.

3.9 Let $y_1, \cdots, y_N$ be samples taken from the zero mean Gaussian distribution. Consider the test of Gaussian distribution variance

$$\begin{cases} H_0 : \sigma^2 = \sigma_0^2 \\ H_1 : \sigma^2 = \sigma_1^2 \end{cases}$$

where σ_0 and σ_1 are known constants and satisfy $\sigma_0^2 < \sigma_1^2$.

(1) Calculate the logarithmic likelihood ratio;

(2) Assume that the threshold Th satisfies

$$\text{LLR}\left(y_1, \cdots, y_N\right) < \text{Th} \Rightarrow H_0,$$
$$\text{LLR}\left(y_1, \cdots, y_N\right) \geqslant \text{Th} \Rightarrow H_1,$$

where, LLR is the abbreviation of logarithmic likelihood function. Prove that $T\left(y_1, \cdots, y_N\right) = \sum_{i=1}^{N} y_i^2$ is a sufficient statistic; And represent the threshold η of the statistic

$$T\left(y_1, \cdots, y_N\right) < \eta \quad \Rightarrow \quad H_0$$
$$T\left(y_1, \cdots, y_N\right) \geqslant \eta \quad \Rightarrow \quad H_1$$

as a function of the threshold Th and constants σ_0, σ_1;

(3) Try to find the expressions of α_0 and β;

(4) Draw the receiver operating characteristic curve when $N = 1$, $\sigma_1^2 = 1.5$ and $\sigma_0^2 = 1$.

3.10 Prove that the decision function of Bayes test can be written in the form of

$$\frac{p\left(y_1, \cdots, y_N \mid H_1\right)}{p\left(y_1, \cdots, y_N \mid H_0\right)} \geqslant \text{Th},$$

i.e., the Bayes test is equivalent to likelihood ratio test.

3.11 Let $y = \sum_{i=1}^{n} x_i$, where $x_i \sim \mathcal{N}(0, \sigma^2)$ is an independent Gaussian variable with the same distribution, and n is a random variable of Poisson distribution

$$P(n = k) = \frac{\lambda^k}{k!}\exp(-\lambda), \quad k = 0, 1, \cdots.$$

Now we need to decide between two hypotheses

$$\begin{cases} H_0 : n > 1 \\ H_1 : n \leqslant 1 \end{cases}.$$

Try to write the expression of Neyman-Paron test with false alarm probability α_0.

3.12 In the above question, if the assumption is replaced by

$$\begin{cases} H_0 : \lambda = \lambda_0 \\ H_1 : \lambda = \lambda_1 \end{cases},$$

please write the expression of Neyman-Pearson test with false alarm probability α_0.

3.13 The method of combining two tests with a certain probability scheme is called randomization decision rule. Taking likelihood ratio test as an example, the threshold values of the two tests are Th_1 and Th_2, respectively. Then the randomization decision refers to the test results with probability η using Th_1 as the threshold and probability $1-\eta$ using Th_2 as the threshold value.

(1) Represent the detection probability of randomization decision by the detection probability of two likelihood ratio tests;
(2) Prove that the operating characteristic curves of the receiver tested by continuous likelihood ratio test are concave.

3.14 Prove that the slope of a receiver operating characteristic curve at a specific point is equal to the threshold of likelihood ratio test.

3.15 The observation data $y_1, \cdots, y_N$ are generated by the following model

$$y_i = \theta + n_i,$$

where θ obeys Gaussian distribution $\mathcal{N}(0, \sigma^2)$, $and n_i$ is an independent identically distributed Gaussian variable $\mathcal{N}(0, \sigma_n^2)$. Try to find the least mean square estimation and the maximum a posteriori estimation of θ.

3.16 The value set of the independent identically distributed variables $y_1, \cdots, y_N$ is $\{0, 1\}$, where

$$P(y_i = 0) = p, P(y_i = 1) = (1-p), \quad i = 0, 1, \cdots, N.$$

The hypothesis testing on parameter p is as follows

$$\begin{cases} H_0 : p = p_0 \\ H_1 : p = p_1 \end{cases}.$$

(1) Try to determine the sufficient statistics for the hypothesis testing;
(2) Write the expression of Neyman-Pearson test with false alarm probability α_0.

3.17 The joint distribution density of random variables $y_1, \cdots, y_N$ is

$$p(y_1, \cdots, y_N \mid H_1) = \sum_{i=1}^{N} p_i \frac{1}{(2\pi\sigma^2)^{N/2}} \exp\left(-\frac{(y_i-m)^2}{2\sigma^2}\right) \prod_{k\neq i}^{N} \exp\left(-\frac{y_k^2}{2\sigma^2}\right),$$

$$p(y_1, \cdots, y_N \mid H_0) = \prod_{i=1}^{N} \frac{1}{\sqrt{2\pi}\sigma} \exp\left(-\frac{y_i^2}{2\sigma^2}\right),$$

where $\sum_{i=1}^{N} p_i = 1$.

(1) Try to find the likelihood ratio test;

(2) In the condition of $N = 2$ and $p_1 = p_2 = \frac{1}{2}$, draw the decision region corresponding to each detection threshold on $y1, y_2$ plane;

(3) Write the expression of false alarm probability and power function. Get the upper and lower bounds of the two expressions by changing their integral domain.

3.18 Let $y_1, \cdots, y_N$ be the samples taken from the Gaussian distribution. Consider the problem of hypothesis testing for Gaussian distribution

$$\begin{cases} H_0 : p(y) = \frac{1}{\sqrt{2\pi}\sigma_0} \exp\left(-\frac{(y-m_0)^2}{2\sigma_0^2}\right) \\ H_1 : p(y) = \frac{1}{\sqrt{2\pi}\sigma_1} \exp\left(-\frac{(y-m_1)^2}{2\sigma_1^2}\right) \end{cases}$$

(1) Compute the likelihood ratio;

(2) Denote $l_\alpha = \sum_{i=1}^{N} y_i$ and $l_\beta = \sum_{i=1}^{N} y_i^2$. Try to draw the decision region in l_α, l_β plane in the condition of $2m_0 = m_1 > 0, 2\sigma_1 = \sigma_0$.

3.19 Let $y_1, \cdots, y_N$ be samples of Gaussian distribution. Consider the Gaussian distribution test

$$\begin{cases} H_0 : p(y) = \frac{1}{\sqrt{2\pi}\sigma} \exp\left(-\frac{y^2}{2\sigma_0}\right) \\ H_1 : p(y) = \frac{1}{\sqrt{2\pi}\sigma} \exp\left(-\frac{(y-m)^2}{2\sigma^2}\right) \end{cases},$$

where $m > 0$ is an unknown non-random parameter. Determine if there is a uniform most power test for this test? If exists, try to construct the test; if not, try to explain why.

3.20 Change the conditions for the previous question:

(1) Change the condition to $m < 0$, repeat the previous problem;

(2) Change the condition to $m \neq 0$, repeat the previous problem.

3.21 Suppose there are N statistically independent random variables $y_1, \cdots, y_N$, and their distribution density under the two hypotheses is

$$\begin{cases} H_0 : p(y) = \frac{1}{\sqrt{2\pi}\sigma_0} \exp\left(-\frac{(y-m_0)^2}{2\sigma_0^2}\right) \\ H_1 : p(y) = \frac{1}{\sqrt{2\pi}\sigma_1} \exp\left(-\frac{(y-m_1)^2}{2\sigma_1^2}\right) \end{cases},$$

where σ_0 is known, and σ_1 is an unknown non random parameter satisfying $\sigma_1 > \sigma_0$ parameters.

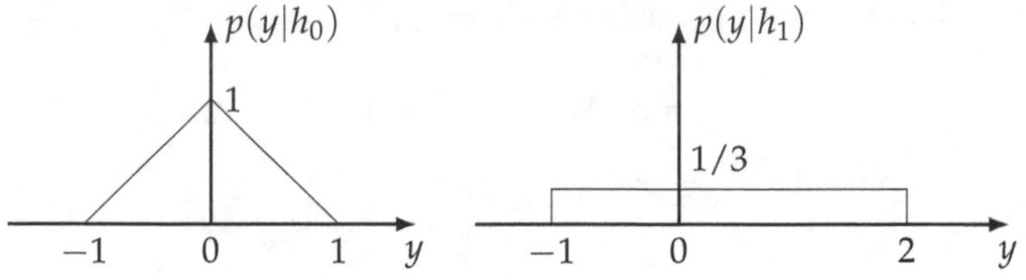

Fig-Exer. The distribution of observation signals under the two hypotheses

(1) Assume that we want the false alarm probability to be α_0, construct the upper bound of the power function;
(2) Is there a uniform most power test? If exists, try to construct it; if not, try to explain why.

3.22 The distribution density of the random variable m is

$$p(m) = \frac{1}{\sqrt{2\pi}\sigma_m} \exp\left(-\frac{m^2}{2\sigma_m^2}\right)$$

Try to determine the Neyman-Pearson test expression when the false alarm probability is α_0.

3.23 In the binary hypothesis testing problem, the distribution of observation signals under the two hypotheses is shown in the figure below. Get the Bayes decision expression. **3.24** Consider the hypothesis testing problem of ternary signals. The hypothesis is

$$\begin{cases} H_0: & y = n \\ H_1: & y = 1 + n\,, \\ H_2: & y = 2 + n \end{cases}$$

where, the noise n follows distribution $p(n) = 1 - |n|, \quad -1 \leqslant n \leqslant 1$. In the case of prior probability $P(H_0) = P(H_1)$, find the minimum total error probability P_E.

3.25 The hypothesis testing problem of continuous time observation data is

$$\begin{cases} H_1: y(t) = 1 + w(t) \\ H_0: y(t) = w(t) \end{cases}$$

where $w(t)$ represents Gaussian white noise with zero mean.

(1) For the given threshold Th, calculate the corresponding detection probability P_D and false alarm probability P_F;
(2) Draw the receiver operating characteristic curve (***ROC***) when the variance of $w(t)$ is $0.5, 1, 2, 4$, respectively.

Appendix 3A: Error Function Table

x	erf(x)	x	erf(x)	x	erf(x)	x	erf(x)	x	erf(x)	x	erf(x)	x	erf(x)	x	erf(x)
0.00	0.00000	0.25	0.27632	0.50	0.52049	0.75	0.71115	1.00	0.84270	1.25	0.92290	1.50	0.96610	1.75	0.98667
0.01	0.01128	0.26	0.28689	0.51	0.52924	0.76	0.71753	1.01	0.84681	1.26	0.92523	1.51	0.96727	1.76	0.98719
0.02	0.02256	0.27	0.29741	0.52	0.53789	0.77	0.72382	1.02	0.85083	1.27	0.92751	1.52	0.96841	1.77	0.98769
0.03	0.03384	0.28	0.30788	0.53	0.54646	0.78	0.73001	1.03	0.85478	1.28	0.92973	1.53	0.96951	1.78	0.98817
0.04	0.04511	0.29	0.31828	0.54	0.55493	0.79	0.73610	1.04	0.85864	1.29	0.93189	1.54	0.97058	1.79	0.98864
0.05	0.05637	0.30	0.32862	0.55	0.56322	0.80	0.74210	1.05	0.86243	1.30	0.93400	1.55	0.97162	1.80	0.98909
0.06	0.06762	0.31	0.33890	0.56	0.57161	0.81	0.74800	1.06	0.86614	1.31	0.93606	1.56	0.97262	1.81	0.98952
0.07	0.07885	0.32	0.34912	0.57	0.57981	0.82	0.75381	1.07	0.86977	1.32	0.93806	1.57	0.97360	1.82	0.98994
0.08	0.09007	0.33	0.35927	0.58	0.58792	0.83	0.75952	1.08	0.87332	1.33	0.94001	1.58	0.97454	1.83	0.99034
0.09	0.10128	0.34	0.36936	0.59	0.59593	0.84	0.76514	1.09	0.87680	1.34	0.94191	1.59	0.97546	1.84	0.99073
0.10	0.11246	0.35	0.37938	0.60	0.60385	0.85	0.77066	1.10	0.88020	1.35	0.94376	1.60	0.97634	1.85	0.99111
0.11	0.12362	0.36	0.38932	0.61	0.61168	0.86	0.77610	1.11	0.88353	1.36	0.94556	1.61	0.97720	1.86	0.99147
0.12	0.13475	0.37	0.39920	0.62	0.61941	0.87	0.78143	1.12	0.88678	1.37	0.94731	1.62	0.97803	1.87	0.99182
0.13	0.14586	0.38	0.40900	0.63	0.62704	0.88	0.78668	1.13	0.88997	1.38	0.94901	1.63	0.97884	1.88	0.99215
0.14	0.15694	0.39	0.41873	0.64	0.63458	0.89	0.79184	1.14	0.89308	1.39	0.95067	1.64	0.97962	1.89	0.99247
0.15	0.16799	0.40	0.42839	0.65	0.64202	0.90	0.79690	1.15	0.89612	1.40	0.95228	1.65	0.98037	1.90	0.99279
0.16	0.17901	0.41	0.43796	0.66	0.64937	0.91	0.80188	1.16	0.89909	1.41	0.95385	1.66	0.98110	1.91	0.99308
0.17	0.18999	0.42	0.44746	0.67	0.65662	0.92	0.80676	1.17	0.90200	1.42	0.95537	1.67	0.98181	1.92	0.99337
0.18	0.20093	0.43	0.45688	0.68	0.66378	0.93	0.81156	1.18	0.90483	1.43	0.95685	1.68	0.98249	1.93	0.99365
0.19	0.21183	0.44	0.46622	0.69	0.67084	0.94	0.81627	1.19	0.90760	1.44	0.95829	1.69	0.98315	1.94	0.99392
0.20	0.22270	0.45	0.47548	0.70	0.67780	0.95	0.82089	1.20	0.91031	1.45	0.95969	1.70	0.98379	1.95	0.99417
0.21	0.23352	0.46	0.48465	0.71	0.68466	0.96	0.82542	1.21	0.91295	1.46	0.96105	1.71	0.98440	1.96	0.99442
0.22	0.24429	0.47	0.49374	0.72	0.69143	0.97	0.82987	1.22	0.91553	1.47	0.96237	1.72	0.98500	1.97	0.99466
0.23	0.25502	0.48	0.50274	0.73	0.69810	0.98	0.83423	1.23	0.91805	1.48	0.96365	1.73	0.98557	1.98	0.99489
0.24	0.26570	0.49	0.51166	0.74	0.70467	0.99	0.83850	1.24	0.92050	1.49	0.96486	1.74	0.98613	1.99	0.99511
0.25	**0.27632**	**0.50**	**0.52049**	**0.75**	**0.71115**	**1.00**	**0.84270**	**1.25**	**0.92290**	**1.50**	**0.96610**	**1.75**	**0.98667**	**2.00**	**0.99532**

4 Modern Spectral Estimation

Power spectrum estimation refers to the estimation of the power spectrum of a stationary stochastic signal using the given set of sample data. The analysis and estimation of the power spectrum is of great significance in practice because it can show the distribution of the energy of the analyzed object with respect to the frequency. For instance, in radar signal processing, the position, the radiation intensity and the velocity of the moving object can be determined by the power spectrum, the width, height, and position of the spectral peak of the radar echo signal. In passive sonar signal processing, the direction (azimuth) of the torpedo can be estimated from the position of the spectral peak. In biomedical engineering, the peak shape and waveform of the power spectral density show the cycle of epileptic seizures. In target recognition, the power spectrum can be used as one of the characteristics of the target.

The smooth periodic graph for estimating power spectral density is a kind of nonparametric method, which is independent of any model parameters. The main problem of this method is that the estimated power spectrum is not matched with the real power spectrum due to the assumption of zero value of the autocorrelation function outside the data observation interval. In general, the asymptotic properties of periodic graphs cannot give a satisfactory approximation of the real power spectrum, so it is a low resolution spectral estimation method.

Different from the periodic graph method, another class of power spectrum estimation methods use parameterized models, referred to as parameterized power spectrum estimation. Since this kind of method can give much higher frequency resolution than the periodic graph method, which is called high resolution spectrum estimation method or modern spectrum estimation method.

Various modern spectrum estimation methods will be discussed in this chapter, which constitutes a very important field in modern signal processing and forms the common basis of many signal processing techniques (e.g. radar signal processing, communication signal processing, sonar signal processing, seismic signal processing, and biomedical signal processing).

4.1 Nonparametric Spectral Estimation

In digital signal processing, a continuous time random process must first be sampled and then processed into a discrete sequence. Therefore, it is necessary to extend the concept of the continuous stochastic process to a discrete form. This process includes the change from continuous function to discrete sequence, from analog system to discrete system and from Fourier integration to Fourier series.

https://doi.org/10.1515/9783110475562-004

4.1.1 Discrete Stochastic Process

A discrete process $x(n)$ is a sequence of real or complex random variables defined for each integer n. Let the sampling interval be T, and for simplicity, $x(nT)$ is often abbreviated as $x(n)$. The autocorrelation function and the autocovariance function of a discrete process $x(n)$ are respectively defined as

$$R_{xx}(n_1, n_2) \stackrel{\text{def}}{=} \mathrm{E}\{x(n_1)x^*(n_2)\}, \tag{4.1.1}$$

$$\begin{aligned} C_{xx}(n_1, n_2) &\stackrel{\text{def}}{=} \mathrm{E}\{[x(n_1) - \mu_x(n_1)][x(n_2) - \mu_x(n_2)]^* \\ &= R_{xx}(n_1, n_2) - \mu_x(n_1)\mu_x^*(n_2), \end{aligned} \tag{4.1.2}$$

where $\mu_x(n) = \mathrm{E}\{x(n)\}$ is the mean of the signal at time n.

The cross correlation function $R_{xy}(n_1, n_2)$ and the cross covariance function $C_{xy}(n_1, n_2)$ of the discrete processes $x(n)$ and $y(n)$ are defined as

$$R_{xy}(n_1, n_2) \stackrel{\text{def}}{=} \mathrm{E}\{x(n_1)y^*(n_2)\}, \tag{4.1.3}$$

$$\begin{aligned} C_{xy}(n_1, n_2) &\stackrel{\text{def}}{=} \mathrm{E}\{[x(n_1) - \mu_x(n_1)][y(n_2) - \mu_y(n_2)]^* \\ &= R_{xy}(n_1, n_2) - \mu_x(n_1)\mu_y^*(n_2), \end{aligned} \tag{4.1.4}$$

The discrete process $x(n)$ is a (generalized) stationary process if the mean is a constant and the autocorrelation function depends only on the time difference $k = n_1 - n_2$, i.e.,

$$R_{xx}(k) = \mathrm{E}\{x(n)x^*(n-k)\} = C_{xx}(k) + |\mu_x|^2. \tag{4.1.5}$$

The two random processes $x(n)$ and $y(n)$ are (generalized) jointly stationary if each of them is stationary and their cross correlation function depends only on the time difference $k = n_1 - n_2$, i.e.,

$$R_{xy}(k) = \mathrm{E}\{x(n)y^*(n-k)\} = C_{xy}(k) + \mu_x\mu_y^*. \tag{4.1.6}$$

The power spectral density of the stationary discrete process $x(n)$ is defined as the Fourier series of the self-covariance function, i.e.,

$$P_{xx}(\omega) \stackrel{\text{def}}{=} \sum_{k=-\infty}^{\infty} C_{xx}(k)e^{-jkT\omega}. \tag{4.1.7}$$

$P_{xx}(\omega)$ is a function of period $\sigma = \frac{\pi}{T}$. Hence, the autocovariance function can be expressed by the power spectral density as

$$C_{xx}(\tau) = \frac{1}{2\sigma}\int_{-\sigma}^{\sigma} P_{xx}(\omega)e^{j\tau T\omega}d\omega. \tag{4.1.8}$$

Similarly, the cross-power spectrum is defined as

$$P_{xy}(\omega) \stackrel{\text{def}}{=} \sum_{k=-\infty}^{\infty} C_{xy}(k)e^{-jkT\omega}. \tag{4.1.9}$$

4.1.2 Non-parametric Power Spectrum Estimation

Assume that the discrete random process has N data samples $x(0), x(1), \cdots, x(N-1)$. Without loss of generality, assume that these data are zero averaged. Spectrum estimation of the discrete signal $x(n)$ can be divided into non-parametric and parametric methods. Non-parametric spectrum estimation also regarded as classical spectrum estimation can be divided into direct and indirect methods.

The direct method first calculates the Fourier transform (i.e. spectrum) of the following N data

$$X_N(\omega) = \sum_{n=0}^{N-1} x(n)e^{-jn\omega}, \tag{4.1.10}$$

Then take the product of the spectrum and its conjugate to get the power spectrum

$$P_x(\omega) = \frac{1}{N}|X_N(\omega)|^2 = \frac{1}{N}\left|\sum_{n=0}^{N-1} x(n)e^{-jn\omega}\right|^2. \tag{4.1.11}$$

The indirect method first estimates the sample autocorrelation function based on N sample data

$$\hat{R}_x(k) = \frac{1}{N}\sum_{n=0}^{N-1} x(n+k)x^*(n),\ k = 0, 1, \cdots, M, \tag{4.1.12}$$

where $1 \ll M < N$, $\hat{R}_x(-k) = \hat{R}_x^*(k)$. The power spectrum can be obtained by calculating the Fourier transform of the sample autocorrelation function

$$P_x(\omega) = \sum_{k=-M}^{M} \hat{R}_x(k)e^{-jk\omega}. \tag{4.1.13}$$

When computing the Fourier transform of Eqs. (4.1.11) and (4.1.13), $x(n)$ and $\hat{R}_x(k)$ are regarded as periodic functions, so the power spectra estimated by direct and indirect methods are often called periodic graphs. The power spectrum estimated by the periodic graph method is biased. To reduce its deviation, a window function is usually needed to smooth the periodogram.

There are two different methods of adding window functions. One is to add the window function $c(n)$ directly to the sample data, and the resulting power spectrum is often called a modified periodogram, defined as

$$P_x(\omega) \stackrel{\text{def}}{=} \frac{1}{NW}\left|\sum_{n=0}^{N-1} x(n)c(n)e^{-jn\omega}\right|^2, \tag{4.1.14}$$

where

$$W = \frac{1}{N}\sum_{n=0}^{N-1}|c(n)|^2 = \frac{1}{2\pi N}\int_{-\pi}^{\pi}|C(\omega)|^2 d\omega, \tag{4.1.15}$$

where $C(\omega)$ is the Fourier transform of the window function $c(n)$. The other method is to add the window function $w(n)$ directly to the sample autocorrelation function, and the resulting power spectrum is called periodogram smoothing, which is proposed by Blackman and Tukey[29], also known as Blackman-Tukey method. The power spectrum is defined as

$$P_{BT}(\omega) \stackrel{\text{def}}{=} \sum_{k=-M}^{M} \hat{R}_x(k)w(k)e^{-jk\omega}. \tag{4.1.16}$$

The window function $c(n)$ directly added to the data is called the data window, while the window function $w(k)$ added to the autocorrelation function is called the lag window, and its Fourier transform is called the spectral window.

The following are some typical window functions.

(1) Hanning window

$$w(n) = \begin{cases} 0.5 - 0.5\cos(\frac{2\pi n}{N-1}), & n = 0, 1, \cdots, N-1, \\ 0, & \text{Others}, \end{cases} \tag{4.1.17}$$

(2) Hamming window

$$w(n) = \begin{cases} 0.54 - 0.46\cos(\frac{2n\pi}{N-1}), & n = 0, 1, \cdots, N-1, \\ 0, & \text{Others}, \end{cases} \tag{4.1.18}$$

(3) Blackman window

$$w(n) = \begin{cases} 0.42 - 0.5\cos(\frac{2n\pi}{N-1}) + 0.08\cos(\frac{4n\pi}{N-1}), & n = 0, 1, \cdots, N-1, \\ 0, & \text{Others}. \end{cases} \tag{4.1.19}$$

Window functions can reduce the deviation of the periodogram and improve the smoothness of the power spectrum curve. However, as a nonparametric spectrum estimation, the periodic graph has inherent defects of low resolution, which can not meet the needs of high resolution power spectrum estimation. In contrast, parametric spectrum estimation can provide much higher frequency resolution than periodogram, so it is often called high resolution spectrum estimation. Parametric spectral estimation is the main topic of this chapter.

4.2 Stationary ARMA Process

The system is called a time invariant system if its parameters do not change with time. Quite a number of stationary random processes can be generated by exciting a linear time invariant system with white noise, and the linear system can be described by a linear difference equation, which is referred to as autoregressive moving average (ARMA) model.

If the discrete stochastic process $\{x(n)\}$ obeys the linear difference equation

$$x(n) + \sum_{i=1}^{p} a_i x(n-i) = e(n) + \sum_{j=1}^{q} b_j e(n-j), \tag{4.2.1}$$

where $e(n)$ is a discrete white noise, it is called an ARMA process. The difference equation shown by Eq. (4.2.1) is called an ARMA model. The coefficients $a_1, \cdots, a_p$ and $b_1, \cdots, b_q$ are the autoregressive (AR) and moving average (MA) parameters, respectively. p and q are the order of AR and MA , respectively. It is clear that the ARMA model describes a linear time invariant system. The ARMA process with AR order p and MA order q is usually abbreviated as ARMA(p, q).

ARMA process can be written in a more compact form

$$A(z)x(n) = B(z)e(n), n = 0, \pm 1, \pm 2 \cdots, \tag{4.2.2}$$

where $A(z)$ and $B(z)$ are called AR and MA polynomials respectively, i.e.,

$$A(z) = 1 + a_1 z^{-1} + \cdots + a_p z^{-p}, \tag{4.2.3}$$

$$B(z) = 1 + b_1 z^{-1} + \cdots + b_q z^{-q}, \tag{4.2.4}$$

where z^{-j} is a backward shift operator, defined as

$$z^{-j}x(n) \stackrel{\text{def}}{=} x(n-j), j = 0, \pm 1, \pm 2 \cdots . \tag{4.2.5}$$

The transfer function of the linear time invariant system described by the ARMA model is defined as

$$H(z) \stackrel{\text{def}}{=} \frac{B(z)}{A(z)} = \sum_{i=-\infty}^{\infty} h_i z^{-i}, \tag{4.2.6}$$

where h_i is the impulse response coefficient of the system. It can be seen that the pole $A(z) = 0$ contribution of the system is autoregressive and the zero $B(z) = 0$ contribution is moving average.

There are two special cases of the ARMA process.

(1) If $B(z) = 1$, ARMA(p, q) is reduced to

$$x(n) + a_1 x(n-1) + \cdots + a_p x(n-p) = e(n), \tag{4.2.7}$$

which is called an AR process with order p, abbreviated as AR(p) process.

(2) If $A(z) = 1$, ARMA(p, q) is reduced to

$$x(n) = e(n) + b_1 e(n-1) + \cdots + b_q e(n-p), \tag{4.2.8}$$

which is called an MA process with order q, abbreviated as MA(q) process.

The important properties of the ARMA process are discussed below.

First, to make the linear time invariant system stable, i.e., bounded input $e(n)$ certainly generates bounded output $x(n)$, then the impulse response h_i of the system must be definitely summable

$$\sum_{i=-\infty}^{\infty} |h_i| < \infty. \tag{4.2.9}$$

This condition is equivalent to that the system transfer function cannot have poles on the unit circle, i.e., $A(z) \neq 0, |z| = 1$.

Secondly, the system model cannot be abbreviated, which requires that the polynomials $A(z)$ and $B(z)$ have no common factors, or $A(z)$ and $B(z)$ are mutually prime.

In addition to stability and mutuality, the linear time invariant system is also required to be physically realizable, namely a causal system.

Definition 4.2.1. *(Causal process) An ARMA process defined by $A(z)x(n) = B(z)e(n)$ is called a causal process, i.e., $x(n)$ is the causal function of $e(n)$, if there is a constant sequence satisfying the following two conditions*

$$\sum_{i=0}^{\infty} |h_i| < \infty, \tag{4.2.10}$$

$$x(n) = \sum_{i=0}^{\infty} h_i e(n-i). \tag{4.2.11}$$

Eq. (4.2.10) is to ensure that the output of the system is bounded at any time, while Eq. (4.2.11) is the real condition for causality. The above two conditions mean that $h_i = 0, i < 0$. It should be noted that causality is not a separate property of output $x(n)$, but a relationship between the output and the input excitation.

The following theorem gives a necessary and sufficient condition for the ARMA process to be a causal process.

Theorem 4.2.1. *Let $\{x(n)\}$ be an ARMA(p, q) process with no common zeros for $A(z)$ and $B(z)$, then x is causal if and only if $A(z) \neq 0$ for all $|z| \geq 1$.*

Proof. Prove sufficiency first ($\Rightarrow$). Assume $A(z) \neq 0$, $|z| \geq 1$. This means that there is an arbitrarily small nonnegative number $\epsilon \geq 0$ such that $\frac{1}{A(z)}$ has a power series expansion

$$\frac{1}{A(z)} = \sum_{i=0}^{\infty} \xi_i z^{-i} = \xi(z), \ |z| > 1 + \epsilon.$$

In other words, when $i \to \infty$, $\xi_i(1 + \epsilon/2)^{-i} \to 0$. Therefore, there exists $K \in (0, +\infty)$ such that

$$|\xi_i| < K(1 + \epsilon/2)^i, \ i = 0, 1, 2, \cdots.$$

Accordingly, $\sum_{i=0}^{\infty} |\xi_i| < \infty$ and $\xi(z)A(z) = 1, \forall |z| \geq 1$. Multiply the two sides of the difference equation $A(z)x(n) = B(z)e(n)$ by the same operator $\xi(z)$, we have

$$x(n) = \xi(z)B(z)e(n) = \frac{B(z)}{A(z)} e(n).$$

Let $H(z) = \xi(z)B(z)$, the desired expression can be obtained

$$x(n) = \sum_{i=0}^{\infty} h_i z^{-i} e(n) = \sum_{i=0}^{\infty} h_i e(n-i).$$

Then prove the necessity ($\Leftarrow$). Assume $\{x(n)\}$ is causal, i,e., $x(n) = \sum_{i=0}^{\infty} h_i e(n-i)$, the sequence satisfying $\sum_{i=0}^{\infty} |h_i| < \infty$ and $H(z) \neq 0$, $|z| \geq 1$. This means $x(n) = H(z)e(n)$. We have

$$B(z)e(n) = A(z)x(n) = A(z)H(z)e(n).$$

Let $\eta(z) = A(z)H(z) = \sum_{i=0}^{\infty} \eta_i z^{-i}$, $|z| \geq 1$, then the above equation can written as

$$\sum_{i=0}^{q} \theta_i e(n-i) = \sum_{i=0}^{\infty} \eta_i e(n-i), |z| \geq 1,$$

multiplying both sides of the above equation by $e(n-k)$ to get the mathematical expectation, since $e(n)$ is a white noise satisfying $E\{e(n-i)e(n-k)\} = \sigma^2\delta(k-i)$, $\eta_i = \theta_i$, $i = 0, 1, \cdots, q$ and $\eta_i = 0$, $i > q$. So we have

$$B(z) = \eta(z) = A(z)H(z), |z| \geq 1. \tag{4.2.12}$$

On the other side,

$$|H(z)| = |\sum_{i=0}^{\infty} h_i z^{-i}| < \sum_{i=0}^{\infty} |h_i||z^{-i}| \leq \sum_{i=0}^{\infty} |h_i|, |z| \geq 1.$$

However, according to the stability condition, h_i is absolutely summable, so from the above equation, we have

$$|H(z)| < \infty, |z| \geq 1. \tag{4.2.13}$$

There are no common zeros for $B(z)$ and $A(z)$, so from Eq. (4.2.12) and Eq. (4.2.13), for $|z| \geq 1$, it is impossible to have $A(z) = 0$. So far, the theorem is proved. □

Theorem 4.2.1 shows that if and only if all the poles of the system are within the unit circle, the output $x(n)$ is a causal function of the input $e(n)$. If all the poles of the system are outside the unit circle, the output is an inverse causal function of the input and the corresponding system is called an anti-causal system. Note that stability requires that the system poles cannot be on the unit circle. The system whose poles are both inside and outside the unit circle is called a noncausal system.

The function of the zero point of the system is discussed below, which determines the reversibility of the system.

Definition 4.2.2. *(Reversible process) An ARMA process defined by $A(z)x(n) = B(z)e(n)$ is called a reversible process, if there is a constant sequence $\{\pi_i\}$ satisfying*

$$\sum_{i=0}^{\infty} |\pi_i| < \infty, \tag{4.2.14}$$

$$e(n) = \sum_{i=0}^{\infty} \pi_i x(n-i). \tag{4.2.15}$$

Like causality, reversibility is not a separate property of the ARMA process $\{x(n)\}$, but a relationship between the output and the input excitation $e(n)$. The following theorem gives the necessary and sufficient conditions for invertibility.

Theorem 4.2.2. *Let $\{x(n)\}$ be an ARMA(p, q) process with no common zeros for $A(z)$ and $B(z)$. The ARMA process is invertible if and only if $B(z) \neq 0$ for all complex z with $|z| \geq 1$. The coefficient π_i of the reversible process Eq. (4.2.15) is determined by*

$$\pi(z) = \sum_{i=0}^{\infty} \pi_i z^{-i} = \frac{A(z)}{B(z)}, \ |z| \geq 1. \tag{4.2.16}$$

Proof. Similar to the proof of Theorem 4.2.1, this proof is left for the reader to exercise. □

Theorem 4.2.2 shows that if and only if all the system zeros are within the unit circle, the input $e(n)$ is the invertible function of the output $x(n)$. If an ARMA(p, q) process is invertible, all the poles of its inverse system $A(z)/B(z)$ are within the unit circle, so it is a causal system. A reversible system is also called a minimum phase system. If the zeros of the system are on or outside the unit circle, it is called the maximum phase system. If the system has zeros inside and outside the unit circle, it is called a non minimum phase system. Note that when the system has zeros on the unit circle, its inverse system is unstable. When all zeros of the system are outside the unit circle, the inverse system is an anti-causal system.

More generally, we discuss the case of $A(z) \neq 0$ when $|z| = 1$. In this case, it is known from the complex number analysis that there exists a radius $r > 1$, making the Laurent series

$$\frac{A(z)}{B(z)} = \sum_{i=-\infty}^{\infty} h_i z^{-i} = H(z), \tag{4.2.17}$$

absolutely converge in the annular region $r^{-1} < |z| < r$. The convergence of the Laurent series plays a key role in the proof of the following theorem.

Theorem 4.2.3. *If $A(z) \neq 0$ for all $|z| = 1$, then ARMA process has unique stationary solution*

$$x(n) = H(z)e(n) = \sum_{i=-\infty}^{\infty} h_i e(n-i), \tag{4.2.18}$$

where the coefficient h_i is determined by Eq. (4.2.17).

Proof. Prove sufficiency first ($\Rightarrow$). If $A(z) \neq 0$ for all $|z| = 1$, from Theorem 4.2.1, there exists $\delta > 1$ making the series $\sum_{i=-\infty}^{\infty} \xi_i z^{-i} = \frac{1}{A(z)} = \xi(z)$ absolutely converge in annular region $\delta^{-1} < |z| < \delta$. Therefore, multiplying both sides of the ARMA model $A(z)x(n) =$

$B(z)e(n)$ by the same operator $\xi(z)$ yields

$$\xi(z)A(z)x(n) = \xi(z)B(z)e(n).$$

Because $\xi(z)A(z) = 1$, the above equation can be written as

$$x(n) = \xi(z)B(z)e(n) = H(z)e(n) = \sum_{i=-\infty}^{\infty} h_i e(n-i),$$

where $H(z) = \xi(z)B(z) = \frac{B(z)}{A(z)}$, i.e., the coefficient h_i of $H(z)$ is determined by Eq. (4.2.17). Then prove the necessity ($\Leftarrow$). Assume the ARMA process has a unique stationary solution Eq. (4.2.18). Applying the operator $A(z)$ to both sides of Eq. (4.2.18) yields

$$A(z)x(n) = A(z)H(z)e(n) = B(z)e(n),$$

Namely, the process with a unique stationary solution is an ARMA process. Since the ARMA process needs to satisfy stability, $A(z) \neq 0$ should always hold for all $|z| = 1$. □

Combined with Theorem 4.2.2 and Theorem 4.2.3, the Wold decomposition theorem describing the relationship between ARMA, MA, and AR processes can be obtained.

Theorem 4.2.4. *(Wold decomposition theorem) Any ARMA or MA process with finite variance can be expressed as a unique AR process with probably infinite order; Similarly, any ARMA or AR process can also be expressed as an MA process with probably infinite order. If $A(z) \neq 0$ for all $|z| = 1|$, then ARMA process has unique stationary solution*

$$x(n) = H(z)e(n) = \sum_{i=-\infty}^{\infty} h_i e(n-i), \tag{4.2.19}$$

where the coefficient h_i is determined by Eq. (4.2.17).

The above theorem plays an important role in practical application. If one of the three models selected is wrong, a reasonable approximation can still be obtained by setting a very high order. Therefore, an ARMA model can be approximated by an AR model with enough high order. Compared with the ARMA model, which needs not only the determination of AR and MA order but also parameter estimation of AR and MA models (among them MA model parameter estimation needs to solve nonlinear equations), the AR model only needs parameter estimation, hence many engineers and technicians often like to use AR model for approximation.

For MA(q) random process, the parameters of the model are exactly the same as the impulse response of the system, i.e.,

$$b_i = h_i, i = 0, 1, \cdots, q, \tag{4.2.20}$$

where $b_0 = h_0 = 1$, because

$$\begin{aligned} x(n) &= e(n) + b_1 e(n-1) + \cdots + b_q e(n-q) = \sum_{i=0}^{\infty} h_i e(n-i) \\ &= e(n) + h_1 e(n-1) + \cdots + h_q e(n-q). \end{aligned}$$

Since there are only $q + 1$ impulse response coefficients, such a system is called a finite impulse response system, or FIR system for short. Here FIR is the abbreviation of the finite impulse response. Hence MA model is also called the FIR model. On the contrary, ARMA and AR systems are called infinite impulse response system, because they have infinite impulse response coefficients.

4.3 Power Spectral Density of Stationary Process

The power spectral density of a stationary ARMA process is widely representative. For example, any rational spectral density, AR process observed in additive white noise and sine wave (more generalized as harmonic) process with line spectrum can be expressed by ARMA spectral density. Due to its wide representativeness and applicability, ARMA spectral analysis has become one of the most important methods in modern spectral analysis.

4.3.1 Power Spectral Density of ARMA Process

Theorem 4.3.1. *Let $\{y(n)\}$ be a discrete-time stationary process with zero mean and power spectral density $P_y(\omega)$. If $x(n)$ is described by*

$$x(n) = \sum_{i=-\infty}^{\infty} h_i y(n-i), \tag{4.3.1}$$

where h_i is absolutely summable, i.e., $\sum_{i=-\infty}^{\infty} |h_i| < \infty$, then $x(n)$ is also a stationary process with zero mean and power spectral density represented as

$$P_x(\omega) = |H(e^{-j\omega})|^2 P_y(\omega), \tag{4.3.2}$$

where $H(e^{-j\omega})$ is a polynomial of $e^{-j\omega}$

$$H(e^{-j\omega}) = \sum_{i=-\infty}^{\infty} h_i z^{-i}\bigg|_{z=e^{j\omega}}. \tag{4.3.3}$$

Proof. Under the condition that h_i can be absolutely summable, when the mathematical expectation is calculated on both sides of the Eq. (4.3.1), the position of the mathematical expectation can be exchanged with the summation operator, so

$$\mathrm{E}\{x(n)\} = \sum_{i=-\infty}^{\infty} h_i \mathrm{E}\{y(n-i)\} = 0.$$

The above equation uses the assumption of $\mathrm{E}\{y(n)\} = 0$. Calculating the autocorrelation function of the ARMA process $\{x(n)\}$ and considering that $y(n)$ is a stationary process, we have

$$\begin{aligned} R_x(n_1, n_2) &= \mathrm{E}\{x(n_1)x^*(n_2)\} \\ &= \sum_{i=-\infty}^{\infty}\sum_{j=-\infty}^{\infty} h_i h_j^* \mathrm{E}\{y(n_1-i)y^*(n_2-j)\} \\ &= \sum_{i=-\infty}^{\infty}\sum_{j=-\infty}^{\infty} h_i h_j^* R_y(n_1-n_2+j-i). \end{aligned}$$

Let $\tau = n_1 - n_2$ denote the time difference, then the above equation can be written as

$$R_x(\tau) = \sum_{i=-\infty}^{\infty}\sum_{j=-\infty}^{\infty} h_i h_j^* R_y(\tau + j - i). \tag{4.3.4}$$

So $\{x(n)\}$ is a stationary process with zero mean.

Compute the complex conjugate of both sides of Eq. (4.3.3), we have

$$H^*(e^{-j\omega}) = \sum_{i=-\infty}^{\infty} h_i^* z^i \Big|_{z=e^{j\omega}}.$$

Using variable substitution $\tau = \tau' + j - i$, the power spectral density $P_x(\omega) = \sum_{\tau=-\infty}^{\infty} C_x(\tau)e^{-j\omega\tau}$ can be rewritten as

$$\begin{aligned} P_x(\omega) &= \sum_{i=-\infty}^{\infty} h_i e^{-j\omega i} \sum_{j=-\infty}^{\infty} h_j e^{j\omega j} \sum_{\tau=-\infty}^{\infty} C_y(\tau' + j - i) e^{-j\omega(\tau'+j-i)} \\ &= H(e^{-j\omega})H^*(e^{-j\omega})P_y(\omega) \\ &= |H(e^{-j\omega})|^2 P_y(\omega), \end{aligned}$$

which is Eq. (4.3.2). □

As an application example of Theorem 4.3.1, the power spectral density of any stationary process ARMA(p, q) is derived below. $e(n)$ is a normal distribution process with zero mean and variance σ^2, denoted by $e(n) \sim \mathcal{N}(0, \sigma^2)$.

Theorem 4.3.2. *Let $x(n)$ be a stationary ARMA(p, q) process, satisfying the following difference equation*

$$x(n) + a_1 x(n-1) + \cdots + a_p x(n-p) = e(n) + b_1 e(n-1) + \cdots + b_q e(n-q), \tag{4.3.5}$$

where $e(n) \sim \mathcal{N}(0, \sigma^2)$, then the power spectral density is

$$P_x(\omega) = \sigma^2 \frac{|B(z)|^2}{|A(z)|^2}\bigg|_{z=e^{j\omega}} = \sigma^2 \frac{|B(e^{j\omega})|^2}{|A(e^{j\omega})|^2}, \tag{4.3.6}$$

where

$$A(z) = 1 + a_1 z^{-1} + \cdots + a_p z^{-p}, \tag{4.3.7}$$

$$B(z) = 1 + b_1 z^{-1} + \cdots + b_q z^{-q}. \tag{4.3.8}$$

Proof. From Theorem 4.2.3, the only stationary solution of Eq. (4.3.5) can be written as $x(n) = \sum_{i=-\infty}^{\infty} h_i e(n-i)$, where $\sum_{i=-\infty}^{\infty} |h_i| < \infty$, and $H(z) = B(z)/A(z)$. $e(n)$ is white noise, its covariance function $C_e(\tau) = \sigma^2\delta(\tau)$, i.e., the power spectral density is a constant σ^2, hence using Theorem 4.3.1 directly yields

$$P_x(\omega) = |H(e^{j\omega})|^2 P_e(\omega) = \sigma^2 \frac{|B(e^{j\omega})|^2}{|A(e^{j\omega})|^2}.$$

□

The power spectral density defined by Eq. (4.3.6) is the ratio of two polynomials, so it is usually called rational spectral density. Theorem 4.3.2 shows an important result: The power spectral density of a discrete ARMA(p, q) process is a rational function of $e^{-j\omega}$. On the contrary, if a stationary process $\{x(n)\}$ has rational spectral density expressed as Eq. (4.3.6), $\{x(n)\}$ can also be proven to be an ARMA(p, q) process described by Eq. (4.3.5). Hence, let $H(z) = B(z)/A(z) = \sum_{i=-\infty}^{\infty} h_i z^{-i}$. Take $\{x(n)\}$ as a process generated by Eq. (4.3.1), then from the known conditions and Theorem 4.3.1, we get $P_y(\omega) = \sigma^2$. So $\{y(n)\}$ is Gaussian white noise $\mathcal{N}(0, \sigma^2)$, i.e., $\{x(n)\}$ can be written as

$$x(n) = \sum_{i=-\infty}^{\infty} h_i e(n-i),\ e(n) \sim \mathcal{N}(0, \sigma^2).$$

Since $H(z) = B(z)/A(z)$, the above equation is equivalent to Eq. (4.3.5).

For the ARMA process shown in Eq. (4.3.5), an important equation can be obtained from Eq. (4.3.4)

$$R_x(\tau) = \sum_{i=-\infty}^{\infty} \sum_{j=-\infty}^{\infty} h_i h_j^* \sigma^2 \delta(\tau + j - i).$$

Notice that the condition of $\delta(\tau+j-i) = 1$ is $j = i-\tau$, while in other cases $\delta(\tau+j-i) \equiv 0$, so $R_x(\tau)$ can be expressed as

$$R_x(\tau) = \sigma^2 \sum_{i=-\infty}^{\infty} h_i h_{i-\tau}^*.$$

The above equation represents the relationship between the autocorrelation function and the impulse response of an ARMA process $\{x(n)\}$, which is of great significance.

In particular, when $\tau = 0$, Eq. (4.3.6) gives the result

$$R_x(0) = \mathrm{E}\{x(n)x^*(n)\} = \sigma^2 \sum_{i=-\infty}^{\infty} |h_i|^2,$$

where $\mathrm{E}\{x(n)x^*(n)\} = \mathrm{E}\{|x(n)|^2\}$ indicating the energy of the ARMA process $\{x(n)\}$. To make the energy of $\{x(n)\}$ limited, from the above equation, the following condition must be satisfied

$$\sum_{i=-\infty}^{\infty} |h_i|^2 < \infty, \tag{4.3.9}$$

which is called the square summability condition of the impulse response.

Here is an introduction of the other two processes which can be expressed by ARMA spectral density Eq. (4.3.6).

1. White Noise Driven AR Process

Assume $\{s(n)\}$ is an AR(p) process satisfying the following difference equation

$$s(n) + a_1 s(n-1) + \cdots + a_p s(n-p) = e(n),\ e(n) \sim \mathcal{N}(0, \sigma^2),$$

and this process is observed in additive white noise, i.e., $x(n) = s(n) + v(n)$, where $v(n)$ and $s(n)$ are independent, and the variance of the white noise is σ_v^2. Let's calculate the power spectral density of the process $\{x(n)\}$.

From Theorem 4.3.2, the power spectral density of signal $s(n)$ is

$$P_s(\omega) = \frac{\sigma^2}{|1 + a_1 e^{-j\omega} + \cdots + a_p e^{-j\omega p}|^2} = \left.\frac{\sigma^2}{|A(z)|^2}\right|_{z=e^{j\omega}}.$$

When $s(n)$ and $v(n)$ are independent of each other, using the definition of covariance function, it is easy to prove

$$C_x(\tau) = C_s(\tau) + C_v(\tau) = C_s(\tau) + \sigma_v^2 \delta(\tau). \tag{4.3.10}$$

Via the definition of power spectral density from discrete process, it is obvious to get the following result

$$P_x(\omega) = P_s(\omega) + P_v(\omega) = \left.\frac{\sigma^2}{|A(z)|^2}\right|_{z=e^{j\omega}} + \sigma_v^2 = \sigma_\omega^2 \left.\frac{|B(z)|^2}{|A(z)|^2}\right|_{z=e^{j\omega}}, \tag{4.3.11}$$

where $\sigma_\omega^2 = \sigma^2 + \sigma_v^2$, and $B(z)B^*(z) = [\sigma^2 + \sigma_v^2 A(z)A^*(z)]/(\sigma^2 + \sigma_v^2)$.

This example shows that an AR(p) process in white noise is an ARMA(p, q) process, with white noise as excitation and variance $\sigma_\omega^2 = \sigma^2 + \sigma_v^2$, i.e., $w(n) \sim \mathcal{N}(0, \sigma^2 + \sigma_v^2)$. Note that Eq. (4.3.10) and Eq. (4.3.11) are applicable to any two independent processes $\{s(n)\}$ and $\{v(n)\}$.

2. Predictable Process

Definition 4.3.1. *If $\{s(n)\}$ is a recursive process without excitation*

$$s(n) + a_1 s(n-1) + \cdots + a_p s(n-p) = 0, \tag{4.3.12}$$

it is called a (fully) predictable process.

The predictable process is also called the degenerate AR process or non incentive AR process. Another equivalent expression of Eq. (4.3.12) is

$$s(n) = -\sum_{i=1}^{p} a_i s(n-i). \tag{4.3.13}$$

If p values $s(1), \cdots, s(p)$ are given, $s(p+1), s(p+2), \cdots$ can be calculated in turn according to Eq. (4.3.13). In fact, as long as the values of any p continuous time of the signal $s(n)$ are given, the signal values at other times can be completely predicted according to these p values. Hence get the name of the predictable process.

Assume $s(n)$ is a predictable process with order p, and it is observed in additive white noise $v(n)$. Let $x(n) = s(n) + v(n)$, where $v(n) \sim \mathcal{N}(0, \sigma^2)$ is independent of $s(n)$. Substituting $s(n) = x(n) - v(n)$ into the Eq. (4.3.12) yields,

$$x(n) + \sum_{i=1}^{p} a_i x(n-i) = v(n) + \sum_{j=1}^{p} b_j v(n-j). \tag{4.3.14}$$

The above equation shows that the predictable process in additive white noise is a special ARMA(p, q) process, with the same MA order and AR order.

The power spectral density of the predictable process is derived below.

First, multiply both sides of Eq. (4.3.12) by $s(n-\tau)$, $\tau \geq 0$, and compute the mathematical expectation, we have

$$R_s(\tau) + a_1 R_s(\tau - 1) + \cdots + a_p R_s(\tau - p) = 0, \forall \tau \geq 0. \tag{4.3.15}$$

Proposition 4.3.1. *Let z_k be the root of the characteristic polynomial $A(z) = 1 + a_1 z^{-1} + \cdots + a_p z^{-p}$, then the autocorrelation function of the predictable process $\{s(n)\}$ can be expressed as*

$$R_s(m) = \sum_{l=1}^{p} c_i z_i^m, \ |z| \leq 1, \tag{4.3.16}$$

where c_i is the constant to be determined.

Proof. Since z_k is the root of $A(z) = 0$, we have

$$A(z_k) = \sum_{i=0}^{p} a_i z_k^{-i} = 0, \tag{4.3.17}$$

where $a_0 = 1$. Construct function $R_s(m)$ according to Eq. (4.3.16), then

$$\sum_{i=0}^{p} a_i R_s(m-i) = \sum_{i=0}^{p} a_i \sum_{k=1}^{p} c_k z_k^{m-i} = \sum_{k=1}^{p} c_k z_k^m \sum_{i=0}^{p} a_i z_k^{-i}.$$

Substituting Eq. (4.3.16) into the right side of the above equation, the result is equal to zero, i.e.,

$$\sum_{i=0}^{p} a_i R_s(m-i) = 0.$$

That is to say, the function $R_s(m)$ satisfies the relation Eq. (4.3.14) that only the autocorrelation function of the predictable process satisfies. Hence, the function $R_s(m)$ defined by Eq. (4.3.16) is really the autocorrelation function of predictable process. □

In particular, if the order p of the predictable process is an even number, and the coefficients satisfy the symmetry condition $a_i = a_{p-i}, i = 0, 1, \cdots, p/2, (a_0 = 1)$, then the p pairs of conjugate roots z_i of the characteristic polynomial $A(z) = 0$ are all on the unit circle. Hence, The autocorrelation function Eq. (4.3.16) of the predictable process with symmetric coefficients can be specifically expressed as

$$R_s(m) = \sum_{i=1}^{p} c_i e^{\mathrm{j}m\omega_i}, \, |\omega_i| < \pi. \tag{4.3.18}$$

Since the mean of the predictable process is zero and its auto-covariance function and autocorrelation function are equal, the power spectral density of the predictable process is

$$P_s(\omega) = \sum_{k=-\infty}^{\infty} R_s(k) e^{-\mathrm{j}k\omega} = \sum_{i=1}^{p} c_i \sum_{k=-\infty}^{\infty} e^{-\mathrm{j}k(\omega-\omega_i)}.$$

Using the well-known discrete time Fourier transform pair $e^{\mathrm{j}\omega_0 k} \leftrightarrow \frac{1}{2\pi}\delta(\omega - \omega_0)$, the above equation can be written as

$$P_s(\omega) = \frac{1}{2\pi} \sum_{i=1}^{p} c_i \delta(\omega - \omega_i). \tag{4.3.19}$$

This shows that the power spectral density of the p-order predictable (real) process is composed of p individual linear spectra, which are called line spectrum.

The power spectral density $P(\omega)$ of any stationary process can be written as the sum of the two parts

$$P(\omega) = P_\alpha(\omega) + P_\beta(\omega), \tag{4.3.20}$$

where $P_\alpha(\omega)$ is the rational power spectral density represented by Eq. (4.3.6), while $P_\beta(\omega)$ is line spectrum represented by Eq. (4.3.19). Eq. (4.3.20) is also called Wold decomposition[171]. Note that it is different from the meaning of the Wold decomposition theorem in Section 4.2. Wold decomposition theorem describes the approximate equivalence among three difference models of a stationary process, while Wold decomposition here refers to the decomposition of the power spectral density of a stationary process.

4.3.2 Power Spectrum Equivalence

Power spectrum describes the distribution of signal power with frequency, so it plays a very important role in many practical projects. However, it is necessary to point out that the self power spectral density has a limitation, that is, the signals obtained from some different ARMA models may have the same power spectrum, which is called power spectrum equivalence.

Let's investigate ARMA(p, q) process $A(z)x(n) = B(z)e(n)$, where $e(n) \sim \mathcal{N}(0, \sigma^2)$, and $A(z) \neq 0$, $|z| = 1$, $B(z) \neq 0$, $|z| = 1$. Suppose the linear system has p poles α_i and q zeros β_i, then the original ARMA model can be rewritten as

$$\prod_{i=1}^{p}(1 - \alpha_i z^{-1})x(n) = \prod_{i=1}^{q}(1 - \beta_i z^{-1})e(n), \, e(n) \sim \mathcal{N}(0, \sigma^2). \tag{4.3.21}$$

Suppose there are p poles of which r poles are within the unit circle and the rest are outside the unit circle. Similarly, s of q zeros are inside the unit circle and the rest are outside the unit circle. Namely, we have

$$\begin{aligned}
&|\alpha_i| < 1, \quad 1 \le i \le r \quad \text{(Causal part)},\\
&|\alpha_i| > 1, \quad r < i \le p \quad \text{(Anti causal part)},\\
&|\beta_i| < 1, \quad 1 \le i \le s \quad \text{(Minimum phase part)},\\
&|\beta_i| > 1, \quad s < i \le q \quad \text{(Maximum phase part)}.
\end{aligned}$$

From Theorem 4.3.2, the power spectral density of the ARMA process $\{x(n)\}$ can be written as

$$P_x(\omega) = \sigma^2 \frac{|B(z)|^2}{|A(z)|^2}\bigg|_{z=e^{j\omega}} = \sigma^2 \frac{\prod_{i=1}^{q} |1-\beta_i e^{-j\omega}|^2}{\prod_{i=1}^{p} |1-\alpha_i e^{-j\omega}|^2}. \tag{4.3.22}$$

A new ARMA process is obtained by inverting all the zeros and poles outside the unit circle into the unit circle

$$\tilde{A}(z)\tilde{x}(n) = \tilde{B}(z)\tilde{e}(n), \tag{4.3.23}$$

where

$$\begin{aligned}
\tilde{A}(z) &= \prod_{i=1}^{r}(1-\alpha_i z^{-1}) \prod_{i=r+1}^{p} (1-\bar{\alpha}_i z^{-1}),\\
\tilde{B}(z) &= \prod_{i=1}^{s}(1-\beta_i z^{-1}) \prod_{i=s+1}^{p} (1-\bar{\beta}_i z^{-1}),
\end{aligned}$$

where $\bar{\alpha}_i = 1/\alpha_i^*, i = r+1, \cdots, p$; $\bar{\beta}_i = 1/\beta_i^*, i = s+1, \cdots, q$. Obviously, the power spectral density of the new ARMA process $\{\tilde{x}(n)\}$ is

$$P_{\tilde{x}}(\omega) = \sigma^2 \frac{|B(z)|^2}{|A(z)|^2}\bigg|_{z=e^{j\omega}} = \sigma^2 \frac{\left|\prod_{i=1}^{s}(1-\beta_i e^{-j\omega}) \prod_{i=s+1}^{p}(1-\bar{\beta}_i e^{-j\omega})\right|^2}{\left|\prod_{i=1}^{r}(1-\alpha_i e^{-j\omega}) \prod_{i=r+1}^{p}(1-\bar{\alpha}_i e^{-j\omega})\right|^2}. \tag{4.3.24}$$

From complex number operation, the following expression can be obtained

$$|1-\bar{\alpha}_i e^{-j\omega}| = |1-\alpha_i^{-1} e^{j\omega}| = |\alpha_i^{-1} e^{j\omega}||\alpha_i e^{-j\omega} - 1| = |\alpha_i^{-1}||1-\alpha_i e^{-j\omega}|.$$

Similarly, get the following equation

$$|1-\bar{\beta}_i e^{-j\omega}| = |\beta_i^{-1}||1-\beta_i e^{-j\omega}|.$$

Substituting the above two results into the Eq. (4.3.24), yields

$$P_{\tilde{x}}(\omega) = \sigma^2 \frac{\prod_{i=s+1}^{q} |\beta_i^{-1}|^2}{\prod_{i=r+1}^{p} |\alpha_i^{-1}|^2} \frac{\left|\prod_{i=1}^{q}(1-\beta_i e^{-j\omega})\right|^2}{\left|\prod_{i=1}^{p}(1-\alpha_i e^{-j\omega})\right|^2} = \frac{\prod_{i=r+1}^{p} |\alpha_i|^2}{\prod_{i=s+1}^{q} |\beta_i|^2} P_x(\omega).$$

It shows that the two ARMA processes $\{x(n)\}$ and $\{\tilde{x}(n)\}$ have exactly the same power spectral density shape, and the only difference is a fixed scale factor.

Example 4.3.1 ARMA process

$$x(n) - 2.5x(n-1) = e(n) + 4e(n-1),\ e(n) \sim \mathcal{N}(0, \sigma_e^2),$$

is an anti causal and maximum phase system because its pole 2.5 and zero −4 are all outside the unit circle. A causal and minimum phase ARMA process is obtained by inverting its pole and zero inside the unit circle

$$\tilde{x}(n) - 0.4\tilde{x}(n-1) = \tilde{e}(n) + 0.25\tilde{e}(n-1),\ \tilde{e}(n) \sim \mathcal{N}(0, \sigma_{\tilde{e}}^2).$$

Then the power spectral density of the above two ARMA processes have the same shape, with the only difference being a fixed scale factor. In particular, if $\sigma_{\tilde{e}}^2 = 2.56\sigma_e^2$, $\{\tilde{x}(n)\}$ and $\{x(n)\}$ have exactly the same power spectral density.

In fact, if any zeros and/or poles of an ARMA model are inversed from the inside to the outside of the unit circle, or from the outside to the inside of the unit circle, the power spectral density with exactly the same shape can be obtained, with only a scale factor being different. This property is called the equivalence of power spectrum. In other words, it is impossible to distinguish whether an ARMA model is a causal minimum phase or a noncausal non minimum phase process by self power spectral density. Therefore, the power spectrum equivalence is also called the multiplicity of the ARMA model. Since the power spectral density is obtained by the discrete Fourier transform of the autocovariance function, the power spectral equivalence means the autocovariance function equivalence. The equivalence or multiplicity tells us that the causality and minimum phase of the ARMA process can not be distinguished or identified by using auto covariance function or power spectral density. In order to ensure the uniqueness of ARMA model identification, it is usually assumed that the ARMA model is a causal and minimum phase when auto covariance function or power spectral density is used as an analysis tool.

When a linear system $H(e^{j\omega})$ is excited by an input $y(n)$, it is known from Theorem 4.3.2 that the output power spectral density of $\{x(n)\}$ is $P_x(\omega) = |H(e^{j\omega})|^2 P_y(\omega)$. It can be seen from this expression that even if both $P_y(\omega)$ and $P_x(\omega)$ are known, only $|H(e^{j\omega})|^2$ can be identified, but not $H(e^{j\omega})$, because the previous analysis has shown that $|H(e^{j\omega})|$ has the same form after taking the conjugate reciprocal of any zeros or poles of the system.

However, if the cross power spectral density is used to identify the system, the result will be different.

Considering the output of the discrete linear time invariant system described by Eq. (4.3.1), From Theorem 4.3.1, when the input $y(n)$ is a generalized stationary process with zero mean, the output $x(n)$ is also a generalized stationary process with zero mean.

From Eq. (4.3.1), we have

$$y(n)x^*(n-\tau) = \sum_{k=-\infty}^{\infty} y(n)y^*(n-\tau-k)h^*(k),$$

$$x(n)x^*(n-\tau) = \sum_{k=-\infty}^{\infty} y(n-k)x^*(n-\tau)h(k).$$

Take the mathematical expectation of the above two equations to get

$$C_{yx}(\tau) = R_{yx}(\tau) = \sum_{k=-\infty}^{\infty} R_{yy}(\tau+k)h^*(k), \tag{4.3.25}$$

$$C_{xx}(\tau) = R_{xx}(\tau) = \sum_{k=-\infty}^{\infty} R_{yx}(\tau-k)h(k). \tag{4.3.26}$$

Taking the discrete Fourier transform of the above two equations, yields

$$P_{yx}(\omega) = P_{yy}(\omega)H^*(e^{j\omega}), \tag{4.3.27}$$

$$P_{xx}(\omega) = P_{yx}(\omega)H(e^{j\omega}). \tag{4.3.28}$$

Obviously, if the mutual power spectral density $P_{yx}(\omega)$ between the input and output and the power spectral density $P_{yy}(\omega)$ of the input are known, or $P_{yx}(\omega)$ and $P_{xx}(\omega)$ are known, the real transfer function $H(e^{j\omega})$ of the system can be identified according to Eq. (4.3.27) and Eq. (4.3.28). Therefore, although the power spectral density can not identify the noncausality and non minimum phase of the system, the cross power spectral density can do so. By the way, using higher-order statistics can identify noncausal non minimum phase systems, which will be discussed in Chapter 6.

4.4 ARMA Spectrum Estimation

The expression Eq. (4.3.6) of the power spectral density for the ARMA process is derived in Section 4.3. The purpose of ARMA spectrum estimation is to calculate the power spectral density of ARMA process $\{x(n)\}$ using N known observation data $x(0), x(1), \cdots, x(N-1)$. Obviously, when Eq. (4.3.6) is used to estimate the spectrum directly, it is necessary to identify the whole ARMA model and the variance of the excitation noise in advance. The identification of the ARMA model involves the determination of the order of AR and MA models, as well as the estimation of AR and MA parameters. The estimation of MA parameters needs to solve nonlinear equations (see section 4.5 for details). Can we avoid this nonlinear operation and only use the linear operation to estimate the power spectral density of the ARMA process? The answer is yes.

4.4.1 Two Linear Methods for ARMA Power Spectrum Estimation

As to the power spectral density $P_x(\omega) = P_x(z)\Big|_{z=e^{j\omega}}$, for simplicity, the ARMA power spectral density is expressed as $P_x(z)$. Therefore, the ARMA power spectral density shown in Eq. (4.3.6) can be written as

$$P_x(z) = \sigma^2 \frac{|B(z)|^2}{|A(z)|^2} = \sigma^2 \frac{B(z)B(z^{-1})}{A(z)A(z^{-1})}, \tag{4.4.1}$$

where $A(z^{-1}) = A^*(z)$; $B(z^{-1}) = B^*(z)$.

1. Cadzow Spectrum Estimator

In 1980, Cadzow[40] proposed to decompose the power spectral density of the ARMA process into the sum of two parts

$$P_x(z) = \sigma^2 \frac{B(z)B(z^{-1})}{A(z)A(z^{-1})} = \frac{N(z)}{A(z)} + \frac{N(z^{-1})}{A(z^{-1})}, \tag{4.4.2}$$

where $N(z)$ is a p-order polynomial, defined as

$$N(z) = \sum_{i=0}^{p} N_i z^{-i}. \tag{4.4.3}$$

The power spectral density $P_x(z)$ is divided into two terms: $N(z)/A(z)$ is a polynomial of z^{-1}, and $N(z^{-1})/A(z^{-1})$ is a polynomial of z.

To satisfy Eq. (4.4.2), obviously, the following equation should hold

$$N(z)A(z^{-1}) + N(z^{-1})A(z) = \sigma^2 B(z)B(z^{-1}). \tag{4.4.4}$$

On the other hand, the power spectral density expressed by the discrete Fourier series of the covariance function can be similarly decomposed

$$P_x(z) = \sum_{k=-\infty}^{\infty} C_x(k)z^{-k} = \sum_{k=0}^{\infty} \rho(k)z^{-k} + \sum_{k=0}^{\infty} \rho(-k)z^{k}, \tag{4.4.5}$$

where $\rho(-k) = \rho(k)$, and

$$\rho(k) = \begin{cases} \frac{1}{2}C_x(k), & k = 0, \\ C_x(k), & \text{Others.} \end{cases} \tag{4.4.6}$$

To ensure the decomposition shown in Eq. (4.4.2) and Eq. (4.4.5) equal, let

$$\frac{N(z)}{A(z)} = \frac{\sum_{i=0}^{p} n_i z^{-i}}{\sum_{i=0}^{p} a_i z^{-i}} = \sum_{k=0}^{\infty} \rho(k)z^{-k}, \tag{4.4.7}$$

Multiply both sides of Eq. (4.4.7) with $\sum_{i=0}^{p} a_i z^{-i}$ and compare the coefficients of the same power term to get

$$n_k = \sum_{i=0}^{p} a_i \rho(k-i),\ k = 0, 1, \cdots, p. \tag{4.4.8}$$

In short, the key of the Cadzow spectrum estimator is to determine the AR order P and estimate the AR parameters, because the coefficients n_k can be calculated directly by Eq. (4.4.8). This method avoids the determination of MA order, the estimation of MA parameters and the exciting white noise variance.

2. Kaveh Spectrum Estimator

The ARMA power spectral density expression Eq. (4.4.1) can be rewritten as

$$P_x(z) = \sigma^2 \frac{B(z)B(z^{-1})}{A(z)A(z^{-1})} = \frac{\sum_{i=-q}^{q} c_k z^{-k}}{A(z)A(z^{-1})} = \sum_{l=-\infty}^{\infty} C_x(l) z^{-l}. \tag{4.4.9}$$

To ensure that the equation holds, the relationship between the coefficients c_k and the MA parameters should satisfy the following equation

$$\sigma^2 B(z)B(z^{-1}) = \sum_{k=-q}^{q} c_k z^{-k}. \tag{4.4.10}$$

It can be seen that the coefficients c_k are symmetrical, i.e., $c_{-k} = c_k$.

From the third equation of Eq. (4.4.9), we can get

$$\sum_{k=-q}^{q} c_k(k) z^{-k} = A(z)A(z^{-1}) \sum_{l=-\infty}^{\infty} C_x(l) z^{-l}. \tag{4.4.11}$$

Note $A(z)A(z^{-1}) = \sum_{i=0}^{p} \sum_{j=0}^{p} a_i a_j^*(k) z^{-i+j}$, and compare the coefficients of the same power term on both sides of Eq. (4.4.11), we can get the calculation formula of coefficients c_k

$$c_k = \sum_{i=0}^{p} \sum_{j=0}^{p} a_i a_j^* C_x(k - i + j), \; k = 0, 1, \cdots, q. \tag{4.4.12}$$

The ARMA spectrum estimator proposed by Kaveh is

$$P_x(\omega) = \left. \frac{\sum_{k=-q}^{q} c_k z^{-k}}{\left|1 + \sum_{i=1}^{p} a_i z^{-i}\right|^2} \right|_{z=e^{j\omega}}. \tag{4.4.13}$$

Obviously, the Kaveh spectrum estimator does not need the white noise variance σ^2 and MA parameters b_i, but needs to know MA order.

4.4.2 Modified Yule-Walker Equation

Both the Cadzow spectrum estimator and Kaveh spectrum estimator need to be known the order and parameters of the AR model. In practice, they can be estimated by observation data. Hence, it is necessary to deduce the linear equations of AR parameters.

According to Theorem 4.2.3, the causal ARMA process $\{x(n)\}$ has a unique stationary solution

$$x(n) = \sum_{i=0}^{\infty} h(i) e(n - i). \tag{4.4.14}$$

The correlation function of $\{x(n)\}$ is

$$\begin{aligned} R_x(\tau) &= \mathrm{E}\{x(n)x(n+\tau)\} \\ &= \mathrm{E}\left\{\left[\sum_{i=0}^{\infty} h(i)e(n-i)\right]\left[\sum_{k=0}^{\infty} h(k)e(n+\tau-k)\right]\right\} \\ &= \sum_{i=0}^{\infty}\sum_{k=0}^{\infty} h(i)h(k)\mathrm{E}\{e(n-i)e(n+\tau-k)\}, \end{aligned} \tag{4.4.15}$$

and $e(n)$ is white noise, so

$$\mathrm{E}\{e(n-i)e(n+\tau-k)\} = \begin{cases} \sigma^2, & k=\tau+i, \\ 0, & \text{Others.} \end{cases}$$

Substituting the above equation into Eq. (4.4.15), we have

$$R_x(\tau) = \sigma^2 \sum_{i=0}^{\infty} h(i)h(i+\tau). \tag{4.4.16}$$

This equation describing the relationship between the correlation function and the impulse response is important and will be often used later.

The impulse response $h(n)$ of a linear system is the output response when the system is excited by the impulse signal $\delta(n)$. Therefore, according to the definition of the ARMA process, there is the following equation directly

$$\sum_{i=0}^{p} a_i h(n-i) = \sum_{k=0}^{q} b_k \delta(n-k) = b_n. \tag{4.4.17}$$

So, using Eq. (4.4.16) and Eq. (4.4.17), it is easy to get

$$\sum_{i=0}^{p} a_i R_x(l-i) = \sigma^2 \sum_{k=0}^{\infty} h(k) \sum_{i=0}^{p} h(k+l-i) = \sigma^2 \sum_{k=0}^{\infty} h(k) b_{k+l}. \tag{4.4.18}$$

Note that for an ARMA(p, q) process, its MA parameters $b_i = 0, i > q$, so Eq. (4.4.18) is always equal to zero if $l > q$, i.e.,

$$R_x(l) + \sum_{i=1}^{p} a_i R_x(l-i) = 0, \forall l > q. \tag{4.4.19}$$

This normal equation is the famous modified Yule-Walker equation, often abbreviated as the MYW equation.

In particular, for an AR(p) process, Eq. (4.4.19) can be simplified as

$$R_x(l) + \sum_{i=1}^{p} a_i R_x(l-i) = 0, \forall l > 0. \tag{4.4.20}$$

This normal equation is called Yule-Walker equation, sometimes called YW equation for short.

Since the modified Yule-Walker equation holds for all $l > q$, is it necessary to solve infinitely many equations to determine the AR parameters $a_1, \cdots, a_p$? This is the problem of unique identifiability of parameters mentioned in Chapter 4. The answer to this problem is given by the following theorem proposed by Gersch[84] in 1970.

Theorem 4.4.1. *(The identifiability of AR parameters) If the polynomials $A(z)$ and $B(z)$ of the ARMA (p, q) model have no cancellable common factors and $a_p \neq 0$, then the AR parameters $a_1, \cdots, a_p$ of this ARMA model can be uniquely determined or identified by the following p modified Yule-Walker equations*

$$\sum_{i=1}^{p} a_i R_x(l-i) = -R_x(l), l = q+1, \cdots, q+p. \tag{4.4.21}$$

Theorem 4.4.1 tells us that when the true AR order p and the autocorrelation function $R_x(\tau)$ of ARMA(p, q) process $\{x(n)\}$ are known, only p modified Yule-Walker equations need to be solved to identify AR parameters. However, in practical application, the AR order and the autocorrelation function are unknown. How can we solve this problem?

Let's still assume that the autocorrelation function $R_x(\tau)$ is known, but AR order p is unknown. In this case, if the original ARMA(p, q) process $\{x(n)\}$ is regarded as an ARMA(p_e, q) process with extended AR order $p_e \geq p$, then the modified Yule-Walker equation still holds when p is replaced by p_e and $l > q_e$ (where $q_e \geq q$). Write it as

$$\boldsymbol{R}_e \boldsymbol{a}_e = \boldsymbol{0}, \tag{4.4.22}$$

where

$$\boldsymbol{R}_e = \begin{bmatrix} R_x(q_e+1) & R_x(q_e) & \cdots & R_x(q_e+1-p_e) \\ R_x(q_e+2) & R_x(q_e+1) & \cdots & R_x(q_e+2-p_e) \\ \vdots & \vdots & \vdots & \vdots \\ R_x(q_e+M) & R_x(q_e+M-1) & \cdots & R_x(q_e+M-p_e) \end{bmatrix}, \tag{4.4.23}$$

$$a_e = [1, a_1, \cdots, a_p, a_{p+1}, \cdots, a_{p_e}]^T. \tag{4.4.24}$$

Here $M \gg p$. Therefore, Eq. (4.4.22) is an overdetermined set of equations. Now the question is whether the rank of matrix $\boldsymbol{R}_e$ is still equal to p? The following proposition answers this question.

Proposition 4.4.1. [41] *If $M \geq p_e$, $p_e \geq p$, $q_e \geq q$, and $q_e - p_e \geq q - p$, then rank$(\boldsymbol{R}_e) = p$.*

This proposition shows that if the autocorrelation function $R_x(\tau)$ is known, only p of the p_e+1 singular values of matrix $\boldsymbol{R}_e$ are not equal to zero, and the rest are all equal to zero. Therefore, AR order p can be determined by singular value decomposition of matrix $\boldsymbol{R}_e$. However, in practical application, not only the AR order is unknown, but also the true autocorrelation function is unknown. Only N observation data $x(1), \cdots, x(N)$

can be used. So we need to calculate the sample autocorrelation function $\hat{R}_x(\tau)$ first, and then use it to replace the element $R_x(\tau)$ in matrix $\boldsymbol{R}_e$. A natural question is: can a time averaged sample autocorrelation function $\hat{R}_x(\tau)$ replace the real autocorrelation function $R_x(\tau)$? The following second-order mean square ergodic theorem definitely answers this question.

Theorem 4.4.2. *(Second-order mean square ergodic theorem) Let $\{x(n)\}$ be a Gaussian wide-sense stationary complex process with zero mean. If the true or total autocorrelation is square summable, i.e.,*

$$lim_{N\to\infty}\frac{1}{N}\sum_{k=0}^{N-1}|R_x(k)|^2 = 0. \tag{4.4.25}$$

Then for any fixed $\tau = 0, \pm1, \pm2, \cdots$, there is

$$lim_{N\to\infty}E\{[\hat{R}_x(\tau) - R_x(\tau)]^2\} = 0, \tag{4.4.26}$$

where the sample autocorrelation function is

$$\hat{R}_x(\tau) = \frac{1}{N}\sum_{n=1}^{N}x(n)x^*(n-\tau). \tag{4.4.27}$$

Proof. See reference [120]. □

4.4.3 Singular Value Decomposition Method for AR Order Determination

The methods to determine the order of an ARMA model can be divided into two categories: information criterion method and linear algebra method.

The most famous information criterion methods are the final prediction error (FPE) method[6] and Akachi information criterion (AIC) method[7], which were proposed by a Japanese mathematical statistician Akachi in 1969 and 1974 respectively.

The final prediction error (FPE) criterion selects the order (p, q) of the ARMA model which makes the information

$$\text{FPE}(p, q) \stackrel{\text{def}}{=} \hat{\sigma}_{wp}^2\left(\frac{N+p+q+1}{N-p-q-1}\right), \tag{4.4.28}$$

minimum, where $\hat{\sigma}_{wp}^2$ is the variance of the linear prediction error, and the calculation formula is

$$\hat{\sigma}_{wp}^2 = \sum_{i=0}^{p}\hat{a}_i\hat{R}_x(q-i). \tag{4.4.29}$$

In the AIC criterion, the selection criterion of ARMA model order (p, q) is to minimize information

$$\text{AIC}(p, q) \stackrel{\text{def}}{=} \ln\sigma_{wp}^2 + 2(p+q)/N, \tag{4.4.30}$$

where N is the data length (also called sample size). Obviously, whether using FPE or AIC criterion, it is necessary to use the least square method to fit the ARMA model of various possible orders in advance, and then according to the "parsimony principle", determine a combination of order (p, q) of the ARMA model as small as possible. When the sample size $N \to \infty$, the information criteria FPE(p, q) and AIC(p, q) are equivalent. Kashyap[121] proved that when $N \to \infty$, the error probability of the AIC criterion choosing the right order does not tend to zero. Hence, the AIC criterion is a statistically inconsistent estimation.

The improved form of AIC is called the BIC criterion. The principle of choosing (p, q) is to minimize the information

$$\mathrm{BIC}(p, q) \overset{\mathrm{def}}{=} \ln\sigma_{wp}^2 + (p+q)\frac{\ln N}{N}. \tag{4.4.31}$$

Rissanen[183] presented another information criterion to choose the order (p, q) of the ARMA model using the minimum description length (MDL), which is called the MDL criterion. The information is defined as

$$\mathrm{MDL}(p, q) \overset{\mathrm{def}}{=} N\ln\hat{\sigma}_{wp}^2 + (p+q)\ln N. \tag{4.4.32}$$

The criterion autoregressive transfer function is also a commonly used information criterion, referred to as CAT criterion, proposed by Parzen[172] in 1974. CAT function is defined as

$$\mathrm{CAT}(p, q) \overset{\mathrm{def}}{=} \left(\frac{1}{N}\sum_{k=1}^{p}\frac{1}{\bar{\sigma}_{k,q}^2}\right) - \frac{1}{\bar{\sigma}_{p,q}^2}, \tag{4.4.33}$$

where

$$\bar{\sigma}_{k,q}^2 = \frac{N}{N-k}\hat{\sigma}_{k,q}^2. \tag{4.4.34}$$

The order (p, q) should selected to minimize CAT(p, q).

MDL information criterion is statistically consistent. The experimental results show that for a short data length, the AR order should be in the range of $N/3 \sim N/2$ to get good results. Ulrych and Clayton[212] proved that for a short data segment, none of the FPE, AIC, and CAT methods works well.

Typical linear algebra order determination methods include determinant test method[56], Gram-Schmidt orthogonal method[48] and singular value decomposition method. The singular value decomposition method is introduced here.

Singular value decomposition (SVD) is mainly used to solve linear equations. The matrix associated with the equations not only represents the characteristics of the desired solution, but also represents the information of dynamic performance. Therefore, it is necessary to study the characteristics of this characteristic matrix. The matrix singular value decomposition in the following theorem can play this role.

Theorem 4.4.3. *Let* $\boldsymbol{A}$ *be an* $m \times n$ *complex matrix, then there exist an* $m \times m$ *unitary matrix* $\boldsymbol{U}$ *and an* $n \times n$ *unitary matrix* $\boldsymbol{V}$*, so that* $\boldsymbol{A}$ *can be decomposed as*

$$\boldsymbol{A} = \boldsymbol{U}\boldsymbol{\Sigma}\boldsymbol{V}^H, \tag{4.4.35}$$

where the superscript H denotes the conjugate transpose of a complex matrix, $\boldsymbol{\Sigma}$ *is an* $m \times n$ *diagonal matrix with nonnegative principal diagonal elements arranged in the following order*

$$\sigma_{11} \geq \sigma_{22} \geq \cdots \geq \sigma_{hh} \geq 0, \tag{4.4.36}$$

where $h = min(m, n)$.

The proof of the above theorem can be found in many works of linear algebra or matrix theory, such as references [107], [93] and [240].

A complex matrix $\boldsymbol{B}$ is called unitary matrix if $\boldsymbol{B}^{-1} = \boldsymbol{B}^{\mathrm{H}}$. In particular, a real unitary matrix is called an orthogonal matrix, i.e., $\boldsymbol{B}^{-1} = \boldsymbol{B}^{\mathrm{T}}$. The elements σ_{kk} on the main diagonal of diagonal matrix $\boldsymbol{\Sigma}$ are called singular values of matrix $\boldsymbol{A}$. The unitary matrices $\boldsymbol{U}$ and $\boldsymbol{V}$ are called left singular vector matrix and right singular vector matrix respectively. The column vectors of $\boldsymbol{u}_i$ and $\boldsymbol{v}_j$ are called left singular vector and right singular vector of matrix $\boldsymbol{U} = [\boldsymbol{u}_1, \cdots, \boldsymbol{u}_m]^T$ and $\boldsymbol{V} = [\boldsymbol{v}_1, \cdots, \boldsymbol{v}_n]^T$, respectively.

Singular values σ_{kk} contain useful information about the properties of the rank of matrix $\boldsymbol{A}$. In practical applications, it is often necessary to find the best approximation $\hat{\boldsymbol{A}}$ of $m \times n$ matrix $\boldsymbol{A}$ in the sense of Frobenious norm.

Keeping the first k singular values of $\boldsymbol{\Sigma}$ unchanged and setting other singular values zero to get a matrix $\boldsymbol{\Sigma}_k$, which is called the rank k approximation of $\boldsymbol{\Sigma}$. Namely, we can use

$$\boldsymbol{A}^{(k)} = \boldsymbol{U}\boldsymbol{\Sigma}_k\boldsymbol{V}^{\mathrm{H}}, \tag{4.4.37}$$

to approximate matrix $\boldsymbol{A}$. The quality of the approximation is measured by the Frobenious norm of the matrix difference $\boldsymbol{A} - \boldsymbol{A}^{(k)}$

$$||\boldsymbol{A} - \boldsymbol{A}^{(k)}||_{\mathrm{F}} = ||\boldsymbol{U}\boldsymbol{\Sigma}\boldsymbol{V}^{\mathrm{H}} - \boldsymbol{U}\boldsymbol{\Sigma}_k\boldsymbol{V}^{\mathrm{H}}||_{\mathrm{F}}. \tag{4.4.38}$$

According to the operation of norm, for $m \times m$ unitary matrix U and $n \times n$ unitary matrix V, their norms are $||\boldsymbol{U}||_{\mathrm{F}} = \sqrt{m}$ and $||\boldsymbol{V}||_{\mathrm{F}} = \sqrt{n}$, respectively, so Eq. (4.4.38) can be simplified as

$$||\boldsymbol{A} - \boldsymbol{A}^{(k)}||_{\mathrm{F}} = ||\boldsymbol{U}||_{\mathrm{F}}||\boldsymbol{\Sigma} - \boldsymbol{\Sigma}_k||_{\mathrm{F}}||\boldsymbol{V}^{\mathrm{H}}||_{\mathrm{F}} = \sqrt{mn}\left(\sum_{i=k+1}^{\min(m,n)} \sigma_{ii}^2\right)^{1/2}.$$

The above equation shows that the accuracy of matrix $\boldsymbol{A}^{(k)}$ approximating $\boldsymbol{A}$ depends on the sum of the squares of the singular values set to zero. Obviously, if k is larger, $||\boldsymbol{A} - \boldsymbol{A}^{(k)}||_{\mathrm{F}}$ is smaller, and when $k = h = \min(m, n)$, $||\boldsymbol{A} - \boldsymbol{A}^{(k)}||_{\mathrm{F}}$ is equal to zero. Naturally, we hope that the Frobenious norm of the approximation error matrix $\boldsymbol{A} - \boldsymbol{A}^{(k)}$ is small enough when k takes a suitable value p, and the Frebenious norm will not decrease significantly when $k > p$. This value p is called the effective rank of matrix $\boldsymbol{A}$. The method of determining the effective rank of $\boldsymbol{A}$ can be realized by the following method. Define a normalized ratio

$$v(k) = \frac{||\boldsymbol{A}^{(k)}||_{\mathrm{F}}}{||\boldsymbol{A}||_{\mathrm{F}}} = \left(\frac{\sigma_{11}^2 + \cdots + \sigma_{kk}^2}{\sigma_{11}^2 + \cdots + \sigma_{hh}^2}\right)^{1/2}, 1 \leq k \leq h, \tag{4.4.39}$$

and pre-determine a threshold very close to 1 (e.g. 0.995). Therefore, when p is the smallest integer while $\nu(k)$ is greater than or equal to the threshold, the first p singular values can be considered as the primary singular values, while all the other singular values are less important singular values, thus p is determined as the effective rank of matrix $\boldsymbol{A}$.

In addition to the normalized ratio $\nu(k)$, the normalized singular value can also be used to determine the effective rank of a matrix. The normalized singular value is defined as

$$\bar{\sigma}_{kk} \stackrel{\text{def}}{=} \sigma_{kk}/\sigma_{11}, 1 \leq k \leq h. \tag{4.4.40}$$

Obviously, $\bar{\sigma}_{11} = 1$. Contrary to the case of using normalized ratio $\nu(k)$, when using normalized singular value to determine the effective rank, a positive number close to zero (e.g. 0.05) is selected as the threshold, and the largest integer k when $\bar{\sigma}_{kk}$ greater than the threshold is taken as the effective rank p of the matrix $\boldsymbol{A}$.

In Eq. (4.4.23), using the sample correlation function instead of the ensemble correlation function yields,

$$\boldsymbol{R}_e = \begin{bmatrix} \hat{R}_x(q_e+1) & \hat{R}_x(q_e) & \cdots & \hat{R}_x(q_e+1-p_e) \\ \hat{R}_x(q_e+2) & \hat{R}_x(q_e+1) & \cdots & \hat{R}_x(q_e+2-p_e) \\ \vdots & \vdots & \vdots & \vdots \\ \hat{R}_x(q_e+M) & \hat{R}_x(q_e+M-1) & \cdots & \hat{R}_x(q_e+M-p_e) \end{bmatrix}. \tag{4.4.41}$$

The AR order of the ARMA model can be estimated by determining the effective rank of the above matrix via the normalized ratio $\nu(k)$ or the normalized singular value $\bar{\sigma}_{kk}$.

4.4.4 Total Least Squares Method for AR Parameter Estimation

Once the AR order p is determined, how to get the estimated values of p AR parameters? The intuitive idea is to use the least square method, but this will bring two problems: first, we must relist the normal equations, so that they contain only p unknown numbers; Second, the least square method for solving $\boldsymbol{Ax} = \boldsymbol{b}$ only considers that $\boldsymbol{b}$ contains errors, but actually the coefficient matrix $\boldsymbol{A}$ (here is the sample correlation matrix) also contains errors. Therefore, a more reasonable method than the least squares should consider the error or disturbance of $\boldsymbol{A}$ and $\boldsymbol{b}$ at the same time. Let the error matrix of $m \times n$ matrix $\boldsymbol{A}$ be $\boldsymbol{E}$ and the error vector of vector $\boldsymbol{b}$ be $\boldsymbol{e}$, that is, consider the least square solution of the matrix equation

$$(\boldsymbol{A} + \boldsymbol{E})\boldsymbol{x} = \boldsymbol{b} + \boldsymbol{e}. \tag{4.4.42}$$

Because the total error is considered, this method is called the total least squares (TLS) method.

Eq. (4.4.42) can be equivalently written as

$$\left([\; -b \;\vdots\; A \;] + [\; -e \;\vdots\; E \;] \right) \begin{bmatrix} 1 \\ \cdots \\ x \end{bmatrix} = \mathbf{0}, \tag{4.4.43}$$

or

$$(B + D)z = \mathbf{0}, \tag{4.4.44}$$

where

$$B = [\; -b \;\vdots\; A \;], D = [\; -e \;\vdots\; E \;], z = \begin{bmatrix} 1 \\ x \end{bmatrix}.$$

So, the TLS method for solving the equations Eq. (4.4.42) can be expressed as solving the vector z such that

$$||D||_F = \min, \tag{4.4.45}$$

where

$$||D||_F = \left(\sum_{i=1}^{m} \sum_{j=1}^{n} d_{ij}^2 \right)^{1/2}, \tag{4.4.46}$$

is the Frobenious norm of the disturbance matrix D. Here the overdetermined equation solution is considered, that is, assuming $m > n + 1$.

The basic idea of the TLS method is to minimize the influence of noise disturbance from A and b. The specific step is to find a disturbance matrix $D \in \mathcal{R}^{m\times(n+1)}$ with minimum norm so that $B + D$ is not full rank (If full rank, there is only trivial solution $z = \mathbf{0}$). The singular value decomposition can achieve this purpose. Let

$$B = U\Sigma V^{\mathrm{H}}, \tag{4.4.47}$$

and the singular values are still in descending order

$$\sigma_{11} \geq \sigma_{22} \geq \cdots \geq \sigma_{n+1,n+1} \geq 0.$$

Let $m \times (n + 1)$ matrix $\hat{B}$ be the best approximation of B and the effective rank of B be p, then from the previous discussion, the best approximation $\hat{B}$ is

$$\hat{B} = U\Sigma_p V^{\mathrm{H}}, \tag{4.4.48}$$

where the first p singular values of Σ_p are the same as the first p singular values of Σ, while the other singular values are zero.

Let $m \times (p + 1)$ matrix $\hat{B}(j, p + j)$ be a submatrix of the $m \times (n + 1)$ best approximation matrix $\hat{B}$, defined as

$$\hat{B}(j, p + j) \stackrel{\text{def}}{=} \text{Submatrix composed of } j\text{-th to } (p + j)\text{-th columns of matrix.} \tag{4.4.49}$$

Obviously, there are $n + 1 - p$ submatrices, namely $\hat{B}(1, p + 1), \cdots, \hat{B}(n + 1 - p, n + 1)$.

The effective rank of matrix B is p, which means that only p undetermined parameters in the unknown parameter vector x are independent. Let these parameters be the first p parameters of $\boldsymbol{x}$, and together with 1, they form a $(p+1\times 1)$-dimensional parameter vector $\boldsymbol{a}=[1,x_1,\cdots,x_p]^{\mathrm{T}}$. So the solution of the original equations Eq. (4.4.44) becomes the solution of $n+1-p$ equations.

$$\hat{\boldsymbol{B}}(k,p+k)\boldsymbol{a}=\boldsymbol{0},\ k=1,\cdots,n+1-p, \tag{4.4.50}$$

or equivalent to the solution of

$$\begin{bmatrix}\hat{\boldsymbol{B}}(1:p+1)\\ \hat{\boldsymbol{B}}(2:p+2)\\ \vdots\\ \hat{\boldsymbol{B}}(n+1-p:n+1)\end{bmatrix}\boldsymbol{a}=\boldsymbol{0}. \tag{4.4.51}$$

It's not hard to prove

$$\hat{\boldsymbol{B}}(k,p+k)=\sum_{j=1}^{p}\sigma_{jj}\boldsymbol{u}_j(\boldsymbol{v}_j^k)^{\mathrm{H}}, \tag{4.4.52}$$

where $\boldsymbol{u}_j$ is the j-th column of the unitary matrix $\boldsymbol{U}$, and $\boldsymbol{v}_j^k$ is a windowed segment of column j of the unitary matrix $\boldsymbol{V}$, defined as

$$\boldsymbol{v}_j^k=[v(k,j),v(k+1,j),\cdots,v(k+p,j)]^{\mathrm{T}}, \tag{4.4.53}$$

where $v(k,j)$ is the element of row k and column j of matrix $\boldsymbol{V}$.

According to the principle of least squares, finding the least square solution of the equations Eq. (4.4.51) is equivalent to minimizing the following cost function

$$\begin{aligned}f(\boldsymbol{a}) &= [\hat{\boldsymbol{B}}(1:p+1)\boldsymbol{a}]^{\mathrm{H}}\hat{\boldsymbol{B}}(1:p+1)\boldsymbol{a}+\cdots\\ &\quad+[\hat{\boldsymbol{B}}(n+1-p:n+1)\boldsymbol{a}]^{\mathrm{H}}\hat{\boldsymbol{B}}(n+1-p:n+1)\boldsymbol{a}\\ &= \boldsymbol{a}^{\mathrm{H}}\left[\sum_{i=1}^{n+1-p}\hat{\boldsymbol{B}}^{\mathrm{H}}(i:p+i)\hat{\boldsymbol{B}}(i:p+i)\right]\boldsymbol{a}.\end{aligned} \tag{4.4.54}$$

Define $(p+1)\times(p+1)$ matrix

$$\boldsymbol{S}^{(p)}=\sum_{i=1}^{n+1-p}\hat{\boldsymbol{B}}^{\mathrm{H}}(i:p+i)\hat{\boldsymbol{B}}(i:p+i). \tag{4.4.55}$$

Substituting Eq. (4.4.52) into the above equation yields

$$\boldsymbol{S}^{(p)}=\sum_{j=1}^{p}\sum_{i=1}^{n+1-p}\sigma_{jj}^2\boldsymbol{v}_j^i(\boldsymbol{v}_j^i)^{\mathrm{H}}. \tag{4.4.56}$$

In addition, substituting Eq. (4.4.55) into Eq. (4.4.54), the cost function can be rewritten as $f(\boldsymbol{a})=\boldsymbol{a}^{\mathrm{H}}\boldsymbol{S}^{(p)}\boldsymbol{a}$. From $\frac{\partial f(\boldsymbol{a})}{\partial \boldsymbol{a}}=0$, we have

$$\boldsymbol{S}^{(p)}\boldsymbol{a}=\alpha\boldsymbol{e}, \tag{4.4.57}$$

where $\boldsymbol{e} = [1, 0, \cdots, 0]^H$. The normalization constant α should be selected so that the first element of the parameter vector $\boldsymbol{a}$ is 1.

It is easy to solve Eq. (4.4.57). If $\boldsymbol{S}^{-(p)}$ is the inverse matrix of $\boldsymbol{S}^{(p)}$, the solution depends only on the first column of the inverse matrix $\boldsymbol{S}^{-(p)}$. It is easy to solve

$$\hat{x}_i = \boldsymbol{S}^{-(p)}(i+1, 1)/\boldsymbol{S}^{-(p)}(1, 1), \ i = 1, \cdots, p. \tag{4.4.58}$$

To sum up, the algorithm for finding the total least squares solution consists of the following steps.

Algorithm 4.4.1. *SVD-TLS Algorithm*

Step 1 Calculate the SVD of the augmented matrix $\boldsymbol{B}$ *and store the singular values and matrix* $\boldsymbol{V}$*;*

Step 2 Determine the effective rank p of the augmented matrix $\boldsymbol{B}$*;*

Step 3 Calculate matrix $\boldsymbol{S}^{(p)}$ *using Eq.* (4.4.56) *and Eq.* (4.4.53)*;*

Step 4 Solve the inverse matrix $\boldsymbol{S}^{-(p)}$ *of* $\boldsymbol{S}^{(p)}$ *and calculate the total least squares estimation of unknown parameters by Eq.* (4.4.58).

Taking the autocorrelation matrix $\boldsymbol{R}_e$ defined in Eq. (4.4.41) as the augmented matrix $\boldsymbol{B}$ in the above algorithm, the AR order p of the ARMA model can be determined, and the overall least squares solution of p AR parameters can be obtained.

Once the AR order p and AR parameters $a_1, \cdots, a_p$ are estimated, the power spectral density of ARMA process $\{x(n)\}$ can be obtained using the Cadzow spectrum estimator or Kaveh spectrum estimator.

4.5 ARMA Model Identification

In some applications, we hope not only to obtain AR order and AR parameters, but also to obtain the estimation of MA order and MA parameters, and finally get the identification of the whole ARMA model.

4.5.1 MA Order Determination

For an ARMA model, the determination of its MA order is very simple in theory, but the effectiveness of the actual algorithms is not as simple as expected.

Modified Yule-Walker equation (4.4.19) holds for all $l > q$. This equation also shows

$$R_x(l) + \sum_{i=1}^{p} a_i R_x(l-i) \neq 0, \, l = q, \tag{4.5.1}$$

otherwise, it will mean that Eq. (4.4.19) is true for all l, that is, the order of the ARMA process is $q-1$, which is contrary to ARMA(p, q) model. In fact, Eq. (4.5.1) may or may not hold for some $l < q$, but it must hold for $l = q$. This shows that the MA order is the maximum integer l that makes the following equation t

$$R_x(l) + \sum_{i=1}^{p} a_i R_x(l-i) \neq 0, \quad l \leq q. \tag{4.5.2}$$

The principle of determining MA order was proposed by Chow[56] in 1972. The problem is: in the case of short data, the sample correlation function $\hat{R}_x(l)$ has a large estimation error and variance. Therefore, the test method of Eq. (4.5.2) lacks numerical stability. The better method is the linear algebra method[251]. Its theoretical basis is that the information of MA order q is contained in a Hankel matrix.

Proposition 4.5.1.[251] *Let $\boldsymbol{R}_1$ be a $(p+1)\times(p+1)$ Hankel matrix, i.e.,*

$$\boldsymbol{R}_1 = \begin{bmatrix} R_x(q) & R_x(q-1) & \cdots & R_x(q-p) \\ R_x(q+1) & R_x(q) & \cdots & R_x(q+1-p) \\ \vdots & \vdots & \vdots & \vdots \\ R_x(q+p) & R_x(q+p-1) & \cdots & R_x(q) \end{bmatrix}. \tag{4.5.3}$$

If $a_p \neq 0$, rank$(\boldsymbol{R}_1) = p+1$.

Furthermore, the rank of matrix $\boldsymbol{R}_{1e}$ with extended order q is considered.

Proposition 4.5.2.[251] *Suppose p has been determined and let $\boldsymbol{R}_{1e}$ be a $(p+1)\times(p+1)$ matrix, defined as*

$$\boldsymbol{R}_{1e} = \begin{bmatrix} R_x(q_e) & R_x(q_e-1) & \cdots & R_x(q_e-p) \\ R_x(q_e+1) & R_x(q_e) & \cdots & R_x(q_e-p+1) \\ \vdots & \vdots & \vdots & \vdots \\ R_x(q_e+p) & R_x(q_e+p-1) & \cdots & R_x(q_e) \end{bmatrix}. \tag{4.5.4}$$

Then when $q_e > q$, rank$(\boldsymbol{R}_{1e}) = p$, and only when $q = p$, rank$(\boldsymbol{R}_{1e}) = p+1$.

Proposition 4.5.2 shows that the real MA order q is implied in the matrix $\boldsymbol{R}_{1e}$. Theoretically, the order q can be determined as follows: start from $Q = q_e > q$, take $Q \leftarrow Q-1$ in turn, and use SVD to determine the rank of $\boldsymbol{R}_{1e}$; When the first turning point of rank from p to $p+1$ occurs in $Q = q$. However, in practical application, due to the use of the sample autocorrelation function, the turning point of order from p jump to $p+1$ is often not obvious. In order to develop a practical algorithm for MA order determination, the overdetermined matrix $\boldsymbol{R}_{2e}$ can be used with elements $R_{2e}[i,j] = \hat{R}_x(q_e+i-j), i = 1, \cdots, M; j = 1, \cdots, p_e+1; M \gg p_e$. Obviously, we have

$$\text{rank}(\boldsymbol{R}_{2e}) = \text{rank}(\boldsymbol{R}_{1e}). \tag{4.5.5}$$

Because $\boldsymbol{R}_{2e}$ contains the whole $\boldsymbol{R}_{1e}$, any k-th ($k \geq p+2$) column (or row) is linearly related to its left p columns (or upper p rows). This does not change the rank of the whole matrix.

The algorithm[251] for determining the MA order of ARMA(p, q) model using SVD is as follows.

Algorithm 4.5.1. *MA order determination algorithm*

Step 1 Step 1 and step 2 of the SVD-TLS algorithm for AR order determination and parameter estimation are used to determine the AR order and take $Q = q_e > q$*;*

Step 2 Let $Q \leftarrow Q - 1$*, and construct the sample autocorrelation function* $\boldsymbol{R}_{2e}$ *to calculate its SVD;*

Step 3 If the $p + 1$*-th singular value has an obvious turning point compared with the last calculation result, select* $q = Q$*; Otherwise, return to Step 2 and continue the above steps until* q *is selected.*

Note 1 Due to $p \geq q$, the simplest and effective way to select the initial value of Q is to take $Q = q_e = p + 1$, so that q can be found as soon as possible.

Note 2 The above MA order determination method is only related to AR order, but independent of AR parameters. In other words, the AR order and MA order can be determined respectively before parameter estimation.

Note 3 The key of the algorithm is to determine the turning point of the $p + 1$-th singular value in Step 3. As a turning point test rule, the ratio can be considered

$$\alpha = \frac{\sigma^{(Q)}_{p+1,p+1}}{\sigma^{(Q+1)}_{p+1,p+1}}, \tag{4.5.6}$$

where $\sigma^{(Q)}_{p+1,p+1}$ is the $p + 1$-th singular value when $\boldsymbol{R}_{2e}$ corresponds to Q value. If the relative change rate of the $p + 1$-th singular value is greater than a given threshold for a Q value, the Q value is accepted as the turning point.

Example 4.5.1 A time series is generated by

$$x(n) = \sqrt{20}\cos(2\pi 0.2n) + \sqrt{2}\cos(2\pi 0.213n) + v(n),$$

where $v(n)$ is Gaussian white noise with mean value of 0 and variance of 1. The signal-to-noise ratio (SNR) of each cosine wave is defined as the ratio of the power of the cosine wave to the average noise power, i.e. variance. Therefore, the cosine waves with frequencies of 0.2 and 0.213 have SNR of 10dB and 0dB respectively. A total of 10 independent experiments were conducted, and the running data length of each time is 300. In each operation, the SVD method gives the AR order estimation result of $p = 4$, and the above SVD method for MA order determination also gives the MA order estimation of $q = 4$, with the calculated minimum ratio $\alpha = 38.94\%$. The following are the singular values corresponding to $\alpha = 38.94\%$.

When $Q = 5$, the singular values of $\boldsymbol{R}_e$ are

$$\sigma_{11} = 102.942,\ \sigma_{22} = 102.349\ \sigma_{33} = 2.622,\ \sigma_{44} = 2.508,$$

$$\sigma_{55} = 0.588,\ \sigma_{66} = 0.517,\ \cdots,\ \sigma_{15,15} = 0.216,$$

It is easy to see that the first four singular values dominate, so the AR order estimation of $p = 4$ is given.

When $Q = 4$, the singular values of $\boldsymbol{R}_{2e}$ are

$$\sigma_{11} = 103.918,\ \sigma_{22} = 102.736\ \sigma_{33} = 2.621,\ \sigma_{44} = 2.510,$$

$$\sigma_{55} = 0.817,\ \sigma_{66} = 0.575,\ \cdots,\ \sigma_{15,15} = 0.142.$$

We can see that σ_{55} has an obvious turning point when $Q = 4$, so $q = 4$ is selected as the result of MA order estimation.

4.5.2 MA Parameter Estimation

When deriving the Kaveh spectrum estimator, Eq. (4.4.10) has been obtained. By comparing the coefficients of the same power term on the left and right sides of this equation, a set of important equations can be obtained

$$\left.\begin{array}{rl} \sigma^2(b_0^2 + b_1^2 + \cdots + b_q^2) & = c_0 \\ \sigma^2(b_0 b_1 + \cdots + b_{q-1} b_q) & = c_1 \\ & \vdots \\ \sigma^2 b_0 b_q & = c_q \end{array}\right\} . \tag{4.5.7}$$

Observing the above nonlinear equations, we can see that there are $q + 2$ unknown parameters $b_0, b_1, \cdots, b_q$ and σ^2, but there are only $q + 1$ equations. To ensure the uniqueness of the solution, it is usually assumed that $\sigma^2 = 1$ or $b_0 = 1$. In fact, the two assumptions are inclusive, because under the assumption of $\sigma^2 = 1$, $\sigma^2 = b_0^2$ can still be obtained by normalizing the MA parameter $b_0 = 1$. For convenience, it is assumed here that $\sigma^2 = 1$.

The Newton-Raphson algorithm[32] for solving nonlinear equations Eq. (4.5.7) is introduced below.

Define the fitting error function

$$f_k = \sum_{i=0}^{q} b_i b_{i+k} - c_k,\ k = 0, 1, \cdots, q, \tag{4.5.8}$$

as well as $(q + 1) \times 1$ vectors

$$\boldsymbol{b} = [b_0, b_1, \cdots, b_q]^{\mathrm{T}}, \tag{4.5.9}$$

$$\boldsymbol{f} = [f_0, f_1, \cdots, f_q]^{\mathrm{T}}, \tag{4.5.10}$$

and $(q + 1) \times (q + 1)$ matrix

$$\boldsymbol{F} = \frac{\partial \boldsymbol{f}}{\partial \boldsymbol{b}^{\mathrm{T}}} = \begin{bmatrix} \frac{\partial f_0}{\partial b_0} & \frac{\partial f_0}{\partial b_1} & \cdots & \frac{\partial f_0}{\partial b_q} \\ \vdots & \vdots & \vdots & \vdots \\ \frac{\partial f_q}{\partial b_0} & \frac{\partial f_q}{\partial b_1} & \cdots & \frac{\partial f_q}{\partial b_q} \end{bmatrix} . \tag{4.5.11}$$

Take the partial derivative of Eq. (4.5.8) and substitute it into Eq. (4.5.11), we have

$$\boldsymbol{F} = \begin{bmatrix} b_0 & b_1 & \cdots & b_q \\ b_1 & \cdots & b_q & \\ \vdots & ⋰ & & \\ b_q & & & 0 \end{bmatrix} + \begin{bmatrix} b_0 & b_1 & \cdots & b_q \\ & b_0 & \cdots & b_{q-1} \\ & & \ddots & \vdots \\ 0 & & & b_0 \end{bmatrix} . \tag{4.5.12}$$

According to the principle of the Newton-Raphson method, if the MA parameter vector obtained in the i-the iteration is $\boldsymbol{b}^{(i)}$, the estimated value of the MA parameter vector in the $i + 1$-the iteration is given by the following equation

$$\boldsymbol{b}^{(i+1)} = \boldsymbol{b}^{(i)} - \boldsymbol{F}^{-(i)}\boldsymbol{f}^{(i)}, \tag{4.5.13}$$

where $\boldsymbol{F}^{-(i)}$ is the inverse of matrix $\boldsymbol{F}^{(i)}$ of $\boldsymbol{F}$ in the i-th iteration.

To sum up, the Newton-Raphson algorithm for estimating MA parameters consists of the following steps.

Algorithm 4.5.2. *Newton-Raphson Algorithm*

Initialization Use Eq. (4.4.12) to calculate the MA spectrum coefficient c_k, $k = 0, 1, \cdots, q$, and let the initial values $b_0^{(0)} = \sqrt{c_0}$, $b_i^{(0)} = 0$, $i = 1, \cdots, q$.

Step 1 Calculate the fitting error function $f_k^{(i)}$, $k = 0, 1, \cdots, q$ from Eq. (4.5.8), and calculate $\boldsymbol{F}^{(i)}$ with Eq. (4.5.12);

Step 2 Update MA parameter estimation vector $\boldsymbol{b}^{(i+1)}$ by Eq. (4.5.13);

Step 3 Test whether the MA parameter estimation vector converges. If it converges, stop the iteration and output the MA parameter estimation result; Otherwise, let $i \leftarrow i + 1$, return to Step 1, and repeat the above steps until the MA parameter estimation converges.

As a rule of terminating the iteration, we can compare the absolute error of the estimated values of each parameter obtained by the previous and subsequent two iterations. However, this method is more suitable for MA parameters with large absolute value, but not for M parameters with small absolute value. A better method is to use relative error[248] to measure whether a parameter converges during iteration, such as

$$\left[\frac{b_k^{(i+1)} - b_k^{(i)}}{b_k^{(i+1)}}\right] \leq \alpha, \tag{4.5.14}$$

where the threshold can be a small number, such as 0.05.

When there is additive AR colored noise, the ARMA power spectrum estimation needs to use the generalized least square algorithm, and the bootstrap method is used to calculate the AR parameters of the ARMA model from the modified Yule-Walker equation under AR colored noise. Readers interested in this algorithm can refer to reference [249].

4.6 Maximum Entropy Spectrum Estimation

The theoretical basis of ARMA spectrum estimation is the modeling of a stochastic process. This section introduces another power spectrum estimation method from the point of view of information theory–maximum entropy method. This method was proposed by Burg[39] in 1967. Due to a series of advantages of the maximum entropy method, it has become an important branch of modern spectrum estimation. Interestingly, the maximum entropy spectrum estimation is equivalent to AR and ARMA spectrum estimation under different conditions.

4.6.1 Burg Maximum Entropy Spectrum Estimation

In information theory, we are often interested in how much information can be obtained after an event $X = x_k$ with probability p_k is observed. This information is represented by the symbol $I(x_k)$ and is defined as

$$I(x_k) \stackrel{\text{def}}{=} I(X = x_k) = \log\frac{1}{p_k} = -\log p_k, \tag{4.6.1}$$

where the base of the logarithm can be selected arbitrarily. When natural logarithm is used, the unit of information is nat. When the base 2 logarithm is used, its unit is bit. In the following description, the base 2 logarithm is used, unless otherwise specified. No matter what logarithm is used, it is easy to prove by Eq. (4.5.13) that the amount of information has the following properties.

Property 1 The event that must occur contains no information, i.e.,

$$I(x_k) = 0, \ \forall p_k = 1. \tag{4.6.2}$$

Property 2 The amount of information is nonnegative, i.e.,

$$I(x_k) \geq 0, \ 0 \leq p_k \leq 1. \tag{4.6.3}$$

This is called the nonnegative property of information, which shows that the occurrence of a random event either brings information or does not bring any information, but it will never cause the loss of information.

Property 3 The smaller the probability of an event, the more information we get from it, i.e.,

$$I(x_k) > I(x_i), \text{ if } \ p_k < p_i. \tag{4.6.4}$$

Consider the discrete random variable X, whose character set of value is $\mathcal{X}$. Let the probability that the random variable X takes value of x_k be $\Pr\{X = x_k\}$, $x_k \in \mathcal{X}$.

Definition 4.6.1. *The average value of information $I(x)$ in the character set $\mathcal{X}$ is called the entropy of the discrete random variable X, denoted as $H(x)$ and defined as*

$$H(X) \stackrel{\text{def}}{=} \mathrm{E}\left\{I(x)\right\} = \sum_{x_k \in \mathcal{X}} p_k I(x_k) = -\sum_{x_k \in \mathcal{X}} p_k \log p_k. \tag{4.6.5}$$

Here, $0\log 0 = 0$ is agreed, which can be easily proved from the limit $\lim_{x\to 0} x\log x = 0$. This shows that adding a zero probability term to the definition of entropy will not have any effect on entropy.

Entropy is a function of the distribution of random variable X, which has nothing to do with the actual value of random variable X, but only with the probability of this value.

If the character set $\mathcal{X}$ consists of $2K+1$ characters, the entropy can be expressed as

$$H(X) = \sum_{k=-K}^{K} p_k I(x_k) = -\sum_{k=-K}^{K} p_k \log p_k. \tag{4.6.6}$$

The entropy is a bounded function, i.e,

$$0 \le H(X) \le \log(2K+1). \tag{4.6.7}$$

The following are the properties of the lower and upper bounds of entropy.

(1) $H(x) = 0$, if and only if $p_k = 1$ for a certain $X = x_k$, so that the probability of X taking other values in set $\mathcal{X}$ is all zero; In other words, the lower bound 0 of entropy corresponds to no uncertainty;
(2) $H(x) = \log(2K+1)$, if and only if $p_k = 1/(2K+1)$ is constant for all k, i.e., all discrete values are equal probability. Therefore, the upper bound of entropy corresponds to the maximum uncertainty.

The proof of property 2 can be found in Reference [94].

Example 4.6.1 Let

$$X = \begin{cases} 1, & \text{with probability } p, \\ 0, & \text{with probability } 1-p. \end{cases}$$

The entropy of X can be calculated by Eq. (4.6.5)

$$H(x) = -p\log p - (1-p)\log(1-p).$$

In particular, if $p = 1/2$, $H(X) = 1$ bit.

Example 4.6.2 Let

$$X = \begin{cases} a, & \text{with probability } 1/2, \\ b, & \text{with probability } 1/8, \\ c, & \text{with probability } 1/4, \\ d, & \text{with probability } 1/8, \end{cases}$$

then the entropy of X is

$$H(x) = -\frac{1}{2}\log\frac{1}{2} - \frac{1}{8}\log\frac{1}{8} - \frac{1}{4}\log\frac{1}{4} - \frac{1}{8}\log\frac{1}{8} = \frac{7}{4} \text{ bit}.$$

If Definition 4.6.2 is extended to continuous random variables, there is the following definition.

Definition 4.6.2. *(Entropy of continuous random variables) Let the distribution density function of continuous random variable x be p(x), then its entropy is defined as*

$$H(x) \stackrel{\text{def}}{=} -\int_{-\infty}^{\infty} p(x)\ln p(x)dx = -\mathrm{E}\left\{\{\ln p(x)\}\right\}. \tag{4.6.8}$$

In 1967, following the definition of entropy of continuous random variables, Burg[39] defined

$$H[P(\omega)] = \frac{1}{2\pi}\int_{-\pi}^{\pi} \ln P(\omega)\mathrm{d}\omega, \tag{4.6.9}$$

as the entropy of power spectrum $P(\omega)$ (spectral entropy for short), and proposed that the spectral entropy should be maximized when estimating the power spectrum using a given $2p+1$ sample autocorrelation functions $\hat{R}_x(k)$, $k = 0, \pm 1, \cdots \pm p$. This is the well-known Burg maximum entropy spectrum estimation method. Of course, the inverse Fourier transform of the estimated power spectrum should also be able to restore the original $2p+1$ sample autocorrelation functions $\hat{R}_x(k)$, $k - 0, \pm 1, \cdots \pm p$. Specifically, the Burg maximum entropy spectrum estimation can be described as: finding the power spectral density $P(\omega)$ so that $P(\omega)$ can maximize the spectral entropy $H[P(\omega)]$ under the constraint

$$\hat{R}_x(m) = \frac{1}{2\pi}\int_{-\pi}^{\pi} P(\omega)\mathrm{e}^{\mathrm{j}\omega m}\mathrm{d}\omega, \quad m = 0, \pm 1, \cdots, \pm p. \tag{4.6.10}$$

This constrained optimization problem can be easily solved by the Lagrange multiplier method.

Construct objective function

$$J[P(\omega)] = \frac{1}{2\pi}\int_{-\pi}^{\pi} \ln P(\omega)\mathrm{d}\omega + \sum_{k=-p}^{p}\lambda_k\left[\hat{R}_x(k) - \frac{1}{2\pi}\int_{-\pi}^{\pi} P(\omega)\mathrm{e}^{\mathrm{j}\omega k}\mathrm{d}\omega\right], \tag{4.6.11}$$

where λ_k is a Lagrange multiplier. Find the partial derivative of $J[P(\omega)]$ relative to $P(\omega)$ and make it equal to 0, then we have

$$P(\omega) = \frac{1}{\sum_{k=-p}^{p}\lambda_k \mathrm{e}^{\mathrm{j}\omega k}}. \tag{4.6.12}$$

Replace the variable $\mu_k = \lambda_{-k}$, then Eq. (4.6.12) can be written as

$$P(\omega) = \frac{1}{\sum_{k=-p}^{p}\mu_k \mathrm{e}^{-\mathrm{j}\omega k}}. \tag{4.6.13}$$

Let

$$W(z) = \sum_{k=-p}^{p} \mu_k z^{-k},$$

then $P(\omega) = 1/W(e^{j\omega})$. Since the power spectral density is non negative, there is

$$W(e^{j\omega}) \geq 0.$$

Theorem 4.6.1. *(Fejer-Riesz Theorem)*[170] *If*

$$W(z) = \sum_{k=-p}^{p} \mu_k z^{-k} \quad or \quad W(e^{j\omega}) \geq 0,$$

we can find a function

$$A(z) = \sum_{i=0}^{p} a(i) z^{-i}, \tag{4.6.14}$$

so that

$$W(e^{j\omega}) = |A(e^{j\omega})|^2. \tag{4.6.15}$$

If the roots of $A(z) = 0$ are all in the unit circle, the function $A(z)$ is uniquely determined.

According to Fejer-Riesz Theorem, if $a(0) = 1$ is assumed, Eq. (4.6.13) can be expressed as

$$P(\omega) = \frac{\sigma^2}{|A(e^{j\omega})|^2}. \tag{4.6.16}$$

The above equation is exactly the previous AR power spectral density. This shows that Burg maximum entropy power spectrum is equivalent to the AR power spectrum.

4.6.2 Levinson Recursion

To realize the maximum entropy spectrum estimation, the order p and coefficients a_i need to be determined. This raises a question: what is the appropriate order? Burg proposes to use the linear prediction method to recursively calculate the coefficients of the predictor with different orders, and then compare the prediction error power of each predictor. The basis of this recursive calculation is the famous Levinson recursion (also known as Levinson-Durbin recursion).

The maximum entropy method uses both forward and backward prediction. The forward prediction is to use the given m data $x(n-m), \cdots, x(n-1)$ to predict the value of $x(n)$, which is called m-order forward linear prediction.

The filter that realizes forward linear prediction is called a forward linear prediction filter or forward linear predictor. The m-order forward linear prediction value of $x(n)$ is denoted as $\hat{x}(n)$ and defined as

$$\hat{x}(n) \stackrel{\text{def}}{=} -\sum_{i=1}^{m} a_m(i) x(n-i), \tag{4.6.17}$$

where $a_m(i)$ represents the i-th coefficient of the m-order forward linear prediction filter.

Similarly, predicting the value of $x(n-m)$ with given m data $x(n-m+1), \cdots, x(n)$ is called m-order backward linear prediction, which is defined as

$$\hat{x}(n-m) \overset{\text{def}}{=} -\sum_{i=1}^{m} a_m^*(i)x(n-m+i), \tag{4.6.18}$$

where $a_m^*(i)$ is the complex conjugate of $a_m(i)$. The filter that realizes backward linear prediction is called a backward linear prediction filter.

The following discusses how to design forward and backward linear prediction filters according to the minimum mean square error (MMSE) criterion.

The forward and backward linear prediction errors are respectively defined as

$$f(n) \overset{\text{def}}{=} x(n) - \hat{x}(n) = \sum_{i=0}^{m} a_m(i)x(n-i), \tag{4.6.19}$$

$$g(n-m) \overset{\text{def}}{=} x(n-m) - \hat{x}(n-m) = \sum_{i=0}^{m} a_m^*(i)x(n-m+i), \tag{4.6.20}$$

where $a_m(0) = 1$. According to the principle of orthogonality, to make the prediction value $\hat{x}(n)$ be the linear mean square estimation of $x(n)$, the forward prediction error $f(n)$ must be orthogonal to the known data $x(n-m), \cdots, x(n-1)$, i.e.,

$$\mathrm{E}\{f(n)x^*(n-k)\} = 0, \quad 1 \le k \le m. \tag{4.6.21}$$

Substituting Eq. (4.6.19) into Eq. (4.6.21), a set of normal equations can be obtained

$$\left.\begin{array}{r} R_x(0)a_m(1) + R_x(-1)a_m(2) + \cdots + R_x(-m+1)a_m(m) = -R_x(1) \\ R_x(1)a_m(1) + R_x(0)a_m(2) + \cdots + R_x(-m+2)a_m(m) = -R_x(2) \\ \vdots \\ R_x(m-1)a_m(1) + R_x(m-2)a_m(2) + \cdots + R_x(0)a_m(m) = -R_x(m) \end{array}\right\}, \tag{4.6.22}$$

where $R_x(k) = \mathrm{E}\{x(n)x^*(n-k)\}$ is the autocorrelation function of $\{x(n)\}$.

Define the mean square error of the forward linear prediction as

$$\begin{aligned} P_m &\overset{\text{def}}{=} \mathrm{E}\{|f(n)|^2\} = \mathrm{E}\{f(n)[x(n) - \hat{x}(n)]^*\} \\ &= \mathrm{E}\{f(n)x^*(n)\} - \sum_{i=1}^{m} a_m^*(i)\mathrm{E}\{f(n)x^*(n-i)\} \\ &= \mathrm{E}\{f(n)x^*(n)\}. \end{aligned} \tag{4.6.23}$$

The summation term $\sum$ in the above equation is zero, which is the result of the direct substitution of Eq. (4.6.21). Expand the right side of Eq. (4.6.23) to obtain the following equation directly

$$P_m = \sum_{i=0}^{m} a_m(i)R_x(-i). \tag{4.6.24}$$

The mean square error of forward linear prediction is the output power of forward prediction error, which is referred to as forward prediction error power for short.

Combine Eq. (4.6.22) and Eq. (4.6.24), we have

$$\begin{bmatrix} R_x(0) & R_x(-1) & \cdots & R_x(-m) \\ R_x(1) & R_x(0) & \cdots & R_x(-m+1) \\ \vdots & \vdots & \vdots & \vdots \\ R_x(m) & R_x(m-1) & \cdots & R_x(0) \end{bmatrix} \begin{bmatrix} 1 \\ a_m(1) \\ \vdots \\ a_m(m) \end{bmatrix} = \begin{bmatrix} P_m \\ 0 \\ \vdots \\ 0 \end{bmatrix}. \tag{4.6.25}$$

By solving Eq. (4.6.25), the coefficients $a_m(1), \cdots, a_m(m)$ of m-order forward prediction filter can be obtained directly, but the corresponding prediction error power is not necessarily small. Therefore, it needs to calculate the coefficients of various possible orders of forward prediction filter and the corresponding forward prediction error power. For different m, it is obviously too time-consuming to solve the filter equations independently. Assuming that the coefficients $a_{m-1}(1), \cdots, a_{m-1}(m-1)$ of $m-1$-order forward prediction filter have been calculated, how to recursively calculate the coefficients $a_m(1), \cdots, a_m(m)$ of m-order forward prediction filter from them?

From Eq. (4.6.25), it is easy to list the equations of $m-1$-order forward prediction filter

$$\begin{bmatrix} R_x(0) & R_x(-1) & \cdots & R_x(-m+1) \\ R_x(1) & R_x(0) & \cdots & R_x(-m) \\ \vdots & \vdots & \vdots & \vdots \\ R_x(m-1) & R_x(m-2) & \cdots & R_x(0) \end{bmatrix} \begin{bmatrix} 1 \\ a_{m-1}(1) \\ \vdots \\ a_{m-1}(m-1) \end{bmatrix} = \begin{bmatrix} P_m \\ 0 \\ \vdots \\ 0 \end{bmatrix}. \tag{4.6.26}$$

Considering the $m-1$-order backward prediction error, from Eq. (4.6.20), we have

$$\begin{aligned} g(n-m+1) &\overset{\text{def}}{=} x(n-m+1) - \hat{x}(n-m+1) \\ &= \sum_{i=0}^{m-1} a_{m-1}^{*}(i)x(n-m+1+i). \end{aligned} \tag{4.6.27}$$

From the principle of orthogonality, In order for $\hat{x}(n-m+1)$ to be a linear mean square estimate of $x(n-m+1)$, the backward prediction error $g(n-m+1)$ should be orthogonal to the known data $x(n-m+2), \cdots, x(n)$, i.e

$$\mathrm{E}\{g(n-m+1)x^{*}(n-m+1+k)\} = 0, \ k = 1, \cdots, m-1. \tag{4.6.28}$$

Combine Eq. (4.6.27) and Eq. (4.6.28) to get

$$\sum_{i=0}^{m-1} a_{m-1}^{*}(i)R_x(i-k) = 0, \ k = 1, \cdots, m-1, \tag{4.6.29}$$

$$P_{m-1} = \mathrm{E}\{|g(n-m+1)|^2\} = \sum_{i=0}^{m-1} a_{m-1}^{*}(i)R_x(i). \tag{4.6.30}$$

Eq. (4.6.29) and Eq. (4.6.30) can be combined together and written in matrix form as

$$\begin{bmatrix} R_x(0) & R_x(-1) & \cdots & R_x(-m+1) \\ R_x(1) & R_x(0) & \cdots & R_x(-m) \\ \vdots & \vdots & \vdots & \vdots \\ R_x(m-1) & R_x(m-2) & \cdots & R_x(0) \end{bmatrix} \begin{bmatrix} a^*_{m-1}(m-1) \\ \vdots \\ a^*_{m-1}(1) \\ 1 \end{bmatrix} = \begin{bmatrix} 0 \\ \vdots \\ 0 \\ P_{m-1} \end{bmatrix}. \tag{4.6.31}$$

The basic idea of using $a_{m-1}(i)$ to recursively calculate $a_m(i)$ is to directly use the sum of $a_{m-1}(i)$ and the correction term as the value of $a_m(i)$. This idea of the recursive formula is widely used in signal processing (such as adaptive filtering), neural networks, and neural computing (such as learning algorithms). In Levinson recursion, the correction term is reflected by the coefficients of the $m-1$-order backward prediction filter, i.e.,

$$a_m(i) = a_{m-1}(i) + K_m a^*_{m-1}(m-i), \ i = 0, 1, \cdots, m, \tag{4.6.32}$$

where K_m is the reflection coefficient.

The following are two edge cases of Eq. (4.6.32).

(1) If $i = 0$, since $a_m(0) = a_{m-1}(0) + K_m a^*_{m-1}(m)$ and $a_{m-1}(m) = 0$, there is $a_m(0) = a_{m-1}(0)$. This shows that if $a_1(0) = 1$, then $a_m(0) = 1$, $m \geq 2$.

(2) If $i = m$, since $a_m(m) = a_{m-1}(m) + K_m a^*_{m-1}(0)$, $a_{m-1}(m) = 0$, and $a_{m-1}(0) = 1$, there is $a_m(m) = K_m$.

To get the recursion of the prediction error P_m, Eq. (4.6.32) is rewritten as

$$\begin{bmatrix} 1 \\ a_m(1) \\ \vdots \\ a_m(m-1) \\ a_m(m) \end{bmatrix} = \begin{bmatrix} 1 \\ a_{m-1}(1) \\ \vdots \\ a_{m-1}(m-1) \\ 0 \end{bmatrix} + K_m \begin{bmatrix} 0 \\ a^*_{m-1}(m-1) \\ \vdots \\ a^*_{m-1}(1) \\ 1 \end{bmatrix}. \tag{4.6.33}$$

Substituting Eq. (4.6.33) into Eq. (4.6.25), we have

$$\begin{bmatrix} R_x(0) & R_x(-1) & \cdots & R_x(-m) \\ R_x(1) & R_x(0) & \cdots & R_x(-m+1) \\ \vdots & \vdots & \vdots & \vdots \\ R_x(m) & R_x(m-1) & \cdots & R_x(0) \end{bmatrix} \times$$

$$\left\{ \begin{bmatrix} 1 \\ a_{m-1}(1) \\ \vdots \\ a_{m-1}(m-1) \\ 0 \end{bmatrix} + K_m \begin{bmatrix} 0 \\ a^*_{m-1}(m-1) \\ \vdots \\ a^*_{m-1}(1) \\ 1 \end{bmatrix} \right\} = \begin{bmatrix} P_m \\ 0 \\ \vdots \\ 0 \end{bmatrix}. \tag{4.6.34}$$

Substituting Eq. (4.6.26) and Eq. (4.6.31) into the above equation yields

$$\begin{bmatrix} P_{m-1} \\ 0 \\ \vdots \\ 0 \\ \hdashline X \end{bmatrix} + K_m \begin{bmatrix} Y \\ \hdashline 0 \\ \vdots \\ 0 \\ P_{m-1} \end{bmatrix} = \begin{bmatrix} P_m \\ 0 \\ \vdots \\ 0 \end{bmatrix},$$

or

$$P_{m-1} + K_m Y = P_m, \tag{4.6.35}$$

$$X + K_m P_{m-1} = 0, \tag{4.6.36}$$

where

$$X = \sum_{i=0}^{m-1} a_{m-1}(i) R_x(m-i), \tag{4.6.37}$$

$$Y = \sum_{i=0}^{m-1} a^{*}_{m-1}(i) R_x(i-m). \tag{4.6.38}$$

Note that $R_x(i-m) = R^{*}_x(m-i) =$, so from Eq. (4.6.37) and Eq. (4.6.38) we have

$$Y = X^{*}, \tag{4.6.39}$$

while from Eq. (4.6.36) we have

$$X = -K_m P_{m-1},$$

so $Y = -K^{*}_m P_{m-1}$. Note that the prediction error power P_{m-1} is a real number. By substituting $Y = -K^{*}_m P_{m-1}$ into Eq. (4.6.35), the recursive equation of the prediction error power is obtained

$$P_m = (1 - |K_m|^2) P_{m-1}. \tag{4.6.40}$$

To summarize the above discussion, we have the following recursive algorithms.

Algorithm 4.6.1. *Levinson recursive algorithm (upward recursion)*

$$a_m(i) = a_{m-1}(i) + K_m a^{*}_{m-1}(m-i), \; i = 1, \cdots, m-1, \tag{4.6.41}$$

$$a_m(m) = K_m, \tag{4.6.42}$$

$$P_m = (1 - |K_m|^2) P_{m-1}. \tag{4.6.43}$$

When m = 0, Eq. (4.6.24) gives the initial value of the prediction error power

$$P_0 = R_x(0) = \frac{1}{N} \sum_{n=1}^{N} |x(n)|^2. \tag{4.6.44}$$

Eq. (4.6.41) to Eq. (4.6.43) are called (upward) Levinson recursive equations, i.e., the coefficients of the second-order prediction filter are recursively derived from the coefficients of the first-order prediction filter, and then the third-order is from the second-order. Sometimes, we are interested in the recursive calculation of $M-1$ order prediction filter from M order prediction filter, and then $m-2$ order from $M-1$ order. Such a recursive method of calculating low-order filter coefficients from high-order filter coefficients is called downward recursion. The following is the downward Levinson recursive algorithm [178].

Algorithm 4.6.2. *Levinson recursive algorithm (downward recursion)*

$$a_m(i) = \frac{1}{1-|K_{m+1}|^2}[a_{m+1}(i) - K_{m+1}a_{m+1}(m-i+1)], \tag{4.6.45}$$

$$K_m = a_m(m), \tag{4.6.46}$$

$$P_m = \frac{1}{1-|K_{m+1}|^2}P_{m+1}, \tag{4.6.47}$$

where $i = 1, \cdots, m$.

In the recursion process of the algorithm, when $|k_0| = 1$, *the recursion stops.*

4.6.3 Burg Algorithm

By observing Levinson recursion, the remaining problem is how to find the recursion formula of the reflection coefficient K_m. This problem has limited the practical application of Levinson recursion until Burg[39] studied the maximum entropy method. The basic idea of the Burg algorithm is to minimize the average power of forward and backward prediction errors.

Burg defined m-order forward and backward prediction errors as

$$f_m(n) = \sum_{i=0}^{m} a_m(i)x(n-i), \tag{4.6.48}$$

$$g_m(n) = \sum_{i=0}^{m} a_m^*(m-i)x(n-i). \tag{4.6.49}$$

Substituting Eq. (4.6.41) into Eq. (4.6.48) and Eq. (4.6.49) respectively, the order recursive equations of forward and backward prediction errors can be obtained as

$$f_m(n) = f_{m-1}(n) + K_m g_{m-1}(n-1), \tag{4.6.50}$$

$$g_m(n) = K_m^* f_{m-1}(n) + g_{m-1}(n-1). \tag{4.6.51}$$

Define the average power of the m-order (forward and backward) prediction errors as

$$P_m = \frac{1}{2}\sum_{n=m}^{N}[|f_m(n)|^2 + |g_m(n)|^2]. \tag{4.6.52}$$

Substituting the order recursion equations Eq. (4.6.50) and Eq. (4.6.51) into Eq. (4.6.52), and letting $\frac{\partial P_m}{\partial K_m} = 0$ yields

$$K_m = \frac{-\sum_{n=m+1}^{N} f_{m-1}(n) g_{m-1}^{*}(n-1)}{\frac{1}{2}\sum_{n=m+1}^{N}[|f_{m-1}(n)|^2 + |g_{m-1}(n-1)|^2]}, \tag{4.6.53}$$

where $m = 1, 2, \cdots$.

To sum up, the Burg algorithm for calculating the coefficients of the forward prediction filter is as follows.

Algorithm 4.6.3. *Burg Algorithm*

Step 1 Calculate the initial value of error power

$$P_0 = \frac{1}{N}\sum_{n=1}^{N} |x(n)|^2,$$

the initial value of the forward and backward prediction error

$$f_0(n) = g_0(n) = x(n),$$

and let $m = 1$;

Step 2 Calculate the reflection coefficient

$$K_m = \frac{-\sum_{n=m+1}^{N} f_{m-1}(n) g_{m-1}^{*}(n-1)}{\frac{1}{2}\sum_{n=m+1}^{N}[|f_{m-1}(n)|^2 + |g_{m-1}(n-1)|^2]};$$

Step 3 Calculate the forward prediction filter coefficients

$$\begin{aligned} a_m(i) &= a_{m-1}(i) + K_m a_{m-1}^{*}(m-i), \; i = 1, \cdots, m-1, \\ a_m(m) &= K_m; \end{aligned}$$

Step 4 Calculate the prediction error power

$$P_m = (1 - |K_m|^2) P_{m-1};$$

Step 5 Calculate the output of the filter

$$\begin{aligned} f_m(n) &= f_{m-1}(n) + K_m g_{m-1}(n-1), \\ g_m(n) &= K_m^{*} f_{m-1}(n) + g_{m-1}(n-1); \end{aligned}$$

Step 6 Let m $\leftarrow$ *m* + 1, *and repeat steps 2 to 5 until the prediction error power* P_m *is no longer significantly reduced.*

4.6.4 Burg Maximum Entropy Spectrum Analysis and ARMA Spectrum Estimation

If a constraint is added, Burg maximum entropy power spectral density is equivalent to ARMA power spectral density. This conclusion was independently proved by Lagunas[131] and Ihara[110] using different methods. Lagunas et al.[130] deduced this equivalence from the perspective of engineering application, using an easy to understand the method, and proposed a specific maximum entropy ARMA spectrum estimation algorithm.

Considering the logarithmic power spectral density $\ln P(\omega)$, the inverse Fourier transform is called the cepstrum coefficient, i.e

$$c_x(k) = \frac{1}{2\pi}\int_{-\pi}^{\pi} \ln P(\omega)\mathrm{e}^{\mathrm{j}\omega k}\mathrm{d}\omega. \tag{4.6.54}$$

Let $2M + 1$ autocorrelation functions $\hat{R}_x(m)$, $m = 0, \pm 1, \cdots, \pm M$ and $2N$ cepstrum coefficients $\hat{c}_x(l)$, $l = \pm 1, \cdots, \pm N$ be known. Now consider finding the power spectral density $P(\omega)$ under the constraints of autocorrelation function matching

$$\hat{R}_x(m) = \frac{1}{2\pi}\int_{-\pi}^{\pi} P(\omega)\mathrm{e}^{\mathrm{j}\omega m}\mathrm{d}\omega, \quad m = 0, \pm 1, \cdots, \pm M, \tag{4.6.55}$$

and cepstrum matching

$$\hat{c}_x(k) = \frac{1}{2\pi}\int_{-\pi}^{\pi} \ln P(\omega)\mathrm{e}^{\mathrm{j}\omega l}\mathrm{d}\omega, \, l = \pm 1, \cdots, \pm N, \tag{4.6.56}$$

to maximize its spectral entropy $H[P(\omega)]$. Note that when $l = 0$, the cepstrum $\hat{c}_x(0)$ defined by Eq. (4.6.56) is exactly the spectral entropy, so $\hat{c}_x(0)$ should not be included in cepstrum matching condition Eq. (4.6.56).

Using the Lagrange multiplier method, construct the cost function

$$\begin{aligned} J[P(\omega)] = {} & \frac{1}{2\pi}\int_{-\pi}^{\pi} \ln P(\omega)\mathrm{d}\omega + \sum_{m=-M}^{M} \lambda_m[\hat{R}_x(m) - \frac{1}{2\pi}\int_{-\pi}^{\pi} P(\omega)\mathrm{e}^{\mathrm{j}\omega m}\mathrm{d}\omega] \\ & + \sum_{l\neq 0, l=-N}^{N} \mu_l[\hat{c}_x(l) - \frac{1}{2\pi}\int_{-\pi}^{\pi} \ln P(\omega)\mathrm{e}^{\mathrm{j}\omega l}\mathrm{d}\omega], \end{aligned} \tag{4.6.57}$$

where λ_m and μ_l are two undetermined Lagrange multipliers. Let $\frac{\partial J[P(\omega)]}{\partial P(\omega)} = 0$, and substitute $\beta_l = -\mu_l$ to obtain the expression of power spectral density

$$P(\omega) = \frac{\sum_{l=-N}^{N} \beta_l \mathrm{e}^{\mathrm{j}\omega l}}{\sum_{m=-M}^{M} \lambda_m \mathrm{e}^{\mathrm{j}\omega m}}, \tag{4.6.58}$$

where $\beta_0 = 1$.

From theorem 4.6.1, that is, Fejer-Riesz theorem, Eq. (4.6.58) can be written as

$$P(\omega) = \frac{|B(z)|^2}{|A(z)|^2} = \left.\frac{|\sum_{i=0}^{N} b_i z^{-i}|^2}{|1 + \sum_{i=1}^{M} a_i z^{-i}|^2}\right|_{z=\mathrm{e}^{\mathrm{j}\omega}}. \tag{4.6.59}$$

Obviously, this is a rational ARMA power spectrum.

It is not difficult to see that the constraint of the autocorrelation function leads to the autoregressive part, which acts as the pole, while the constraint of the cepstrum contributes to the moving average part, which acts as the zero.

In addition, Lagunas et al.[130] also introduced the specific algorithm of maximum entropy ARMA spectrum estimation, which is omitted here.

The above maximum entropy method uses both autocorrelation function matching and cepstrum matching to obtain the maximum entropy ARMA power spectrum, while Burg maximum entropy method only uses autocorrelation function matching, which is equivalent to the AR power spectrum.

In the spectral entropy defined by Burg, we can see that it is significantly different from the entropy of continuous random variables defined by Eq. (4.6.8). If the entropy of the power spectrum is defined strictly following Eq. (4.6.8), it should be

$$H_2[P(\omega)] = -\frac{1}{2\pi}\int_{-\pi}^{\pi} P(\omega)\ln P(\omega)\mathrm{d}\omega, \tag{4.6.60}$$

which is called configuration entropy, proposed by Frieden[80].

Traditionally, the power spectrum estimation using spectral entropy maximization is called the first type of maximum entropy method (MEM-1 for short), and the power spectrum estimation using configuration entropy (which is negative entropy) minimization is called the second type of maximum entropy method (MEM-2 for short).

By using the Lagrange multiplier method, the cost function is constructed

$$\begin{aligned} J[P(\omega)] &= -\frac{1}{2\pi}\int_{-\pi}^{\pi} \ln P(\omega)\mathrm{d}\omega \\ &\quad + \sum_{m=-M}^{M} \lambda_m \left[\frac{1}{2\pi}\int_{-\pi}^{\pi} P(\omega)\mathrm{e}^{\mathrm{j}\omega m}\mathrm{d}\omega - \hat{R}_x(m)\right]. \end{aligned} \tag{4.6.61}$$

Let $\frac{\partial J[P(\omega)]}{\partial P(\omega)} = 0$ to get

$$\ln P(\omega) = -1 + \sum_{m=-M}^{M} \lambda_m \mathrm{e}^{\mathrm{j}\omega m}, \tag{4.6.62}$$

where λ_m is the undetermined Lagrange multiplier. Via variable substitution $c_m = \lambda_{-m}$, $m = \pm 1, \cdots, \pm M$ and $c_0 = -1 + \lambda_0$, Eq. (4.6.62) can be rewritten as

$$\ln P(\omega) = \sum_{m=-M}^{M} c_m \mathrm{e}^{-\mathrm{j}\omega m}.$$

So the maximum entropy power spectral density is

$$P(\omega) = \exp\left(\sum_{m=-M}^{M} c_m \mathrm{e}^{-\mathrm{j}\omega m}\right). \tag{4.6.63}$$

Obviously, the key to calculate MEM-2 power spectral density is to estimate the coefficient c_m. In general, c_m is called complex cepstrum.

Substituting Eq. (4.6.63) into the autocorrelation function matching Eq. (4.6.55), there is

$$\frac{1}{2\pi}\int_{-\pi}^{\pi}\exp\left(\sum_{k=-M}^{M} c_k e^{-j\omega k}\right)e^{j\omega m}d\omega = \hat{R}_x(m),\quad m = 0, \pm1, \cdots, \pm M. \tag{4.6.64}$$

This is a nonlinear equation system. Using the Newton-Raphson method, the complex cepstrum coefficient c_k can be obtained from the nonlinear equations according to the given $2M + 1$ sample autocorrelation functions. The biggest disadvantage of MEM-2 lies in its nonlinear calculation.

Nadeu et al.[161] observed through simulation experiments that for the ARMA model whose poles are not close to the unit circle, the spectral estimation performance of MEM-2 is better than MEM-1, but when the poles are close to the unit circle, MEM-1 is better than MEM-2. This result is explained theoretically in reference [218].

The analysis in this section shows that two different maximum entropy power spectrum estimation methods can be obtained by using different definitions of power spectrum entropy: using spectrum entropy definition, two different spectral estimators equivalent to AR and ARMA spectral estimation are obtained under different constraints.

4.7 Pisarenko Harmonic Decomposition Method

The harmonic process is often encountered in many signal processing applications, and it is necessary to determine the frequency and power of these harmonics (collectively referred to as harmonic recovery). The key task of harmonic recovery is the estimation of the number and frequency of harmonics. This section introduces the Pisarenko harmonic decomposition method for harmonic frequency estimation, which lays the theoretical foundation for harmonic recovery.

4.7.1 Pisarenko Harmonic Decomposition

In the Pisarenko harmonic decomposition method, the process composed of p real sine waves is considered

$$x(n) = \sum_{i=1}^{P} A_i \sin(2\pi f_i n + \theta_i). \tag{4.7.1}$$

When phase θ_i is a constant, the above harmonic process is a non-stationary deterministic process. To ensure the stationarity of the harmonic process, it is usually assumed that the phase is a random number uniformly distributed in the range of $[-\pi, \pi]$. At this time, the harmonic process is a random process.

The harmonic process can be described by a difference equation. Consider a single sine wave, for simple calculation, let $x(n) = \sin(2\pi fn + \theta)$. From the trigonometric function identity, we have

$$\sin(2\pi fn + \theta) + \sin[2\pi f(n-2) + \theta] = 2\cos(2\pi f)\sin[2\pi f(n-1) + \theta].$$

If $x(n) = \sin(2\pi fn + \theta)$ is substituted into the above formula, the second-order difference equation is obtained

$$x(n) - 2\cos(2\pi f)x(n-1) + x(n-2) = 0.$$

Calculating the Z transform of the above equation to obtain

$$[1 - 2\cos(2\pi f)z^{-1} + z^{-2}]X(z) = 0.$$

Then the characteristic polynomial is obtained

$$1 - 2\cos(2\pi f)z^{-1} + z^{-2} = 0,$$

which has a pair of conjugate complex roots, i.e.,

$$z = \cos(2\pi f) \pm \mathrm{j}\sin(2\pi f) = \mathrm{e}^{\pm \mathrm{j}2\pi f}.$$

Note that the modulus of the conjugate roots are 1, i.e., $|z_1| = |z_2| = 1$, and the frequency of the sine wave can be determined by the roots, i.e.,

$$f_i = \arctan[\mathrm{Im}(z_i)/\mathrm{Re}(z_i)]/2\pi. \tag{4.7.2}$$

Usually, only positive frequencies are taken. Obviously, if p real sine wave signals have no repetition frequency, these p frequencies should be determined by the roots of the characteristic polynomial

$$\prod_{i=1}^{p}(z - z_i)(z - z_i^*) = \sum_{i=0}^{2p} a_i z^{2p-i} = 0,$$

or

$$1 + a_1 z^{-1} + \cdots + a_{2p-1} z^{-(2p-1)} + z^{-2p} = 0. \tag{4.7.3}$$

It is easy to know that the modules of these roots are all equal to 1. Because all roots exist in the form of conjugate pairs, the coefficients of characteristic polynomials have features of symmetry, i.e.,

$$a_i = a_{2p-1},\ i = 0, 1, \cdots, p. \tag{4.7.4}$$

The difference equation corresponding to Eq. (4.7.3) is

$$x(n) + \sum_{i=1}^{2p} a_i x(n-i) = 0, \tag{4.7.5}$$

which is an AR process without excitation and has exactly the same form as the difference equation of the predictable process introduced in Section 4.3.

Note that the sine wave process is generally observed in the additive white noise $w(n)$, that is, the observation process can be expressed as

$$y(n) = x(n) + w(n) = \sum_{i=1}^{p} A_i \sin(2\pi f_i n + \theta_i) + w(n), \tag{4.7.6}$$

where $w(n) \sim \mathcal{N}(0, \sigma_w^2)$ is the Gaussian white noise, statistically independent of the sine wave signal $x(n)$. By substituting $x(n) = y(n) - w(n)$ into Eq. (4.7.5), the difference equation satisfied by the sine wave process in white noise is obtained

$$y(n) + \sum_{i=1}^{2p} a_i y(n - I) = w(n) + \sum_{i=1}^{2p} a_i w(n - i). \tag{4.7.7}$$

This is a special ARMA process, as not only the AR order is equal to the MA order, but also the AR parameters are exactly the same as the MA parameters.

Now, the normal equation satisfied by the AR parameters of this special ARMA process is derived. So define the following vectors

$$\left.\begin{array}{lcl} \boldsymbol{y} & = & [y(n), y(n-1), \cdots, y(n-2p)]^T \\ \boldsymbol{a} & = & [1, a_1, \cdots, a_{2p}]^T \\ \boldsymbol{w} & = & [w(n), w(n-1), \cdots, w(n-2p)]^T \end{array}\right\}. \tag{4.7.8}$$

Thus, Eq. (4.7.7) can be written as

$$\boldsymbol{y}^T \boldsymbol{a} = \boldsymbol{w}^T \boldsymbol{a}. \tag{4.7.9}$$

Left multiply Eq. (4.7.9) by vector $\boldsymbol{y}$ and take the mathematical expectation to obtain

$$\mathrm{E}\{\boldsymbol{y}\boldsymbol{y}^T\}\boldsymbol{a} = \mathrm{E}\{\boldsymbol{y}\boldsymbol{w}^T\}\boldsymbol{a}, \tag{4.7.10}$$

Let $R_y(k) = \mathrm{E}\{y(n+k)y(n)\}$ to obtain

$$\begin{aligned} \mathrm{E}\{\boldsymbol{y}\boldsymbol{y}^T\} &= \begin{bmatrix} R_y(0) & R_y(-1) & \cdots & R_y(-2p) \\ R_y(1) & R_y(0) & \cdots & R_y(-2p+1) \\ \vdots & \vdots & \vdots & \vdots \\ R_y(2p) & R_y(2p-1) & \cdots & R_y(0) \end{bmatrix} \stackrel{\text{def}}{=} \boldsymbol{R}_y, \\ \mathrm{E}\{\boldsymbol{y}\boldsymbol{w}^T\} &= \mathrm{E}\{(\boldsymbol{x}+\boldsymbol{w})\boldsymbol{w}^T\} = \mathrm{E}\{\boldsymbol{w}\boldsymbol{w}^T\} = \sigma^2 \boldsymbol{I}, \end{aligned}$$

where the assumption that $x(n)$ and $w(n)$ are statistically independent is used. By substituting the above two equations into Eq. (4.7.10), an important normal equation is obtained

$$\boldsymbol{R}_y \boldsymbol{a} = \sigma_w^2 \boldsymbol{a}, \tag{4.7.11}$$

which shows that σ_w^2 is the eigenvalue of the autocorrelation matrix $\boldsymbol{R}_y$ of the observation process $\{y(n)\}$, and the coefficient vector $\boldsymbol{a}$ of the characteristic polynomial is the eigenvector corresponding to this eigenvalue. This is the theoretical basis of the Pisarenko harmonic decomposition method, which enlightens us that the harmonic recovery problem can be transformed into the eigenvalue decomposition of the autocorrelation matrix $\boldsymbol{R}_y$.

When performing Pisarenko decomposition method, it usually starts from an $m \times m$ $(m > 2p)$ autocorrelation matrix $\boldsymbol{R}$. If the minimum eigenvalue of the autocorrelation matrix has multiple degrees, the coefficient vector $\boldsymbol{a}$ will have multiple solutions. The solution is to reduce the dimension of the autocorrelation matrix until it has exactly one minimum eigenvalue. The problem is that the dimension of the autocorrelation matrix is small and the sample autocorrelation functions used are not enough, which will seriously affect the estimation accuracy of the coefficient vector. Although Pisarenko harmonic decomposition establishes the relationship between the coefficient vector of the characteristic polynomial and the eigenvector of the autocorrelation matrix for the first time in theory, it is not an effective harmonic recovery algorithm from the practical effect. In contrast, the ARMA modeling method introduced below is a very effective harmonic recovery method.

4.7.2 ARMA Modeling Method for Harmonic Recovery

As mentioned above, when the harmonic signal is observed in additive white noise, the observation process is a special ARMA random process with exactly the same AR parameters and MA parameters. Because $A(z)$ and $B(z) = A(z)$ have common factors, the modified Yule-Walker equation cannot be directly applied. Now, from another point of view, we establish the normal equation of this special ARMA process.

Multiply $x(n-k)$ on both sides of the difference Eq. (4.7.5) of the AR model without excitation and take the mathematical expectation, then there is

$$R_x(k) + \sum_{i=1}^{2p} a_i R_x(k-i) = 0, \ \forall k. \tag{4.7.12}$$

Note that the harmonic signal $x(n)$ and the additive white noise $w(n)$ are statistically independent, so $R_y(k) = R_x(k) + R_v(k) = R_x(k) + \sigma_w^2 \delta(k)$. Substituting this relation into Eq. (4.7.12), we can get

$$R_y(k) + \sum_{i=1}^{2p} a_i R_y(k-i) = \sigma_w^2 \sum_{i=0}^{2p} a_i \delta(k-i).$$

Obviously, when $k > 2p$, the impulse function $\delta(\cdot)$ in the summation term on the right side of the above equation is equal to 0, so the above equation can be simplified to

$$R_y(k) + \sum_{i=1}^{2p} a_i R_y(k-i) = 0, \ k > 2p. \tag{4.7.13}$$

This is the normal equation obeyed by the special ARMA process of Eq. (4.7.7), which is consistent with the modified Yule-Walker equation of ARMA($2p$, $2p$) process in form.

Similar to the modified Yule-Walker equation, the normal equation (4.7.13) can also form an overdetermined system of equations, which can be solved by the SVD-TLS algorithm.

Algorithm 4.7.1. *ARMA modeling algorithm for harmonic recovery*

Step 1 Using the sample autocorrelation function $\hat{R}_y(k)$ of the observation data, the extended order autocorrelation matrix of the normal equation (4.7.13) *is constructed*

$$\boldsymbol{R}_e = \begin{bmatrix} \hat{R}_y(p_e+1) & \hat{R}_y(p_e) & \cdots & \hat{R}_y(1) \\ \hat{R}_y(p_e+2) & \hat{R}_y(p_e+1) & \cdots & \hat{R}_y(2) \\ \vdots & \vdots & \vdots & \vdots \\ \hat{R}_y(p_e+M) & \hat{R}_y(p_e+M-1) & \cdots & \hat{R}_y(M) \end{bmatrix}, \tag{4.7.14}$$

where $p_e > 2p$, and $M \gg p$;

Step 2 Matrix $\boldsymbol{R}_e$ is regarded as augmented matrix $\boldsymbol{B}$, and SVD-TLS algorithm is used to determine the overall least squares estimation of AR order $2p$ and the coefficient vector $\boldsymbol{a}$;

Step 3 Calculate the conjugate root pairs (z_i, z_i^*), $i = 1, \cdots, p$ of the characteristic polynomial

$$A(z) = 1 + \sum_{i=1}^{2p} a_i z^{-i}; \tag{4.7.15}$$

Step 4 Calculate the frequency of each harmonic using Eq. (4.7.2).

The above algorithm has very good numerical stability due to the use of singular value decomposition and the total least squares method, and the estimation of AR order and parameters also has very high accuracy. For short, it is an effective algorithm for harmonic recovery.

Finally, it should be pointed out that the above results are still applicable when the harmonic signal is complex harmonic, except that the order of the difference equation is not $2p$ but p, and the roots of the characteristic polynomial are no longer conjugate root pairs.

4.8 Extended Prony Method

As early as 1975, Prony proposed a mathematical model using the linear combination of exponential functions to describe uniformly spaced sampling data. Therefore, the traditional Prony method is not a power spectrum estimation technology in the general sense.

After appropriate expansion, the Prony method can be used to estimate rational power spectral density. This section introduces this extended Prony method[123].

The mathematical model adopted by the extended Prony method is a set of p exponential functions with arbitrary amplitude, phase, frequency, and attenuation factor. Its discrete time function is

$$\hat{x}(n) = \sum_{i=1}^{p} b_i z_i^n, \quad n = 0, 1, \cdots, N-1, \tag{4.8.1}$$

and $\hat{x}(n)$ is used as the approximation of $x(n)$. In Eq. (4.8.1), b_i and z_i are assumed as complex numbers, i.e.,

$$b_i = A_i \exp(\mathrm{j}\theta_i), \tag{4.8.2}$$

$$z_i = \exp[(\alpha_i + \mathrm{j}2\pi f_i)\Delta t], \tag{4.8.3}$$

where A_i is the amplitude; θ_i is the phase (in radians); α_i attenuation factor; f_i represents the oscillation frequency; Δt represents the sampling interval. For convenience, let $\Delta t = 1$.

Construct cost function as

$$\epsilon = \sum_{n=0}^{N-1} |x(n) - \hat{x}(n)|^2. \tag{4.8.4}$$

If the square sum of error is minimized, the parameter quadruple $(A_i, \theta_i, \alpha_i, f_i)$ can be solved. However, this requires solving nonlinear equations. Generally, finding a nonlinear solution is an iterative process. For example, see reference [153]. Here, only the linear estimation of parameter quaternion is discussed.

The key of the Prony method is to realize that the fitting of Eq. (4.8.1) is a homogeneous solution of a linear difference equation with constant coefficients. To derive the linear difference equation, define the characteristic polynomial first as

$$\psi(z) = \prod_{i=1}^{p}(z - z_i) = \sum_{i=0}^{p} a_i z^{p-i}, \tag{4.8.5}$$

where $a_0 = 1$. From Eq. (4.8.1), we have

$$\hat{x}(n-k) = \sum_{i=1}^{p} b_i z_i^{n-k}, \quad 0 \le n-k \le N-1.$$

Multiply both sides by a_k and sum to get

$$\sum_{k=0}^{p} a_k \hat{x}(n-k) = \sum_{i=1}^{p} b_i \sum_{k=0}^{p} a_k z_i^{n-k}, \quad p \le n \le N-1.$$

Substitute $z_i^{n-k} = z_i^{n-p} z_i^{p-k}$ to obtain

$$\sum_{k=0}^{p} a_k \hat{x}(n-k) = \sum_{i=1}^{p} b_i z_i^{n-p} \sum_{k=0}^{p} a_k z_i^{p-k} = 0. \tag{4.8.6}$$

Eq. (4.8.6) is equal to 0 because the second summation term is exactly the characteristic polynomial $\psi(z_i) = 0$ of Eq. (4.8.5) at root z_i.

Eq. (4.8.6) means that $\hat{x}(n)$ satisfies the recursive difference equation

$$\hat{x}(n) = -\sum_{i=1}^{p} a_i \hat{x}(n-i), \quad n = 0, 1, \cdots, N-1. \tag{4.8.7}$$

To establish Prony method, the error between the actual measured data $x(n)$ and its approximate value $\hat{x}(n)$ is defined as $e(n)$, i.e.,

$$x(n) = \hat{x}(n) + e(n), \quad n = 0, 1, \cdots, N-1. \tag{4.8.8}$$

Substituting Eq. (4.8.7) into Eq. (4.8.8) yields

$$x(n) = -\sum_{i=1}^{p} a_i x(n-i) + \sum_{i=0}^{p} a_i e(n-i), \quad n = 0, 1, \cdots, N-1. \tag{4.8.9}$$

The difference Eq. (4.8.9) shows that the exponential process in white noise is a special ARMA(p, q) process, which has the same AR and MA parameters, and the excitation noise is the original additive white noise $e(n)$. This is very similar to the complex ARMA(p, q) process in the Pisarenko harmonic decomposition method, except that the root of the characteristic polynomial $\psi(z)$ in the extended Prony method is not constrained by the unit module root (i.e. no attenuation harmonic).

Now, the criterion for the least squares estimation of parameters $a_1, \cdots, a_p$ is to minimize the error square sum $\sum_{n=p}^{N-1} |e(n)|^2$. However, this will lead to a set of nonlinear equations. The linear method of estimating $a_1, \cdots, a_p$ is to define

$$\epsilon(n) = \sum_{i=0}^{p} a_i e(n-i), \quad n = p, \cdots, N-1, \tag{4.8.10}$$

and rewrite Eq. (4.8.9) as

$$x(n) = -\sum_{i=1}^{p} a_i x(n-i) + \epsilon(n), \quad n = 0, 1, \cdots, N-1. \tag{4.8.11}$$

If minimizing $\sum_{n=p}^{N-1} |\epsilon(n)|^2$ instead of $\sum_{n=p}^{N-1} |e(n)|^2$, we can get a set of linear matrix equations as

$$\begin{bmatrix} x(p) & x(p-1) & \cdots & x(0) \\ x(p+1) & x(p) & \cdots & x(1) \\ \vdots & \vdots & \vdots & \vdots \\ x(N-1) & x(N-2) & \cdots & x(N-p-1) \end{bmatrix} \begin{bmatrix} 1 \\ a_1 \\ \vdots \\ a_p \end{bmatrix} = \begin{bmatrix} \epsilon(p) \\ \epsilon(p+1) \\ \vdots \\ \epsilon(N-1) \end{bmatrix}, \tag{4.8.12}$$

or

$$\boldsymbol{Xa} = \boldsymbol{\epsilon}, \tag{4.8.13}$$

for simple expression. The linear least squares method for solving Eq. (4.8.13) is called the extended Prony method.

To minimize the cost function

$$J(\boldsymbol{a}) = \sum_{n=p}^{N-1} |\epsilon(n)|^2 = \sum_{n=p}^{N-1} \left| \sum_{j=0}^{p} a_j x(n-j) \right|^2, \tag{4.8.14}$$

let $\frac{\partial J(\boldsymbol{a})}{\partial a_i} = 0, i = 1, \cdots, p$, then we have

$$\sum_{j=0}^{p} a_j \left[\sum_{n=p}^{N-1} x(n-j)x^*(n-i) \right] = 0, \ i = 1, \cdots, p. \tag{4.8.15}$$

The corresponding minimum error energy is

$$\epsilon_p = \sum_{j=0}^{p} a_j \left[\sum_{n=p}^{N-1} x(n-j)x^*(n) \right]. \tag{4.8.16}$$

Define

$$r(i,j) = \sum_{n=p}^{N-1} x(n-j)x^*(n-i), \ i,j = 0, 1, \cdots, p. \tag{4.8.17}$$

Then Eqs. (4.8.15) and (4.8.16) can be combined together to get the normal equation form of the Prony method

$$\begin{bmatrix} r(0,0) & r(0,1) & \cdots & r(0,p) \\ r(1,0) & r(1,1) & \cdots & r(1,p) \\ \vdots & \vdots & \vdots & \vdots \\ r(p,0) & r(p,1) & \cdots & r(p,p) \end{bmatrix} \begin{bmatrix} 1 \\ a_1 \\ \vdots \\ a_p \end{bmatrix} = \begin{bmatrix} \epsilon_p \\ 0 \\ \vdots \\ 0 \end{bmatrix}. \tag{4.8.18}$$

By solving the above equation, the estimation of the coefficients $a_1, \cdots, a_p$ and minimum error energy ϵ_p can be obtained.

Once the coefficients $a_1, \cdots, a_p$ have been obtained, the roots $z_i, i = 1, \cdots, p$, of the characteristic polynomial

$$1 + a_1 z^{-1} + \cdots + a_p z^{-p}, \tag{4.8.19}$$

can be found. Sometimes, z_i is called the Prony pole.

From now on, Eq. (4.8.1) of the exponential model is simplified to a linear equation with unknown parameter b_i, expressed in matrix form as

$$\boldsymbol{Zb} = \hat{\boldsymbol{x}}, \tag{4.8.20}$$

where

$$\boldsymbol{Z} = \begin{bmatrix} 1 & 1 & \cdots & 1 \\ z_1 & z_2 & \cdots & z_p \\ \vdots & \vdots & \vdots & \vdots \\ z_1^{N-1} & z_2^{N-1} & \cdots & z_p^{N-1} \end{bmatrix}, \tag{4.8.21}$$

$$\boldsymbol{b} = [b_1, b_2, \cdots, b_p]^{\mathrm{T}}, \tag{4.8.22}$$

$$\hat{\boldsymbol{x}} = [\hat{x}(0), \hat{x}(1), \cdots, \hat{x}(N-1)]^{\mathrm{T}}. \tag{4.8.23}$$

Here, $\boldsymbol{Z}$ is an $N \times p$ Vandermonde matrix. Because z_i is different, the columns of Vandermonde matrix $\boldsymbol{Z}$ are linearly independent, that is, it is full column rank. Therefore, the least square solution of Eq. (4.8.20) is

$$\boldsymbol{b} = (\boldsymbol{Z}^{\mathrm{H}}\boldsymbol{Z})^{-1}\boldsymbol{Z}^{\mathrm{H}}\hat{\boldsymbol{x}}. \tag{4.8.24}$$

It is easy to prove

$$\boldsymbol{Z}^{textH}\boldsymbol{Z} = \begin{bmatrix} y_{11} & y_{12} & \cdots & y_{1p} \\ y_{21} & y_{22} & \cdots & y_{2p} \\ \vdots & \vdots & \vdots & \vdots \\ y_{p1} & y_{p2} & \cdots & y_{pp} \end{bmatrix}, \tag{4.8.25}$$

where

$$y_{ij} = \frac{(z_i^* z_j)^N - 1}{(z_i^* z_j) - 1}. \tag{4.8.26}$$

To summarize the above discussion, the extended Prony method can be described as follows.

Algorithm 4.8.1. *Extended Prony Algorithm for Harmonic Recovery*

Step 1 Calculate the sample function $r(i, j)$ using Eq. (4.8.17), and construct the extended order matrix

$$\boldsymbol{R}_e = \begin{bmatrix} r(1,0) & r(1,1) & \cdots & r(1,p_e) \\ r(2,0) & r(2,1) & \cdots & r(2,p_e) \\ \vdots & \vdots & \vdots & \vdots \\ r(p_e,0) & r(p_e,1) & \cdots & r(p_e,p_e) \end{bmatrix}, \quad p_e \gg p; \tag{4.8.27}$$

Step 2 Use Algorithm 4.4.1 (SVD-TLS Algorithm) to determine the effective rank p of matrix $\boldsymbol{R}_e$ and the total least squares estimation of coefficients $a_1, \cdots, a_p$;

Step 3 Find the roots $z_1, \cdots, z_p$ of the characteristic polynomial $a_1, \cdots, a_p$, and use Eq. (4.8.7) to calculate $\hat{x}(n)$, $n = 1, \cdots, N-1$, where $\hat{x}(0) = x(0)$;
Step 4 Calculate the parameters $b_1, \cdots, b_p$ using Eq. (4.8.24)~ Eq. (4.8.26);
Step 5 Use the following equation to calculate the amplitude A_i, phase θ_i, frequency f_i and attenuation factor α_i:

$$\left.\begin{aligned} A_i &= |b_i| \\ \theta_i &= arctan[Im(b_i)/Re(b_i)]/(2\pi\Delta t) \\ \alpha_i &= \ln|z_i|/\Delta t \\ f_i &= arctan[Im(z_i)/Re(z_i)]/(2\pi\Delta t) \end{aligned}\right\}, \quad i = 1, \cdots, p. \tag{4.8.28}$$

With a little generalization, the extended Prony method can be used for power spectrum estimation. From $\hat{x}(n)$, $n = 0, 1, \cdots, N-1$, which is calculated in Step 3, the spectrum can be obtained

$$\hat{X}(f) = \mathcal{F}[\hat{x}(n)] = \sum_{i=1}^{p} A_i \exp(j\theta_i) \frac{2\alpha_i}{\alpha_i^2 + [2\pi(f - f_i)]^2}. \tag{4.8.29}$$

where $\mathcal{F}[\hat{x}(n)]$ represents the Fourier transform of $\hat{x}(n)$. Thus, the Prony power spectrum can be calculated as

$$P_{Prony}(f) = |\hat{X}(f)|^2. \tag{4.8.30}$$

It should be noted that in some cases, the additive noise will seriously affect the estimation accuracy of Prony pole z_i, and the noise will also make the calculation of the attenuation factor have relatively large errors.

Prony harmonic decomposition method is superior to Pisarenko harmonic decomposition method in the following aspects:

(1) Prony method does not need to estimate the sample autocorrelation function;
(2) The estimated variance of frequency and power (or amplitude) given by the Prony method is relatively small;
(3) Prony method only needs to solve two sets of homogeneous linear equations and one time factorization, while the Pisarenko method requires solving characteristic equations.

If p sine wave signals are real, non attenuated, and observed in noise, there is a special variant[103] of the Prony method. At this time, the signal model in Eq. (4.8.1) becomes

$$\hat{x}(n) = \sum_{i=1}^{p} (b_i z_i^n + b_i^* z_i^{*n}) = \sum_{i=1}^{p} A_i \cos(2\pi f_i n + \theta_i), \tag{4.8.31}$$

where $b_i = 0.5A_i e^{j\theta_i}$; $z_i = e^{j2\pi f_i}$. The corresponding characteristic polynomial becomes

$$\psi(z) = \prod_{i=1}^{p} (z - z_i)(z - z_i)^* = \sum_{i=0}^{2p} a_i z^{2p-i} = 0, \tag{4.8.32}$$

where $a_0 = 1$, and a_i is the real coefficient. Since z_i are the unit module roots, they appear in the form of conjugate pairs. Therefore, z_i in Eq. (4.8.32) is still valid after being replaced by z_i^{-1}, i.e.,

$$z^{2p}\psi(z^{-1}) = z^{2p}\sum_{i=0}^{2p} a_i z^{i-2p} = \sum_{k=0}^{2p} a_k z^k = 0. \tag{4.8.33}$$

By comparing Eq. (4.8.32) with Eq. (4.8.33), it can be concluded that: $a_i = a_{2p-i}$, $i = 0, 1, \cdots, p$, and $a_0 = a_{2p} = 1$.

In this case, the corresponding form of Eq. (4.8.11) is

$$\epsilon(n) = \sum_{i=0}^{p} \bar{a}_i[x(n+i) + x(n-i)], \tag{4.8.34}$$

where $\bar{a}_i = a_i, i = 0, 1, \cdots, p-1$, while $\bar{a}_p = \frac{1}{2}a_p$. The factor $\bar{a}_p$ is halved because the factor $\bar{a}_p$ is calculated twice in Eq. (4.8.34).

Let

$$\sum_{n=p}^{N-1} |\epsilon(n)|^2 = \sum_{n=p}^{N-1} \left| \sum_{i=0}^{p} [x(n+i) + x(n-i)] \right|^2 . \tag{4.8.35}$$

By minimizing the square sum of errors in the above equation, a normal equation similar to Eq. (4.8.18) is obtained

$$\sum_{i=0}^{p} \bar{a}_i r(i,j) = 0, \tag{4.8.36}$$

except that the function $r(i,j)$ takes a different form from Eq. (4.8.17), i.e.,

$$r(i,j) = \sum_{n=p}^{N-1} [x(n+j) + x(n-j)][x^*(n+i) + x^*(n-i)]. \tag{4.8.37}$$

The above method of estimating coefficients $\bar{a}_i$ is called Prony spectral line estimation.

Algorithm 4.8.2. *Prony spectral line estimation algorithm*

Step 1 Use Eq. (4.8.37) to calculate function $r(i,j)$, $i, j = 0, 1, \cdots, p_e$, where $p_e \gg p$; Construct matrix $\mathbf{R}_e$ by Eq. (4.8.27);

Step 2 Determine the effective rank p and coefficients $\bar{a}_1, \cdots, \bar{a}_p$ of $\mathbf{R}_e$ using the SVD-TLS algorithm. Let $a_{2p-i} = \bar{a}_i$, $i = 1, \cdots, p-1$ and $a_p = 2\bar{a}_p$;

Step 3 Find the conjugate root pairs (z_i, z_i^*), $i = 1, \cdots, p$ of the characteristic polynomial $1 + a_1 z^{-1} + \cdots + a_{2p} z^{-2p} = 0$;

Step 4 Calculate the frequency of p harmonics

$$f_i = \arctan[Im(z_i)/Re(z_i)]/(2\pi\Delta t).$$

Summary

This chapter introduces some main methods of modern power spectrum estimation from different aspects:

(1) ARMA spectrum estimation is modern spectrum estimation based on the signal difference model;
(2) Burg maximum entropy spectrum estimation is modern spectrum estimation derived from information theory, which is equivalent to AR and ARMA spectrum estimation respectively under different constraints;
(3) Pisarenko harmonic decomposition is a spectrum estimation method with a harmonic signal as a specific object, which transforms the estimation of harmonic frequency into eigenvalue decomposition of signal correlation matrix;
(4) Extended Prony method uses the complex harmonic model to fit complex signals.

Exercises

4.1 Let $\{x(t)\}$ and $\{y(t)\}$ be stationary stochastic processes satisfying the following difference equations

$$x(t) - \alpha x(t-1) = w(t), \quad \{w(t)\} \sim \mathcal{N}(0, \sigma^2),$$
$$y(t) - \alpha y(t-1) = x(t) + u(t), \quad \{u(t)\} \sim \mathcal{N}(0, \sigma^2),$$

where $|\alpha| < 1$; $\{w(t)\}$ and $\{u(t)\}$ are uncorrelated. Calculate the power spectrum of $\{y(t)\}$.

4.2 Assume that the input signal $\{x(t)\}$ is Gaussian white noise with zero mean and power spectrum $P_X(f) = N_0$ and the impulse response of the linear system is

$$h(t) = \begin{cases} e^{-t}, t \geq 0, \\ 0, \text{ others.} \end{cases}$$

Solve the power spectrum and covariance function of the output $y(t) = x(t) * h(t)$.

4.3 It is known that the transfer function of a wireless channel is described by

$$H(f) = Ke^{-j2\pi f\tau_0}, \quad \tau_0 = r/c,$$

where r is the propagation distance and c is the speed of light. Such channel is called a nondispersive channel. Assume that the transmitted signal is

$$x(t) = A\cos(2\pi f_c t + \Phi),$$

where Φ is a random variable uniformly distributed in $[-\pi, \pi]$. Let $y(t)$ be the signal received by the receiver after the transmitted signal $x(t)$ passes through the nondispersive channel. There is Gaussian white noise $n(t)$ at the receiver, with zero mean and the power spectral density N_0. The additive noise $n(t)$ is independent of the transmitted signal $x(t)$. Find the power spectrum $P_y(f)$ of the received signal $y(t)$ of the receiver and the cross power spectrum $P_{xy}(f)$ between the transmitted signal and the received signal.

4.4 The power spectral density of a random signal is

$$P(\omega) = \frac{1.25 + \cos\omega}{1.0625 + 0.5\cos\omega}.$$

If this is regarded as the power spectrum of the output of the linear causal and minimum phase system $H(z)$ excited by white noise with unit power spectrum, solve this linear system $H(z)$.

4.5 The power spectrum of a random signal $x(n)$ is a rational expression of ω

$$P(\omega) = \frac{\omega^2 + 4}{\omega^2 + 1}.$$

If signal $x(n)$ is regarded as the output of linear causal and minimum phase system $H(z)$ excited by a white noise of unit power spectrum, try to determine this system.

4.6 The discrete-time second-order AR process is described by the difference equation

$$x(n) = a_1 x(n-1) + a_2 x(n-2) + w(n),$$

where $w(n)$ is a white noise with zero mean and variance σ_w^2. Prove that the power spectrum of $x(n)$ is

$$P_x(f) = \frac{\sigma_w^2}{1 + a_1^2 + a_2^2 - 2a_1(1-a_2)\cos(2\pi f) - 2a_2\cos(4\pi f)},$$

4.7 The second-order moving average process is defined by

$$x(n) = w(n) + b_1 w(n-1) + b_2 w(n-2), \quad \{w(n)\} \sim \mathcal{N}(0, \sigma^2),$$

where $N(0, \sigma^2)$ represents normal distribution with zero mean and variance σ^2. Find the power spectrum of $x(n)$.

4.8 The difference model of an MA random process is $y(t) = w(t) + 1.5w(y-1) - w(t-2)$, where $\{w(t)\}$ is a Gaussian white noise process with zero mean and variance $\sigma_w^2 = 1$. Find another equivalent MA model of $y(t)$.

4.9 Let $x(t)$ be an unknown random process with zero mean, and the first three values of its autocorrelation function are $R_x(0) = 2$, $R_x(1) = 0$, and $R_x(2) = -1$. In this case, can an ARMA(1,1) model be used to fit it?

4.10 The error power is defined as

$$P_m(r_m) = \frac{1}{2}\mathrm{E}\{|e_{m-1}^f(n) + r_m e_{m-1}^b(n-1)|^2 + |r_m^* e_{m-1}^f(n-1) + e_{m-1}^b(n-1)|^2\}.$$

(1) Calculate $\min_{r_m}[P_m(r_m)]$;

(2) Prove

$$r_m = \frac{-2\sum_{n=m+1}^{N} e_{m-1}^{f}(n)e_{m-1}^{b*}(n-1)}{\sum_{n=m+1}^{N}[|e_{m-1}^{f}(n)|^2 + |e_{m-1}^{b}(n-1)|^2]}, \quad m = 1, 2, \cdots.$$

(3) Prove $|r_m| \le 1$ is constant for $m = 1, 2, \cdots$.

4.11 If the forward and backward prediction errors are defined as

$$e_m^f(n) = \sum_{k=0}^{m} a_m(k)x(n-k),$$

$$e_m^b(n) = \sum_{k=0}^{m} a_{m-k}^{*}(k)x(n-k),$$

Using Burg recurrence equation to prove

(1) $e_m^f(n) = e_{m-1}^f(n) + r_m e_{m-1}^b(n-1)$;

(2) $e_m^b(n) = r_m^{*} e_{m-1}^f(n) + e_{m-1}^b(n-1)$.

4.12 The observation is

$$x(n) = \begin{cases} 1, & n = 0, 1, \cdots, N-1. \\ 0, & \text{others.} \end{cases}$$

Now use the Prony method to model $x(n)$ so that $x(n)$ is the unit impulse response of a linear time invariant filter $H(z)$ with only one pole and one zero. Find the expression of the filter transfer function $H(z)$ and solve $H(z)$ when $n = 21$.

4.13 An observation data vector $\boldsymbol{x} = [1, \alpha, \alpha^2, \cdots, \alpha^N]^{\mathrm{T}}$ is known, where $|\alpha| < 1$. Suppose the Prony method is used to fit the data, and the filter transfer function is $H(z) = \frac{b_0}{1+a_1 z^{-1}}$, find the coefficients a_1 and b_0, and write the specific form of $H(z)$.

4.14 Consider a code division multiple access (CDMA) system in wireless communication, which has K users[207]. User 1 is the desired user. A receiver receives the signals transmitted by all the K users, and the vector form of the received signal is given by

$$\boldsymbol{y}(n) = \sum_{k=1}^{K} \boldsymbol{y}_k(n) = \boldsymbol{h}_1 w_1(n) + \boldsymbol{H}\boldsymbol{w}(n) + \boldsymbol{v}(n),$$

where $w_1(n)$ is the bit signal transmitted by the desired user to be detected, and $\boldsymbol{h}_1$ is the equivalent characteristic waveform vector of the desired user, which is known; $\boldsymbol{H}$ and $\boldsymbol{w}(n)$ are the matrix and interference bit vector composed of characteristic waveform vectors of all the other users (interference users for short). It is assumed that the additive noise of the channel is Gaussian white noise with each noise componentthe of zero mean and the variance σ^2.

(1) Design a minimum variance receiver $\boldsymbol{f}$ to minimize the mean square error between the receiver output $\hat{w}_1(n) = \boldsymbol{f}^{\mathrm{T}}\boldsymbol{y}(n)$ and $w_1(n)$ while meeting the constraint $\boldsymbol{f}^{\mathrm{T}}\boldsymbol{h}_1 = 1$. Find the expression of the minimum variance receiver $\boldsymbol{f}$.

(2) If the desired user's equivalent characteristic waveform vector $\boldsymbol{h}_1$ is $\boldsymbol{h}_1 = \boldsymbol{C}_1\boldsymbol{g}_1$, where

$$\boldsymbol{C}_1 = \begin{bmatrix} c_1(0) & \cdots & 0 \\ \vdots & \vdots & c_1(0) \\ c_1(P-1) & \cdots & \vdots \\ 0 & \cdots & c_1(P-1) \end{bmatrix}, \quad \boldsymbol{g}_1 = \begin{bmatrix} g_1(0) \\ \cdots \\ g_1(L) \end{bmatrix},$$

and $c_1(0), \cdots, c_1(P-1)$ in the above equation is the spreading code of the desired user, and $g_1(l)$ represents the parameter of the l-th transmission path. Design a minimum variance distortionless response (MVDR) beamformer $\boldsymbol{g}$ and prove it to be the generalized eigenvector corresponding to the minimum generalized eigenvalue of the matrix beam $(\boldsymbol{C}_1^{\mathrm{H}}\boldsymbol{R}_y\boldsymbol{C}_1, \boldsymbol{C}_1^{\mathrm{H}}\boldsymbol{C}_1)$.

4.15 Consider the following generalization of Pisarenko harmonic decomposition of M real harmonic signals[127]. Let the dimension of the noise subspace be greater than 1, so the elements of each column vector of matrix $\boldsymbol{V}_n$ spanning noise subspace satisfy

$$\sum_{k=0}^{2M} v_k \mathrm{e}^{\mathrm{j}\omega_i k} = \sum_{k=0}^{2M} v_k \mathrm{e}^{-\mathrm{j}\omega_i k} = 0, \quad 1 \le i \le M.$$

Let $\bar{\boldsymbol{p}} = \boldsymbol{V}_n\boldsymbol{\alpha}$ represent a nondegenerate linear combination of the column vectors of $\boldsymbol{V}_n$. The so-called nondegenerate means that the polynomial $p(z)$ composed of the elements of vector $\bar{\boldsymbol{p}} = [\bar{p}_0, \bar{p}_1, \cdots, \bar{p}_{2M}]^{\mathrm{T}}$ has at least $2M$ order, i.e., $p(z) = \bar{p}_0 + \bar{p}_1 z + \cdots + \bar{p}_{2M} z^{2M}$, $\bar{p}_{2M} \neq 0$. Therefore, this polynomial also satisfies the above equation. This means that all harmonic frequencies can be obtained from the $2M$ roots of the polynomial $p(z)$ on the unit circle. Now we want to select the coefficient vector $\boldsymbol{\alpha}$ satisfying the conditions $p_0 = 1$ and $\sum_{k=1}^{K} p_k^2 = \min$.

(1) Let $\boldsymbol{v}^{\mathrm{T}}$ be the first row of matrix $\boldsymbol{V}_n$, and $\boldsymbol{V}$ be a matrix composed of all the other rows of $\boldsymbol{V}_n$. If $\boldsymbol{p}$ is a vector composed of all the elements of $\bar{\boldsymbol{p}}$ except for the first element, try to prove

$$\boldsymbol{\alpha} = \arg\min \boldsymbol{\alpha}^{\mathrm{T}}\boldsymbol{V}^{\mathrm{T}}\boldsymbol{V}\boldsymbol{\alpha},$$

with constraint $\boldsymbol{V}^{\mathrm{T}}\boldsymbol{\alpha} = 1$.

(2) Using the Lagrange multiplier method to prove that the solution of the constrained optimization problem is

$$\boldsymbol{\alpha} = \frac{(\boldsymbol{V}^{\mathrm{T}}\boldsymbol{V})^{-1}\boldsymbol{v}}{\boldsymbol{v}^{\mathrm{T}}(\boldsymbol{V}^{\mathrm{T}}\boldsymbol{V})^{-1}\boldsymbol{v}},$$

$$\boldsymbol{p} = \frac{\boldsymbol{V}(\boldsymbol{V}^{\mathrm{T}}\boldsymbol{V})^{-1}\boldsymbol{v}}{\boldsymbol{v}^{\mathrm{T}}(\boldsymbol{V}^{\mathrm{T}}\boldsymbol{V})^{-1}\boldsymbol{v}}.$$

4.16 Each element $e_i(t)$ of the additive noise vector $\boldsymbol{e}(t) = [e_1(t), \cdots, e_m(t)]^{\mathrm{T}}$ is a zero mean complex white noise and has the same variance σ^2. Assume that these complex white noises are statistically uncorrelated with each other. Prove that the noise vector satisfies the condition

$$\mathrm{E}\{\boldsymbol{e}(n)\} = \mathbf{0}, \quad \mathrm{E}\{\boldsymbol{e}(n)\boldsymbol{e}^{\mathrm{T}}(n)\} = \sigma^2\boldsymbol{I}, \quad \mathrm{E}\{\boldsymbol{e}(n)\boldsymbol{e}^{\mathrm{T}}(n)\} = \boldsymbol{O},$$

where $\boldsymbol{I}$ and $\boldsymbol{o}$ are the identity matrix and zero matrix, respectively.

4.17 Assume that the simulated observation data are generated by

$$x(n) = \sqrt{20}\sin(2\pi 0.2n) + \sqrt{2}\sin(2\pi 0.213n) + w(n),$$

where $w(n)$ is a white Gaussian noise with zero mean and variance 1, and $n = 1, \cdots, 128$. Using the TLS and SVD-TLS methods to estimate the AR parameters of the ARMA model of the observation data and estimate the frequencies of the sine waves. In computer simulation, when running the least square TLS method, the AR order is set to 4 and 6 respectively; When running the SVD-TLS method, it is assumed that the AR order is unknown. The computer simulation will run at least 20 times independently. It is required to complete the computer simulation experiment report, which mainly includes:

(1) The basic theory and method of harmonic recovery;
(2) The statistical results (mean and deviation) of the AR parameters and the sine wave frequency estimates;
(3) The discussion of the advantages and precautions of using SVD to determine the effective rank of the sample autocorrelation matrix.

5 Adaptive Filter

Filter is a device that extracts signals from noisy observation data in the form of physical hardware or computer software. The filter can realize the basic tasks of information processing such as filtering, smoothing and prediction. If the output of the filter is a linear function of the input, it is called a linear filter; Otherwise, it is called a nonlinear filter. If the impulse response of the filter is infinite, it is called infinite impulse response (IIR) filter , and the filter with limited impulse response is called finite impulse response (FIR) filter . If the filter is implemented in the time domain, frequency domain, or spatial domain, it is called time domain filter, frequency domain filter or spatial domain filter respectively. The courses "Signals and Systems" and "Digital Signal Processing" mainly discuss filters in the frequency domain. This chapter focuses on time domain filters.

In real-time signal processing, it is often hoped that the filter can track and adapt to the dynamic changes of the system or environment when realizing tasks such as filtering, smoothing, or prediction, which requires that the parameters of the (time domain) filter can be changed or updated with time simply, because the complex operation does not meet the needs of real-time fast processing. In other words, the parameters of the filter should be updated adaptively in a recursive way. Such filters are collectively referred to as adaptive filters.

This chapter will first discuss two common optimal filters – the matched filter and Wiener filter, and then focus on various adaptive implementation algorithms of Kalman filter and Wiener filter. For any kind of adaptive filter, the adaptive algorithm itself is important, but the statistical performance of the algorithm, especially the ability to track the dynamic changes of the system or environment, and its application in practice are also important. They constitute the main content of this chapter.

5.1 Matched Filter

Roughly speaking, a filter is a signal extractor. Its function is to extract the original signal from the signal polluted by noise. Of course, the signal extraction should meet some optimization criteria. There are two optimal design criteria for continuous time filters. One criterion is to make the output of the filter reach the maximum signal-to-noise ratio, called matched filter. The other is to minimize the mean square estimation error of the output filter, called Wiener filter. This section introduces the matched filter and Section 5.2 discusses continuous time Wiener filter.

https://doi.org/10.1515/9783110475562-005

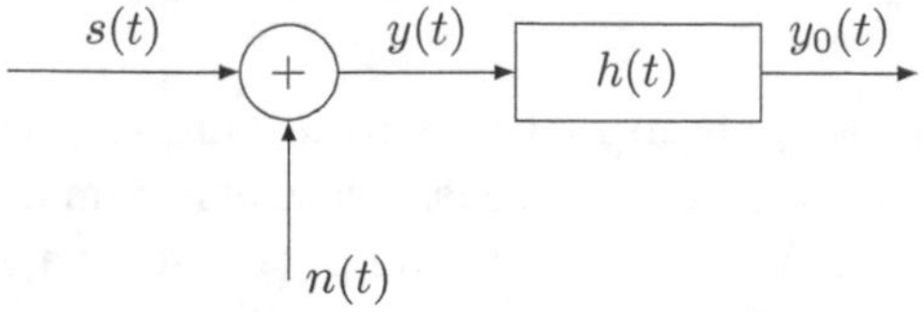

Fig. 5.1.1: Schematic illustration for the linear continuous time filter

5.1.1 Matched Filter

Consider the received or observed signal

$$y(t) = s(t) + n(t), \quad -\infty < t < \infty, \tag{5.1.1}$$

where $s(t)$ is a known signal; $n(t)$ is stationary noise with zero mean. Note that the additive noise $n(t)$ can be white or colored.

Let $h(t)$ be the time invariant impulse response function of the filter. Our goal is to design the impulse response function $h(t)$ of the filter to maximize the signal-to-noise ratio of the filter output. Fig 5.1.1 shows the structure of the linear continuous time filter. From the figure, the output of the filter can be expressed as

$$\begin{aligned} y_O(t) &= \int_{-\infty}^{\infty} h(t-\tau)y(\tau)\mathrm{d}\tau \\ &= \int_{-\infty}^{\infty} h(t-\tau)s(\tau)\mathrm{d}\tau + \int_{-\infty}^{\infty} h(t-\tau)n(\tau)\mathrm{d}\tau \\ &\overset{\text{def}}{=} s_O(t) + n_O(t), \end{aligned} \tag{5.1.2}$$

where

$$s_O(t) \overset{\text{def}}{=} \int_{-\infty}^{\infty} h(t-\tau)s(\tau)\mathrm{d}\tau, \tag{5.1.3}$$

$$n_O(t) \overset{\text{def}}{=} \int_{-\infty}^{\infty} h(t-\tau)n(\tau)\mathrm{d}\tau, \tag{5.1.4}$$

$s_O(t)$ and $n_O(t)$ are respectively the signal component and noise component in the filter output. It can be seen from the above definition that the signal component and the noise component are actually the output of the signal and additive noise after passing through the filter, respectively.

The output signal-to-noise ratio of the filter at time t is defined as

$$\left(\frac{S}{N}\right)^2 \overset{\text{def}}{=} \frac{\text{Output instantaneous signal power at } t = T_0}{\text{Average power of output noise}} = \frac{s_0^2(T_0)}{\mathrm{E}\{n_0^2(t)\}}. \tag{5.1.5}$$

Apply the Parseval theorem to Eq. (5.1.3) to get

$$\int_{-\infty}^{\infty} x^*(\tau)y(\tau)\mathrm{d}\tau = \frac{1}{2\pi}\int_{-\infty}^{\infty} X^*(\omega)Y(\omega)\mathrm{d}\omega, \tag{5.1.6}$$

then the output signal can be rewritten as

$$s_0(t) = \frac{1}{2\pi}\int_{-\infty}^{\infty} H(\omega)S(\omega)\mathrm{e}^{\mathrm{j}\omega t}\mathrm{d}\omega, \tag{5.1.7}$$

where

$$H(\omega) = \int_{-\infty}^{\infty} h(t)\mathrm{e}^{-\mathrm{j}\omega t}\mathrm{d}t, \tag{5.1.8}$$

$$S(\omega) = \int_{-\infty}^{\infty} s(t)\mathrm{e}^{-\mathrm{j}\omega t}\mathrm{d}t, \tag{5.1.9}$$

and they are the transfer function of the filter and the signal spectrum.

The instantaneous power of the output signal at $t = T_0$ can be obtained from Eq. (5.1.7) as

$$s_0^2(T_0) = \left|\frac{1}{2\pi}\int_{-\infty}^{\infty} H(\omega)S(\omega)\mathrm{e}^{\mathrm{j}\omega T_0}\mathrm{d}\omega\right|^2, \tag{5.1.10}$$

and the average power of output noise can be obtained from Eq. (5.1.4) as

$$\mathrm{E}\{n_0^2(t)\} = \mathrm{E}\left\{\left[\int_{-\infty}^{\infty} h(t-\tau)n(\tau)\mathrm{d}\tau\right]^2\right\}. \tag{5.1.11}$$

Let $P_n(\omega)$ be the power spectral density of the additive noise $n(t)$, then the power spectral density of the output noise is

$$P_{n_0}(\omega) = |H(\omega)|^2 P_n(\omega), \tag{5.1.12}$$

and the average power of the output noise can be written as

$$\mathrm{E}\{n_0^2(t)\} = \frac{1}{2\pi}\int_{-\infty}^{\infty} P_{n_0}(\omega)\mathrm{d}\omega = \frac{1}{2\pi}\int_{-\infty}^{\infty} |H(\omega)|^2 P_n(\omega)\mathrm{d}\omega. \tag{5.1.13}$$

Substituting Eqs. (5.1.10) and (5.1.13) into the output signal-to-noise ratio defined by Eq. (5.1.5), yields

$$\left(\frac{S}{N}\right)^2 = \frac{\left|\frac{1}{2\pi}\int_{-\infty}^{\infty} H(\omega)S(\omega)\mathrm{e}^{\mathrm{j}\omega T_0}\mathrm{d}\omega\right|^2}{\frac{1}{2\pi}\int_{-\infty}^{\infty} |H(\omega)|^2 P_n(\omega)\mathrm{d}\omega}$$

$$= \frac{1}{2\pi}\frac{\left|\int_{-\infty}^{\infty}\left(H(\omega)\sqrt{P_n(\omega)}\right)\left(\frac{S(\omega)}{\sqrt{P_n(\omega)}}\right)\mathrm{e}^{\mathrm{j}\omega T_0}\mathrm{d}\omega\right|^2}{\int_{-\infty}^{\infty} |H(\omega)|^2 P_n(\omega)\mathrm{d}\omega}. \tag{5.1.14}$$

Reviewing Cauchy-Schwartz inequality

$$\left|\int_{-\infty}^{\infty} f(x)g(x)\mathrm{d}x\right|^2 \le \left(\int_{-\infty}^{\infty} |f(x)|^2\mathrm{d}x\right)\left(\int_{-\infty}^{\infty} |g(x)|^2\mathrm{d}x\right), \tag{5.1.15}$$

the equal sign holds if and only if $f(x) = cg^*(x)$, where c is an arbitrary complex constant. Without loss of generality, taking $c = 1$ in the following.

In Eq. (5.1.15), let

$$f(x) = H(\omega)\sqrt{P_n(\omega)} \quad \text{and} \quad g(x) = \frac{S(\omega)}{\sqrt{P_n(\omega)}}\mathrm{e}^{\mathrm{j}\omega T_0}.$$

Apply these substitutions to Eq. (5.1.14) to get

$$\left(\frac{S}{N}\right)^2 \le \frac{1}{2\pi}\frac{\int_{-\infty}^{\infty} |H(\omega)|^2 P_n(\omega)\mathrm{d}\omega \int_{-\infty}^{\infty}\frac{|S(\omega)|^2}{\sqrt{P_n(\omega)}}\mathrm{d}\omega}{\int_{-\infty}^{\infty} |H(\omega)|^2 P_n(\omega)\mathrm{d}\omega}, \tag{5.1.16}$$

or

$$\left(\frac{S}{N}\right)^2 \le \frac{1}{2\pi}\int_{-\infty}^{\infty}\frac{|S(\omega)|^2}{P_n(\omega)}\mathrm{d}\omega, \tag{5.1.17}$$

for simplification. Denote the filter transfer function when the equal sign in Eq. (5.1.17) is true as $H_{\text{opt}}(\omega)$. From the condition for equality of Cauchy-Schwartz inequality, we have

$$H_{\text{opt}}(\omega)\sqrt{P_n(\omega)} = \left[\frac{S(\omega)}{\sqrt{P_n(\omega)}}\right]^* \mathrm{e}^{-\mathrm{j}\omega T_0} = \frac{S^*(\omega)}{\sqrt{P_n^*(\omega)}}\mathrm{e}^{-\mathrm{j}\omega T_0},$$

that is,

$$H_{\text{opt}}(\omega) = \frac{S(-\omega)}{P_n(\omega)}\mathrm{e}^{-\mathrm{j}\omega T_0}, \tag{5.1.18}$$

where $S(-\omega) = S^*(\omega)$.

When the transfer function of the filter takes the form of Eq. (5.1.18), the equal sign of Eq. (5.1.17) holds, i.e., the maximum signal-to-noise ratio of the filter output is

$$\text{SNR}_{\max} = \frac{1}{2\pi}\int_{-\infty}^{\infty}\frac{|S(\omega)|^2}{P_n(\omega)}\mathrm{d}\omega. \tag{5.1.19}$$

In the sense of maximizing the output signal-to-noise ratio, the filter defined in Eq. (5.1.18) is the optimal linear filter. Therefore, the transfer function $H_{\text{opt}}(\omega)$ shown in Eq. (5.1.18) is the transfer function of the optimal linear filter.

Discussing the following two cases of additive noise will help us to understand the linear optimal filter more deeply.

1. Optimal Filter in the Case of White Noise – Matched Filter

When the additive noise $n(t)$ is white noise with zero mean and unit variance, because its power spectral density $P_n(\omega) = 1$, Eq. (5.1.18) is simplified as

$$H_0(\omega) = S(-\omega)e^{-j\omega T_0}. \tag{5.1.20}$$

From the above equation, there is $|H_0(\omega)| = |S^*(\omega)| = |S(\omega)|$. In other words, when the filter reaches the maximum output signal-to-noise ratio, the amplitude frequency characteristic $|H(\omega)|$ of the filter is equal to the amplitude frequency characteristic $|S(\omega)|$ of the signal $s(t)$, or they match. Therefore, the linear filter $H_0(\omega)$ that maximizes the signal-to-noise ratio in the case of white noise is often called a matched filter.

Perform inverse Fourier transform on both sides of Eq. (5.1.20) to obtain the impulse response of matched filter $H_0(\omega)$

$$h_0(t) = \int_{-\infty}^{\infty} S(-\omega)e^{-j\omega T_0}e^{j\omega t}d\omega.$$

By variable substitution $\omega' = -\omega$, the above formula becomes

$$h_0(t) = \int_{-\infty}^{\infty} S(\omega')e^{j\omega'(T_0-t)}d\omega' = s(T_0 - t). \tag{5.1.21}$$

That is, the impulse response $h_0(t)$ of the matched filter is a mirror signal of the signal $s(t)$.

2. Optimal Filter in the Case of Colored Nise – Generalized Matched Filter

Let $w(t)$ be a filter with transfer function

$$W(\omega) = \frac{1}{\sqrt{P_n(\omega)}}. \tag{5.1.22}$$

Then, when the colored noise $n(t)$ is filted by $W(\omega)$, the power spectral density of the output signal $\tilde{n}(t)$ is

$$P_{\tilde{n}}(\omega) = |W(\omega)|^2 P_n(\omega) = 1. \tag{5.1.23}$$

Therefore, the filter $W(\omega)$ shown in Eq. (5.1.22) is a whitening filter for colored noise. Consequently, Eq. (5.1.18) can be written as

$$H_{\text{opt}}(\omega) = \frac{S^*(\omega)}{P_n(\omega)}e^{-j\omega T_0} = W(\omega)[S^*(\omega)W^*(\omega)e^{-j\omega T_0}]. \tag{5.1.24}$$

Let $\tilde{S}(\omega) = S(\omega)W(\omega)$, and perform inverse Fourier transform on both sides to obtain $\tilde{s}(t) = s(t) * w(t)$, that is, $\tilde{s}(t)$ is the filtering result of the original signal $s(t)$ by using the whitening filter $w(t)$, and $H_0(\omega) = \tilde{S}^*(\omega)e^{-j\omega T_0}$ can be regarded as the filter extracting the signal from the filtered observation process $\tilde{y}(t)$, where

$$\tilde{y}(t) = y(t) * w(t) = [s(t) + n(t)] * w(t) = s(t) * w(t) + n(t) * w(t) = \tilde{s}(t) + \tilde{n}(t).$$

The difference is that $\tilde{n}(t) = n(t) * w(t)$ has become white noise, so $H_0(\omega)$ is a matched filter. Therefore, the linear filter $H_{\text{opt}}(\omega)$ that maximizes the signal-to-noise ratio in the case of colored noise is formed by cascading the whitening filter $W(\omega)$ and the matched filter $H_0(\omega)$. In view of this, $H_{\text{opt}}(\omega)$ is often called generalized matched filter, and its working principle is shown in Fig. 5.1.2.

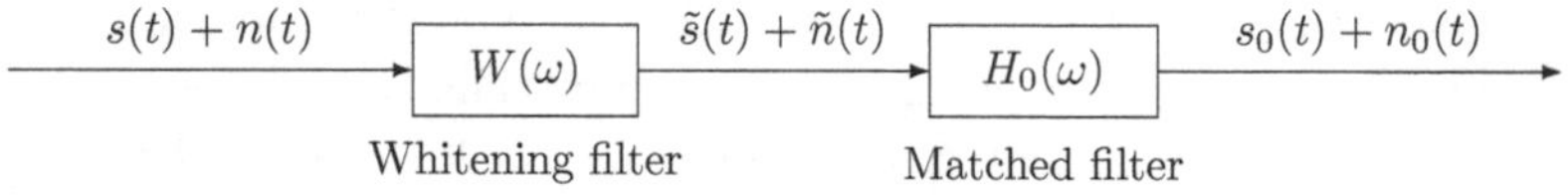

Fig. 5.1.2: Working principle of the generalized matched filter

Example 5.1.1 Given that the original signal is a harmonic process

$$s(t) = A\cos(2\pi f_c t),\ f_c = \frac{1}{T},$$

while the additive noise is colored noise, and its power spectrum is

$$P_n(f) = \frac{1}{1 + 4\pi^2 f^2}.$$

Solve the impulse response of the optimal linear filter with the maximum signal-to-noise ratio.

Solution. Since

$$s(t) = A\cos(2\pi f_c t) = \frac{A}{2}[e^{j2\pi f_c t} + e^{-j2\pi f_c t}],$$

the spectrum of the harmonic signal $s(t)$ is

$$S(f) = \frac{A}{2}\int_{-\infty}^{\infty}[e^{-j2\pi(f-f_c)t} + e^{-j2\pi(f+f_c)t}]dt = \frac{A}{2}[\delta(f + f_c) + \delta(f - f_c)].$$

From the above equation, we have

$$S(-f) = \frac{A}{2}[\delta(-f - f_c) + \delta(-f + f_c)] = \frac{A}{2}[\delta(f + f_c) + \delta(f - f_c)].$$

Here we use the properties of δ function $\delta(-x) = \delta(x)$. Therefore, the optimal linear filter $H_{\text{opt}}(f)$ that maximizes the signal-to-noise ratio is given by Eq. (5.1.18), and its impulse response is determined by Eq. (5.1.21) as

$$\begin{aligned} h_{\text{opt}}(t) &= \frac{A}{2}\int_{-\infty}^{\infty} \frac{[\delta(f+f_c)+\delta(f-f_c)]}{(1+4\pi^2 f^2)^{-1}} \mathrm{e}^{-\mathrm{j}2\pi f(T-t)}\mathrm{d}f \\ &= \frac{A}{2}(1+4\pi^2 f_c^2)[\mathrm{e}^{\mathrm{j}2\pi f_c(T-t)} + \mathrm{e}^{-\mathrm{j}2\pi f_c(T-t)}] \\ &= A(1+4\pi^2 f_c^2)\cos[2\pi f_c(T-t)]. \end{aligned}$$

□

5.1.2 Properties of Matched Filter

Since matched filter has important applications in many engineering problems, it is necessary to understand its important properties.

Property 1 Among all linear filters, the output signal-to-noise ratio of the matched filter is the largest, and $\text{SNR}_{\max} = \frac{E_s}{N_0/2}$, which is independent of the waveform of the input signal and the distribution characteristics of the additive noise.

Property 2 The instantaneous power of the output signal of the matched filter reaches the maximum at $t = T_0$.

Property 3 The time T_0 when the output signal-to-noise ratio of the matched filter reaches the maximum should be selected to be equal to the duration T of the original signal $s(t)$.

Property 4 The matched filter is adaptive to the delay signal with the same waveform but different amplitude.

Property 5 Matched filter has no adaptability to frequency shift signal.

Let $s_2(t)$ be the frequency shift signal of $s(t)$, that is, $S_2(\omega) - S(\omega + \omega_\alpha)$. For example, $S(\omega)$ represents the frequency spectrum of radar fixed target echo signal, $S_2(\omega)$ represents the frequency spectrum of moving target echo with radial velocity, and ω_α is called Doppler frequency shift. From Eq. (5.1.18), the transfer function of the matched filter corresponding to signal $s_2(t)$ is

$$H_2(\omega) = S^*(\omega + \omega_\alpha)\mathrm{e}^{-\mathrm{j}\omega T_0}.$$

Let $\omega' = \omega + \omega_\alpha$, then

$$H_2(\omega') = S^*(\omega')\mathrm{e}^{-\mathrm{j}\omega' T_0 + \mathrm{j}\omega_\alpha T_0} = H(\omega')\mathrm{e}^{\mathrm{j}\omega_\alpha T_0}.$$

It can be seen that the transfer functions of the matched filter of the original signal $s(t)$ and the frequency shift signal $s_2(t)$ are different, i.e., the matched filter has no adaptability to the frequency shift signal.

Note 1 If $T_0 < T$, the obtained matched filter will not be physically realizable. At this time, if the matched filter is approximated by a physically realizable filter, its output signal-to-noise ratio at $T_0 = T$ will not be maximum.

Note 2 If the signal becomes very small from time T_0, i.e., the matched filter can be designed using the signal up to time T_0, which is a quasi-optimal linear filter.

5.1.3 Implementation of Matched Filter

If the precise structure of signal $s(t)$ is known, the impulse response of the matched filter can be directly determined by Eq. (5.1.21), so as to realize matched filtering. However, in many practical applications, only the signal power spectrum $P_s(\omega) = |S(\omega)|^2$ can be known. In such cases, it is necessary to separate the spectral expression $S(\omega)$ of the signal from the power spectrum, and then use Eq. (5.1.20) to design the transfer function of the matched filter.

In addition, when designing a whitening filter with colored noise, it is often the case that only the power spectrum $P_n(\omega) = |N(\omega)|^2$ of the noise is known. In order to design the whitening filter $W(\omega) = \frac{1}{N(\omega)} = \frac{1}{\sqrt{P_n(\omega)}}$, it is also necessary to decompose the noise power spectrum to obtain the noise spectrum $N(\omega) = \sqrt{P_n(\omega)}$. The process of obtaining spectrum from power spectrum is called factorization of power spectrum, also called spectral decomposition for short.

For any stationary signal $x(t)$, its power spectral density $P_x(\omega) = |X(\omega)|^2$ is generally a rational function, expressed as

$$P_x(\omega) = \sigma^2 \frac{(\omega + z_1) \cdots (\omega + z_n)}{(\omega + p_1) \cdots (\omega + p_m)}, \tag{5.1.25}$$

where $z_i, i = 1, \cdots, n$, and $p_j, j = 1, \cdots, m$, are respectively called the zeros and poles of power spectrum. Usually, $n < m$ is assumed. In addition, any zero and the poles are irreducible.

The power spectrum is a nonnegative real and even function, i.e.,

$$P_x(\omega) = P_x^{\star}(\omega). \tag{5.1.26}$$

It can be seen that the zeros and poles of $P_x(\omega)$ appear in conjugate pairs certainly. Therefore, the power spectrum can always be written as

$$P_x(\omega) = \left[\alpha \frac{(\mathrm{j}\omega + \alpha_1) \cdots (\mathrm{j}\omega + \alpha_q)}{(\mathrm{j}\omega + \beta_1) \cdots (\mathrm{j}\omega + \beta_q)}\right] \left[\alpha \frac{(-\mathrm{j}\omega + \alpha_1) \cdots (-\mathrm{j}\omega + \alpha_q)}{(-\mathrm{j}\omega + \beta_1) \cdots (-\mathrm{j}\omega + \beta_q)}\right]. \tag{5.1.27}$$

Denote $P_x^+(\omega)$ as the factor formed by the zeros and poles of $P_x(\omega)$ on the left half plane, and $P_x^-(\omega)$ as factor formed by the zeros and poles on the right half plane, and divide the zeros and poles of $P_x(\omega)$ on the axis in half to $P_x^+(\omega)$ and $P_x^-(\omega)$. In this way, the power spectrum $P_x(\omega)$ can be factorized into

$$P_x(\omega) = P_x^+(\omega) P_x^-(\omega), \tag{5.1.28}$$

which is called spectrum decomposition.

To make the matched filter physically realizable, just take

$$S(\omega) = P_s^+(\omega), \tag{5.1.29}$$

and substitute it into Eq. (5.1.20). Similarly, by selecting

$$W(\omega) = \frac{1}{P_n^+(\omega)}, \tag{5.1.30}$$

a physically realizable whitening filter can be obtained.

5.2 Continuous Time Wiener Filter

In the matched filter, the receiver must know and store the precise structure or power spectrum of the signal, and the integral interval must be synchronized with the interval in which the signal takes non-zero value. Unfortunately, sometimes it is difficult to know the structure or power spectrum of the signal alone, and once the signal propagation delay, phase drift, or frequency drift occurs in the transmission process, the synchronization between the integral interval and the signal interval will also lead to error. In these cases, the application of matched filter is difficult to obtain satisfactory results, or even impossible. It is necessary to find other linear optimal filters.

Since the observation data $y(t) = s(t) + n(t)$, using filter $H(\omega)$, the estimation of signal $s(t)$ can be obtained

$$\hat{s}(t) = \int_{-\infty}^{\infty} h(t-\tau)y(\tau)\mathrm{d}\tau = \int_{-\infty}^{\infty} h(\tau)y(t-\tau)\mathrm{d}\tau. \tag{5.2.1}$$

Recall that it has been pointed out in Chapter 2 (Parameter Estimation Theory) that the estimation error $s(t) - \hat{s}(t)$ is a random variable and is not suitable for evaluating the performance of a parameter estimator or filter. Whereas different from the estimation error, the mean square error is a deterministic quantity and is one of the main measures of filter performance.

Consider minimizing the mean square error

$$J = \mathrm{E}\{[s(t) - \hat{s}(t)]^2\} = \mathrm{E}\left\{\left[s(t) - \int_{-\infty}^{\infty} h(\tau)y(t-\tau)\mathrm{d}\tau\right]^2\right\}, \tag{5.2.2}$$

which is the minimum mean square error (MMSE) criterion. Thus, the impulse response of the linear optimal filter can be expressed as

$$h_{\mathrm{opt}}(t) = \arg\min_{h(t)} \mathrm{E}\left\{\left[s(t) - \int_{-\infty}^{\infty} h(\tau)y(t-\tau)\mathrm{d}\tau\right]^2\right\}, \tag{5.2.3}$$

and its Fourier transform is the frequency response of the linear optimal filter.

The linear optimal filter

$$H_{\text{opt}}(\omega) = \frac{P_{sy}(\omega)}{P_{yy}(\omega)}, \tag{5.2.4}$$

is a noncausal Wiener filter, since the filter impulse response $h_{\text{opt}}(t)$ takes value in time interval $(-\infty, +\infty)$. Note that the noncausal Wiener filter is physically unrealizable.

Any noncausal linear system can be divided into causal and anticausal parts. The causal part is physically realizable, and the anticausal part is physically unrealizable. Therefore, it is considered that the causal part is separated from a noncausal Wiener filter to obtain a physically realizable causal Wiener filter.

Given

$$H(\omega) = \frac{P_{sy}(\omega)}{P_{yy}(\omega)}. \tag{5.2.5}$$

In general, it is difficult to separate the causal part $H_{\text{opt}}(\omega) = \sum_{k=0}^{\infty} h(k)\mathrm{e}^{-\mathrm{j}\omega k}$ from $H(\omega) = \sum_{k=-\infty}^{\infty} h(k)\mathrm{e}^{-\mathrm{j}\omega k}$. However, if the power spectrum $P_{yy}(\omega)$ is a rational function of ω, the causal Wiener filter $H_{\text{opt}}(\omega)$ can be easily obtained.

Firstly, the rational power spectrum $P_{yy}(\omega)$ is decomposed into

$$P_{yy}(\omega) = A_{yy}^{+}(\omega)A_{yy}^{-}(\omega), \tag{5.2.6}$$

where, the zeros and poles of $A_{yy}^{+}(\omega)$ are all located in the left half plane, while the zeros and poles of $A_{yy}^{-}(\omega)$ are all located in the right half plane, and the zeros and poles located on the ω axis are divided into $A_{yy}^{+}(\omega)$ and $A_{yy}^{-}(\omega)$ in half.

Then, further decomposition can be performed to obtain

$$\frac{P_{sy}(\omega)}{A_{yy}^{-}(\omega)} = B^{+}(\omega) + B^{-}(\omega), \tag{5.2.7}$$

where the zeros and poles of $B^{+}(\omega)$ are all located in the left half plane, while the zeros and poles of $B^{-}(\omega)$ are all located in the right half plane, and the zeros and poles located on the ω axis are also divided into $B^{+}(\omega)$ and $B^{-}(\omega)$ in half.

Finally, rewrite Eq. (5.2.5) as

$$H(\omega) = \frac{P_{sy}(\omega)}{A_{yy}^{+}(\omega)A_{yy}^{-}(\omega)} = \frac{1}{A_{yy}^{+}(\omega)}\frac{P_{sy}(\omega)}{A_{yy}^{-}(\omega)} = \frac{1}{A_{yy}^{+}(\omega)}[B^{+}(\omega) + B^{-}(\omega)], \tag{5.2.8}$$

then

$$H_{\text{opt}}(\omega) = \frac{B^{+}(\omega)}{A_{yy}^{+}(\omega)}, \tag{5.2.9}$$

contains only the zeros and poles of the left half plane, so obviously it is physically realizable.

To summarize the above discussions, when $P_{yy}(\omega)$ is a rational power spectrum, the design algorithm of causal Wiener filter is as follows.

Algorithm 5.2.1. *Causal Wiener filter design algorithm 1*

Step 1 Perform spectral decomposition of $P_{yy}(\omega)$ according to Eq. (5.2.6);
Step 2 Calculate Eq. (5.2.7);
Step 3 Get the transfer function $H_{opt}(\omega)$ of causal Wiener filter using Eq. (5.2.9).

If $z = e^{j\omega}$ and the power spectral density is written as $P_{sy}(z)$ and $P_{yy}(z)$, the above algorithm can be easily extended as follows.

Algorithm 5.2.2. *Causal Wiener filter design algorithm 2*
Step 1 Perform spectral decomposition of $P_{yy}(z)$ to obtain

$$P_{yy}(z) = A^{+}_{yy}(z)A^{-}_{yy}(z), \tag{5.2.10}$$

where the zeros and poles of $A^{+}_{yy}(z)$ are all located inside the unit circle, while the zeros and poles of $A^{-}_{yy}(z)$ are all located outside the unit circle.
Step 2 Calculate

$$\frac{P_{sy}(z)}{A^{-}_{yy}(z)} = B^{+}(z) + B^{-}(z), \tag{5.2.11}$$

where the zeros and poles of $B^{+}(z)$ are all located inside the unit circle, while the zeros and poles of $B^{-}(z)$ are all located outside the unit circle.
Step 3 Determine the transfer function $H_{opt}(z)$ of the causal Wiener filter by

$$H_{opt}(z) = \frac{B^{+}(z)}{A^{+}_{yy}(z)}. \tag{5.2.12}$$

It should be noted that the decomposition of Eq. (5.2.10) is the factorization of the power spectrum, while the decomposition of Eq. (5.2.11) is the decomposition of the positive and negative frequency parts of the power spectrum.

5.3 Optimal Filtering Theory and Wiener Filter

The continuous time Wiener filter was discussed in Section 5.2. In digital signal processing, it is desirable to obtain discrete time filters so that they can be implemented with digital hardware or computer software. Therefore, it is necessary to discuss the optimal filtering of discrete time signals.

5.3.1 Linear Optimal Filter

Consider the linear discrete time filter shown in Fig. 5.3.1. The input of the filter includes infinite time series $u(0), u(1), \cdots$, and the impulse response of the filter is an also infinite sequence as $w_0, w_1, \cdots$. Let $y(n)$ represent the output of the filter at discrete time n, hoping that it is an estimate of the desired response $d(n)$.

The estimation error $e(n)$ is defined as the difference between the expected response $d(n)$ and the filter output $y(n)$, i.e., $e(n) = d(n) - y(n)$. The requirement for the filter is

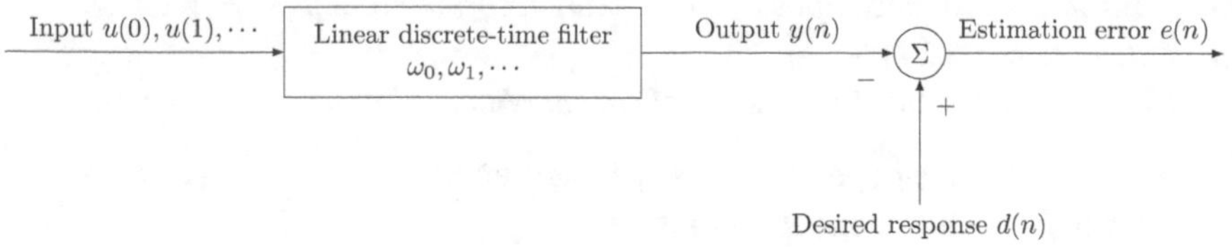

Fig. 5.3.1: Schematic illustration for the linear discrete time filter

to make the estimation error as small as possible in a certain statistical sense. For this purpose, the filter is subject to the following constraints:
(1) The filter is linear (on the one hand, it is to prevent the signal from distortion after passing through the filter, on the other hand, it is to facilitate the mathematical analysis of the filter);
(2) The filter is discrete in time domain, which will enable the filter to be implemented by digital hardware or software.

According to whether the impulse response is finite or infinite, the linear discrete time filter is divided into finite impulse response (FIR) filter and infinite impulse response (IIR) filter. Since FIR filter is a special case of IIR filter, IIR filter is discussed here.

The filter whose estimation error is as small as possible in a statistical sense is called the optimal filter in this statistical sense. So, how to design the criteria of statistical optimization? The most common criterion is to minimize a cost function.

There are many forms of the cost function, and the most typical forms are:
(1) The mean square value of the estimation error;
(2) The expected value of the absolute estimation error;
(3) The expected value of the third or higher power of the absolute estimation error.

The statistical optimization criterion that minimizes the mean square value of estimation error is called the minimum mean square error criterion, i.e., the MMSE criterion. It is the most widely used optimization criterion in the design of filters, estimators, detectors, and so on.

To sum up the above discussion, the optimal design problem of linear discrete time filter can be expressed as follows.

The coefficient w_k of a linear discrete time filter is designed so that the output $y(n)$ gives the estimation of the expected response $d(n)$ given the input sample set $u(0), u(1), \cdots$, and the mean square value $\mathrm{E}\{|e(n)|^2\}$ of the estimation error $e(n) = d(n) - y(n)$ can be minimized.

5.3.2 Orthogonality Principle

Consider the optimal design of the linear discrete-time filter shown in Fig.5.3.1. The output $y(n)$ of the filter in discrete time n is the linear convolution sum of the input $u(k)$ and the filter impulse response w_k^*, that is

$$y(n) = \sum_{k=0}^{\infty} w_k^* u(n-k), \quad n = 1, 2, \cdots . \tag{5.3.1}$$

Assume that both the filter input and the expected response are a single instance of the generalized stationary random process. Since the estimation of the expected response $d(n)$ is always accompanied by errors, which can be defined as

$$e(n) = d(n) - y(n), \tag{5.3.2}$$

the MMSE criterion is used to design the optimal filter. Therefore, the cost function is defined as the mean square error

$$J(n) = \mathrm{E}\{|e(n)|^2\} = \mathrm{E}\{e(n)e^*(n)\}. \tag{5.3.3}$$

For complex input data, the tap weight coefficient w_k of the filter is generally complex-valued. Assuming that there are an infinite number of tap weight coefficients w_k, this filter is an IIR filter. The tap weight coefficient may be divided into real and imaginary parts as

$$w_k = a_k + \mathrm{j}b_k, \; k = 0, 1, 2, \cdots . \tag{5.3.4}$$

Define the gradient operator as

$$\nabla_k = \frac{\partial}{\partial a_k} + \mathrm{j}\frac{\partial}{\partial b_k}, \quad k = 0, 1, 2, \cdots , \tag{5.3.5}$$

we have

$$\nabla_k J(n) \overset{\text{def}}{=} \frac{\partial J(n)}{\partial w_k} = \frac{\partial J(n)}{\partial a_k} + \mathrm{j}\frac{\partial J(n)}{\partial b_k}, \quad k = 0, 1, 2, \cdots . \tag{5.3.6}$$

To minimize the cost function J, all elements of gradient $\nabla_k J(n)$ must be equal to zero at the same time, i.e.,

$$\nabla_k J(n) = 0, \; k = 0, 1, 2, \cdots . \tag{5.3.7}$$

Under this set of conditions, the filter is optimal in the sense of minimum mean square error.

From Eqs. (5.3.2) and (5.3.3), it is easy to obtain

$$\nabla_k J(n) = \mathrm{E}\left\{ \frac{\partial e(n)}{\partial a_k} e^*(n) + \frac{\partial e^*(n)}{\partial a_k} e(n) + \mathrm{j}\frac{\partial e(n)}{\partial b_k} e^*(n) + \mathrm{j}\frac{\partial e^*(n)}{\partial b_k} e(n) \right\} . \tag{5.3.8}$$

Using Eqs. (5.3.2) and (5.3.5), the partial derivative can be expressed as

$$\left.\begin{aligned} \frac{\partial e(n)}{\partial a_k} &= -u(n-k) \\ \frac{\partial e(n)}{\partial b_k} &= \mathrm{j}u(n-k) \\ \frac{\partial e^*(n)}{\partial a_k} &= -u^*(n-k) \\ \frac{\partial e^*(n)}{\partial b_k} &= -\mathrm{j}u^*(n-k) \end{aligned}\right\} . \tag{5.3.9}$$

Substituting Eq. (5.3.9) into Eq. (5.3.8), there is

$$\nabla_k J(n) = -2\mathrm{E}\left\{u(n-k)e^*(n)\right\}. \tag{5.3.10}$$

Let $e_{\mathrm{opt}}(n)$ represent the estimation error of the filter under the optimal condition. From Eq. (5.3.10), $e_{\mathrm{opt}}(n)$ should satisfy $\nabla_k J = -2\mathrm{E}\left\{u(n-k)e^*_{\mathrm{opt}}(n)\right\} = 0$, or equivalently

$$\mathrm{E}\left\{u(n-k)e^*_{\mathrm{opt}}(n)\right\} = 0, \; k = 0, 1, 2, \cdots. \tag{5.3.11}$$

Eq. (5.3.11) indicates that the necessary and sufficient condition for minimizing the cost function J is that the estimation error $e_{\mathrm{opt}}(n)$ is orthogonal to the input $u(0), \cdots, u(n)$. This is the famous orthogonality principle, which is often used as a theorem. It is one of the most important theorems in linear optimal filtering theory. At the same time, it also provides a mathematical basis for the test method to test whether the filter works under the optimal condition.

On the other hand, it is easy to verify

$$\mathrm{E}\left\{y(n)e^*(n)\right\} = \mathrm{E}\left\{\sum_{k=0}^{\infty} w_k^* u(n-k)e^*(n)\right\} = \sum_{k=0}^{\infty} w_k^* \mathrm{E}\left\{u(n-k)e^*(n)\right\}. \tag{5.3.12}$$

Let $y_{\mathrm{opt}}(n)$ represent the output of the optimal filter in the sense of minimum mean square. According to Eqs. (5.3.11) and (5.3.12), the orthogonality principle can be equivalently written as

$$\mathrm{E}\left\{y_{\mathrm{opt}}(n)e^*_{\mathrm{opt}}(n)\right\} = 0. \tag{5.3.13}$$

The above equation means that when the filter works under the optimal condition, the estimation of the desired response $y_{\mathrm{opt}}(n)$ is defined by the filter output, and the corresponding estimation error $e_{\mathrm{opt}}(n)$ are orthogonal to each other. This is called the lemma of the orthogonality principle.

5.3.3 Wiener Filter

The necessary and sufficient condition for the filter to be in the optimal working state is derived and shown in Eq. (5.3.11). Take Eq. (5.3.2) into consideration, Eq. (5.3.11) can be rewritten as

$$\mathrm{E}\left\{u(n-k)\left[d^*(n) - \sum_{i=0}^{\infty} w_{\mathrm{opt},i} u^*(n-i)\right]\right\} = 0, \; k = 1, 2, \cdots, \tag{5.3.14}$$

where $w_{\mathrm{opt},i}$ represents the i-th coefficient of the impulse response of the optimal filter. Expand the above equation and rearrange it to obtain

$$\sum_{i=0}^{\infty} w_{\mathrm{opt},i} \mathrm{E}\left\{u(n-k)u^*(n-i)\right\} = \mathrm{E}\left\{u(n-k)d^*(n)\right\}, \; k = 1, 2, \cdots. \tag{5.3.15}$$

The two mathematical expectation terms in Eq. (5.3.15) have the following physical meanings respectively:

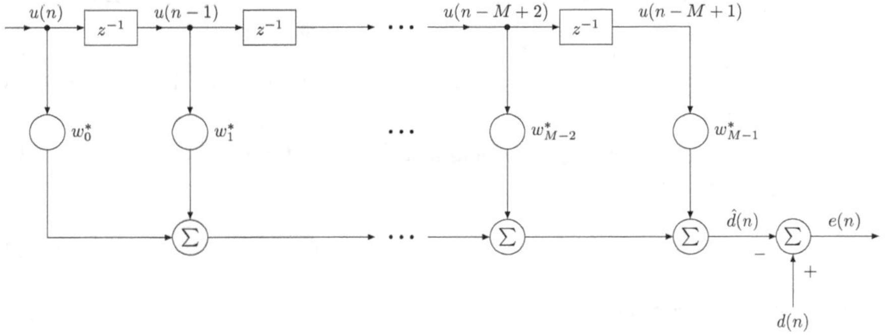

Fig. 5.3.2: FIR filter

(1) The mathematical expectation term $\mathrm{E}\left\{u(n-k)u^*(n-i)\right\}$ represents the autocorrelation function $R_{uu}(i-k)$ of the filter input at lag $i-k$, i.e.,

$$R_{uu}(i-k)=\mathrm{E}\left\{u(n-k)u^*(n-i)\right\}. \tag{5.3.16}$$

(2) The mathematical expectation term $\mathrm{E}\left\{u(n-k)d^*(n)\right\}$ is equal to the cross-correlation function $R_{u,d}(-k)$ between the filter input $u(n-k)$ and the expected response $d(n)$ at lag $-k$, i.e.,

$$R_{u,d}(-k)=\mathrm{E}\left\{u(n-k)d^*(n)\right\}. \tag{5.3.17}$$

Using Eqs. (5.3.16) and (5.3.17), Eq. (5.3.15) can be written in a concise form as

$$\sum_{i=0}^{\infty} w_{\mathrm{opt},i}R_{u,u}(i-k)=R_{u,d}(-k), \quad k=1,2,\cdots. \tag{5.3.18}$$

This is the famous Wiener-Hopf (difference) equation, which defines the conditions that the optimal filter coefficients must obey. In principle, if the autocorrelation function $R_{u,u}(\tau)$ of the filter input and the cross-correlation function $R_{u,d}(\tau)$ between the input and the expected response can be estimated, the coefficients of the optimal filter can be obtained by solving the Wiener-Hopf equation, so as to complete the optimal filter design. However, for the IIR filter design, it is unrealistic to solve Wiener-Hopf equation, because infinite equations need to be solved.

If the filter has a finite number of impulse response coefficients, the filter design will be greatly simplified. This kind of filter is FIR filter, also known as transversal filter. Fig.5.3.2 shows the principle of an FIR filter.

As shown in Fig.5.3.2, the filter impulse response is defined by M tap weight coefficients $w_0, w_1, \cdots, w_{M-1}$. Thus, the filter output is

$$y(n)=\sum_{i=0}^{M-1} w_i^* u(n-i), \; n=0,1,\cdots, \tag{5.3.19}$$

and consequently the Wiener-Hopf equation (5.3.18) is simplified to M homogeneous equations

$$\sum_{i=0}^{M-1} w_{\text{opt},i} R_{u,u}(i-k) = R_{u,d}(-k), \; k = 0, 1, \cdots, M-1, \tag{5.3.20}$$

where $w_{\text{opt},i}$ represents the optimal tap weight coefficient of the transversal filter. Define an $M \times 1$ input vector

$$\boldsymbol{u}(n) = [u(n), u(n-1), \cdots, u(n-M+1)]^{\mathrm{T}}, \tag{5.3.21}$$

to get its autocorrelation matrix as

$$\boldsymbol{R} = \mathrm{E}\{\boldsymbol{u}(n)\boldsymbol{u}^{\mathrm{H}}(n)\} = \begin{bmatrix} R_{u,u}(0) & R_{u,u}(1) & \cdots & R_{u,u}(M-1) \\ R^*_{u,u}(1) & R_{u,u}(0) & \cdots & R_{u,u}(M-2) \\ \vdots & \vdots & \vdots & \vdots \\ R^*_{u,u}(M-1) & R^*_{u,u}(M-2) & \cdots & R_{u,u}(0) \end{bmatrix}. \tag{5.3.22}$$

Similarly, the cross-correlation vector between the input and the desired response is

$$\boldsymbol{r} = \mathrm{E}\{\boldsymbol{u}(n)d^*(n)\} = [R_{u,d}(0), R_{u,d}(-1), \cdots, R_{u,d}(-M+1)]^{\mathrm{T}}. \tag{5.3.23}$$

Using Eqs. (5.3.21) ~ (5.3.23), Wiener-Hopf equation (5.3.20) can be written in a compact matrix form as

$$\boldsymbol{R}\boldsymbol{w}_{\text{opt}} = \boldsymbol{r}, \tag{5.3.24}$$

where $\boldsymbol{w}_{\text{opt}} = [w_{\text{opt},0}, w_{\text{opt},1}, \cdots, w_{\text{opt},M-1}]^{\mathrm{T}}$ represents the $M \times 1$ optimal tap weight vector of the transversal filter.

From matrix equation (5.3.24), the solution of the optimal tap weight vector can be obtained as

$$\boldsymbol{w}_{\text{opt}} = \boldsymbol{R}^{-1}\boldsymbol{r}. \tag{5.3.25}$$

The discrete-time transversal filter satisfying this relationship is called Wiener filter, which is optimal under the criterion of minimum mean square error. In fact, the discrete-time filter in many signal processing problems has the form of Wiener filter. By the way, the optimal filtering theory was first established by Wiener for continuous time signals.

Two main conclusions about the Wiener filter can be drawn from Eq. (5.3.25).

(1) The calculation of the optimal tap weight vector of Wiener filter requires the following statistics to be known: ① the autocorrelation matrix $\boldsymbol{R}$ of the input vector $\boldsymbol{u}(n)$; ② the cross correlation vector $\boldsymbol{r}$ between the input vector $\boldsymbol{u}(n)$ and the expected response $d(n)$.

(2) Wiener filter is actually the solution of the optimal filtering problem of unconstrained optimization.

5.4 Kalman Filter

In section 5.3, the linear optimal filter in the presence of expected response is analyzed, and Wiener filter is obtained. A natural question is: if the expected response is unknown, how to perform linear optimal filtering? This question will be answered in this section based on the state space model. The linear optimal filter based on the state space model is proposed by Kalman, which is called Kalman filter.

Kalman filtering theory is the extension of Wiener filtering theory. It was first used for parameter estimation of stochastic processes, and then it was widely used in various optimal filtering and optimal control problems. Kalman filter has the following characteristics: (1) its mathematical formula is described by the concept of state space; (2) Its solution is calculated recursively, i.e., unlike the Wiener filter, the Kalman filter is an adaptive filter. It is worth noting that the Kalman filter provides a unified framework for deriving a large class of adaptive filters called recursive least squares filter. The widely used recursive least squares algorithm in practice is a special case of the Kalman filter.

5.4.1 Kalman Filtering Problem

A discrete-time dynamic system is considered, which is represented by the process equation describing the state vector and the observation equation describing the observation vector.

(1) The process equation is

$$\boldsymbol{x}(n+1) = \boldsymbol{F}(n+1, n)\boldsymbol{x}(n) + \boldsymbol{v}_1(n), \tag{5.4.1}$$

where $M \times 1$ vector $\boldsymbol{x}(n)$ represents the state vector of the system at discrete time n and is unobservable; $M \times M$ matrix $\boldsymbol{F}(n+1, n)$ is called the state transition matrix describing the transition of the dynamic system from the state of time n to the state of $n+1$, which should be known; The $M \times 1$ vector $\boldsymbol{v}_1(n)$ is the process noise vector, which describes the additive noise or error in the state transition.

(2) The observation equation is

$$\boldsymbol{y}(n) = \boldsymbol{C}(n)\boldsymbol{x}(n) + \boldsymbol{v}_2(n), \tag{5.4.2}$$

where $\boldsymbol{y}(n)$ represents the $N \times 1$ observation vector of the dynamic system at time n; $N \times M$ matrix $\boldsymbol{C}(n)$ is called the observation matrix (describing that the state becomes observable data through its function), which is required to be known; $\boldsymbol{v}_2(n)$ represents the observation noise vector, and its dimension is the same as that of the observation vector.

The process equation is also called the state equation. For the convenience of analysis, it is usually assumed that both process noise $\boldsymbol{v}_1(n)$ and observation noise $\boldsymbol{v}_2(n)$ are white noise processes with zero mean, and their correlation matrices are

$$\mathrm{E}\{\boldsymbol{v}_1(n)\boldsymbol{v}_1^{\mathrm{H}}(k)\} = \begin{cases} \boldsymbol{Q}_1(n), & n = k, \\ \boldsymbol{0}, & n \neq k, \end{cases} \tag{5.4.3}$$

$$\mathrm{E}\{\boldsymbol{v}_2(n)\boldsymbol{v}_2^{\mathrm{H}}(k)\} = \begin{cases} \boldsymbol{Q}_2(n), & n = k, \\ \boldsymbol{0}, & n \neq k. \end{cases} \tag{5.4.4}$$

It is also assumed that the initial value $\boldsymbol{x}(0)$ of the state are not correlated with $\boldsymbol{v}_1(n)$ and $\boldsymbol{v}_2(n)$ (where $n > 0$) , and the noise vectors $\boldsymbol{v}_1(n)$ and $\boldsymbol{v}_2(n)$ are not correlated with each other either, that is

$$\mathrm{E}\{\boldsymbol{v}_1(n)\boldsymbol{v}_2^{\mathrm{H}}(k)\} = \boldsymbol{0}, \ \forall n, k. \tag{5.4.5}$$

The problem of the Kalman filter can be described as: using the observed data vector $\boldsymbol{y}(1), \cdots, \boldsymbol{y}(n)$, for $n \geq 1$, find the least squares estimation of each entry of the state vector $\boldsymbol{x}(i)$. According to the different values of i and n, the Kalman filtering problem can be further divided into the following three types:

(1) Filtering ($i = n$): Extracting the information at time n by using the measurement data at time n and before;
(2) Smoothing ($1 \leq i < n$): Different from filtering, the information to be extracted is not necessarily at time n, but generally the information of a certain time before n. And the measurement data after time n can also be used. in other words, the time to obtain the result of interest usually lags behind the time to obtain the measurement data. Since not only the measurement data at time n and before, but also the measurement data after time n can be used, the smoothing result is more accurate than the filtering result in a sense.
(3) Prediction ($i > n$): Determine the information of time $n + \tau$ (where $\tau > 0$) using the measurement data of time n and before, so it is a prediction result of the actual information of time $n + \tau$.

5.4.2 Innovation Process

Consider the one-step prediction problem: given the observation value $\boldsymbol{y}(1), \cdots, \boldsymbol{y}(n)$, find the least squares estimation of the observation vector $\boldsymbol{y}(n)$ and denote it as $\hat{\boldsymbol{y}}_1(n) \stackrel{\text{def}}{=} \hat{\boldsymbol{y}}(n|\boldsymbol{y}(1), \cdots, \boldsymbol{y}(n-1))$. Such a one-step prediction problem is easy to solve via the innovation method, proposed by Kailath in 1968[116, 117].

1. Properties of the Innovation Process

The innovation process of $\boldsymbol{y}(n)$ is defined as

$$\boldsymbol{\alpha}(n) = \boldsymbol{y}(n) - \hat{\boldsymbol{y}}_1(n), \quad n = 1, 2, \cdots, \tag{5.4.6}$$

where $N \times 1$ vector $\boldsymbol{\alpha}(n)$ represents the new information of the observation data $\boldsymbol{y}(n)$, called innovation for short.

The innovation $\boldsymbol{\alpha}(n)$ has the following properties.

Property 1 The innovation $\boldsymbol{\alpha}(n)$ at time n is orthogonal to all past observations $\boldsymbol{y}(1), \cdots, \boldsymbol{y}(n-1)$, that is,

$$\mathrm{E}\{\boldsymbol{\alpha}(n)\boldsymbol{y}^{\mathrm{H}}(k)\} = \mathbf{0}, \ 1 \leq k \leq n-1, \tag{5.4.7}$$

where $\mathbf{0}$ represents a zero matrix (i.e., a matrix whose elements are all zero).

Property 2 The innovation process consists of a random vector sequence $\{\boldsymbol{\alpha}(n)\}$ which is orthogonal to each other, i.e.,

$$\mathrm{E}\{\boldsymbol{\alpha}(n)\boldsymbol{\alpha}^{\mathrm{H}}(k)\} = \mathbf{0}, \ 1 \leq k \leq n-1. \tag{5.4.8}$$

Property 3 The random vector sequence $\{\boldsymbol{y}(1), \cdots, \boldsymbol{y}(n)\}$ representing the observation data corresponds to the random vector sequence $\{\boldsymbol{\alpha}(1), \cdots, \boldsymbol{\alpha}(n)\}$ representing the innovation process one by one, i.e.,

$$\{\boldsymbol{y}(1), \cdots, \boldsymbol{y}(n)\} \Leftrightarrow \{\boldsymbol{\alpha}(1), \cdots, \boldsymbol{\alpha}(n)\}, \tag{5.4.9}$$

The above properties show that the innovation $\boldsymbol{\alpha}(n)$ at time n is a random process that is not related to the observation data $\boldsymbol{y}(1), \cdots, \boldsymbol{y}(n-1)$ before time n and has the property of white noise, but it can provide new information about $\boldsymbol{y}(n)$. This is just the physical meaning of innovation.

2. Calculation of Innovation Process

Let the correlation matrix of the innovation process be

$$\boldsymbol{R}(n) = \mathrm{E}\{\boldsymbol{\alpha}(n)\boldsymbol{\alpha}^{\mathrm{H}}(n)\}. \tag{5.4.10}$$

In the Kalman filter, the one-step prediction $\hat{\boldsymbol{y}}_1(n)$ of the observation data vector is not estimated directly, but the one-step prediction of the state vector is calculated first

$$\hat{\boldsymbol{x}}_1(n) \overset{\text{def}}{=} \boldsymbol{x}(n|\boldsymbol{y}(1), \cdots, \boldsymbol{y}(n-1)), \tag{5.4.11}$$

and then obtain

$$\hat{\boldsymbol{y}}_1(n) = \boldsymbol{C}(n)\hat{\boldsymbol{x}}_1(n). \tag{5.4.12}$$

By substituting the above equation into Eq. (5.4.6), the innovation process can be rewritten as

$$\boldsymbol{\alpha}(n) = \boldsymbol{y}(n) - \boldsymbol{C}(n)\hat{\boldsymbol{x}}_1(n) = \boldsymbol{C}(n)[\boldsymbol{x}(n) - \hat{\boldsymbol{x}}_1(n)] + \boldsymbol{v}_2(n). \tag{5.4.13}$$

This is the actual calculation formula of the innovation process, provided that the one-step prediction $\hat{\boldsymbol{x}}_1(n)$ of the state vector has been obtained.

Define the one-step prediction error of the state vector

$$\boldsymbol{\epsilon}(n, n-1) \overset{\text{def}}{=} \boldsymbol{x}(n) - \hat{\boldsymbol{x}}_1(n), \tag{5.4.14}$$

and substituting the above equation into Eq. (5.4.13) yields

$$\boldsymbol{\alpha}(n) = \boldsymbol{C}(n)\boldsymbol{\epsilon}(n, n-1) + \boldsymbol{v}_2(n). \tag{5.4.15}$$

Substitute Eq. (5.4.14) into the correlation matrix defined in Eq. (5.4.10) of the innovation process, and note that the observation matrix $\boldsymbol{C}(n)$ is a known deterministic matrix, so we have

$$\begin{aligned}\boldsymbol{R}(n) &= \boldsymbol{C}(n)\mathrm{E}\{\boldsymbol{\epsilon}(n,n-1)\boldsymbol{\epsilon}^{\mathrm{H}}(n,n-1)\}\boldsymbol{C}^{\mathrm{H}}(n)+\mathrm{E}\{\boldsymbol{v}_2(n)\boldsymbol{v}_2^{\mathrm{H}}(n)\}\\ &= \boldsymbol{C}(n)\boldsymbol{K}(n,n-1)\boldsymbol{C}^{\mathrm{H}}(n)+\boldsymbol{Q}_2(n),\end{aligned} \tag{5.4.16}$$

where $\boldsymbol{Q}_2(n)$ is the correlation matrix of the observation noise $\boldsymbol{v}_2(n)$, and

$$\boldsymbol{K}(n,n-1)=\mathrm{E}\{\boldsymbol{\epsilon}(n,n-1)\boldsymbol{\epsilon}^{\mathrm{H}}(n,n-1)\} \tag{5.4.17}$$

represents the correlation matrix of (one-step) prediction state error.

5.4.3 Kalman Filtering Algorithm

With the knowledge and information about the innovation process, we can discuss the core problems of the Kalman filtering algorithm: How to use the innovation process to predict state vector? The most natural method is to use the linear combination of the innovation process sequence $\boldsymbol{\alpha}(1),\cdots,\boldsymbol{\alpha}(n)$ to directly construct the one-step prediction of the state vector

$$\hat{\boldsymbol{x}}_1(n+1)\stackrel{\text{def}}{=}\hat{\boldsymbol{x}}(n+1|\boldsymbol{y}(1),\cdots,\boldsymbol{y}(n))=\sum_{k=1}^{n}\boldsymbol{W}_1(k)\boldsymbol{\alpha}(k), \tag{5.4.18}$$

where $\boldsymbol{W}_1(k)$ represents the weight matrix corresponding to the one-step prediction, and k is discrete time. The question now becomes how to determine the weight matrix.

According to the principle of orthogonality, the estimation error $\boldsymbol{\epsilon}(n+1,n)=\boldsymbol{x}(n+1)-\hat{\boldsymbol{x}}_1(n+1)$ of the optimal prediction should be orthogonal to the known values, so we have

$$\mathrm{E}\{\boldsymbol{\epsilon}(n+1,n)\boldsymbol{\alpha}^{\mathrm{H}}(k)\}=\mathrm{E}\{[\boldsymbol{x}(n+1)-\hat{\boldsymbol{x}}_1(n+1)]\boldsymbol{\alpha}^{\mathrm{H}}(k)\}=\boldsymbol{0},\ k=1,\cdots,n. \tag{5.4.19}$$

Substituting Eq. (5.4.18) into Eq. (5.4.19), and using the orthogonality of the innovation process, we obtain

$$\mathrm{E}\{\boldsymbol{x}(n+1)\boldsymbol{\alpha}^{\mathrm{H}}(k)\}=\boldsymbol{W}_1(k)\mathrm{E}\{\boldsymbol{\alpha}(k)\boldsymbol{\alpha}^{\mathrm{H}}(k)\}=\boldsymbol{W}_1(k)\boldsymbol{R}(k).$$

Thus, the expression of the weight matrix can be obtained

$$\boldsymbol{W}_1(k)=\mathrm{E}\{\boldsymbol{x}(n+1)\boldsymbol{\alpha}^{\mathrm{H}}(k)\}\boldsymbol{R}^{-1}(k), \tag{5.4.20}$$

and substituting Eq. (5.4.20) into Eq. (5.4.18) further, the minimum mean square estimation of one-step prediction of the state vector can be expressed as

$$\begin{aligned}\hat{\boldsymbol{x}}_1(n+1)=&\sum_{k=1}^{n-1}\mathrm{E}\{\boldsymbol{x}(n+1)\boldsymbol{\alpha}^{\mathrm{H}}(k)\}\boldsymbol{R}^{-1}(k)\boldsymbol{\alpha}(k)\\ &+\mathrm{E}\{\boldsymbol{x}(n+1)\boldsymbol{\alpha}^{\mathrm{H}}(n)\}\boldsymbol{R}^{-1}(n)\boldsymbol{\alpha}(n).\end{aligned} \tag{5.4.21}$$

Notice $\mathrm{E}\{\boldsymbol{v}_1(n)\boldsymbol{\alpha}(k)\} = \mathbf{0}$, $k = 0, 1, \cdots, n$, and use the state equation (5.4.1), it is easy to find that

$$\mathrm{E}\{\boldsymbol{x}(n+1)\boldsymbol{\alpha}^{\mathrm{H}}(k)\} = \mathrm{E}\{[\boldsymbol{F}(n+1,n)\boldsymbol{x}(n) + \boldsymbol{v}_1(n)]\boldsymbol{\alpha}^{\mathrm{H}}(k)\} \tag{5.4.22}$$

$$= \boldsymbol{F}(n+1,n)\mathrm{E}\{\boldsymbol{x}(n)\boldsymbol{\alpha}^{\mathrm{H}}(k)\}, \tag{5.4.23}$$

holds for $k = 0, 1, \cdots, n$.

By substituting Eq. (5.4.23) into the first term (summation term) on the right of Eq. (5.4.21), it can be reduced to

$$\sum_{k=1}^{n-1}\mathrm{E}\{\boldsymbol{x}(n+1)\boldsymbol{\alpha}^{\mathrm{H}}(k)\}\boldsymbol{R}^{-1}(k)\boldsymbol{\alpha}(k) = \boldsymbol{F}(n+1,n)\sum_{k=1}^{n-1}\mathrm{E}\{\boldsymbol{x}(n)\boldsymbol{\alpha}^{\mathrm{H}}(k)\}\boldsymbol{R}^{-1}(k)\boldsymbol{\alpha}(k)$$

$$= \boldsymbol{F}(n+1,n)\hat{\boldsymbol{x}}_1(n). \tag{5.4.24}$$

On the other hand, if

$$\boldsymbol{G}(n) \stackrel{\mathrm{def}}{=} \mathrm{E}\{\boldsymbol{x}(n+1)\boldsymbol{\alpha}^{\mathrm{H}}(n)\}\boldsymbol{R}^{-1}(n), \tag{5.4.25}$$

is defined and Eqs. (5.4.24) and (5.4.25) are substituted into Eq. (5.4.21), the update equation of one-step prediction of the state vector can be obtained as

$$\hat{\boldsymbol{x}}_1(n+1) = \boldsymbol{F}(n+1,n)\hat{\boldsymbol{x}}_1(n) + \boldsymbol{G}(n)\boldsymbol{\alpha}(n). \tag{5.4.26}$$

Eq. (5.4.26) plays a key role in Kalman filtering algorithm because it shows that the one-step prediction of the state vector at time $n+1$ can be divided into nonadaptive (i.e. determinate) part $\boldsymbol{F}(n+1,n)\hat{\boldsymbol{x}}_1(n)$ and adaptive (i.e. corrective) part $\boldsymbol{G}(n)\boldsymbol{\alpha}(n)$. Therefore, $\boldsymbol{G}(n)$ is called Kalman gain (matrix).

The following is the Kalman adaptive filtering algorithm based on one-step prediction.

Algorithm 5.4.1. *Kalman adaptive filtering algorithm*
Initial conditions:

$$\hat{\boldsymbol{x}}_1(1) = E\{\boldsymbol{x}(1)\},$$
$$\boldsymbol{K}(1,0) = E\{[\boldsymbol{x}(1) - \bar{\boldsymbol{x}}(1)][\boldsymbol{x}(1) - \bar{\boldsymbol{x}}(1)]^H\},$$

where $\bar{\boldsymbol{x}}(1) = E\{\boldsymbol{x}(1)\}$.

Input observation vector process:

Observation vector sequence $\{\boldsymbol{y}(1), \cdots, \boldsymbol{y}(n)\}$.

Known parameters:

State transition matrix $\boldsymbol{F}(n+1,n)$,
Observation matrix $\mathbf{C}(n)$,
Correlation matrix $\mathbf{Q}_1(n)$ *of the process noise vector,*
Correlation matrix $\mathbf{Q}_2(n)$ *of the observation noise vector,*

Calculate: $n = 1, 2, 3, \cdots$

$$\begin{aligned} \boldsymbol{G} &= \boldsymbol{F}(n+1,n)\boldsymbol{K}(n,n-1)\boldsymbol{C}^H(n)\left[\boldsymbol{C}(n)\boldsymbol{K}(n,n-1)\boldsymbol{C}^H(n)+\boldsymbol{Q}_2(n)\right]^{-1}, \\ \boldsymbol{\alpha}(n) &= \boldsymbol{y}(n)-\boldsymbol{C}(n)\hat{\boldsymbol{x}}_1(n), \\ \hat{\boldsymbol{x}}_1(n+1) &= \boldsymbol{F}(n+1,n)\hat{\boldsymbol{x}}_1(n)+\boldsymbol{G}(n)\boldsymbol{\alpha}(n), \\ \boldsymbol{P}(n) &= \boldsymbol{K}(n,n-1)-\boldsymbol{F}^{-1}(n+1,n)\boldsymbol{G}(n)\boldsymbol{C}(n)\boldsymbol{K}(n,n-1), \\ \boldsymbol{K}(n,n-1) &= \boldsymbol{F}(n+1,n)\boldsymbol{P}(n)\boldsymbol{F}^H(n+1,n)+\boldsymbol{Q}_1(n). \end{aligned}$$

The estimation performance of the Kalman filter is that it minimizes the trace of the correlation matrix $\boldsymbol{P}(n)$ of the filtered state estimation error. This means that the Kalman filter is a linear minimum variance estimate of the state vector $\boldsymbol{x}(n)$[100].

5.5 LMS Adaptive Algorithms

Different from the Kalman filtering algorithm based on the state space model, another kind of adaptive algorithm is based on the (gradient) descent algorithm in optimization theory. There are two main implementation forms of the descent algorithm. One is the adaptive gradient algorithm, the other is the adaptive Gauss-Newton algorithm.

The adaptive gradient algorithm includes the least mean squares algorithm, its variants and improvements (collectively referred to as LMS adaptive algorithms). And adaptive Gauss-Newton algorithm includes the recursive least square algorithm and its variants and improvements.

This section describes LMS adaptive algorithms.

5.5.1 Descent Algorithm

The most common criterion for filter design is to minimize the mean square error $\mathrm{E}\{|e(n)|^2\}$ between the actual output $y(n) = \boldsymbol{u}^{\mathrm{T}}(n)\boldsymbol{w}^* = \boldsymbol{w}^{\mathrm{H}}\boldsymbol{u}(n)$ of the filter and the expected response $d(n)$, which is the famous minimum mean square error (MMSE) criterion.

Fig. 5.5.1 shows a schematic diagram of an adaptive FIR filter.

Let

$$e(n) = d(n) - \boldsymbol{w}^{\mathrm{H}}\boldsymbol{u}(n), \tag{5.5.1}$$

represent the estimation error of the filter at time n, and define the mean square error

$$J(n) \stackrel{\text{def}}{=} \mathrm{E}\{|e(n)|^2\} = \mathrm{E}\left\{\left|d(n) - \boldsymbol{w}^{\mathrm{H}}\boldsymbol{u}(n)\right|^2\right\}, \tag{5.5.2}$$

as the cost function.

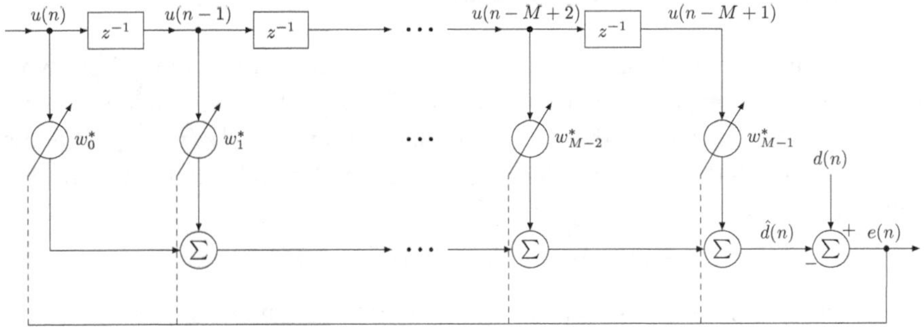

Fig. 5.5.1: Schematic diagram of an adaptive FIR filter

The gradient of the cost function relative to the filter tap weight vector $\boldsymbol{w}$ is

$$\begin{aligned}\nabla_k J(n) &= -2\mathrm{E}\{u(n-k)e^*(n)\}\\ &= -2\mathrm{E}\left\{u(n-k)[d(n)-\boldsymbol{w}^{\mathrm{H}}\boldsymbol{u}(n)]^*\right\},\ k=0,1,\cdots,M-1.\end{aligned} \tag{5.5.3}$$

Let $w_i = a_i + jb_i$, $i = 0, 1, \cdots, M-1$, and define the gradient vector

$$\begin{aligned}\boldsymbol{\nabla} J(n) &\stackrel{\mathrm{def}}{=} [\nabla_0 J(n), \nabla_1 J(n), \cdots, \nabla_{M-1} J(n)]^{\mathrm{T}}\\ &= \begin{bmatrix} \frac{\partial J(n)}{\partial a_0(n)} + \mathrm{j}\frac{\partial J(n)}{\partial b_0(n)}\\ \frac{\partial J(n)}{\partial a_1(n)} + \mathrm{j}\frac{\partial J(n)}{\partial b_1(n)}\\ \vdots\\ \frac{\partial J(n)}{\partial a_{M-1}(n)} + \mathrm{j}\frac{\partial J(n)}{\partial b_{M-1}(n)}\end{bmatrix},\end{aligned} \tag{5.5.4}$$

the input vector and the tap weight vector

$$\boldsymbol{u}(n) = [u(n), u(n-1), \cdots, u(n-M+1)]^{\mathrm{T}}, \tag{5.5.5}$$

$$\boldsymbol{w}(n) = [w_0(n), w_1(n), \cdots, w_{M-1}(n)]^{\mathrm{T}}, \tag{5.5.6}$$

then Eq. (5.5.4) can be written in vector form

$$\boldsymbol{\nabla} J(n) = -2\mathrm{E}\{\boldsymbol{u}(n)[d^*(n) - \boldsymbol{u}^{\mathrm{H}}(n)\boldsymbol{w}(n)]\} = -2\boldsymbol{r} + 2\boldsymbol{R}\boldsymbol{w}(n), \tag{5.5.7}$$

where

$$\boldsymbol{R} = \mathrm{E}\{\boldsymbol{u}(n)\boldsymbol{u}^{\mathrm{H}}(n)\}, \tag{5.5.8}$$

$$\boldsymbol{r} = \mathrm{E}\{\boldsymbol{u}(n)d^*(n)\}. \tag{5.5.9}$$

The most widely used adaptive algorithm is "descent algorithm"

$$\boldsymbol{w}(n) = \boldsymbol{w}(n-1) + \mu(n)\boldsymbol{v}(n), \tag{5.5.10}$$

where $\boldsymbol{w}(n)$ is the weight vector of the n-th iteration (i.e. time n), $\mu(n)$ is the update step of the n-th iteration, and $\boldsymbol{v}(n)$ is the update direction (vector) of the n-th iteration.

5.5.2 LMS Algorithm and Its Basic Variants

The most commonly used descent algorithm is the gradient descent method, which is often called the steepest descent method. In this algorithm, the update direction vector $\boldsymbol{v}(n)$ is taken as the negative gradient of the cost function $J[\boldsymbol{w}(n-1)]$ of iteration $n-1$, i.e., the unified form of the steepest descent method (also known as gradient algorithm) is

$$\boldsymbol{w}(n) = \boldsymbol{w}(n-1) - \frac{1}{2}\mu(n)\nabla J(n-1), \tag{5.5.11}$$

where the coefficient 1/2 is to make the resulted update formula simpler.

By substituting Eq. (5.5.7) into Eq. (5.5.11), the update formula of tap weight vector $\boldsymbol{w}(n)$ can be obtained

$$\boldsymbol{w}(n) = \boldsymbol{w}(n-1) + \mu(n)[\boldsymbol{r} - \boldsymbol{R}\boldsymbol{w}(n-1)], \quad n = 1, 2, \cdots. \tag{5.5.12}$$

The update formula (5.5.12) shows that

(1) $\boldsymbol{r} - \boldsymbol{R}\boldsymbol{w}(n-1)$ is the error vector, which represents the correction amount of $\boldsymbol{w}(n)$ in each step.

(2) The parameter $\mu(n)$ is multiplied by the correction amount, which is a parameter controlling the actual correction amount of $\boldsymbol{w}(n)$ at each step, so $\mu(n)$ is called the "step parameter" at time n. This parameter determines the convergence speed of the update algorithm as Eq. (5.5.12).

(3) When the adaptive algorithm tends to converge, there is $\boldsymbol{r} - \boldsymbol{R}\boldsymbol{w}(n-1) \to 0$ (if $n \to \infty$), i.e.,

$$\lim_{n\to\infty} \boldsymbol{w}(n-1) = \boldsymbol{R}^{-1}\boldsymbol{r}$$

that is, the tap weight vector converges to Wiener filter.

When the mathematical expectation terms $\mathrm{E}\{\boldsymbol{u}(n)d^*(n)\}$ and $\mathrm{E}\{\boldsymbol{u}(n)\boldsymbol{u}^{\mathrm{H}}(n)\}$ in Eq. (5.5.7) are replaced by their instantaneous values $\boldsymbol{u}(n)d^*(n)$ and $\boldsymbol{u}(n)\boldsymbol{u}^{\mathrm{H}}(n)$ respectively, the estimated value of the real gradient vector is obtained

$$\hat{\nabla} J(n) = -2[\boldsymbol{u}(n)d^*(n) - \boldsymbol{u}(n)\boldsymbol{u}^{\mathrm{H}}(n)\boldsymbol{w}(n)], \tag{5.5.13}$$

which is commonly referred to as the instantaneous gradient.

If the real gradient vector $\nabla J(n-1)$ in the gradient algorithm shown by Eq. (5.5.11) is replaced by the instantaneous gradient vector $\hat{\nabla} J(n-1)$, the instantaneous gradient algorithm is obtained

$$\begin{aligned} \boldsymbol{w}(n) &= \boldsymbol{w}(n-1) + \mu(n)\boldsymbol{u}(n)[d(n) - \boldsymbol{u}^{\mathrm{T}}(n)\boldsymbol{w}^*(n-1)]^* \\ &= \boldsymbol{w}(n-1) + \mu(n)e^*(n)\boldsymbol{u}(n), \end{aligned} \tag{5.5.14}$$

where

$$e(n) = d(n) - \boldsymbol{u}^{\mathrm{T}}(n)\boldsymbol{w}^*(n-1) = d(n) - \boldsymbol{w}^{\mathrm{H}}(n-1)\boldsymbol{u}(n). \tag{5.5.15}$$

Note that although $e(n)$ and $\epsilon(n)$ defined in Eq. (5.5.3) represent the estimation error of the filter at time n, they are different in that $e(n)$ is determined by $\boldsymbol{w}(n-1)$ and $\epsilon(n)$ is determined by $\boldsymbol{w}(n)$. In order to distinguish, $e(n)$ is often called a priori estimation error and $\epsilon(n)$ is a posteriori estimation error.

The algorithm shown in Eq. (5.5.14) is the famous least mean square error adaptive algorithm, referred to as LMS algorithm, which was proposed by Widrow in the early 1960s[227].

It is easy to verify that the instantaneous gradient vector is an unbiased estimation of the real gradient vector because

$$\begin{aligned} \mathrm{E}\{\hat{\nabla}J(n)\} &= -2\mathrm{E}\{\boldsymbol{u}(n)[d^*(n) - \boldsymbol{u}^{\mathrm{H}}(n)\boldsymbol{w}(n-1)]\} \\ &= -2[\boldsymbol{r} - \boldsymbol{R}\boldsymbol{w}(n-1)] = \nabla J(n). \end{aligned} \tag{5.5.16}$$

For the convenience of readers, the LMS adaptive algorithm and its several basic variants are summarized as follows.

Algorithm 5.5.1. *LMS adaptive algorithm and its basic variants*
Step 1 Initialization $\boldsymbol{w}(0) = \boldsymbol{0}$;
Step 2 Update $n = 1, 2, \cdots$

$$\begin{aligned} e(n) &= d(n) - \boldsymbol{w}^H(n-1)\boldsymbol{u}(n), \\ \boldsymbol{w}(n) &= \boldsymbol{w}(n-1) + \mu(n)\boldsymbol{u}(n)e^*(n), \end{aligned}$$

Here are some notes about the LMS algorithm.

Note 1 If $\mu(n) = \text{constant}$, it is called basic LMS algorithm.

Note 2 If $\mu(n) = \frac{\alpha}{\beta + \boldsymbol{u}^{\mathrm{H}}(n)\boldsymbol{u}(n)}$, where $\alpha \in (0, 2)$, $\beta \ge 0$, the normalized LMS algorithm is obtained, which is an improvement of the basic LMS algorithm.

Note 3 In the power normalized LMS algorithm, $\mu(n) = \frac{\alpha}{\sigma_u^2(n)}$ is taken, where $\sigma_u^2(n)$ represents the variance of $u(n)$, which can be calculated recursively by $\sigma_u^2(n)\lambda\sigma_u^2(n-1) + e^2(n)$. Here $\lambda \in (0, 1]$ is the forgetting factor, which is determined by $0 < \alpha < \frac{2}{M}$, and M is the order of the filter.

Note 4 When the desired signal is unknown, $d(n)$ in Step 2 can be directly replaced by the actual output $y(n)$ of the filter.

5.5.3 Decorrelation LMS Algorithm

In the LMS algorithm, there is an implicit assumption of independence that the input vector $\boldsymbol{u}(1), \cdots, \boldsymbol{u}(n)$ of the transverse filter is a sequence of vectors that are statistically independent of each other. When they do not meet the condition of statistical independence, the performance of the basic LMS algorithm may degrade, especially the convergence speed will be slow. Therefore, in this case, it is necessary to decorrelate the input vectors at each time to keep them as statistically independent as possible.

This operation is called decorrelation. A large number of studies show that ([90] and other related references) decorrelation can effectively accelerate the convergence rate of the LMS algorithm.

1. Time Domain Decorrelation LMS Algorithm

Define the correlation coefficient between $\boldsymbol{u}(n)$ and $\boldsymbol{u}(n-1)$ at time n as

$$a(n) \stackrel{\text{def}}{=} \frac{\boldsymbol{u}^{\mathrm{H}}(n-1)\boldsymbol{u}(n)}{\boldsymbol{u}^{\mathrm{H}}(n-1)\boldsymbol{u}(n-1)}. \tag{5.5.17}$$

According to the definition, if $a(n) = 1$, $\boldsymbol{u}(n)$ is the coherent signal of $\boldsymbol{u}(n-1)$; If $a(n) = 0$, $\boldsymbol{u}(n)$ is not related to $\boldsymbol{u}(n-1)$; When $0 < a(n) < 1$, $\boldsymbol{u}(n)$ is correlated with $\boldsymbol{u}(n-1)$, and the greater $a(n)$, the stronger the correlation between them.

Obviously, $a(n)\boldsymbol{u}(n-1)$ represents the part of $\boldsymbol{u}(n)$ related to $\boldsymbol{u}(n-1)$. If this part is subtracted from $\boldsymbol{u}(n)$, the subtraction operation is equivalent to "decorrelation". Now, use the result of decorrelation as the update direction vector

$$\boldsymbol{v}(n) = \boldsymbol{u}(n) - a(n)\boldsymbol{u}(n-1), \tag{5.5.18}$$

From this point of view, it is more appropriate to call $a(n)$ as the decorrelation coefficient.

On the other hand, the step parameter $\mu(n)$ should be the solution satisfying the minimization problem

$$\mu(n) = \arg\min_{\mu} J[\mathbf{w}(n-1) + \mu\mathbf{v}(n)]. \tag{5.5.19}$$

Thus

$$\mu(n) = \frac{e(n)}{\boldsymbol{u}^{\mathrm{H}}(n)\boldsymbol{v}(n)}. \tag{5.5.20}$$

Based on the above results, the decorrelation LMS algorithm[75] can be summarized as follows.

Algorithm 5.5.2. *Decorrelation LMS algorithm*
Step 1 Initialization: $\mathbf{w}(0) = \mathbf{0}$*;*
Step 2 Update: $n = 1, 2, \cdots$

$$\begin{aligned}
e(n) &= d(n) - \boldsymbol{w}^{H}(n-1)\boldsymbol{u}(n),\\
a(n) &= \frac{\boldsymbol{u}^{H}(n-1)\boldsymbol{u}(n)}{\boldsymbol{u}^{H}(n-1)\boldsymbol{u}(n-1)},\\
\boldsymbol{v}(n) &= \boldsymbol{u}(n) - a(n)\boldsymbol{u}(n-1),\\
\mu(n) &= \frac{\rho e(n)}{\boldsymbol{u}^{H}(n)\boldsymbol{v}(n)},\\
\boldsymbol{w}(n) &= \boldsymbol{w}(n-1) + \mu(n)\boldsymbol{v}(n).
\end{aligned}$$

In the above algorithm, the parameter ρ is called the trimming factor.

The decorrelation LMS algorithm can be regarded as an adaptive auxiliary variable method, in which the auxiliary variable is given by $\mathbf{v}(n) = \mathbf{u}(n) - a(n)\mathbf{u}(n-1)$. Roughly speaking, the selection principle of the auxiliary variable is that it should be strongly related to the lagging input and output, but not to interference. Readers interested in the auxiliary variable methods and their adaptive algorithms can refer to [242].

Further, the auxiliary variable in the above algorithm can be replaced by the error vector of a forward predictor. Let $\boldsymbol{a}(n)$ be the weight vector of the M-order forward predictor, and calculate the forward prediction error

$$e^{\mathrm{f}}(n) = u(n) + \sum_{i=1}^{M} a_i(n)u(n-i) = u(n) + \boldsymbol{a}^{\mathrm{H}}(n)\mathbf{u}(n-1), \tag{5.5.21}$$

where $\mathbf{u}(n-1) = [u(n-1), \cdots, u(n-M)]^{\mathrm{T}}$ and $\boldsymbol{a}(n) = [a_1(n), \cdots, a_M(n)]^{\mathrm{T}}$.

The forward prediction error vector is used as the auxiliary variable, that is, the update direction vector

$$\mathbf{v}(n) = \mathbf{e}^{\mathrm{f}}(n) = [e^{\mathrm{f}}(n), e^{\mathrm{f}}(n-1), \cdots, e^{\mathrm{f}}(n-M+1)]^{\mathrm{T}}. \tag{5.5.22}$$

If the forward predictor is used to filter the instantaneous estimation error $e(n) = y(n) - \mathbf{w}a^{\mathrm{H}}(n-1)\mathbf{u}(n)$, the filtered LMS algorithm is obtained as follows[151].

Algorithm 5.5.3. *Filtered LMS algorithm*
Step 1 Initialization: $\mathbf{w}(0) = \mathbf{0}$*;*
Step 2 Update: $n = 1, 2, \cdots$

Given the estimation of a forward predictor $\boldsymbol{a}(n)$*,*

$$\begin{aligned}
e(n) &= d(n) - \mathbf{w}^H(n-1)\mathbf{u}(n),\\
\mathbf{e}(n) &= [e(n), e(n-1), \cdots, e(n-M+1)]^T,\\
e^f(n) &= u(n) + \boldsymbol{a}^T(n)\mathbf{u}(n-1),\\
\mathbf{e}^f(n) &= [e^f(n), e^f(n-1), \cdots, e^f(n-M+1)]^T,\\
\tilde{e}(n) &= e(n) + \boldsymbol{a}^H(n)\mathbf{e}(n) \quad (\textit{filtering}),\\
\mathbf{w}(n) &= \mathbf{w}(n-1) + \mu\mathbf{e}^f(n)\tilde{e}(n).
\end{aligned}$$

2. Transform Domain Decorrelation LMS Algorithm

The early work to improve the performance of the LMS algorithm is to use unitary transformation for the input data vector $\mathbf{u}(n)$. For some types of input signals, the algorithm of unitary transformation can improve the convergence rate, but the computational complexity is equivalent to that of the LMS algorithm. These algorithms and their variants are collectively referred to as transform domain adaptive filtering algorithms[22].

The unitary transform can use discrete Fourier transform (DFT), discrete cosine transform (DCT), and discrete Hartley transform (DHT), which can effectively improve the convergence rate of the LMS algorithm.

Let $\boldsymbol{S}$ be a $M \times M$ unitary transformation matrix, i.e.,

$$\boldsymbol{S}\boldsymbol{S}^{\mathrm{H}} = \beta \boldsymbol{I}, \tag{5.5.23}$$

where $\beta > 0$ is a fixed scalar.

Being transformed by the unitary matrix $\boldsymbol{S}$, the input data vector $\boldsymbol{u}(n)$ becomes $\boldsymbol{x}(n)$ as

$$\boldsymbol{x}(n) = \boldsymbol{S}\boldsymbol{u}(n). \tag{5.5.24}$$

Accordingly, the unitarily transformed weight vector $\boldsymbol{w}(n-1)$ becomes

$$\hat{\boldsymbol{w}}(n-1) = \frac{1}{\beta}\boldsymbol{S}\boldsymbol{w}(n-1), \tag{5.5.25}$$

which is the weight vector of the transform domain adaptive filter that needs to update.

Therefore, the original prediction error $e(n) = d(n) - \boldsymbol{w}^{\mathrm{H}}(n-1)\boldsymbol{u}(n)$ can be represented by the transformed input data vector $\boldsymbol{x}(n)$ and the filter weight vector $\hat{\boldsymbol{w}}(n-1)$ as,

$$e(n) = d(n) - \hat{\boldsymbol{w}}^{\mathrm{H}}(n-1)\boldsymbol{x}(n). \tag{5.5.26}$$

Comparing the input data vectors $\boldsymbol{u}(n)$ and $\boldsymbol{x}(n)$ before and after transformation, it can be seen that the elements of the original data vector are the shift form of $u(n-i+1)$, and they have strong correlation, while the elements of $\boldsymbol{x}(n) = [x_1(n), x_2(n), \cdots, x_M(n)]^{\mathrm{T}}$ are equivalent to the signal of M channels. It can be expected that they have a weaker correlation than the original signal $\boldsymbol{u}(n)$. In other words, a certain degree of decorrelation is achieved in the transform domain through unitary transformation. From the perspective of filter, the original single channel M-order FIR transverse filter becomes an equivalent multi-channel filter, and the original input signal $\boldsymbol{u}(n)$ is equivalent to passing through a filter bank containing M filters.

Summarizing the above analysis, it is easy to get the transform domain LMS algorithm as follows.

Algorithm 5.5.4. *Transform domain LMS algorithm*
Step 1 Initialization: $\hat{\boldsymbol{w}}(0) = \boldsymbol{0}$;
Step 2 Given a unitary transformation matrix $\boldsymbol{S}$,
Update: $n = 1, 2, \cdots$

$$\begin{aligned}
\boldsymbol{x}(n) &= \boldsymbol{S}\boldsymbol{u}(n),\\
e(n) &= d(n) - \hat{\boldsymbol{w}}^{H}(n-1)\boldsymbol{x}(n),\\
\hat{\boldsymbol{w}}(n) &= \hat{\boldsymbol{w}}(n-1) + \mu(n)\boldsymbol{x}(n)e(n).
\end{aligned}$$

In particular, if the unitary transform adopts DFT, $\boldsymbol{u}$ becomes the Fourier transform of the input data vector $\boldsymbol{u}(n)$ within a sliding window. This shows that the estimated weight vector $\hat{\boldsymbol{w}}(n)$ is the frequency response of the time domain filter $\boldsymbol{w}(n)$. Therefore, the adaptation, in this case, occurs in the frequency domain, and the resulted filter is called frequency domain adaptive filter accordingly.

5.5.4 Selection of the Learning Rate Parameter

The step size parameter μ in the LMS algorithm determines the update amount of tap weight vector in each iteration, which is the key parameter affecting the convergence rate of the algorithm. Since the purpose of the LMS algorithm is to make the tap weight vector approach the Wiener filter in the updating process, the updating process of the weight vector can be regarded as a learning process, and μ determines the speed of the learning process of LMS algorithm. In this sense, the step parameter μ is also called the learning rate parameter. The following discusses the selection of the learning rate parameter from the perspective of LMS algorithm convergence.

The convergence of the basic LMS algorithm can be divided into mean convergence and mean square convergence[100].

1. Mean Convergence

From Eq. (5.5.14), the condition that the convergence of the basic LMS algorithm must meet is

$$\mathrm{E}\{e(n)\} \to 0 \quad \text{if } n \to \infty.$$

Or equivalent to that $\hat{\boldsymbol{w}}(n)$ converging to the optimal Wiener filter, i.e.,

$$\lim_{n\to\infty} \mathrm{E}\{\hat{\boldsymbol{w}}(n)\} = \boldsymbol{w}_{opt}. \tag{5.5.27}$$

This is called mean convergence.

To ensure the convergence of the weight vector mean of the LMS algorithm, the learning rate parameter $\mu(n)$ must satisfy

$$0 < \mu < \frac{2}{\lambda_{\mathrm{maxt}}}. \tag{5.5.28}$$

2. Mean Square Convergence

LMS algorithm is called mean square convergence. If the number of iterations n tends to infinity, the mean square value of the error signal $\epsilon(n) = d(n) - \boldsymbol{w}^{\mathrm{H}}(n)\boldsymbol{u}(n)$ converges to a constant, i.e.,

$$\lim_{n\to\infty} \mathrm{E}\left\{|e(n)|^2\right\} = c, \tag{5.5.29}$$

where c is a positive constant.

The condition that the weight vector mean square convergence of the LMS algorithm needs to meet is

$$0 < \mu < \frac{2}{\mathrm{tr}(\boldsymbol{R})}, \tag{5.5.30}$$

or

$$0 < \mu < \frac{2}{\text{Total input energy}}, \tag{5.5.31}$$

where $\mathrm{tr}(\boldsymbol{R})$ is the rank of the autocorrelation function matrix of the filter output vector.

For the transverse filter shown in Fig. 5.5.1, the denominator of Eq. (5.5.31) is M times the input energy $\mathrm{E}\left\{|u(n)|^2\right\}$.

3. Adaptive Learning Rate Parameter

The conditions that the learning rate parameters should meet are obtained from the perspective of mean convergence and mean square convergence of the LMS algorithm. In the LMS algorithm, the simplest selection of the learning rate parameter is to take $\mu(n)$ as a constant, i.e., $\mu(n) = \mu$. μ is determined by Eq. (5.5.28), Eq. (5.5.30) or Eq. (5.5.31). However, this method will cause a contradiction between convergence and steady-state performance: a large learning rate can improve the convergence rate of the filter, but the steady-state performance will be reduced; On the contrary, when a small learning rate is used to improve the steady-state performance, the convergence will slow down. Therefore, the selection of learning rate should take into account the steady-state performance and convergence rate. A simple and effective method is to use different learning rate parameters at different iteration times, that is, time-varying learning rate[184], and the simplest time-varying learning rate is

$$\mu(n) = \frac{c}{n}, \tag{5.5.32}$$

where c is a constant. This choice is often called the simulated annealing rule. It should be noted that if parameter c is large, the LMS algorithm will fall into divergence after several iterations.

A better approach is to use a large learning rate in the transient stage, namely, the transition stage, and a small learning rate in the steady state. This selection of learning rate parameters is called gear-shifting approach[227]. For example, the "constant plus time-varying" learning rate is a typical gear-shifting method. Here are two typical examples.

The first example is to use the so-called "search first, then convergence" rule[70]

$$\mu(n) = \frac{\mu_0}{1 + (n/\tau)}, \tag{5.5.33}$$

where μ_0 is a fixed learning rate parameter, and τ represents a search time constant. As can be seen from Eq. (5.5.33), this rule uses an approximately fixed learning rate μ_0 in the iteration time of $n \le \tau$; When the iteration time n is greater than the search time parameter τ, the learning rate decreases with time, and the decrease speed is faster and faster.

The second example is the "fixed first, then exponential decay" rule[232]

$$\mu(n) = \begin{cases} \mu_0, & n \le N_0, \\ \mu_0 e^{-N_d(n-N_0)}, & n > N_0, \end{cases} \tag{5.5.34}$$

where μ_0 and N_d are positive constants, respectively; N_0 is a positive integer.

The above time-varying learning rate is predetermined and has no direct relationship with the actual running state of the LMS algorithm. If the time-varying learning rate is controlled by the actual running state of the LMS algorithm, this kind of time-varying learning rate is called adaptive learning rate, also known as learning of learning rules, which was proposed by Amari in 1967[11]. Many methods have been proposed to select adaptive learning rate, and three examples are introduced here.

(1) Harries et al. controlled the learning rate by testing the polarity of adjacent sample values of LMS algorithm estimation error[99]. If there are m_0 adjacent sign changes in the estimation error, the learning rate is appropriately reduced; If there are m_1 adjacent identical signs, the learning rate is appropriately increased.
(2) Kwong and Johston proposed to adjust the learning rate according to the square of the prediction error[129].
(3) The above methods require the user to select some additional constants and initial learning rates, which are based on language rules such as “large learning rate in the initial stage and small learning rate in the steady-state stage”, and convert these language rules into mathematical models to adjust the learning rate parameters. Naturally, the tuning of the learning rate can also be realized by using fuzzy system theory and language models to form the so-called fuzzy step adjustment. Readers interested in this method can refer to [83].

5.5.5 Statistical Performance Analysis of LMS Algorithm

The basic LMS adaptive algorithm and the selection of its learning rate are introduced above. Next, the statistical performance of the LMS algorithm is analyzed by using the independence theory.

The independence theory of the LMS algorithm was first proposed by Widrow et al.[226] and Mazo[150], which was based on the following independence assumption.
(1) The input vectors $\boldsymbol{u}(1), \boldsymbol{u}(2), \cdots, \boldsymbol{u}(n)$ are statistically independent of each other;
(2) At time n, the input vector $\boldsymbol{u}(n)$ is statistically independent of the expected responses $d(1), \cdots, d(n-1)$ at all past times;
(3) At time n, the expected response $d(n)$ is related to the input vector $\boldsymbol{u}(n)$, but statistically independent of all the input vectors at the past time;
(4) The input vector $\boldsymbol{u}(n)$ and the expected response $d(n)$ constitute a random variable of joint Gaussian distribution for all n.

Let $\boldsymbol{w}_{\text{opt}}$ represent the optimal Wiener filter, then the weight error vector is defined as

$$\boldsymbol{\epsilon}(n) \stackrel{\text{def}}{=} \boldsymbol{w}(n) - \boldsymbol{w}_{\text{opt}}. \tag{5.5.35}$$

Thus, the estimation error $e(n) \stackrel{\text{def}}{=} d(n) - \boldsymbol{w}^{\mathrm{H}}(n)\boldsymbol{u}(n)$ generated by LMS algorithm can be rewritten as

$$e(n) = d(n) - \boldsymbol{w}_{\text{opt}}^{\mathrm{H}}\boldsymbol{u}(n) - \boldsymbol{\epsilon}^{\mathrm{H}}(n)\boldsymbol{u}(n) = e_{\text{opt}}(n) - \boldsymbol{\epsilon}^{\mathrm{H}}(n)\boldsymbol{u}(n), \tag{5.5.36}$$

where $e_{\text{opt}}(n)$ is the estimation error of the optimal Wiener filter. The mean square value of the estimation error of the tap weight vector $\boldsymbol{w}(n)$, shortened to the mean square error, is written as

$$\xi(n) = \text{MSE}(\boldsymbol{w}(n)) = \text{E}\{|e(n)|^2\}. \tag{5.5.37}$$

Using the independence hypothesis, it is easy to obtain

$$\begin{aligned}\xi(n) &= \mathrm{E}\left\{\left[e_{\mathrm{opt}}(n) - \boldsymbol{\epsilon}^{\mathrm{H}}(n)\boldsymbol{u}(n)\right]\left[e^{*}_{\mathrm{opt}}(n) - \boldsymbol{u}^{\mathrm{H}}(n)\boldsymbol{\epsilon}(n)\right]\right\} \\ &= \xi_{\min}(n) + \mathrm{E}\left\{\boldsymbol{\epsilon}^{\mathrm{H}}(n)\boldsymbol{u}(n)\boldsymbol{u}^{\mathrm{H}}(n)\boldsymbol{\epsilon}(n)\right\}, \qquad (5.5.38)\end{aligned}$$

where

$$\xi_{\min}(n) = \mathrm{E}\left\{|e_{\mathrm{opt}}(n)|^{2}\right\} = \mathrm{E}\left\{e_{\mathrm{opt}}(n)e^{*}_{\mathrm{opt}}(n)\right\}, \qquad (5.5.39)$$

is the minimum mean square error generated by the optimal Wiener filter.

Calculate the second term on the right of Eq. (5.5.38) to yield

$$\begin{aligned}\mathrm{E}\left\{\boldsymbol{\epsilon}^{\mathrm{H}}(n)\boldsymbol{u}(n)\boldsymbol{u}^{\mathrm{H}}(n)\boldsymbol{\epsilon}(n)\right\} &= \mathrm{E}\left\{\mathrm{tr}\left(\boldsymbol{\epsilon}^{\mathrm{H}}(n)\boldsymbol{u}(n)\boldsymbol{u}^{\mathrm{H}}(n)\boldsymbol{\epsilon}(n)\right)\right\} \\ &= \mathrm{tr}\left(\mathrm{E}\left\{\boldsymbol{u}(n)\boldsymbol{u}^{\mathrm{H}}(n)\boldsymbol{\epsilon}(n)\boldsymbol{\epsilon}^{\mathrm{H}}(n)\right\}\right). \qquad (5.5.40)\end{aligned}$$

The above equation uses the properties of matrix trace $\mathrm{tr}(\boldsymbol{AB}) = \mathrm{tr}(\boldsymbol{BA})$ and assumes that $\boldsymbol{\epsilon}(n)$ and $\boldsymbol{u}(n)$ are statistically independent.

Using the independence hypothesis, Eq. (5.5.40) can also be written as

$$\mathrm{E}\left\{\boldsymbol{\epsilon}^{\mathrm{H}}(n)\boldsymbol{u}(n)\boldsymbol{u}^{\mathrm{H}}(n)\boldsymbol{\epsilon}(n)\right\} = \mathrm{tr}\left(\mathrm{E}\left\{\boldsymbol{u}(n)\boldsymbol{u}^{\mathrm{H}}(n)\right\}\mathrm{E}\left\{\boldsymbol{\epsilon}(n)\boldsymbol{\epsilon}^{\mathrm{H}}(n)\right\}\right) = \mathrm{tr}[\boldsymbol{RK}(n)], \qquad (5.5.41)$$

where $\boldsymbol{R} = E\left\{\boldsymbol{u}(n)\boldsymbol{u}^{\mathrm{H}}(n)\right\}$ is the correlation matrix of the input vector, and $\boldsymbol{K}(n) = E\left\{\boldsymbol{\epsilon}(n)\boldsymbol{\epsilon}^{\mathrm{H}}(n)\right\}$ represents the correlation matrix of the filter weight error vector at time n, abbreviated as the weight error correlation matrix.

By further substituting Eq. (5.5.41) into Eq. (5.5.38), the mean square error in the LMS algorithm can be expressed as

$$\xi(n) = \xi_{\min} + \mathrm{tr}[\boldsymbol{RK}(n)]. \qquad (5.5.42)$$

The difference between the mean square error $\xi(n)$ generated by the adaptive algorithm at time n and the minimum mean square error $\xi_{\min}$ generated by the optimal Wiener filter is called the residual mean square error of the adaptive algorithm at time n, which is recorded as $\xi_{\mathrm{ex}}(n)$, i.e.,

$$\xi_{\mathrm{ex}}(n) = \xi(n) - \xi_{\min} = \mathrm{tr}[\boldsymbol{RK}(n)]. \qquad (5.5.43)$$

When $n \to \infty$, the limit of the residual mean square error is called steady-state residual mean square error (or asymptotic residual mean square error), denoted as

$$\xi_{\mathrm{ex}} = \xi_{\mathrm{ex}}(\infty) = \lim_{n\to\infty} \mathrm{tr}[\boldsymbol{RK}(n)]. \qquad (5.5.44)$$

Finally, this subsection considers a special choice of the expected response $d(n)$, namely, $d \equiv 0$. At this time, the cost function of the minimum mean square error criterion becomes

$$J(n) = \mathrm{E}\left\{|\boldsymbol{w}^{\mathrm{H}}\boldsymbol{u}(n)|^{2}\right\}. \qquad (5.5.45)$$

Since the right side of the above equation represents the filter output energy, the minimization of the above equation is called the minimum output energy (MOE) criterion.

Define the mean output energy of the filter tap weight vector $\boldsymbol{w}(n)$ at time n as

$$\eta(n) \stackrel{\text{def}}{=} \text{MOE}((n)) = \text{E}\left\{\left|\boldsymbol{w}^{\text{H}}\boldsymbol{u}(n)\right|^2\right\}. \tag{5.5.46}$$

Since the weight error vector $\boldsymbol{\epsilon}(n) = \boldsymbol{w}(n) - \boldsymbol{w}_{\text{opt}}$ and $\boldsymbol{w}_{\text{opt}}$ are statistically independent of the filter input vector $\boldsymbol{u}(n)$, from Eq. (5.5.46) it is obtained

$$\begin{aligned}\eta(n) &= \text{E}\left\{\left|\left[\boldsymbol{w}_{\text{opt}} + \boldsymbol{\epsilon}(n)\right]^{\text{H}}\boldsymbol{u}(n)\right|^2\right\} \\ &= \text{E}\left\{\left|\boldsymbol{w}_{\text{opt}}^{\text{H}}\boldsymbol{u}(n)\right|^2\right\} + \text{E}\left\{\boldsymbol{\epsilon}^{\text{H}}(n)\boldsymbol{u}(n)\boldsymbol{u}^{\text{H}}(n)\boldsymbol{\epsilon}(n)\right\} \\ &= \eta_{\text{min}} + \text{E}\left\{\boldsymbol{\epsilon}^{\text{H}}(n)\boldsymbol{u}(n)\boldsymbol{u}^{\text{H}}(n)\boldsymbol{\epsilon}(n)\right\},\end{aligned} \tag{5.5.47}$$

where $\eta_{\text{min}} = \text{E}\left\{\left|\boldsymbol{w}_{\text{opt}}^{\text{H}}\boldsymbol{u}(n)\right|^2\right\}$ represents the output energy of the optimal filter, which is the minimum output energy that the adaptive filter can achieve.

The design criterion that makes the filter reach the minimum output energy is called the MOE criterion

Define the residual output energy as

$$\eta_{\text{ex}}(n) \stackrel{\text{def}}{=} \eta(n) - \eta_{\text{min}}, \tag{5.5.48}$$

and from Eq. (5.5.41), we have

$$\eta_{\text{ex}}(n) = \text{E}\left\{\boldsymbol{\epsilon}^{\text{H}}(n)\boldsymbol{u}(n)\boldsymbol{u}^{\text{H}}(n)\boldsymbol{\epsilon}(n)\right\} = \text{tr}[\boldsymbol{RK}(n)]. \tag{5.5.49}$$

Compare Eq. (5.5.49) and Eq. (5.5.43) to obtain

$$\eta_{\text{ex}}(\infty) = \xi_{\text{ex}}(\infty). \tag{5.5.50}$$

That is, the steady-state residual output energy of the filter is equivalent to the steady-state residual output mean square error.

The above analysis shows that although the filter tap weight vectors designed according to the MMSE criterion and MOE criterion may be different, their steady-state residual mean square error and steady-state residual output energy are equivalent. In particular, the variation curve of the actually measured residual mean square error $\xi_{\text{ex}}(n)$ relative to the iteration time n is called the learning curve of the LMS algorithm. It is a curve that decreases with time, from which we can see the convergence performance of the LMS algorithm (the speed of convergence and the value of steady-state residual mean square error).

5.5.6 Tracking Performance of LMS Algorithm

sis of the statistical performance of the LMS algorithm is carried out under the basic assumption that the Wiener filter is fixed. Therefore, these statistical performances are the "average performance" of standard LMS algorithms, which are suitable for stationary environments.

In the non-stationary environment, the parameters of the system are time-varying, so the parameters of the Wiener filter should also be time-varying to track the dynamic changes of the system. The index to evaluate the adaptability of the LMS algorithm to a non-stationary environment is its tracking performance. According to the speed of parameters changing with time, time-varying systems can be divided into fast time-varying and slow time-varying. Only the slow time-varying environment is discussed here.

An unknown dynamic system can be modeled by a transverse filter whose tap weight vector, i.e. impulse response vector $\boldsymbol{w}_{\text{opt}}(n)$, follows a first-order Markov process

$$\boldsymbol{w}_{\text{opt}}(n+1) = a\boldsymbol{w}_{\text{opt}}(n) + \boldsymbol{\omega}(n), \tag{5.5.51}$$

where a is a fixed parameter. For slow time-varying systems, a is a positive number very close to 1, $\boldsymbol{\omega}(n)$ is the process noise with zero mean and correlation matrix Q.

The transverse filter output $\boldsymbol{w}_{\text{opt}}^{\text{H}}(n)\boldsymbol{u}(n)$ approximates the desired response, and its approximation error $v(n)$ is called the measurement noise. Therefore, the desired response of the transverse filter can be expressed as

$$d(n) = \boldsymbol{w}_{\text{opt}}^{\text{H}}(n)\boldsymbol{u}(n) + v(n). \tag{5.5.52}$$

The input, process, and measurement noises of the filter are assumed as follows:

(1) The process noise vector $\boldsymbol{\omega}(n)$ is independent of the input vector $\boldsymbol{u}(n)$ and the measurement noise vector $\boldsymbol{v}(n)$;

(2) The input vector $\boldsymbol{u}(n)$ and the measurement noise $v(n)$ are independent of each other;

(3) The measurement noise $v(n)$ is white noise with zero mean and finite variance $\sigma_v^2 < \infty$.

To describe the fast and slow change of the model, Macchi[144, 145] defines the degree of nonstationarity of the time-varying system as the ratio of the average noise power caused by the process noise vector $\boldsymbol{\omega}(n)$ to the average noise power caused by the measurement noise, i.e.,

$$\alpha \overset{\text{def}}{=} \left(\frac{\mathrm{E}\left\{ \left| \boldsymbol{\omega}^{\text{H}}(n)\boldsymbol{u}(n) \right|^2 \right\}}{\mathrm{E}\left\{ |v(n)|^2 \right\}} \right)^{1/2}. \tag{5.5.53}$$

Note that the degree of nonstationarity α is only a characteristic description of time-varying systems, and it does not describe adaptive filters.

Using the statistical independence between the process noise vector $\boldsymbol{\omega}(n)$ and the input vector $\boldsymbol{u}(n)$, and noting that for scalar $\boldsymbol{x}^{\text{H}}\boldsymbol{y}$, $\mathrm{E}\left\{\boldsymbol{x}^{\text{H}}\boldsymbol{y}\right\} = \mathrm{tr}\left[\mathrm{E}\left\{\boldsymbol{x}\boldsymbol{y}^{\text{H}}\right\}\right]$, it is easy to obtain that the numerator of Eq. (5.5.53) is[100]

$$
\begin{aligned}
\mathrm{E}\left\{\left|\boldsymbol{\omega}^{\mathrm{H}}(n)\boldsymbol{u}(n)\right|^2\right\} &= \mathrm{E}\left\{\boldsymbol{\omega}^{\mathrm{H}}(n)\boldsymbol{u}(n)\boldsymbol{u}^{\mathrm{H}}(n)\boldsymbol{\omega}(n)\right\} \\
&= \mathrm{tr}\left[\mathrm{E}\left\{\boldsymbol{\omega}^{\mathrm{H}}(n)\boldsymbol{u}(n)\boldsymbol{u}^{\mathrm{H}}(n)\boldsymbol{\omega}(n)\right\}\right] \\
&= \mathrm{E}\left\{\mathrm{tr}\left[\boldsymbol{\omega}(n)\boldsymbol{\omega}^{\mathrm{H}}(n)\boldsymbol{u}(n)\boldsymbol{u}^{\mathrm{H}}(n)\right]\right\} \\
&= \mathrm{tr}\left[\mathrm{E}\left\{\boldsymbol{\omega}(n)\boldsymbol{\omega}^{\mathrm{H}}(n)\right\}\mathrm{E}\left\{\boldsymbol{u}(n)\boldsymbol{u}^{\mathrm{H}}(n)\right\}\right] \\
&= \mathrm{tr}(\boldsymbol{QR}),
\end{aligned}
\tag{5.5.54}
$$

where $\boldsymbol{Q} = \mathrm{E}\left\{\boldsymbol{\omega}(n)\boldsymbol{\omega}^{\mathrm{H}}(n)\right\}$ is the correlation matrix of the process noise vector $\boldsymbol{\omega}(n)$; $\boldsymbol{R} = \mathrm{E}\left\{\boldsymbol{u}(n)\boldsymbol{u}^{\mathrm{H}}(n)\right\}$ represents the correlation matrix of the input vector $\boldsymbol{u}(n)$.

On the other hand, the denominator of Eq. (5.5.53) is the variance σ_v^2 of the measurement noise $v(n)$ with zero mean. By substituting this result and Eq. (5.5.54) into Eq. (5.5.53), the degree of nonstationary can be simplified as

$$
\alpha = \frac{1}{\sigma_v}(\mathrm{tr}[\boldsymbol{QR}])^{1/2} = \frac{1}{\sigma_v}(\mathrm{tr}[\boldsymbol{RQ}])^{1/2}, \tag{5.5.55}
$$

where $\mathrm{tr}[\boldsymbol{QR}] = \mathrm{tr}[\boldsymbol{RQ}]$ since the matrix products $\boldsymbol{QR}$ and $\boldsymbol{RQ}$ have the same diagonal elements.

In addition to the convergence rate introduced earlier, misadjustment is another important index to measure the performance of the adaptive filter. The misadjustment of adaptive filter is defined as the ratio of filter steady-state residual mean square error J_{ex} to filter minimum mean square error $J_{\min}$, i.e.,

$$
\mathcal{M} \overset{\mathrm{def}}{=} \frac{J_{\mathrm{ex}}}{J_{\min}}, \tag{5.5.56}
$$

In the above equation, the steady-state residual mean square error is defined as the difference between the actual mean square error and the minimum mean square error of the filter output, i.e., $J_{\mathrm{out}} - J_{\min}$.

Obviously, when $J_{\mathrm{ex}} = 0$, the filter output reaches the minimum output mean square error exactly, which is the optimal filter in the sense of minimum mean square error. At this time, $\mathcal{M} = 0$, i.e., there is no misadjustment in the filter. It can be seen that the misadjustment $\mathcal{M}$ is actually a measure of the deviation of the filter from the optimal filter. As long as the remaining output energy is not equal to zero, the filter is said to be maladjusted. It is generally expected that the smaller the misadjustment, the better the adaptive filter, which depends on the design of the filter and the environment in which the filter is located (for example, the degree of nonstationarity of the signal that the filter wants to track).

The relationship between the degree of nonstationary α and the misadjustment of the adaptive filter is analyzed below. For a filter designed by the MMSE criterion, its minimum mean square error $J_{\min}$ is equal to the variance of the measurement noise, i.e.,

$$J_{\min} = \sigma_V^2. \tag{5.5.57}$$

On the other hand, it can be seen from the Markov model defined by Eq. (5.5.51) that the process noise vector $\boldsymbol{\omega}(n)$ is actually the filter weight error vector, i.e., $\boldsymbol{\omega}(n) \approx \boldsymbol{w}_{\text{opt}}(n+1) - \boldsymbol{w}_{\text{opt}}(n) = \boldsymbol{\epsilon}(n)$, because the coefficient α in Eq. (5.5.51) is very close to 1. This shows that at time n, the correlation matrix of the process noise vector $\boldsymbol{\omega}(n)$ satisfying $\boldsymbol{Q}(n) = \mathrm{E}\{\boldsymbol{\omega}(n)\boldsymbol{\omega}^{\mathrm{H}}(n)\} \approx \mathrm{E}\{\boldsymbol{\epsilon}(n)\boldsymbol{\epsilon}^{\mathrm{H}}(n)\} = \boldsymbol{K}(n)$. Substituting this relation into Eq. (5.5.49), we can find the residual output energy of the adaptive filter at time n to be

$$\eta_{\text{ex}}(n) \approx \text{tr}(\boldsymbol{RQ}).$$

If the adaptive filter has no misadjustment, its steady-state residual output energy $J_{\text{ex}} = \eta_{\text{ex}}(\infty) = \text{tr}(\boldsymbol{RQ})$. That is, the minimum mean square error filter is equivalent to the minimum energy filter. If the adaptive filter has misadjustment, its steady-state residual output energy $J_{\text{ex}} > \text{tr}(\boldsymbol{RQ})$. Therefore, there is

$$J_{\text{ex}} \geq \text{tr}(\boldsymbol{RQ}). \tag{5.5.58}$$

Substituting Eqs. (5.5.57) and (5.5.58) into Eq. (5.5.56), there is

$$\mathcal{M} \geq \frac{\text{tr}(\boldsymbol{RQ})}{\sigma_V^2} = \alpha^2. \tag{5.5.59}$$

In other words, the misadjustment $\mathcal{M}$ of the adaptive filter is the upper bound of the square value of the degree of nonstationarity of the time-varying system.

The following conclusions can be drawn from the above analysis[100].

(1) For slow time-varying systems, because the degree of nonstationary α is small, the adaptive filter can track the changes of time-varying systems.

(2) If the time-varying system changes so fast that the degree of nonstationarity α is greater than 1, in this case, the misadjustment $\mathcal{M}$ caused by the adaptive filter is also greater than 1, i.e., the misadjustment will exceed 100%. This means that the adaptive filter will not be able to track the changes of this fast time-varying system anymore.

5.6 RLS Adaptive Algorithm

This section will discuss the adaptive implementation of the least square method. Its purpose is to design an adaptive transverse filter so that when the filter tap weight coefficient at time $n-1$ is known, the filter tap weight coefficient at time n can be obtained through simple update. Such an adaptive least squares algorithm is called recursive least squares (RLS) algorithm.

5.6.1 RLS Algorithm

Different from the general least squares method, an exponentially weighted least squares method is considered here. As the name suggests, in this method, the exponentially weighted sum of squares of errors is used as the cost function, i.e.,

$$J(n) = \sum_{i=0}^{n} \lambda^{n-i} |\epsilon(i)|^2, \tag{5.6.1}$$

where the weighting factor $0 < \lambda < 1$ is called forgetting factor. Its function is to add a larger weight to the error closer to time n and a smaller weight to the error farther from time n. In other words, λ has a certain forgetting effect on the error of each time, so it is called a forgetting factor. In this sense, $\lambda \equiv 1$ means that the errors at all times are treated equally, that is, there is no forgetting function or infinite memory function. At this time, the exponentially weighted least squares method degenerates into a general least squares method. On the contrary, if $\lambda = 0$, only the error of the current moment works, while the error of the past moment is completely forgotten and has no effect. In the nonstationary environment, to track the changing system, these two extreme forgetting factor values are not appropriate.

The estimation error in Eq. (5.6.1) is defined as

$$\epsilon(i) = d(i) - \boldsymbol{w}^{\mathrm{H}}(n)\boldsymbol{u}(i), \tag{5.6.2}$$

where $d(i)$ represents the expected response at time i. When the desired response cannot be known, the actual output of the filter can be taken directly as the desired response $d(i)$. Note that the tap weight vector in Eq. (5.6.2) is the weight vector $\boldsymbol{w}(n)$ at time n rather than the weight vector $\boldsymbol{w}(i)$ at time i. The reason is as follows: in the adaptive update process, the filter is always getting better and better, which means that the absolute value $|\epsilon(i)| = |d(i) - \boldsymbol{w}^{\mathrm{H}}(n)\boldsymbol{u}(i)|$ of the estimation error is always smaller than $|e(i)| = |d(i) - \boldsymbol{w}^{\mathrm{H}}(i)\boldsymbol{u}(i)|$ for any time $i \le n$. Therefore, the cost function $J(n)$ composed of $\epsilon(i)$ is always smaller than the cost function $\tilde{J}(n)$ composed of $e(i)$, thus the cost function $J(n)$ is more reasonable than $\tilde{J}(n)$. According to the definition, $\epsilon(i)$ is called the posterior estimation error of the filter at time i, and $e(i)$ is called the prior estimation error at time i. Therefore, the complete expression of the sum of squares of the weighted errors is

$$J(n) = \sum_{i=0}^{n} \lambda^{n-i} \left| d(i) - \boldsymbol{w}^{\mathrm{H}}(n)\boldsymbol{u}(i) \right|^2, \tag{5.6.3}$$

which is a function of $\boldsymbol{w}(n)$. From $\frac{\partial J(n)}{\partial \boldsymbol{w}} = 0$, it is easy to get $\boldsymbol{R}(n)\boldsymbol{w}(n) = \boldsymbol{r}(n)$, and its solution is

$$\boldsymbol{w}(n) = \boldsymbol{R}^{-1}(n)\boldsymbol{r}(n), \tag{5.6.4}$$

where

$$R(n) = \sum_{i=0}^{n} \lambda^{n-i} u(i) u^{H}(i), \tag{5.6.5}$$

$$r(n) = \sum_{i=0}^{n} \lambda^{n-i} u(i) d^{*}(i). \tag{5.6.6}$$

Again, Eq. (5.6.2) shows that the solution $w(n)$ of the exponentially weighted least squares problem is a Wiener filter. Consider its adaptive update below.

According to the definitions of Eq. (5.6.5) and Eq. (5.6.6), it is easy to obtain the following recursive estimation equations

$$R(n) = \lambda R(n-1) + u(n) u^{H}(n), \tag{5.6.7}$$

$$r(n) = \lambda r(n-1) + u(n) d^{*}(n). \tag{5.6.8}$$

Using the famous matrix inverse lemma for Eq. (5.6.7), the recursive expression of the inverse matrix $P(n) = R^{-1}(n)$ can be obtained

$$\begin{aligned} P(n) &= \frac{1}{\lambda}\left[P(n-1) - \frac{P(n-1)u(n)u^{H}(n)P(n-1)}{\lambda + u^{H}(n)P(n-1)u(n)}\right] \\ &= \frac{1}{\lambda}\left[P(n-1) - k(n)u^{H}(n)P(n-1)\right], \end{aligned} \tag{5.6.9}$$

where $k(n)$ is called the gain vector, defined as

$$k(n) = \frac{P(n-1)u(n)}{\lambda + u^{H}(n)P(n-1)u(n)}. \tag{5.6.10}$$

By using Eq. (5.6.9), it is not difficult to prove

$$\begin{aligned} P(n)u(n) &= \frac{1}{\lambda}\left[P(n-1)u(n) - k(n)u^{H}(n)P(n-1)u(n)\right] \\ &= \frac{1}{\lambda}\left\{\left[\lambda + u^{H}(n)P(n-1)u(n)\right]k(n) - k(n)u^{H}(n)P(n-1)u(n)\right\} \\ &= k(n). \end{aligned} \tag{5.6.11}$$

Meanwhile, from Eq. (5.6.4) it can be obtained

$$\begin{aligned} w(n) &= R^{-1}(n)r(n) = P(n)r(n) \\ &= \frac{1}{\lambda}\left[P(n-1) - k(n)u^{H}(n)P(n-1)\right]\left[\lambda r(n-1) + d^{*}(n)u(n)\right] \\ &= P(n-1)r(n-1) + \frac{1}{\lambda}d^{*}(n)\left[P(n-1)u(n) - k(n)u^{H}(n)P(n-1)u(n)\right] \\ &- k(n)u^{H}(n)P(n-1)r(n-1). \end{aligned}$$

Substitute Eq. (5.6.11) into the above equation to obtain

$$w(n) = w(n-1) + d^{*}(n)k(n) - k(n)u^{H}(n)w(n-1)$$

After simplification, we get

$$\boldsymbol{w}(n) = \boldsymbol{w}(n-1) + \boldsymbol{k}(n)e^*(n), \tag{5.6.12}$$

where

$$e(n) = d(n) - \boldsymbol{u}^{\mathrm{T}}(n)\boldsymbol{w}^*(n-1) = d(n) - \boldsymbol{w}^{\mathrm{H}}(n-1)\boldsymbol{u}(n), \tag{5.6.13}$$

is a priori estimation error.

To sum up, the RLS direct algorithm can be obtained as follows.

Algorithm 5.6.1. *RLS direct algorithm*
Step 1 Initialization: $\boldsymbol{w}(0) = \boldsymbol{0}$, $\boldsymbol{P}(0) = \delta^{-1}\boldsymbol{I}$, *where* δ *is a very small number.*
Step 2 Update: $n = 1, 2, \cdots$

$$\begin{aligned} e(n) &= d(n) - \boldsymbol{w}^H(n-1)\boldsymbol{u}(n), \\ \boldsymbol{k}(n) &= \frac{\boldsymbol{P}(n-1)\boldsymbol{u}(n)}{\lambda + \boldsymbol{u}^H(n)\boldsymbol{P}(n-1)\boldsymbol{u}(n)}, \\ \boldsymbol{P}(n) &= \frac{1}{\lambda}\left[\boldsymbol{P}(n-1) - \boldsymbol{k}(n)\boldsymbol{u}^H(n)\boldsymbol{P}(n-1)\right], \\ \boldsymbol{w}(n) &= \boldsymbol{w}(n-1) + \boldsymbol{k}(n)e^*(n). \end{aligned}$$

The application of RLS algorithm requires the initial value $\boldsymbol{P}(0) = \boldsymbol{R}^{-1}(0)$. In the non-stationary environment, the initial value is

$$\boldsymbol{P}(0) = \boldsymbol{R}^{-1}(0) = \left(\sum_{i=-n_0}^{0} \lambda^{-i}\boldsymbol{u}(i)\boldsymbol{u}^{\mathrm{H}}(i)\right)^{-1}. \tag{5.6.14}$$

Therefore, the correlation matrix Eq. (5.6.6) becomes

$$\boldsymbol{R}(n) = \sum_{i=1}^{n} \lambda^{n-i}\boldsymbol{u}(i)\boldsymbol{u}^{\mathrm{H}}(i) + \boldsymbol{R}(0), \tag{5.6.15}$$

Due to the forgetting effect of λ, it is natural to hope that $\boldsymbol{R}(0)$ plays a small role in Eq. (5.6.15). Considering this, a small identity matrix is used to approximate $\boldsymbol{R}(0)$, i.e.,

$$\boldsymbol{R}(0) = \delta\boldsymbol{I}, \tag{5.6.16}$$

with δ being a small positive number and $\boldsymbol{I}$. being identity matrix. Then, the initial value of $\boldsymbol{P}(0)$ is

$$\boldsymbol{P}(0) = \delta^{-1}\boldsymbol{I}, \tag{5.6.17}$$

with δ being the same as in Eq. (5.6.16).This is why in Algorithm 5.6.1, the initial value $\boldsymbol{P}(0) = \delta^{-1}\boldsymbol{I}$ (where δ is very small).

The smaller the value of δ, the smaller the proportion of the initial value of the correlation matrix $\boldsymbol{R}(0)$ in the calculation of $\boldsymbol{R}(n)$, which is desirable; otherwise, the role of $\boldsymbol{R}(0)$ will be highlighted, which should be avoided. The typical value of δ is

$\delta = 0.01$ or less. Generally, when $\delta = 0.01$ and $\delta = 10^{-4}$ are taken, there is no obvious difference in the results given by the RLS algorithm, but taking $\delta = 1$ will seriously affect the convergence speed and convergence results of the RLS algorithm, which must be paid attention to when applying RLS algorithm.

5.6.2 Comparison between RLS Algorithm and Kalman Filtering Algorithm

A special non excitation dynamic model is considered

$$\boldsymbol{x}(n+1) = \lambda^{-1/2}\boldsymbol{x}(n), \tag{5.6.18}$$

$$y(n) = \boldsymbol{u}^{\mathrm{H}}(n)\boldsymbol{x}(n) + v(n), \tag{5.6.19}$$

where $\boldsymbol{x}(n)$ is the state vector of the model; $y(n)$ is a scalar observation or reference signal; $\boldsymbol{u}^{\mathrm{H}}(n)$ is the observation matrix; $v(n)$ represents a scalar white noise process with zero mean and unit variance. The model parameter λ is a real positive constant.

From Eq. (5.6.18), it is easy to get

$$\boldsymbol{x}(n) = \lambda^{-n/2}\boldsymbol{x}(0), \tag{5.6.20}$$

where $\boldsymbol{x}(0)$ is the initial value of the state vector. Substitute Eq. (5.6.20) into Eq. (5.6.19) and use the common term $\boldsymbol{x}(0)$ to represent the observations at each time to get

$$\left.\begin{aligned} y(0) &= \boldsymbol{u}^{\mathrm{H}}(0)\boldsymbol{x}(0) + v(0) \\ y(1) &= \lambda^{-1/2}\boldsymbol{u}^{\mathrm{H}}(1)\boldsymbol{x}(0) + v(1) \\ &\vdots \\ y(n) &= \lambda^{-n/2}\boldsymbol{u}^{\mathrm{H}}(n)\boldsymbol{x}(0) + v(n) \end{aligned}\right\}, \tag{5.6.21}$$

or equivalently

$$\left.\begin{aligned} y(0) &= \boldsymbol{u}^{\mathrm{H}}(0)\boldsymbol{x}(0) + v(0) \\ \lambda^{1/2}y(1) &= \boldsymbol{u}^{\mathrm{H}}(1)\boldsymbol{x}(0) + \lambda^{1/2}v(1) \\ &\vdots \\ \lambda^{n/2}y(n) &= \boldsymbol{u}^{\mathrm{H}}(n)\boldsymbol{x}(0) + \lambda^{n/2}v(n) \end{aligned}\right\}, \tag{5.6.22}$$

From the viewpoint of Kalman filtering, the equation set (5.6.22) represents the random characteristics of the non excitation dynamic model.

Different from the Kalman filter using a stochastic model, the RLS algorithm adopts a deterministic model, i.e., the desired signal (also known as reference signal) can be expressed as a linear regression model

$$\left.\begin{aligned} d^*(0) &= \boldsymbol{u}^{\mathrm{H}}(0)\boldsymbol{w}_o + e_o^*(0) \\ d^*(1) &= \boldsymbol{u}^{\mathrm{H}}(1)\boldsymbol{w}_o + e_o^*(1) \\ &\vdots \\ d^*(n) &= \boldsymbol{u}^{\mathrm{H}}(n)\boldsymbol{w}_o + e_o^*(n) \end{aligned}\right\}, \tag{5.6.23}$$

where $\boldsymbol{w}_o$ represents the unknown parameter vector of the model, $\boldsymbol{u}(n)$ is the input vector, and $e_o(n)$ is the observation noise.

If the initial value of the state vector in the Kalman filter is equal to the tap weight vector in the RLS algorithm, i.e.,

$$\boldsymbol{x}(0) = \boldsymbol{w}_o, \tag{5.6.24}$$

it is easy to see that the equivalence condition between the determination model in Eq. (5.6.23) of the RLS algorithm and the special random model in Eq. (5.6.22) of the Kalman filtering algorithm is the following one-to-one correspondence

$$y(n) = \lambda^{-n/2} d^*(n), \tag{5.6.25}$$

$$v(n) = \lambda^{-n/2} e_o^*(n), \tag{5.6.26}$$

with the left and right sides of the above equations being the parameters of the state space model and the linear regression model, respectively.

Summarizing the above analysis, we can draw the following conclusion: the deterministic linear regression model of the RLS adaptive algorithm is a special non excitation state space model of the Kalman filtering algorithm. This equivalence relationship was established by Sayed and Kailaith in 1994[190].

Table 5.6.1 summarizes the corresponding relationship of each variable between the Kalman filter algorithm and RLS algorithm[100].

Tab. 5.6.1: The association between the parameters of Kalman filter algorithm and that of RLS algorithm

Kalman algorithm		RLS algorithm	
Parameter	**Variable**	**Variable**	**Parameter**
Initial state vector	$\boldsymbol{x}(0)$	$\boldsymbol{w}(0)$	Tap weight vector
State vector	$\boldsymbol{x}(n)$	$\lambda^{-n/2}\boldsymbol{w}_0$	Exponentially weighted tap weight vector
Reference signal	$y(n)$	$\lambda^{-n/2}d^*(n)$	Expected response
Observation noise	$v(n)$	$\lambda^{-n/2}e_o^*(n)$	Measurement error
One-step predicted state vector	$\hat{\boldsymbol{x}}(n+1\|y_1,\cdots,y_n)$	$\lambda^{-n/2}\hat{\boldsymbol{w}}(n)$	Estimation of tap weight vector
Correlation matrix of state prediction error	$\boldsymbol{K}(n)$	$\lambda^{-1}\boldsymbol{P}(n)$	Inverse of input vector correlation matrix
Kalman gain	$\boldsymbol{g}(n)$	$\lambda^{-1/2}\boldsymbol{k}(n)$	Gain vector
Innovation	$\alpha(n)$	$\lambda^{-n/2}\xi^*(n)$	A priori estimation error
Initial condition	$\hat{\boldsymbol{x}}(1) = \boldsymbol{0}$	$\hat{\boldsymbol{w}}(0) = \boldsymbol{0}$	Initial condition
	$\boldsymbol{K}(0)$	$\delta^{-1}\boldsymbol{P}(0)$	

5.6.3 Statistical Performance Analysis of RLS Algorithm

Since the measurement error $e_{\text{opt}}(n) = d(n) - \boldsymbol{w}_{\text{opt}}^{\text{H}}\boldsymbol{u}(n)$ of the Wiener filter has minimum mean square value, the desired response $d(n)$ can be written as

$$d(n) = e_{\text{opt}}(n) + \boldsymbol{w}_{\text{opt}}^{\text{H}}\boldsymbol{u}(n), \tag{5.6.27}$$

which is often called the linear regression model of the expected response $d(n)$, and the $M \times 1$ weight vector $\boldsymbol{w}_{\text{opt}}$ represents the regression parameter vector of the model.

From Eq. (5.6.13) and Eq. (5.6.27), by eliminating $d(n)$, the a priori estimation error can be expressed as

$$\begin{aligned} e(n) &= e_{\text{opt}}(n) - \left[\boldsymbol{w}(n-1) - \boldsymbol{w}_{\text{opt}}\right]^{\text{H}}\boldsymbol{u}(n) \\ &= e_{\text{opt}}(n) - \boldsymbol{\epsilon}^{\text{H}}(n-1)\boldsymbol{u}(n), \end{aligned} \tag{5.6.28}$$

where

$$\boldsymbol{\epsilon}(n-1) = \boldsymbol{w}(n-1) - \boldsymbol{w}_{\text{opt}}, \tag{5.6.29}$$

represents the difference between the actual tap weight vector at time $n-1$ and the tap weight vector of the optimal Wiener filter, shortened to the weight error vector.

Consider the mean square value of the a priori estimation error, i.e., the mean square estimation error

$$\xi(n) = \text{MSE}(\boldsymbol{w}(n)) = \text{E}\{|e(n)|^2\}. \tag{5.6.30}$$

Substitute Eq. (5.6.28) into Eq. (5.6.30) and sort it out to obtain

$$\begin{aligned} \xi(n) = {} & \text{E}\left\{|e_{\text{opt}}(n)|^2\right\} + \text{E}\left\{\boldsymbol{u}^{\text{H}}(n)\boldsymbol{\epsilon}(n-1)\boldsymbol{\epsilon}^{\text{H}}(n-1)\boldsymbol{u}(n)\right\} \\ & - \text{E}\left\{e_{\text{opt}}(n)\boldsymbol{u}^{\text{H}}(n)\boldsymbol{\epsilon}(n-1)\right\} - \text{E}\left\{\boldsymbol{\epsilon}^{\text{H}}(n-1)\boldsymbol{u}(n)e_{\text{opt}}^{*}(n)\right\}. \end{aligned} \tag{5.6.31}$$

The values of each item on the right side of Eq. (5.6.31) are analyzed below.

(1) The first term on the right of Eq. (5.6.31) represents the mean square error of the optimal Wiener filter, which is the minimum mean square error that all filters would have, denoted as

$$\xi_{\min} = \text{E}\left\{|e_{\text{opt}}(n)|^2\right\}. \tag{5.6.32}$$

(2) Calculate the second term on the right of Eq. (5.6.31) and get

$$\begin{aligned} \text{E}\left\{\boldsymbol{u}^{\text{H}}(n)\boldsymbol{\epsilon}(n-1)\boldsymbol{\epsilon}^{\text{H}}(n-1)\boldsymbol{u}(n)\right\} &= \text{E}\left\{\text{tr}\left[\boldsymbol{u}^{\text{H}}(n)\boldsymbol{\epsilon}(n-1)\boldsymbol{\epsilon}^{\text{H}}(n-1)\boldsymbol{u}(n)\right]\right\} \\ &= \text{E}\left\{\text{tr}\left[\boldsymbol{u}(n)\boldsymbol{u}^{\text{H}}(n)\boldsymbol{\epsilon}(n-1)\boldsymbol{\epsilon}^{\text{H}}(n-1)\right]\right\} \\ &= \text{tr}\left[\text{E}\left\{\boldsymbol{u}(n)\boldsymbol{u}^{\text{H}}(n)\boldsymbol{\epsilon}(n-1)\boldsymbol{\epsilon}^{\text{H}}(n-1)\right\}\right] \\ &= \text{tr}\left[\text{E}\left\{\boldsymbol{u}(n)\boldsymbol{u}^{\text{H}}(n)\right\}\text{E}\left\{\boldsymbol{\epsilon}(n-1)\boldsymbol{\epsilon}^{\text{H}}(n-1)\right\}\right] \\ &= \text{tr}\left[\boldsymbol{R}\boldsymbol{K}(n-1)\right], \end{aligned} \tag{5.6.33}$$

where $\boldsymbol{R} = \mathrm{E}\left\{\boldsymbol{u}(n)\boldsymbol{u}^{\mathrm{H}}(n)\right\}$ is the correlation matrix of the filter input, and $\boldsymbol{K}(n-1) = \mathrm{E}\left\{\boldsymbol{\epsilon}(n-1)\boldsymbol{\epsilon}^{\mathrm{H}}(n-1)\right\}$ is the weight error correlation matrix at time $n-1$.

(3) Since the weight error vector $\boldsymbol{\epsilon}(n-1)$ at time $n-1$ is statistically independent of the input vector $\boldsymbol{u}(n)$ and the measurement error $e_{\mathrm{opt}}(n)$ at time n, the third term on the right of Eq. (5.6.31) is

$$\mathrm{E}\left\{e_{\mathrm{opt}}(n)\boldsymbol{u}^{\mathrm{H}}(n)\boldsymbol{\epsilon}(n-1)\right\} = \mathrm{E}\left\{e_{\mathrm{opt}}(n)\boldsymbol{u}^{\mathrm{H}}(n)\right\}\mathrm{E}\left\{\boldsymbol{\epsilon}(n-1)\right\},$$

According to the orthogonality principle, the measurement error $e_{\mathrm{opt}}(n)$ is orthogonal to all elements of the input vector $\boldsymbol{u}(n)$, i.e., $\mathrm{E}\left\{e_{\mathrm{opt}}(n)\boldsymbol{u}^{\mathrm{H}}(n)\right\} = \boldsymbol{0}$, so we can get

$$\mathrm{E}\left\{e_{\mathrm{opt}}(n)\boldsymbol{u}^{\mathrm{H}}(n)\boldsymbol{\epsilon}(n-1)\right\} = 0. \tag{5.6.34}$$

(4) Similarly, we have

$$\mathrm{E}\left\{\boldsymbol{\epsilon}^{\mathrm{H}}(n-1)\boldsymbol{u}(n)e^{*}_{\mathrm{opt}}(n)\right\} = \mathrm{E}\left\{\boldsymbol{\epsilon}^{\mathrm{H}}(n-1)\right\}\mathrm{E}\left\{\boldsymbol{u}(n)e^{*}_{\mathrm{opt}}(n)\right\} = 0. \tag{5.6.35}$$

Substitute Eq. (5.6.32) ~ Eq. (5.6.35) into Eq. (5.6.31) to obtain $\xi(n) = \xi_{\min} + \mathrm{tr}\left[\boldsymbol{RK}(n-1)\right]$, from which the residual mean square error

$$\xi_{\mathrm{ex}}(n) = \xi(n) - \xi_{\min} = \mathrm{tr}\left[\boldsymbol{RK}(n-1)\right], \tag{5.6.36}$$

and the steady-state or asymptotic residual mean square error

$$\xi_{\mathrm{ex}}(\infty) = \lim_{n\to\infty}\mathrm{tr}\left[\boldsymbol{RK}(n-1)\right], \tag{5.6.37}$$

can be obtained.

The curve of the actual measured residual mean square error $\xi_{\mathrm{ex}}(n)$ relative to the iteration time n is called the learning curve of the RLS algorithm. In general, it is a curve that decreases with time, indicating the convergence performance, i.e., convergence rate and the steady-state residual mean square error of the RLS algorithm.

5.6.4 Fast RLS Algorithm

It has been proved that the Kalman gain vector in the RLS direct algorithm can be updated in a fast way so that the RLS algorithm can be implemented quickly[43, 141]. The key of fast RLS algorithm is to make proper use of the shift invariant property of the data matrix. For this purpose, consider the data vector $\boldsymbol{x}_M(n) = [u(n), u(n-1), \cdots, u(n-M+1)]^T$ after the data vector $\boldsymbol{x}_{M+1}(n) = [u(n), u(n-1), \cdots, u(n-M)]^T$ is increased by one order. Obviously, it has two different block forms

$$\boldsymbol{x}_{M+1}(n) = \begin{bmatrix}\boldsymbol{x}_M(n)\\ u(n-M)\end{bmatrix} = \begin{bmatrix}u(n)\\ \boldsymbol{x}_{M-1}(n-1)\end{bmatrix}. \tag{5.6.38}$$

Using these two block forms, the appropriate blocking of the increased order autocorrelation matrix $\boldsymbol{R}_{M+1}(n)$ can be obtained. According to these blocks, the Kalman gain vector $\boldsymbol{c}_M(n)$ at time n can be obtained from $\boldsymbol{c}_{M-1}(n-1)$ by the increased order vector $\boldsymbol{c}_{M+1}(n)$. The overall renewal mechanism can be expressed as

$$\begin{array}{ccccc} & \boldsymbol{a}_M(n-1) & & \boldsymbol{b}_M(n-1) & \\ & \downarrow & & \downarrow & \\ \boldsymbol{c}_{M-1}(n-1) & \rightarrow & \boldsymbol{c}_{M+1}(n) & \rightarrow & \boldsymbol{c}_M(n)\ , \\ & \downarrow & & \downarrow & \\ & \boldsymbol{a}_M(n) & & \boldsymbol{b}_M(n) & \end{array} \tag{5.6.39}$$

where the auxiliary vectors $\boldsymbol{a}_M(n)$ and $\boldsymbol{b}_M(n)$ represent the forward and backward least squares predictors, which are FIR transversal filters after letting $y(n) = u(n+1)$ and $y(n) = u(n-M)$ in Eq. (5.5.1). Here, the stabilized fast RLS algorithm[90] is given.

Algorithm 5.6.2. *Stabilized fast RLS algorithm*
Initialization:

$$\begin{aligned} &\boldsymbol{w}_M(0) = \boldsymbol{0},\ \boldsymbol{c}_M(0) = \boldsymbol{0},\ \boldsymbol{a}_M(0) = [1, 0, \cdots, 0]^T, \\ &\boldsymbol{b}_M(0) = [0, \cdots, 0, 1]^T,\ \alpha_M(0) = 1,\ \alpha_M^f(0) = \lambda^M \alpha_M^b(0),\ \alpha_M^b(0) = \delta > 0, \end{aligned}$$

Calculate: $n = 1, 2, \cdots$

$$\begin{aligned} e_M^f(n) &= u(n) + \boldsymbol{a}_M^T(n-1)\boldsymbol{x}_M(n-1), \\ \epsilon_M^f(n) &= e_M^f(n)/\alpha_M(n-1), \\ \boldsymbol{a}_M(n) &= \boldsymbol{a}_M(n-1) - \boldsymbol{c}_M(n-1)\epsilon_M^f(n), \\ \alpha_M^f(n) &= \lambda\alpha_M^f(n-1) + e_M^f(n)\epsilon_M^f(n), \\ k_{M+1}(n) &= \lambda^{-1}\alpha_M^f(n-1)e_M^f(n), \\ \boldsymbol{c}_{M+1}(n) &= \begin{bmatrix} 0 \\ \boldsymbol{c}_M(n) \end{bmatrix} + \begin{bmatrix} 1 \\ \boldsymbol{a}_M(n-1) \end{bmatrix} k_{M+1}(n-1), \\ \boldsymbol{d}_{M+1} &= \begin{bmatrix} \boldsymbol{d}_M(n) \\ d_{M+1}(n) \end{bmatrix}, \\ e_M^b(n) &= \lambda\alpha_M^b(n-1)d_{M+1}(n), \\ \tilde{e}_M^b(n) &= u(n-M) + \boldsymbol{b}_M^T(n-1)x_M(n), \\ \Delta^b(n) &= \tilde{e}_M^b(n) - e_M^b(n), \\ \hat{e}_{M,i}^b(n) &= \tilde{e}_M^b(n) + \sigma_i\Delta^b(n),\ i = 1, 2, 3, \cdots, \\ \boldsymbol{c}_M(n) &= \boldsymbol{d}_M(n) - \boldsymbol{b}_M(n-1)d_{M+1}(n), \\ \alpha_{M+1}(n) &= \alpha_M(n-1) - k_{M+1}(n)e_M^f(n), \\ \alpha_M(n) &= \alpha_{M+1}(n) + d_{M+1}(n)\hat{e}_{M,1}^b(n), \\ \tilde{\alpha}_M(n+1) &= 1 + \boldsymbol{c}_M^T(n)x_M(n), \end{aligned}$$

$$
\begin{aligned}
\Delta^{\alpha}(n) &= \tilde{\alpha}_M(n) + \sigma\Delta^{\alpha}(n),\\
\epsilon^b_{M,i}(n) &= \hat{e}^b_{M,i}(n)/\hat{\alpha}_M(n),\ i = 2, 3, \cdots,\\
\boldsymbol{b}_M(n) &= \boldsymbol{b}_M(n-1) - \boldsymbol{c}_M(n)\epsilon^b_{M,2}(n),\\
\alpha^b_M(n) &= \lambda\alpha^b_M(n-1) + \hat{e}^b_{M,3}(n)\epsilon^b_{M,3}(n),\\
e_M(n) &= y(n) - \boldsymbol{w}^T(n-1)\boldsymbol{x}_M(n),\\
\epsilon_M(n) &= e_M(n)/\hat{\alpha}_M(n),\\
\mu(n) &= \alpha_M(n)/[1 - \rho\hat{\alpha}_M(n)],\\
\boldsymbol{w}_M(n) &= \boldsymbol{w}_M(n-1) + \mu(n)\boldsymbol{c}_M(n)\epsilon_M(n).
\end{aligned}
$$

The relevant calculation equations of the above algorithm reflect the update relationship shown in Eq. (5.6.39). For example, the relationship between $\boldsymbol{c}_{M-1}(n-1)$ and $\boldsymbol{a}_M(n-1)$ on the left of Eq. (5.6.39) to synthesize $\boldsymbol{a}_M(n)$ is reflected in $\boldsymbol{a}_M(n) = \boldsymbol{a}_M(n-1) - \boldsymbol{c}_M(n-1)\epsilon^f_M(n)$, while the relationship between $\boldsymbol{c}_M(n)$ and $\boldsymbol{a}_M(n-1)$ to synthesize $\boldsymbol{c}_{M+1}(n)$ is reflected in

$$
\boldsymbol{c}_{M+1}(n) = \begin{bmatrix} 0 \\ \boldsymbol{c}_M(n) \end{bmatrix} + \begin{bmatrix} 1 \\ \boldsymbol{a}_M(n-1) \end{bmatrix} k_{M+1}(n-1).
$$

5.7 Adaptive Line Enhancer and Notch Filter

Adaptive spectral line enhancer was first proposed by Widrow et al. in 1975 when studying the adaptive noise cancellation[224]. The purpose is to separate sine wave from broadband noise to extract sine wave signal. On the contrary, if the sine wave signal is the noise or interference to be suppressed (for example, in biomedical instruments, 50Hz AC is called the power line interference), the adaptive filter for this purpose is called notch filter. Now, the adaptive spectral line enhancer and notch filter have been widely used in instantaneous frequency estimation, spectral analysis, narrowband detection, speech coding, narrowband interference suppression, interference detection, adaptive carrier recovery of digital data receiver, etc. Please refer to reference [238].

5.7.1 Transfer Functions of Line Enhancer and Notch Filter

Consider the following observed signal

$$
x(n) = s(n) + v(n) = \sum_{i=1}^{p} A_i \sin(\omega_i n + \theta_i) + v(n), \tag{5.7.1}
$$

where A_i, ω_i, θ_i are the amplitude, frequency, and initial phase of the i-th sine wave signal, respectively; $v(n)$ is additive broadband noise and it can be colored.

Now we hope to design a filter, when $x(n)$ passes through this filter, the output contains only p sine wave signals $s(n)$ without any other signal or noise. Because the power spectrum of p sine wave signals is p discrete spectral lines, this filter that only extracts sine wave signals is called line enhancer.

Let $H(\omega)$ be the transfer function of the line enhancer. To extract p sine waves and reject all other signals and noise, the transfer function $H(\omega)$ must meet the following condition

$$H_{\text{LE}}(\omega) = \begin{cases} 1, & \text{if } \omega = \omega_1, \cdots, \omega_p, \\ 0, & \text{others.} \end{cases} \tag{5.7.2}$$

Conversely, if the transfer function of the filter is

$$H_{\text{notch}}(\omega) = \begin{cases} 0, & \text{if } \omega = \omega_1, \cdots, \omega_p, \\ 1, & \text{others,} \end{cases} \tag{5.7.3}$$

then the filter will suppress p sine wave signals and let $v(n)$ pass completely. The function of this filter is equivalent to the trap of sine waves, called a notch filter.

Obviously, the relationship between the transfer functions of the line enhancer and notch filter is

$$H_{\text{LE}}(\omega) = 1 - H_{\text{notch}}(\omega), \tag{5.7.4}$$

Fig. 5.7.1 (a) and (b)show the curves of the transfer functions of the line enhancer and

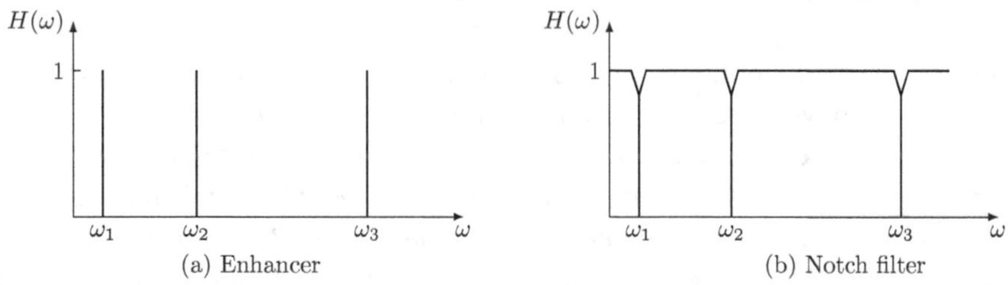

Fig. 5.7.1: Curves of the transfer functions of the line enhancer and the notch filter

the notch filter for three sine wave signals, respectively.

An adaptive line enhancer or notch filter is an adaptive filter, and the transfer function satisfies Eq. (5.7.2) or Eq. (5.7.3). In fact, the adaptive line enhancer can be easily realized by an adaptive notch filter, as shown in Fig. 5.7.2.

As shown in Fig. 5.7.2, the observation signal $x(n) = s(n) + v(n)$ suppresses the sine wave signal through the adaptive notch filter to generate the optimal estimation $\hat{v}(n)$ of $v(n)$, and then subtracts it from the observation signal to generate the estimation $\hat{s}(n) = s(n)+v(n)-\hat{v}(n)$ of the sine wave signal. If the notch filter is ideal, then $\hat{v}(n) = v(n)$ so that $\hat{s}(n) = s(n)$. The adaptive line enhancer constructed by notch filter is called notch adaptive line enhancer for short.

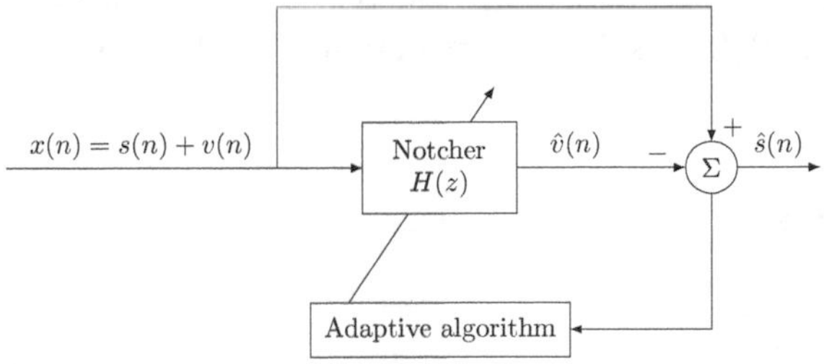

Fig. 5.7.2: Adaptive line enhancer realized by adaptive notch filter

5.7.2 Adaptive Notch Filter based on Lattice IIR Filter

Based on the adaptive infinite impulse response (IIR) filter, the adaptive notch filter and adaptive line enhancer can be realized. The adaptive line enhancer proposed by Rao and Kung only needs to adaptively adjust $2p$ weight coefficients for p sine waves[182]. In 1985, Nehorai proposed another notch adaptive line enhancer[162]. By limiting the zero point of the notch filter to the unit circle, only p weight coefficients of the filter need to be adjusted for p sine waves. The idea of using an IIR notch filter is attractive because it can reject interference signals. In addition, the filter length required is much smaller than that of the adaptive line enhancer using an FIR filter.

To enhance a sine wave signal $s(n) = re^{j\omega n}$, the transfer function of the notch filter is determined by

$$H(z) = \frac{(1 - re^{j\omega}z^{-1})(1 - re^{-j\omega}z^{-1})}{(1 - \alpha re^{j\omega}z^{-1})(1 - \alpha re^{-j\omega}z^{-1})}, \tag{5.7.5}$$

$$= \frac{1 + \omega_1 z^{-1} + \omega_2 z^{-2}}{1 + \alpha\omega_1 z^{-1} + \alpha^2\omega_2 z^{-2}}, \tag{5.7.6}$$

where $\omega_1 = -2r\cos\omega$, $\omega_2 = r^2$, and α is a parameter determining the bandwidth of the notch filter.

From Eq. (5.7.5), when $z = re^{\pm j\omega}$ and $\alpha \neq 1$, $H(z) = 0$. On the other hand, it can also be seen that when $z \neq re^{\pm j\omega}$ and $\alpha \to 1$, $H(z) \approx 1$. Therefore, as long as $\alpha \to 1$ is selected, the notch effect can be approximately realized, and the closer α is to 1, the more ideal the notch effect of $H(z)$ is. The adaptive algorithm of the line enhancer is to adjust the weight coefficients ω_1 and ω_2 to minimize the mean square value of the estimation error $\hat{v}(n)$, which can be realized by the Gaussian-Newton algorithm (such as LMS algorithm). However, the Gauss-Newton algorithm is sensitive to some initial conditions.

To improve the defect of direct IIR notch filter, Cho et al.[53] proposed to use lattice IIR notch filter to realize notch transfer function in the line enhancer. The structure of this lattice IIR filter is shown in Fig. 5.7.3, which is formed by cascading two lattice filters. The input of the upper lattice filter $H_1(z)$ is $x(n)$ and the output is $s_0(n)$; While the input of the lower lattice filter $H_2(z)$ is $s_0(n)$ and the output is $s_2(n)$.

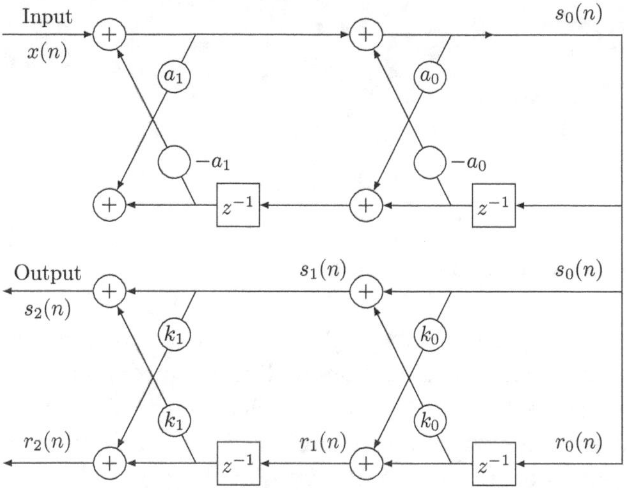

Fig. 5.7.3: Structure of a lattice IIR filter

From Fig. 5.7.3, it is easy to write the input-output equations of the lattice filters $H_1(z)$ and $H_2(z)$ as follows

$$\begin{aligned} s_0(n) + a_0(1 + a_1)s_0(n-1) + a_1 s_0(n-2) &= x(n), \\ s_0(n) + k_0(1 + k_1)s_0(n-1) + k_1 s_0(n-2) &= s_2(n), \end{aligned}$$

or written in Z-transform expression

$$[1 + a_0(1 + a_1)z^{-1} + a_1 z^{-2}]S_0(z) = X(z),$$

$$[1 + k_0(1 + k_1)z^{-1} + k_1 z^{-2}]S_0(z) = S_2(z).$$

Therefore, the transfer functions of the two lattice filters are defined as

$$H_1(z) \stackrel{\text{def}}{=} \frac{S_0(z)}{X(z)} = \frac{1}{1 + a_0(1 + a_1)z^{-1} + a_1 z^{-2}}, \tag{5.7.7}$$

$$H_2(z) \stackrel{\text{def}}{=} \frac{S_2(z)}{S_0(z)} = 1 + k_0(1 + k_1)z^{-1} + k_1 z^{-2}. \tag{5.7.8}$$

Thus, the transfer function of the whole lattice filter is

$$H(z) \stackrel{\text{def}}{=} \frac{S_2(z)}{X(z)} = \frac{S_2(z)}{S_0(z)}\frac{S_0(z)}{X(z)} = \frac{1 + k_0(1 + k_1)z^{-1} + k_1 z^{-2}}{1 + a_0(1 + a_1)z^{-1} + a_1 z^{-2}}. \tag{5.7.9}$$

From Fig. 5.7.3, it can be seen that the upper lattice filter $H_1(z)$ contributes to the pole part of the whole lattice filter, which is equivalent to an AR model; The lower lattice filter $H_2(z)$ contributes to the zero part of the whole lattice filter, which is a lattice FIR filter. Therefore, the whole lattice filter has an infinite number of impulse responses, which is a lattice IIR filter.

Since Eq. (5.7.9) must meet the condition Eq. (5.7.6) of the notch filter, there is

$$a_0(1 + a_1) = \alpha k_0(1 + k_1) \text{ and } a_1 = \alpha^2 k_1, \tag{5.7.10}$$

then we have

$$a_0 = \frac{\alpha k_0(1 + k_1)}{1 + \alpha^2 k_1}, \tag{5.7.11}$$

Eq. (5.7.10) and Eq. (5.7.11) show that the weight coefficients a_0 and a_1 are determined by the weight coefficients k_0 and k_1. Since α is close to 1, there is an approximate relationship

$$\begin{aligned} a_1 &= \alpha^2 k_1 \approx \alpha k_1, \\ a_0(1 + a_1) &= \alpha k_0(1 + k_1) \approx k_0(1 + \alpha k_1) \approx k_0(1 + a_1), \end{aligned}$$

or written as

$$a_1 \approx \alpha k_1, \tag{5.7.12}$$

$$a_0 \approx k_0. \tag{5.7.13}$$

This shows that it is only needed to deduce the adaptive update equations of k_0 and k_1.

Since $H_1(z)$ is a pole model, to ensure the stability of this filter, the pole of $H_1(z)$ must be located inside the unit circle, i.e., the modulus $|a_0|$ and $|a_1|$ of the weight coefficients must be less than 1. Therefore, the modulus $|k_0|$ and $|k_1|$ of the weight coefficients of $H_2(z)$ must also be less than 1.

The weight coefficients k_0 and k_1 can be adjusted adaptively by using the adaptive algorithm of lattice FIR filter[146]:

$$k_m(n) = -\frac{C_m(n)}{D_m(n)}, \tag{5.7.14}$$

$$C_m(n) = \lambda C_m(n-1) + s_m(n) r_m(n-1), \tag{5.7.15}$$

$$D_m(n) = \lambda D_m(n-1) + \frac{1}{2}\left[s_m^2(n) + r_m^2(n-1)\right], \tag{5.7.16}$$

where $m = 0, 1$; the forgetting factor $0 < \lambda \le 1$; $s_m(n)$ and $r_m(n)$ are the forward and backward residuals of the m-th stage of the lower lattice filter in Fig. 5.7.3, respectively.

The lattice IIR filter shown in Fig. 5.7.3 can enhance only one sine wave signal. For enhancing p sine wave signals, p IIR lattice line enhancers need to be cascaded. Each lattice filter can be adaptively adjusted by using the algorithm shown in Eq. (5.7.14) ~ Eq. (5.7.16).

5.8 Generalized Sidelobe Canceller

Let $\boldsymbol{c}_0$ be a vector representing the feature of the desired source $s_0(t)$, shortened to the feature vector. For example, in array signal processing, $\boldsymbol{c}_0 = \boldsymbol{a}(\phi_0)$ is the direction vector of the desired source. As another example, in the code division multiple access system of wireless communication, $\boldsymbol{c}_0$ is the signature vector of the desired user. Now we want to design a narrowband beamformer $\boldsymbol{w}$ to extract the desired source $s_0(t)$, i.e., $\boldsymbol{w}$ is required to meet the following linear constraint

$$\boldsymbol{w}^{\mathrm{H}}\boldsymbol{c}_0 = g, \tag{5.8.1}$$

where $g \neq 0$ is a constant.

Suppose there are p signals $s_1(t), \cdots, s_p(t)$, and the vectors representing their characteristics are $\boldsymbol{c}_1, \cdots, \boldsymbol{c}_p$. If one of them is to be extracted and all the other signals are to be suppressed, the single linear constraint should be extended to p linear constraints, i.e.,

$$\boldsymbol{C}^{\mathrm{H}}\boldsymbol{w} = \boldsymbol{g}, \tag{5.8.2}$$

where $\boldsymbol{C}$ is called the constraint matrix, represented as $\boldsymbol{C} = [\boldsymbol{c}_1, \cdots, \boldsymbol{c}_p]$. The column vector $\boldsymbol{g}$ is called a gain vector, and its elements determine whether the corresponding signal is extracted or suppressed. Take two linear constraints as an example

$$[\boldsymbol{c}_1, \boldsymbol{c}_2]^{\mathrm{H}}\boldsymbol{w} = \begin{bmatrix} 1 \\ 0 \end{bmatrix}, \tag{5.8.3}$$

which indicates that the source $s_1(t)$ will be extracted and the source $s_2(t)$ will be suppressed. Using the terminology of array signal processing, the first constraint $\boldsymbol{c}_1^{\mathrm{H}}\boldsymbol{w} = 1$ represents the main lobe of the array, and the second constraint $\boldsymbol{c}_2^{\mathrm{H}}\boldsymbol{w} = \alpha$ $(\alpha < 1)$ represents the side lobe of the array. When $\alpha = 0$, the side lobe is cancelled. Given this, the filter satisfying Eq. (5.8.3) is often called a sidelobe canceller. So, how to realize the adaptive sidelobe canceller?

Suppose that M array elements are used to receive L sources, i.e., in Eq. (5.8.2), there are L linear constraints, that is, C is an $M \times L$ constraint matrix, while the sidelobe canceller $\boldsymbol{w}$ has M tap coefficients, that is, W is $M \times 1$-dimensional, and the gain vector $\boldsymbol{g} = [1, 0, \cdots, 0]^{\mathrm{T}}$ is $\times 1$-dimensional. Therefore, if $\boldsymbol{g} = [1, 0, \cdots, 0]^{\mathrm{T}}$, the sidelobe canceller satisfying Eq. (5.8.2) only retains the desired signal $s_1(t)$ and eliminates the other $L - 1$ (interference) signals, i.e., all sidelobes are suppressed.

Generally, the number of array elements M is greater than the number of signals L. Let the columns of $M \times (M - L)$ matrix $\boldsymbol{C}_a$ be linearly independent vectors, forming a set of basis vectors. Assume that the space expanded by these basis vectors is an orthogonal complement of the space expanded by the columns of the constraint matrix $\boldsymbol{C}$. According to the definition of orthogonal complement, a matrix $\boldsymbol{C}$ and its orthogonal complement matrix $\boldsymbol{C}_a$ are orthogonal to each other, then we have

$$\boldsymbol{C}^{\mathrm{H}}\boldsymbol{C}_a = \boldsymbol{0}_{L\times(M-L)}, \tag{5.8.4}$$

$$\boldsymbol{C}_a^{\mathrm{H}}\boldsymbol{C} = \boldsymbol{0}_{(M-L)\times L}, \tag{5.8.5}$$

where $\boldsymbol{0}_{L\times(M-L)}$ and $\boldsymbol{0}_{(M-L)\times L}$ are all zero matrices, and the subscript indicates their dimension.

Using the constraint matrix $\boldsymbol{C}$ and its orthogonal complement $\boldsymbol{C}_a$ as submatrices, a matrix is synthesized

$$\boldsymbol{U} = [\boldsymbol{C}, \boldsymbol{C}_a], \tag{5.8.6}$$

Using the matrix $\boldsymbol{U}$, the $M\times 1$ weight vector $\boldsymbol{w}$ of the beamformer can be defined as

$$\boldsymbol{w} = \boldsymbol{U}\boldsymbol{q} \text{ or } \boldsymbol{q} = \boldsymbol{U}^{-1}\boldsymbol{w}. \tag{5.8.7}$$

Partition the vector $\boldsymbol{q}$ into blocks as

$$\boldsymbol{q} = \begin{bmatrix} \boldsymbol{v} \\ -\boldsymbol{w}_a \end{bmatrix}, \tag{5.8.8}$$

where $\boldsymbol{v}$ is an $L\times 1$ vector, and $\boldsymbol{w}_a$ is an $(M-L)\times 1$ vector. Further substituting Eqs. (5.8.6) and (5.8.8) into Eq. (5.8.7) yields

$$\boldsymbol{w} = [\boldsymbol{C}, \boldsymbol{C}_a]\begin{bmatrix} \boldsymbol{v} \\ -\boldsymbol{w}_a \end{bmatrix} = \boldsymbol{C}\boldsymbol{v} - \boldsymbol{C}_a\boldsymbol{w}_a. \tag{5.8.9}$$

Use $\boldsymbol{C}^{\mathrm{H}}$ to premultiply both sides of Eq. (5.8.9), and then substitute it into Eq. (5.8.2) to get

$$\boldsymbol{g} = \boldsymbol{C}^{\mathrm{H}}\boldsymbol{C}\boldsymbol{v} - \boldsymbol{C}^{\mathrm{H}}\boldsymbol{C}_a\boldsymbol{w}_a.$$

Using the orthogonality of $\boldsymbol{C}$ and $\boldsymbol{C}_a$ (i.e., $\boldsymbol{C}^{\mathrm{H}}\boldsymbol{C}_a$ is equal to zero matrix $\boldsymbol{0}$), the above equation can be abbreviated as $\boldsymbol{g} = \boldsymbol{C}^{\mathrm{H}}\boldsymbol{C}\boldsymbol{v}$. Solve this equation to obtain

$$\boldsymbol{v} = (\boldsymbol{C}^{\mathrm{H}}\boldsymbol{C})^{-1}\boldsymbol{g}. \tag{5.8.10}$$

Substitute Eq. (5.8.10) into Eq. (5.8.9) to obtain

$$\boldsymbol{w} = \boldsymbol{C}(\boldsymbol{C}^{\mathrm{H}}\boldsymbol{C})^{-1}\boldsymbol{g} - \boldsymbol{C}_a\boldsymbol{w}_a. \tag{5.8.11}$$

Define

$$\boldsymbol{w}_0 \stackrel{\text{def}}{=} \boldsymbol{C}(\boldsymbol{C}^{\mathrm{H}}\boldsymbol{C})^{-1}\boldsymbol{g}, \tag{5.8.12}$$

then Eq. (5.8.11) can be represented as

$$\boldsymbol{w} = \boldsymbol{w}_0 - \boldsymbol{C}_a\boldsymbol{w}_a. \tag{5.8.13}$$

Eq. (5.8.13) shows that the sidelobe canceller $\boldsymbol{w}$ defined in Eq. (5.8.2) can be divided into two parts:

(1) The filter $\boldsymbol{w}_0$ defined by Eq. (5.8.12) is a fixed part of the sidelobe canceller and is determined by the constraint matrix $\boldsymbol{C}$ and the gain vector $\boldsymbol{g}$;
(2) $\boldsymbol{C}_a\boldsymbol{w}_a$ represents the adaptive part of the sidelobe canceller.

Since the filter $\boldsymbol{w}_0$ and the orthogonal complement matrix $\boldsymbol{C}_a$ of the constraint matrix are known invariants after the constraint matrix $\boldsymbol{C}$ and the gain vector $\boldsymbol{g}$ are given, the adaptive update of the sidelobe canceller $\boldsymbol{w}$ is converted to the update of the adaptive filter $\boldsymbol{w}_a$. Given this, the sidelobe canceller defined in Eq. (5.8.13) is often called a generalized sidelobe canceller.

The following is a further physical explanation of the generalized sidelobe canceller.

(1) Substituting Eq. (5.8.13) into the constraint Eq. (5.8.2) of the sidelobe canceller, we get

$$\boldsymbol{C}^{\mathrm{H}}\boldsymbol{w}_0 - \boldsymbol{C}^{\mathrm{H}}\boldsymbol{C}_a\boldsymbol{w}_a = \boldsymbol{g}.$$

Since $\boldsymbol{C}^{\mathrm{H}}\boldsymbol{C}_a = \boldsymbol{0}$, the above equation can be simplified as

$$\boldsymbol{C}^{\mathrm{H}}\boldsymbol{w}_0 = \boldsymbol{g}, \tag{5.8.14}$$

which shows that the filter $\boldsymbol{w}_0$ is actually a fixed sidelobe canceller satisfying the constraint Eq. (5.8.2).

(2) The decomposition shown in Eq. (5.8.13) is a typical orthogonal decomposition due to

$$< \boldsymbol{C}_a\boldsymbol{w}_a, \boldsymbol{w}_0\rangle = \boldsymbol{w}_a^{\mathrm{H}}\boldsymbol{C}_a^{\mathrm{H}}\boldsymbol{C}(\boldsymbol{C}^{\mathrm{H}}\boldsymbol{C})^{-1}\boldsymbol{g} = 0,$$

where the orthogonality of $\boldsymbol{C}_a$ and $\boldsymbol{C}$ is used, i.e., $\boldsymbol{C}_a^{\mathrm{H}}\boldsymbol{C} = \boldsymbol{0}$.

The term "generalized sidelobe canceller" was first introduced by Griffiths and Jim[96] and is further discussed in literature [215], [16] and [100].

Generalized sidelobe canceller has important applications in array signal processing and multi-user detection of wireless communication. For example, refer to literature [241]. In the next section, its application in blind multiuser detection will be introduced.

5.9 Blind Adaptive Multiuser Detection

Taking the code division multiple access (CDMA) system in wireless communication as an example, this section introduces how to use the generalized sidelobe canceller and adaptive filtering algorithm to realize blind multiuser detection of CDMA, and compares the statistical performance of LMS, RLS, and Kalman filtering algorithms to track the desired user signal in this application.

5.9.1 Canonical Representation of Blind Multiuser Detection

A direct sequence code division multiple access (DS-CDMA) system is discussed in this subsection. It has K users and the wireless channel is an additive Gaussian white noise channel. After a series of processing (chip filtering, chip rate sampling), the discrete-time output of the receiver during a symbol interval can be represented by the following signal model

$$y(n) = \sum_{k=1}^{K} A_k b_k(n) s_k(n) + \sigma v(n), \quad n = 0, 1, \cdots, T_s - 1, \tag{5.9.1}$$

where $v(n)$ is the channel noise; A_k, $b_k(n)$ and $s_k(n)$ are the received amplitude, information character sequence and characteristic waveform of the k-th user, respectively; σ is a constant. Now, it is assumed that the information characters of each user are independently and equally selected from $\{-1, +1\}$, and the characteristic waveform has unit energy, i.e.,

$$\sum_{n=0}^{T_s-1} |s_k(n)|^2 = 1.$$

The support interval of the characteristic waveform is $[0, T_s]$, where $T_s = NT_c$ is the symbol interval; N and T_c are the spread spectrum gain and chip interval, respectively.

The blind multiuser detection problem is to estimate the information characters $b_d(0), b_d(1), \cdots, b_d(N-1)$ transmitted by the desired user when only the received signal $y(0), \cdots, y(N-1)$ within one symbol interval and the characteristic waveform $s_d(0), s_d(1), \cdots, s_d(N-1)$ of the desired user are known. Here, the term "blind" means that we do not know any information about other users. Without losing generality, user 1 is assumed to be the desired user.

Define

$$\mathbf{y}(n) = [y(0), y(1), \cdots, y(N-1)]^{\mathrm{T}}, \tag{5.9.2}$$

$$\mathbf{v}(n) = [v(0), v(1), \cdots, v(N-1)]^{\mathrm{T}}, \tag{5.9.3}$$

as the received signal vector and noise vector respectively, and the characteristic waveform vector of user k as

$$s_k(n) = [s_k(0), s_k(1), \cdots, s_k(N-1)]^{\mathrm{T}}. \tag{5.9.4}$$

Then Eq. (5.9.1) can be written in vector form

$$\mathbf{y}(n) = A_1 b_1(n) \mathbf{s}_1 + \sum_{k=2}^{K} A_k b_k(n) \mathbf{s}_k + \sigma \mathbf{v}(n), \tag{5.9.5}$$

where the first term on the right side is the desired user signal, the second term is the sum of the interference signals of all other users (collectively referred to as interference users), and the third term represents channel noise.

Now, the multiuser detector $\boldsymbol{c}_1$ is designed for the desired user, and the detector output is $\boldsymbol{c}_1^{\mathrm{T}}\boldsymbol{y}(n) =< \boldsymbol{c}_1, \boldsymbol{y}\rangle$. Therefore, the detection result of the information symbol of the desired user within the n-th symbol interval is

$$\hat{b}_1(n) = \operatorname{sgn}(\langle \boldsymbol{c}_1, \boldsymbol{y}\rangle) = \operatorname{sgn}(\boldsymbol{c}_1^{\mathrm{T}}(n)\boldsymbol{y}(n)). \tag{5.9.6}$$

Blind multiuser detector $\boldsymbol{c}_1$ has two canonical representations:

Canonical representation 1:

$$\boldsymbol{c}_1(n) = \boldsymbol{s}_1 + \boldsymbol{x}_1(n). \tag{5.9.7}$$

Canonical representation 2:

$$\boldsymbol{c}_1(n) = \boldsymbol{s}_1 - \boldsymbol{C}_{1,\mathrm{null}}\boldsymbol{w}_1. \tag{5.9.8}$$

Both canonical representations decompose the adaptive multiuser detector into the sum of the fixed part $\boldsymbol{s}_1$ and the other adaptive part, and the two parts are orthogonal (orthogonal decomposition), i.e., the two parts respectively satisfy the following equations

$$\langle \boldsymbol{s}_1, \boldsymbol{x}_1\rangle = 0, \tag{5.9.9}$$

$$\langle \boldsymbol{s}_1, \boldsymbol{C}_{1,\mathrm{null}}\boldsymbol{w}_1\rangle = 0. \tag{5.9.10}$$

Canonical representation 1 was proposed by Honig et al.[106]. The constraint condition Eq. (5.9.7) can be equivalently expressed as

$$\langle \boldsymbol{c}_1, \boldsymbol{s}_1\rangle = \langle \boldsymbol{s}_1, \boldsymbol{s}_1\rangle = 1. \tag{5.9.11}$$

Since $\langle \boldsymbol{c}_1, \boldsymbol{s}_1\rangle = 1$, $\boldsymbol{c}_1(n)$ is a standardized multiuser detector, which is the meaning of canonical representation.

Canonical representation 2 is obtained by Kapoor et al.[118], under the framework of generalized sidelobe canceller and the constraint $\langle \boldsymbol{c}_1, \boldsymbol{s}_1\rangle = 1$. In this canonical representation, the matrix $\boldsymbol{C}_{1,\mathrm{null}}$ is expanded into the zero space of the desired user characteristic waveform vector $\boldsymbol{s}_1 \geqslant 1$, i.e., $\langle \boldsymbol{s}_1, \boldsymbol{C}_{1,\mathrm{null}}\boldsymbol{w}_1\rangle = 0$. It is easy to see that the canonical representation 1 and 2 are equivalent.

5.9.2 LMS and RLS Algorithms for Blind Multiuser Detection

Consider using canonical representation 1 to derive LMS and RLS algorithms for blind multiuser detection.

1. LMS Algorithm

Considering the blind multiuser detector $\boldsymbol{c}_1(n)$ described by canonical representation 1, the mean output energy and mean square error of its output signal $\langle \boldsymbol{c}_1, \boldsymbol{y}\rangle$

are

$$\mathrm{MOE}(\boldsymbol{c}_1) = \mathrm{E}\left\{\langle \boldsymbol{c}_1, \boldsymbol{y}\rangle^2\right\} = \mathrm{E}\left\{(\boldsymbol{c}_1^{\mathrm{T}}(n)\boldsymbol{y}(n))^2\right\}, \tag{5.9.12}$$

$$\mathrm{MSE}(\boldsymbol{c}_1) = \mathrm{E}\left\{(A_1 b_1 - \langle \boldsymbol{c}_1, \boldsymbol{y}\rangle)^2\right\}. \tag{5.9.13}$$

Define

$$e(n) = \langle \boldsymbol{c}_1, \boldsymbol{y}\rangle = \boldsymbol{c}_1^{\mathrm{T}}(n)\boldsymbol{y}(n), \tag{5.9.14}$$

then the mean of $e(n)$ is zero and the variance [106, 187] is

$$\mathrm{cov}\{e(n)\} = \mathrm{E}\{e^2(n)\} = A_1^2 + \mathrm{MSE}(\boldsymbol{c}_1(n)). \tag{5.9.15}$$

Find the unconstrained gradient of the MOE with respect to $\boldsymbol{c}_1(n)$, and obtain

$$\nabla \mathrm{MOE} = 2\mathrm{E}\{\langle \boldsymbol{y}, \boldsymbol{s}_1 + \boldsymbol{x}_1\rangle\}\boldsymbol{y}. \tag{5.9.16}$$

Therefore, the random gradient adaptive algorithm of the adaptive part $\boldsymbol{x}_1(i)$ of the blind multiuser detector $\boldsymbol{c}_1(n)$ is

$$\boldsymbol{x}_1(i) = \boldsymbol{x}_1(i-1) - \mu\hat{\nabla}\mathrm{MOE}, \tag{5.9.17}$$

where $\hat{\nabla}\mathrm{MOE}$ is the estimation of $\nabla\mathrm{MOE}$. The instantaneous gradient is adopted, i.e., the mathematical expectation in Eq. (5.9.16) is directly replaced by its instantaneous value to obtain

$$\hat{\nabla}\mathrm{MOE} = 2\langle \boldsymbol{y}, \boldsymbol{s}_1 + \boldsymbol{x}_1\rangle\boldsymbol{y}. \tag{5.9.18}$$

It is easy to prove

$$[\boldsymbol{y} - \langle \boldsymbol{y}, \boldsymbol{s}_1\rangle\boldsymbol{s}_1]^{\mathrm{T}}\boldsymbol{s}_1 = \boldsymbol{y}^{\mathrm{T}}\boldsymbol{s}_1\boldsymbol{s}_1^{\mathrm{T}}\boldsymbol{s}_1 = 0.$$

Since each user characteristic waveform has unit energy, that is, $\boldsymbol{s}_1^{\mathrm{T}}\boldsymbol{s}_1 = 1$, the above equation is equivalent to

$$[\boldsymbol{y} - \langle \boldsymbol{y}, \boldsymbol{s}_1\rangle\boldsymbol{s}_1] \perp \boldsymbol{s}_1,$$

indicating that the component orthogonal to $\boldsymbol{s}_1$ in $\boldsymbol{y}$ is

$$\boldsymbol{y} - \langle \boldsymbol{y}, \boldsymbol{s}_1\rangle\boldsymbol{s}_1. \tag{5.9.19}$$

Therefore, from Eqs. (5.9.18) and (5.9.19), the projection gradient (namely, the component orthogonal to $\boldsymbol{s}_1$ in the gradient) is

$$2\langle \boldsymbol{y}, \boldsymbol{s}_1 + \boldsymbol{x}_1\rangle[\boldsymbol{y} - \langle \boldsymbol{y}, \boldsymbol{s}_1\rangle\boldsymbol{s}_1]. \tag{5.9.20}$$

Let the matched filter output responses of $\boldsymbol{s}_1$ and $\boldsymbol{s}_1 + \boldsymbol{x}_1$ be

$$Z_{\mathrm{MF}}(i) = \langle \boldsymbol{y}(i), \boldsymbol{s}_1\rangle, \tag{5.9.21}$$

$$Z(i) = \langle \boldsymbol{y}(i), \boldsymbol{s}_1 + \boldsymbol{x}_1(i-1)\rangle, \tag{5.9.22}$$

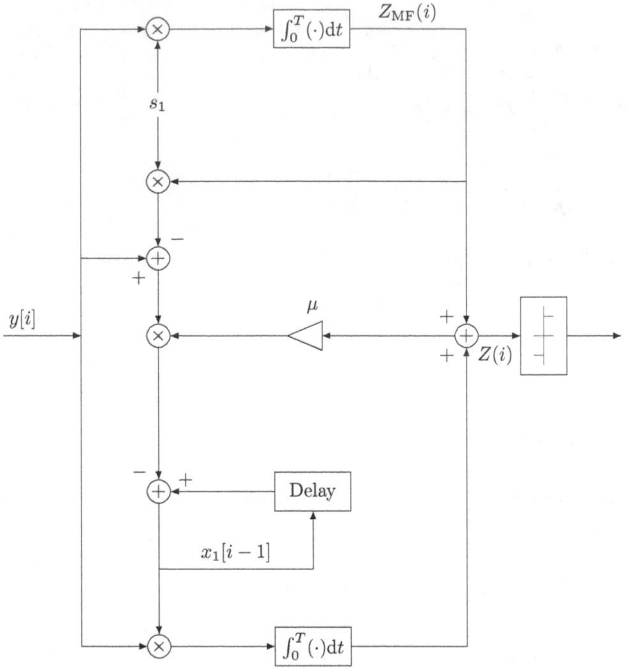

Fig. 5.9.1: Schematic illustration of LMS algorithm for blind multiuser detection

respectively. Substituting Eqs. (5.9.21) and (5.9.22) into Eq. (5.9.17), the update equation of the random gradient adaptive algorithm is obtained

$$\mathbf{x}_1(i) = \mathbf{x}_1(i-1) - \mu Z(i)[\mathbf{y}(i) - Z_{\mathrm{MF}}(i)\mathbf{s}_1]. \tag{5.9.23}$$

The implementation of the algorithm is shown in Fig. 5.9.1. This is the LMS algorithm for blind multiuser detection proposed by Honig et al.

When there is no interference characteristic waveform information, the initial condition of Eq. (5.9.23) can usually be selected to be $\boldsymbol{x}_1(0) = 0$.

2. RLS Algorithm

Different from the LMS algorithm proposed by Honig et al., which minimizes the MOE of the blind detector, poor and Wang[175] proposed to minimize the exponentially weighted output energy of the blind detector, i.e.,

$$\min \sum_{i=1}^{n} \lambda^{n-i} \left[\mathbf{c}_1^{\mathrm{T}}(n)\mathbf{y}(i)\right]^2 \quad \text{subject to } \mathbf{s}_1^{\mathrm{T}}\mathbf{c}_1(n) = 1, \tag{5.9.24}$$

where $0 < \lambda < 1$ is a forgetting factor.

It can be proved that the optimal detector satisfying Eq. (5.9.24) is

$$\boldsymbol{c}_1(n) = \frac{\boldsymbol{R}^{-1}(n)\boldsymbol{s}_1}{\boldsymbol{s}_1^{\mathrm{T}}\boldsymbol{R}^{-1}(n)\boldsymbol{s}_1}, \tag{5.9.25}$$

where

$$\boldsymbol{R}(n) = \sum_{i=1}^{n} \lambda^{n-i}\boldsymbol{y}(i)\boldsymbol{y}^{\mathrm{T}}(i) \tag{5.9.26}$$

is the autocorrelation matrix of the observed signal. From the matrix inversion lemma, the update formula of $\boldsymbol{R}^{-1}(n)$ can be obtained, and the RLS algorithm for updating blind multiuser detector $\boldsymbol{c}_1(n)$ is as follows

$$\boldsymbol{k}(n) = \frac{\boldsymbol{R}^{-1}(n-1)\boldsymbol{y}(n)}{\lambda + \boldsymbol{y}^{\mathrm{T}}(n)\boldsymbol{R}^{-1}(n-1)\boldsymbol{y}(n)}, \tag{5.9.27}$$

$$\boldsymbol{h}(n) = \boldsymbol{R}^{-1}(n)\boldsymbol{s}_1 = \frac{1}{\lambda}[\boldsymbol{h}(n-1) - \boldsymbol{k}(n)\boldsymbol{y}^{\mathrm{T}}(n)\boldsymbol{h}(n-1)], \tag{5.9.28}$$

$$\boldsymbol{c}_1(n) = \frac{1}{\boldsymbol{s}_1\boldsymbol{h}(n)}\boldsymbol{h}(n), \tag{5.9.29}$$

$$\boldsymbol{R}^{-1}(n) = \frac{1}{\lambda}[\boldsymbol{R}^{-1}(n-1) - \boldsymbol{k}(n)\boldsymbol{y}^{\mathrm{T}}(n)\boldsymbol{R}^{-1}(n-1)]. \tag{5.9.30}$$

This is the RLS algorithm for blind adaptive multiuser detection proposed by poor and Wang[175].

5.9.3 Kalman Adaptive Algorithm for Blind Multiuser Detection

Now consider using canonical representation 2 to design a Kalman adaptive algorithm for blind multiuser detection. Given the eigenvector $\boldsymbol{s}_1$ of user 1, $\boldsymbol{C}_{1,\text{null}}$ can be easily obtained by Gram-Schmidt orthogonalization or singular value decomposition.

For a time invariant CDMA system, an important fact is that the optimal detector or tap weight vector $\boldsymbol{c}_{\text{opt}1}(n)$ is also time invariant, i.e., $\boldsymbol{c}_{\text{opt}1}(n+1) = \boldsymbol{c}_{\text{opt}1}(n)$. Let $\boldsymbol{w}_{\text{opt}1}$ be the adaptive part in canonical representation 2 of $\boldsymbol{c}_1$, then there is the following state equation for the state variable $\boldsymbol{w}_{\text{opt}1}$

$$\boldsymbol{w}_{\text{opt}1}(n+1) = \boldsymbol{w}_{\text{opt}1}(n). \tag{5.9.31}$$

Meanwhile, take canonical representation 2 into consideration and substituting Eq. (5.9.8) into Eq. (5.9.14) gives

$$e(n) = \boldsymbol{s}_1^{\mathrm{T}}\boldsymbol{y}(n) - \boldsymbol{y}^{\mathrm{T}}(n)\boldsymbol{C}_{1,\text{null}}\boldsymbol{w}_1(n). \tag{5.9.32}$$

Let $\tilde{y}(n) = \boldsymbol{s}_1^{\mathrm{T}}\boldsymbol{y}(n)$ and $\boldsymbol{d}^{\mathrm{T}}(n) = \boldsymbol{y}^{\mathrm{T}}(n)\boldsymbol{C}_{1,\text{null}}$. If $\boldsymbol{w}_1$ reaches $\boldsymbol{w}_{\text{opt}1}$, Eq. (5.9.32) can be written as a measurement equation

$$\tilde{y}(n) = \boldsymbol{d}^{\mathrm{T}}(n)\boldsymbol{w}_{\text{opt}1}(n) + e_{\text{opt}}(n). \tag{5.9.33}$$

The state equation (5.9.31) and measurement equation (5.9.33) together constitute the dynamic system equation of user 1, which is the basis of Kalman filtering. The Kalman filtering problem in blind multiuser detection can be described as: given the measurement matrix $\boldsymbol{d}^{\mathrm{T}}(n)$, use the observation data $\tilde{y}(n)$ to calculate the minimum mean square error estimation of each coefficient of the state vector $\boldsymbol{w}_{\mathrm{opt1}}$ for each $n \geq 1$.

From Eq. (5.9.15), the variance of the optimal detection error is obtained

$$\xi_{\min} = \operatorname{cov}\{e_{\mathrm{opt}}(n)\} = \mathrm{E}\{e_{\mathrm{opt}}^2(n)\} = A_1^2 + \epsilon_{\min}, \tag{5.9.34}$$

where $\xi_{\min} = \mathrm{MSE}(\boldsymbol{c}_{\mathrm{opt1}}(n))$ represents the minimum mean square error when the tap weight vector $\boldsymbol{c}_1$ is optimal, so $\xi_{\min} = \mathrm{MOE}(\boldsymbol{c}_{\mathrm{opt1}}(n))$ represents the minimum average output energy of the dynamic system of user 1.

For synchronization model Eq. (5.9.1), it is easy to prove that $\mathrm{E}\{e_{\mathrm{opt}}(n)e_{\mathrm{opt}}(l)\} = 0, n \neq l$, because $E\{y(n)y(l) = 0\}, n \neq l$. This shows that $e_{\mathrm{opt}}(n)$ is a white noise with zero mean and variance $\xi_{\min}$ in the case of synchronization.

Compared with the first-order dynamic system model described by Eq. (5.4.1) and Eq. (5.4.2), the linear first-order state space model determined by Eq. (5.9.31) and Eq. (5.9.33) has the following characteristics:

(1) The state vector is $\boldsymbol{w}_{\mathrm{opt1}}$, the state transition matrix $\boldsymbol{F}(n+1, n)$ is an $N \times N$ identity matrix, and the process noise is a zero vector;
(2) The observation vector becomes a scalar $\tilde{y}(n) = \boldsymbol{s}_1^{\mathrm{T}}\boldsymbol{y}(n)$ and the observation matrix becomes a vector $\boldsymbol{d}^{\mathrm{T}}(n) = \boldsymbol{y}^{\mathrm{T}}(n)\boldsymbol{C}_{1,\mathrm{null}}$.

Comparing Eq. (5.4.1) and Eq. (5.4.2) of the standard dynamic system model with Eq. (5.9.31) and Eq. (5.9.33) of the dynamic system of user 1, it is easy to extend the standard Kalman filtering algorithm to the following Kalman adaptive algorithm for blind multiuser detection.

Algorithm 5.9.1. *Kalman adaptive algorithm for blind multiuser detection*
Initial condition: $\boldsymbol{K}(1, 0) = \boldsymbol{I}$
Iterative calculation: $n = 1, 2, 3, \cdots$

$$\boldsymbol{g}(n) = \boldsymbol{K}(n, n-1)\boldsymbol{d}(n)\left\{\boldsymbol{d}^H(n)\boldsymbol{K}(n, n-1)\boldsymbol{d}(n) + \xi_{min}\right\}^{-1}, \tag{5.9.35}$$

$$\boldsymbol{K}(n+1, n) = \boldsymbol{K}(n, n-1) - \boldsymbol{g}(n)\boldsymbol{d}^H(n)\boldsymbol{K}(n, n-1), \tag{5.9.36}$$

$$\hat{\boldsymbol{w}}_{opt1}(n) = \hat{\boldsymbol{w}}_{opt1}(n-1) + \boldsymbol{g}(n)\left\{y(n) - \boldsymbol{d}^H(n)\hat{\boldsymbol{w}}_{opt1}(n-1)\right\}, \tag{5.9.37}$$

$$\boldsymbol{c}_1(n) = \boldsymbol{s}_1 - \boldsymbol{C}_{1,null}\hat{\boldsymbol{w}}_{opt1}(n), \tag{5.9.38}$$

where $\hat{\boldsymbol{w}}_{opt1}(n), \boldsymbol{g}(n)$ *and* $\boldsymbol{d}(n)$ *are* $(N-1) \times 1$ *matrices;* $\boldsymbol{K}(n+1, n)$ *is an* $(N-1) \times (N-1)$ *matrix.*

The Kalman adaptive filtering algorithm for blind multiuser detection was proposed by Zhang and Wei[250] in 2002. To optimize the Kalman filter, the initial state is required to be a Gaussian random vector. Therefore, the initial prediction estimation $\hat{\boldsymbol{w}}_{\text{opt1}}(0) = \mathrm{E}\{\hat{\boldsymbol{w}}_{\text{opt1}}(n)\} = \boldsymbol{0}$ can be selected, and its correlation matrix is

$$\boldsymbol{K}(1,0) = \mathrm{E}\left\{[\boldsymbol{w}_{\text{opt1}}(0) - \mathrm{E}\{\boldsymbol{w}_{\text{opt1}}(0)\}][\boldsymbol{w}_{\text{opt1}}(0) - \mathrm{E}\{\boldsymbol{w}_{\text{opt1}}(0)\}]^{\mathrm{T}}\right\} = \boldsymbol{I}.$$

Eq. (5.9.37) is important because it shows that the correction term is equal to the innovation process $e(n) = \tilde{y}(n) - \boldsymbol{d}^{\mathrm{T}}(n)\hat{\boldsymbol{w}}(n-1)$ multiplied by the Kalman gain $\boldsymbol{g}(n)$. Although the calculation of $\boldsymbol{g}(n)$ in Eq. (5.9.35) requires that the minimum mean square error $\xi_{\min}$ is known or estimated, this requirement does not play much role. There are two reasons:

(1) The Kalman gain vector $\boldsymbol{g}(n)$ is only a time-varying step for updating $\hat{\boldsymbol{w}}_{\text{opt1}}(n)$;
(2) According to Ref.[106], after the inter symbol interference is suppressed, the signal-to-interference ratio (SIR) of the desired user output is defined as

$$\mathrm{SIR} = 10\log\frac{\langle \boldsymbol{c}_{\text{opt1}}, \boldsymbol{s}_1\rangle^2}{c_{\min}} = -10\log\epsilon_{\min}(\mathrm{dB}).$$

The above equation is based on Eq. (5.9.11). Since the maximum SIR of the desired user output is usually expected to be greater than 10dB, the minimum mean square error $\xi_{\min}$ is generally less than 0.1. It is noted that the received signal amplitude A_1 of the desired user in the n-th symbol interval is usually large and satisfies $A_1^2 \gg 0.1$, so $\hat{\xi}_{\min} \approx A_1^2$ can be directly taken as the estimation of the unknown parameter $\xi_{\min}$ in Eq. (5.9.34).

Although the above discussion takes the stationary wireless channel as the assumption, the obtained Kalman algorithm is also applicable to the slow time-varying channels. According to Ref. [100], a slow time-varying dynamic system can be modeled by a transverse filter whose tap weight vector $\boldsymbol{w}_{\text{opt1}}$ follows a first-order Markov process

$$\boldsymbol{w}_{\text{opt1}}(n+1) = a\boldsymbol{w}_{\text{opt1}}(n) + \boldsymbol{v}_1(n), \tag{5.9.39}$$

where a is a fixed model parameter; $\boldsymbol{v}_1(n)$ is the process noise with zero mean and correlation matrix $\boldsymbol{Q}_1$. Therefore, Eq. (5.9.36) and Eq. (5.9.37) in the Kalman algorithm equations should be replaced by

$$\boldsymbol{K}(n+1,n) = \boldsymbol{K}(n,n-1) - \boldsymbol{g}(n)\boldsymbol{d}^{\mathrm{H}}(n)\boldsymbol{K}(n,n-1) + \boldsymbol{Q}_1, \tag{5.9.40}$$

$$\hat{\boldsymbol{w}}_{\text{opt1}}(n) = \hat{\boldsymbol{w}}_{\text{opt1}}(n-1) + \boldsymbol{g}(n)[\tilde{y}(n) - \boldsymbol{d}^{\mathrm{T}}(n)\hat{\boldsymbol{w}}_{\text{opt1}}(n-1)], \tag{5.9.41}$$

respectively.

For a slow time-varying CDMA system, it can be assumed that the parameter a is very close to 1, and each element of the process noise correlation matrix $\boldsymbol{Q}_1$ takes a small value. Therefore, although the Kalman algorithm needs the unknown parameters such as $\xi_{\min}$, a and $\boldsymbol{Q}_1$ in a slow time-varying CDMA system, the estimated values $\hat{\xi}_{\min} \approx A_1^2$, $a \approx 1$ and $\boldsymbol{Q}_1 \approx \boldsymbol{0}$ (zero matrix) can be used. That is, the Kalman algorithm is suitable for slow time-varying CDMA systems.

The following is a comparison of the computational complexity of LMS, RLS, and Kalman algorithms (the amount of calculation to update the tap weight vector $\mathbf{c}_1(n)$ in each symbol interval).

LMS algorithm[106]: $4N$ times multiplication and $6N$ times addition;
RLS algorithm[175]: $4N^2 + 7N$ times multiplication and $3N^2 + 4N$ times addition;
Kalman algorithm[250]: $4N^2 - 3N$ times multiplication and $4N^2 - 3N$ times addition.

Theorem 5.9.1. [250] *When n is large enough, the mean output energy $\xi(n)$ of the Kalman filtering algorithm is*

$$\xi(n) \leq \xi_{min}(1 + n^{-1}N), \tag{5.9.42}$$

for a stationary CDMA system.

Theoretical analysis in reference [250] shows that when the blind adaptive multiuser detector converges, the steady-state residual mean output energy of the above three algorithms is

$$\xi_{\min}(\infty) = \begin{cases} \xi_{\min} \frac{\frac{\mu}{2}\operatorname{tr}(\boldsymbol{R}_{vy})}{1-\frac{\mu}{2}\operatorname{tr}(\boldsymbol{R}_{vy})}, & \text{LMS algorithm}, \\ \frac{1-\lambda}{\lambda}(N-1)\xi_{\min}, & \text{RLS algorithm}, \\ 0, & \text{Kalman algorithm}, \end{cases} \tag{5.9.43}$$

where $\boldsymbol{R}_{vy} = \mathrm{E}\{(\boldsymbol{I} - \mathbf{s}_1\mathbf{s}_1^{\mathrm{T}})\mathbf{y}(n)\mathbf{y}^{\mathrm{T}}(n)\}$ is the cross-correlation matrix of vectors $\mathbf{v}(n) = (\boldsymbol{I} - \mathbf{s}_1\mathbf{s}_1^{\mathrm{T}})\mathbf{y}(n)$ and $\mathbf{y}(n)$; μ and λ are the step size of LMS algorithm and the forgetting factor of RLS algorithm, respectively.

Notes From Eq. (5.9.42), for Kalman algorithm, when the number of iterations $n = 2N$, the mean output energy $\xi(n) \leq 1.5\xi_{\min}$; when $n = 16N$, $\xi(n) \leq 1.0625\xi_{\min}$. This shows that the output energy $\xi(n)$ of the Kalman filtering algorithm quickly tends to the minimum mean output energy $\xi_{\min}$ with the increase of n. Therefore, the convergence performance of the average output energy $\xi(n)$ of the three algorithms is as follows:

(1) As shown in Eq. (5.9.42), the convergence of the Kalman filtering algorithm only depends on the spread spectrum gain N, which is independent of data correlation matrix $\boldsymbol{R}$;
(2) The convergence of the LMS algorithm depends on the eigenvalue distribution of data correlation matrix $\boldsymbol{R}$;
(3) The convergence of RLS algorithm depends on the trace of matrix product $\boldsymbol{RM}(n-1)$, where $\boldsymbol{M}(n) = \mathrm{E}\{[\mathbf{c}_1(n) - \mathbf{c}_{\mathrm{opt1}}][\mathbf{c}_1(n) - \mathbf{c}_{\mathrm{opt1}}]^{\mathrm{T}}\}$.

When using the LMS algorithm, the step size μ must meet the stability condition of the output mean square error convergence[106]

$$\mu < \frac{2}{\sum_{k=1}^{K} A_k^2 + N\sigma^2}. \tag{5.9.44}$$

To compare the multiple access interference suppression ability of different algorithms, the time average SIR of n-step iteration is often used as an index

$$\mathrm{SIR}(n) = 10\log\frac{\sum_{i=1}^{M}(\boldsymbol{c}_{1l}^{\mathrm{T}}(n)\boldsymbol{s}_1)^2}{\sum_{l=1}^{M}}\boldsymbol{c}_{1l}^{\mathrm{T}}(n)\left(\boldsymbol{y}_l(n) - b_{1,l}(n)\boldsymbol{s}_1\right)^2, \tag{5.9.45}$$

where M is the number of independent experiments, the subscript l indicates the l-th experiment, and the variance of the background noise is σ^2. Assume that the SNR of user k is $\mathrm{SNR} = 10\log(E_k/\sigma^2)$, where $E_k = A_k^2$ is the bit energy of user k.

Summary

The main content of this chapter is the optimal design and adaptive implementation of filters. Firstly, different filters are introduced from three perspectives:

(1) Based on the principle of maximum SNR, the matched filter is discussed;
(2) Based on the minimum mean square error criterion, the Wiener filter is derived;
(3) Based on the state space model, the Kalman filter and its adaptive algorithm are introduced.

Then, for the adaptive implementation of the Wiener filter, LMS and RLS adaptive algorithms are introduced. To overcome the disadvantage of slow convergence of the transverse filter, the LMS lattice filter with symmetric structure and LS lattice filter with the asymmetric structure are introduced in this chapter.

Finally, as the application of adaptive filters, adaptive line enhancer and notch filter, generalized sidelobe canceller, and blind adaptive multiuser detector are introduced, respectively.

Exercises

5.1 The harmonic signal is

$$s(t) = A\cos(2\pi f_c t), \quad 0 \le t \le T, f_c = \frac{1}{T}.$$

The observation sample is $y(t) = s(t) + w(t)$, where $w(t)$ is a Gaussian white noise with zero mean and variance σ^2. Find the output of the matched filter when $t = T$ and its mean and variance.

5.2 Suppose that the transmitter transmits signals $s_1(t)$ and $s_2(t)$ in turn,

$$\begin{aligned} s_1(t) &= A\cos(2\pi f_c t), \quad 0 \le t \le T, f_c = \frac{1}{T}, \\ s_2(t) &= A\sin(2\pi f_c t), \quad 0 \le t \le T. \end{aligned}$$

The attenuation in the signal transmission process is ignored, and the observation noise at the receiver is white noise $w(t)$, with zero mean and variance σ^2. A matched filter is designed at the receiver to extract signal $s_1(t)$ and calculate the output of the matched filter when $t = T$.

5.3 Set the signal to be

$$s(t) = \begin{cases} e^{-t}, & t > 0, \\ 0, & t < 0, \end{cases}$$

and the noise $n(t)$ to be a white noise with zero mean and variance 1. Solve the impulse response $h_0(t)$ of the matched filter.

5.4 The autocorrelation function

$$R_{ss}(\tau) = A\frac{\sin^2(\alpha\tau)}{\tau^2},$$

of signal $s(t)$ and the autocorrelation function $R_{vv}(\tau) = N\delta(\tau)$ of the additive noise are known. The signal is not correlated with the noise, i.e., $R_{sv}(\tau) = \mathrm{E}\{s(t)v(t-\tau)\} = 0, \forall\tau$. Find the non causal Wiener filter $H(\omega)$ for estimating $s(t)$ with the observation data $x(t) = s(t) + v(t)$.

5.5 Let $s(t)$ be a stationary random process, and

$$R_{ss}(\tau) = \mathrm{E}\{s(t)s(t-\tau)\} = \frac{1}{2}e^{-|\tau|},$$

$$R_{nn}(\tau) = \mathrm{E}\{n(t)n(t-\tau)\} = \begin{cases} 1, & \tau = 0, \\ 0, & \tau \neq 0. \end{cases}$$

The signal is uncorrelated with the noise, i.e., $\mathrm{E}\{s(t)n(t-\tau)\} = 0, \forall\tau$. Find the transfer function expression of the causal Wiener filter.

5.6 Let $y(t) = s(t) + n(t)$. Given

$$P_{ss}(\omega) = \frac{N_0}{\alpha^2 + \omega^2}, \quad P_{nn}(\omega) = N \text{ and } P_{sn}(\omega) = 0,$$

where $\alpha > 0$. Find the transfer function of the causal Wiener filter.

5.7 The discrete-time signal $s(n)$ is a first-order AR process, and its correlation function is $R_s(k) = \alpha^{|k|}, 0 < \alpha < 1$. Let the observation data be $x(n) = s(n) + v(n)$, where $s(n)$ and $v(n)$ are uncorrelated. $v(n)$ is a white noise with zero mean and variance of σ_v^2. Design its Wiener filter $H(z)$.

5.8 Assume that the filter acting on the observation data $y(t) = s(t) + n(t)$ has a transfer function

$$a(t, u) = \frac{m(1+u)^{m-1}}{(1+t)^m}, \quad 0 \le u \le t, m > 0.$$

Find the innovation process of $y(t)$, and design a filter $b(t, s)$ acting on the innovation.

5.9 Suppose the power spectrum of the observation signal $y(t) = s(t) + n(t)$ is

$$P_{yy}(\omega) = \frac{\omega^2 + 25}{(\omega^2 + 1)(\omega^2 + 4)}.$$

Solve the innovation process $w(t)$ and the expression of linear mean square estimation $\hat{s}(t)$ of signal $s(t)$.

5.10 Let $x(t)$ be a time invariant scalar random variable, which is observed in additive white Gaussian noise $v(t)$, that is, $y(t) = x(t) + v(t)$ is the observed data. If Kalman filter is used to adaptively estimate $x(t)$, try to design a Kalman filter:

(1) Construct a discrete-time state space equation;

(2) Find the update equation of the state variable $x(k)$.

5.11 Kalman filter estimation of AR(1) process. The state variable obeys AR(1) model $x(n) = 0.8x(n-1) + w(n)$, where $w(n)$ is a white noise with zero mean and variance $\sigma_w^2 = 0.36$. The observation equation is $y(n) = x(n) + v(n)$, where $v(n)$ is a white noise independent of $w(n)$, with zero mean and variance $\sigma_v^2 = 1$. The Kalman filter is used to estimate the state variable. Find the specific expression of $\hat{x}(n)$.

5.12 The state transition equation and observation equation of a time-varying system are

$$\boldsymbol{x}(n+1) = \begin{bmatrix} 1/2 & 1/8 \\ 1/8 & 1/2 \end{bmatrix} \boldsymbol{x}(n) + \boldsymbol{v}_1(n),$$

and

$$\boldsymbol{y}(n) = \boldsymbol{x}(n) + \boldsymbol{v}_2(n),$$

respectively, where

$$\begin{aligned}
\mathrm{E}\{\boldsymbol{v}_1(n)\} &= 0, \\
\mathrm{E}\{\boldsymbol{v}_1(n)\boldsymbol{v}_1^{\mathrm{T}}(k)\} &= \begin{cases} \sigma_1^2\boldsymbol{I}, & n = k, \\ \boldsymbol{0}, & n \neq k, \end{cases} \\
\mathrm{E}\{\boldsymbol{v}_2(n)\boldsymbol{v}_2^{\mathrm{T}}(k)\} &= \begin{cases} \sigma_2^2\boldsymbol{I}, & n = k, \\ \boldsymbol{0}, & n \neq k. \end{cases} \\
\mathrm{E}\{\boldsymbol{v}_1(n)\boldsymbol{v}_2^{\mathrm{T}}(k)\} &= \boldsymbol{0}, \quad \forall n, k, \\
\mathrm{E}\{\boldsymbol{x}(1)\boldsymbol{x}^{\mathrm{T}}(1)\} &= \boldsymbol{I},
\end{aligned}$$

where $\boldsymbol{0}$ and $\boldsymbol{I}$ are zero matrix and identity matrix, respectively. Find the update equation of $\boldsymbol{x}(n)$.

5.13 The following figure shows a two-dimensional radar tracking schematic diagram. In the figure, the target aircraft flies in the x direction at a constant speed V, the radar is at the origin O, the distance between the aircraft and the radar is r, and the azimuth is θ. To make the radar track the aircraft, Kalman filter is applied to adaptively estimate the aircraft distance r, the aircraft velocity $\dot{r}$, azimuth θ and angular velocity $\dot{\theta}$ on the radar line of sight. For the problem of two-dimensional radar tracking, try to construct a discrete-time state space equation.

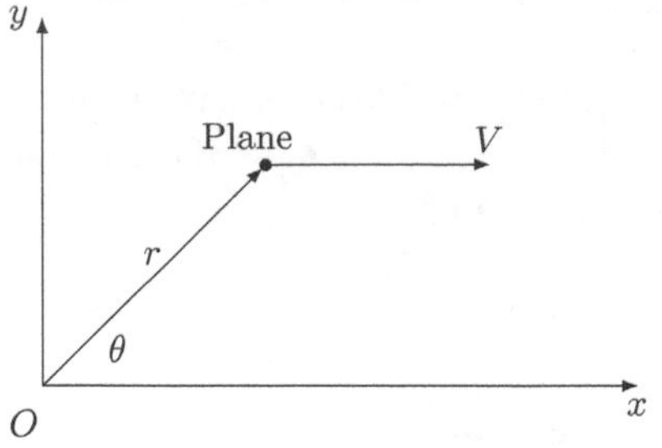

Fig-Exer. Geometry illustration of radar tracking

5.14 The time-varying real ARMA process is described by the difference equation

$$y(n) + \sum_{i=1}^{p} a_i(n)y(n-i) = \sum_{i=1}^{q} a_{p+i}v(n-i) + v(n),$$

where $a_1(n), \cdots, a_p(n), a_{p+1}(n), \cdots, a_{p+q}(n)$ are parameters of the ARMA model, $v(n)$ is the input, and $y(n)$ is the output. Assume that the input process $v(n)$ is Gaussian white noise with variance σ^2. The parameters of the ARMA model obey a random disturbance model

$$a_k(n+1) = a_k(n) + w_k(n), \ k = 1, \cdots, p, p+1, \cdots, p+q,$$

where $w_k(n)$ is a Gaussian white noise with zero mean, which is independent of $w_j(n), j \neq k$ and $v(n)$. Define the $(p+q) \times 1$ state vector

$$\boldsymbol{x}(n) = [a_1(n), \cdots, a_p(n), a_{p+1}(n), \cdots, a_{p+q}(n)]^{\mathrm{T}}.$$

And define the measurement matrix

$$\boldsymbol{C}(n) = [-y(n-1), \cdots, -y(n-p), v(n-1), \cdots, v(n-q)],$$

which is essentially a row vector here. According to the above conditions, solve the following problems:

(1) Establish the state space equation of the time-varying ARMA process;
(2) Give the Kalman adaptive filtering algorithm for updating the state vector $\boldsymbol{x}(n+1)$;
(3) How to set the initial values?

5.15 In the case of stationary system, if the state transition matrix $\boldsymbol{F}(n+1, n)$ is the identity matrix and the state noise vector is zero. Prove that the predicted state error correlation matrix $\boldsymbol{K}(n+1, n)$ is equal to the filtered state error correlation matrix $\boldsymbol{K}(n)$.

5.16 In wireless communication, the FIR filter with known impulse response is often used as the wireless channel model. If the channel output, i.e., the signal $y(n)$ received by the receiver, is given by

$$y(n) = \boldsymbol{h}^{\mathrm{T}}\boldsymbol{x}(n) + w(n),$$

where $\boldsymbol{h}$ is an $M \times 1$ vector, representing the channel impulse response; $\boldsymbol{x}(n)$ is an $M \times 1$ vector, representing the current value and $M - 1$ previous transmission values of the channel input; $w(n)$ is a Gaussian white noise with zero mean and variance σ_w^2. At time n, the channel input $u(n)$ is composed of binaries $\{-1, +1\}$, which is statistically independent of $w(n)$. Therefore, the state equation can be written as[133]

$$\boldsymbol{x}(n+1) = \boldsymbol{A}\boldsymbol{x}(n) + \boldsymbol{e}_1 v(n),$$

where $v(n)$ is a Gaussian white noise with zero mean and variance σ_v^2, which is independent of $w(n)$. The matrix $\boldsymbol{A}$ is an $M \times M$ matrix, with elements

$$a_{ij} = \begin{cases} 1, & i = j+1, \\ 0, & \text{others.} \end{cases}$$

and $\boldsymbol{e}_1$ is an $M \times 1$ vector, with elements

$$e_i = \begin{cases} 1, & i = 1, \\ 0, & \text{others.} \end{cases}$$

When the channel model and the observation $y(n)$ with noise are known, an equalizer is constructed by using Kalman filter, which can give an estimated value of the channel input $u(n)$ at a certain delay time $(n + D)$, where $0 \le D \le M - 1$. Prove that the equalizer is an infinite impulse response filter, and its coefficients are determined by two groups of different parameters:

(1) $M \times 1$ channel impulse response vector;

(2) Kalman gain vector (which is an M dimensional column vector).

5.17 Consider a code division multiple access (CDMA) system with K users. Assuming that user 1 is the desired user, its characteristic waveform vector $\boldsymbol{s}_1$ is known and satisfies the unit energy condition $\langle \boldsymbol{s}_1, \boldsymbol{s}_1 \rangle = \boldsymbol{s}_1^{\mathrm{T}}\boldsymbol{s}_1 = 1$. The observation data vector of the receiver is $\boldsymbol{y}(n)$, which contains the linear mixture of K user signals. To detect the desired user's signal, we want to design a multiuser detector $\boldsymbol{c}_1$ to minimize the output energy of the detector. If the multiuser detector obeys the constraint $\boldsymbol{c}_1 = \boldsymbol{s}_1 + \boldsymbol{U}_i\boldsymbol{w}$, where $\boldsymbol{U}_i$ is called the interference subspace, i.e., its columns are expanded into interference subspace.

(1) Find the LMS adaptive algorithm of the linear detector $\boldsymbol{c}_1$;

(2) How to compute the interference subspace $\boldsymbol{U}_i$?

5.18 If the j-order least squares backward prediction error vector is given by

$$\boldsymbol{P}_{0,j-1}^{\perp}(n)z^{-j}\boldsymbol{x}(n) = \boldsymbol{e}_j^b(n).$$

Prove

$$P_{1,j}^{\perp}(n)z^{-j-1}\boldsymbol{x}(n) = z^{-1}\boldsymbol{e}_j^b(n).$$

5.19 Given a time signal $\boldsymbol{v}(n) = [v(1), v(2), v(3), \cdots, v(n)]^{\mathrm{T}}$. Calculate
(1) the data vectors $\boldsymbol{v}(2)$ and $\boldsymbol{v}(3)$;
(2) the vectors $z^{-1}\boldsymbol{v}(2)$ and $z^{-2}\boldsymbol{v}(2)$;
(3) the vectors $z^{-1}\boldsymbol{v}(3)$ and $z^{-2}\boldsymbol{v}(3)$.

If $\boldsymbol{u}(n) = z^{-1}\boldsymbol{v}(n)$, calculate
(4) the projection matrix $\boldsymbol{P}_u(2)$ and $\boldsymbol{P}_u(3)$;
(5) the least square prediction of $\boldsymbol{v}(n)$ by $\boldsymbol{u}(n)$. This prediction is called one-step forward prediction of $\boldsymbol{v}(n)$;
(6) the forward prediction error vectors $\boldsymbol{e}_1^f(2)$ and $\boldsymbol{e}_1^f(3)$.

5.20 It is known that the forward and backward prediction residuals are respectively

$$\begin{aligned}
\epsilon_m^f(n) &= \langle \boldsymbol{x}(n), \boldsymbol{P}_{1,m}^{\perp}(n)\boldsymbol{x}(n)\rangle, \\
\epsilon_m^b(n) &= \langle z^{-m}\boldsymbol{x}(n), \boldsymbol{P}_{0,m-1}^{\perp}(n)z^{-m}\boldsymbol{x}(n)\rangle,
\end{aligned}$$

and the partial correlation coefficient $\Delta_{m+1}^f(n) = \langle \boldsymbol{e}_m^f(n), z^{-1}\boldsymbol{e}_m^b(n)\rangle$. Prove

$$\begin{aligned}
\epsilon_{m+1}^f(n) &= \epsilon_m^f(n) - \frac{\Delta_{m+1}^2(n)}{\epsilon_m^b(n-1)}, \\
\epsilon_{m+1}^b(n) &= \epsilon_m^b(n-1) - \frac{\Delta_{m+1}^2(n)}{\epsilon_m^f(n)}.
\end{aligned}$$

6 Higher-Order Statistical Analysis

The signal processing methods used in the previous chapters use the second-order statistics (time-domain is the correlation function and frequency-domain is the power spectrum) as the mathematical analysis tool. The correlation function and the power spectrum have some shortcomings. For example, they have equivalence or multiplicity and can not identify the nonminimum phase systems; they are sensitive to additive noise and can only deal with the observation with additive white noise. To overcome these shortcomings, third-order or higher-order statistics must be used. Here all these statistics are called higher-order statistics. The signal analysis based on higher-order statistics is referred to as higher-order statistical analysis of the signal, also known as non-Gaussian signal processing. Second-order statistical analysis can only extract the main information of the signal, i.e., the profile, while higher-order statistical analysis can provide detailed information about the signal. Therefore, higher-order statistics is an indispensable mathematical tool in signal processing.

As early as the 1960s, higher-order statistics had been studied by mathematicians. However, this research did not gain comparative development because no appropriate applications were found at that time. It was not until the late 1980s that the experts in signal processing ignited the fire of this research and it rapidly developed into an important branch of modern signal processing. This chapter will systematically introduce the theory, method, and some typical applications of the higher-order statistical analysis.

6.1 Moments and Cumulants

The most commonly used higher-order statistics are higher-order cumulants and higher-order spectra.

6.1.1 Definition of Higher-order Moments and Cumulants

The characteristic function method is one of the main analysis tools of probability theory and mathematical statistics. Using the characteristic function, it is easy to get the definition of the higher-order moments and higher-order cumulants.

Consider a continuous random variable x, if the probability density function of x is $f(x)$, and $g(x)$ is an arbitrary function, then the expectation of $g(x)$ can be defined as

$$E\{g(x)\} \stackrel{\text{def}}{=} \int_{-\infty}^{+\infty} f(x)g(x)dx. \tag{6.1.1}$$

https://doi.org/10.1515/9783110475562-006

Especially, when $g(x) = e^{j\omega x}$, we have

$$\Phi(\omega) \stackrel{\text{def}}{=} E\{e^{j\omega x}\} = \int_{-\infty}^{+\infty} f(x)e^{j\omega x}dx, \tag{6.1.2}$$

and this function is called the first characteristic function. In other words, the first characteristic function is the inverse Fourier transform of the characteristic function $f(x)$. Since the probability density function $f(x) \geq 0$, the first characteristic function $\Phi(x)$ has a maximum value at the origin, i.e.,

$$|\Phi(\omega)| \leq \Phi(0) = 1. \tag{6.1.3}$$

Taking the kth-order derivative of the first characteristic function, it is obtained that

$$\Phi^k(\omega) = \frac{d^k\Phi(\omega)}{d\omega^k} = j^k E\{x^k e^{j\omega x}\}. \tag{6.1.4}$$

Given a random variable x, its kth-order (original) moment m_k and central moment μ_k are defined as

$$m_k \stackrel{\text{def}}{=} E\{x^k\} = \int_{-\infty}^{+\infty} x^k f(x)dx, \tag{6.1.5}$$

$$\mu_k \stackrel{\text{def}}{=} E\{(x-\eta)^k\} = \int_{-\infty}^{+\infty} (x-\eta)^k f(x)dx, \tag{6.1.6}$$

where $\eta = E\{x\}$ represents the first-order moment (i.e. mean) of the random variable x. For a random variable x with zero mean, the kth-order original moment m_k and central moment μ_k are equivalent. In the following, let the mean of random variables and signals be zero.

By making $\omega = 0$ in Eq. (6.1.4), the kth-order moment of x can be obtained

$$m_k = E\{x^k\} = (-j)^k \left.\frac{d^k\Phi(\omega)}{d\omega^k}\right|_{\omega=0} = (-j)^k\Phi^k(0). \tag{6.1.7}$$

Since the kth-order moment $E\{x^k\}$ of x can be generated using the first characteristic function, the first characteristic function is usually called the moment-generating function.

The natural logarithm of the first characteristic function is called the second characteristic function, denoted by

$$\Psi(\omega) \stackrel{\text{def}}{=} \ln\Phi(\omega). \tag{6.1.8}$$

Similar to the definition of Eq. (6.1.7) for kth-order moments, one can also define the kth-order cumulants of a random variable x as

$$c_{kx} = (-j)^k \left.\frac{d^k \ln\Psi(\omega)}{d\omega^k}\right|_{\omega=0} = (-j)^k\Psi^k(0). \tag{6.1.9}$$

Therefore, the second characteristic function is called the cumulant-generating function.

The above discussion on single random variable x can be easily extended to multiple random variables. Let $x_1, \cdots, x_k$ be k continuous random variables and the joint probability density function is $f(x_1, \cdots, x_k)$, then the first joint characteristic function for these k random variables is defined as

$$\begin{aligned}\Phi(\omega_1, \cdots, \omega_k) &\overset{\text{def}}{=} E\{e^{j(\omega_1 x_1 + \cdots + \omega_k x_k)}\} \\ &= \int_{-\infty}^{\infty} \cdots \int_{-\infty}^{\infty} f(x_1, \cdots, x_k) e^{j(\omega_1 x_1 + \cdots + \omega_k x_k)} dx_1 \cdots dx_k. \end{aligned} \tag{6.1.10}$$

Taking the $(r = r_1 + \cdots + r_k)$th-order partial derivatives of $\Phi(\omega_1, \cdots, \omega_k)$ with respect to $\omega_1, \cdots, \omega_k$, we have

$$\frac{\partial^r \Phi(\omega_1, \cdots, \omega_k)}{\partial \omega_1^{r_1} \cdots \partial \omega_k^{r_k}} = (-j)^r E\{x_1^{r_1} \cdots x_k^{r_k} e^{j(\omega_1 x_1 + \cdots + \omega_k x_k)}\}. \tag{6.1.11}$$

Therefore, the rth-order joint moment of k random variables is

$$m_{r_1 \cdots r_k} \overset{\text{def}}{=} E\{x_1^{r_1} \cdots x_k^{r_k}\} = (-j)^r \left. \frac{\partial^r \Phi(\omega_1, \cdots, \omega_k)}{\partial \omega_1^{r_1} \cdots \partial \omega_k^{r_k}} \right|_{\omega_1 = \cdots = \omega_k = 0}. \tag{6.1.12}$$

Similarly, the second joint characteristic function is defined as

$$\Psi(\omega_1, \cdots, \omega_k) = \ln \Phi(\omega_1, \cdots, \omega_k). \tag{6.1.13}$$

The rth-order joint cumulants of the random variables $x_1, \cdots, x_k$ are defined as

$$c_{r_1 \cdots r_k} \overset{\text{def}}{=} \text{cum}(x_1^{r_1} \cdots x_k^{r_k}) = (-j)^r \left. \frac{\partial^r \ln \Phi(\omega_1, \cdots, \omega_k)}{\partial \omega_1^{r_1} \cdots \partial \omega_k^{r_k}} \right|_{\omega_1 = \cdots = \omega_k = 0}. \tag{6.1.14}$$

In practice, it is often taken as $r_1 = \cdots = r_k = 1$, from which the kth-order moments and kth-order cumulants of the k random variables are obtained as

$$m_{1 \cdots 1} \overset{\text{def}}{=} E\{x_1 \cdots x_k\} = (-j)^k \left. \frac{\partial^k \Phi(\omega_1, \cdots, \omega_k)}{\partial \omega_1 \cdots \partial \omega_k} \right|_{\omega_1 = \cdots = \omega_k = 0}, \tag{6.1.15}$$

$$c_{1 \cdots 1} \overset{\text{def}}{=} \text{cum}(x_1 \cdots x_k) = (-j)^k \left. \frac{\partial^k \ln \Phi(\omega_1, \cdots, \omega_k)}{\partial \omega_1 \cdots \partial \omega_k} \right|_{\omega_1 = \cdots = \omega_k = 0}. \tag{6.1.16}$$

Consider a stationary continuous random signal $x(t)$. Let $x_1 = x(t), x_2 = x(t + \tau_1), \cdots, x_k = x(t + \tau_{k-1})$, then $m_{kx}(\tau_1, \cdots, \tau_{k-1}) = m_{1 \cdots 1}$ is called the kth-order moment of the random signal $x(t)$. Hence, from Eq. (6.1.15) we get

$$m_{kx}(\tau_1, \cdots, \tau_{k-1}) = E\{x(t)x(t + \tau_1) \cdots x(t + \tau_{k-1})\}. \tag{6.1.17}$$

Similarly, higher-order cumulants of random signal $x(t)$ can be expressed as

$$c_{kx}(\tau_1, \cdots, \tau_{k-1}) = \text{cum}[x(t), x(t + \tau_1), \cdots, x(t + \tau_{k-1})]. \tag{6.1.18}$$

The above equation is only a formal definition of higher-order cumulants and does not give the concrete expression of cumulants. In fact, cumulants can be expressed by moments, and we will discuss this later.

Especially, the third- and fourth-order cumulants are most commonly used in higher-order statistical analysis.

6.1.2 Higher-order Moments and Cumulants of Gaussian Signal

Let x be a Gaussian random signal with zero mean and variance σ^2, or it can be denoted by distribution symbol as $x \sim \mathcal{N}(0, \sigma^2)$. Since the probability density function of x is

$$f(x) = \frac{1}{\sqrt{2\pi}\sigma} \exp\left(-\frac{x^2}{2\sigma^2}\right), \tag{6.1.19}$$

then the moment-generating function of Gaussian random signal x is given by

$$\begin{aligned} \Phi(\omega) &= \int_{-\infty}^{\infty} f(x) e^{j\omega x} dx \\ &= \frac{1}{\sqrt{2\pi}\sigma} \int_{-\infty}^{\infty} \exp\left(-\frac{x^2}{2\sigma^2} + j\omega x\right) dx. \end{aligned} \tag{6.1.20}$$

In the following integral formula

$$\int_{-\infty}^{\infty} \exp(-Ax^2 \pm 2Bx - C) dx = \sqrt{\frac{\pi}{A}} \exp\left(-\frac{AC - B^2}{A}\right), \tag{6.1.21}$$

let $A = \frac{1}{2\sigma^2}$, $B = \frac{j\omega}{2}$, $C = 0$, then from Eq. (6.1.20) and (6.1.21) we can get

$$\Phi(\omega) = e^{-\sigma^2\omega^2/2}. \tag{6.1.22}$$

Computing the derivatives of $\Phi(\omega)$, we obtain

$$\begin{aligned} \Phi'(\omega) &= -\sigma^2 \omega e^{-\sigma^2\omega^2/2}, \\ \Phi''(\omega) &= (\sigma^4\omega^2 - \sigma^2) e^{-\sigma^2\omega^2/2}, \\ \Phi^3(\omega) &= (3\sigma^4\omega - \sigma^6\omega^3) e^{-\sigma^2\omega^2/2}, \\ \Phi^4(\omega) &= (3\sigma^4 - 6\sigma^6\omega^2 + \sigma^8\omega^4) e^{-\sigma^2\omega^2/2}. \end{aligned}$$

Substituting these values into Eq. (6.1.7), we can get the moments of Gaussian random variable

$$m_1 = 0, m_2 = \sigma^2, m_3 = 0, m_4 = 3\sigma^4.$$

By extension, for any integer k, the moments of Gaussian random variable can be written uniformly

$$m_k = \begin{cases} 0, & k = \text{odd}, \\ 1 \cdot 3 \cdots (k-1)\sigma^k, & k = \text{even}. \end{cases} \quad (6.1.23)$$

From Eq. (6.1.22), the cumulant-generating function of Gaussian random variable x can be obtained directly

$$\Psi(\omega) = \ln \Phi(\omega) = \frac{\sigma^2\omega^2}{2},$$

its derivatives are $\Psi'(\omega) = -\sigma^2\omega$, $\Psi''(\omega) = -\sigma^2$, and $\Psi^{(k)}(\omega) \equiv 0$, $k = 3, 4, \cdots$. Substituting these values into Eq. (6.1.9), we can obtain the cumulants of Gaussian random variable, i.e., $c_1 = 0$, $c_2 = \sigma^2$ and $c_k = 0$ for $k = 3, 4, \cdots$.

The above results of moments and cumulants of Gaussian random variables can be easily generalized as follows: the second-order moment and the second-order cumulant of any zero mean Gaussian stochastic process are identical, which are equal to the variance σ^2; its odd-order moments are always zero, but its even-order moments are not zero; higher-order (third-order and above) cumulants are equal to zero. In this sense, the higher-order cumulants are said to be "blind" to the Gaussian stochastic processes.

6.1.3 Transformation Relationships between Moments and Cumulants

Let $\{x_1, \cdots, x_k\}$ be a set of k random variables, and its indices set is $I = \{1, 2, \cdots, k\}$. Now consider dividing set I into several subsets such that none of these subsets is empty, no two subsets have the same elements and these subsets have no order. Such division is called nonintersecting and nonempty division of set I. That is, the division is an unordered collection of nonintersecting nonempty subsets I_p such that $\bigcup I_p = I$. Here $\bigcup I_p$ denotes the union of all the subsets. Let $m_x(I)$ and $c_x(I)$ denote the kth-order moment and kth-order cumulant of a random signal $x(t)$, respectively. Moreover, $m_x(I_p)$ and $c_x(I_p)$ represent the moments and cumulants of indices set I_p. For example, $I_p = \{1, 3\}$, then $m_x(I_p) = E\{x(t)x(t+\tau_2\}$ and $c_x(I_p) = cum\{x(t), x(t+\tau_2\}$.

Using moments,the cumulants can be described as

$$c_x(I) = \sum_{\bigcup_{p=1}^{q} I_p = I} (-1)^{q-1}(q-1)! \prod_{p=1}^{q} m_x(I_p), \quad (6.1.24)$$

this relation is known as the moment-to-cumulant transform formula or M-C formula for short.

Similarly, moments can also be expressed as cumulants

$$m_x(I) = \sum_{\bigcup_{p=1}^{q} I_p = I} \prod_{p=1}^{q} c_x(I_p), \quad (6.1.25)$$

this equation is called cumulant-to-moment transform formula or C-M formula for short.

Next, taking the third-order cumulant as an example, we will discuss how to get the formula of cumulant expressed by moment using the M-C formula.

(1) If $q = 1$, i.e., set $I = \{1, 2, 3\}$ decomposes into one subset, then $I_1 = \{1, 2, 3\}$.
(2) If $q = 2$, i.e., set $I = \{1, 2, 3\}$ decomposes into two subsets, then we have three kinds of decomposition: $I_1 = \{1\}, I_2 = \{2, 3\}$; $I_1 = \{2\}, I_2 = \{3, 1\}$ and $I_1 = \{3\}$, $I_2 = \{1, 2\}$.
(3) If $q = 3$, i.e., set $I = \{1, 2, 3\}$ decomposes into three subsets, then we have only one kind of decomposition: $I_1 = \{1\}, I_2 = \{2\}$ and $I_3 = \{3\}$.

Substituting the above decomposition into Eq. (6.1.24), we can get

$$\begin{aligned} c_{3x}(\tau_1, \tau_2) =& E\{x(t)x(t+\tau_1)x(t+\tau_2)\} - E\{x(t)\}E\{x(t+\tau_1)x(t+\tau_2)\} \\ & - E\{x(t+\tau_1)\}E\{x(t+\tau_2)x(t)\} - E\{x(t+\tau_2)\}E\{x(t)x(t+\tau_1\} \\ & + 2E\{x(t)\}E\{x(t+\tau_1)\}E\{x(t+\tau_2)\}. \end{aligned} \tag{6.1.26}$$

If let the mean of stationary random real signal $x(t)$ is $\mu_x = E\{x(t)\}$ and the correlation function is $R_x = E\{x(t)x(t+\tau)\}$, then Eq. (6.1.26) can be represented as

$$\begin{aligned} c_{3x}(\tau_1, \tau_2) =& E\{x(t)x(t+\tau_1)x(t+\tau_2)\} - \mu_x R_x(\tau_2 - \tau_1) \\ & - \mu_x R_x(\tau_2) - \mu_x R_x(\tau_1) + 2\mu_x^3, \end{aligned} \tag{6.1.27}$$

this form is very complicated. Similarly, the fourth-order cumulant can be expressed by the first-, second-, and third-order moments, but its form will be more complex. However, these expressions can be greatly simplified when $x(t)$ is a random signal with zero mean. For convenience, the second-, third- and fourth-order cumulants of a zero mean random real signal $x(t)$ are summarized as follows:

$$c_{2x}(\tau) = E\{x(t)x(t+\tau)\} = R_x(\tau), \tag{6.1.28}$$

$$c_{3x}(\tau_1, \tau_2) = E\{x(t)x(t+\tau_1)x(t+\tau_2)\}, \tag{6.1.29}$$

$$\begin{aligned} c_{4x}(\tau_1, \tau_2, \tau_3) =& E\{x(t)x(t+\tau_1)x(t+\tau_2)x(t+\tau_3)\} - R_x(\tau_1)R_x(\tau_3 - \tau_2) \\ & - R_x(\tau_2)R_x(\tau_3 - \tau_1) - R_x(\tau_3)R_x(\tau_2 - \tau_1). \end{aligned} \tag{6.1.30}$$

In real applications, it is necessary to estimate the cumulants of each order based on the known data samples. In order to obtain the consistent sample estimation of kth-order cumulants, it is usually necessary to assume that the non-Gaussian signal $x(t)$ is $2k$th-order absolutely summable, i.e.,

$$\sum_{\tau_1=-\infty}^{\infty} \cdots \sum_{\tau_{m-1}=-\infty}^{\infty} |c_{mx}(\tau_1, \cdots, \tau_{m-1})| < \infty, m = 1, \cdots, 2k. \tag{6.1.31}$$

When $x(t)$ satisfies this condition, the higher-order cumulants can be estimated using the data samples $x(1), \cdots, x(N)$

$$\hat{c}_{3x}(\tau_1, \tau_2) = \frac{1}{N}\sum_{n=1}^{N} x(n)x(n+\tau_1)x(n+\tau_2), \tag{6.1.32}$$

$$\hat{m}_{4x}(\tau_1, \tau_2, \tau_3) = \frac{1}{N}\sum_{n=1}^{N} x(n)x(n+\tau_1)x(n+\tau_2)x(n+\tau_3), \tag{6.1.33}$$

$$\begin{aligned}\hat{c}_{4x}(\tau_1, \tau_2, \tau_3) = {} & \hat{m}_{4x}(\tau_1, \tau_2, \tau_3) - \hat{R}_x(\tau_1)\hat{R}_x(\tau_3 - \tau_2) \\ & - \hat{R}_x(\tau_2)\hat{R}_x(\tau_3 - \tau_1) - \hat{R}_x(\tau_3)\hat{R}_x(\tau_2 - \tau_1),\end{aligned} \tag{6.1.34}$$

where

$$\hat{R}_x(\tau) = \frac{1}{N}\sum_{n=1}^{N} x(n)x(n+\tau), \qquad \hat{R}_x(-\tau) = R_x(\tau). \tag{6.1.35}$$

In the above formulas, let $x(n) = 0$ when $n \le 0$ or $n > N$.

6.2 Properties of Moments and Cumulants

Now we discuss the important properties of moment and cumulant, and further reveal the difference between moment and cumulant. In particular, the properties of cumulant will be frequently cited later. For the convenience of narration, $\text{mom}(x_1, \cdots, x_k)$ and $\text{cumm}(x_1, \cdots, x_k)$ are respectively used to represent the moment and cumulant of k random variables $x_1, \cdots, x_k$.

Property 1 Let λ_i be constant, x_i be random variable, $i = 1, \cdots, k$, then

$$\text{mom}(\lambda_1 x_1, \cdots, \lambda_k x_k) = \prod_{i=1}^{k} \lambda_i \text{mom}(x_1, \cdots, x_k), \tag{6.2.1}$$

$$\text{cum}(\lambda_1 x_1, \cdots, \lambda_k x_k) = \prod_{i=1}^{k} \lambda_i \text{cum}(x_1, \cdots, x_k). \tag{6.2.2}$$

Proof. From the definition of moment and the assumption that $\lambda_1, \cdots, \lambda_k$ are constants, it is immediately seen that Eq. (6.2.1) holds. In order to prove Eq. (6.2.2) holds, notice that the variable sets $y = \{\lambda_1 x_1, \cdots, \lambda_k x_k\}$ and $x = \{x_1, \cdots, x_k\}$ have the same indicator set, i.e., $I_y = I_x$. According to the M-C formula Eq. (6.1.24), we have

$$c_y(I_y) = \sum_{\bigcup_{q=1}^{p} I_p = I} (-1)^{q-1}(q-1)! \prod_{p=1}^{q} m_y(I_p),$$

where

$$\prod_{p=1}^{q} m_y(I_p) = \prod_{p=1}^{q} \lambda_p \prod_{p=1}^{q} m_x(I_p).$$

Here, the definition of the moment and the properties of expectation are used. Thus,

$$c_y(I_y) = \prod_{p=1}^{q} \lambda_p c_x(I_x). \tag{6.2.3}$$

Since $I_y = I_x$, Eq. (6.2.3) is the equivalent representation of Eq. (6.2.2). □

Property 2 Moments and cumulants are symmetric in their arguments, i.e.,

$$\text{mom}(x_1, \cdots, x_k) = \text{mom}(x_{i_1}, \cdots, x_{i_k}), \tag{6.2.4}$$

$$\text{cum}(x_1, \cdots, x_k) = \text{cum}(x_{i_1}, \cdots, x_{i_k}), \tag{6.2.5}$$

where $(i_1, \cdots, i_k)$ is a permutation of $(1, \cdots, k)$.

Proof. Since $\text{mom}(x_1, \cdots, x_k) = E\{x_1 \cdots x_k\}$, then exchanging the positions of the arguments has no effect on the moment. Obviously, Eq. (6.2.4) holds. On the other hand, from the M-C formula Eq. (6.1.24), we know that the division of set I_x satisfies $\bigcup I_p = I$ is an unordered collection of nonintersecting nonempty subsets. Thus, the order of the cumulant arguments is independent of the cumulant value, and the result is that the cumulant is symmetric in their arguments. □

Property 3 Moments and cumulants are additive in their arguments, i.e.,

$$\text{mom}(x_1 + y_1, x_2, \cdots, x_k) = \text{mom}(x_1, x_2, \cdots, x_k) + \text{mom}(y_1, x_2, \cdots, x_k), \tag{6.2.6}$$

$$\text{cum}(x_1 + y_1, x_2, \cdots, x_k) = \text{cum}(x_1, x_2, \cdots, x_k) + \text{cum}(y_1, x_2, \cdots, x_k). \tag{6.2.7}$$

This property means that the cumulants of the sum are equal to the sum of cumulants, terminology "cumulant" is named because of this.

Proof. Noticing

$$\text{mom}(x_1 + y_1, x_2, \cdots, x_k) = E\{(x_1 + y_1)x_2 \cdots x_k\} = E\{x_1 x_2 \cdots x_k\} + E\{y_1 x_2 \cdots x_k\}$$

is the equivalent form of Eq. (6.2.6). Let $\boldsymbol{z} = (x_1 + y_1, x_2, \cdots, x_k)$, $\boldsymbol{x} = (x_1, x_2, \cdots, x_k)$ and $\boldsymbol{v} = (y_1, x_2, \cdots, x_k)$. Since $m_z(I_p)$ is the expectation of the product of the elements in the subdivision I_p, and $x_1 + y_1$ only appears in the form of single power, thus

$$\prod_{p=1}^{q} m_z(I_p) = \prod_{p=1}^{q} m_x(I_p) + \prod_{p=1}^{q} m_v(I_p).$$

Substituting the above equation into the M-C formula (6.1.24), we can get the result of Eq. (6.2.7). □

Property 4 If the random variables $\{x_i\}$ are independent of the random variables $\{y_i\}$, then the cumulants are "semi-invariant", i.e.,

$$\text{cum}(x_1 + y_1, \cdots, x_k + y_k) = \text{cum}(x_1, \cdots, x_k) + \text{cum}(y_1, \cdots, y_k). \tag{6.2.8}$$

But the higher-order moments are generally not semi-invariance, i.e.,

$$\text{mom}(x_1 + y_1, \cdots, x_k + y_k) \neq \text{mom}(x_1, \cdots, x_k) + \text{mom}(y_1, \cdots, y_k). \tag{6.2.9}$$

This property gives another name of cumulant – semi-invariant.

Proof. Let $\boldsymbol{z} = (x_1+y_1, \cdots, x_k+y_k) = \boldsymbol{x}+\boldsymbol{y}$, where $\boldsymbol{x} = (x_1, \cdots, x_k)$ and $\boldsymbol{y} = (y_1, \cdots, y_k)$. According to the statistically independent of $\{x_i\}$ and $\{y_i\}$, it is immediately seen that

$$\begin{aligned}
\Psi_z(\omega_1, \cdots, \omega_k) &= \ln E\left\{e^{j[\omega_1(x_1+y_1)+\cdots+\omega_k(x_k+y_k)]}\right\} \\
&= \ln E\left\{e^{j(\omega_1 x_1+\cdots+\omega_k x_k)}\right\} + \ln E\left\{e^{j(\omega_1 y_1+\cdots+\omega_k y_k)}\right\} \\
&= \Psi_x(\omega_1, \cdots, \omega_k) + \Psi_y(\omega_1, \cdots, \omega_k).
\end{aligned}$$

From the above equation and the definition of cumulant, Eq. (6.2.9) holds. □

Property 5 If a subset of k variables $\{x_1, \cdots, x_k\}$ is independent of the rest, then

$$\text{cum}(x_1, \cdots, x_k) = 0, \tag{6.2.10}$$

$$\text{mom}(x_1, \cdots, x_k) \neq 0. \tag{6.2.11}$$

Proof. According to property 2, cumulants are symmetric in their arguments. Therefore, without loss of generality, assume that $\{x_1, \cdots, x_i\}$ is independent of $\{x_{i+1}, \cdots, x_k\}$, hence

$$\begin{aligned}
\Psi_x(\omega_1, \cdots, \omega_k) &= \ln E\left\{e^{j(\omega_1 x_1+\cdots+\omega_i x_i)}\right\} + \ln E\left\{e^{j(\omega_{i+1} x_{i+1}+\cdots+\omega_k x_k)}\right\} \\
&= \Psi_x(\omega_1, \cdots, \omega_i) + \Psi_x(\omega_{i+1}, \cdots, \omega_k),
\end{aligned} \tag{6.2.12}$$

and

$$\Phi_x(\omega_1, \cdots, \omega_k) = \Phi_x(\omega_1, \cdots, \omega_i)\Phi_x(\omega_{i+1}, \cdots, \omega_k). \tag{6.2.13}$$

From Eq. (6.2.12), it follows that

$$\frac{\partial^k \Psi_x(\omega_1, \cdots, \omega_k)}{\partial\omega_1 \cdots \partial\omega_k} = \frac{\partial^k \Psi_x(\omega_1, \cdots, \omega_i)}{\partial\omega_1 \cdots \partial\omega_k} + \frac{\partial^k \Psi_x(\omega_{i+1}, \cdots, \omega_k)}{\partial\omega_1 \cdots \partial\omega_k} = 0 + 0 = 0. \tag{6.2.14}$$

This is because $\Psi_x(\omega_1, \cdots, \omega_i)$ does not contain variables $\omega_{i+1}, \cdots, \omega_k$, while $\Psi_x(\omega_{i+1}, \cdots, \omega_k)$ does not contain variables $\omega_1, \cdots, \omega_i$. Thus, their kth-order partial derivatives with respect to $\omega_1, \cdots, \omega_k$ are equal to zero, respectively. According to the defintion of cumulant (6.1.15) and Eq. (6.2.14), we know immediately that Eq. (6.2.10) is true.

From Eq. (6.2.13), it follows that

$$\frac{\partial^k \Phi_x(\omega_1, \cdots, \omega_k)}{\partial \omega_1 \cdots \partial \omega_k} = \frac{\partial^k}{\partial \omega_1 \cdots \partial \omega_k}[\Phi_x(\omega_1, \cdots, \omega_i)\Phi_x(\omega_{i+1}, \cdots, \omega_k)].$$

Because $\Phi_x(\omega_1, \cdots, \omega_i)\Phi_x(\omega_{i+1}, \cdots, \omega_k)$ contains variables $\omega_1, \cdots, \omega_k$, the above partial derivative is not zero, that is, Eq. (6.2.11) holds. □

Property 6 If α is constant, then

$$\text{cum}(x_1 + \alpha, x_2, \cdots, x_k) = \text{cum}(x_1, x_2, \cdots, x_k), \tag{6.2.15}$$

$$\text{mom}(x_1 + \alpha, x_2, \cdots, x_k) \neq \text{mom}(x_1, x_2, \cdots, x_k). \tag{6.2.16}$$

Proof. Using Property 3 and Property 5, it follows that

$$\begin{aligned}\text{cum}(x_1 + \alpha, x_2 \cdots, x_k) &= \text{cum}(x_1, x_2, \cdots, x_k) + \text{cum}(\alpha, x_2, \cdots, x_k)\\ &= \text{cum}(x_1, x_2, \cdots, x_k) + 0.\end{aligned}$$

This is Eq. (6.2.15). But

$$\begin{aligned}\text{mom}(x_1 + \alpha, x_2 \cdots, x_k) &= \text{mom}(x_1, x_2, \cdots, x_k) + \text{mom}(\alpha, x_2, \cdots, x_k)\\ &= \text{mom}(x_1, x_2, \cdots, x_k) + \alpha E\{x_2 \cdots x_k\},\end{aligned}$$

is not equal to $\text{mom}(x_1, x_2 \cdots, x_k)$, that is, Eq. (6.2.16) holds. □

The above properties of cumulants are often used in the following sections. Here, we give three important examples to illustrate the important applications of cumulant properties.

1. The Symmetric Forms of the Third-order Cumulant

Property 2 shows that the kth-order cumulant has $k!$ symmetric forms. Taking the third-order cumulant as an example, there are six symmetric forms

$$\begin{aligned}c_{3x}(m, n) &= c_{3x}(n, m) = c_{3x}(-n, m - n) = c_{3x}(n - m, -m)\\ &= c_{3x}(m - n, -n) = c_{3x}(-m, n - m).\end{aligned} \tag{6.2.17}$$

2. Independently Identically Distributed Stochastic Process

As the name implies, an independently identically distributed (IID) random process is a kind of random variable whose values are independent at any time and obey the same distribution. According to Property 5, the cumulant of an independently identically distributed process $\{e(t)\}$ is

$$\begin{aligned}c_{ke}(\tau_1, \cdots, \tau_{k-1}) &= \text{cum}\{e(t), e(t + \tau_1), \cdots, e(t + \tau_{k-1})\}\\ &= \begin{cases} \gamma_{ke}, & \tau_1 = \cdots = \tau_{k-1} = 0 \\ 0, & others \end{cases}\\ &= \gamma_{ke}\delta(\tau_1, \cdots, \tau_{k-1}),\end{aligned} \tag{6.2.18}$$

where $\delta(\tau_1, \cdots, \tau_{k-1})$ is a $(k-1)$-dimemsional δ function, and given by

$$\delta(\tau_1, \cdots, \tau_{k-1}) = \begin{cases} 1, & \tau_1 = \cdots = \tau_{k-1} = 0, \\ 0, & others. \end{cases} \tag{6.2.19}$$

Eq. (6.2.18) shows that the kth-order cumulant of independently identically distributed stochastic processes is a $(k-1)$-dimensional δ function. In particular, Eq. (6.2.18) degenerates into $R_e(\tau) = \sigma_e^2\delta(\tau)$ when $k = 2$, and this is known as white noise. Just as the power spectrum of white noise is constant and has the property of white light, since the kth-order cumulant of an independently identically distributed process satisfying Eq. (6.2.17) for $k \geq 2$ is a $(k-1)$-dimensional δ function, its $(k-1)$-dimensional Fourier transform (called higher-order spectrum) is also a constant. That is, the higher-order spectrum of independently identically distributed non-Gaussian noise is multidimensional flat, so it is called higher-order white noise.

It is necessary to point out that the kth-order moment of an independently identically distributed stochastic process is not a δ function. Taking the fourth-order moment as an example, it is easy to know

$$c_{ke}(0, \tau, \tau) = E\{e^2(t)e^2(t+\tau)\} = E\{e^2(t)\}E\{e^2(t+\tau)\} = \sigma_e^4$$

i.e., the fourth-order moment of an independently identically distributed stochastic process is not a δ function, hence the fourth-order moment spectrum is not multidimensional flat.

3. Blindness to Colored Gaussian Noise

Consider a random signal $x(t)$ observed in colored Gaussian noise $v(t)$, if $v(t)$ and $x(t)$ are statistically independent, we can know from property 4 that the cumulant of the observation process $y(t) = x(t) + v(t)$ is

$$c_{ky}(\tau_1, \cdots, \tau_{k-1}) = c_{kx}(\tau_1, \cdots, \tau_{k-1}) + c_{kv}(\tau_1, \cdots, \tau_{k-1}).$$

However, since the higher-order cumulants of any colored Gaussian noise are equal to zero, the above formula can be simplified as

$$c_{ky}(\tau_1, \cdots, \tau_{k-1}) = c_{kx}(\tau_1, \cdots, \tau_{k-1}), \quad k > 2.$$

This shows that when a non-Gaussian signal is observed in additive colored Gaussian noise, the higher-order cumulants of the observation process are equivalent to that of the non-Gaussian signal, that is, the higher-order cumulants are blind or immune to colored Gaussian noise. However, according to Eq. (6.2.9), the higher-order moments of the observation process are not necessarily equal to the higher-order moments of the non-Gaussian signal, that is, the higher-order moments are sensitive to Gaussian noise.

The above important application answers an important question: why do we usually use higher-order cumulants instead of higher-order moments in higher-order statistical analysis to analyze and process non-Gaussian signals?

For a stationary process $\{x(t)\}$ with zero mean, its kth-order cumulant can be defined as [171]

$$c_{ky}(\tau_1, \cdots, \tau_{k-1}) = E\{x(t)x(t+\tau_1)\cdots x(t+\tau_{k-1})\} - E\{g(t)g(t+\tau_1)\cdots g(t+\tau_{k-1})\}, \tag{6.2.20}$$

where $\{g(t)\}$ is a Gaussian stochastic process and it has the same correlation function and power spectrum as $\{x(t)\}$, i.e.,

$$E\{g(t)g(t+\tau)\} = E\{x(t)x(t+\tau)\}. \tag{6.2.21}$$

Formula (6.2.21) is an engineering definition, which is more intuitive and easy to understand. In particular, it provides a measure of the deviation of a stochastic process $\{x(t)\}$ from normal or Gaussian.

6.3 Higher-order Spectra

For stationary random signal $x(t)$ with zero mean, the power spectral density is defined as the Fourier transform of the autocorrelation function. Similarly, higher-order moment spectra and higher-order cumulant spectra can be defined.

6.3.1 Higher-order Moment Spectra and Higher-order Cumulant Spectra

When defining the power spectrum, the autocorrelation function should be absolutely summable. Similarly, in order to ensure the existence of the Fourier transform of higher-order moments and higher-order cumulants, it is also required that the higher-order moments and higher-order cumulants are absolutely summable.

Definition 6.3.1. *If the higher-order moment $m_{kx}(\tau_1, \cdots, \tau_{k-1})$ is absolutely summable, i.e.,*

$$\sum_{\tau_1=-\infty}^{\infty} \cdots \sum_{\tau_{k-1}=-\infty}^{\infty} |m_{kx}(\tau_1, \cdots, \tau_{k-1})| < \infty, \tag{6.3.1}$$

then the kth-order moment spectrum is defined as the $(k-1)$-dimensional discrete Fourier transform of the kth-order moment, namely

$$M_{kx}(\omega_1, \cdots, \omega_{k-1}) = \sum_{\tau_1=-\infty}^{\infty} \cdots \sum_{\tau_{k-1}=-\infty}^{\infty} m_{kx}(\tau_1, \cdots, \tau_{k-1}) e^{-j(\omega_1\tau_1+\cdots+\omega_{k-1}\tau_{k-1})}. \tag{6.3.2}$$

Definition 6.3.2. *Suppose that the higher-order cumulant $c_{kx}(\tau_1, \cdots, \tau_{k-1})$ is absolutely summable, i.e.,*

$$\sum_{\tau_1=-\infty}^{\infty} \cdots \sum_{\tau_{k-1}=-\infty}^{\infty} |c_{kx}(\tau_1, \cdots, \tau_{k-1})| < \infty, \tag{6.3.3}$$

then the kth-order cumulant spectrum is defined as the $(k-1)$-dimensional discrete Fourier transform of the kth-order cumulant, namely

$$S_{kx}(\omega_1, \cdots, \omega_{k-1}) = \sum_{\tau_1=-\infty}^{\infty} \cdots \sum_{\tau_{k-1}=-\infty}^{\infty} c_{kx}(\tau_1, \cdots, \tau_{k-1}) e^{-j(\omega_1\tau_1+\cdots+\omega_{k-1}\tau_{k-1})}. \tag{6.3.4}$$

Higher-order moments, higher-order cumulants, higher-order moment spectra, and higher-order cumulant spectra are the main four higher-order statistics. In general, higher-order cumulants and higher-order cumulant spectra are often used, while higher-order moments and higher-order moment spectra are rarely used. For this reason, higher-order cumulant spectra are often referred to as higher-order spectra, although higher-order spectra are the combination of higher-order moment spectra and higher-order cumulant spectra.

Higher-order spectra is also called multispectrum, which means the spectra of multiple frequencies. In particular, the third-order spectrum $S_{3x}(\omega_1, \omega_2)$ is called bispectrum, and the fourth-order spectrum $S_{4x}(\omega_1, \omega_2, \omega_3)$ is called trispectrum, because they are energy spectra of two and three frequencies respectively. Generally, $B_x(\omega_1, \omega_2)$ and $T_x(\omega_1, \omega_2, \omega_3)$ are used to represent bispectrum and trispectrum. Next, we focus on the properties and defined region of the bispectrum.

Bispectrum has the following properties.

(1) Bispectrum is usually complex, i.e.,

$$B_x(\omega_1, \omega_2) = |B_x(\omega_1, \omega_2)| e^{j\phi_B(\omega_1, \omega_2)}, \tag{6.3.5}$$

where $|B_x(\omega_1, \omega_2)|$ and $\phi_B(\omega_1, \omega_2)$ represent the amplitude and phase of the bispectrum, respectively.

(2) Bispectrum is a biperiodic function, and both periods are 2π, i.e.,

$$B_x(\omega_1, \omega_2) = B_x(\omega_1 + 2\pi, \omega_2 + 2\pi). \tag{6.3.6}$$

(3) The bispectrum has symmetry,

$$\begin{aligned} B_x(\omega_1, \omega_2) &= B_x(\omega_2, \omega_1) = B_x^{\star}(-\omega_1, -\omega_2) \\ &= B_x^{\star}(-\omega_2, -\omega_1) = B_x(-\omega_1 - \omega_2, \omega_2) \\ &= B_x(\omega_1, -\omega 1 - \omega_2) = B_x(-\omega_1 - \omega_2, \omega_1) \\ &= B_x(\omega_2, -\omega_1 - \omega_2). \end{aligned} \tag{6.3.7}$$

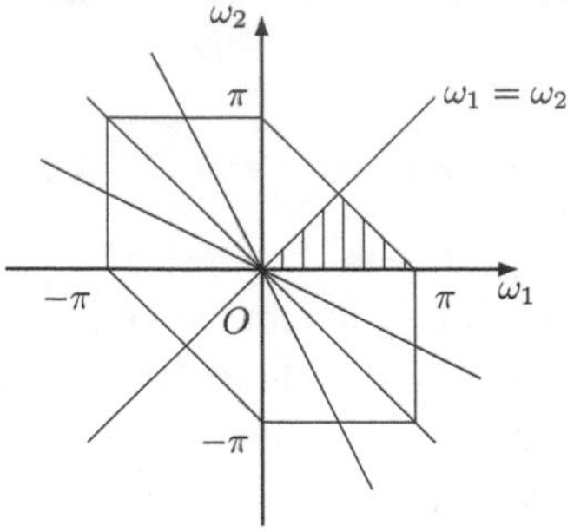

Fig. 6.3.1: The symmetrical regions of the bispectrum

As an example, we prove that $B_x(\omega_1, \omega_2) = B_x(-\omega_1 - \omega_2, \omega_2)$. From property 2 of cumulants, we can get

$$\begin{aligned} B_x(\omega_1, \omega_2) &= \sum_{m=-\infty}^{\infty} \sum_{n=-\infty}^{\infty} c_{3x}(m, n) e^{-j(m\omega_1 + n\omega_2)} \\ &= \sum_{m=-\infty}^{\infty} \sum_{m=-\infty}^{\infty} c_{3x}(-m, n-m) e^{-j[m(\omega_1+\omega_2)+(n-m)\omega_2]} \\ &= \sum_{\tau_1=-\infty}^{\infty} \sum_{\tau_2=-\infty}^{\infty} c_{3x}(\tau_1, \tau_2) e^{-j[(-\omega_1-\omega_2)\tau_1+\omega_2\tau_2]} \\ &= B_x(-\omega_1 - \omega_2, \omega_2). \end{aligned}$$

As shown in Fig. 6.3.1, the defined region of the bispectrum can be divided into 12 sectors. Therefore, from the symmetry of the bispectrum, it is known that all bispectra can be completely described by knowing the bispectrum within the triangle $\omega_2 \geq 0$, $\omega_1 \geq \omega_2$, $\omega_1 + \omega_2 \leq \pi$ (as shown in the shaded area). Because all the bispectra in the other sectors can be obtained from the bispectrum in the triangle using the symmetry. In Ref [173], Pflug et al. pointed out that the trispectrum of a real signal has 96 symmetrical regions.

6.3.2 Bispectrum Estimation

Two nonparametric methods of bispectrum estimation can be obtained by extending the two periodogram methods (direct method and indirect method) of power spectrum estimation.

Let $x(0), x(1), \cdots, x(N-1)$ be the observation sample with zero mean, and its sampling frequency is f_s.

Algorithm 6.3.1. *Direct algorithm of bispectrum estimation*

Step 1 Divided the data into K segments, each segment contains M observation samples, which are recorded as $x^{(k)}(0), x^{(k)}(1), \cdots, x^{(k)}(M-1)$, *where* $k = 1, \cdots, K$. *Note that overlap between two adjacent data segments is allowed here.*

Step 2 Calculate the discrete Fourier transform (DFT) coefficients

$$X^{(k)}(\lambda) = \frac{1}{M} \sum_{n=0}^{M-1} x^{(k)}(n) e^{-j2\pi n\lambda/M} \tag{6.3.8}$$

where $\lambda = 0, 1, \cdots, M/2$; $k = 1, \cdots, K$.

Step 3 Calculate the triple correlation of DFT coefficients

$$\hat{b}_k(\lambda_1, \lambda_2) = \frac{1}{\triangle_0^2} \sum_{i_1=-L_1}^{L_1} \sum_{i_2=-L_1}^{L_1} X^{(k)}(\lambda_1 + i_1) X^{(k)}(\lambda_2 + i_2) X^{(k)}(-\lambda_1 - \lambda_2 - i_1 - i_2)$$

$$k = 1, \cdots, K; 0 \le \lambda_2 \le \lambda_1, \lambda_1 + \lambda_2 \le f_s/2$$

where $\triangle_0 = f_s/N_0$, N_0 *and* L_1 *should be selected to satisfy* $M = (2L_1 + 1)N_0$.

Step 4 The mean value of K bispectral estimation gives the bispectral estimation of the given data $x(0), x(1), \cdots, x(N-1)$

$$\hat{B}_D(\omega_1, \omega_2) = \frac{1}{K} \sum_{k=1}^{K} \hat{b}_k(\omega_1, \omega_2), \tag{6.3.9}$$

where $\omega_1 = \frac{2\pi f_s}{N_0} \lambda_1$, $\omega_2 = \frac{2\pi f_s}{N_0} \lambda_2$.

Algorithm 6.3.2. *Indirect algorithm of bispectrum estimation*

Step 1 Divided the data into K segments and each segment contains M observation samples.

Step 2 Let $x^{(k)}(0), x^{(k)}(1), \cdots, x^{(k)}(M-1)$ *be the k-th segment data, then estimate the third-order cumulant of each data segment*

$$c^{(k)}(i,j) = \frac{1}{M} \sum_{n=-M_1}^{M_2} x^{(k)}(n) x^{(k)}(n+i) x^{(k)}(n+j), \quad k = 1, \cdots, K, \tag{6.3.10}$$

where $M_1 = max(0, -i-j)$ *and* $M_2 = min(M-1, M-1-i, M-1-j)$.

Step 3 Take the average of the third-order cumulants of all segments as the third-order cumulant estimation of the whole observation data, namely

$$\hat{c}(i,j) = \frac{1}{K} \sum_{k=1}^{K} c^{(k)}(i,j). \tag{6.3.11}$$

Step 4 Calculate the bispectrum estimation

$$\hat{B}_{IN}(\omega_1, \omega_2) = \sum_{i=-L}^{L} \sum_{l=-L}^{L} \hat{c}(i,l) w(i,l) e^{-j(\omega_1 i + \omega_2 l)}, \tag{6.3.12}$$

where $L < M-1$ *and* $w(i,j)$ *is the two-dimensional lag window function.*

The two-dimensional window function $w(m,n)$ of bispectral estimation was first derived and discussed in reference [189]. It is proved that the two-dimensional window function must satisfy the following four conditions:

(1) $w(m,n) = w(n,m) = w(-m,n-m) = w(m-n,-n)$;
(2) If (m,n) is outside the support region of the cumulant estimation $\hat{c}_{3x}(m,n)$, then $w(m,n) = 0$;
(3) $w(0,0) = 1$ (normalization condition);
(4) $W(\omega_1,\omega_2) \geq 0$ for $\forall(\omega_1,\omega_2)$.

It is easy to see that constraint (1) can guarantee that $c_{3x}(m,n)w(m,n)$ has the same symmetry as the third-order cumulant $c_{3x}(m,n)$. It is worth pointing out that the two-dimensional window function $w(m,n)$ satisfying the above four constraints can be constructed by using the one-dimensional lag window function $d(m)$, that is

$$w(m,n) = d(m)d(n)d(n-m), \tag{6.3.13}$$

where the one-dimensional lag window $d(m)$ should satisfy the following four conditions:

$$d(m) = d(-m) \tag{6.3.14}$$

$$d(m) = 0, \quad m > L \tag{6.3.15}$$

$$d(0) = 1 \tag{6.3.16}$$

$$D(\omega) \geq 0, \quad \forall\omega \tag{6.3.17}$$

where $D(\omega)$ is the Fourier transform of $d(n)$.

It is easy to prove that the following three window functions satisfy the above constraints.

(1) Optimum window

$$d_{\text{opt}}(m) = \begin{cases} \frac{1}{\pi}|sin\frac{\pi m}{L}| \left(1 - \frac{|m|}{L}\right) cos\frac{\pi m}{L}, & |m| \leq L \\ 0, & |m| > L \end{cases} \tag{6.3.18}$$

(2) Parzen window

$$d_{\text{Parzen}}(m) = \begin{cases} 1 - 6\left(\frac{|m|}{L}\right)^2 + 6\left(\frac{|m|}{L}\right)^3, & |m| \leq L/2 \\ 2\left(1 - \frac{|m|}{L}\right)^3, & L/2 < |m| \leq L \\ 0, & |m| > L \end{cases} \tag{6.3.19}$$

(3) Uniform window in the spectral domain

$$W_{\text{uniform}}(\omega_1,\omega_2) = \begin{cases} \frac{4\pi}{3\Omega_0}, & |\omega| \leq \Omega_0 \\ 0, & |\omega| > \Omega_0 \end{cases} \tag{6.3.20}$$

where $|\omega| = \max(|\omega_1|, |\omega_2|, |\omega_1 + \omega_2|)$, $\Omega_0 = a_0/L$ and a_0 is a constant.

In order to evaluate the above three window functions, the bispectral bias spectrum

$$J = \frac{1}{(2\pi)^2} \int_{-\pi}^{\pi} \int_{-\pi}^{\pi} (\omega_1 - \omega_2)^2 W(\omega_1, \omega_2) d\omega_1 d\omega_2, \tag{6.3.21}$$

and approximate normalized bispectral variance

$$V = \sum_{m=-L}^{L} \sum_{n=-L}^{L} |w(m, n)|^2 \tag{6.3.22}$$

are defined in reference [189]. In fact, V denotes the energy of the window function.

Table 6.3.1 lists the performance evaluation results of the three window functions.

Tab. 6.3.1: Performance of three bispectral estimation window functions

Window function	Deviation supremum (J)	Variance (V)
Optimum window	$J_{\text{opt}} = \frac{6\pi^2}{L^2}$	$V_{\text{opt}} : 0.05L^2$
Parzen window	$J_{\text{Parzen}} = \frac{72}{L^2}$	$V_{\text{Parzen}} : 0.037L^2$
Uniform window	$J_{\text{uniform}} = \frac{5}{6}\left(\frac{a_0}{L}\right)^2$	$V_{\text{uniform}} : \frac{4\pi}{3}\left(\frac{L}{a_0}\right)^2$

It can be seen from the table that $J_{\text{uniform}} \approx 3.7J_{\text{opt}}$ when $V_{\text{uniform}} = V_{\text{opt}}$, that is, the upper bound on the deviation of the uniform window is significantly larger than that of the optimal window. Comparing the optimal window with the Parzen window, we also know that $J_{\text{Parzen}} = 1.215J_{\text{opt}}$ and $V_{\text{Parzen}} = 0.74V_{\text{opt}}$. The optimal window function outperforms the other two window functions in the sense of having the smallest upper bound on the deviation.

In reference [163], it is proved that the bispectral estimations are asymptotically unbiased and consistent, and they obey the asymptotically complex normal distribution. For sufficiently large M and N, both indirect and direct methods give an asymptotically unbiased bispectral estimation, i.e.,

$$E\{\hat{B}_{\text{IN}}(\omega_1, \omega_2)\} \approx E\{\hat{B}_{\text{D}}(\omega_1, \omega_2)\} \approx B(\omega_1, \omega_2). \tag{6.3.23}$$

And the indirect and the direct methods have the asymptotic variance respectively

$$\begin{aligned} \text{var}\{\text{Re}[\hat{B}_{\text{IN}}(\omega_1, \omega_2)]\} &= \text{var}\{\text{Im}[\hat{B}_{\text{IN}}(\omega_1, \omega_2)]\} \\ &\approx \frac{V}{(2L+1)^2 K} P(\omega_1)P(\omega_2)P(\omega_1 + \omega_2), \end{aligned} \tag{6.3.24}$$

$$\begin{aligned} \text{var}\{\text{Re}[\hat{B}_{\text{D}}(\omega_1, \omega_2)]\} &= \text{var}\{\text{Im}[\hat{B}_{\text{D}}(\omega_1, \omega_2)]\} \\ &\approx \frac{1}{KM_1} P(\omega_1)P(\omega_2)P(\omega_1 + \omega_2), \end{aligned} \tag{6.3.25}$$

where V is defined in Eq. (6.3.22), $P(\omega)$ represents the true power spectral density of signal $\{x(n)\}$. Note that when the window function is not used in formula (6.3.12) of the indirect bispectrum estimation, then $V/(2L+1)^2 = 1$. If the direct method is not smoothing in frequency domain (i.e. $M_1 = 1$), then Eq. (6.3.24) and Eq. (6.3.25) are equivalent.

6.4 Non-Gaussian Signal and Linear System

Signals whose probability density distribution is nonnormal distribution are called non-Gaussian signals. The higher-order cumulants of the Gaussian signal are equal to zero, but there must be some higher-order cumulants of the non-Gaussian signals that are not equal to zero. This section provides a further discussion on the distinction between Gaussian and non-Gaussian signals.

6.4.1 Sub-Gaussian and Super-Gaussian Signal

In the high-order statistical analysis of signal, it is often of interest to a special slice of the high-order statistics of real signal $x(t)$. Consider the special slices $c_{3x}(0,0) = E\{x^3(t)\}$ and $c_{4x}(0,0,0) = E\{x^4(t)\} - 3E^2\{x^2(t)\}$ of higher-order cumulants when all delays are equal to zero, where $E^2\{x^2(t)\}$ is the square of the expectation $E\{x^2(t)\}$. From these two special slices, two important terms can be derived.

Definition 6.4.1. *The skewness of real signal $x(t)$ is defined as*

$$S_x \overset{def}{=} E\{x^3(t)\}, \tag{6.4.1}$$

kurtosis is defined as

$$K_x \overset{def}{=} E\{x^4(t)\} - 3E^2\{x^2(t)\}, \tag{6.4.2}$$

and

$$K_x \overset{def}{=} \frac{E\{x^4(t)\}}{E^2\{x^2(t)\}} - 3 \tag{6.4.3}$$

is called the return to zero kurtosis.

For any signal, if its skewness is equal to zero, then its third-order cumulant is equal to zero. If the skewness is equal to zero, it means that the signal obeys symmetric distribution, while if the skewness is not equal to zero, it must obey asymmetric distribution. In other words, skewness is actually a measure of how skewed the distribution of a signal deviates from the symmetric distribution.

There is another definition of kurtosis.

Definition 6.4.2. *The normalized kurtosis of the real signal is defined as*

$$K_x \overset{def}{=} \frac{E\{x^4(t)\}}{E^2\{x^2(t)\}}. \tag{6.4.4}$$

Kurtosis can be used not only to distinguish Gaussian and non-Gaussian signals but also to further classify non-Gaussian signals into sub-Gaussian and super-Gaussian signals.

(1) Signal classification based on the return to zero kurtosis
Gaussian signal: a signal with kurtosis equal to zero;
Sub-Gaussian signal: a signal with kurtosis less than zero;
Super-Gaussian signal: a signal with kurtosis greater than zero.

(2) Signal classification based on the normalized kurtosis
Gaussian signal: real signal with normalized kurtosis equal to 3 or complex signal with normalized kurtosis equal to 2;
Sub-Gaussian signal: real signal with normalized kurtosis less than 3 or complex signal with normalized kurtosis less than 2;
Super-Gaussian signal: real signal with normalized kurtosis greater than 3 or complex signal with normalized kurtosis greater than 2.

It can be seen that the kurtosis of the sub-Gaussian signal is lower than that of the Gaussian signal, and the kurtosis of the super-Gaussian signal is higher than that of the Gaussian signal. This is why they are called sub-Gaussian and super-Gaussian signals, respectively. Most of the digitally modulated signals used in wireless communication are sub-Gaussian signals.

6.4.2 Non-Gaussian Signal Passing Through Linear System

Consider the single-input single-output linear time-invariant system depicted in Fig. 6.4.1 which is excited by discrete-time non-Gaussian noise $e(n)$.

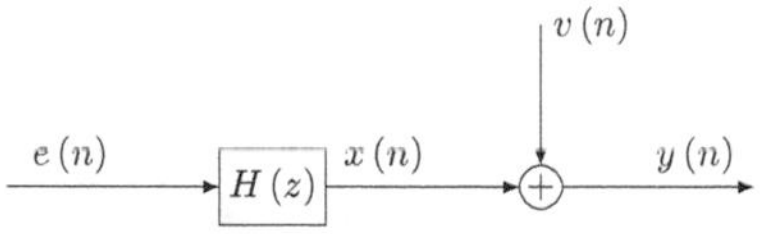

Fig. 6.4.1: A single-input single-output linear time-invariant system

It is assumed that the additive noise $v(n)$ is colored Gaussian noise and is statistically independent of $e(n)$. Thus, it is statistically independent of the system output $x(n)$. Since the higher-order cumulants of any Gaussian stochastic process are equal to zero, we have

$$\begin{aligned} c_{ky}(\tau_1, \cdots, \tau_{k-1}) &= c_{kx}(\tau_1, \cdots, \tau_{k-1}) + c_{kv}(\tau_1, \cdots, \tau_{k-1}) \\ &= c_{kx}(\tau_1, \cdots, \tau_{k-1}). \end{aligned}$$

On the other hand, since the output $x(n)$ is equal to the convolution of the input $e(n)$ and the system impulse response, i.e.,

$$x(n) = e(n) * h(n) = \sum_{i=-\infty}^{\infty} h(i)e(n-i). \tag{6.4.5}$$

Using this result and the definition of cumulant, and applying property 1 and property 5 repeatedly, we can obtain

$$\begin{aligned} c_{kx}(\tau_1, \cdots, \tau_{k-1}) &= \text{cum}[x(n), x(n+\tau_1), \cdots, x(n+\tau_{k-1})] \\ &= \text{cum}\left(\sum_{i_1=-\infty}^{\infty} h(i_1)e(n-i_1), \cdots, \sum_{i_k=-\infty}^{\infty} h(i_k)e(n+\tau_{k-1}-i_k) \right) \\ &= \sum_{i_1=-\infty}^{\infty} \cdots \sum_{i_k=-\infty}^{\infty} h(i_1)\cdots h(i_k)\text{cum}[e(n-i_1), \cdots, e(n+\tau_{k-1}-i_k)] \end{aligned}$$

Using the defintion of cumulant $c_{ke}(\tau_1, \cdots, \tau_{k-1}) = \text{cum}[e(n), e(n+\tau_1), \cdots, e(n+\tau_{k-1})]$, the above equation can be expressed as

$$c_{kx}(\tau_1, \cdots, \tau_{k-1}) = \sum_{i_1=-\infty}^{\infty} \cdots \sum_{i_k=-\infty}^{\infty} h(i_1)\cdots h(i_k)c_{ke}(\tau_1+i_1-i_2, \cdots, \tau_{k-1}+i_1-i_k). \tag{6.4.6}$$

This formula describes the relationship between the cumulant of the output and the cumulant of the input noise and the system impulse response.

Taking the $(k-1)$-dimensional Fourier transform and Z-transform of Eq. (6.4.6), two other important formulas can be obtained

$$S_{kx}(\omega_1, \cdots, \omega_{k-1}) = S_{ke}(\omega_1, \cdots, \omega_{k-1})H(\omega_1)\cdots H(\omega_{k-1})H(-\omega_1 - \cdots - \omega_{k-1}) \tag{6.4.7}$$

$$S_{kx}(z_1, \cdots, z_{k-1}) = S_{ke}(z_1, \cdots, z_{k-1})H(z_1)\cdots H(z_{k-1})H(-z_1 - \cdots - z_{k-1}) \tag{6.4.8}$$

where $H(\omega) = \sum_{i=-\infty}^{\infty} h(i)e^{-j\omega i}$ and $H(z) = \sum_{i=-\infty}^{\infty} h(i)z^{-i}$ represent system transfer function and its Z-transform form, respectively. Eq. (6.4.7) describes the relationship between the higher-order spectra of the system output signal and the higher-order spectra of the input signal and the system transfer function, while Eq. (6.4.8) is the Z-transform of Eq. (6.4.7).

Eqs. (6.4.6) to (6.4.8) were first obtained by Bartlett [20], but he only considered the special case of $k = 2, 3, 4$ at that time. Later, Bringer and Rosenblatt generalized these three formulas to any kth-order [34]. Therefore, Eqs. (6.4.6) to (6.4.8) are often referred to as the Bartlett-Brllinger-Rosenblatt formula or the BBR formula for short.

In particular, when the input $e(n)$ of the system is independently identically distributed higher-order white noise, Eqs. (6.4.6) to (6.4.8) can be simplified as

$$c_{kx}(\tau_1,\cdots,\tau_{k-1}) = \gamma_{ke}\sum_{i=-\infty}^{\infty} h(i)h(i+\tau_1)\cdots h(i+\tau_{k-1}), \tag{6.4.9}$$

$$S_{kx}(\omega_1,\cdots,\omega_{k-1}) = \gamma_{ke}H(\omega_1)\cdots H(\omega_{k-1})H(-\omega_1-\cdots-\omega_{k-1}), \tag{6.4.10}$$

$$S_{kx}(z_1,\cdots,z_{k-1}) = \gamma_{ke}H(z_1)\cdots H(z_{k-1})H(z_1^{-1}\cdots z_{k-1}^{-1}). \tag{6.4.11}$$

For convenience, the BBR formula of $k = 2, 3, 4$ are summarized as follows:

(1) BBR formula for the cumulants

$$c_{2x}(\tau) = R_x(\tau) = \sigma_e^2\sum_{i=-\infty}^{\infty} h(i)h(i+\tau), \tag{6.4.12}$$

$$c_{3x}(\tau_1,\tau_2) = \gamma_{3e}\sum_{i=-\infty}^{\infty} h(i)h(i+\tau_1)h(i+\tau_2), \tag{6.4.13}$$

$$c_{4x}(\tau_1,\tau_2,\tau_3) = \gamma_{4e}\sum_{i=-\infty}^{\infty} h(i)h(i+\tau_1)h(i+\tau_2)h(i+\tau_3). \tag{6.4.14}$$

(2) BBR formula for spectrum, bispectrum ,and trispectrum

$$P_x(\omega) = \sigma_e^2 H(\omega)H^*(\omega) = \sigma_e^2|H(\omega)|^2, \tag{6.4.15}$$

$$B_x(\omega_1,\omega_2) = \gamma_{3e}H(\omega_1)H(\omega_2)H(-\omega_1-\omega_2), \tag{6.4.16}$$

$$T_x(\omega_1,\omega_2,\omega_3) = \gamma_{3e}H(\omega_1)H(\omega_2)H(\omega_3)H(-\omega_1-\omega_2-\omega_3). \tag{6.4.17}$$

The above BBR formula will be used frequently in the future. As an example, we consider the special slice $c_{3x}(m) = c_{3x}(m, m)$ of the third-order cumulant, which is often called the diagonal slice.

According to the BBR formula (6.4.9) for cumulant, the third-order cumulant of the diagonal slice can be written as

$$c_{3x}(m) = \gamma_{ke}\sum_{i=-\infty}^{\infty} h(i)h^2(i+m), \tag{6.4.18}$$

its Z-transform is

$$\begin{aligned} C(z) &= \gamma_{3e}\sum_{m=-\infty}^{\infty}\left[\sum_{i=-\infty}^{\infty} h(i)h^2(i+m)\right]z^{-m} \\ &= \gamma_{3e}\sum_{m=-\infty}^{\infty} h(i)z^i\sum_{k=-\infty}^{\infty} h^2(k)z^{-k} \\ &= \gamma_{3e}H(z^{-1})H_2(z), \end{aligned} \tag{6.4.19}$$

where

$$H(z^{-1}) = \sum_{i=-\infty}^{\infty} h(i)z^{i}, \tag{6.4.20}$$

$$H_2(z) = \sum_{k=-\infty}^{\infty} h^2(k)z^{-k} = H(z) * H(z). \tag{6.4.21}$$

Note that $h^2(k) = h(k)h(k)$ is a product, and the Z-transform of the product $a(k)b(k)$ corresponds to the convolution $A(z) * B(z)$.

Since the power spectrum $P(z) = \sigma_e^2 H(z)H(z^{-1})$, multiply both sides of Eq. (6.4.19) by $\sigma_e^2 H(z)$, we can get

$$H_2(z)P(z) = \frac{\sigma_e^2}{\gamma_{3e}} H(z)C(z). \tag{6.4.22}$$

The Z-transform $C(z)$ of the third-order cumulant diagonal slice $c_{3x}(m)$ is called the $1\frac{1}{2}$-D spectrum, and the relationship formula (6.4.22) between it and the power spectrum $P(z)$ is derived in reference [87]. This relationship plays an important role in the q-slice method of FIR system identification.

6.5 FIR System Identification

The finite impulse response (FIR) filter plays an important role in wireless communication, radar, and other signal processing. The output of the FIR system is equivalent to a MA stochastic process. In modern spectral estimation, the relationship between the autocorrelation function and MA parameters is a set of nonlinear equations, and it is known by autocorrelation equivalence that only the minimum phase FIR system can be identified by the autocorrelation function. Compared with the FIR system identification based on autocorrelation function, the FIR system identification with higher-order cumulants is not only linear but also suitable for the nonminimum phase system identification.

6.5.1 RC Algorithm

The method that applies correlation (R) functions and cumulants (C) to identify the FIR system is called the RC algorithm.

Consider a stationary non-Gaussian MA(q) stochastic process

$$x(n) = \sum_{i=0}^{q} b(i)e(n-i), \quad e(n) \sim \text{IID}(0, \sigma_e^2, \gamma_{ke}), \tag{6.5.1}$$

where $b(0) = 1$, $b(q) \neq 0$, and $e(n) \sim \text{IID}(0, \sigma_e^2, \gamma_{ke})$ means that $e(n)$ is an IID process with zero mean, variance σ_e^2 and kth-order cumulant γ_{ke}. Without loss of generality, suppose $\gamma_{ke} \neq 0$ for a $k > 2$.

Consider two different FIR systems, their outputs are described by difference equations

$$\text{FIR system one}: x(n) = e(n) + 0.3e(n-1) - 0.4e(n-2), \qquad (6.5.2)$$

$$\text{FIR system two}: x'(n) = e(n) - 1.2e(n-1) - 1.6e(n-2). \qquad (6.5.3)$$

The characteristic polynomial of system one is

$$1 + 0.3z^{-1} - 0.4z^{-2} = (1 - 0.5z^{-1})(1 + 0.8z^{-1}),$$

its zero points are $z_1 = 0.5$ and $z_2 = -0.8$; the characteristic polynomial of system two is

$$1 - 1.2z^{-1} - 1.6z^{-2} = (1 - 2z^{-1})(1 + 0.8z^{-1}),$$

its zero points are $z_1 = 2$ and $z_2 = -0.8$. Obviously, system one is a minimum phase system and system two is a nonminimum phase system. They have a same zero point and the other zero point is reciprocal.

If $\sigma_e^2 = 1$, then the autocorrelation function of signal $x(n)$ and $x'(n)$ can be calculated using the BBR formula (6.4.12)

$$R_x(0) = b^2(0) + b^2(1) + b^2(2) = 1.25, \quad R_{x'}(0) = 5.0;$$
$$R_x(1) = b(0)b(1) + b(1)b(2) = 0.18, \quad R_{x'}(1) = 0.72;$$
$$R_x(2) = b(0)b(2) = -0.4, \quad R_{x'}(2) = -1.6;$$
$$R_x(\tau) = 0, \quad \forall\tau > 2, \quad R_{x'}(\tau) = 0, \quad \forall\tau > 2.$$

This shows that the autocorrelation functions of this two stochastic processes differ only by a fixed scale factor, i.e., $R_{x'}(\tau) = 4R_x(\tau)$, $\forall\tau$. Since their autocorrelation functions have exactly the same shape, it would be impossible to distinguish between these two different systems using the autocorrelation functions.

The case of cumulant is quite different. From the BBR formula1 (6.4.13) of the cumulant, it is not difficult to calculate the third-order cumulants of $x(n)$ and $x'(n)$ (for convenience, let $\gamma_{3e} = 1$ here) as

$$c_{3x}(0,0) = b^3(0) + b^3(1) + b^3(2) = 0.963, \quad c_{3x'}(0,0) = -4.878;$$
$$c_{3x}(0,1) = b^2(0)b(1) + b^2(1)b(2) = 1.264, \quad c_{3x'}(0,1) = -3.504;$$
$$c_{3x}(0,2) = b^2(0)b(2) = -0.4, \quad c_{3x'}(0,2) = -1.6.$$

It can be seen that the third-order cumulants of signal $x(n)$ and $x'(n)$ are completely different. This shows that the third-order cumulants can be used to distinguish these two different systems.

1.GM Agorithm

Taking the inverse Z-transform on both sides of Eq. (6.4.22), then its time-domain expression can be obtained

$$\sum_{i=-\infty}^{\infty} b^2(i)R_x(m-i) = \varepsilon_3 \sum_{i=-\infty}^{\infty} b(i)c_{3x}(m-i, m-i), \qquad (6.5.4)$$

where $\varepsilon_3 = \sigma_e^2/\gamma_{3e}$.

For the FIR system, Eq. (6.5.4) can be written as

$$\sum_{i=0}^{q} b^2(i)R_x(m-i) = \varepsilon_3 \sum_{i=0}^{q} b(i)c_{3x}(m-i, m-i), \quad -q \le m \le 2q. \tag{6.5.5}$$

The fourth-order result corresponding to the above equation is

$$\sum_{i=0}^{q} b^3(i)R_x(m-i) = \varepsilon_4 \sum_{i=0}^{q} b(i)c_{4x}(m-i, m-i), \quad -q \le m \le 2q, \tag{6.5.6}$$

where $\varepsilon_4 = \sigma_e^2/\gamma_{4e}$

Eqs. (6.5.5) and (6.5.6) are established by Giannakis and Mendel and are called GM equations [87]. The linear algebraic method for solving this equation is called the GM algorithm.

For the GM algorithm, we must pay attention to several problems [154]:

(1) GM algorithm treats $b^2(i)$ and $b^3(i)$ as independent parameters, however, they are not. Therefore, this "over parameterization" method is suboptimal.

(2) Eq. (6.5.5) has $3q + 1$ equations with $2q + 1$ unknown parameters $b(1), \cdots, b(q)$, $b^2(1), \cdots, b^2(q)$ and ε_3, which is overdetermined equation. However, it is possible that the rank of the coefficient matrix may not equal $2q + 1$, hence more slices may have to be used to determine the $2q + 1$ unknown parameters. Exactly how many slices and which slices are needed to ensure the identifiability of the parameters is an open question.

(3) As an RC method, due to the use of correlation function, Eq. (6.5.5) is only applicable to the special case without additive noise, where $R_y(\tau) = R_x(\tau)$. When the additive noise is white noise, in which case $R_y(m) = R_x(m) + \sigma_e^2\delta(m)$, the lag m cannot include the values $0, 1, \cdots, q$ to avoid the effects of noise. This leads to an underdetermined equation of GM algorithm

$$\sum_{i=0}^{q} b^2(i)R_x(m-i) = \varepsilon_3 \sum_{i=0}^{q} b(i)c_{3y}(m-i, m-i), \quad -q \le m \le -1; \quad q+1 \le m \le 2q.$$

It has $2q$ equations with $2q + 1$ unknowns. The above equation can be rearranged as

$$\sum_{i=1}^{q} b(i)c_{3y}(m-i, m-i) - \sum_{i=0}^{q} [\varepsilon b^2(i)]R_y(m-i) = -c_{3y}(m, m),$$
$$-q \le m \le -1; \quad q+1 \le m \le 2q, \tag{6.5.7}$$

where $\varepsilon = \gamma_{3e}/\sigma_e^2$.

2. Tugnait Algorithm

In order to make the GM algorithm suitable for additive white noise, a new equation must be added. Therefore, using the BBR formula we can obtain

$$\begin{aligned}\sum_{i=0}^{q} b(i)c_{3y}(i-\tau,q) &= \sum_{i=0}^{q} b(i)\sum_{k=0}^{q} \gamma_{3e}h(k)h(k+i-\tau)h(k+q)\\ &= \gamma_{3e}\sum_{i=0}^{q} b(i)h(0)h(i-\tau)h(q).\end{aligned} \tag{6.5.8}$$

Notice that $h(q) = b(q)$, $h(0) = 1$ and the second-order of BBR formula

$$\sum_{i=0}^{q} b(i)h(i-\tau) = \sum_{i=0}^{q} h(i)h(i-\tau) = \sigma_e^{-2}R_x(\tau) = \sigma_e^{-2}[R_y(\tau) - \sigma_e^2\delta(\tau)],$$

then Eq. (6.5.8) becomes

$$\sum_{i=0}^{q} b(i)c_{3y}(i-m,q) = [\varepsilon b(q)][R_y(m) - \sigma_e^2\delta(m)]. \tag{6.5.9}$$

Obviously, in order to avoid the influence of the white noise $v(n)$, the above formula cannot contain $m = 0$. Rearranging the above formula yields

$$\sum_{i=0}^{q} b(i)c_{3y}(i-m,q) - [\varepsilon b(q)]R_y(m) = -c_{3y}(-m,q), \quad 1 \le m \le q. \tag{6.5.10}$$

Concatenating Eqs. (6.5.7) and (6.5.10), and solving the unknown parameter $b(1), \cdots, b(q)$, $\varepsilon b(q)$ and $\varepsilon b^2(1), \cdots, \varepsilon b^2(q)$, the RC algorithm of Tugnait is formed[209]. In this algorithm, the equations are overdetermined, it has $4q$ equations with $2q+2$ unknowns, and it has been proved that these parameters are uniquely identifiable. Incidentally, this algorithm is obtained by rearranging and modifying another RC algorithm of Tugnait[208].

3.Combined Cumulant Slice Method

In addition to the two typical RC algorithms mentioned above, there is another variant of the RC algorithm, which is called the combined cumulant slice method, proposed by Fonollasa and Vidal[79].

Using $b(i) = h(i)$, the BBR formula for FIR system can be written as

$$c_{kx}(\tau_1, \cdots, \tau_{k-1}) = \gamma_{ke}\sum_{j=0}^{q}\prod_{l=0}^{k-1} b(j+\tau_l), \quad \tau_0 = 0, \quad k \ge 2. \tag{6.5.11}$$

If let $\tau_1 = i$ be a variable and $\tau_2, \cdots, \tau_{k-1}$ be fixed, then the 1-D slice cumulant can be expressed as the correlation between parameters $b(j)$ and $b(i; \tau_2, \cdots, \tau_{k-1})$, i.e.,

$$c_{kx}(i, \tau_2, \cdots, \tau_{k-1}) = \sum_{j=0}^{q} b(j+i)b(j; \tau_2, \cdots, \tau_{k-1}), \tag{6.5.12}$$

where the causal sequence $b(i;\tau_2,\cdots,\tau_{k-1})$ is defined as

$$b(i;\tau_2,\cdots,\tau_{k-1}) = \gamma_{ke} b(i) \prod_{j=2}^{k-1} b(i+\tau_j). \tag{6.5.13}$$

Concatenating Eqs. (6.5.12) and (6.5.13), then the linear combination of any slice

$$C_w(i) = w_2 c_{2x}(i) + \sum_{j=-q}^{q} w_3(j) c_{3x}(i,j) + \sum_{j=-q}^{q}\sum_{l=-q}^{j} w_4(j,l) c_{4x}(i,j,l) + \cdots \tag{6.5.14}$$

can be expressed as the correlation of $b(i)$ and $g_w(i)$, i.e.,

$$C_w(i) = \sum_{n=0}^{\infty} b(n+i) g_w(n), \tag{6.5.15}$$

where $g_w(n)$ is a causal sequence

$$g_w(n) = w_2 b(n) + \sum_{j=-q}^{q} w_3(j) b(n;j) + \sum_{j=-q}^{q}\sum_{l=-q}^{j} w_4(j,l) b(n;j,l) + \cdots. \tag{6.5.16}$$

It can be regarded as the weighted coefficient of MA parameter $b(i)$.

Eq. (6.5.15) shows that for a MA model, any w-slice can be expressed as the correlation of two finite causal sequences $b(n)$ and $g_w(n)$. Therefore, if we choose the weight coefficient

$$g_w(n) = \delta(n) = \begin{cases} 1, & n = 0 \\ 0, & n \neq 0 \end{cases},$$

we can develop an FIR system identification method, which is called w-slice method[79].

6.5.2 Cumulant Algorithm

The main deficiency of RC and w-slice algorithm is that they can only be applied to additive white noise (Gaussian or non-Gaussian). Obviously, in order to completely suppress the colored Gaussian noise in theory, it is necessary to avoid using the autocorrelation function and only use higher-order cumulants. Such an algorithm has been proposed in Ref.[253].

It is assumed that the non-Gaussian MA process $\{x(n)\}$ is observed in an additive colored Gaussian noise $v(n)$ which is independent of $x(n)$, i.e. the observed data $y(n) = x(n) + v(n)$. At this point, $c_{ky}(\tau_1,\cdots,\tau_{k-1}) = c_{kx}(\tau_1,\cdots,\tau_{k-1})$. Without loss of generality, it is also assumed that $h(0) = 1$.

It is noted that for a MA(q) process, $h(i) = 0 (i < 0$, or $i > q)$ always hold, so the BBR formula can be simplified as

$$c_{ky}(\tau_1,\cdots,\tau_{k-1}) = c_{kx}(\tau_1,\cdots,\tau_{k-1}) = \gamma \sum_{i=0}^{q} h(i)h(i+\tau_1)\cdots h(i+\tau_{k-1}). \tag{6.5.17}$$

Consider a special slice $\tau_1 = \tau, \tau_2 = \cdots = \tau_{k-1} = 0$, then

$$c_{ky}(\tau, 0, \cdots, 0) = \gamma_{ke} \sum_{i=0}^{q} h^{k-1}(i) h(i+\tau). \tag{6.5.18}$$

Using $c_{ky}(m, n) = c_{ky}(m, n, 0, \cdots, 0)$ and substituting $b(i+\tau) = h(i+\tau)$ into the above equation, we can get

$$c_{ky}(\tau, 0) = \gamma_{ke} \sum_{i=0}^{q} h^{k-1}(i) b(i+\tau) = \gamma_{ke} \sum_{i=0}^{q} b(j) h^{k-1}(j-\tau), \forall \tau, \tag{6.5.19}$$

where $b(i) = h(i) = 0$ ($i < 0$, or $i > q$) is used.

On the other hand, from Eq. (6.5.17) we have

$$c_{ky}(q, 0) = \gamma_{ke} h(q), \tag{6.5.20}$$

$$c_{ky}(q, n) = \gamma_{ke} h(n) h(q). \tag{6.5.21}$$

In order to ensure the uniqueness of MA(q) process, it is generally assumed that $b(0) \neq 0$ and $b(q) \neq 0$. From thess assumption and Eqs. (6.5.20) and (6.5.21), it is immediately seen that $c_{ke}(q, 0) \neq 0$ and $c_{ky}(q, q) \neq 0$.

Combining Eqs. (6.5.20) and (6.5.21), we can obtain an important formula

$$h(n) = \frac{c_{ky}(q, n)}{c_{ky}(q, 0)} = b(n). \tag{6.5.22}$$

Because of the form that the above equation has, it is customary to call it the $C(q, n)$ formula. This formula is proposed by Giannakis[85] (Chinese scholar Qiansheng Cheng[180] has obtained the same results almost independently at the same time).

$C(q, n)$ formula shows that the parameters of the MA model can be calculated directly according to the cumulants. However, due to the large error and variance in the case of short data, such a direct algorithm is not practical for short data. However, using it can help us to obtain a set of linear normal equations for MA parameter estimation from Eq. (6.5.19). Therefore, substituting Eq. (6.5.22) into Eq. (6.5.19) and organizing it, we can obtain

$$\gamma_{ke} \sum_{i=0}^{q} b(i) c_{ky}^{k-1}(q, i-\tau) = c_{ky}(\tau, 0) c_{ky}^{k-1}(q, 0), \quad \forall \tau. \tag{6.5.23}$$

This equation is called the first normal equation of MA parameter estimation.

Similarily, if subsituting $b(i) = h(i)$ into Eq. (6.5.18) and keeping $h(i+\tau)$ unchanged, then we have

$$c_{ke}(\tau, 0) = \gamma_{ke} \sum_{i=0}^{q} b^{k-1}(i) h(i+\tau), \quad \forall \tau. \tag{6.5.24}$$

Subsituting Eq. (6.5.22) into Eq. (6.5.24), the second normal equation can be obtained

$$\gamma_{ke} \sum_{i=0}^{q} b^{k-1}(i) c_{ky}(q, i+\tau) = c_{ky}(\tau, 0) c_{ky}(q, 0). \tag{6.5.25}$$

Normal equations Eqs. (6.5.23) and (6.5.25) are formally similar to the modified Yule-Walker equation of the ARMA model. Theoretically, solving these two normal equations separately yields an estimation of the parameters $\gamma_{ke}, \gamma_{ke}b(1), \cdots, \gamma_{ke}b(q)$ or $\gamma_{ke}, \gamma_{ke}b^{k-1}(1), \cdots, \gamma_{ke}b^{k-1}(q)$. However, if the estimated value of γ_{ke} is very small, it can easily lead to ill-conditioned problems. Fortunately, this problem can be easily overcome. The method is to let $\tau = q$ in Eq. (6.5.25) to obtain

$$\gamma_{ke} = \frac{c_{ky}^2(q,0)}{c_{ky}(q,q)}. \tag{6.5.26}$$

Then substituting Eq. (6.5.26) into the normal equation (6.5.23), we can get the normal equation without γ_{ke}

$$\sum_{i=0}^{q} b(i)c_{ky}^{k-1}(q,i-\tau) = c_{ky}(\tau,0)c_{ky}^{k-3}(q,0)c_{ky}(q,q), \quad \tau = -q,\cdots,0,\cdots,q. \tag{6.5.27}$$

This normal equation is the basis of the method for cumulant estimation of FIR system identification in reference [253].

The parametric identifiability of this normal equation is analyzed below. Defining

$$\boldsymbol{C}_1 = \begin{bmatrix} c_{ky}^{k-1}(q,q) & & & & 0 \\ c_{ky}^{k-1}(q,q-1) & c_{ky}^{k-1}(q,q) & & & \\ \vdots & \vdots & \ddots & & \\ c_{ky}^{k-1}(q,1) & c_{ky}^{k-1}(q,2) & \cdots & c_{ky}^{k-1}(q,q) & 0 \end{bmatrix}, \tag{6.5.28}$$

$$\boldsymbol{C}_2 = \begin{bmatrix} c_{ky}^{k-1}(q,0) & c_{ky}^{k-1}(q,1) & \cdots & c_{ky}^{k-1}(q,q) \\ & c_{ky}^{k-1}(q,0) & \cdots & c_{ky}^{k-1}(q,q-1) \\ & & \ddots & \vdots \\ 0 & & & c_{ky}^{k-1}(q,0) \end{bmatrix}, \tag{6.5.29}$$

$$\boldsymbol{b}_1 = [b(0), b(1), \cdots, b(q)]^{\mathrm{T}}, \tag{6.5.30}$$

and

$$c_1 = \begin{bmatrix} c_{ky}(-q,0)c_{ky}^{k-3}(q,0)c_{ky}(q,q) \\ c_{ky}(-q+1,0)c_{ky}^{k-3}(q,0)c_{ky}(q,q) \\ \vdots \\ c_{ky}(-1,0)c_{ky}^{k-3}(q,0)c_{ky}(q,q) \end{bmatrix}, \tag{6.5.31}$$

$$c_2 = \begin{bmatrix} c_{ky}(0,0)c_{ky}^{k-3}(q,0)c_{ky}(q,q) \\ c_{ky}(1,0)c_{ky}^{k-3}(q,0)c_{ky}(q,q) \\ \vdots \\ c_{ky}^{k-2}(q,0)c_{ky}(q,q) \end{bmatrix}, \tag{6.5.32}$$

then the normal equation Eq. (6.5.27) can be simplified as

$$\begin{bmatrix} \boldsymbol{C}_1 \\ \boldsymbol{C}_2 \end{bmatrix} \boldsymbol{b}_1 = \begin{bmatrix} c_1 \\ c_2 \end{bmatrix} . \tag{6.5.33}$$

Solving the matrix equation Eq. (6.5.33), the estimation of the FIR parameters $b(0), b(1), \cdots, b(q)$ can be obtained. The question is, can this method guarantee the unique identifiability of the FIR system parameters? The following theorem gives a positive answer to this question.

Theorem 6.5.1. *Assume that the true cumulants $c_{ky}(q, \tau_1)$, $0 \le \tau_1 \le q$ and $c_{ky}(\tau_2, 0)$, $-q \le \tau_2 \le q$ are known, then the $q+1$ parameters of FIR system are uniquely determined by the solution of Eq. (6.5.33).*

Proof. Since $c_{ky}(q, 0) \neq 0$, then the determinant of the $(q+1) \times (q+1)$ upper triangular matrix $\boldsymbol{C}_2$ is

$$\det(\boldsymbol{C}_2) = \prod_{i=1}^{q+1} c_2(i, i) = c_{ky}^{(k-1)(q+1)}(q, 0) \neq 0.$$

Hence

$$\operatorname{rank} \begin{bmatrix} \boldsymbol{C}_1 \\ \boldsymbol{C}_2 \end{bmatrix} = \operatorname{rank}(\boldsymbol{C}_2) = q + 1.$$

This shows that the matrix equation Eq. (6.5.33) with $q+1$ unknowns has a unique least square solution $\boldsymbol{b}_1 = (\boldsymbol{C}^{\mathrm{T}}\boldsymbol{C})^{-1}\boldsymbol{C}^{\mathrm{T}}c$, where $\boldsymbol{C} = [\boldsymbol{C}_1^{\mathrm{T}}, \boldsymbol{C}_2^{\mathrm{T}}]^{\mathrm{T}}$ and $c = [c_1^{\mathrm{T}}, c_2^{\mathrm{T}}]^{\mathrm{T}}$. □

Similarly, by substituting Eq. (6.5.26) into Eq. (6.5.25), we can obtain another cumulant algorithm for FIR system parameters estimation

$$\sum_{i=0}^{q} b^{k-1}(i) c_{ky}(q, i+\tau) = c_{ky}(\tau, 0) c_{ky}(q, q) / c_{ky}(q, 0), \quad \tau = -q, \cdots, 0, \cdots, q. \tag{6.5.34}$$

This algorithm also ensures the unique identifiability of the unknown parameters.

Theorem 6.5.2. *Assume that the true cumulants $c_{ky}(q, \tau_1)$, $0 \le \tau_1 \le q$ and $c_{ky}(\tau_2, 0)$, $-q \le \tau_2 \le q$ are known, then the $q+1$ unknowns $b^{k-1}(0), b^{k-1}(1), \cdots, b^{k-1}(q)$ can be recovered uniquely by the solution of Eq. (6.5.34).*

Proof. This proof is exactly similar to the proof of Theorem 6.5.1, which is omitted here. □

For the above two algorithms, the following notes are available.

(1) In Eq. (6.5.27), $\hat{b}(i)/\hat{b}(0)$ are taken as the final estimation of $b(i)$, $i = 1, \cdots, q$ to satisfy the normalized condition $b(0) = 1$. Similarly, in Eq. (6.5.34), $\hat{b}^{k-1}(i)/\hat{b}^{k-1}(0)$ are taken as the final estimations of $b^{k-1}(i)$, $i = 1, \cdots, q$. However, since we are only interested in the estimation of $b(i)$, when $k = 4$ (i.e., using the fourth-order

cumulant), we can directly take $\hat{b}(i) = \sqrt[3]{\hat{b}^3(i)\hat{b}^3(0)}$, $i = 1, \cdots, q$; when $k = 3$, the sign of the estimation $\hat{b}(i)$ is taken as that of $b(n) = c_{3y}(q, n)/c_{3x}(q, 0)$ given by $C(q, n)$ method, and the amplitude of $\hat{b}(i)$ is taken as $|\hat{b}(i)| = \sqrt{\hat{b}^2(i)/\hat{b}^2(0)}$.

(2) It is better to implement these two algorithms simultaneously. If the $|\hat{b}(0)|$ obtained by Algorithm 1 is much smaller than 1, then the estimation given by Algorithm 1 is considered to be poor. Similarly, if the $|\hat{b}^{k-1}(0)|$ obtained by Algorithm 2 is much smaller than 1, then the estimation given by Algorithm 2 is considered to be poor. In this case, the estimation given by another algorithm should be taken. In this way, the performance of the MA parameter estimation can be improved.

(3) The GM algorithm and the Tugnait method are overparametrized in the sense that they estimate not only $b(i)$ but also $b^2(i)$. Thus, they are suboptimal. In contrast, the cumulant method is well parametrized in the sense that only $b(i)$ or $b^{k-1}(i)$ are estimated. In addition, the GM algorithm, Tugnait method, and w-slice method are only applicable to the additive white noise, while the cumulant method can theoretically completely suppress the additive colored Gaussian noise.

In addition to the four linear normal equation methods that can be used to estimate the MA parameters, there are several closed-form recursive estimation methods proposed by Ref.[87], [202],[208], [209] and [254], respectively (not presented here due to space limitations). By the way, it is noted that the first four recursions use both correlation function and higher-order cumulants and thus are only applicable to additive white noise, while the last recursion only uses higher-order cumulants.

6.5.3 MA Order Determination

The above discussion only deals with the parameter estimation of the MA model, while implicitly assuming that the MA order is known. In practice, this order is required to be determined in advance before parameter estimation.

It can be seen from BBR formula (6.5.18) and $C(q, n)$ formula (6.5.22) of special slice cumulant

$$c_{ky}(q, 0) = c_{ky}(q, 0, \cdots, 0) \neq 0, \tag{6.5.35}$$

$$c_{ky}(q, n) = c_{ky}(q, n, \cdots, n) = 0, \quad \forall n > q. \tag{6.5.36}$$

The above equation implies that the MA order q should be the smallest positive integer n that satisfies Eq. (6.5.36). This is the order determination method in Ref.[88]. The problem is that for a group of short data observed in colored Gaussian noise, the sample cumulant $\hat{c}_{ky}(q, n)$ tends to exhibit large error and variance, which makes the test of Eq. (6.5.36) difficult to manipulate.

In Ref.[252], a singular value decomposition method for determining the order q of MA is proposed, which has good numerical stability. The basic idea of this method is to turn the estimation of the MA order into a problem of determining the matrix rank, the key of which is to construct the following $(q+1)\times(q+1)$-dimensional cumulant matrix

$$\boldsymbol{C}_{\text{MA}} = \begin{bmatrix} c_{ky}(0,0) & c_{ky}(1,0) & \cdots & c_{ky}(q,0) \\ c_{ky}(1,0) & \cdots & c_{ky}(q,0) & \\ \vdots & \iddots & & \\ c_{ky}(q,0) & & & 0 \end{bmatrix}. \tag{6.5.37}$$

Since the diagonal elements $c_{ky}(q,0) \neq 0$, $\boldsymbol{C}_{\text{MA}}$ is clearly a full rank matrix, i.e., there is

$$\text{rank}(\boldsymbol{C}_{\text{MA}}) = q+1. \tag{6.5.38}$$

Although the order estimation of MA now becomes the determination of the matrix rank, the cumulant matrix contains the unknown order q. To make this approach practical, consider the extended cumulant matrix

$$\boldsymbol{C}_{\text{MA,e}} = \begin{bmatrix} c_{ky}(0,0) & c_{ky}(1,0) & \cdots & c_{ky}(q_e,0) \\ c_{ky}(1,0) & \cdots & c_{ky}(q_e,0) & \\ \vdots & \iddots & & \\ c_{ky}(q_e,0) & & & 0 \end{bmatrix}, \tag{6.5.39}$$

where $q_e > q$. Since $c_{ky}(m,0) = 0, \forall m > q$, it is easy to verify

$$\text{rank}(\boldsymbol{C}_{\text{MA,e}}) = \text{rank}(\boldsymbol{C}_{\text{MA}}) = q+1. \tag{6.5.40}$$

In practice, the elements of the cumulant matrix $\boldsymbol{C}_{\text{MA,e}}$ are replaced by the sample cumulants, and then q can be determined from the effective rank of $\boldsymbol{C}_{\text{MA,e}}$ (which is equal to $q+1$) which can be obtained by using the singular value decomposition.

On the other hand, it is easy to know from the upper triangular structure of the cumulant matrix $\boldsymbol{C}_{\text{MA,e}}$ that the determination of its effective rank is equivalent to the judgment that the product of diagonal elements is not equal to zero, i.e., q is the maximum integer m such that

$$c_{ky}^{m+1}(m,0) \neq 0, \quad m = 1,2,\cdots. \tag{6.5.41}$$

Obviously, from the point of numerical performance, testing for Eq. (6.5.41) is more numerically robust than testing for Eq. (6.5.36).

To summarize the above discussion, the MA order can be determined either by the effective rank of the extended cumulant matrix $C_{\text{MA,e}}$ or estimated from Eq. (6.5.41). These two linear algebraic methods have stable numerical performance.

6.6 Identification of Causal ARMA Models

Compared with the finite impulse response (FIR) system introduced in the previous section, the infinite impulse response (IIR) system is more representative. For the IIR system, the ARMA model is more reasonable than the MA model and AR model from the point of view of parameter parsimony, because the latter two models use too many parameters. This section discusses the identification of the causal ARMA model.

6.6.1 Identification of AR Parameters

Consider the following ARMA model

$$\sum_{i=0}^{p} a(i)x(n-i) = \sum_{i=0}^{q} b(i)e(n-i), \tag{6.6.1}$$

where the ARMA(p, q) random process $\{x(n)\}$ is observed in additive noise $v(n)$, i.e.,

$$y(n) = x(n) + v(n). \tag{6.6.2}$$

Without loss of generality, the following conditions are assumed to hold.

(AS1)The system transfer function $H(z) = B(z)/A(z) = \sum_{i=0}^{\infty} h(i)z^{-i}$ is free of pole-zero cancellations, i.e., $a(p) \neq 0$ and $b(q) \neq 0$.

(AS2)The input $e(n)$ is a non-Gaussian white noise with finite nonzero cumulant γ_{ke}.

(AS3)The observe noise $v(n)$ is a colored Gaussian noise and is independent of $e(n)$ and $x(n)$.

Condition (AS1) means that the system is causal (the impulse response is constant zero when time is negative) and the ARMA(p, q) model cannot be further simplified. Note that no constraint is placed on the zeros of the system, which means that they can lie inside and outside the unit circle. Moreover, if the inverse system of the ARMA model is not used, the zeros are also allowed to be on the unit circle.

Let $c_{kx}(m, n) = c_{kx}(m, n, 0, \cdots, 0)$, then under the conditions (AS1) $\sim$ (AS3), it is easy to obtain the following equation by the BBR formula

$$\begin{aligned}\sum_{i=0}^{p} a(i)c_{kx}(m-i, n) &= \gamma_{ke} \sum_{j=0}^{\infty} h^{k-2}(j)h(j+n) \sum_{i=0}^{p} a(i)h(j+m-i) \\ &= \gamma_{ke} \sum_{j=0}^{\infty} h^{k-2}(j)h(j+n)b(j+m),\end{aligned} \tag{6.6.3}$$

where the definition of the impulse response

$$\sum_{i=0}^{p} a(i)h(n-i) = \sum_{j=0}^{q} b(j)\delta(n-j) = b(n) \tag{6.6.4}$$

is used.

Since $a(p) \neq 0$ and $b(q) \neq 0$ are assumed, and $b(j) \equiv 0 (j > q)$, then an important set of normal equations

$$\sum_{i=0}^{p} a(i) c_{kx}(m-i, n) = 0, \quad m > q, \quad \forall n \tag{6.6.5}$$

can be obtained from Eq. (6.6.3). This is the modified Yule-Walker equation expressed in terms of higher-order cumulants or the MYW equation for short. Note that the MYW equation can also be written in other forms, such as[204]

$$\sum_{i=0}^{p} a(i) c_{kx}(i-m, n) = 0, \quad m > q, \quad \forall n. \tag{6.6.6}$$

This normal equation can be derived by imitating the derivation of Eq. (6.6.5). If these two MYW equations use the same range of (m, n), they are essentially equivalent.

An important question is how to take the appropriate values of m and n in the MYW equation to ensure that the solution of the AR parameter is unique. This problem is called the identifiability of AR parameters based on higher-order cumulants. It may be useful to look at an example first[205].

Example 6.6.1. *Consider the following causal maximum phase system*

$$H(z) = \frac{(z - \alpha_1^{-2})(z - \alpha_1^{-1}\alpha_2^{-1})}{(z - \alpha_1)(z - \alpha_2)}, \tag{6.6.7}$$

where $\alpha_1 \neq \alpha_2$. Assuming that the third-order cumulant is used and takes $n \neq 0$, then the MYW equation is

$$c_{kx}(-m, n) - \alpha_2 c_{kx}(1-m, n) = 0, \quad m > 1, n \neq 0. \tag{6.6.8}$$

This equation shows that if $n = 0$ is excluded, no matter how to choose the combination of m and n, it is impossible to identify the pole α_1. However, if $n = 0$ is included, the MYW equation will be different from Eq. (6.6.8) *and will be able to identify the poles α_1 and α_2.*

This example shows that the combination of m and n needs to be chosen carefully and cannot be chosen arbitrarily. So, what combination will ensure the unique identifiable of the AR parameters? The answer to this question is given in the following theorem.

Theorem 6.6.1. [86, 88] *Under the conditions (AS1) ∼ (AS3), the AR parameters of the ARMA model* (6.6.1) *can be identified uniquely as the least square solution of the following equation*

$$\sum_{i=0}^{p} a(i) c_{ky}(m-i, n) = 0, \quad m = q+1, \cdots, q+p; n = q-p, \cdots, q. \tag{6.6.9}$$

In principle, the determination of the AR order p is to find the rank of the cumulant matrix $\boldsymbol{C}$. However, since C itself uses the unknown orders p and q, the structure of $\boldsymbol{C}$

must be modified so that the new cumulants no longer contain the unknown p and q, but it still has rank p. Considering the connection with the total least square method of AR parameter estimation, the result of the modification can be described as follows.

Theorem 6.6.2. [88] *Define the $M_2(N_2 - N_1 + 1) \times M_2$ extended cumulant matrix*

$$\boldsymbol{C}_e = \begin{bmatrix} c_{kx}(M_1, N_1) & \cdots & c_{kx}(M_1 + M_2 - 1, N_1) \\ \vdots & \vdots & \vdots \\ c_{kx}(M_1, N_2) & \cdots & c_{kx}(M_1 + M_2 - 1, N_2) \\ \vdots & \vdots & \vdots \\ c_{kx}(M_1 + M_2, N_1) & \cdots & c_{kx}(M_1 + 2M_2 - 1, N_1) \\ \vdots & \vdots & \vdots \\ c_{kx}(M_1 + M_2, N_2) & \cdots & c_{kx}(M_1 + 2M_2 - 1, N_2) \end{bmatrix}, \tag{6.6.10}$$

where $M_1 \geq q + 1 - p$, $M_2 \geq p$, $N_1 \leq q - p$ and $N_2 \geq q$. Matrix $\boldsymbol{C}_e$ has rank p if and only if the ARMA(p, q) model is free of pole-zero cancellations.

This theorem suggests that if the true cumulant $c_{kx}(m, n)$ of the signal is replaced by the sample cumulant $\hat{c}_{kx}(m, n)$ of the observed data, the effective rank of the sample cumulant matrix $\boldsymbol{C}_e$ will be equal to p.

Collating the above results, the singular value decomposition-total least squares (SVD-TLS) algorithm for AR order determination and AR parameter estimation of the causal ARMA model can be obtained as follows.

Algorithm 6.6.1. *Cumulant-based SVD-TLS algorithm*

Step 1 Define the $M_2(N_2 - N_1 + 1) \times M_2$ extended cumulant matrix

$$\boldsymbol{C}_e = \begin{bmatrix} \hat{c}_{kx}(M_1 + M_2 - 1, N_1) & \cdots & \hat{c}_{kx}(M_1, N_1) \\ \vdots & \vdots & \vdots \\ \hat{c}_{kx}(M_1 + M_2 - 1, N_2) & \cdots & \hat{c}_{kx}(M_1, N_2) \\ \vdots & \vdots & \vdots \\ \hat{c}_{kx}(M_1 + 2M_2 - 1, N_1) & \cdots & \hat{c}_{kx}(M_1 + M_2, N_1) \\ \vdots & \vdots & \vdots \\ \hat{c}_{kx}(M_1 + 2M_2 - 1, N_2) & \cdots & \hat{c}_{kx}(M_1 + M_2, N_2) \end{bmatrix}, \tag{6.6.11}$$

calculate its singular value decomposition $\boldsymbol{C}_e = \boldsymbol{U\Sigma V}^T$ and store the matrix $\boldsymbol{V}$.

Step 2 Determine the effective rank of matrix $\boldsymbol{C}_e$ and give the AR order estimate p.

Step 3 Compute the $(p + 1) \times (p + 1)$ matrix

$$\boldsymbol{S}^p = \sum_{j=1}^{p} \sum_{i=1}^{M_2 - p} \sigma_j^2 \boldsymbol{v}_j^i (\boldsymbol{v}_j^i)^T, \tag{6.6.12}$$

where $\boldsymbol{v}_j^i = [v(i, j), v(i + 1, j), \cdots, v(i + p, j)]^T$ and $v(i, j)$ is ith row and jth column element of the matrix $\boldsymbol{V}$.

Step 4 Find the inverse matrix $\mathbf{S}^{(-p)}$ *of the matrix* $\mathbf{S}^{(p)}$*, then the total least square estimation of the AR parameters are given by*

$$\hat{a}(i) = \mathbf{S}^{-(p)}(i+1,1)/S^{-(p)}(1,1). \tag{6.6.13}$$

Here, $\mathbf{S}^{-(p)}(i,1)$ *is the ith row and first column element of the inverse matrix* $\mathbf{S}^{-(p)}$.

Applying the SVD-TLS algorithm to solve the MYW equation can significantly improve the estimation accuracy of the AR parameters, which is especially important for system identification using higher-order cumulants because the sample higher-order cumulants for short data inherently have relatively large estimation errors.

6.6.2 MA order Determination

If the AR part of the causal ARMA process has been identified, then using the available AR order and AR parameters, we can get the following equation by "filtering" the original observed process

$$\tilde{y}(n) = \sum_{i=0}^{p} a(i)y(n-i) \tag{6.6.14}$$

and $\{\tilde{y}(n)\}$ is called the "residual time series". Substitute Eq. (6.6.2) into Eq.(6.6.14) yields

$$\tilde{y}(n) = \sum_{i=0}^{p} a(i)x(n-i) + \sum_{i=0}^{p} a(i)v(n-i) = \sum_{j=0}^{q} b(j)e(n-j) + \tilde{v}(n). \tag{6.6.15}$$

Here, in order to obtain the first term of the second equation, Eq. (6.6.1) is substituted, and

$$\tilde{v}(n) = \sum_{i=0}^{p} a(i)v(n-i) \tag{6.6.16}$$

is still colored Gaussian noise.

Eq. (6.6.15) shows that the residual time series $\{\tilde{y}(n)\}$ is a MA(q) process observed in colored Gaussian noise. Therefore, MA parameter estimates for the causal ARMA process can be obtained by simply applying the RC method or the cumulant method for FIR system parameter estimation introduced in Section 6.5 to the residual time series.

The method to transform the MA parameter estimation of the causal ARMA process into a pure FIR system parameter identification of the residual time series is called the residual time series method and is proposed in the [88]. Moreover, the method to estimate the MA order of the ARMA model is also proposed. Here, the maximum integer m satisfying

$$c_{k\tilde{y}}(m,0,\cdots,0) \neq 0 \tag{6.6.17}$$

is used as the MA order of the ARMA model. More generally, all the order determination methods applicable to pure MA models can be applied to the residual time series to obtain the estimation of the MA order of the ARMA model.

Now consider the method of determining the MA order of the ARMA model directly without constructing residual time series. Therefore, the fitting error function is defined as

$$f_k(m,n) = \sum_{i=0}^{p} a(i)c_{ky}(m-i,n) = \sum_{i=0}^{p} a(i)c_{kx}(m-i,n). \tag{6.6.18}$$

Here $c_{ky}(m,n) = c_{kx}(m,n)$ is used. Substituting Eq. (6.6.3) into Eq. (6.6.18), it is immediately obtain

$$f_k(m,n) = \gamma_{ke} \sum_{j=0}^{\infty} h^{k-2}(j)h(j+n)b(j+m). \tag{6.6.19}$$

Obviously,

$$f_k(q,0) = \gamma_{ke}h^{k-1}(0)b(q) = \gamma_{ke}b(q) \neq 0 \tag{6.6.20}$$

because $h(0) = 1$ and $b(q) \neq 0$.

On the other hand, using the fitting error function, the MYW equation can be expressed as

$$f_k(m,n) = 0, \quad m > q, \forall n. \tag{6.6.21}$$

Eqs. (6.6.20) and (6.6.21) inspire a method to determine the MA order of ARMA model, i.e., q is the largest integer m that makes

$$f_k(m) \stackrel{\text{def}}{=} f_k(m,0) = \sum_{i=0}^{p} a(i)c_{ky}(m-i,0,\cdots,0) \neq 0 \tag{6.6.22}$$

hold. Although this order determination method is theoretically attractive, it is difficult to adopt in practice because there will be a large error in the estimated value $\hat{f}_k(m)$ when the data are relatively short.

To overcome this difficulty, a simple and effective method is proposed in Ref.[252]. The basic idea is to introduce a fitted error matrix and to transform the determination of the MA order into the determination of the effective rank of this matrix, which can be achieved using a stable method such as singular value decomposition.

The basic idea is that the introduction of a fitted error matrix transforms the determination of the MA order into the determination of the effective rank of this matrix, and the latter can be achieved using a numerically stable method such as singular value decomposition.

Construct the fitted error matrix

$$\boldsymbol{F} = \begin{bmatrix} f_k(0) & f_k(1) & \cdots & f_k(q) \\ f_k(1) & \cdots & f_k(q) & \\ \vdots & ⋰ & & \\ f_k(q) & & & 0 \end{bmatrix}. \tag{6.6.23}$$

This is a Henkel matrix, and it is also an upper triangular matrix. Clearly

$$\det(\boldsymbol{F}) = \prod_{i=1}^{q+1} f(i, q+2-i) = f_k^{q+1}(q) \neq 0. \tag{6.6.24}$$

That is to say, rank($\boldsymbol{F}$) = $q + 1$. Since the MA order q is unknown, it is necessary to modify the fitted error matrix F so that its elements no longer contain the unknown q, but still have the rank $q + 1$. Therefore, construct the extended fitted error matrix

$$\boldsymbol{F}_e = \begin{bmatrix} f_k(0) & f_k(1) & \cdots & f_k(q_e) \\ f_k(1) & \cdots & f_k(q_e) & \\ \vdots & \iddots & & \\ f_k(q_e) & & & 0 \end{bmatrix}, \quad q_e > q. \tag{6.6.25}$$

Using Eqs. (6.6.20) and (6.6.21), it is easy to verify that

$$\text{rank}(\boldsymbol{F}_e) = \text{rank}(\boldsymbol{F}) = q + 1. \tag{6.6.26}$$

Although q is unknown, it is a not difficult to choose one q_e such that $q_e > q$.

Eq. (6.6.26) shows that the effective rank of $\boldsymbol{F}_e$ can be determined using the singular value decomposition (SVD) if the elements of the matrix $\boldsymbol{F}_e$ are replaced by their estimations $\hat{f}_k(m)$. On the other hand, by the upper triangular structure of the matrix $\boldsymbol{F}_e$, it is known that the effective rank determination of $\boldsymbol{F}_e$ is equivalent to testing the following inequality

$$\hat{f}_k^{m+1}(m) \neq 0. \tag{6.6.27}$$

The largest integer m that makes the above inequality hold approximately is the estimation of the MA order q. The test of $\hat{f}_k^{m+1}(m) \neq 0$ is called as the product of diagonal entries (PODE) test. Clearly, PODE test inequality Eq. (6.6.27) is more reasonable than directly test inequality Eq. (6.6.22), since the former can be expected to provide better numerical stability than that of the latter.

Ref.[252] found that when using the SVD or PODE test to determine the MA order of the ARMA model alone, the model order determined may be overdetermined or underdetermined. Therefore, it is suggested a combination of the SVD and PODE test. The specific operation is: firstly, the order M determined by SVD is taken as the reference. If $f_k^{M+1}(M)$ and $f_k^{M+2}(M+1)$ differ greatly from zero, then this M is underdetermined, and we should use Eq. (6.6.27) to take the PODE test for order $M' = M + 1$; conversely, if $f_k^{M+1}(M)$ and $f_k^{M+2}(M+1)$ obviously are close to zero, then this M is overdetermined, and we should use Eq. (6.6.27) to take the PODE test for order $M' = M - 1$. A perfect estimate of M should satisfy that $f_k^{M+1}(M)$ differs greatly from zero and $f_k^{M+2}(M+1)$ is close to zero.

6.6.3 Estimation of MA Parameters

It has been pointed out earlier that the estimation of the MA parameters of the ARMA model can be obtained by applying either the RC method of pure MA parameter estimation or the higher-order cumulant method to the residual time series. This estimation problem will be discussed in the following.

1. Residual Time Series Cumulant Method

In fact, the MA parameters of the ARMA model can be estimated successively without computing the residual time series. Consider the cumulants of the residual time series

$$c_{k\tilde{y}}(\tau_1, \cdots, \tau_{k-1}) = \text{cum}[\tilde{y}(n), \tilde{y}(n+\tau_1), \cdots, \tilde{y}(n+\tau_{k-1})]. \tag{6.6.28}$$

Substituting the residual time series defined in Eq.(6.6.14) into the above equation and using the definition and properties of the cumulants, we have

$$\begin{aligned} c_{k\tilde{y}}(\tau_1, \cdots, \tau_{k-1}) &= \text{cum}\left[\sum_{i_1=0}^{p} a(i_1)y(n-i_1), \cdots, \sum_{i_k=0}^{p} a(i_k)y(n-i_k)\right] \\ &= \sum_{i_1=0}^{p} \cdots \sum_{i_k=0}^{p} a(i_1)\cdots a(i_k)c_{ky}(\tau_1+i_1-i_2, \cdots, \tau_{k-1}+i_1-i_k). \end{aligned} \tag{6.6.29}$$

In particularly, the k-th order ($k = 2, 3, 4$) are

$$R_{\tilde{y}}(m) = \sum_{i_1=0}^{p}\sum_{j=0}^{p} a(i)a(j)R_y(m+i-j), \tag{6.6.30}$$

$$c_{3\tilde{y}}(m, n) = \sum_{i_1=0}^{p}\sum_{i_2=0}^{p}\sum_{i_3=0}^{p} a(i_1)a(i_2)a(i_3)c_{ky}(m+i_1-i_2, n+i_1-i_3), \tag{6.6.31}$$

$$c_{4\tilde{y}}(m, n, l) = \sum_{i_1=0}^{p}\cdots\sum_{i_4=0}^{p} a(i_1)\cdots a(i_4)c_{ky}(m+i_1-i_2, n+i_1-i_3, l+i_1-i_4). \tag{6.6.32}$$

The above method that can directly obtain the cumulants of the residual time series without computing the residual time series itself is proposed in Ref.[247]. Although this method calculates the third- and fourth-order cumulants of the residual time series involving triple and quadruple summation, respectively, compared with the (direct) residual time series method which first generates the residual time series and then calculates the estimates of the third-order and fourth-order cumulants, the computational complexity of Eqs. (6.6.31) and (6.6.32) is much smaller.

2. q-slice Algorithm

Let $m = q$ in Eq.(6.6.19), then we have

$$f_k(q, n) = \gamma_{ke} \sum_{i=0}^{\infty} h^{k-2}(i)h(i+n)b(i+q) = \gamma_{ke}h(n)b(q). \tag{6.6.33}$$

Dividing the above formula by Eq. (6.6.20), yields

$$h(n) = \frac{f_k(q, n)}{f_k(q, 0)} = \frac{\sum_{i=0}^{p} a(i)c_{ky}(q-i, n)}{\sum_{i=0}^{p} a(i)c_{ky}(q-i, 0)}. \tag{6.6.34}$$

Since for a fixed n, only the 1-D slices of cumulants $c_{ky}(q, n), \cdots, c_{ky}(q-p, n)$ are used in the numerator of Eq. (6.6.34), while q alices are required to calculate the q impulse response coefficients $h(1), \cdots, h(n)$. Consequently, the above estimation method for the impulse response of the ARMA system is called the "q-slice" algorithm. By the way, it is noted that Eq. (6.5.22) of $C(q, n)$ for estimating the impulse response of the FIR system given in Section 6.5 is a special case of Eq. (6.6.34) of q-slice but with AR order $p = 0$.

Once the impulse response $h(n)$, $n = 1, \cdots, q$ of the ARMA model has been directly calculated or estimated from Eq. (6.6.34), the familiar equation

$$b(n) = \sum_{i=0}^{p} a(i)h(n-i), \quad n = 1, \cdots, q \tag{6.6.35}$$

can be used to calculate the MA parameters directly.

Interestingly, the algorithm for estimating AR parameters and impulse response can be obtained by modifying the q-slice algorithm. This algorithm was proposed by Swami and Mendel[203]. Let $\varepsilon = -f_k(q, 0)$, then Eq. (6.6.34) can be modified as

$$\sum_{i=0}^{p} a(i)c_{ky}(q-i, n) + \varepsilon h(n) = -c_{ky}(q, n). \tag{6.6.36}$$

Concatenating the above equation for $n = 0, 1, \cdots, Q (Q \geq q)$, we can obtain

$$\begin{bmatrix} c_{ky}(q-1, 0) & \cdots & c_{ky}(q-p, 0) \\ c_{ky}(q-1, 1) & \cdots & c_{ky}(q-p, 1) \\ \vdots & \vdots & \vdots \\ c_{ky}(q-1, Q) & \cdots & c_{ky}(q-p, Q) \end{bmatrix} \begin{bmatrix} a(1) \\ a(2) \\ \vdots \\ a(p) \end{bmatrix} + \begin{bmatrix} \varepsilon \\ \varepsilon h(1) \\ \vdots \\ \varepsilon h(Q) \end{bmatrix} = - \begin{bmatrix} c_{ky}(q, 0) \\ c_{ky}(q, 1) \\ \vdots \\ c_{ky}(q, Q) \end{bmatrix} \tag{6.6.37}$$

which may be compactly written as

$$\boldsymbol{C}_1 \boldsymbol{a} + \varepsilon \boldsymbol{h} = -\boldsymbol{c}_1. \tag{6.6.38}$$

Concatenating Eq. (6.6.38) with the MYW equation to yield

$$\begin{bmatrix} \boldsymbol{C} & \boldsymbol{0} \\ \boldsymbol{C}_1 & \boldsymbol{I} \end{bmatrix} \begin{bmatrix} \boldsymbol{a} \\ \varepsilon \boldsymbol{h} \end{bmatrix} = - \begin{bmatrix} \boldsymbol{c} \\ \boldsymbol{c}_1 \end{bmatrix}. \tag{6.6.39}$$

Eq. (6.6.39) represents an overdetermined system of linear equations from which the unknown AR parameters $a(1), \cdots, a(p)$ and the implus response parameters $h(1), \cdots, h(Q)$ can be simultaneously estimated.

3. The Closed Solution

The higher-order spectra of a random process $\{x(n)\}$ defined in Eq. (6.4.11) can be rewritten as the Z-transform form

$$\begin{aligned} S_x(z_1, \cdots, z_{k-1}) &= \gamma_{ke} H(z_1) \cdots H(z_{k-1}) H(z_1^{-1} \cdots z_{k-1}^{-1}) \\ &= \gamma_{ke} \frac{\beta_k(z_1, \cdots, z_{k-1})}{\alpha_k(z_1, \cdots, z_{k-1})} \end{aligned} \tag{6.6.40}$$

or

$$S_x(z_1, \cdots, z_{k-1}) = \sum_{i_1=-\infty}^{\infty} \cdots \sum_{i_{k-1}=-\infty}^{\infty} c_{ky}(i_1, \cdots, i_{k-1}) z_1^{-i_1} \cdots z_{k-1}^{-i_{k-1}}. \tag{6.6.41}$$

In Eq. (6.6.40),

$$\begin{aligned} \alpha_k(z_1, \cdots, z_{k-1}) &= A(z_1) \cdots A(z_{k-1}) A(z_1^{-1} \cdots z_{k-1}^{-1}) \\ &= \sum_{i_1=-p}^{p} \cdots \sum_{i_{k-1}=-p}^{p} \alpha_k(i_1, \cdots, i_{k-1}) z_1^{-i_1} \cdots z_{k-1}^{-i_{k-1}}, \end{aligned} \tag{6.6.42}$$

$$\begin{aligned} \beta_k(z_1, \cdots, z_{k-1}) &= B(z_1) \cdots B(z_{k-1}) B(z_1^{-1} \cdots z_{k-1}^{-1}) \\ &= \sum_{i_1=-p}^{p} \cdots \sum_{i_{k-1}=-p}^{p} \beta_k(i_1, \cdots, i_{k-1}) z_1^{-i_1} \cdots z_{k-1}^{-i_{k-1}}. \end{aligned} \tag{6.6.43}$$

Comparing the coefficients of the same power term on both sides of the above two equations respectively, it is easy to obtain

$$\alpha_k(z_1, \cdots, z_{k-1}) = \sum_{j=0}^{p} a(j) a(j + i_1) \cdots a(j + i_{k-1}), \tag{6.6.44}$$

$$\beta_k(z_1, \cdots, z_{k-1}) = \sum_{j=0}^{q} b(j) b(j + i_1) \cdots b(j + i_{k-1}). \tag{6.6.45}$$

According to their specific forms, $\alpha_k(z_1, \cdots, z_{k-1})$ and $\beta_k(z_1, \cdots, z_{k-1})$ are called the kth-order correlation coefficients of the AR and MA parameters [89], respectively.

If letting $i_1 = p$, $i_2 = \cdots = i_{k-1} = 0$, then from Eq. (6.6.44) we can get

$$a(p) = \alpha_k(p, 0, \cdots, 0).$$

If letting $i_1 = p$, $i_2 = i$ and $i_3 = \cdots = i_{k-1} = 0$, then Eq. (6.6.44) gives the result

$$a(p) a(i) = \alpha_k(p, i, 0, \cdots, 0).$$

From the above two equations, it is immediately seen that

$$a(i) = \frac{\alpha_k(p, i, 0, \cdots, 0)}{\alpha_k(p, 0, \cdots, 0)}, \quad i = 1, \cdots, p. \tag{6.6.46}$$

Similarly, from Eq. (6.6.45), it follows that

$$b(i) = \frac{\beta_k(q, i, 0, \cdots, 0)}{\beta_k(q, 0, \cdots, 0)}, \quad i = 1, \cdots, q. \tag{6.6.47}$$

This shows that the if $\beta_k(q, i, 0, \cdots, 0)$ is known, the MA parameters can be calculated directly from Eq. (6.6.47). Therefore, the key to the direct calculation of the MA parameters is how to obtain the kth-order correlation coefficients $\beta_k(q, i, 0, \cdots, 0)$ for the MA parameters.

From Eqs. (6.6.40) and (6.6.41), it is easy to know that

$$\sum_{i_1=-p}^{p} \cdots \sum_{i_{k-1}=-p}^{p} \alpha_k(i_1, \cdots, i_{k-1}) c_{ky}(\tau_1 - i_1, \cdots, \tau_{k-1} - i_{k-1})$$
$$= \begin{cases} \gamma_{ke}\beta_k(\tau_1, \cdots, \tau_{k-1}), & \tau_i \in [-q, q] \\ 0, & \text{others} \end{cases}. \tag{6.6.48}$$

Combining Eqs. (6.6.47) and (6.6.48), another formula for directly calculating the MA parameters can be obtained

$$b(m) = \frac{\beta_k(q, m, 0, \cdots, 0)}{\beta_k(q, 0, \cdots, 0)} = \frac{\gamma_{ke}\beta_k(q, m, 0, \cdots, 0)}{\gamma_{ke}\beta_k(q, 0, \cdots, 0)}$$
$$= \frac{\sum_{i_1=-p}^{p} \cdots \sum_{i_{k-1}=-p}^{p} \alpha_k(i_1, \cdots, i_{k-1}) c_{ky}(q - i_1, m - i_2, -i_3, \cdots, -i_{k-1})}{\sum_{i_1=-p}^{p} \cdots \sum_{i_{k-1}=-p}^{p} \alpha_k(i_1, \cdots, i_{k-1}) c_{ky}(q - i_1, -i_2, -i_3, \cdots, -i_{k-1})}, \tag{6.6.49}$$

where $m = 1, \cdots, q$.

After the AR parameter is estimated, its kth-order correlation $\alpha_k(i_1, \cdots, i_{k-1})$ can be calculated by its definition Eq. (6.6.44), so Eq. (6.6.49) is easy to calculate. It should be pointed out that although the direct calculation formulation of MA parameters is theoretically appealing and again reflects the superiority of higher-order cumulants in the identification of ARMA model systems, it is rarely used directly in practice. This is mainly because the higher-order cumulants have relatively large estimation error and variance in the case of short data, which will seriously affect the estimation error and variance of MA parameters obtained by direct calculation.

Finally, it is noted that there are other closed-form recursive solutions for the estimation of the MA parameters of the ARMA process, and the interested reader is referred to Ref.[254].

6.7 Harmonic Retrieval in Colored Noise

In Chapter 4, we have discussed various methods of harmonic retrieval in additive white noise, and the second-order statistics (correlation function) are used. This section analyzes the problem of harmonic retrieval in colored noise. In order to suppress the influence of colored noise, higher-order statistics are used.

6.7.1 Cumulant Definition for Complex Signal

The discussion in the previous sections is limited to the higher-order statistics of real signals. Since the complex harmonic process is a complex signal, it is necessary to introduce the definition of the cumulants of the complex signal. Brillinger and Rosenblatt[35] pointed out that "in particular, for each partition of k (divided into j conjugate elements and $k-j$ unconjugated elements), there can be a corresponding kth-order spectral density". Therefore, there are 2^k possible expressions for the kth-order spectrum.

Assuming that the complex signal of interest is a complex harmonic process

$$x(n) = \sum_{i=1}^{p} \alpha_i e^{j(\omega_i n + \phi_i)}, \tag{6.7.1}$$

where the ϕ_i's are independent identically distributed random variables uniformly distributed over $[-\pi, \pi]$. Since ϕ_i's are independent of each other, we conclude from the properties of the cumulant, that the cumulant of the signal $x(n)$ is the sum of the cumulants due to the individual harmonics. Therefore, when discussing the definition of cumulants for complex harmonic processes, it is sufficient to consider only the definition of a single cumulant.

Let $s = e^{j\phi}$, where ϕ is uniformly distributed over $[0, 2\pi)$ ($\phi \sim U[0, 2\pi)$). For $m \neq 0$, we have $E\{e^{jm\phi}\} = 0$. Hence, it is easy to have

$$\text{cum}(s, s) = 0, \quad \text{cum}(s^*, s) = E\{|s|^2\} = 1, \tag{6.7.2}$$

$$\text{cum}(s, s, s) = \text{cum}(s^*, s, s) = 0, \tag{6.7.3}$$

$$\text{cum}(s, s, s, s) = \text{cum}(s^*, s, s, s) = 0, \tag{6.7.4}$$

$$\text{cum}(s^*, s^*, s, s) = E\{|s|^4\} - |E\{s^2\}|^2 - 2E\{|s|^2\}E\{|s|^2\} = -1. \tag{6.7.5}$$

From the above four equations, the following important conclusions can be drawn.

(1) There are two definitions for the second-order cumulant of the complex harmonic processes, of which only definition $\text{cum}(s^*, s)$ gives nonzero results;
(2) No matter how the cumulant is defined, the third-order cumulant of the harmonic signal is identically zero;
(3) There are several different ways of defining the fourth-order cumulant of the harmonic process, only the definition $\text{cum}(s^*, s^*, s, s)$ yields nonzero values.

Of course, the fourth-order cumulant $\text{cum}(s^*, s, s^*, s)$ and $\text{cum}(s, s, s^*, s^*)$ also give nonzero cumulants, we conclude from the symmetry of the cumulants that these definitions are equivalent to $\text{cum}(s^*, s^*, s, s)$. Without loss of generality, the conjugated elements will be arranged in the front and the unconjugated elements will be arranged in the back. Accordingly, the definition of the fourth-order cumulant of the complex signals can be derived.

Definition 6.7.1. *The fourth-order cumulant of the complex process $\{x(n)\}$ with zero mean is defined as*

$$c_{4x}(m_1, m_2, m_3) = cum[x^*(n), x^*(n+m_1), x(n+m_2), x(n+m_3)]. \tag{6.7.6}$$

Note that, there are two special harmonic processes whose higher-order cumulants must take other definition forms. These two special harmonic processes are the quadratic phase-coupled harmonic process and the cubically phase-coupled harmonic process, respectively.

1.Quadratic Phase-coupled Harmonic Process

Definition 6.7.2. *If the harmonic process $x(n) = \sum_{i=1}^{3} \alpha_i e^{j(\omega_i n+\phi_i)}$ satisfy that $\phi_3 = \phi_1 + \phi_2$ and $\omega_3 = \omega_1 + \omega_2$, then this harmonic process is called quadratic phase-coupled process.*

Definition 6.7.3. *The third-order cumulant of the quadratic phase-coupled harmonic process is defined as*

$$c_{3x}(\tau_1, \tau_2) \stackrel{def}{=} cum[x^*(n), x(n+\tau_1), x(n+\tau_2)]. \tag{6.7.7}$$

It is easy to verify that the above definition gives a nonzero cumulant

$$c_{3x}(\tau_1, \tau_2) = \alpha_1\alpha_2\alpha_3^* \left[e^{j(\omega_1\tau_1+\omega_2\tau_2)} + e^{j(\omega_2\tau_1+\omega_1\tau_2)}\right], \tag{6.7.8}$$

while the other definition $cum[x(n), x(n+m_1), x(n+m_2)]$ and $cum[x^*(n), x^*(n+m_1), x(n+m_2)]$ are identically zero.

Taking the 2-D Fourier transform of Eq. (6.7.8), it is easy to see that the bispectrum of the quadratic phase-coupled harmonic process is

$$B_x(\lambda_1, \lambda_2) = \alpha_1\alpha_2\alpha_3^*[\delta(\lambda_1 - \omega_1, \lambda_2 - \omega_2) + \delta(\lambda_1 - \omega_2, \lambda_2 - \omega_1), \tag{6.7.9}$$

where

$$\delta(i, j) = \begin{cases} 1, & i = j = 0 \\ 0, & \text{others} \end{cases} \tag{6.7.10}$$

is a 2-D δ function.

Eq. (6.7.9) shows that the bispectrum consists of a pair of impulses at (ω_1, ω_2) and (ω_2, ω_1), and is identically zero at other frequency. This is an important property of the quadratic phase-coupled harmonic process, which can be used to test whether a harmonic process is a quadratic phase coupled process.

2. Cubically Phase-coupled Harmonic Process

Definition 6.7.4. *If the harmonic process $x(n) = \sum_{i=1}^{4} \alpha_i e^{j(\omega_i n+\phi_i)}$ satisfy that $\phi_4 = \phi_1 + \phi_2 + \phi_3$ and $\omega_4 = \omega_1 + \omega_2 + \omega_3$, then this harmonic process is called cubically phase-coupled process.*

Note that, for the cubically phase-coupled harmonic process, the fourth-order cumulant given by definition 6.7.1 is identically zero. Hence, the following definition of cumulant must be used.

Definition 6 .7.5. *The fourth-order cumulant of the cubically phase-coupled harmonic process is defined as*

$$c_{4x}(\tau_1,\tau_2,\tau_3) \stackrel{def}{=} cum[x^*(n), x(n+\tau_1), x(n+\tau_2), x(n+\tau_3)]. \tag{6.7.11}$$

According to Eq. (6.7.11), the fourth-order cumulant of the cubically phase-coupled harmonic process can be derived as

$$\begin{aligned} c_{4x}(\tau_1,\tau_2,\tau_3) = \alpha_1\alpha_2\alpha_3\alpha_4^* \Big[& e^{j(\omega_1\tau_1+\omega_2\tau_2+\omega_3\tau_3)} + e^{j(\omega_2\tau_1+\omega_1\tau_2+\omega_3\tau_3)} \\ & + e^{j(\omega_1\tau_1+\omega_3\tau_2+\omega_2\tau_3)} + e^{j(\omega_2\tau_1+\omega_3\tau_2+\omega_1\tau_3)} \\ & + e^{j(\omega_3\tau_1+\omega_1\tau_2+\omega_2\tau_3)} + e^{j(\omega_3\tau_1+\omega_2\tau_2+\omega_1\tau_3)} \Big]. \end{aligned} \tag{6.7.12}$$

Taking the 3-D Fourier transform of the above equation, then the trispectrum of the cubically phase-coupled harmonic process is obtained

$$\begin{aligned} T_x(\lambda_1,\lambda_2,\lambda_3) = \alpha_1\alpha_2\alpha_3\alpha_4^*[& \delta(\lambda_1-\omega_1,\lambda_2-\omega_2,\lambda_3-\omega_3) \\ & + \delta(\lambda_1-\omega_2,\lambda_2-\omega_1,\lambda_3-\omega_3) + \delta(\lambda_1-\omega_1,\lambda_2-\omega_3,\lambda_3-\omega_2) \\ & + \delta(\lambda_1-\omega_2,\lambda_2-\omega_3,\lambda_3-\omega_1) + \delta(\lambda_1-\omega_3,\lambda_2-\omega_1,\lambda_3-\omega_2) \\ & + \delta(\lambda_1-\omega_3,\lambda_2-\omega_2,\lambda_3-\omega_1), \end{aligned} \tag{6.7.13}$$

i.e., the trispectrum consists of impulses at $(\omega_1, \omega_2, \omega_3)$ and its permutations. This important property can be used to test whether a harmonic process is a cubically phase coupled process.

From the discussion above, it can be seen that the definition of cumulants of complex processes is not unique, and different cumulant definitions should be used depending on the circumstances. In the following discussion, it will be assumed that the harmonic process is not a phase-coupled process.

6.7.2 Cumulants of Harmonic Process

Regarding the cumulants of the harmonic process $\{x(n)\}$, Swami and Mendel proved the following propositions[204].

Proposition 6.7.1. *For the harmonic process* $\{x(n)\}$ *shown in Eq.* (6.7.1), *its autocorrelation and fourth-order cumulant are given by*

$$R_x(\tau) = \sum_{k=1}^{p} |\alpha_k|^2 e^{j\omega\tau}, \tag{6.7.14}$$

$$c_{4x}(\tau_1, \tau_2, \tau_3) = -\sum_{k=1}^{p} |\alpha_k|^4 e^{j\omega_k(-\tau_1+\tau_2+\tau_3)}. \tag{6.7.15}$$

Proposition 6.7.2. *Consider a real harmonic signal*

$$x(n) = \sum_{k=1}^{p} \alpha_k cos(\omega_k n + \phi_k), \tag{6.7.16}$$

if ϕ_k*'s are independent and uniformly distributed over* $[-\pi, \pi)$, *and* $\alpha_k > 0$, *then the autocorrelation and the fourth-order cumulant of* $x(n)$ *are given by*

$$R_x(\tau) = \frac{1}{2}\sum_{k=1}^{p} \alpha_k^2 cos(\omega_k \tau), \tag{6.7.17}$$

$$c_{4x}(\tau_1, \tau_2, \tau_3) = -\frac{1}{8}\sum_{k=1}^{p} \alpha_k^4 [cos(\tau_1 - \tau_2 - \tau_3) + cos(\tau_2 - \tau_3 - \tau_1)]$$
$$+ cos(\tau_3 - \tau_1 - \tau_2). \tag{6.7.18}$$

The above two propositions are important because they describe the relationship between the cumulants of complex-valued and real-valued harmonic signals and the harmonic parameters, respectively. In particular, let us examine the 1-D special slice of the cumulants - the diagonal slice.

Let $\tau_1 = \tau_2 = \tau_3 = \tau$ in Eq. (6.7.15), then for the complex harmonic process, there is

$$c_{4x}(\tau) \stackrel{\text{def}}{=} c_{4x}(\tau, \tau, \tau) = -\sum_{k=1}^{p} |\alpha_k|^4 e^{j\omega_k \tau}. \tag{6.7.19}$$

Similarly, let $\tau_1 = \tau_2 = \tau_3 = \tau$ in Eq. (6.7.18), then for the real harmonic process, there is

$$c_{4x}(\tau) \stackrel{\text{def}}{=} c_{4x}(\tau, \tau, \tau) = -\frac{1}{8}\sum_{k=1}^{p} \alpha_k^4 cos(\omega_k \tau). \tag{6.7.20}$$

Eqs. (6.7.19)) and (6.7.20) are of special significance because they clearly show that the diagonal slices of the fourth-order cumulants of complex and real harmonic processes retain all the pertinent information needed to recover all the parameters (the number of harmonics p, amplitude α_k and frequency ω_k) of the harmonic process.

Compare Eq. (6.7.19) with Eq. (6.7.14), and Eq. (6.7.20) with Eq. (6.7.17), it is immediately obtain the following important result.

Corollary 6.7.1. *The 1-D slice of the fourth-order cumulant of the complex harmonic process $x(n)$ in Eq. (6.7.1) is identical (ignoring the negative sign) with the autocorrelation of the signal*

$$\tilde{x}(n) = \sum_{k=1}^{p} |\alpha_k|^2 e^{j(\omega_k n + \phi_k)}. \tag{6.7.21}$$

Additionally, the diagonal slice of the fourth-order cumulant of the real harmonic process $x(n)$ in Eq. (6.7.16) is identical (ignoring the scale factor $-3/4$) with the autocorrelation of the signal

$$\tilde{x}(n) = \sum_{k=1}^{p} \alpha_k^2 \cos(\omega_k n + \phi_k). \tag{6.7.22}$$

6.7.3 Harmonic Retrieval in Colored Gaussian Noise

Corollary 6.7.1 enlightens us that the harmonic retrieval methods based on autocorrelation functions (e.g., ARMA modeling method, etc.) can become cumulant-based harmonic retrieval methods after the autocorrelation functions are replaced by the diagonal slices of the fourth-order cumulant.

As a typical example, we consider the ARMA modeling method of the complex harmonic retrieval in colored Gaussian noise. The MYW equation of the complex harmonic signal based on the autocorrelation function is

$$\sum_{i=0}^{p} a(i) R_x(m-i) = 0, \quad m > p \quad \text{(complex harmonic)}. \tag{6.7.23}$$

Multiplying both sides of the above equation by -1, and taking into account $c_{4x}(\tau) = -R_x(\tau)$, hence the above equation becomes the MYW equation for the complex harmonic signal based on the fourth-order cumulant, i.e., we have

$$\sum_{i=0}^{p} a(i) c_{4x}(m-i) = 0, \quad m > p \quad \text{(complex harmonic)}. \tag{6.7.24}$$

Obviously, the unknown AR order p (i.e., the number of complex harmonic signals) of the ARMA model can be determined using the singular value decomposition (SVD) for Eq. (6.7.24), and then the SVD-TLS is used to obtain the total least-squares solution of the AR parameters. Finally, the roots of the characteristic polynomial

$$1 + a(1)z^{-1} + \cdots + a(p)z^{-p} = 0 \quad \text{(complex harmonic)} \tag{6.7.25}$$

are given as estimations of the harmonic frequencies.

For the real harmonic signal, the MYW equation based on the autocorrelation is

$$\sum_{i=0}^{2p} a(i) R_x(m-i) = 0, \quad m > 2p \quad \text{(real harmonic)}. \tag{6.7.26}$$

Multiplying both sides of the above equation by $-3/4$, and substituting $c_{4x}(\tau) = -\frac{3}{4}R_x(\tau)$, then we can get the MYW equation based on the cumulants

$$\sum_{i=0}^{2p} a(i)c_{4x}(m-i) = 0, \quad m > 2p \quad \text{(real harmonic)}. \tag{6.7.27}$$

In the case of colored Gaussian noise $v(n)$, since the fourth-order cumulant of the observed data $y(n) = x(n) + v(n)$ is identical to the fourth-order cumulant of the harmonic signal $x(n)$, the corresponding method for harmonic retrieval directly using the diagonal slice of the fourth-order cumulant of the observed data can be obtained by simply replacing $c_{4x}(\tau)$ with $c_{4y}(\tau)$ in the above equations.

6.7.4 Harmonic Retrieval in Colored Non-Gaussian Noise

Consider a real harmonic process $\{x(n)\}$, it observed in a colored non-Gaussian ARMA noise

$$V(n) + \sum_{i=1}^{n_b} b(i)v(n-i) = \sum_{j=0}^{n_d} d(j)w(n-j), \tag{6.7.28}$$

where n_b and n_d are the AR order and MA order of the ARMA noise process, respectively. $w(n)$ is an independently identically distributed process that takes real values, and the third-order cumulant $y_{3w} \neq 0$.

Let

$$B(q) = \sum_{i=0}^{n_b} b(i)q^{-i} \quad and \quad D(q) = \sum_{i=0}^{n_d} d(i)q^{-i}, \tag{6.7.29}$$

where q^{-1} is the backward shift operator, i.e., $q^{-i}x(n) = x(n-i)$. Using Eq. (6.7.29), Eq. (6.7.28) can be simplified as

$$B(q)v(n) = D(q)w(n). \tag{6.7.30}$$

Assume that the additive non-Gaussian noise $v(n)$ is independent of the harmonic signal $x(n)$, and they are of zero mean. Since the third-order cumulant of the harmonic signal $x(n)$ is identically zero, the third-order cumulant of the observed data $y(n) = x(n) + v(n)$ is identically to that of the non-Gaussian noise $v(n)$, that is

$$c_{3y}(\tau_1, \tau_2) = c_{3v}(\tau_1, \tau_2). \tag{6.7.31}$$

This shows that the AR parameters $b(i), i = 1, \cdots, b(n_b)$ of the non-Gaussian ARMA noise model can be identified using the third-order cumulant $c_{3y}(\tau_1, \tau_2)$ of the observed data.

Once the AR parameters of the noise process are identified, then the observed process can be filtered to obtain

$$\tilde{y}(n) = B(q)y(n) = B(q)[x(n) + v(n)] = \tilde{x}(n) + \tilde{v}(n), \tag{6.7.32}$$

where

$$\tilde{x}(n) = B(q)x(n) = \sum_{i=0}^{n_b} b(i)x(n-i) \tag{6.7.33}$$

$$\tilde{v}(n) = B(q)v(n) = D(q)w(n) = \sum_{j=0}^{n_d} d(j)w(n-j) \tag{6.7.34}$$

represent the filtered harmonic signal and the filtered colored non-Gaussian noise, respectively. Note that since $x(n)$ and $v(n)$ are statistically independent, their linear transformations $\tilde{x}(n)$ and $\tilde{v}(n)$ are also statistically independent.

After filtering, the original additive non-Gaussian noise is transformed from ARMA to the MA process shown in Eq. (6.7.34). Using the truncated tail property of the autocorrelation and the higher-order cumulants of the MA process, we have

$$R_{\tilde{v}}(\tau) = 0, \quad \tau > n_d, \tag{6.7.35}$$

$$c_{k\tilde{v}}(\tau_1, \cdots, \tau_{k-1}) = 0, \quad \tau > n_d;\ k > 2. \tag{6.7.36}$$

The question is, is it possible to retrieval the harmonic signal from the filtered observation process $\tilde{y}(n)$? The answer is yes: using Eqs. (6.7.35) and (6.7.36), two methods for harmonic retrieval in colored non-Gaussian noise can be obtained.

1. The Hybrid Method

For the filtered harmonic signal $\tilde{x}(n)$, by Eq. (6.7.33) we can obtain

$$\begin{aligned} E\{\tilde{x}(n)\tilde{x}(n-\tau)\} &= E\left\{\sum_{i=0}^{n_b} b(i)x(n-i) \cdot \sum_{j=0}^{n_b} b(j)x(n-\tau-j)\right\} \\ &= \sum_{i=0}^{n_b}\sum_{j=0}^{n_b} b(i)b(j)E\{x(n-i)x(n-\tau-j)\}, \end{aligned}$$

namely,

$$R_{\tilde{x}}(\tau) = \sum_{i=0}^{n_b}\sum_{j=0}^{n_b} b(i)b(j)R_x(\tau+j-i). \tag{6.7.37}$$

Let $a(1), \cdots, a(2p)$ be the coefficents of the characteristic polynomial $1 + a(1)z^{-1} + \cdots + a(2p)z^{-2p} = 0$ of the harmonic signal $x(n)$. Interestingly, from Eq. (6.7.37), it is easy to obtain

$$\sum_{k=0}^{2p} R_{\tilde{x}}(m-k) = \sum_{i=0}^{n_b}\sum_{j=0}^{n_b} b(i)b(j)\sum_{k=0}^{2p} a(k)R_x(m-k+j-i) = 0, \quad \forall m. \tag{6.7.38}$$

Since the harmonic signal $x(n)$ is a completely predictable process, it obeys the normal equation $\sum_{k=0}^{2p} a(k)R_x(m-k) = 0, \forall m$.

From the statistical independence of $\tilde{x}(n)$ and $\tilde{v}(n)$, we know that $R_{\tilde{y}}(m) = R_{\tilde{x}}(m) + R_{\tilde{v}}(m)$. Substituting this relationship into Eq. (6.7.38), we have

$$\sum_{i=0}^{2p} a(i)R_{\tilde{y}}(m-i) = \sum_{i=0}^{2p} a(i)R_{\tilde{v}}(m-i), \forall m. \tag{6.7.39}$$

Substituting Eq. (6.7.35) into Eq. (6.7.39) yields

$$\sum_{i=0}^{2p} a(i)R_{\tilde{y}}(m-i) = 0, \quad m > 2p + n_d. \tag{6.7.40}$$

It is easy to prove that when $L \geq (p_e + 1)$, the rank of $L \times (p_e + 1)$ matrix

$$\boldsymbol{R}_e = \begin{bmatrix} R_{\tilde{y}}(q_e+1) & R_{\tilde{y}}(q_e) & \cdots & R_{\tilde{y}}(q_e-p_e+1) \\ R_{\tilde{y}}(q_e+2) & R_{\tilde{y}}(q_e+1) & \cdots & R_{\tilde{y}}(q_e-p_e+2) \\ \vdots & \vdots & \vdots & \vdots \\ R_{\tilde{y}}(q_e+L) & R_{\tilde{y}}(q_e+L-1) & \cdots & R_{\tilde{y}}(q_e-p_e+L) \end{bmatrix} \tag{6.7.41}$$

is equal to $2p$, if $p_e \geq 2p$ and $q_e \geq p_e + n_d$ are taken.

Based on the above discussion, a hybrid approach to harmonic retrieval in non-Gaussian ARMA(n_b, n_d) noise can be obtained[245].

Algorithm 6.7.1. *The hybrid approach to harmonic retrieval in non-Gaussian noise*

Step 1 Use the SVD-TLS method and the third-order cumulant of the observed process $y(n)$ to estimate the AR order n_b and the AR parameters $b(1), \cdots, b(n_b)$ of the non-Gaussian ARMA noise.

Step 2 Use the estimated AR parameters $b(1), \cdots, b(n_b)$ and Eq. (6.7.32) to filter the observed data $y(n)$, obtain the filtered observed process $\tilde{y}(n)$.

Step 3 Replace the true autocorrelation $R_{\tilde{y}}(m)$ by the sample autocorrelation $\hat{R}_{\tilde{y}}(m)$ of $\tilde{y}(n)$ in Eq. (6.7.41) and take $p_e \geq 2p$ and $q_e \geq p_e + n_d$ to determine the effective rank of $\hat{\boldsymbol{R}}_e$, say $2p$. Then apply the SVD-TLS method to estimate the AR parameters $a(1), \cdots, a(2p)$ of the harmonic process $x(n)$.

Step 4 Find the roots z_i of the characteristic polynomial $A(z) = 1 + a(1)z^{-1} + \cdots + a(2p)z^{-2p} = 0$ and compute the frequencies of the harmonics (only take the positive frequencies) as follows:

$$\omega_i = \frac{1}{2\pi}\arctan[Im(z_i)/Re(z_i)]. \tag{6.7.42}$$

2. Prefiltering-based ESPRIT Method

In the following, another method to harmonic retrieval in non-Gaussian ARMA noise (prefiltering-based ESPRIT method[244]) is introduced. The core of the ESPRIT method is how to construct the appropriate matrix pencil. Therefore, let us consider two cross-covariance matrices of the filtered observation process $\{\tilde{y}(n)\}$. Denoting

$$\tilde{y}_1(n) = \tilde{y}(n+m+n_d) \tag{6.7.43}$$

$$\tilde{y}_2(n) = \tilde{y}(n+m+n_d+1) \tag{6.7.44}$$

and taking $m > d$ (for the complex harmonic process $d = p$ and for the real harmonic process $d = 2p$).

Construct two $m \times 1$ vectors

$$\begin{aligned}\tilde{\boldsymbol{y}}_1 &= [\tilde{y}_1(n), \cdots, \tilde{y}_1(n+1), \cdots, \tilde{y}_1(n+m-1)]^{\mathrm{T}} \\ &= [\tilde{y}_1(n+m+n_d), \cdots, \tilde{y}_1(n+2m+n_d-1)]^{\mathrm{T}}, \end{aligned} \tag{6.7.45}$$

$$\begin{aligned}\tilde{\boldsymbol{y}}_2 &= [\tilde{y}_2(n), \cdots, \tilde{y}_2(n+1), \cdots, \tilde{y}_2(n+m-1)]^{\mathrm{T}} \\ &= [\tilde{y}_2(n+m+n_d+1), \cdots, \tilde{y}_2(n+2m+n_d)]^{\mathrm{T}}. \end{aligned} \tag{6.7.46}$$

Define

$$\tilde{\boldsymbol{y}}(n) = [\tilde{y}(n), \tilde{y}(n+1), \cdots, \tilde{y}(n+m-1)]^{\mathrm{T}}, \tag{6.7.47}$$

$$\tilde{\boldsymbol{x}}(n) = [\tilde{x}(n), \tilde{x}(n+1), \cdots, \tilde{x}(n+m-1)]^{\mathrm{T}}, \tag{6.7.48}$$

$$\tilde{\boldsymbol{v}}(n) = [\tilde{v}(n), \tilde{v}(n+1), \cdots, \tilde{v}(n+m-1)]^{\mathrm{T}}, \tag{6.7.49}$$

then Eq. (6.7.32) can be written in vector form

$$\tilde{\boldsymbol{y}}(n) = \tilde{\boldsymbol{x}}(n) + \tilde{\boldsymbol{v}}(n), \tag{6.7.50}$$

and

$$\tilde{\boldsymbol{y}}_1(n) = \tilde{\boldsymbol{x}}(n+m+n_d) + \tilde{\boldsymbol{v}}(n+m+n_d), \tag{6.7.51}$$

$$\tilde{\boldsymbol{y}}_2(n) = \tilde{\boldsymbol{x}}(n+m+n_d+1) + \tilde{\boldsymbol{v}}(n+m+n_d+1). \tag{6.7.52}$$

For the complex harmonic signal $x(n) = \sum_{i=1}^{p} \alpha_i e^{j(\omega_i n + \phi_i)}$, its vector form $\boldsymbol{x}(n) = [x(n), x(n+1), \cdots, x(n+m-1)]^{\mathrm{T}}$ can be expressed as

$$\boldsymbol{x}(n) = \boldsymbol{A}\boldsymbol{s}_1(n), \tag{6.7.53}$$

where

$$\boldsymbol{s}_1(n) = [\alpha_1 e^{j(\omega_1 n+\phi_1)}, \cdots, \alpha_p e^{j(\omega_p n+\phi_p)}]^{\mathrm{T}}, \tag{6.7.54}$$

$$\boldsymbol{A} = \begin{bmatrix} 1 & 1 & \cdots & 1 \\ e^{j\omega_1} & e^{j\omega_2} & \cdots & e^{j\omega_p} \\ \vdots & \vdots & \vdots & \vdots \\ e^{j(m-1)\omega_1} & e^{j(m-1)\omega_2} & \cdots & e^{j(m-1)\omega_p} \end{bmatrix}. \tag{6.7.55}$$

Similarly, the vector $\boldsymbol{x}(n-k) = [x(n-k), x(n-k+1), \cdots, x(n-k+m-1)]^{\mathrm{T}}$ can be written as

$$\boldsymbol{x}(n-k) = \boldsymbol{A}\boldsymbol{\Phi}^{-k}\boldsymbol{s}_1(n), \tag{6.7.56}$$

where $\boldsymbol{\Phi}^{-k}$ is the $-k$th-order power of the diagonal matrix $\boldsymbol{\Phi}$ that is defined by

$$\boldsymbol{\Phi} = \mathrm{diag}(e^{j\omega_1}, \cdots, e^{j\omega_p}). \tag{6.7.57}$$

Using Eq. (6.7.56) it is easy to obtain

$$
\begin{aligned}
\tilde{\mathbf{s}}(n) &= \left[\sum_{i=0}^{p} a(i)x(n-i), \cdots, \sum_{i=0}^{p} a(i)x(n+m-1-i)\right]^{\mathrm{T}} \\
&= \sum_{i=0}^{p} a(i)[x(n-i), \cdots, x(n+m-1-i)]^{\mathrm{T}} = \sum_{i=0}^{p} a(i)\mathbf{x}(n-i) \\
&= \sum_{i=0}^{p} a(i)\boldsymbol{A}\boldsymbol{\Phi}^{-i}\mathbf{s}_1(n) = \boldsymbol{A}\left[\boldsymbol{I} + \sum_{i=0}^{p} a(i)\boldsymbol{\Phi}\right]\mathbf{s}_1(n).
\end{aligned} \tag{6.7.58}
$$

Similarly, we have

$$
\tilde{\mathbf{s}}(n+k) = \sum_{i=0}^{p} a(i)\boldsymbol{A}\boldsymbol{\Phi}^{k-i}\mathbf{s}_1(n) = \boldsymbol{A}\boldsymbol{\Phi}^{k}\left[\boldsymbol{I} + \sum_{i=1}^{p} a(i)\boldsymbol{\Phi}^{-i}\right]\mathbf{s}_1(n). \tag{6.7.59}
$$

Let

$$
\boldsymbol{\Phi}_1 = \boldsymbol{I} + \sum_{i=1}^{p} a(i)\boldsymbol{\Phi}^{-i}, \tag{6.7.60}
$$

then Eqs. (6.7.58) and (6.7.59) can be simplified as

$$
\tilde{\mathbf{s}}(n) = \boldsymbol{A}\boldsymbol{\Phi}_1\mathbf{s}_1(n), \tag{6.7.61}
$$

$$
\tilde{\mathbf{s}}(n+k) = \boldsymbol{A}\boldsymbol{\Phi}^{k}\boldsymbol{\Phi}_1\mathbf{s}_1(n). \tag{6.7.62}
$$

Using Eqs. (6.7.61) and (6.7.62), Eq. (6.7.50)∼ (6.7.52) can be rewritten as

$$
\tilde{\mathbf{y}}(n) = \boldsymbol{A}\boldsymbol{\Phi}_1\mathbf{s}_1(n) + \tilde{\mathbf{v}}(n), \tag{6.7.63}
$$

$$
\tilde{\mathbf{y}}_1(n) = \boldsymbol{A}\boldsymbol{\Phi}^{m+n_d}\boldsymbol{\Phi}_1\mathbf{s}_1(n) + \tilde{\mathbf{v}}(n+m+n_d), \tag{6.7.64}
$$

$$
\tilde{\mathbf{y}}_2(n) = \boldsymbol{A}\boldsymbol{\Phi}^{m+n_d+1}\boldsymbol{\Phi}_1\mathbf{s}_1(n) + \tilde{\mathbf{v}}(n+m+n_d+1). \tag{6.7.65}
$$

Proposition 6.7.3. [244] *Let* $\boldsymbol{R}_{\tilde{y},\tilde{y}_i}$ *be the cross-covariance matrix of the vector processes* $\tilde{\mathbf{y}}$ *and* $\tilde{\mathbf{y}}_i$ *(*$i = 1, 2$*), namely,* $\boldsymbol{R}_{\tilde{y},\tilde{y}_i} = E\{\tilde{\mathbf{y}}\tilde{\mathbf{y}}_i^H\}$*, where the superscript H denotes the conjugate transposition of the vector. If let*

$$
\boldsymbol{S} = E\{\mathbf{s}_1\mathbf{s}_1^H\}, \tag{6.7.66}
$$

then we have

$$
\boldsymbol{R}_{\tilde{y},\tilde{y}_1} = \boldsymbol{A}\boldsymbol{\Phi}_1\boldsymbol{S}\boldsymbol{\Phi}_1^H(\boldsymbol{\Phi}^{m+n_d})^H\boldsymbol{A}^H, \tag{6.7.67}
$$

$$
\boldsymbol{R}_{\tilde{y},\tilde{y}_2} = \boldsymbol{A}\boldsymbol{\Phi}_1\boldsymbol{S}\boldsymbol{\Phi}_1^H(\boldsymbol{\Phi}^{m+n_d+1})^H\boldsymbol{A}^H. \tag{6.7.68}
$$

It is easy to prove that the structures of the matrices $\boldsymbol{R}_{\tilde{y},\tilde{y}_1}$ and $\boldsymbol{R}_{\tilde{y},\tilde{y}_2}$ in Proposition 6.7.3 are

$$\boldsymbol{R}_{\tilde{y},\tilde{y}_1} = \begin{bmatrix} R_{\tilde{y}}(m+n_d) & R_{\tilde{y}}(m+n_d+1) & \cdots & R_{\tilde{y}}(2m+n_d-1) \\ R_{\tilde{y}}(m+n_d-1) & R_{\tilde{y}}(m+n_d) & \cdots & R_{\tilde{y}}(2m+n_d-2) \\ \vdots & \vdots & \vdots & \vdots \\ R_{\tilde{y}}(n_d+1) & R_{\tilde{y}}(n_d+2) & \cdots & R_{\tilde{y}}(n_d+m) \end{bmatrix}, \quad (6.7.69)$$

$$\boldsymbol{R}_{\tilde{y},\tilde{y}_2} = \begin{bmatrix} R_{\tilde{y}}(m+n_d+1) & R_{\tilde{y}}(m+n_d+2) & \cdots & R_{\tilde{y}}(2m+n_d) \\ R_{\tilde{y}}(m+n_d) & R_{\tilde{y}}(m+n_d+1) & \cdots & R_{\tilde{y}}(2m+n_d-1) \\ \vdots & \vdots & \vdots & \vdots \\ R_{\tilde{y}}(n_d+2) & R_{\tilde{y}}(n_d+3) & \cdots & R_{\tilde{y}}(n_d+m+1) \end{bmatrix}. \quad (6.7.70)$$

Theorem 6.7.1. [244] *Define $\boldsymbol{\Gamma}$ as the generalized eigenvalue matrix associated with the matrix pencil $\{\boldsymbol{R}_{\tilde{y},\tilde{y}_1}, \boldsymbol{R}_{\tilde{y},\tilde{y}_2}\}$, then the following results are true*

(1) The $m \times d$ Vandermonde matrix $\boldsymbol{A}$ defined in Eq. (6.7.55) is full rank, and the $d \times d$ diagonal matrix $\boldsymbol{S}$ is nonsingular, i.e., $\mathrm{rank}(\boldsymbol{A}) = \mathrm{rank}(\boldsymbol{S}) = d$;

(2) $m \times m$ matrix $\boldsymbol{\Gamma}$ is related to the diagonal matrix $\boldsymbol{\Phi}$ by

$$\boldsymbol{\Gamma} = \begin{bmatrix} \boldsymbol{\Phi} & \boldsymbol{O} \\ \boldsymbol{O} & \boldsymbol{O} \end{bmatrix} \quad (6.7.71)$$

where $\boldsymbol{O}$ is the zero matrix and the diagonal entries of $\boldsymbol{\Phi}$ may be permutated.

Theorem 6.7.1 leads to the prefiltering-based ESPRIT method for the harmonic retrieval in non-Gaussian noise [244].

Algorithm 6.7.2. *Prefiltering-based ESPRIT Method*

Step 1 Cumulant estimation: Estimate the third-order cumulant $c_{3y}(\tau_1, \tau_2)$ from the observed data $y(1), \cdots, y(N)$.

Step 2 AR modeling of the noise: Use SVD-TLS method and the third-order cumulant $c_{3y}(\tau_1, \tau_2)$ to estimate the AR order n_b and AR parameters $b(1), \cdots, b(n_b)$ of the non-Gaussian ARMA noise.

Step 3 Prefiltering: Use the estimated AR parameters $b(1), \cdots, b(n_b)$ and Eq. (6.7.32) to filter the observed data $y(n)$ and obtain the filtered observed process $\tilde{y}(n)$.

Step 4 Cross-covariance matrices compuatation: Estimate the autocorrelation $R_{\tilde{y}}(\tau)$ of the filtered process and construct the cross-covariance matrices $\boldsymbol{R}_{\tilde{y},\tilde{y}_1}$ and $\boldsymbol{R}_{\tilde{y},\tilde{y}_2}$ using Eqs. (6.7.69) and (6.7.70).

Step 5 TLS-ESPRIT algorithm: Compute the SVD $\boldsymbol{R}_{\tilde{y},\tilde{y}_1} = \boldsymbol{U}\boldsymbol{\Sigma}\boldsymbol{V}^H$ of $\boldsymbol{R}_{\tilde{y},\tilde{y}_1}$ to determine its effective rank, yielding the estimation of the number of harmonics d. The left and right singular matrices and singular value matrices corresponding to the d large singular values are $\boldsymbol{U}$, $\boldsymbol{V}$, and $\boldsymbol{\Sigma}$, respectively. Then, Compute the d non-zero generalized eigenvalues, say, $\gamma_1, \cdots, \gamma_d$, of the matrix pencil $\{\boldsymbol{\Sigma}_1, \boldsymbol{U}_1^H \boldsymbol{R}_{\tilde{y},\tilde{y}_2} \boldsymbol{V}_1\}$. The harmonic *frequencies are given by*

$$\omega_i = \frac{1}{2\pi} \arctan[Im(y_i)/Re(y_i)], \quad i = 1, \cdots, d. \tag{6.7.72}$$

The method for harmonic retrieval in mixed colored Gaussian and non-Gaussian noises can be found in reference [243].

6.8 The Adaptive Filtering of Non-Gaussian Signal

Some typical adaptive filtering algorithms are presented in Chapter 5. The application of these algorithms assumes that the additive noise is white since the cost functions used (i.e., the mean-squared error or the weighted sum of squared errors) are second-order statistics. Therefore, in order to perform adaptive filtering of non-Gaussian signals in the presence of additive colored noise, it is necessary to use the higher-order statistics and modify the form of the cost function.

Assuming adaptive filtering of the observed data signal $x(n)$ using m weight coefficients $w_1, \cdots, w_m$, the principle of determining the optimal weight coefficients by MMSE criterion is to minimize the mean square error

$$J_1 \stackrel{\text{def}}{=} E\{|e(n)|^2\} = E\left\{\left|x(n) - \sum_{i=1}^{m} w_i x(n-i)\right|^2\right\}, \tag{6.8.1}$$

where

$$e(n) = x(n) - \sum_{i=1}^{m} w_i x(n-i) \tag{6.8.2}$$

is the error signal. The solution to this optimization problem is the Weiner filter

$$\boldsymbol{w}_{opt} = \boldsymbol{R}^{-1}\boldsymbol{r}, \tag{6.8.3}$$

where $\boldsymbol{R} = E\{\boldsymbol{x}(n)\boldsymbol{x}^{\mathrm{H}}(n)\}$ and $\boldsymbol{r} = E\{\boldsymbol{x}(n)x(n)\}$, and they are sensitive to additive noises.

If $x(n)$ is the output of a linear system $H(z)$ drived by an independently identically distributed noise $v(n) \sim \mathrm{IID}(0, \sigma_v^2, \gamma_{3v})$, then the power spectrum is $P_x(\omega) = \sigma_v^2 H(\omega)H^*(\omega)$ and the bispectrum is $B(\omega_1, \omega_2) = \gamma_{3v}H(\omega_1)H(\omega_2)H^*(\omega_1 + \omega_2)$. Consider a special bispectrum slice with $\omega_1 = \omega$ and $\omega_2 = \omega$, it is clearly that

$$P_x(\omega) = \frac{B(\omega, 0)}{\gamma_{3v}H(0)/\sigma_v^2} = \alpha B(\omega, 0), \tag{6.8.4}$$

where $\alpha = \sigma_v^2/[\gamma_{3v}H(0)]$ is a constant.

Taking the inverse Fourier transform of Eq. (6.8.4) yields

$$R_x(\tau) = \alpha \sum_{m=-\infty}^{\infty} c_{3x}(\tau, m). \tag{6.8.5}$$

This relationship inspires a cost function immune to the additive colored Gaussian noise, i.e., the correlation function $R_x(\tau)$ is replaced by Eq. (6.8.5). Eq. (6.8.5) can be equivalently written as

$$E\{x(n)x(n+\tau)\} = \alpha \sum_{m=-\infty}^{\infty} E\{x(n)x(n+\tau)x(n+m)\}. \tag{6.8.6}$$

In particular, if $\tau = 0$, then

$$E\{|x(n)|^2)\} = \alpha \sum_{m=-\infty}^{\infty} E\{x(n+m)|x(n)|^2)\}. \tag{6.8.7}$$

Similarly, define

$$E\{|e(n)|^2)\} = \alpha \sum_{m=-\infty}^{\infty} E\{e(n+m)|e(n)|^2)\}, \tag{6.8.8}$$

then the cost function J_1 defined in Eq. (6.8.1) can be modified as

$$J_2 \stackrel{\text{def}}{=} \sum_{m=-\infty}^{\infty} E\{x(n+m)|e(n)|^2)\}, \tag{6.8.9}$$

where the error signal $e(n)$ is difined by Eq. (6.8.2).

The cost function Eq. (6.8.9) is proposed by Delopoulos and Giannakis [74], they also prove that $J_2 = \alpha J_1$, where $\alpha = \sigma_v^2/[\gamma_{3v}H(0)]$.

Unlike the J_1 criterion, the criterion J_2 uses the third-order cumulant of $x(n)$. Therefore, not only does it theoretically suppresses the colored Gaussian noise completely, but also suppresses the colored non-Gaussian noise of symmetric distribution. Because the third-order cumulant of this colored non-Gaussian noise is identically zero.

6.9 Time Delay Estimation

Time delay estimation is an important problem in the applications such as sonar, radar, biomedicine, and geophysics. For example, target localization in sonar and radar, stratigraphic structures (such as dams, etc.), and other issues need to determine the time delay between the received signals of two sensors or the time delay between the received signals relative to the transmitted signals. These two kinds of time delays are referred to as time delays.

6.9.1 The Generalized Correlation Mehtod

Consider two spatially separated sensors whose observations $x(n)$ and $y(n)$ satisfy

$$x(n) = s(n) + w_1(n), \tag{6.9.1}$$

$$y(n) = s(n-D) + w_2(n), \tag{6.9.2}$$

where $s(n)$ is real, and $s(n-D)$ denotes the delayed signal of $s(n)$ where the time delay is D and may also include the amplitude factor α. $w_1(n)$ and $w_2(n)$ are the observed noise of the two sensors, respectively, which are statistically independent of each other and uncorrelated with the signal $s(n)$. The problem of time delay estimation is to estimate the time delay parameter D by using the observed data $x(n)$ and $y(n)$ (where $n = 1, \cdots, N$).

Essentially, time delay estimation is to find the time difference (lag) of the maximum similarity between two signals. In signal processing, "finding the similarity between the two" can be translated as "finding the cross correlation function between them". For this purpose, we examine the cross correlation function

$$\begin{aligned} R_{xy}(\tau) &\stackrel{\text{def}}{=} E\{x(n)y(n+\tau)\} \\ &= E\{[s(n)+w_1(n)][s(n+\tau-D)+w_2(n+\tau)]\} \\ &= R_{ss}(\tau-D) \end{aligned} \tag{6.9.3}$$

between $x(n)$ and $y(n)$, where $R_{ss}(\tau) = E\{s(n)s(n+\tau)\}$ is the autocorrelation of the signal $s(n)$. Since the autocorrelation has the property that $R_{ss}(\tau) \le R_{ss}(0)$, the cross correlation takes the maximum value at $\tau = D$. In other words, the lag τ at which the cross correlation takes its maximum value gives an estimation of the time delay D.

Since the estimated cross correlation functions using $x(n)$ and $y(n)$ may have relatively large biases, in order to obtain better time delay estimation, it is necessary to smooth the estimated cross correlation function. For example, using

$$R_{xy}(\tau) = \mathcal{F}^{-1}[P_{xy}(\omega)W(\omega)] = R_{xy}(\tau) * w(\tau) \tag{6.9.4}$$

to estimate the time delay parameter D, where $*$ denotes convolution, and $P_{xy}(\omega) = \mathcal{F}[R_{xy}(\tau)]$ is the Fourier transform of the cross correlation $R_{xy}(\tau)$, i.e., the cross power spectrum of $x(n)$ and $y(n)$. This method, proposed by Knapp and Carter, is known as the generalized correlation method[125].

The key of the generalized correlation method is the choice of the smoothed function $w(n)$. The following are some typical window functions.

1. The Smoothed Coherence Transform Window

Knapp and Carter proposed to use the window function[125]

$$W(\omega) = \frac{1}{\sqrt{P_x(\omega)P_y(\omega)}} = H_1(\omega)H_2(\omega), \tag{6.9.5}$$

where $P_x(\omega)$ and $P_y(\omega)$ are the power spectra of $x(n)$ and $y(n)$ respectively, and

$$H_1(\omega) = \frac{1}{\sqrt{P_x(\omega)}}, \tag{6.9.6}$$

$$H_2(\omega) = \frac{1}{\sqrt{P_y(\omega)}}. \tag{6.9.7}$$

Since $y(n) = s(n - D)$ is the coherent signal of $x(n) = s(n)$ in the noiseless case, $H_1(\omega)H_2(\omega)$ can be regarded as a coherent transformation, which is exactly the reason for calling the smoothed window function $W(\omega)$ a smoothed coherent transformation window.

2.The Maximum Likelihood Window or Hannan-Thompson Window

Chan et al.[49] proposed that the window function takes

$$W(\omega) = \frac{z(\omega)}{|P_{xy}(\omega)|} = \frac{1}{|P_{xy}(\omega)|}\frac{|\gamma_{xy}(\omega)|}{1 - |\gamma_{xy}(\omega)|^2}, \tag{6.9.8}$$

where

$$|\gamma_{xy}(\omega)|^2 = \frac{|P_{xy}(\omega)|^2}{P_x(\omega)P_y(\omega)} \tag{6.9.9}$$

is the correlation coefficient of the magnitude square, which takes values between 0 and 1. Chan et al. proved that for uncorrelated Gaussian processes $x(n)$ and $y(n)$ with zero mean, the function

$$z(\omega) = \frac{|\gamma_{xy}(\omega)|}{1 - |\gamma_{xy}(\omega)|^2} \propto \frac{1}{\text{the phase variance of } P_{xy}(\omega)}, \tag{6.9.10}$$

where $a \propto b$ denotes that a is proportional to b.

In the generalized cross correlation method using the maximum likelihood window, the lag τ corresponding to the maximum magnitude of the cross correlation function

$$R_{xy}(\tau) = \mathcal{F}^{-1}\left[\frac{P_{xy}(\omega)}{|P_{xy}(\omega)|}z(\omega)\right] \tag{6.9.11}$$

is used as the estimation of the time delay D.

6.9.2 Higher-Order Statistics Method

In many practical applications (such as passive and active sonar), signal $s(n)$ is often a non-Gaussian process, while the additive noise $w_1(n)$ and $w_2(n)$ are Gaussian processes[188, 220]. In such cases, it is more reasonable to use the higher-order cumulants for time delay estimation, since the Gaussian noise can be theoretically completely suppressed.

Let $x(n)$ and $y(n)$ be the observed data with zero mean, then the third-order cumulant of $x(n)$ is

$$c_{3x}(\tau_1, \tau_2) \stackrel{\text{def}}{=} E\{x(n)x(n + \tau_1)x(n + \tau_2)\} = c_{3s}(\tau_1, \tau_2), \tag{6.9.12}$$

where

$$c_{3s}(\tau_1, \tau_2) \stackrel{\text{def}}{=} E\{s(n)s(n + \tau_1)s(n + \tau_2)\}. \tag{6.9.13}$$

Define the cross third-order cumulant of the signal $x(n)$ and $y(n)$ be

$$c_{xyx}(\tau_1, \tau_2) \stackrel{\text{def}}{=} E\{x(n)y(n + \tau_1)x(n + \tau_2)\} = c_{3s}(\tau_1 - D, \tau_2). \tag{6.9.14}$$

Taking the 2-D Fiourier transform of Eqs. (6.9.12) and (6.9.14) yields the bispectrum

$$P_{3x}(\omega_1, \omega_2) = P_{3s}(\omega_1, \omega_2) \tag{6.9.15}$$

and the cross bispectrum

$$P_{xyx}(\omega_1, \omega_2) = P_{3s}(\omega_1, \omega_2) e^{j\omega_1 D}. \tag{6.9.16}$$

In addition, we have

$$P_{3x}(\omega_1, \omega_2) = |P_{3x}(\omega_1, \omega_2)| e^{j\phi_{3x}(\omega_1, \omega_2)}, \tag{6.9.17}$$

$$P_{xyx}(\omega_1, \omega_2) = |P_{xyx}(\omega_1, \omega_2)| e^{j\phi_{xyx}(\omega_1, \omega_2)}, \tag{6.9.18}$$

where $\phi_{3x}(\omega_1, \omega_2)$ and $\phi_{xyx}(\omega_1, \omega_2)$ represent the phase of the bispectrum $P_{3x}(\omega_1, \omega_2)$ and $P_{xyx}(\omega_1, \omega_2)$, respectively.

The following are several higher-order statistics methods for time delay estimation.

1. Non-parametric Bispectrum Method 1[188, 206]

Define the following function

$$I(\omega_1, \omega_2) = \frac{P_{xyx}(\omega_1, \omega_2)}{P_{3x}(\omega_1, \omega_2)}. \tag{6.9.19}$$

Substituting Eqs. (6.9.15) and (6.9.16) into Eq. (6.9.19), then the function $I(\omega_1, \omega_2)$ can be rewritten as

$$I(\omega_1, \omega_2) = e^{j\omega_1 D}. \tag{6.9.20}$$

Therefore

$$T_1(\tau) = \int_{-\infty}^{\infty}\int_{-\infty}^{\infty} I_1(\omega_1, \omega_2) e^{-j\omega_1 \tau} d\omega_1 d\omega_2 = \int_{-\infty}^{\infty} d\omega_2 \int_{-\infty}^{\infty} e^{j\omega_1 (D-\tau)} d\omega_1 \tag{6.9.21}$$

peaks at $\tau = D$.

2. Non-parametric Bispectrum Method 2[165]

Using the difference between the phase of the cross bispectrum $P_{xyx}(\omega_1, \omega_2)$ and the phase of the bispectrum $P_{3x}(\omega_1, \omega_2)$, a new phase

$$\phi(\omega_1, \omega_2) = \phi_{xyx}(\omega_1, \omega_2) - \phi_{3x}(\omega_1, \omega_2) \tag{6.9.22}$$

can be defined. Then, using this new phase, a new function

$$I_2(\omega_1, \omega_2) = e^{j\phi(\omega_1, \omega_2)} \tag{6.9.23}$$

can be constructed. From Eq. (6.9.15) to Eq. (6.9.18), it follows that $I_2(\omega_1, \omega_2) = e^{j\omega_1 D}$, hence the function

$$T_2(\tau) = \int_{-\infty}^{\infty}\int_{-\infty}^{\infty} I_2(\omega_1, \omega_2) e^{-j\omega_1 \tau} d\omega_1 d\omega_2 = \int_{-\infty}^{\infty} d\omega_2 \int_{-\infty}^{\infty} e^{j\omega_1 (D-\tau)} d\omega_1 \tag{6.9.24}$$

also peaks at $\tau = D$.

Algorithm 6.9.1. *Non-parametric bispectrum method for time delay estimation*

Step 1 Divide the N data samples into K segments, each segment contains M data and there is 50% data overlap between two adjacent segments, i.e., $N = KM/2$. The data in the kth segment are denoted as $x^{(k)}(n)$ and $y^{(k)}(n)$, where $k = 1, \cdots, K$; $n = 0, 1, \cdots, M-1$.

Step 2 Compute the discrete Fourier transform for each segment of data

$$X^{(k)}(\omega) = \sum_{n=0}^{M-1} x^{(k)}(n) e^{-j\frac{n\omega}{M}}, \tag{6.9.25}$$

$$Y^{(k)}(\omega) = \sum_{n=0}^{M-1} y^{(k)}(n) e^{-j\frac{n\omega}{M}}, \tag{6.9.26}$$

where $k = 1, \cdots, K$.

Step 3 Estimate the bispectrum and the cross bispectrum for each segment separately

$$P_{3x}^{(k)}(\omega_1, \omega_2) = X^{(k)}(\omega_1) X^{(k)}(\omega_2) X^{(k)*}(\omega_1 + \omega_2), \tag{6.9.27}$$

$$P_{xyx}^{(k)}(\omega_1, \omega_2) = X^{(k)}(\omega_1) Y^{(k)}(\omega_2) X^{(k)*}(\omega_1 + \omega_2), \tag{6.9.28}$$

where $k = 1, \cdots, K$, and $X^{(k)*}(\omega)$ denotes the complex conjugate of $X^{(k)}(\omega)$.

Step 4 Smoothing the K bispectrum and the cross bispectrum to obtain the bispectrum and the cross bispectrum of N data

$$\hat{P}_{3x}(\omega_1, \omega_2) = \frac{1}{K} \sum_{k=1}^{K} P_{3x}^{(k)}(\omega_1, \omega_2), \tag{6.9.29}$$

$$\hat{P}_{xyx}(\omega_1, \omega_2) = \frac{1}{K} \sum_{k=1}^{K} P_{xyx}^{(k)}(\omega_1, \omega_2). \tag{6.9.30}$$

Step 5 Compute the phases of the bispectrum and the cross bispectrum

$$\hat{\phi}_{3x}(\omega_1, \omega_2) = \arctan \left\{ \frac{Im[\hat{P}_{3x}(\omega_1, \omega_2)]}{Re[\hat{P}_{3x}(\omega_1, \omega_2)]} \right\}, \tag{6.9.31}$$

$$\hat{\phi}_{xyx}(\omega_1, \omega_2) = \arctan \left\{ \frac{Im[\hat{P}_{xyx}(\omega_1, \omega_2)]}{Re[\hat{P}_{xyx}(\omega_1, \omega_2)]} \right\}. \tag{6.9.32}$$

Step 6 Compute

$$\hat{\phi}(\omega_1, \omega_2) = \hat{\phi}_{xyx}(\omega_1, \omega_2) - \hat{\phi}_{3x}(\omega_1, \omega_2) \tag{6.9.33}$$

and construct

$$\hat{I}_2(\omega_1, \omega_2) = e^{j\hat{\phi}(\omega_1, \omega_2)}. \tag{6.9.34}$$

Step 7 Compute

$$\hat{T}_2(\tau) = \sum_{\omega_1=0}^{M-1} \sum_{\omega_2=0}^{M-1} \hat{I}_2(\omega_1, \omega_2) e^{-j\omega_1 \tau}. \tag{6.9.35}$$

Step 8 Choose the τ when $\hat{T}_2(\tau)$ takes its maximum as the estimation $\hat{D}$ of the time delay.

Obviously, by replacing Eq. (6.9.34) in the above algorithm with Eq. (6.9.19), the algorithm for estimating the time delay parameter by Sasaki et al. is obtained[188].

3. The Cross Bicepstrum Method

The above methods are not direct estimations of the time delay parameters and belong to the category of non-direct methods. A direct method proposed in Ref.[164], which uses both the cross bicepstrum and the cross bispectrum, is presented below.

From Eqs. (6.9.19) and (6.9.20), Eq. (6.9.19) can be rewritten in Z-transform form as

$$\frac{P_{3x}(z_1, z_2)}{P_{xyx}(z_1, z_2)} = z_1^{-D}, \tag{6.9.36}$$

where $z_1^{-D} = e^{j\omega_1 D}$, and

$$P_{3x}(\omega_1, \omega_2) = P_{3x}(z_1, z_2)|_{z_1=e^{-j\omega_1}, z_2=e^{-j\omega_2}}, \tag{6.9.37}$$

$$P_{xyx}(\omega_1, \omega_2) = P_{xyx}(z_1, z_2)|_{z_1=e^{-j\omega_1}, z_2=e^{-j\omega_2}}. \tag{6.9.38}$$

If we take the complex logarithm of Eq. (6.9.36), then we have

$$\ln[P_{xyx}(z_1, z_2)] - \ln[P_{3x}(z_1, z_2)] = -D\ln[z_1], \tag{6.9.39}$$

where $\ln[P_{3x}(z_1, z_2)]$ and $\ln[P_{xyx}(z_1, z_2)]$ are called the bicepstrum of $x(n)$ and the cross bicepstrum of $x(n)$ and $y(n)$, respectively. Taking the partial derivative of Eq. (6.9.39) with respect to z_1, we can obtain

$$\frac{1}{P_{xyx}(z_1, z_2)} \frac{\partial P_{xyx}(z_1, z_2)}{\partial z_1} - \frac{1}{P_{3x}(z_1, z_2)} \frac{\partial P_{3x}(z_1, z_2)}{\partial z_1} = \frac{D}{z_1}, \tag{6.9.40}$$

and the corresponding time-domain representation is

$$c_{3x}(m, n) \star [m \cdot c_{xyx}(m, n)] - c_{xyx}(m, n) \star [m \cdot m_{3x}(m, n)] = Dc_{xyx}(m, n) \star c_{3x}(m, n).$$

Taking the 2-D Fourier transform on both sides of the above equation with respect to variables m and n yields

$$D(\omega_1, \omega_2) = \frac{\mathfrak{F}_2[m \cdot c_{xyx}(m, n)]}{\mathfrak{F}_2[c_{xyx}(m, n)]} - \frac{\mathfrak{F}_2[m \cdot c_{3x}(m, n)]}{\mathfrak{F}_2[c_{3x}(m, n)]}, \tag{6.9.41}$$

where

$$\mathfrak{F}_2[m \cdot c_{xyx}(m, n)] = \int_{-\infty}^{\infty} m \cdot c_{xyx}(m, n) e^{-j(\omega_1 m + \omega_2 n)} dm dn,$$

$$\mathfrak{F}_2[m \cdot c_{3x}(m, n)] = \int_{-\infty}^{\infty} m \cdot c_{3x}(m, n) e^{-j(\omega_1 m + \omega_2 n)} dm dn.$$

The peak of $D(\omega_1, \omega_2)$ gives the estimation of the time delay parameter D. Note that in order to reduce the estimation error and variance, a segmented smoothing method can be adopted: $\hat{D}^{(k)}(\omega_1, \omega_2)$ for the K segments of data is first found, then

$$\hat{D}(\omega_1, \omega_2) = \frac{1}{K}\sum_{k=1}^{K} \hat{D}^{(k)}(\omega_1, \omega_2) \tag{6.9.42}$$

is calculated, and finally its peak value is used as the estimation of the time delay D.

4. The Fourth-order Statistics Method

The time delay estimation problem can be solved by optimization methods. Chan et al. pointed out [50] that the generalized cross correlation method without weighting, i.e., adding a rectangular window function, is equivalent to selecting the time delay D to minimize the cost function

$$J_2(D) = E\{[x(n-D) - y(n)]^2\}. \tag{6.9.43}$$

Tugnait generalizes the above cost function to fourth-order statistics and proposes that using the cost function [210]

$$J_4(D) = E\{[x(n-D) - y(n)]^4\} - 3(E\{[x(n-D) - y(n)]^2\})^2, \tag{6.9.44}$$

the desired time delay D_0 is defined as the solution to the following optimization problem:

(1) If the kurtosis of signal $x(n)$

$$\gamma_{4s} = E\{s^4(n)\} - 3\sigma_s^4 \tag{6.9.45}$$

is greater than zero, then D_0 is the solution that minimizes $J_4(D)$;

(2) If the kurtosis γ_{4s} is smaller than zero, then D_0 is the solution that maximizes $J_4(D)$.

In practice, the cost function is

$$\hat{J}_4(D) = \frac{1}{N}\sum_{n=1}^{N}[x(n-D) - y(n)]^4 - 3\left(\frac{1}{N}\sum_{n=1}^{N}[x(n-D) - y(n)]^2\right)^2, \tag{6.9.46}$$

where kurtosis γ_{4s} can be estimated by [210]

$$\hat{\gamma}_{4s} = \frac{1}{2}(A_x + A_y), \tag{6.9.47}$$

where

$$A_x = \frac{1}{N}\sum_{n=1}^{N} x^4(n) - 3\left[\frac{1}{N}\sum_{n=1}^{N} x^2(n)\right]^2, \tag{6.9.48}$$

$$A_y = \frac{1}{N}\sum_{n=1}^{N} y^4(n) - 3\left[\frac{1}{N}\sum_{n=1}^{N} y^2(n)\right]^2. \tag{6.9.49}$$

There are still some better methods for time delay estimation, which will not be discussed here due to space. The reader can refer to references [166] and [69] for more details.

6.10 Application of Bispectrum in Signal Classification

As the name suggests, signal classification is to classify the signals with unknown attributes into several types; object identification is to classify and recognize the target signals with unknown attributes. Automatic signal classification and object identification may be subdivided into three phases: (1)data acquisition; (2) data representation; (3)data classification. In the data representation phase, some salient features are extracted from the collected data. Such a representation is called a feature vector. In the classification phase, the feature vector of an unknown signal is compared with those of known signals stored in the database to determine the type.

Because the phase information is provided, higher-order statistics (especially bispectra) were widely used in feature extraction. The direct use of bispectra results in a complex 2-D matching, which limits the application of the direct bispectrum in real-time object identification. To overcome this difficulty, it is necessary to transform the bispectra into a 1-D function or other characteristic function that facilitates real-time applications. This is the main content that will be presented in this section.

6.10.1 The Integrated Bispectra

In many important applications (such as radar target recognition), it is usually required that the extracted signal features have time translation invariance, scale variant, and phase preserving (that is, the phase information cannot be destroyed) for the following reasons:

(1) Aircraft, especially fighter aircraft, are usually in maneuvering flight. If the features of the target aircraft change with the attitude of the aircraft, i.e., the features have time translation variance, then this is undoubtedly very unfavorable to the radar target recognition.
(2) Different aircraft have different geometries (especially length and wing width). If the signal features contain the aircraft scale information, it will be helpful for radar target recognition.
(3) The phase information reflects the radiation and scattering characteristics of the aircraft to the electromagnetic waves, which are directly related to the skin and critical parts of the aircraft (such as engines, radomes, air vents, etc.).

The bispectrum has exactly the above three properties because the cumulant and multispectra retain the amplitude and phase information of the signal and are independent of time. In addition, the cumulant and multispectra can suppress any colored Gaussian noise. However, since the bispectrum is a 2-D function, directly using all the bispectrum of the signal as the signal features will lead to 2-D template matching, which is computationally intensive and cannot meet the requirements of real-time object identification. Obviously, one way to overcome this difficulty is to transform the 2-D bispectrum into a 1-D function. There are three methods to achieve this transformation of the bispectrum. The transformed bispectrum is referred to as the integrated bispectrum.

1. Radially Integrated Bispectra (RIB)

Chandran and Elgar[51] were the first to propose using the phase of radially integrated bispectra (PRIB)

$$\mathrm{PRIB}(a) = \arctan\left(\frac{I_i(a)}{I_r(a)}\right) \tag{6.10.1}$$

as the signal features, where

$$I(a) = I_i(a) + jI_r(a) = \int_0^{1/(1+a)} B(f_1, af_1)df_1 \tag{6.10.2}$$

is the integration of the bispectra along radial lines passing through the origin in bifrequency space. Here $0 < a \le 1$ and $j = \sqrt{-1}$ is an imaginary number.

$I(a)$ defined by Eq. (6.10.2) is called radially integrated bispectra, and $\mathrm{PRIB}(a)$ is the phase of the integrated bispectra. In the PRIB method, the intraclass mean and intraclass variance of $\mathrm{PRIB}(a)$ need to be computed for each class of know signal over the entire training data set. Moreover, the interclass mean and interclass variance between two classes of the signal also need to be computed. Then, the K sets of phases of integrated bispectra $\mathrm{PRIB}(a)$ that maximize

$$\text{Interclass separation} = \frac{\text{the mean-square difference between PRIB}(a)\text{ for two classes}}{\text{the sum of the two intraclass standard deviations}} \tag{6.10.3}$$

are chosen as the characteristic parameters combinations $P(1), \cdots, P(K)$. These selected integration paths a and the corresponding integrated bispectra phase $\mathrm{PRIB}(a)$ are stored as the feature parameters of the signal and used as model features. In the test phase, the values of $\mathrm{PRIB}(a)$ on these integration paths are computed for the test samples and compared to the values of $\mathrm{PRIB}(a)$ of the various types of signals on the model. Finally, the known signal with the greatest similarity is selected as the class of the test sample.

Since only the phase of the radially integrated bispectrum is taken, the amplitude information of the bispectrum is lost, i.e., $\mathrm{PRIB}(a)$ loses the scale variance. Obviously, if the radially integrated bispectrum

$$\mathrm{RIB}(a) = I(a) = \int_0^{1/(1+a)} B(f_1, af_1)df_1 \tag{6.10.4}$$

is taken directly as the feature parameter, RIB(a) will be time-shift invariant, scale invariant and phase preserving at the same time.

2. Axially Integrated Bispectra (AIB)

Another integrated bispectra method was developed by Tugnait [211] and the integrated paths are parallel to the ω_1 or ω_2 axes.

Let $B(\omega_1, \omega_2)$ be the bispectrum of the signal $x(t)$, and denoted as

$$y(t) \stackrel{\text{def}}{=} x^2(t) - E\{x^2(t)\} \quad \text{and} \quad \tilde{y}(t) \stackrel{\text{def}}{=} x^2(t), \tag{6.10.5}$$

then the axially integrated bispectra (AIB) is defined as

$$\text{AIB}(\omega) = P_{\tilde{y}x}(\omega) = \frac{1}{2\pi}\int_{-\infty}^{\infty} B(\omega, \omega_2)d\omega_2 = \frac{1}{2\pi}\int_{-\infty}^{\infty} B(\omega_1, \omega)d\omega_1, \tag{6.10.6}$$

where

$$P_{\tilde{y}x}(\omega) \stackrel{\text{def}}{=} \int_{-\infty}^{\infty} E\{\tilde{y}(t)x(t+\tau)\}e^{-j\omega\tau}d\tau. \tag{6.10.7}$$

The estimation variance of AIB(ω) is equal to the estimation variance of the power spectrum thus much less than that of the bispectrum. In addition, AIB(ω) retains the amplitude information of the bispectrum and thus has the scale variance. However, a shortcoming of the AIB method is that the AIB loses most of the phase information of the bispectrum. From Eqs. (6.10.6) and (6.10.7), it is easy to see that

$$\text{AIB}(\omega) = \int_{-\infty}^{\infty} c_{3x}(0, \tau)e^{-j\omega\tau}d\tau, \tag{6.10.8}$$

since $E\{\tilde{y}(t)x(t)\} = E\{x(t)x(t)x(t+\tau)\} = c_{3x}(0, \tau)$, that is AIB($\omega$) is only the Fourier transform of the cumulant slice $c_{3x}(0, \tau)$, while the phase information of the bispectra corresponding to the other slices $c_{3x}(m, \tau)$, $m \neq 0$ is lost.

3. Circularly Integrated Bispectra (CIB)

The third integrated bispectra, called circularly integrated bispectra (CIB), was proposed by Liao and Bao [139]. Unlike the RIB and AIB, the integral paths of CIB are a set of concentric circles with the origin as the center, i.e.,

$$\text{CIB}(a) = \int B_p(a, \theta)d\theta, \tag{6.10.9}$$

where $B_p(a, \theta)$ is the polar representation of $B(\omega_1, \omega_2)$, i.e., $B_p(a, \theta) = B(\omega_1, \omega_2)$ with $\omega_1 = a\cos\theta$ and $\omega_2 = a\sin\theta$.

Since $B_p(a, k\pi/2)$ with integer k provides no phase information, and $B_p(a, \theta)$ with k near 2π provides little phase information. These bispectra should not be integrated. Therefore, the weighted circularly integrated bispectra (WCIB)

$$\text{WCIB}(a) = \int w(\theta)B_p(a, \theta)d\theta \tag{6.10.10}$$

is used instead of the CIB, where $w(\theta)$ takes a very small value when $\theta \approx k\pi/2k$ is an integer.

Summarizing the above discussion, it can be seen that all the above integrated bispectra can be considered as some form of axially integrated bispectra:

(1) The AIB are the axially integrated bispectra of $B(\omega_1, \omega_2)$ and the integrated paths are parallel to the ω_1 or ω_2 axes.
(2) The RIB are the axially integrated bispectra of the polar representation $B_p(a, \theta)$ and the integrated paths are parallel to the a axis.
(3) The PRIB are the phases of the RIB.
(4) The CIB is the axially integrated bispectra of the polar representation $B_p(a, \theta)$ and the integrated paths are parallel to the θ axis.

The integrated bispectra transform the 2-D bispectra function into a 1-D function, which facilitates the implementation of real-time target recognition. However, these methods have the following common drawbacks.

(1) The computer implementation of the integrated bispectra is usually the result of summing the integrals over a path. Obviously, the integrated bispectra of a certain path is selected as the signal feature, which means that the sum of all bispectra in the path plays an important role in target recognition. However, this does not mean that every bispectrum on that path plays an important role in target recognition. In other words, there may be some bispectra points, which have little effect on target recognition and belong to ordinary bispectrum.
(2) If there is a cross term in the original observed signal (the range profile of high-resolution radar is a typical example), the cross term will be more serious when the third-order cumulant is obtained through the cubic correlation function. Therefore, the cross-terms are generally more serious in the obtained bispectra estimation. Since the cross-terms are randomly distributed, they will be difficult to avoid in the chosen integration path. Usually, the cross-terms are detrimental to the target identification.

6.10.2 Selected Bispectra

In order to overcome the drawbacks of the integrated bispectra mentioned above, the selected bispectra method is proposed in Ref.[246]. The so-called selected bispectra are that only those bispectra with the most discriminant power are selected as signal feature parameters. Obviously, this can avoid either trivial bispectra or the cross terms.

In order to select the powerful bispectrum set as the feature parameter set, a discriminant measure $m(\omega)$ is required to judge the role of a bispectrum value in signal type recognition. Fisher's class separability is one such well-known measure.

Consider interclass separation of class i and class j using the bispectra. For simplicity, denote $\omega = (\omega_1, \omega_2)$ and $B(\omega) = B(\omega_1, \omega_2)$. Suppose the training set consists of bispectrum samples $\{B_k^{(i)}(\omega)\}_{k=1,\cdots,N_i}$ and $\{B_k^{(j)}(\omega)\}_{k=1,\cdots,N_j}$, where the subscript k stands for bispectra computed from the kth set of observed data, and the superscript i and j represent the type of the signal, and N_i and N_j are the set number of observed data of the ith and jth class signals, respectively. Taking three classes of signals as example, we need compute $m^{(12)}(\omega)$, $m^{(23)}(\omega)$ and $m^{(13)}(\omega)$, respectively.

The Fisher class separability measure between the ith and jth classes is difined by

$$m^{(i,j)}(\omega) = \frac{\sum_{l=i,j} p^{(l)} \left[\text{mean}_k \left(B_k^l(\omega)\right) - \text{mean}_l \left[\text{mean}_k \left(B_k^l(\omega)\right)\right]\right]^2}{\sum_{l=i,j} p^{(l)} \text{var}_k \left(B_k^l(\omega)\right)}, \quad i \neq j, \tag{6.10.11}$$

where $p^{(l)}$ is the prior probability of the random variable $B^{(l)} = B_k^{(l)}(\omega)$, $\text{mean}_k(B_k^{(l)}(\omega))$ and $\text{var}_k(B_k^{(l)}(\omega))$ represent the mean and variance of all the sample bispectra at the frequency $\omega = (\omega_1, \omega_2)$ of the lth class, and $\text{mean}_l[\text{mean}_k(B_k^{(l)}(\omega))]$ represents the total centroid of all the sample bispectra at the frequency ω over all the classes. In general, the prior probability $p^{(l)}$ can be equal for every class. Therefore, $p^{(l)}$ can be withdrawn from Eq. (6.10.11).

The larger $m^{(i,j)}(\omega)$ is, the stronger the separability between class i and j. Therefore, the frequency set $\{\omega(h), h = 1, \cdots, Q\}$ with Q largest Fisher separability are chosen as the feature frequencies. The center frequency is called the selected frequencies on the bifrequency plane. The bispectra at these selected frequencies are called the selected bispectra.

Given the kth observation of the lth class of signal $x_k^{(l)}, \cdots, x_k^{(l)}(N)$, where $l = 1, \cdots, c$ and $k = 1, \cdots, N_l$. The off-line training algorithm[246] is shown in the following.

Algorithm 6.10.1. *Off-Line Training Algorithm*
Step 1 Calculate the Fourier transform $X_k^{(l)}(\omega)$ for all the observed data.
Step 2 Compute the bispectra

$$B_k^{(l)}(\omega) = B_k^{(l)}(\omega_1, \omega_2) = X_k^{(l)}(\omega_1)X_k^{(l)}(\omega_2)X_k^{(l)}(-\omega_1 - \omega_2).$$

Step 3 Use Eq. (6.10.11) to compute the Fisher class separability measure $m^{(ij)}(\omega)$ for all class combinations (i, j), and requeue the M largest measures such that

$$m^{(ij)}(v_1) \geq m^{(ij)}(v_2) \geq \cdots m^{(ij)}(v_M).$$

Step 4 Calculate the normalized Fisher class separability measure

$$\bar{m}^{(ij)}(v_p) = \sqrt{\frac{m^{(ij)}(v_p)}{\sum_{k=1}^{M} [m^{(ij)}(v_k)]^2}}, \quad p = 1, \cdots, M, \tag{6.10.12}$$

determine the "effective" number of selected bispectra for inter-class (i, j), and denote it by $H^{(ij)}$. The corresponding frequencies $\{\omega^{(ij)}(p), p = 1, \cdots, H^{(ij)}\}$ are called the "effective" frequencies. The repeated frequency for different combinations (i, j) remains only one.

Step 5 Arrange the obtained effective frequencies $\omega^{(ij)}(p), p = 1, \cdots, H^{(ij)}$ into the sequence $\omega(q), q = 1, \cdots Q$, where $Q = \sum_{i,j} H^{(ij)}$, and arrange the corresponding selected bispectra of the kth record in class l into sequence $\{B_k^{(l)}(q), q = 1, \cdots, Q\}$, $k = 1, \cdots, N_l$.The feature vector of the first type of signal is $\boldsymbol{s}_1, \cdots, \boldsymbol{s}_{N_1}$, the feature vector of the second type of signal is $\boldsymbol{s}_{N_1+1}, \cdots, \boldsymbol{s}_{N_1+N_2}$ and the feature vector of the cth type of signal is $\boldsymbol{s}_{N_1+\cdots+N_{c-1}+1}, \cdots, \boldsymbol{s}_{N_1+\cdots+N_c}$. All these feature vectors have the strongest Fisher class separability.

Step 6 Use the selected bispectra to train the radial-basis function (RBF) neural network as a classifier.

If the selected bispectra in Algorithm 6.10.1 are replaced by the PRIB, RIB, AIB, and CIB, we can get the corresponding integrated bispectra training algorithm.

Consider an RBF network for three types signals classification. Let $\boldsymbol{s}_1, \cdots, \boldsymbol{s}_{N_1}$ be the feature vector of the first type of signal, $\boldsymbol{s}_{N_1+1}, \cdots, \boldsymbol{s}_{N_1+N_2}$ be the feature vector of the second type of signal, and $\boldsymbol{s}_{N_1+N_2+1}, \cdots, \boldsymbol{s}_{N_1+N_2+N_3}$ be the feature vector of the third type of signal. Let $\boldsymbol{H} = [h_{ij}]_{(N_1+N_2+N_3)(N_1+N_2+N_3)}$ represent the hidden node output matrix, where

$$h_{ij} = \exp\left(-\frac{||\boldsymbol{s}_i - \boldsymbol{s}_j||^2}{\sigma^2}\right), \tag{6.10.13}$$

and the variance σ^2 of the Gaussian kernel function is the total variance of all feature vectors $\boldsymbol{s}_i$, $i = 1, \cdots, N_1 + N_2 + N_3$. Hence, the weight matrix of the RBF neural network is given by

$$\boldsymbol{W} = (\boldsymbol{H}^{\mathrm{H}}\boldsymbol{H})^{-1}\boldsymbol{H}^{\mathrm{H}}\boldsymbol{O}, \tag{6.10.14}$$

where $\boldsymbol{O}$ is the $(N_1 + N_2 + N_3) \times 3$ desired output matrix given by

$$\boldsymbol{O} = \begin{bmatrix} 1 & \cdots & 1 & 0 & \cdots & 0 & 0 & \cdots & 0 \\ 0 & \cdots & 0 & 1 & \cdots & 1 & 0 & \cdots & 0 \\ 0 & \cdots & 0 & 0 & \cdots & 0 & 1 & \cdots & 1 \end{bmatrix}, \tag{6.10.15}$$

and the number of elements 1 in rows 1, 2 and 3 are N_1, N_2 and N_3, respectively.

Once the RBF neural network as the classifier is trained, it is sufficient to store the weight matrix $\boldsymbol{W}$ of the neural network.

Let $\boldsymbol{x} = [B(1), \cdots, B(Q)]^{\mathrm{T}}$ be the bispectra computed from a set of observed data, and they are the sample bispectra corresponding to the Q selected frequencies. Then, input this vector to the trained RBF neural network, we can obtain the neural network's hidden node output vector

$$h_i = \exp\left(-\frac{||\boldsymbol{x} - \boldsymbol{s}_i||^2}{\sigma_i^2}\right), \tag{6.10.16}$$

where the variance σ_i^2 in the Gaussian kernel function is the variance of the feature vector $\boldsymbol{s}_i$ determined in the training phase. The output vector of the RBF neural network is given by

$$\boldsymbol{o} = \boldsymbol{W}^{\mathrm{T}}\boldsymbol{h}, \tag{6.10.17}$$

which gives the classification result of the observed signal.

The experimental results of the range profiles of the high-resolution radar in a microwave anechoic chamber and the measured range profiles of the far-field high-resolution radar show that the integrated bispectrum is an effective signal classification tool[246].

Summary

This chapter introduces the theory, methods, and applications of the higher-order statistical analysis and processing of non-Gaussian signals. Firstly, the definitions of moments, cumulants, and higher-order spectra are introduced and their mathematical properties are discussed respectively. Then, the theory and methods of nonminimum phase system identification, harmonic retrieval, and adaptive filtering based on higher-order cumulants are introduced. Finally, time-delay estimation and radar target recognition are used as examples to introduce how to use the cumulant and high-order spectra to solve practical engineering problems.

It can be said that the higher-order statistical analysis of signals is actually a generalization and deepening of the familiar statistical analysis of random signals based on correlation functions and power spectra. The purpose of the higher-order statistical signal analysis is to analyze the higher-level statistical information of the signal.

Exercises

6.1 Let $x_1, x_2, \cdots, x_n$ be n independent Gaussian random variables with mean $E\{x_i\} = \mu$ and variance $\text{var}(x_i) = E\{[x_i - \mu]^2\} = \sigma^2$. Let $\bar{x} = \frac{1}{N}\sum_{i=1}^{n} x_i$. Find the probability density distribution of the sample mean $\bar{x}$.

6.2 Let $\boldsymbol{x}$ be a Gaussian random variable with mean (vector) $\boldsymbol{v}$ and covariance matrix $\boldsymbol{R}$. Prove that the moment generating function of $\boldsymbol{x}$ is

$$\Phi(\boldsymbol{x}) = \exp\left(j\boldsymbol{\omega}^{\mathrm{T}}\boldsymbol{v} - \frac{1}{2}\boldsymbol{\omega}^{\mathrm{T}}\boldsymbol{R}\boldsymbol{\omega}\right).$$

6.3 If the k-th order moment of a finite energy signal $x(n)$ is defined as

$$m_{kx}(\tau_1, \cdots, \tau_{k-1}) \stackrel{def}{=} \sum_{n=-\infty}^{\infty} x(n)x(n+\tau_1)\cdots x(n+\tau_{k-1}).$$

A sequence is given by $x(n) = a^n u(n)$, where $-1 < a < 1$, and $u(n)$ is a unit step function. Find the moment m_{1x}, $m_{2x}(\tau)$, $m_{3x}(\tau_1, \tau_2)$ and $m_{3x}(\tau, \tau, \tau)$ of $x(n)$.

6.4 Let x have a translational exponential distribution $f(x) = e^{-(x+1)}$, $x \geq -1$ (x translates -1 in order to make the mean of x equal to zero). Find $c_{kx} = \text{cum}(x, \cdots, x)$.

6.5 Let the probability density function of x be $f(x) = 0.5e^{-|x|}$, $-\infty < x < \infty$. Find c_{kx}, $k \geq 2$.

6.6 Let x be the square of m independent identically distributed Gaussian random variables with zero mean and unit variance, then the distribution of x is called χ^2 distribution with m degrees of freedom, denoted as χ_m^2. Find the k-order cumulant c_{kx} of the random variable x.

6.7 Given two statistically independent Gaussian stochastic processes $\{x(t)\}$, $\{y(t)\}$ and

$$f_X(x) = \frac{1}{\sqrt{2\pi\sigma_1^2}} \exp\left(-\frac{(x-\mu_1)^2}{2\sigma_1^2}\right),$$

$$f_Y(y) = \frac{1}{\sqrt{2\pi\sigma_2^2}} \exp\left(-\frac{(y-\mu_2)^2}{2\sigma_2^2}\right).$$

Let $z(t) = x(t) + y(t)$. Prove that $\{z(t)\}$ is a Gaussian stochastic process. This conclusion shows that the sum of any two Gaussian stochastic processes is still a Gaussian stochastic process.

6.8 Let $\{e(n)\}$ be a non-Gaussian stationary process, and assume that $\{e(n)\}$ passes through a linear invariant stable system with impulse response $\{h_i\}$, producing an output sequence of $\{y(n)\}$.

(1) Express the cumulant of $\{y(n)\}$ by the cumulant of $\{e(n)\}$;
(2) Find the multispectrum of $\{y(n)\}$ by the multispectrum of $\{e(n)\}$ and the impulse response coefficient $\{h_i\}$.

6.9 $z(n) = x(n)\cos(\omega_c n) + y(n)\sin(\omega_c n)$, where $x(n)$ and $y(n)$ are independent stationary processes, and $E\{x(n)\} = E\{y(n)\} = 0$, $c_{2x}(\tau) = E\{x(n)x(n+\tau)\} = c_{2y}(\tau)$, and $c_{3x}(\tau_1, \tau_2) = E\{x(n)x(n+\tau_1)x(n+\tau_2)\} = c_{3y}(\tau_1, \tau_2)$. Is $z(n)$ a stationary random process?

6.10 Prove that the symmetry of the bispectra $B(\omega_1, \omega_2) = B^*(-\omega_2, -\omega_1) = B(\omega_2, -\omega_1 - \omega_2)$.

6.11 Given that the impulse response $\{h_i\}$ of a linear system satisfies the absolute summability condition $\sum_{i=-\infty}^{\infty} |h_i| < \infty$, proved that the multispectrum existence and continuity of the linear process generated by the independent identically distributed $e(n)$ excitation of the linear system.

6.12 Let $H(e^{j\omega})$ satisfies $H(e^{j\omega}) \neq 0, \forall\omega$, and $\sum_{i=-\infty}^{\infty} |ih_i| < \infty$. Assume that the k-th order cumulant of the excited independent identically distributed processes are not equal to zero, where $k > 2$. Prove that $H(e^{j\omega})$ can be obtained from the k-th order spectrum $S(\omega_1, \cdots, \omega_{k-1})$ differing by at most an unknown complex constant scale factor $Ae^{j\omega m}$, where A is a real number (positive or negative) and m is an integer.

6.13 Prove the multispectra formula

$$S_{kx}(\omega_1, \cdots, \omega_{k-1}) = \gamma_{ke} H(\omega_1) \cdots H(\omega_{k-1}) H(-\omega_1 - \cdots - \omega_{k-1})$$

6.14 Assume $H(e^{j\omega}) \neq 0, \forall \omega$, and $\sum_{i=-\infty}^{\infty} |i \cdot h_i| < \infty$. Prove that the phase $\phi(\omega)$ of $H(e^{j\omega})$ is continuously derivable with respect to ω.

6.15 Let $\{y(n)\}$ be a non-Gaussian MA(q) process with the excitation process $\gamma_{3e} \neq 0$. Find the range of τ_1, τ_2 of $c_{3y}(\tau_1, \tau_2)$ and draw the support region of τ_1, τ_2 in the (τ_1, τ_2) plane.

6.16 Let x be a random process with independently and identically distributed values at each moment, and $\tilde{x} = x * h_j$ be the convolution of x with the impulse response $\{h_i\}$ of a linear time-invariant system. Prove that if $E\{|\tilde{x}|^2\} = E\{|x|^2\}$, then the following relation holds:

(1) $|K(\tilde{x})| \leq |K(x)|$;

(2) $|K(\tilde{x})| \leq |K(x)|$ if and only if $\boldsymbol{s} = [s_1, s_2, \cdots]^{\mathrm{T}}$ is a vector with only one nonzero element (its magnitude is 1).

6.17 Consider MA(1) stochastic process $x(n) = w(n) - w(n-1)$, $n = 0, \pm 1, \pm 2, \cdots$, where $\{w(n)\}$ be an independent identically distributed stochastic processes, and $E\{e(n)\} = 0$, $E\{w^2(n)\} = 1$, $E\{w^3(n)\} = 1$. Find the power spectrum and bispectrum of $\{x(n)\}$.

6.18 Let $P_{1x}(\omega) = \sum_{\tau=-\infty}^{\infty} c_{4x}(\tau, 0, 0)e^{-j\omega\tau}$ be the power spectrum of the special fourth-order cumulant $c_{4x}(\tau, 0, 0)$. Consider a complex harmonic process $x(n) = \alpha e^{j\omega_0 n}$ where ω_0 is a constant, α is a random variable, and $E\{\alpha\} = 0$, $E\{\alpha^2\} = Q$, $E\{\alpha^3\} = 0$, and $E\{\alpha^4\} = \mu$. If the fourth-order cumulant of the complex harmonic process is defined as

$$c_{4x}(\tau_1, \tau_2, \tau_3) = \mathrm{cum}[x(n), x^*(n+\tau_1), x(n+\tau_2), x^*(n+\tau_3)].$$

Prove that $P_{1x}(\omega) = \frac{\gamma}{Q} S_x(\omega)$, where $\gamma = \mu - 3Q^2$ and $S_x(\omega)$ is the power spectrum of $\{x(n)\}$.

6.19 The impulse response of the first-order FIR system is $h(n) = \delta(n) - \alpha\delta(n-1)$. Let $\{x(n)\}$ be the output sequence obtained by exciting the FIR system using the independent identically distributed process $\{e(n)\}$, where $E\{e(n)\} = 0$, $E\{e^2(n)\} = \sigma_e^2$ and $\gamma_{3e} = E\{e^3(n)\} \neq 0$. Prove that the following relationship between the special slice $S_{3x}(\omega, 0)$ of the bispectrum and the power spectrum $S_x(\omega)$ of $\{x(n)\}$

$$S_{3x}(\omega, 0) = \frac{\gamma_{3e}}{\sigma_e^2} H(0) S_x(\omega)$$

where $H(0)$ is the value of the frequency transfer function $H(\omega)$ of the FIR system at zero frequency.

6.20 Consider a real harmonic signal $x(n) = \sum_{k=1}^{p} A_k \cos(\omega_k n + \phi_k)$, where ϕ_k is an independent uniform distribution $U[-\pi, \pi)$ and $A_k > 0$. Prove that the fourth-order cumulant of $\{x(n)\}$ is

$$c_{4x}(\tau_1, \tau_2, \tau_3) = -\frac{1}{8} \sum_{k=1}^{p} \alpha_k^4 [\cos(\tau_1 - \tau_2 - \tau_3) + \cos(\tau_2 - \tau_3 - \tau_1) + \cos(\tau_3 - \tau_1 - \tau_2)]$$

6.21 Let $s(n) = \sum_{i=-\infty}^{\infty} f(i)e(n-i)$ be the signal, while $x(n) = s(n) + w_1(n)$ and $y(n) = s(n-D) + w_2(n)$ are the observed processes, respectively. Define the criterion function

$$J_1(d) = \frac{|\text{cum}[x(n-d), x(n-d), y(n), y(n)]|}{\sqrt{|\text{CUM}_4[x(n)]||\text{CUM}_4[y(n)]|}}$$

where $\text{CUM}_4[x(n)] = \text{cum}[x(n), x(n), x(n), x(n)]$. Prove that the following results hold.
(1) $0 \le J_1(d) \le 1, \forall d$;
(2) $J_1(d) = 1$ if and only if $d = D$.

Notes: The criterion $J_1(d)$ gives an alternative fourth-order cumulant method for time delay estimation.

6.22 Signal $s(n)$ and observation process $x(n)$, $y(n)$ are the same as in the above question, and the criterion function f is defined as

$$J_2(d) = \frac{|\text{CUM}_4[x(n-d) + y(n)]|}{16\sqrt{|\text{CUM}_4[x(n)]||\text{CUM}_4[y(n)]|}}$$

where $\text{CUM}_4[x(n)] = \text{cum}[x(n), x(n), x(n), x(n)]$. Prove that the following results hold.
(1) $0 \le J_2(d) \le 1, \forall d$;
(2) $J_2(d) = 1$ if and only if $d = D$.

Notes: The criterion $J_2(d)$ gives an alternative fourth-order cumulant method for time delay estimation.

6.23 A non-minimum phase MA model is given by

$$x(n) = w(n) + 0.9w(n-1) + 0.385w(n-2) - 0.771w(n-3)$$

where $w(n)$ is an independent identically distributed stochastic process with zero mean, variance of 1, and third-order cumulant $\gamma_{3w} = 1$. The observed data is $y(n) = x(n) + v(n)$, where $v(n)$ is a Gaussian colored noise with zero mean and adjustable variance. Adjust the variance of $v(n)$ to obtain 0 dB and 10 dB signal-to-noise ratios, respectively, and use the Giannakis-Mendel algorithm and cumulant algorithm to estimate the MA parameters, respectively. For each algorithm, 50 computer simulation experiments are run independently under different signal-to-noise ratios, and the parameter estimation results of the two algorithms were tried to be counted.

7 Linear Time-Frequency Transform

In the previous chapters, the signal analysis is either in the time-domain or in the frequency-domain, which constitutes the time-domain analysis or frequency-domain analysis method of the signal. The main mathematical tool used is the Fourier transform, which is only suitable for stationary signals whose statistics do not vary with time. However, the real signals often have a statistic which is a function of time, and such kind of signal is collectively referred to as a non-stationary signal. Many artificial and natural signals are nonstationary, such as temperature and blood pressure.

Although the adaptive filtering algorithms such as Kalman filtering and RLS algorithm are also applicable to non-stationary signals, they are limited to the tracking of slow time-varying signals and can not get the statistics of the general time-varying signals (such as power spectrum, etc). In other words, these signal processing methods cannot meet the special requirements of non-stationary signal analysis. Therefore, it is necessary to discuss the analysis and processing methods of non-stationary signals. Since the statistical properties of nonstationary signals vary with time, the main interest in non-stationary signals should naturally focus on their local statistical properties. For non-stationary signals, the Fourier transform is no longer an effective mathematical analysis tool, because it is a global transformation of the signal, while the analysis of the local performance of the signal must rely on the local transformation of the signal. On the other hand, the local performance of the signal can only be accurately described by using the two-dimensional joint representation of the time domain and frequency domain. In this sense, the two-dimensional analysis of nonstationary signals is often referred to as time-frequency signal analysis.

The time-frequency analysis of non-stationary signals can be divided into two categories: linear and non-linear transformations. In this chapter, the linear transformation of the time-frequency signal is discussed, while the nonlinear transformation of the time-frequency signal, i.e., the quadratic time-frequency distribution, is left to the next chapter.

7.1 Local Transformation of Signals

The Fourier transform $\mathcal{F}$ and the inverse Fourier transform $\mathcal{F}^{-1}$ serve as a bridge to establish a one-to-one correspondence between the signal $s(t)$ and its spectrum $S(f)$

$$S(f) = \mathcal{F}[s(t)] = \int_{-\infty}^{\infty} s(t)e^{-j2\pi ft}dt \quad \text{(Fourier transform)}, \tag{7.1.1}$$

$$s(t) = \mathcal{F}^{-1}[S(f)] = \int_{-\infty}^{\infty} S(f)e^{j2\pi ft}df \quad \text{(inverse Fourier transform)}. \tag{7.1.2}$$

https://doi.org/10.1515/9783110475562-007

This Fourier transform pair can also be expressed in terms of angular frequency as

$$S(\omega) = \int_{-\infty}^{\infty} s(t)e^{-j\omega t}dt \quad \text{(Fourier transform)}, \tag{7.1.3}$$

$$s(t) = \frac{1}{2\pi}\int_{-\infty}^{\infty} S(\omega)e^{j\omega t}d\omega \quad \text{(inverse Fourier transform)}. \tag{7.1.4}$$

Eqs. (7.1.1) and (7.1.2) are called the frequency-domain representation and the time-domain representation of signals respectively, which constitute two ways to observe a signal. The Fourier transform decomposes the signal into different frequency components as a whole and lacks local information, i.e. it does not tell us when a certain frequency component occurs. However, this information is very important for non-stationary signal analysis.

When discussing a linear transformation, it is often convenient to write it in the form of an inner product between the transformed function and the transformed kernel function. Therefore, the inner product of the complex functions $f(x)$ and $g(x)$ is defined as

$$\langle f(x), g(x)\rangle \stackrel{\text{def}}{=} \int_{-\infty}^{\infty} f(x)g^*(x)dx. \tag{7.1.5}$$

Then the Fourier transform pair of Eqs. (7.1.1) and (7.1.2) can be concisely expressed in inner product form as

$$S(f) = \langle s(t), e^{j2\pi ft}\rangle \quad \text{(Fourier transform)}, \tag{7.1.6}$$

$$s(t) = \langle S(f), e^{-j2\pi ft}\rangle \quad \text{(inverse Fourier transform)}. \tag{7.1.7}$$

Obviously, the kernel function of the Fourier transform is an exponential function.

It can be seen from Eqs. (7.1.6) and (7.1.7) that the time lengths of both the original function $s(t)$ and the kernel function $e^{j2\pi ft}$ of the Fourier transform are taken as $(-\infty, \infty)$, while the original function $S(f)$ and the kernel function $e^{-j2\pi ft}$ of the inverse Fourier transform are also taken on the whole frequency axis. In this sense, the Fourier transform is essentially a global transformation of the signal $s(t)$, while the inverse Fourier transform is a global transformation of the spectrum $S(f)$. Although the Fourier transform and its inverse transform are powerful tools for signal analysis, as Gabor pointed out in his classic paper "Theory of Communication" in 1946[82] :

So far, the basis of communication theory consists of two methods of signal analysis: one describing the signal as a function of time and the other describing the signal as a function of frequency (Fourier analysis). Both methods are idealized,······. Yet, in our everyday experience, especially our hearing, has always been a signal described in terms of both time and frequency.

In order to describe a signal using both time and frequency, it is natural to use a joint time-frequency representation $S(t, f)$ for a non-stationary signal $s(t)$. So, how to establish the transform relationship between $s(t)$ and $S(t, f)$? Obviously, we cannot use the Fourier transform, which is a global transform, but should instead use the local transform of the signal.

Since any signal transform can be written as an inner product between that signal and some selected kernel function, it is easy to associate that the signal local transformation can be constructed using two basic forms:

$$\text{Local transform of } s(t) = \langle s(t) \text{taken locally, kernel function infinitely long}\rangle, \quad (7.1.8)$$

or

$$\text{Local transform of } s(t) = \langle s(t) \text{ taken in its entirety, kernel function localized}\rangle. \quad (7.1.9)$$

The following are a few typical examples of local transform of signals.

(1) Short-time Fourier transform

$$\text{STFT}(t, f) = \int_{-\infty}^{\infty} [s(u)g^*(u-t)]e^{-j2\pi fu}du = \langle s(u)g^*(u-t), e^{-j2\pi fu}\rangle, \quad (7.1.10)$$

where $g(t)$ is a narrow window function.

(2) Wigner-Ville time-frequency distribution

$$\begin{aligned} P(t, f) &= \int_{-\infty}^{\infty} z(t+\frac{\tau}{2})z^*(t-\frac{\tau}{2})e^{-j2\pi f\tau}d\tau \\ &= \left\langle z(t+\frac{\tau}{2})z^*(t-\frac{\tau}{2}), e^{-j2\pi f\tau}\right\rangle. \end{aligned} \quad (7.1.11)$$

(3) Wavelet transform

$$\text{WT}(a, b) = \frac{1}{\sqrt{a}}\int_{-\infty}^{\infty} s(t)h^*\left(\frac{t-b}{a}\right)dt = \int_{-\infty}^{\infty} s(t)h_{ab}^*(t)dt = \langle s(t), h_{ab}(t)\rangle, \quad (7.1.12)$$

where the kernel function of the transform

$$h_{ab}(t) = \frac{1}{\sqrt{a}}h\left(\frac{t-b}{a}\right) \quad (7.1.13)$$

is called the wavelet basis function.

(4) Gabor transform

$$a_{mn} = \int_{-\infty}^{\infty} s(t)\gamma^*(t-mT)e^{-j2\pi(nF)t}dt \quad (7.1.14)$$

$$= \int_{-\infty}^{\infty} s(t)\gamma_{mn}^*(t)dt = \langle s(t), \gamma_{mn}(t)\rangle, \quad (7.1.15)$$

where

$$\gamma_{mn}(t) = \gamma(t - mT)e^{j2\pi(nF)t} \tag{7.1.16}$$

is referred to as the Gabor basis function.

It is easy to see that the short-time Fourier transform and the Wigner-Ville distribution belong to the first local transform shown in Eq. (7.1.8), while the wavelet transform and the Gabor transform belong to the second signal local transform shown in Eq. (7.1.9).

In addition to the above four typical forms, there are various other local transform forms, such as the Radon-Wigner transform, fractional Fourier transform, and so on.

The above four local transformations are conventionally called the time-frequency representation of the signal. According to whether the superposition principle or linear principle is satisfied, the time-frequency representation is divided into two categories: linear time-frequency representation and nonlinear time-frequency representation. Specifically, if signal $s(t)$ is a linear combination of several components, and the time-frequency representation $T_s(t, f)$ of $s(t)$ is the same linear combination of the time-frequency representation of each signal component, this $T_s(t, f)$ is called linear time-frequency representation; otherwise, it is called nonlinear time-frequency representation. Take the signal of two components as an example, if

$$s(t) = c_1 s_1(t) + c_2 s_2(t) \to T_s(t, f) = c_1 T_{s_1}(t, f) + c_2 T_{s_2}(t, f), \tag{7.1.17}$$

then $T_s(t, f)$ is linear time-frequency representation.

Short-time Fourier transform, wavelet transform and Gabor transform are linear transforms or linear time-frequency representations of time-frequency signal analysis, while the Wigner-Ville distribution is a nonlinear transform (quadratic transform) of time-frequency signal analysis, which is a nonlinear time-frequency representation. It can be said that the generalized or modified Fourier transform is used for the time-frequency analysis of non-stationary signals.

7.2 Analytic Signal and Instantaneous Physical Quantity

In the analysis and processing of non-stationary signals, the actual signal is often real, but it needs to be transformed into a complex signal for mathematical representation and analysis. In particular, some important instantaneous physical quantities and time-frequency representations are defined directly using the complex signal form of the real signal to be analyzed. So why do we need such a transformation?

When signal $s(t)$ is real, its spectrum

$$S(f) = \int_{-\infty}^{\infty} s(t)e^{-j2\pi ft}dt \tag{7.2.1}$$

has conjugate symmetry because

$$S^*(f) = \int_{-\infty}^{\infty} s(t)e^{j2\pi ft}dt = S(-f). \tag{7.2.2}$$

From the perspective of effective information utilization, the negative frequency spectrum of the real signal is completely redundant, because it can be obtained from the positive frequency spectrum. By removing the negative frequency spectrum part of the real signal and keeping only the positive frequency spectrum part, the signal occupies half of the frequency band, which is beneficial for wireless communication (called single-sideband communication), etc. The signal that retains only the positive frequency spectrum part, its spectrum no longer has conjugate symmetry, and the corresponding time-domain signal should be complex.

The most common way to represent complex variables is to use both real and imaginary components. The same is true for the complex signal, which must be represented using both real and imaginary parts. Of course, the two-way signals will bring trouble in the transmission, so the real signal is always used in the actual signal transmission, while the complex signal is used in the processing of the received signal. In the following, two commonly used complex signals are discussed: analytic signal and baseband signal.

7.2.1 Analytic Signal

The simplest way to represent the complex signal $z(t)$ is to use the given real signal $s(t)$ as its real part and additionally construct a "virtual signal" $\hat{s}(t)$ as its imaginary part, i.e.,

$$z(t) = s(t) + j\hat{s}(t). \tag{7.2.3}$$

The simplest way to construct the virtual signal $\hat{s}(t)$ is to use the original real signal $s(t)$ to excite a filter and use its output as the virtual signal. Let the impulse response of the filter be $h(t)$, then

$$\hat{s}(t) = s(t) * h(t) = \int_{-\infty}^{\infty} s(t-u)h(u)du, \tag{7.2.4}$$

that is, the complex signal can be expressed as

$$z(t) = s(t) + js(t) * h(t), \tag{7.2.5}$$

where * represents the convolution operation. Taking the Fourier transform on both sides of the above equation, then the spectrum expression can be obtained

$$Z(f) = S(f) + jS(f)H(f) = S(f)[1 + jH(f)]. \tag{7.2.6}$$

For the special case of narrow-band signal, the positive frequency part of the signal spectrum is often retained, while the negative frequency part is removed (to keep the total energy of the signal unchanged, the spectrum value of the positive frequency needs to be doubled). This means that the spectrum of complex signal $z(t)$ should have the form

$$Z(f) = \begin{cases} 2S(f), & f > 0 \\ S(f), & f = 0. \\ 0, & f < 0 \end{cases} \tag{7.2.7}$$

Comparing Eqs. (7.2.6) and (7.2.7), it is easy to see that we only need to select the transfer function of the filter to satisfy

$$H(f) = -j\mathrm{sgn}(f) = \begin{cases} -j, & f > 0 \\ 0, & f = 0, \\ j, & f < 0 \end{cases} \tag{7.2.8}$$

where

$$\mathrm{sgn}(f) = \begin{cases} +1, & f > 0 \\ 0, & f = 0 \\ -1, & f < 0 \end{cases} \tag{7.2.9}$$

is the sign function.

Taking the inverse Fourier transform on both sides of Eq. (7.2.8), the impulse response of the filter is obtained

$$h(t) = \int_{-\infty}^{\infty} H(f)e^{j2\pi ft}df = \frac{1}{\pi t}. \tag{7.2.10}$$

Substituting Eq. (7.2.10) into Eq. (7.2.4) creates

$$\hat{s}(t) = \mathcal{H}[s(t)] = s(t) \star \frac{1}{\pi t} = \frac{1}{\pi}\int_{-\infty}^{\infty} \frac{s(\tau)}{t-\tau}d\tau, \tag{7.2.11}$$

where t and τ are real variables, and $\mathcal{H}[s(t)]$ denotes the Hilbert transform of the real signal $s(t)$. Since the impulse response $h(t)$ shown in Eq. (7.2.10) is to make the real signal $s(t)$ become its Hilbert transform, $h(t)$ or $H(f) = \mathcal{H}[h(t)]$ is called the Hilbert transformer, also known as the Hilbert filter.

If the Hilbert transform $\hat{s}(t)$ is known, the original real signal

$$s(t) = \frac{-1}{\pi t} \star \hat{s}(t) = \frac{-1}{\pi}\int_{-\infty}^{\infty} \frac{\hat{s}(\tau)}{t-\tau}d\tau \tag{7.2.12}$$

can also be recovered from it.

Eq. (7.2.8) shows that Hilbert filter $H(f)$ is an all-pass filter, since $|H(f)| = 1, \forall f \neq 0$, see Fig. 7.2.1(a), while the phase characteristics of the Hilbert filter $H(f)$ is shown in Fig.7.2.1(b).

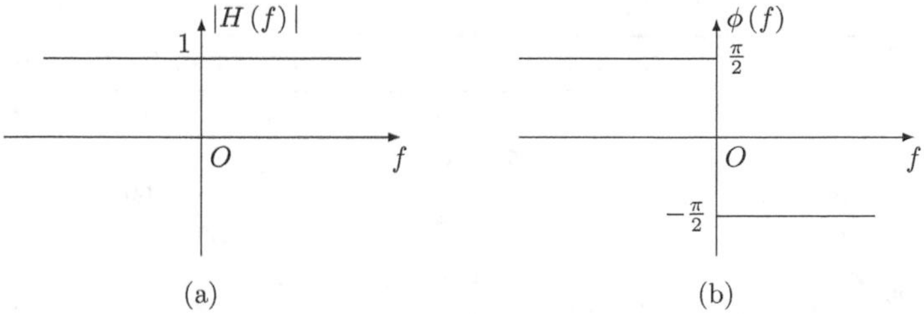

Fig. 7.2.1: The transfer function of the Hilbert filter

Definition 7.2.1. *(Analytic signal) The analytic signal corresponding to the real signal $s(t)$ is denoted as $s_A(t)$ and defined as $s_A(t) = A[s(t)]$, where $A[s(t)] = s(t) + j\mathcal{H}[s(t)]$ is the operator that constitutes the analytic signal and $\hat{s}(t) = \mathcal{H}[s(t)]$ is the Hilbert transform of $s(t)$.*

The Hilbert transform has the following properties.

Property 1 After the signal $s(t)$ passes through the Hilbert transformer, the amplitude of the signal spectrum does not change.

Property 2 $s(t) = -\mathcal{H}[\hat{s}(t)]$.

Property 3 $s(t) = -\mathcal{H}^2[\hat{s}(t)]$, where $\mathcal{H}^2[\hat{s}(t)] = \mathcal{H}\{\mathcal{H}[s(t)]\}$.

It is easy to verify that the Hilbert transform also has the following linearity, time-shift invariance, and scale invariance:

$$x(t) = as_1(t) + bs_2(t) \quad \Rightarrow \quad \hat{x}(t) = a\hat{s}_1(t) + b\hat{s}_2(t), \tag{7.2.13}$$

$$x(t) = s(t-a) \quad \Rightarrow \quad \hat{x}(t) = \hat{s}(t-a), \tag{7.2.14}$$

$$x(t) = s(at),\ a > 0 \quad \Rightarrow \quad \hat{x}(t) = \hat{s}(at), \tag{7.2.15}$$

$$x(t) = s(-at) \quad \Rightarrow \quad \hat{x}(t) = -\hat{s}(-at). \tag{7.2.16}$$

Table 7.2.1 lists some typical signals and their Hilbert transforms[178].

7.2.2 Baseband Signal

For information systems such as communication and radar, the commonly used signal is the real narrowband signal, i.e.,

$$s(t) = a(t)\cos[2\pi f_c t + \phi(t)] = \frac{1}{2}a(t)\left(e^{j[2\pi f_c t+\phi(t)]} + e^{-j[2\pi f_c t+\phi(t)]}\right), \tag{7.2.17}$$

where f_c is the carrier frequency. The positive and negative frequency components of the narrowband signal are clearly separated, and the negative frequency component is easy

Tab. 7.2.1: Hilbert transform pairs

Typical Signals	Signal Representation	Hilbert Transform
Constant signal	a	zero
Sinusoidal signal	$\sin(\omega t)$	$-\cos(\omega t)$
Cosine signal	$\cos(\omega t)$	$\sin(\omega t)$
Exponential signal	$e^{j\omega t}$	$-j\text{sgn}(\omega)e^{j\omega t}$
Square wave pulse signal	$p_a(t) = \begin{cases} 1 & \|t\| \le a \\ 0, & \text{others} \end{cases}$	$\frac{1}{\pi}\ln\left\|\frac{t+a}{t-a}\right\|$
Bipolar pulse signal	$p_a(t)\text{sgn}(t)$	$-\frac{1}{\pi}\ln\left\|1-\frac{a^2}{t^2}\right\|$
Double triangle signal	$tp_a(t)\text{sgn}(t)$	$-\frac{1}{\pi}\ln\left\|1-\frac{a^2}{t^2}\right\|$
Triangle signal	$\text{Tri}(t) = \begin{cases} 1-\|t/a\| & \|t\| \le a \\ 0, & \|t\| > a \end{cases}$	$-\frac{1}{\pi}\left[\ln\left\|\frac{t-a}{t+a}\right\| + \frac{t}{a}\left\|\frac{t^2}{t^2-a^2}\right\|\right]$
Cauchy pulse signal	$\frac{a}{a^2+t^2}$	$\frac{t}{a^2+t^2}$
Gaussian pulse signal	$e^{-\pi t^2}$	$\frac{1}{\pi}\int_0^\infty e^{-\frac{1}{4\pi}\omega^2}\sin(\omega t)d\omega$
Symmetric exponential signal	$e^{-a\|t\|}$	$\frac{1}{\pi}\int_0^\infty \frac{2a}{a^2-\omega^2}\sin(\omega t)d\omega$
Sinc signal	$\frac{\sin(at)}{at}$	$\frac{\sin^2(at/2)}{(at/2)} = \frac{1-\cos(at)}{at}$
Asymmetric exponential signall	$\text{sgn}(t)e^{-a\|t\|}$	$-\frac{1}{\pi}\int_0^\infty \frac{2a}{a^2-\omega^2}\cos(\omega t)d\omega$

to be filtered out. If the positive frequency component is retained and the amplitude is doubled, then the analytic signal can be obtained as

$$s_A(t) = a(t)e^{j\phi(t)}e^{j2\pi f_c t}, \tag{7.2.18}$$

where $e^{j2\pi f_c t}$ is a complex number, which is used as the carrier of information and does not contain useful information. Multiplying both sides of the above equation with $e^{-j2\pi f_c t}$, the signal frequency can be shifted down f_C to zero carrier frequency, and the new signal is obtained as

$$s_B(t) = a(t)e^{j\phi(t)}. \tag{7.2.19}$$

This kind of zero carrier frequency signal is called a baseband signal, or zero intermediate frequency signal.

The above discussion shows that the single side spectrum can be obtained after complex signal processing. Any frequency conversion processing only the carrier frequency f_C is shifted and the envelope information remains unchanged. Different from the complex signal, the real signal cannot shift the carrier frequency very low, otherwise, the mixture of the positive spectrum and negative spectrum will distort the envelope.

Comparing Eq. (7.2.19) and Eq. (7.2.18), it can be seen that there exists a relationship

$$s_A(t) = s_B(t)e^{j2\pi f_0 t} \tag{7.2.20}$$

between the analytic signal and the baseband signal. This shows that the baseband signal $s_B(t)$ is the complex envelope of the analytic signal $s_A(t)$, which is a complex signal like $s_A(t)$.

It should be noted that the baseband signal $s_B(t)$ has a zero medium frequency, it contains both positive and negative frequency components; however, since it is a complex signal, its spectrum does not have conjugate symmetry properties. Therefore, if the negative frequency component is removed from the baseband signal, it will result in the loss of useful information. On the other hand, it is easy to see that the baseband signal of Eq. (7.2.19) is only a frequency-shifted form of the analytic signal of Eq. (7.2.18). Therefore, on many occasions (such as time-frequency analysis, etc.), it is appropriate to use the baseband signal as the analytic signal. Especially in communication signal processing, it is more convenient to use the baseband signal than the analytic signal, because the baseband signal does not contain a carrier, and the analytic signal contains a carrier, and the role of the carrier is only as a carrier of the information signal, does not contain any useful information.

7.2.3 Instantaneous Frequency and Group Delay

The difference between the highest and lowest frequencies of the signal, $B = f_{max} - f_{min}$, is called the bandwidth of the signal, and the duration of the signal, T, is called the time width of the signal.

All actual signals have a time starting point and a time ending point. The time width T has the same role in the time domain as the bandwidth B has in the frequency domain. It is often desirable to know how the energy of the signal is distributed over the time interval $0 < t < T$. This is the so-called frequency characteristic of the signal.

In order to describe the time-varying frequency characteristics of the nonstationary signal, instantaneous physical quantities often play an important role. Instantaneous frequency and group delay are two physical quantities.

"Frequency" is one of the most commonly used technical terms in engineering, physics, and even daily life. In the analysis and processing of stationary signals, frequency refers to the parameters of the Fourier transform, i.e., the circular frequency f or the angular frequency ω, which are independent of time. However, for nonstationary signals, Fourier frequency is no longer an appropriate physical quantity. There are two reasons: (1) nonstationary signals are no longer simply analyzed using the Fourier transform; (2) the frequency of nonstationary signals changes with time. Therefore, another concept of frequency is needed, which is instantaneous frequency.

From the physical point of view, signals can be divided into two categories: single-component and multicomponent signals. Single component signals have only one frequency at any time, which is called the instantaneous frequency of the signal. Multi-component signals have several different instantaneous frequencies at certain moments. There were two different definitions of instantaneous frequency, which were given by Carson and Fry[47] and Gabor[82], respectively. Later, Ville unified these two different definitions[112], and defined the instantaneous frequency of a signal $s(t) =$

$a(t)\cos(\phi(t))$ with instantaneous phase $\phi(t)$ as

$$f_i(t) = \frac{1}{2\pi}\frac{d}{dt}[\arg z(t)], \tag{7.2.21}$$

where $z(t)$ is the analytic signal of the real signal $s(t)$, and $\arg[z(t)]$ is the phase of the analytic signal $z(t)$. That is, the instantaneous frequency is defined as the derivative of the phase $\arg[z(t)]$ of the analytic signal $z(t)$. Eq. (7.2.21) has a very clear physical meaning: since the analytic signal $z(t)$ represents a vector in the complex plane, the instantaneous frequency represents the rotational speed of the argument of this vector (in terms of the number of cycles per unit time, which should be multiplied by 2π if measured in radians). Ville further noticed that since the instantaneous frequency is time-varying, there should exist an instantaneous spectrum corresponding to the instantaneous frequency, and the average frequency of this instantaneous spectrum is the instantaneous frequency.

Let E represents the total energy of the signal $z(t)$, i.e.,

$$E = \int_{-\infty}^{\infty} |z(t)|^2\, dt = \int_{-\infty}^{\infty} |Z(f)|^2\, df. \tag{7.2.22}$$

Thus, the normalized functions $|z(t)|^2/E$ and $|Z(f)|^2/E$ can be considered as the energy density functions of the signal $z(t)$ in the time-domain and frequency-domain, respectively. At this point, the concept of the moment in probability theory can be used to quantitatively describe the performance of the signal. For example, the first-order moment can be used to define the average frequency of the signal spectrum

$$\bar{f} = \frac{1}{E}\int_{-\infty}^{\infty} f|Z(f)|^2\, df = \frac{\int_{-\infty}^{\infty} f|Z(f)|^2\, df}{\int_{-\infty}^{\infty} |Z(f)|^2\, df} \tag{7.2.23}$$

and the time average of the instantaneous frequency

$$\bar{f}_i = \frac{1}{E}\int_{-\infty}^{\infty} f_i(t)|z(t)|^2\, dt = \frac{\int_{-\infty}^{\infty} f_i(t)|z(t)|^2\, dt}{\int_{-\infty}^{\infty} |z(t)|^2\, dt}. \tag{7.2.24}$$

Using Gabor's average measure[82], Ville[112] proved that the average frequency of the signal spectrum is equal to the time average of the instantaneous frequency, i.e., $\bar{f} = \bar{f}_i$.

The instantaneous frequency of Eq. (7.2.21) can also be written in differential form

$$f_i(t) = \lim_{\triangle t \to 0} \frac{1}{4\pi\triangle t}\{\arg[z(t+\triangle t)] - \arg[z(t-\triangle t)]\}. \tag{7.2.25}$$

Let the discrete sampling frequency be f_s, then using Eq. (7.2.25), the instantaneous frequency of the discrete-time signal $s(n)$ can be defined as

$$f_i(n) = \frac{f_s}{4\pi}\{\arg[z(n+1)] - \arg[z(n-1)]\}. \tag{7.2.26}$$

The instantaneous physical quantity corresponding to the time-domain signal $z(t)$ is the instantaneous frequency, while the instantaneous physical quantity corresponding to the frequency-domain signal $Z(f)$ is called the group delay $\tau_g(f)$. The group delay represents the (group) delay of each component of the frequency spectrum $Z(f)$ with frequency f, defined as

$$\tau_g(f) = -\frac{1}{2\pi}\frac{d}{df}\arg[Z(f)], \tag{7.2.27}$$

where $\arg[Z(f)]$ is the phase spectrum of the signal $z(t)$. If $Z(f) = A(f)e^{j\theta(f)}$, then $\arg[Z(f)] = \theta(f)$.

Similar to Eq. (7.2.25), the group delay can also be defined as

$$\tau_g(f) = \lim_{\triangle f\to 0}\frac{1}{4\pi\triangle f}\{\arg[Z(f+\triangle f)] - \arg[Z(f-\triangle f)]\}. \tag{7.2.28}$$

And the group delay of the discrete-time signal $z(n)$ is defined as

$$\tau_g(k) = \frac{1}{4\pi}\{\arg[Z(k+1)] - \arg[Z(k-1)]\}. \tag{7.2.29}$$

Like the instantaneous frequency, the group delay has its physical explanation. If the signal has a linear phase and its initial phase is zero, then the signal is delayed without distortion, and its delay time is the negative slope of the linear phase characteristic,i.e., Eq. (7.2.29). Although the general signal does not have linear phase characteristics, the phase characteristics within a very narrow frequency band around a certain frequency can still be approximated as linear, so it is reasonable to use the slope of its phase characteristic as the group delay of these components.

7.2.4 Exclusion Principle

Since the non-stationary signal analysis uses a joint time-frequency representation, is it possible to obtain the desired time resolution and frequency resolution at the same time? The answer to this question is no.

Let $s(t)$ be a zero-mean signal with finite energy, and $h(t)$ be a window function. The average time $\bar{t}_s$ and average frequency $\bar{\omega}_s$ of the signal $s(t)$ are defined as

$$\bar{t}_s \stackrel{\text{def}}{=} \int_{-\infty}^{\infty} t|s(t)|^2 dt, \tag{7.2.30}$$

$$\bar{\omega}_s \stackrel{\text{def}}{=} \int_{-\infty}^{\infty} \omega|S(\omega)|^2 d\omega, \tag{7.2.31}$$

where $S(\omega)$ is the Fourier transform of $s(t)$.

Similarly, the average time and average frequency of the window function $h(t)$ are defined as

$$\bar{t}_h \stackrel{\text{def}}{=} \int_{-\infty}^{\infty} t|h(t)|^2 dt, \tag{7.2.32}$$

$$\bar{\omega}_h \stackrel{\text{def}}{=} \int_{-\infty}^{\infty} \omega|H(\omega)|^2 d\omega. \tag{7.2.33}$$

The time width T_s and bandwidth B_s of signal $s(t)$ are defined as[82]

$$T_s^2 \stackrel{\text{def}}{=} \int_{-\infty}^{\infty} (t - \bar{t}_s)^2 |s(t)|^2 dt, \tag{7.2.34}$$

$$B_s^2 \stackrel{\text{def}}{=} \int_{-\infty}^{\infty} (\omega - \bar{\omega}_s)^2 |S(\omega)|^2 d\omega. \tag{7.2.35}$$

Time width and bandwidth can also be defined as

$$T_s^2 \stackrel{\text{def}}{=} \frac{\int_{-\infty}^{\infty} t^2 |s(t)|^2 dt}{\int_{-\infty}^{\infty} |s(t)|^2 dt}, \tag{7.2.36}$$

$$B_s^2 \stackrel{\text{def}}{=} \frac{\int_{-\infty}^{\infty} \omega^2 |S(\omega)|^2 d\omega}{\int_{-\infty}^{\infty} |S(\omega)|^2 d\omega}, \tag{7.2.37}$$

which are called the effective time width and effective bandwidth of signal $s(t)$, respectively.

A signal whose energy is approximately distributed within the time width $[-T/2, T/2]$ and bandwidth $[-B/2, B/2]$ is called a finite energy signal.

Consider the relationship between the variation of time width and bandwidth. Let the energy of the signal $s(t)$ be entirely within the time width $[-T/2, T/2]$, that is, the signal has a time width T in the strict sense. Let us see what happens when we stretch $s(t)$ along the time axis without changing the amplitude of the signal. Let $s_k(t) = s(kt)$ represent the stretched signal, where k is the stretching ratio ($k < 1$ corresponds to the compression of the signal in the time region and $k > 1$ corresponds to the stretching of the signal in time region). From the definition of the time width T_s, the time width of the stretched signal is k times that of the original signal, i.e., $T_{s_k} = kT_s$. In addition, calculating the Fourier transform of the stretched signal, we can get $S_k(\omega) = \frac{1}{k}S(\frac{\omega}{k})$, $k > 0$. From the definition of the bandwidth B_s, the bandwidth of the stretched signal is $\frac{1}{k}$ times that of the original signal, i.e., $B_{s_k} = \frac{1}{k}B_s$. Obviously, the product of the time width and bandwidth of the stretched signal is the same as that of the original signal, i.e., $T_{s_k}B_{s_k} = T_sB_s$. This conclusion shows that it is possible to have the relation $T_sB_s =$ constant for any signal. This basic relationship between the time width and bandwidth of a signal can be described in mathematical terms as follows.

Exclusion Principle: The product of time width and bandwidth of any signal $s(t)$ with finite energy or window function $h(t)$ satisfies the inequality

$$\text{Time width -bandwith product} = T_s B_s = \triangle t_s \triangle \omega_s \geq \frac{1}{2} \text{ or } T_h B_h = \triangle t_h \triangle \omega_h \geq \frac{1}{2}. \tag{7.2.38}$$

The exclusion principle is also known as the uncertainty principle or Heisenberg inequality.

The $\triangle t$ and $\triangle \omega$ in Eq. (7.2.38) are called the time resolution and frequency resolution, respectively. As the name implies, the time resolution and frequency resolution are the ability of the signal to differentiate between two time points and two frequency points, respectively. The exclusion principle suggests that time width and bandwidth (i.e., time resolution and frequency resolution) are contradictory quantities, and it is impossible to obtain arbitrarily high time resolution and frequency resolution at the same time. Two extreme examples are: the time width of the impulsive signal $s(t) = \delta(t)$ is zero and has the highest time resolution; while its bandwidth is infinite (its spectrum is equal to 1) and has no frequency resolution; the bandwidth of the unit DC signal $s(t) = 1$ is zero (its spectrum is an impulsive function) and has the highest frequency resolution, but its time width is infinite and its time resolution is zero. Only when the signal is a Gaussian function $e^{-\pi t^2}$, inequality (7.2.38) takes the equal sign.

The window function plays an important role in the non-stationary signal processing: whether the window function has a high time resolution and frequency resolution is related to the non-stationary characteristics of the signal to be analyzed. According to the above analysis, if the impulse function is used as the window function, it is equivalent to only taking the value of the non-stationary signal at the time t for analysis. The time resolution is the highest, but the frequency resolution is completely lost. On the contrary, if the unit DC signal is taken as the window function, that is, the infinitely long signal is analyzed like the Fourier transform, the frequency resolution is the highest, but there is no time resolution at all. This indicates that for non-stationary signals, the window function of the local transform must be chosen with an appropriate compromise between the time resolution and frequency resolution of the signal. It is worth emphasizing that for the local processing of a non-stationary signal with a window, the signal within the window function must be stationary, that is, the window width must be compatible with the local stationarity of the non-stationary signal. Therefore, the frequency resolution obtained by non-stationary signal analysis is related to the local stationary length of the signal. It is impossible to obtain high frequency resolution directly for a short non-stationary signal.

The above relationship between the window function and the local stationary length tells us that the time-frequency analysis is suitable for non-stationary signals with relatively large local stationary length; if the local stationary length is small, the time-frequency analysis is less effective. This point is important to note when performing time-frequency signal analysis. The relationship between the window width selection and the resolution of time-frequency analysis will be further discussed later.

7.3 Short-Time Fourier Transform

Although the instantaneous frequency and group delay are two useful physical quantities to describe the local characteristics of non-stationary signals, they are not applicable to multi-component signals. For example, if $z(t) = A(t)e^{j\phi(t)} = \sum_{i=1}^{p} z_i(t)$ is a p-component signal, we cannot obtain the instantaneous frequencies of each component signals from the derivative of the phase $\phi(t)$. In order to obtain the instantaneous frequencies of each component, an intuitive way is to introduce the concept of "local spectrum": a very narrow window function is used to extract the signal and calculate its Fourier transform. Since this spectrum is the spectrum of the signal in a narrow interval of the window function, excluding the spectrum of the signal outside the window function, it is appropriate to call it the local spectrum of the signal. Fourier transform using a narrow window function is customarily called the short-time Fourier transform, which is a form of the windowed Fourier transform. Windowed Fourier transform was first proposed by Gabor in 1946[82].

7.3.1 The Continuous Short-Time Fourier Transform

Let $g(t)$ be a window function with a short time width that slides along the time axis. Therefore, the continuous short-time Fourier transform (STFT) of signal $z(t)$ is defined as

$$\mathrm{STFT}_z(t,f) = \int_{-\infty}^{\infty} [z(u)g^*(u-t)]e^{-j2\pi fu}du, \tag{7.3.1}$$

where the superscript * denotes the complex conjugate. Obviously, if we take the infinite (global) rectangular window function $g(t) = 1, \forall t$, the short-time Fourier transform will degenerate into the traditional Fourier transform.

Since the signal $z(u)$ multiplied by a very short window function $g(u-t)$ is equivalent to taking a slice of the signal around the analysis time point t, $\mathrm{STFT}(t,f)$ can be understood as the Fourier transform of the signal $z(u)$ around the "analysis time" t (called the "local spectrum").

The continuous Fourier transform has the following basic properties.

Property 1 STFT is a linear time-frequency representation.

Property 2 STFT has frequency-shift invariance

$$\tilde{z}(t) = z(t)e^{j2\pi f_0 t} \rightarrow \mathrm{STFT}_{\tilde{z}}(t,f) = \mathrm{STFT}_z(t,f-f_0), \tag{7.3.2}$$

and does not have time-shift invariance

$$\tilde{z}(t) = z(t - t_0) \to \text{STFT}_{\tilde{z}}(t, f) = \text{STFT}_z(t - t_0, f)e^{-j2\pi t_0 f}, \tag{7.3.3}$$

i.e., does not satisfy $\text{STFT}_{\tilde{z}}(t, f) = \text{STFT}_z(t - t_0, f)$.

In signal processing, the traditional Fourier transform is called Fourier analysis, while the inverse Fourier transform is called Fourier synthesis, because the inverse Fourier transform uses the Fourier spectrum to reconstruct or synthesize the original signal. Similarly, STFT is also divided into analysis and synthesis. Obviously, in order to make STFT as a valuable tool for non-stationary signals, the signal $z(t)$ should be completely reconstructed by $\text{STFT}_z(t, f)$. The reconstruction equation is

$$p(u) = \int_{-\infty}^{\infty}\int_{-\infty}^{\infty} \text{STFT}_z(t, f)\gamma(u - t)e^{j2\pi fu}dtdf. \tag{7.3.4}$$

Substituting Eq. (7.3.1) into Eq. (7.3.4), it is easy to prove that

$$\begin{aligned} p(u) &= \int_{-\infty}^{\infty}\int_{-\infty}^{\infty} \left[\int_{-\infty}^{\infty} e^{-j2\pi f(t'-u)}df\right] z(t')g^*(t' - t)\gamma(u - t)dt'dt \\ &= \int_{-\infty}^{\infty}\int_{-\infty}^{\infty} z(t')g^*(t' - t)\gamma(u - t)\delta(t' - u)dt'dt. \end{aligned}$$

The well-known integral result $\int_{-\infty}^{\infty} e^{-j2\pi f(t'-u)}df = \delta(t' - u)$ is used here. Using the properties of the δ function, we immediately have

$$p(u) = z(u)\int_{-\infty}^{\infty} g^*(u - t)\gamma(u - t)dt = z(u)\int_{-\infty}^{\infty} g^*(t)\gamma(t)dt. \tag{7.3.5}$$

When the reconstructed result $p(u)$ is always equal to the original signal $z(t)$, such a reconstruction is called "complete reconstruction". It can be seen from the above equation that in order to achieve complete reconstruction that is, in order to make $p(u) = z(u)$, the window function $g(t)$ and $\gamma(t)$ must satisfy the condition

$$\int_{-\infty}^{\infty} g^*(t)\gamma(t)dt = 1, \tag{7.3.6}$$

which is known as the STFT complete reconstruction condition.

The complete reconstruction condition is a very wide condition, and for a given analytic window function $g(t)$, the synthesis window function $\gamma(t)$ satisfying condition Eq. (7.3.6) can have infinite choices. Then, how to choose an appropriate synthesis window function $\gamma(t)$? Here are three simplest options: (1) $\gamma(t) = g(t)$; (2) $\gamma(t) = \delta(t)$; (3) $\gamma(t) = 1$.

The most interested is the first choice $\gamma(t) = g(t)$, and the corresponding complete reconstruction condition Eq. (7.3.6) becomes

$$\int_{-\infty}^{\infty} |g(t)|^2 dt = 1. \tag{7.3.7}$$

This formula is called energy normalization. In this case, Eq. (7.3.4) can be written as

$$z(t) = \int_{-\infty}^{\infty}\int_{-\infty}^{\infty} \mathrm{STFT}_z(t', f')g(t - t')e^{j2\pi f' t'} dt' df'. \tag{7.3.8}$$

The above formula can be regarded as the generalized inverse short-time Fourier transform. Different from Fourier transform and inverse Fourier transform which are both 1-D transforms, the short-time Fourier transform Eq. (7.3.1) is a 1-D transform, and the generalized inverse short-time Fourier transform Eq. (7.3.8) is 2-D transform.

In summary, STFT can be regarded as the time-frequency analysis of non-stationary signals, while generalized inverse STFT is the time-frequency synthesis of non-stationary signals. This is why $g(t)$ and $\gamma(t)$ are called the analytic window function and synthesis window function, respectively.

The function $\mathrm{STFT}_z(t, f)$ can be regarded as the inner product of the signal $z(t)$ and the time shift-frequency modulation form $g_{t,f}(u)$ of the window function $g(u)$, i.e.,

$$\mathrm{STFT}_z(t, f) = \langle z, g_{t,f} \rangle, \tag{7.3.9}$$

where $\langle z, g_{t,f} \rangle = \int_{-\infty}^{\infty} z(u)g^*_{t,f}(u)du$ and

$$g_{t,f}(u) = g(u - t)e^{j2\pi fu}. \tag{7.3.10}$$

In principle, the analytic window function $g(t)$ can be chosen arbitrarily in the square integrable space, i.e. $L^2(R)$ space. However, in practical application, it is often desirable to choose a window function $g(t)$ that is a "narrow" time function, so that the integration of Eq. (7.3.1) is affected only by the value of $z(t)$ and its nearby values. Naturally, it is also desirable that the Fourier transform $G(f)$ of $g(t)$ is also a "narrow" function. In order to see the necessity of this requirement, it is useful to recall the following convolution theorem: the product $z(t)g(t)$ of two functions in the time-domain is equivalent to their convolution $Z(f) * G(f)$ in the frequency-domain. If the Fourier transform $G(f)$ of $g(t)$ is wide, the Fourier transform $Z(f)$ of the signal will be affected by $G(f)$ over a wide range of frequency after convolution. This is exactly what we want to be avoided. Unfortunately, according to the previous exclusion principle, the effective time-width τ_{eff} and bandwidth ω_{eff} of the window function $g(t)$ cannot be arbitrarily small, because their product obeys the Heisenberg inequality $\tau_{\mathrm{eff}}\omega_{\mathrm{eff}} \geq 0.5$ and $\tau_{\mathrm{eff}}\omega_{\mathrm{eff}} = 0.5$ when the window function takes a Gaussian function, i.e., $g(t) = e^{-\pi t^2}$. That is, the Gaussian window function has the best (i.e., the smallest) time-width-bandwidth product. In order to make the window function also has unit energy, it is often taken as

$$g^0(t) = 2^{1/4}e^{-\pi t^2}. \tag{7.3.11}$$

The resulting basis function $g^0_{t,f}(t') = g^0(t' - t)e^{j2\pi ft'}$ is called the "standard coherent state" in physics, while in engineering, it was introduced by Gabor when he proposed the windowed Fourier transform. Therefore, $g^0(t)$ is often referred to as Gabor atom and $g^0_{t,f}(t')$ is called a Gabor basis function. The Gabor basis function $g^0_{t,f}$ is highly concentrated in the time-frequency plane around the time-frequency point (t, f).

The practical purpose of the proposed STFT is mainly to understand the local frequency characteristics of the signal. The "local spectrum" has been mentioned repeatedly above. What is the connection between the "local spectrum" and the "global spectrum" based on the (global) Fourier transform? From Eq. (7.3.1), it can be seen that $\text{STFT}_z(t, f)$ at certain time t, that is the Fourier transform of $z(t')g^*(t' - t)$, is not only determined by the signal in the window function near time t, but also related to the window function $g(t)$ itself. Taking a single frequency signal with frequency f_0 as an example, the global spectrum based on the Fourier transform is an impulsive function $\delta(f_0)$ located at f_0. If such a non-time-varying signal is described by time-frequency representation (time as the horizontal axis, frequency as the vertical axis), the "local spectrum" of the signal in the time-frequency plane should be a horizontal impulse line function at f_0, i.e., the slice at any time t is the same impulse spectrum. However, this is not the case in practice, because the "local spectrum" obtained according to Eq. (7.3.1) is equal to $G(f - f_0)e^{j2\pi ft}$, where $G(f)$ represents the spectrum of the analytic window function $g(t)$. Therefore, the local characteristics of the single-frequency signal are expressed in the phase factor $e^{j2\pi ft}$, and the local spectrum is broadened by the spectrum $G(f)$ of the analytic window function. The narrower the window is, the wider the spectrum $G(f)$ is, and the wider the local spectrum of the single-frequency signal is. This indicates that the introduction of the analytical spectrum will reduce the resolution of the local spectrum. In order to maintain the resolution of the local spectrum, the analytic window should be wide, but when the window width exceeds the local stationary length of the non-stationary signal, the signal within the window function will be non-stationary, which in turn will cause the adjacent spectra to be mixed and thus not represent the local spectrum correctly. In other words, the window width should be appropriate to the local stationary length of the signal.

7.3.2 The Discrete Short-Time Fourier Transform

The continuous short-time Fourier transform is discussed above. For any practical application, $\text{STFT}_z(t, f)$ needs to be discretized, i.e., $\text{STFT}_z(t, f)$ is sampled at the equidistant time-frequency grid point (mT, nF), where $T > 0$ and $F > 0$ are the sampling periods of the time and frequency variables, respectively, while m and n are integers. For simplicity, the symbol $\text{STFT}(m, n) = \text{STFT}(mT, nF)$ is introduced. Thus, for a discrete signal

$z(k)$, it is easy to obtain the discretized form

$$\mathrm{STFT}(m,n) = \sum_{k=-\infty}^{\infty} z(k) g^*(kT - mT) e^{-j2\pi(nF)k} \tag{7.3.12}$$

of the short-time Fourier transform formula Eq. (7.3.1) and the discretized form

$$z(k) = \sum_{m=-\infty}^{\infty} \sum_{n=-\infty}^{\infty} \mathrm{STFT}(m,n) \gamma(kT - mT) e^{-j2\pi(nF)k} \tag{7.3.13}$$

of the generalized inverse short-time Fourier transform formula Eq. (7.3.4).

Eqs. (7.3.2) and (7.3.13) are called the discrete short-time Fourier transform and the inverse discrete short-time Fourier transform, respectively.

It should be noted that corresponding to the complete reconstruction constraint Eq. (7.3.6), the time sampling period T, the frequency sampling period F, the discrete analytic window $g(k)$ and the discrete synthesis window $\gamma(k)$ should satisfy the "complete reconstruction condition" in the discrete case

$$\frac{1}{F} \sum_{m=-\infty}^{\infty} g\left(kT + n\frac{1}{F} - mT\right) \gamma^*(kT - mT) = \delta(n), \quad \forall k. \tag{7.3.14}$$

Obviously, the above condition is more rigorous than the complete reconstruction condition $\int_{-\infty}^{\infty} g(t)\gamma^*(t)dt = 1$ in the continuous case. In particular, if $\gamma(k) = g(k)$ is chosen, the discrete short-time Fourier transform is

$$z(k) = \sum_{m=-\infty}^{\infty} \sum_{n=-\infty}^{\infty} \mathrm{STFT}(m,n) g(kT - mT) e^{j2\pi(nF)k}. \tag{7.3.15}$$

STFT has an important application in speech signal processing because the typical example of signal frequency components changing rapidly with time and being complex is human speech. In order to analyze speech signals, Koenig et al.[181] and Potter et al.[177] proposed the (acoustic) spectrogram methods as early as half a century ago. The spectrogram is defined as the square of the modulus of the short-time Fourier transform of the signal, i.e.,

$$\mathrm{SPEC}(t,\omega) = |\mathrm{STFT}(t,\omega)|^2. \tag{7.3.16}$$

The mean time of the signal $z(t)$, the window function $g(t)$, and the spectrogram are defined as

$$\bar{t}_z \stackrel{\text{def}}{=} \int_{-\infty}^{\infty} t|z(t)|^2 dt, \tag{7.3.17}$$

$$\bar{t}_g \stackrel{\text{def}}{=} \int_{-\infty}^{\infty} t|g(t)|^2 dt, \tag{7.3.18}$$

$$\bar{t}_{\mathrm{SPEC}} \stackrel{\text{def}}{=} \int_{-\infty}^{\infty}\int_{-\infty}^{\infty} t|\mathrm{SPEC}(t,\omega)|^2 dt d\omega. \tag{7.3.19}$$

The average frequency is defined as

$$\bar{\omega}_z \stackrel{\text{def}}{=} \int_{-\infty}^{\infty} \omega |Z(\omega)|^2 d\omega, \tag{7.3.20}$$

$$\bar{\omega}_g \stackrel{\text{def}}{=} \int_{-\infty}^{\infty} \omega |G(\omega)|^2 d\omega, \tag{7.3.21}$$

$$\bar{\omega}_{\text{SPEC}} \stackrel{\text{def}}{=} \int_{-\infty}^{\infty}\int_{-\infty}^{\infty} \omega |\text{SPEC}(t,\omega)|^2 dt d\omega. \tag{7.3.22}$$

Using these physical quantities, the time-width of the signal, window function, and spectrogram can be defined as

$$T_z^2 \stackrel{\text{def}}{=} \int_{-\infty}^{\infty} (t-\bar{t}_z)^2 |z(t)|^2 dt, \tag{7.3.23}$$

$$T_g^2 \stackrel{\text{def}}{=} \int_{-\infty}^{\infty} (t-\bar{t}_g)^2 |g(t)|^2 dt, \tag{7.3.24}$$

$$T_{\text{SPEC}}^2 \stackrel{\text{def}}{=} \int_{-\infty}^{\infty}\int_{-\infty}^{\infty} (t-\bar{t}_{\text{SPEC}})^2 |\text{SPEC}(t,\omega)|^2 dt d\omega. \tag{7.3.25}$$

and the bandwidth are

$$B_z^2 \stackrel{\text{def}}{=} \int_{-\infty}^{\infty} (\omega-\bar{\omega}_z)^2 |Z(\omega)|^2 d\omega, \tag{7.3.26}$$

$$B_g^2 \stackrel{\text{def}}{=} \int_{-\infty}^{\infty} (\omega-\bar{\omega}_g)^2 |G(\omega)|^2 d\omega, \tag{7.3.27}$$

$$B_{\text{SPEC}}^2 \stackrel{\text{def}}{=} \int_{-\infty}^{\infty}\int_{-\infty}^{\infty} (\omega-\bar{\omega}_{\text{SPEC}})^2 |\text{SPEC}(t,\omega)|^2 dt d\omega. \tag{7.3.28}$$

By direct calculation, it is easy to verify that the following relationships exist between the spectrogram SPEC and the signal $z(t)$, the window function $g(t)$

$$\bar{t}_{\text{SPEC}} = \bar{t}_z - \bar{t}_g, \tag{7.3.29}$$

$$\bar{\omega}_{\text{SPEC}} = \bar{\omega}_z + \bar{\omega}_g, \tag{7.3.30}$$

$$T_{\text{SPEC}}^2 = T_z^2 + T_g^2, \tag{7.3.31}$$

$$B_{\text{SPEC}}^2 = B_z^2 + B_g^2. \tag{7.3.32}$$

The last two equations are the relationship between the time-width and bandwidth of the spectrogram, the signal, and the window function.

STFT can be considered as a windowed Fourier transform with a very short window function. When the window function takes other forms, other types of window Fourier transforms can be obtained, such as the Gabor transform, which will be described in the next section.

7.4 Gabor Transform

Using series as the expansion form of signal or function is an important signal processing method. According to whether the basis function is orthogonal or not, series expansion can be divided into orthogonal series expansion and non-orthogonal series expansion. Fourier series in Fourier analysis is a typical orthogonal series expansion. This section introduces a non-orthogonal expansion of the signal — Gabor expansion, which was proposed by Gabor in 1949[82]. The integral expression of the Gabor expansion coefficient is called Gabor transform. Now, Gabor expansion and Gabor transform are recognized as one of the best methods of signal representation, especially image representation, in communication and signal processing.

7.4.1 The Continuous Gabor Transform

Let $\phi(t)$ be the real continuous time signal of interest and sample the signal at time interval T. The time and frequency joint function Φ of signal $\phi(t)$ is introduced, which is defined as

$$\Phi(t,f) = \sum_{m=-\infty}^{\infty} \phi(t+mT)e^{-j2\pi fmT} \tag{7.4.1}$$

and is called the complex spectrogram of signal $\phi(t)$. Assume that $g(t)$ is a window function added to signal $\phi(t)$ and

$$G(t,f) = \sum_{m=-\infty}^{\infty} g(t+mT)e^{-j2\pi fmT} \tag{7.4.2}$$

is defined as the complex spectrogram of the window function $g(t)$.

How to use the complex spectrogram $G(t,f)$ of the window function to represent the complex spectrogram $\Phi(t,f)$ of the signal? A simple way is to take

$$\Phi(t,f) = A(t,f)G(t,f), \tag{7.4.3}$$

where $A(t,f)$ is defined as

$$A(t,f) = \sum_{m=-\infty}^{\infty}\sum_{n=-\infty}^{\infty} a_{mn}e^{-j2\pi(mTf-nFt)}. \tag{7.4.4}$$

Here F represents the frequency sampling interval of signal $\phi(t)$.

Substituting Eqs. (7.4.1), (7.4.2) and (7.4.4) into Eq. (7.4.3), then comparing the coefficients of the same power on the left and right sides to obtain

$$\phi(t) = \sum_{m=-\infty}^{\infty} \sum_{n=-\infty}^{\infty} a_{mn} g_{mn}(t), \tag{7.4.5}$$

where

$$g_{mn}(t) = g(t - mT)e^{j2\pi nFt}. \tag{7.4.6}$$

Eq. (7.4.5) is the expansion form of signal $\phi(t)$ proposed by Gabor half a century ago[82], which is now customarily called the continuous Gabor expansion of the (continuous) signal $\phi(t)$. The coefficient a_{mn} is called the Gabor expansion coefficient, and $g_{mn}(t)$ is called the (m, n)-th order Gabor basis function or Gabor atom.

Since the Gabor basis function $g_{mn}(t)$ is constructed only by the translation and modulation of the generating function $g(t)$, if $g(t)$ is a non-orthogonal function, the Gabor basis function $g_{mn}(t)$ is also non-orthogonal. Therefore, the Gabor expansion is a non-orthogonal series expansion. In mathematics, a non-orthogonal series expansion of function is called atomic expansion; in physics, a non-orthogonal expansion is a series expansion concerning a discrete set of coherent states. This is why Gabor basis functions are also called Gabor atoms.

The sampling that satisfies $TF = 1$ is called critical sampling, and the corresponding Gabor expansion is called critical sampling Gabor expansion. In addition, there are two other kinds of Gabor expansions:

(1) undersampling Gabor expansion: $TF > 1$;
(2) oversampling Gabor expansion: $TF < 1$.

It has been shown that undersampling Gabor expansion leads to numerical instability[71], so it is not a practical method and will not be discussed in this book.

The Gabor expansion and Gabor transform in the case of critical sampling and oversampling are discussed below.

1. The Critical Sampling Gabor Expansion

Although the critical sampling Gabor expansion was introduced as early as 1946, there is no good method to determine the coefficients of Gabor expansion, so it has been sleeping for more than 30 years. It was only in 1981 that Bastiaans proposed a simple and effective method[21], which made Gabor expansion develop rapidly.

This method of Bastiaans is called Bastiaan analytic method and its basic idea is to introduce an auxiliary function $\Gamma(t, f)$ under the assumption that $G(t, f)$ can be divided, which is $1/T$ times the conjugate reciprocal of $G(t, f)$, i.e.,

$$\Gamma(t, f)G^*(t, f) = \frac{1}{T} = F, \tag{7.4.7}$$

where

$$\Gamma(t, f) = \sum_{m=-\infty}^{\infty} \gamma(t + mT)e^{-j2\pi fmT}. \tag{7.4.8}$$

Substituting Eq. (7.4.3) into Eq. (7.4.7) yields

$$\frac{1}{T}A(t,f) = \Phi(t,f)\Gamma(t,f). \tag{7.4.9}$$

Then substitute Eqs. (7.4.3), (7.4.4) and (7.4.8) into Eq. (7.4.9) and compare the coefficients of the same power on the left and right sides of this equation, which gives an important formula

$$a_{mn} = \int_{-\infty}^{\infty} \phi(t)\gamma^{*}(t - mT)e^{-j2\pi nFt}dt = \int_{-\infty}^{\infty} \phi(t)\gamma_{mn}^{*}(t)dt, \tag{7.4.10}$$

where

$$\gamma_{mn}(t) = \gamma(t - mT)e^{-j2\pi nFt}. \tag{7.4.11}$$

Eq. (7.4.10) is called the Gabor transform of the signal $\phi(t)$. It shows that when the signal $\phi(t)$ and the auxiliary function $\gamma(t)$ are given, the Gabor expansion coefficient a_{mn} can be obtained using the Gabor transform.

To sum up, the two important issues that need to be addressed when using Eq. (7.4.5) for the Gabor expansion of the signal $\phi(t)$ are

(1) Choosing the window function $g(t)$ in order to construct the Gabor basis function $g_{mn}(t)$ using Eq. (7.4.6);
(2) Selecting the auxiliary function $\gamma(t)$ and calculating the Gabor transform Eq. (7.4.10) to obtain the Gabor expansion coefficients a_{mn}.

Obviously, the key to Gabor expansion is the choice of window function $g(t)$ and auxiliary function $\gamma(t)$.

In the following, the relationship between these two functions is discussed. First, the relationship between $\gamma_{mn}(t)$ and $g_{mn}(t)$ is investigated. Therefore, substituting Eq. (7.4.10) into Eq. (7.4.5) to obtain

$$\begin{aligned}\phi(t) &= \sum_{m=-\infty}^{\infty}\sum_{n=-\infty}^{\infty}\int_{-\infty}^{\infty} \phi(t')\gamma_{mn}^{*}(t')g_{mn}(t)dt' \\ &= \int_{-\infty}^{\infty} \phi(t') \sum_{m=-\infty}^{\infty}\sum_{n=-\infty}^{\infty} g_{mn}(t)\gamma_{mn}^{*}(t')dt'.\end{aligned}$$

This is the reconstruction formula for the signal. If the above equation holds for all time t, then the signal $\phi(t)$ is said to be completely reconstructed. In this case, it is required that $g_{mn}(t)$ and $\gamma_{mn}(t)$ satisfy

$$\sum_{m=-\infty}^{\infty}\sum_{n=-\infty}^{\infty} g_{mn}(t)\gamma_{mn}^{*}(t') = \delta(t - t'). \tag{7.4.12}$$

This is the complete reconstruction formula of Gabor expansion.

Although Eq. (7.4.12) is important, it is inconvenient to use. More practical is the relationship between $g(t)$ and $\gamma(t)$. It can be shown that they should satisfy the relation

$$\int_{-\infty}^{\infty} g(t)\gamma^{*}(t-mT)e^{-j2\pi nFt} = \delta(m)\delta(n). \tag{7.4.13}$$

This relation is called the biorthogonal relation between the window function $g(t)$ and the auxiliary window function $\gamma(t)$. The so-called biorthogonal means that $\gamma(t)$ is orthogonal to $g(t)$ as long as one of m, n in the (m, n)th-order Gabor expansion is not zero. Thus, the auxiliary function $\gamma(t)$ is often called the biorthogonal function of the window function $g(t)$.

In summary, after choosing a suitable Gabor basis function $g(t)$, the analytic method for determining the Gabor expansion coefficients can be carried out in two steps:

(1) Solving the biorthogonal Eq. (7.4.13) to obtain the auxiliary function $\gamma(t)$;
(2) Calculating the Gabor transform Eq. (7.4.10) to obtain the Gabor expansion coefficient a_{mn}.

It can be seen that the introduction of the auxiliary function $\Gamma(t, f)$ makes the determination of the Gabor expansion coefficient a_{mn} very simple, thus solving a difficult problem that has long plagued Gabor expansion.

Interestingly, the biorthogonal relation Eq. (7.4.13) still holds after exchanging the function $g(t)$ and $\gamma(t)$. By extension, the functions $g(t)$ and $\gamma(t)$ in each of the relevant equations obtained from the above discussion can be interchanged. In other words, Gabor expansion Eq. (7.4.5) and Gabor transform Eq. (7.4.10) of signal $\phi(t)$ can also take the following dual forms

$$\phi(t) = \sum_{m=-\infty}^{\infty}\sum_{n=-\infty}^{\infty} a_{mn}\gamma(t-mT)e^{j2\pi nFt} \tag{7.4.14}$$

$$= \sum_{m=-\infty}^{\infty}\sum_{n=-\infty}^{\infty} a_{mn}\gamma_{mn}(t) \tag{7.4.15}$$

and

$$a_{mn} = \int_{-\infty}^{\infty} \phi(t)g^{*}(t-mT)e^{-j2\pi nFt}dt = \int_{-\infty}^{\infty} \phi(t)g^{*}_{mn}(t)dt. \tag{7.4.16}$$

Therefore, $\gamma(t)$ is often called the dual function of $g(t)$. Obviously, $\gamma_{mn}(t)$ and the Gabor basis function $g_{mn}(t)$ are dual, so $\gamma_{mn}(t)$ is also called dual Gabor basis function.

Here are some examples of window function $g(t)$ and its dual function $\gamma(t)$:

(1) Rectangular window function

$$g(t) = \left(\frac{1}{T}\right)^{1/2} p\left(2\frac{t}{T}\right), \quad (7.4.17)$$

$$\gamma(t) = \left(\frac{1}{T}\right)^{1/2} p\left(2\frac{t}{T}\right). \quad (7.4.18)$$

(2) Generalized rectangular window function

$$g(t) = \left(\frac{1}{T}\right)^{1/2} p\left(2\frac{t}{T}\right) f(t), \quad (7.4.19)$$

$$\gamma(t) = \left(\frac{1}{T}\right)^{1/2} p\left(2\frac{t}{T}\right) \frac{1}{f^*(t)}, \quad (7.4.20)$$

where $f(t)$ is an arbitrary function of t.

(3) Gaussian window function

$$g(t) = \left(\frac{\sqrt{2}}{T}\right)^{1/2} e^{-\pi(t/T)^2}, \quad (7.4.21)$$

$$\gamma(t) = \left(\frac{1}{\sqrt{2}T}\right)^{1/2} e^{\pi(t/T)^2} \sum_{n+\frac{1}{2}\geq\frac{1}{T}} (-1)^n e^{-\pi(n+t/T)^2}. \quad (7.4.22)$$

The typical Gabor basis function is

$$g_{mn}(t) = g_T(t - mT)e^{j2\pi nFt}, \quad (7.4.23)$$

where $g_T(t)$ is a Gaussian function, that is,

$$g_T(t) = e^{-\pi(t/T)^2}. \quad (7.4.24)$$

2. The Oversampling Gabor Expansion

For the oversampling case, let the time sampling interval be T_1, the frequency sampling interval be F_1, and $T_1F_1 < 1$. The formulas for the oversampling Gabor expansion and Gabor transform have the same form as those of the critical sampling Gabor expansion and Gabor transform

$$\phi(t) = \sum_{m=-\infty}^{\infty} \sum_{n=-\infty}^{\infty} a_{mn} g_{mn}(t), \quad (7.4.25)$$

$$a_{mn} = \int_{-\infty}^{\infty} \phi(t)\gamma^*(t)dt. \quad (7.4.26)$$

And the Gabor basis function $g_{mn}(t)$ and the dual Gabor basis function $\gamma(t)$ are defined as

$$g_{mn}(t) = g(t - mT_1)e^{j2\pi nF_1 t}, \quad (7.4.27)$$

$$\gamma_{mn}(t) = \gamma(t - mT_1)e^{j2\pi nF_1 t}. \quad (7.4.28)$$

The main difference between the oversampling and critical sampling is that the relationship between the Gabor basis function $g(t)$ and its dual function $\gamma(t)$ needs to be modified. Specifically, the biorthogonal formula Eq.(7.4.13) for critical sampling needs to be modified to[222]

$$\int_{-\infty}^{\infty} g(t)\gamma^*(t - mT_0)e^{-j2\pi nF_0 t} = \frac{T_1}{T_0}\delta(m)\delta(n), \quad T_0 = \frac{1}{F_1}, \quad F_0 = \frac{1}{T_1}. \tag{7.4.29}$$

The complete reconstruction formula Eq.(7.4.12) for critical sampling needs to be modified to

$$\sum_{m=-\infty}^{\infty} g(t - mT_1)\gamma^*(t - mT_1 + nT_0) = \frac{1}{T_0}\delta(n). \tag{7.4.30}$$

Since the right side of Eq. (7.4.29) is multiplied by a non-1 factor, Eq. (7.4.29) is often called a quasi-biorthogonal formula. Similarly, Eq. (7.4.30) is called the quasi-orthogonal formula.

It is worth pointing out that the critical sampling Gabor expansion and Gabor transform do not contain redundancy, which is reflected in the fact that when $g(t)$ is given, the dual function $\gamma(t)$ satisfying the complete reconstruction condition Eq. (7.4.12) is uniquely determined. However, the oversampling Gabor expansion and Gabor transform bring redundancy, because for a given $g(t)$, the dual function $\gamma(t)$ satisfying the full reconstruction condition Eq. (7.4.30) has multiple possible solutions.

Defining the matrix

$$\boldsymbol{W}(t) = \{w_{ij}(t)\} \text{ and } \tilde{\boldsymbol{W}}(t) = \{\tilde{w}_{ij}(t)\}, \tag{7.4.31}$$

where

$$w_{ij}(t) = g[t + (iT_1 - jT_0)] \text{ and } \tilde{w}_{ij}(t) = T_1\gamma^*[t - (iT_0 - jT_1)], \tag{7.4.32}$$

where $i, j = -\infty, \cdots, \infty$. Note that $\boldsymbol{W}(t)$ and $\tilde{\boldsymbol{W}}(t)$ are infinite dimensional matrices. It is easy to verify that the complete reconstruction condition Eq. (7.4.30) can be written as

$$\boldsymbol{W}(t)\tilde{\boldsymbol{W}}(t) = \boldsymbol{I}, \tag{7.4.33}$$

where $\boldsymbol{I}$ is an identity matrix.

The minimum norm solution of matrix equation (7.4.33) is given by

$$\tilde{\boldsymbol{W}}(t) = \boldsymbol{W}^T(t)[\boldsymbol{W}(t)\boldsymbol{W}^T(t)]^{-1}. \tag{7.4.34}$$

The auxiliary function $\gamma(t)$ corresponding to the matrix $\tilde{\boldsymbol{W}}(t)$ is called the optimal biorthogonal function of $g(t)$.

Defining the Zak transform of $g(t)$ as

$$\text{Zak}[g(t)] = \hat{g}(t, f) = \sum_{k=-\infty}^{\infty} g(t - k)e^{-j2\pi kf}, \tag{7.4.35}$$

it can be shown[222] that the biorthogonal function $\gamma(t)$ of $g(t)$ can be calculated as

$$\gamma(t) = 2\pi \int_0^1 \frac{df}{\tilde{g}^*(t,f)}. \tag{7.4.36}$$

Once the $\gamma(t)$ is obtained, the Gabor transform can be calculated directly using Eq. (7.4.26).

In the Gabor expansion of a continuous time signal $\phi(t)$, the Gabor basis function $g_{mn}(t)$ is usually required to obey the energy normalization condition

$$\int_{-\infty}^{\infty} |g_{mn}(t)|^2 dt = 1. \tag{7.4.37}$$

It is necessary to compare the similarities and differences between Gabor transform and STFT. Comparing Eq. (7.4.10) with Eq. (7.3.1), we know that the Gabor transform and STFT are similar in form, but there are the following essential differences between them:

(1) The window function $g(t)$ of STFT must be a narrow window, while the window function $\gamma(t)$ of Gabor transform has no such limitation. Therefore, the Gabor transform can be regarded as a windowed Fourier transform, which has a wider application than STFT;
(2) $\text{STFT}(t,f)$ is a time-frequency 2-D representation of the signal, while the Gabor transform coefficient a_{mn} is a time-shift-frequency-modulated 2-D representation of the signal, because it can be seen from Eq. (7.4.10) that the parameter m is equivalent to the time-shift mT units of the signal $\phi(t)$, while the role of n is reflected in the frequency modulation of the signal $\phi(t)$ using the exponential function $e^{j2\pi nFt}$.

7.4.2 The Discrete Gabor Transform

A sampling of time variables leads to periodicity in the frequency domain, while sampling of frequency variables leads to periodicity in the time domain. Because time and frequency need to be discretized at the same time, the discrete form of the Gabor transform (referred to as the discrete Gabor transform) is only applicable to discrete-time periodic signals. In the following, the discrete-time periodic signal is denoted by $\tilde{\phi}(k)$, and the discrete Gabor expansion coefficient and window function of the periodic signal are denoted by $\tilde{a}_{mn}$ and $\tilde{g}(k)$, respectively.

Let the period of the discrete periodic signal $\tilde{\phi}(k)$ be L, i.e., $\tilde{\phi}(k) = \tilde{\phi}(k + L)$, and its discrete Gabor expansion is defined as[222]

$$\tilde{\phi}(k) = \sum_{m=0}^{M-1} \sum_{n=0}^{N-1} \tilde{a}_{mn} \tilde{g}(k - m\triangle_M) e^{j2\pi nk\triangle_N}, \tag{7.4.38}$$

where the Gabor expansion coefficient is

$$\tilde{a}_{mn} = \sum_{k=0}^{L-1} \tilde{\phi}(k)\tilde{\gamma}^*(k - m\triangle_M)e^{-j2\pi nk\triangle_N}, \tag{7.4.39}$$

where $\triangle_M$ and $\triangle_N$ are the time and frequency sampling intervals, respectively, while M and N are the number of samples sampled at time and frequency, respectively.

The oversampling rate is defined as

$$\alpha = \frac{L}{\triangle_M \triangle_N} \tag{7.4.40}$$

and requires $M\triangle_M = N\triangle_N = L$. Substituting this relationship into Eq. (7.4.40), and the definition of the oversampling rate can be rewritten as

$$\alpha = \frac{\text{Number of Gabor expansion coefficients } MN}{\text{Number of signal samples } L}. \tag{7.4.41}$$

When $\alpha = 1$, the discrete Gabor transform is critical sampling, and the number of Gabor expansion coefficients is equal to the number of signal samples. If $\alpha > 1$, then the discrete Gabor transform is oversampling, i.e., the number of Gabor expansion coefficients is more than the number of signal samples. In other words, the Gabor expansion contains redundancy at this point.

In the following, the discrete Gabor expansion and discrete Gabor transform for the critical sampling and oversampling cases are described, respectively.

1. The Discrete Gabor Transform for the Critical Sampling Case

In the critical sampling case, choosing M to satisfy

$$L = MN, \tag{7.4.42}$$

then the discrete Gabor expansion and Gabor transform become

$$\tilde{\phi}(k) = \sum_{m=0}^{M-1}\sum_{n=0}^{N-1} \tilde{a}_{mn}\tilde{g}_{mn}(k), \tag{7.4.43}$$

$$\tilde{a}_{mn} = \sum_{k=0}^{L-1} \tilde{\phi}(k)\tilde{\gamma}^*_{mn}(k), \tag{7.4.44}$$

where

$$\tilde{g}_{mn}(k) = \tilde{g}(k - mN)e^{j2\pi nk/N}, \tag{7.4.45}$$

$$\tilde{\gamma}_{mn}(k) = \tilde{\gamma}(k - mN)e^{j2\pi nk/N}, \tag{7.4.46}$$

and $\tilde{g}(k)$ is a periodic Gabor basis function with period L, i.e.,

$$\tilde{g}(k) = \sum_{l} \tilde{g}(k + lL) = \tilde{g}(k + L). \tag{7.4.47}$$

And $\tilde{\gamma}(k)$ is also a periodic sequence that satisfies the biorthogonality condition

$$\sum_{k=0}^{L-1}[\tilde{g}(k+mN)e^{-j2\pi nk/N}]\tilde{\gamma}^*(k) = \sum_{k=0}^{L-1}[\tilde{g}^*(k+mN)e^{j2\pi nk/N}]\tilde{\gamma}(k) = \delta(m)\delta(n), \quad (7.4.48)$$

where $0 \le m \le M-1$ and $0 \le n \le N-1$.

For a predetermined L, there may be multiple choices of M and N satisfying the decomposition $L = MN$. Therefore, the discrete Gabor expansion and Gabor transform in the critical sampling case are generally non-uniquely defined, in contrast to the fact that the continuous Gabor expansion and Gabor transform are uniquely determined.

The biorthogonal condition Eq. (7.4.48) can be written in matrix form

$$\boldsymbol{W}\boldsymbol{\gamma} = \boldsymbol{e}_1, \quad (7.4.49)$$

where

$$\boldsymbol{W} = \begin{bmatrix} \boldsymbol{W}^{(0)} & \boldsymbol{W}^{(1)} & \cdots & \boldsymbol{W}^{(M-1)} \\ \boldsymbol{W}^{(1)} & \boldsymbol{W}^{(2)} & \cdots & \boldsymbol{W}^{(0)} \\ \vdots & \vdots & \vdots & \vdots \\ \boldsymbol{W}^{(M-1)} & \boldsymbol{W}^{(0)} & \cdots & \boldsymbol{W}^{(M-2)} \end{bmatrix}, \quad (7.4.50)$$

$$\boldsymbol{\gamma} = [\tilde{\gamma}(0), \tilde{\gamma}(1), \cdots, \tilde{\gamma}(L-1)]^T, \quad (7.4.51)$$

$$\boldsymbol{e}_1 = [1, 0, \cdots, 0]^T, \quad (7.4.52)$$

while $\boldsymbol{W}^{(i)}$ is a $N \times N$ matrix

$$\boldsymbol{W}^{(i)} = \begin{bmatrix} \tilde{g}^*(iN)w^0 & \tilde{g}^*(iN+1)w^0 & \cdots & \tilde{g}^*(iN+N-1)w^0 \\ \tilde{g}^*(iN)w^0 & \tilde{g}^*(iN+1)w^1 & \cdots & \tilde{g}^*(iN+N-1)w^{N-1} \\ \vdots & \vdots & \vdots & \vdots \\ \tilde{g}^*(iN)w^0 & \tilde{g}^*(iN+1)w^{N-1} & \cdots & \tilde{g}^*(iN+N-1)w^1 \end{bmatrix}, \quad w = e^{j2\pi/N}. \quad (7.4.53)$$

The least-squares solution of Eq. (7.4.49) is

$$\boldsymbol{\gamma} = \boldsymbol{W}^{-1}\boldsymbol{e}_1. \quad (7.4.54)$$

Once $\boldsymbol{\gamma}$ is obtained, $\tilde{\gamma}(0), \cdots, \tilde{\gamma}(N-1)$ can be found. Then, using Eqs. (7.4.46) and (7.4.44), $\tilde{\gamma}_{mn}(t)$ and the Gabor expansion coefficient $\tilde{a}_{mn}$ can be estimated successively.

2. The Discrete Gabor Transform for the Oversampling Case

In the case of oversampling ($MN > L$), the period L of the discrete-time periodic function $\tilde{\phi}(k)$ is decomposed into

$$L = \bar{N}M = N\bar{M}, \quad (7.4.55)$$

where $\bar{N}, N, M, \bar{M}$ are positive integers, $\bar{N} < N$ and $\bar{M} < M$. In this case, the Gabor expansion of the periodic signal is

$$\tilde{\phi}(k) = \sum_{m=0}^{M-1}\sum_{n=0}^{N-1} \tilde{a}_{mn}\tilde{g}_{mn}(k). \quad (7.4.56)$$

And the Gabor expansion coefficient is determined by the discrete Gabor transform

$$\tilde{a}_{mn} = \sum_{k=0}^{L-1} \tilde{\phi}(k)\tilde{\gamma}^*_{mn}(k), \tag{7.4.57}$$

where

$$\tilde{g}_{mn}(k) = \tilde{g}(k - m\bar{N})e^{j2\pi nk/N}, \tag{7.4.58}$$

$$\tilde{\gamma}_{mn}(k) = \tilde{\gamma}(k - m\bar{N})e^{j2\pi nk/N}. \tag{7.4.59}$$

The discrete sequence $\tilde{g}(k)$ is defined by Eq. (7.4.47), and $\tilde{\gamma}$ obeys the approximate biorthogonal condition

$$\sum_{k=0}^{L-1}[\tilde{g}^*(k + mN)e^{j2\pi nk/\bar{N}}]\tilde{\gamma}(k) = \frac{L}{MN}\delta(m)\delta(n), \tag{7.4.60}$$

or is written in matrix form

$$\boldsymbol{W}\gamma = \boldsymbol{b}, \tag{7.4.61}$$

where

$$\boldsymbol{W} = \begin{bmatrix} \boldsymbol{W}^{(0)} & \boldsymbol{W}^{(1)} & \cdots & \boldsymbol{W}^{(\bar{M}-1)} \\ \boldsymbol{W}^{(1)} & \boldsymbol{W}^{(2)} & \cdots & \boldsymbol{W}^{(0)} \\ \vdots & \vdots & \vdots & \vdots \\ \boldsymbol{W}^{(\bar{M}-1)} & \boldsymbol{W}^{(0)} & \cdots & \boldsymbol{W}^{(\bar{M}-2)} \end{bmatrix}, \tag{7.4.62}$$

$$\gamma = [\tilde{\gamma}(0), \tilde{\gamma}(1), \cdots, \tilde{\gamma}(L-1)]^T, \tag{7.4.63}$$

$$\boldsymbol{b} = [L/(MN), 0, \cdots, 0]^T. \tag{7.4.64}$$

The matrix equation Eq. (7.4.61) is an underdetermined equation with infinite solutions and its minimum norm solution

$$\gamma = \boldsymbol{W}^H(\boldsymbol{W}\boldsymbol{W}^H)^{-1}\boldsymbol{b} \tag{7.4.65}$$

is uniquely determined. In this case, the window function sequence $\tilde{\gamma}(0), \tilde{\gamma}(1), \cdots, \tilde{\gamma}(L-1)$ has the minimum energy.

It is necessary to point out that in many applications (e.g., signal feature extraction and classification), the Gabor expansion coefficients a_{mn} can be used as features of the signal. This kind of application only uses the Gabor transform and does not need to perform Gabor expansion on the signal. In these cases, it is only necessary to select a basic window function, and it is not necessary to determine its dual function.

7.5 Fractional Fourier transform

Short-Time Fourier Transform and Gabor transform belong to the windowed Fourier transform. This section introduces another generalized form of windowed Fourier transform – fractional Fourier transform.

The fractional power theory of the Fourier transform was first established by Namias in 1980[213], and this generalized Fourier transform is called fractional Fourier transform (FRFT). Later, McBride and Kerr[152] made a more rigorous mathematical definition of fractional Fourier transform, which made it have some important properties.

7.5.1 Definition and Properties of Fractional Fourier Transform

The functions $g(t)$ and $G(\omega)$ are called (symmetric) Fourier transform pairs if

$$G(\omega) = \frac{1}{\sqrt{2\pi}} \int_{-\infty}^{\infty} g(t)e^{-j\omega t}dt, \tag{7.5.1}$$

$$g(t) = \frac{1}{\sqrt{2\pi}} \int_{-\infty}^{\infty} G(\omega)e^{j\omega t}d\omega. \tag{7.5.2}$$

Let $\mathfrak{F}$ and $\mathfrak{F}^{-1}$ denote the Fourier transform operator and the inverse Fourier transform operator, i.e., $G = \mathfrak{F}g$ and $g = \mathfrak{F}^{-1}G$.

If n is an integer and the integer power $\mathfrak{F}^n$ of the Fourier transform represents the n-th order Fourier transform of function $g(t)$, it is easy to conclude that:

(1) The 1st order Fourier transform of function $g(t)$ is its spectrum $G(\omega)$, i.e., $\mathfrak{F}^1 g(t) = G(\omega)$;
(2) The 2nd order Fourier transform of function $g(t)$ is $g(-t)$, because $\mathfrak{F}^2 g(t) = \mathfrak{F}[\mathfrak{F}g(t)] = \mathfrak{F}G(\omega) = g(-t)$;
(3) The 3rd order Fourier transform of function $g(t)$ is $G(-\omega)$, because $\mathfrak{F}^3 g(t) = \mathfrak{F}[\mathfrak{F}^2 g(t)] = \mathfrak{F}g(-t) = G(-\omega)$;
(4) The 4nd order Fourier transform of function $g(t)$ is $g(t)$ itself, i.e., it is equivalent to the zero order Fourier transform, because $\mathfrak{F}^4 g(t) = \mathfrak{F}[\mathfrak{F}^3 g(t)] = \mathfrak{F}G(-\omega) = g(t) = \mathfrak{F}^0 g(t)$.

In the two-dimensional time-frequency plane, the 1st order Fourier transform is equivalent to rotating the time axis counterclockwise by $\frac{\pi}{2}$, the 2nd order Fourier transform is equivalent to rotating the time axis counterclockwise by $2 \cdot \frac{\pi}{2}$, etc. More generally, the n-th order Fourier transform is equivalent to rotating the time axis by $n \cdot \frac{\pi}{2}$.

If let $\alpha = n \cdot \pi/2$ and use the rotation operator $R^\alpha = R^{n\cdot\pi/2} = \mathfrak{F}^n$ to represent the n-th order Fourier transform, then the rotation operator has the following properties:

(1) Zero rotation: the zero rotation operator $R^0 = I$ is an identity operator;
(2) Equivalence with Fourier transform: $R^{\pi/2} = \mathfrak{F}^1$;
(3) Additivity of rotations: $R^{\alpha+\beta} = R^\alpha R^\beta$;
(4) 2π rotation (identity operator): $R^{2\pi} = I$.

Consider an interesting and important question: as shown in Fig. 7.5.1, if the rotation angle $\alpha = p \cdot \frac{\pi}{2}$, where p is a positive fraction, what kind of linear transformation can be obtained?

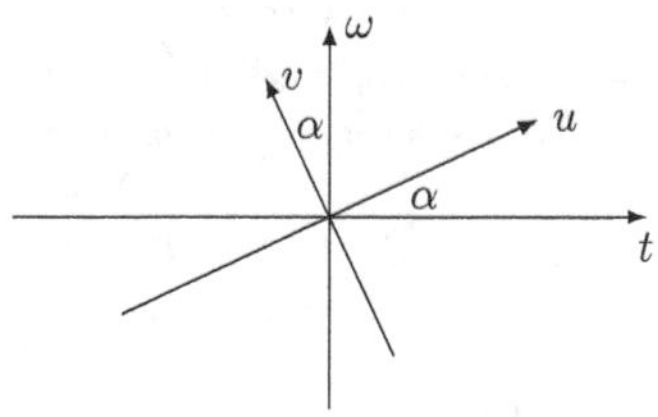

Fig. 7.5.1: (t, ω)-plane rotated to (u, v)-plane

Definition 7.5.1. *(Continuous fractional Fourier transform)*[38] *Let $\alpha = p \cdot \frac{\pi}{2}$ where $p \in \mathbb{R}$. The p-th order Fourier transform $\mathfrak{F}^p$ of function or signal $x(t)$ is a linear integral transform that maps $x(t)$ to a function*

$$X_p(u) = \mathfrak{F}^p(u) = \int_{-\infty}^{\infty} K_p(t, u)x(t)dt, \tag{7.5.3}$$

where the transform kernel function is

$$K_p(t, u) = \begin{cases} c_\alpha exp\left[j\left((u^2 + t^2)cot\alpha - 2\frac{ut}{sin\alpha}\right)\right], & if\ \alpha \neq n\pi \\ \delta(t - u), & if\ \alpha = 2n\pi \\ \delta(t + u), & if\ \alpha = (2n + 1)\pi \end{cases} \tag{7.5.4}$$

with coefficients

$$c_\alpha = \sqrt{1 - jcot\alpha} = \frac{e^{-j[\pi sgn(sin\alpha)/4 - \alpha/2]}}{\sqrt{|sin\alpha|}}. \tag{7.5.5}$$

The fractional Fourier transform $X_p(u)$ of a non-stationary signal $x(t)$ has some typical properties, as shown in Tab.7.5.1[8].

Properties 1 and 2 are the time-shift and frequency-shift characteristics of the fractional Fourier transform, respectively. Properties 3 and 4 are called differential and integral properties of fractional Fourier transform respectively. Property 7 reflects the odd and even properties of the fractional Fourier transform: If $x(t)$ is an even function of t, then $X_p(u)$ is an even function of u; If $x(t)$ is an odd function of t, then $X_p(u)$ is an odd function of u. Property 8 describes the scale property of the fractional Fourier transform. Moreover, Table 7.5.2 lists the fractional Fourier transforms of some common signals.

Assuming that the time-frequency plane coordinate system (t, ω) is transformed into a new coordinate system (u, v) after rotating by an angle $\alpha = p\pi/2$, then there are the

Tab. 7.5.1: The typical properties of the fractional Fourier transform

Property	Signal	Fractional Fourier transform with angle $\alpha = p\pi/2$
1	$x(t-\tau)$	$X_p(u-\tau\cos\alpha)\exp\left[j\left(\frac{\tau^2}{2}\sin\alpha\cos\alpha - u\tau\sin\alpha\right)\right]$
2	$x(t)e^{jvt}$	$X_p(u-v\sin\alpha)\exp\left[-j\left(\frac{v^2}{2}\sin\alpha\cos\alpha + uv\cos\alpha\right)\right]$
3	$x(t)'$	$X_p'\cos\alpha + juX_p(u)\sin\alpha$
4	$\int_a^t x(t')dt'$	if $\alpha - \pi/2$ is not an integer multiple of π, $\sec\alpha\exp\left(-j\frac{u^2}{2}\tan\alpha\right)\int_a^u X_p(z)\exp\left(j\frac{z^2}{2}\tan\alpha\right)dz$ If $\alpha - \pi/2$ is an integer multiple of π, it obeys the property of traditional FT.
5	$tx(t)$	$X_p\cos\alpha) + jX_p'(u)\sin\alpha$
6	$x(t)/t$	$-j\sec\alpha\exp\left(j\frac{u^2}{2}\cos\alpha\right)\int_{-\infty}^u x(z)\exp\left(-j\frac{z^2}{2}\cos\alpha\right)dz$ if α is not an integer multiple of π
7	$x(-t)$	$X_p(-u)$
8	$x(ct)$	$\sqrt{\frac{1-j\cot\alpha}{c^2-j\cot\alpha}}\exp\left[j\frac{u^2}{2}\cos\alpha\left(1-\frac{\cos^2\psi}{\cos^2\alpha}\right)\right]X_p\left(u\frac{\sin\psi}{c\sin\alpha}\right)$ where $\psi = \arctan(c^2\tan\alpha) = q\pi/2$

Tab. 7.5.2: The fractional Fourier transforms of some common signals

Signal	Fractional Fourier transform with angle $\alpha = p\pi/2$
$\delta(t-\tau)$	$\sqrt{\frac{1-j\cot\alpha}{2\pi}}\exp\left[j\left(\frac{\tau^2+u^2}{2}\cot\alpha - u\tau\csc\alpha\right)\right]$, if $\alpha - \pi/2$ is not an integer multiple of π
1	$\sqrt{1+j\tan\alpha}\exp\left[-j(\frac{u^2}{2}\tan\alpha\right]$, if $\alpha - \pi/2$ is not an integer multiple of π
$\exp(jvt)$	$\sqrt{1+j\tan\alpha}\exp\left[j(\frac{v^2+u^2}{2}\tan\alpha + uv\sec\alpha)\right]$, if $\alpha - \pi/2$ is not an integer multiple of π
$\exp(jct^2/2)$	$\sqrt{\frac{1+j\tan\alpha}{1+c\tan\alpha}}\exp\left(j\frac{u^2}{2}\frac{c-\tan\alpha}{1+c\tan\alpha}\right)$, if $\alpha - \arctan c - \pi/2$ is not an integer multiple of π
$\exp(-t^2/2)$	$\exp(-u^2/2)$
$H_n(t)\exp(\frac{-t^2}{2})$	$\exp(-jn\alpha)H_n(u)\exp(-u^2/2)$, where H_n is a Hermitian polynomial
$\exp(-ct^2/2)$	$\sqrt{\frac{1-j\cot\alpha}{c-j\cot\alpha}}\exp\left(j\frac{u^2}{2}\frac{(c^2-1)\cot\alpha}{c^2+\cot^2\alpha}\right)\exp\left(-\frac{u^2}{2}\frac{c\csc^2\alpha}{c^2+\cot^2\alpha}\right)$

following relationships between the new coordinate system and the original coordinate system:

$$\begin{cases} u = t\cos\alpha + \omega\sin\alpha \\ v = -t\sin\alpha + \omega\cos\alpha \end{cases} \tag{7.5.6}$$

and

$$\begin{cases} t = u\cos\alpha - v\sin\alpha \\ \omega = u\sin\alpha + v\cos\alpha \end{cases} \tag{7.5.7}$$

The fractional Fourier transform $Z_p(u)$ of nonstationary signal $z(t)$ can be calculated by the following three steps.

(1) Find the Wigner-Ville distribution of signal $z(t)$

$$W_z(t,\omega) = \int_{-\infty}^{\infty} z(t+\frac{\tau}{2})z^*(t-\frac{\tau}{2})e^{-j\omega\tau}d\tau. \tag{7.5.8}$$

(2) Obtain the fractional Winger-Ville distribution of the rotated α by transforming equation Eq. (7.5.7) of the coordinate system

$$W_P(u, v) = W_Z(ucos\alpha - \omega \sin\alpha, \omega\cos\alpha + t\sin\alpha). \tag{7.5.9}$$

(3) Obtain the fractional Fourier transform of nonstationary signal $z(t)$ from the 1-D inverse Fourier transform of fractional Winger-Ville distribution $W_P(u, v)$ with respect to variable v

$$Z_P(u) = \int_{-\infty}^{\infty} W_P(u, v)e^{juv}dv. \tag{7.5.10}$$

7.5.2 Calculation of Fractional Fourier Transform

In practical applications, the continuous fractional Fourier transform must be transformed into a discrete fractional Fourier transform to facilitate computer calculations.

To better understand the definition of the discrete fractional Fourier transform, it is necessary to review the definition of the discrete Fourier transform of a data vector.

Definition 7.5.2. *(Discrete Fourier transform) Let* $\mathbf{F}$ *be an* $N \times N$*-D Fourier matrix whose elements* $\mathbf{F}_{i,n} = W^{in}/\sqrt{N}$ *and* $W = e^{-j2\pi/N}$*. The discrete Fourier transform of a discrete data vector* $\mathbf{x} = [x(0), x(1), \cdots, x(N-1)]^T$ *is defined as* $N \times 1$ *vector* $\mathbf{X} = \mathbf{F}\mathbf{x} = [X(0), X(1), \cdots, X(N-1)]^T$ *whose element is*

$$X(i) = \sum_{n=0}^{N-1} F_{i,n}x(n) = \frac{1}{\sqrt{N}}\sum_{n=0}^{N-1} x(n)e^{-j2\pi i/N}, \quad i = 0, 1, \cdots, N-1. \tag{7.5.11}$$

The discrete fractional Fourier transform is an extension of the discrete Fourier transform.

Definition 7.5.3. *(Discrete Fractional Fourier transform) The p-th order discrete fractional Fourier transform of the discrete data vector* $\mathbf{x} = [x(0), x(1), \cdots, x(N-1)]^T$ *is defined as an* $N \times 1$*-D vector* $\mathbf{X}_P = \mathbf{F}^\alpha\mathbf{x} = [X_P(0), X_P(1), \cdots, X_P(N-1)]^T$ *whose element is*

$$X_P(i) = \sum_{n=0}^{N-1} F_{i,n}^\alpha x(n), \tag{7.5.12}$$

where $F_{i,n}^\alpha$ *is the i-th row and n-th column element of the* $N \times N$*-D fractional Fourier matrix* $\mathbf{F}^\alpha = \mathbf{E}\mathbf{\Sigma}^\alpha\mathbf{E}^T$*, and* $\mathbf{E}$ *is the eigenvector of the Fourier matrix* $\mathbf{F} = \mathbf{E}\mathbf{\Sigma}\mathbf{E}^T$*.*

The N-th order discrete Fourier transform has a fast algorithm FFT, and its computational complexity is $NlogN$. The discrete fractional Fourier transform does not have a fast algorithm like the FFT, but only a fast approximate implementation.

There are two Matlab algorithms for fast approximation of the discrete fractional Fourier transform.

(1) One uses the subfunction fracF of MATLAB. The relevant computational method was proposed by Ozaktas et al in 1996[168] or published by Ozaktas et al.in reference[169]. The subfunction fracF can be downloaded from the website [128].
(2) The other one uses MATLAB to code fracft, which is a part of the time-frequency analysis software package developed by O'Neill and available from the Mathworks website [167].

It should be noted that the MATLAB subroutine, fracF, requires that the signal length N be odd.

7.6 Wavelet Transform

The short-time Fourier transform and Gabor transform belong to the "windowed Fourier transform", that is, they analyze the signal with a fixed sliding window. With the sliding of the window function, the local frequency characteristics of the signal can be characterized. Obviously, this kind of sliding window processing with equal width in the time domain is not suitable for all signals. For example, an obvious feature of an artificial seismic exploration signal is that it should have a high frequency resolution at the low frequency end of the signal and a low frequency resolution at the high frequency end. From the perspective of the time-frequency exclusion principle, the high-frequency component of this kind of signal should have a high time resolution, while the low frequency component can have a lower time resolution. In fact, not only artificial seismic exploration signals but also many natural signals (such as voice, images, etc.) have similar characteristics. It is easy to associate that the linear time-frequency analysis of such non-stationary signals should have different resolutions at different locations in the time-frequency plane, i.e., it should be a multi-resolution analysis method. Wavelet transform is such a multi-resolution analysis method, and its purpose is to "see not only the forest (the general of the signal), but also the trees (the details of the signal)", so the wavelet transform is often called the mathematical microscope of the signal.

7.6.1 Physical Considerations of Wavelets

STFT and Gabor transform can be expressed uniformly by the inner product $< s(t), g_{mn}(t) >$ of the signal $s(t)$ and the basis function $g_{mn} = g(t - mT)e^{j2\pi nFT}$. Once the window function $g(t)$, the time sampling interval T, and the frequency sampling interval F are selected, STFT and Gabor transform use a fixed window function to perform sliding window processing on the non-stationary signal. Obviously, no matter how m and n change, the envelope of the basis function remains the same, that is,

$|g_{mn}(t)| = |g(t)|, \forall m, n$. Since the basis function has a fixed time sampling interval T and frequency sampling interval F, the window function and the two transforms have equal time width in the time domain and equal bandwidth in the frequency domain. That is to say, STFT and Gabor transform have the same resolution in the time-frequency plane.

Different from STFT and Gabor transform with fixed time interval and fixed frequency interval, the continuous wavelet transform of a square integrable function $s(t)$ is defined as

$$\mathrm{WT}_s(a,b) = \frac{1}{\sqrt{a}} \int_{-\infty}^{\infty} s(t)\psi^*\left(\frac{t-b}{a}\right) dt = \langle s(t), \psi_{a,b}(t)\rangle, \quad a > 0, \tag{7.6.1}$$

that is, the wavelet basis function $\psi_{ab}(t) = \frac{1}{\sqrt{a}}\psi\left(\frac{t-b}{a}\right)$ (multiplied by factor $\frac{1}{\sqrt{a}}$ is introduced to normalize the transform result) is the result of time translation b and scale scaling a of the window function $\psi(t)$. The constants a and b are called scale parameter and translation parameter respectively. The envelope of the wavelet basis function $\psi_{ab}(t)$ varies with a due to the scale parameter a. Specifically, for a given window function $\psi(t)$, if the scale parameter $a > 1$, the basis function is equivalent to stretching the window function to increase the window width; while $a < 1$ is equivalent to compressing the window function so that the window function shrinks.

The effect of the scale parameter in the frequency-domain can be explained by Fourier transform $\Psi(\omega)$ of the window function $\psi(t)$. According to the scaling property of the Fourier transform, the parameter $a > 1$ is equivalent to compressing the frequency characteristic of the window function and reducing the frequency bandwidth; $a < 1$ is equivalent to stretching the frequency characteristic of the window function and increasing the frequency bandwidth. In contrast, the effect of the translation parameter b is only to make the wavelet basis function slide.

From the time-frequency grid division, a large scale parameter a corresponds to the low frequency and has high frequency resolution and low time resolution. On the contrary, a small scale parameter a corresponds to the high frequency and has low frequency resolution and high time resolution. Both the windowed Fourier transform and wavelet transform can be regarded as band-pass filters, and their bandwidths are shown in Fig. 7.6.1(a) and (b), respectively. The multiresolution characteristics of the wavelet transform can be seen in Fig. 7.6.1(b).

In order to have multiresolution properties, the wavelet transform should satisfy the following conditions.

(1) Admissible condition

In physical terms, a wavelet is "a small segment wave". Therefore, wavelet $\psi(t)$ is required to satisfy the admissible condition

$$\int_{-\infty}^{\infty} \psi(t)dt = 0. \tag{7.6.2}$$

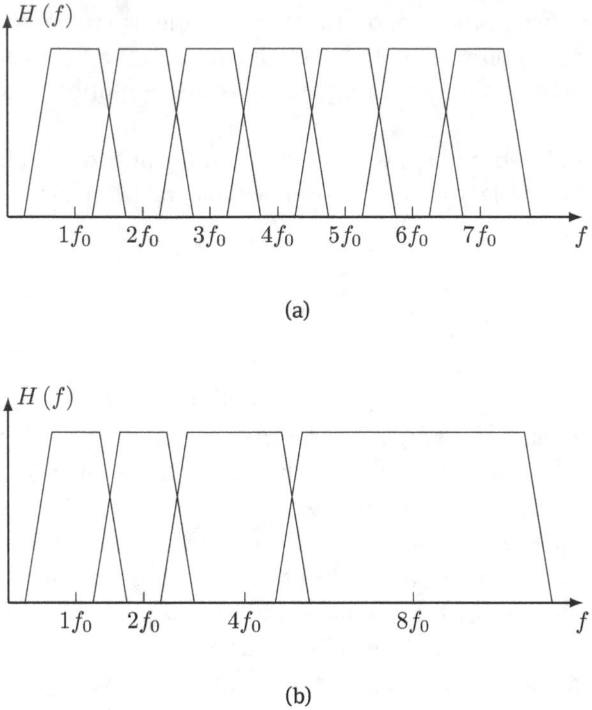

Fig. 7.6.1: (a) Bandwidth of windowed Fourier transform bandpass filter;(b) Bandwidth of wavelet transform bandpass filter.

This condition will make the function $\psi(t)$ conform to the waveform characteristic "a small segment wave", which is the minimum condition that a wavelet must have. The wavelet that satisfies the admissible condition is called an admissible wavelet.

(2) Normalization condition
The wavelet $\psi(t)$ should have unit energy, i.e.,

$$E_\psi = \int_{-\infty}^{\infty} |\psi(t)|^2 dt = 1. \tag{7.6.3}$$

7.6.2 The Continuous Wavelet Transform

The admissible condition and normalization condition are requirements for wavelets from physical considerations. From the point of view of signal transformation, more strict mathematical conditions are also required for the wavelet transform defined by Eq. (7.6.1).

1. Complete Reconstruction Condition

The Fourier transform $\Psi(\omega)$ of the base wavelet $\psi(t)$ must satisfy the condition

$$\int_{-\infty}^{\infty} \frac{|\Psi(\omega)|^2}{|\omega|} d\omega < \infty, \tag{7.6.4}$$

which is called the complete reconstruction condition or the constant resolution condition.

2. Stability Condition

Since the wavelet $\psi_{a.b}(t)$ generated by the base wavelet $\psi(t)$ acts as an observation window for the signal being analyzed in the wavelet transform, $\psi(t)$ should also satisfy the constraint condition

$$\int_{-\infty}^{\infty} |\psi(t)| dt < \infty \tag{7.6.5}$$

of the general window function, i.e., $\Psi(\omega)$ must be a continuous function. This means that in order to satisfy the complete reconstruction condition Eq. (7.6.4), $\Psi(\omega)$ must be equal to zero at the origin, i.e., $\Psi(0) = \int_{-\infty}^{\infty} \psi(t)dt = 0$, which is exactly the admissible condition Eq. (7.6.2) mentioned above that any wavelets must obey.

In order to implement the signal reconstruction to be numerically stable, in addition to the complete reconstruction condition, the Fourier transform of wavelet $\psi(t)$ is required to satisfy the following "stability condition"

$$A \le \sum_{j=-\infty}^{\infty} |\Psi(2^j \omega)|^2 \le B, \tag{7.6.6}$$

where $0 < A \le B < \infty$.

The continuous wavelet transform has the following important properties.

Property 1 (Linear) The wavelet transform of a multi-component signal is equal to the sum of the wavelet transform of each component.

Property 2 (Translation invariance) If $f(t) \leftrightarrow WT_f(a, b)$, then $f(t-\tau) \leftrightarrow WT_f(a, b-\tau)$.

Property 3 (Stretch covariance) If $f(t) \leftrightarrow WT_f(a, b)$, then $f(ct) \leftrightarrow \frac{1}{\sqrt{c}} WT_f(ca, cb)$, where $c > 0$.

Property 4 (Self-similarity) The continuous wavelet transforms corresponding to different scaling parameters a and translation parameters b are self-similar.

Property 5 (Redundancy) There is redundancy of information representation in the continuous wavelet transform.

Property 1 comes directly from the fact that wavelet transform can be written as an inner product, and the inner product has linear property. Property 2 is easy to verify according to the definition of the wavelet transform. The proof of property 3 is as follows: let

$x(t) = f(ct)$, then there is

$$\begin{aligned} \mathrm{WT}_x(a,b) &= \frac{1}{\sqrt{a}} \int_{-\infty}^{\infty} x(t)\psi^*\left(\frac{t-b}{a}\right) dt \\ &= \frac{1}{\sqrt{c}\sqrt{ca}} \int_{-\infty}^{\infty} f(ct)\psi^*\left(\frac{ct-cb}{ca}\right) d(ct) \\ &= \frac{1}{\sqrt{c}} \mathrm{WT}_f(ca, cb), \end{aligned}$$

which is property 3.

Since the wavelet family $\psi_{a,b}(t)$ is obtained by the same wavelet $\psi(t)$ through translation and stretch, and the continuous wavelet transform has translation invariance and stretch covariance, the continuous wavelet transform at different grid points (a, b) has self-similarity, that is, the property 4 holds.

Essentially, the continuous wavelet transform maps one-dimensional signal $f(t)$ equidistant to a two-dimensional scale-time (a, b) plane, and its freedom degree increases obviously so that the wavelet transform contains redundancy, that is, the property 5 holds. Redundancy is also a direct reflection of self-similarity, which is mainly reflected in the following two aspects:

(1) The reconstruction formula for recovering the original signal by the continuous wavelet transform is not unique. That is, there is no one-to-one correspondence between the wavelet transform of signal $s(t)$ and the inverse wavelet transform, while there is a one-to-one correspondence between the Fourier transform and the inverse Fourier transform.

(2) There are many possible choices for the kernel function of the wavelet transform, i.e., the wavelet family functions $\psi_{a,b}(t)$ (for example, they can be non-orthogonal wavelet, orthogonal wavelet or biorthogonal wavelet, or even allowed to be linearly related to each other, as described later).

The correlation of the wavelet transform between different grid points (a, b) increases the difficulty of analyzing and interpreting the results of the wavelet transform. Therefore, the redundancy of wavelet transform should be as small as possible, which is one of the main problems of wavelet analysis.

7.6.3 Discretization of Continuous Wavelet Transform

When using wavelet transform to reconstruct the signal, it is necessary to discretize the wavelet and use the discrete wavelet transform. Different from the traditional time discretization, the continuous wavelet $\psi_{a,b}(t)$ and the continuous wavelet transform $\mathrm{WT}_f(a, b)$ are discretized for the continuous scale parameter a and the continuous translational parameter b, rather than for the time variable t.

Usually, the discretization formulas for the scale parameter a and the translation parameter b are taken as $a = a_0^j$ and $b = ka_0^j b_0$, respectively. The corresponding discrete wavelet $\psi_{j,k}(t)$ is

$$\psi_{j,k}(t) = a_0^{-j/2}\psi(a_0^{-j}t - kb_0). \tag{7.6.7}$$

And the discrete wavelet transform $\mathrm{WT}_f(a_0^j, ka_0^j b_0)$ is abbreviated as $\mathrm{WT}_f(j,k)$ and called

$$c_{j,k} \stackrel{\text{def}}{=} \mathrm{WT}_f(j,k) = \int_{-\infty}^{\infty} f(t)\psi_{j,k}^*(t)dt = \langle f, \psi_{j,k}\rangle \tag{7.6.8}$$

as the discrete wavelet (transform) coefficient.

The purpose of using wavelet transform is to reconstruct the signal. How to select the scale parameter a and the translation parameter b to ensure the accuracy of the reconstructed signal? Qualitatively, the grid points should be as dense as possible (i.e., a_0 and b_0 should be as small as possible), because if the grid points are sparser, the fewer wavelet functions $\psi_{j,k}(t)$ and discrete wavelet coefficients $c_{j,k}$ are used, and the accuracy of the reconstructed signal will be lower. This implies that there exists a threshold for the grid parameters.

In order to make the wavelet transform have variable time and frequency resolution and to adapt to the non-stationary characteristics of the signal to be analyzed, it is natural to change the value of a and b to make the wavelet transform have the "zoom" function. In other words, a dynamic sampling grid is used in practice. The most commonly used is the binary dynamic sampling grid with $a_0 = 2$, $b_0 = 1$. The scale of each grid is 2^j, and the translation is $2^j k$. In particular, when the discretization parameters $a_0 = 2$ and $b_0 = 1$, the discretized wavelet

$$\psi_{j,k}(t) = 2^{j/2}\psi(2^j t - k), \quad j, k \in Z \tag{7.6.9}$$

is called a binary wavelet basis function, where Z denotes the integer domain.

The binary wavelet has a zoom effect on the analysis of the signal. It is assumed that a magnification 2^j is chosen at the beginning, which corresponds to a certain part of the observed signal. If one wants to further view the smaller details of the signal, one needs to increase the magnification i.e. decrease the value of j. Conversely, if one wants to understand the coarser of the signal, one can decrease the magnification i.e. increase the value of j. In this sense, the wavelet transform is called a mathematical microscope.

7.7 Wavelet Analysis and Frame Theory

Fourier analysis is a powerful mathematical tool for stationary signal analysis. Similarly, wavelet analysis is a powerful mathematical tool for non-stationary signal analysis.

7.7.1 Wavelet Analysis

Fourier signal analysis consists of "Fourier (integral) transform" and "Fourier series": the former transforms the continuous signal $f(t)$ by Fourier transform to obtain the spectrum $F(\omega)$ of the signal, while the latter expands by Fourier series to obtain the reconstruction formula of the original signal $f(t)$. The kernel function $e^{-j\omega t}$ of the Fourier transform is called the basis function, and the kernel function $e^{j\omega t}$ of the series expansion is often called the dual basis function.

Like the Fourier analysis of stationary signals, the wavelet analysis of non-stationary signals also consists of two important mathematical entities, the "wavelet (integral) transform" and the "wavelet series". The kernel function $\psi(t)$ of the wavelet transform is called wavelet, while the kernel function $\tilde{\psi}(t)$, which reconstructs the wavelet series of the original signal $f(t)$, is called dual wavelet. The strict mathematical definition of dual wavelet is as follows.

Definition 7.7.1. *(Dual wavelet) If the wavelet $\psi(t)$ satisfies the stability condition Eq. (7.6.6), then there exists a dual wavelet $\tilde{\psi}(t)$ whose Fourier transform $\Psi(\omega)$ is given by the Fourier transform of the wavelet*

$$\tilde{\Psi}(\omega) = \frac{\Psi^*(\omega)}{\sum_{j=-\infty}^{\infty} |\Psi(2^j\omega)|^2}. \tag{7.7.1}$$

In Fourier analysis, any square integrable real function $f(t) \in L^2(R)$ has a Fourier series expression

$$f(t) = \sum_{k=-\infty}^{\infty} c_k e^{jk\omega t}, \tag{7.7.2}$$

in which the expansion parameter

$$c_k = \frac{1}{2\pi}\int_0^{2\pi} f(t)e^{-jk\omega t}dt \tag{7.7.3}$$

is called the Fourier coefficient of the real function f, which is square summable

$$\sum_{k=-\infty}^{\infty} |c_k|^2 < \infty. \tag{7.7.4}$$

Similarly, wavelet analysis can also be defined: any square integrable real function $f(t) \in L^2(R)$ has a wavelet series expression

$$f(t) = \sum_{j=-\infty}^{\infty}\sum_{k=-\infty}^{\infty} c_{j,k}\tilde{\psi}_{j,k}(t), \tag{7.7.5}$$

where the wavelet coefficients $\{c_{j,k}\}$ are defined by Eq. (7.6.8), which is a sequence of square summable, i.e.,

$$\sum_{j=-\infty}^{\infty}\sum_{k=-\infty}^{\infty} |c_{j,k}|^2 < \infty. \tag{7.7.6}$$

The basis function $\tilde{\psi}_{j,k}(t)$ of wavelet level expansion equation Eq. (7.7.5) is called the dual basis of the wavelet basis function $\psi_{j,k}(t)$ and defined as

$$\tilde{\psi}_{j,k}(t) = 2^{j/2}\tilde{\psi}(2^j t - k), \quad j, k \in Z, \tag{7.7.7}$$

where $\tilde{\psi}(t)$ is the dual wavelet of wavelet $\psi(t)$ (Definition 7.7.1). In particular, when the wavelet and its dual wavelet are equal, i.e., $\tilde{\psi}(t) = \psi(t)$, the wavelet basis function and its dual wavelet basis function are also equal, i.e., $\tilde{\psi}_{j,k}(t) = \psi_{j,k}(t)$.

In Fourier analysis and wavelet analysis, the basis function plays an important role.

Definition 7.7.2. *(Hilbert Base)*[223] *Let H be a complete inner product space, i.e., Hilbert space, and the family of discrete sequences $\{\phi_n(t) : n \in Z\}$ (where Z is the integer domain) be called standard orthogonal basis or Hilbert basis within H if the following three conditions are satisfied:*
(1) Orthogonality condition: if $m, n \in Z$ and $m \neq n$, then $< \phi_m, \phi_n >= 0$;
(2) Normalization condition: for each $n \in Z$ with $||\phi_n|| = 1$;
(3) Completeness condition: if $f \in H$ and $\langle f, \phi_n \rangle = 0, \forall n \in Z$, then $f = 0$.

If the family of discrete sequences $\{\phi_n(t)\}$ satisfies only the first and third conditions, it is called an orthogonal basis. A set that satisfies only the first two conditions, but not necessarily the third one, is called a standard orthogonal system. If only the first condition is satisfied, the set is called an orthogonal system.

A Hilbert space is said to be a separable Hilbert space if the number of basis functions of the space is countable. Another expression for the completeness of a separable Hilbert space is called denseness.

Definition 7.7.3. *(Denseness) The family of discrete sequences $\{\phi_n : n \in Z\}$ is dense in H, if for each $f \in H$ and $\epsilon > 0$, we can find a sufficiently large integer N and constant $c_{-N}, c_{-N+1}, \cdots, c_{N-1}, c_N$ such that $||f - \sum_{k=-N}^{N} c_k\phi_k|| < \epsilon$. In other words, any function $f \in H$ can be approximated by a finite number of linear combinations of the function family $\{\phi_n : n \in Z\}$, then $\{\phi_n : n \in Z\}$ is said to be dense in H.*

A standard orthogonal system $\{\phi_n\}$ is dense when and only when it is complete. In other words, a dense standard orthogonal system is a standard orthogonal basis.

The Fourier basis e^{jkt} of Fourier analysis is a standard orthogonal basis, which is chosen uniquely. As mentioned earlier, the redundancy of the wavelet transform increases the difficulty of analyzing and interpreting the results of the wavelet transform, so the redundancy of the wavelet transform should be as small as possible. This means that the linear correlation between wavelets should be reduced. In other words, it is desired that the wavelet family $\psi_{j,k}(t)$ has linear independence. Considering the accuracy of the signal reconstruction, the orthogonal basis is the most ideal basis function for signal reconstruction, so it is more desirable that the wavelets are orthogonal. However, other important aspects must be considered in the selection of wavelets.

As Sweldens points out[217], in order to make the wavelet transform to be a useful signal processing tool, wavelets must satisfy the following three basic requirements:

(1) Wavele is the building block of general function: wavelet can be used as the basis function to expand the general function in wavelet series.
(2) Wavelet has time-frequency aggregation: usually, most of the energy of the wavelet is required to be gathered in a finite interval. Ideally, outside this interval, the energy of the wavelet function $\psi(t)$ should be equal to zero, that is, the wavelet should be a compactly supported function in the frequency domain. However, it is known from the exclusion principle that a function that is compactly supported in the frequency domain has an infinite support region in the time domain. Therefore, the wavelet function should be compactly supported in the time domain and be able to decay rapidly in the frequency domain.
(3) Wavelet has a fast transform algorithm: in order to make the wavelet function easy to realize by computer, we hope that the wavelet transform has a fast algorithm as the Fourier transform.

It can be said that the implementation of these three basic requirements constitutes the core of the wavelet transform. Let us first look at the time-frequency aggregation of the wavelets.

The decay of a wavelet function to high frequency corresponds to the smoothness of the wavelet. The smoother the wavelet, the faster it decays to a higher frequency. If the decay is exponential, the wavelet will be infinitely derivable.

The decay of a wavelet function to low frequency corresponds to the order of the vanishing moment of that wavelet (the definition is given later).

Therefore, with the smoothness and vanishing moment of the wavelet function, the “frequency aggregation” of the wavelet can be ensured and the desired time-frequency aggregation can be obtained.

If the function $f(t)$ has $N-1$th-order continuous derivative and its Nth-order derivative in the neighborhood of the point t_0 are finite, then Taylor theorem of complex function theory tells us that for every point t in the neighborhood, it is possible to find $t_1 = t_1(t)$ in that neighborhood such that

$$f(t) = f(t_0) + \sum_{k=1}^{N-1} \frac{f^{(k)}(t_0)}{k!}(t-t_0)^k + \frac{f^{(N)}(t_1)}{N!}(t-t_0)^N. \tag{7.7.8}$$

If that neighborhood is very small and the Nth-order derivative cannot be too large, then the unknown residual term $\frac{f^{(N)}(t_1)}{N!}(t-t_0)^N$ will be small, i.e., the function $f(t)$ can be adequately approximated by $f(t_0) + \sum_{k=1}^{N-1} \frac{f^{(k)}(t_0)}{k!}(t-t_0)^k$.

Definition 7.7.4. *The wavelet $\psi(t)$ is said to have Nth-order vanishing moment if*

$$\int (t-t_0)^k \psi(t)dt = 0, \quad k = 0, 1, \cdots, N-, 1, \tag{7.7.9}$$

$$\int (t-t_0)^N \psi(t)dt \neq 0. \tag{7.7.10}$$

The vanishing moment determines the smoothness of the function. If $\psi(t)$ has Nth-order vanishing moments at time zero $t_0 = 0$, then its Fourier transform $\Psi(\omega)$ is Nth-order differentiable at frequency zero $\omega = 0$ and $\Psi^k(0) = 0$, where $k = 0, 1, \cdots, N-1$.

Now assume that the signal $f(t)$ has Nth-order continuous derivatives in the neighborhood of t_0 and $|f^{(N)}(t)| \leq M < \infty$ is bounded in that neighborhood. Let $\psi(t)$ be a real wavelet whose support region is $[-R, R]$ and has Nth-order vanishing moments at the zero point. If we use $\psi_a(t) = a\psi(at - t_0)$ to generate the wavelet family function $\{\psi_a(t)\}$ and use Eqs. (7.7.8) and (7.7.9), we can get

$$< f, \psi_a >= \int_{-\infty}^{\infty} f(t)a\psi(at-t_0)dt = f(t_0) + \frac{1}{N!} = \int_{-\infty}^{\infty} f^{(N)}(t_1)(t-t-0)^N a\psi(at-t_0)dt.$$

The above equation can be written as a triangle inequality

$$|\langle f, \psi_a\rangle - f(t_0)| \leq \frac{2M}{N!}\left(\frac{R}{a}\right)^N. \tag{7.7.11}$$

Eq. (7.7.11) shows the following important facts:

(1) The accuracy $|\langle f, \psi_a\rangle - f(t_0)|$ of wavelet transform approximating the original signal $f(t)$ depends on the support region R and scale constant a of the wavelet function $\psi(t)$. The support of a function refers to the closed interval of the definition domain of that function. If its support area is a finite closed interval (this support is called compact support), then the function is said to be a compact support function. If R is finitely large that $\psi(t)$ is a compactly supported function, then Eq. (7.7.11) becomes $|\langle f, \psi_a\rangle - f(t_0)| \to 0$ when the scale parameter $a \to \infty$.
(2) When the scale parameter $a > R$, if the value of N is larger, the higher the accuracy $|\langle f, \psi_a\rangle - f(t_0)|$ of the wavelet transform approximating the original signal $f(t)$.

Therefore, from the perspective of function approximation, wavelet $\psi(t)$ is required to have compact support and Nth-order vanishing moment, and the smaller R and (or) the larger the N, the higher the accuracy of the wavelet transform approximation signal. On the other hand, compactly supported wavelets have good time-local properties and are beneficial for algorithm implementation. However, from the exclusion principle, it is known that the time-local characteristics and the frequency-local characteristics are a pair of contradictions. Considering the frequency resolution, it is desired that the wavelet has a larger time support region.

According to the equation $\mathrm{WT}(a, b) =< f, \psi_{a,b} >$ of the wavelet transform, in order to make $\mathrm{WT}(a, b)$ to keep the phase of the signal $f(t)$ without distortion, the wavelet

$\psi_{ab}(t)$ should have linear phase. The function $g(t)$ is called a symmetric function, if $g(t+T) = g(t-T)$ or $g(t+\frac{T}{2}) = g(t-\frac{T}{2})$ for some integer or semi-integer T (i.e., $T/2$ is an integer); if $g(t+T) = -g(t-T)$ or $g(t+\frac{T}{2}) = -g(t-\frac{T}{2})$, $g(t)$ is said to be an antisymmetric function. The following proposition shows that the linear phase property of the wavelet is determined by its symmetry or antisymmetry.

Proposition 7.7.1. *If the function $g(t)$ is symmetric or antisymmetric with respect to some integer or semi-integer T, then the phase response of $g(t)$ is linear.*

Proof. See reference [223]. □

As block functions for general signals, the basis functions can be non-orthogonal, orthogonal, and biorthogonal.

In practical applications, wavelets are usually expected to have the following properties[113]:

(1) Compact support: If the scaling function and wavelet are compactly supported, the filter H and G are finite impulse response filters, and their summation in the orthogonal fast wavelet transform is the summation of finite terms. This is beneficial for implementation. If they are not compactly supported, a fast decay is desirable.
(2) Symmetry: If the scaling function and wavelet are symmetric, then the filters have a generalized linear phase. If the filters don't have a linear phase, phase distortion will occur after the signal passes through the filters. Therefore, the linear phase requirement of the filter is very important in signal processing applications.
(3) Smoothness: Smoothness is very important in compression applications. Compression is usually achieved by setting the small coefficients $c_{j,k}$ to zero and leaving out the corresponding component $c_{j,k}\psi_{j,k}$ from the original function. If the original function represents an image and the wavelet is not smooth, the error in the compressed image can easily be caught by the human eye. More smoothness corresponds to better frequency localization of the filters.
(4) Orthogonality: In any linear expansion or approximation of signal, the orthogonal basis is the best basis function. Therefore, when an orthogonal scaling function is used, it can provide the best signal approximation.

7.7.2 Frame Theory

The so-called non-orthogonal expansion is to construct non-orthogonal basis functions by using the basic operations of translation and modulation of a single non-orthogonal function and then use these basis functions to expand the signal in series. In fact, we are familiar with this kind of non-orthogonal expansion, because the Gabor expansion is a typical example.

The non-orthogonal expansion has the following advantages in wavelet analysis:

(1) Orthogonal wavelet is a complex function, and any "good" function can be used as the basis wavelet of non-orthogonal expansion.
(2) In some cases of interest, orthogonal bases for coherent states do not even exist, so it is natural to look for non-orthogonal expansions.
(3) The non-orthogonal expansion can obtain higher numerical stability than the orthogonal expansion.

In wavelet analysis, non-orthogonal expansions often use linear independent bases, and the concept of linear independent bases is closely related to the frame.

Definition 7.7.5. *(Frame) The set $\{\psi_{mn}\}$ of sequences in the square summable space, i.e., $l^2(Z^2)$-space, forms a frame if there exist two positive constants A and B ($0 < A \le B < \infty$) such that the following equation holds for all $f(t) \in l^2(R)$*

$$A||f||^2 \le \sum_{-\infty}^{\infty}\sum_{-\infty}^{\infty} |\langle f, \psi_{mn}\rangle|^2 \le B||f||^2, \tag{7.7.12}$$

where $\langle f, \psi_{mn}\rangle$ represents the inner product of the function $f(t)$ and $\psi_{mn}(t)$

$$< f, \psi_{mn} >= \int_{-\infty}^{\infty} f(t)\psi^*_{mn}(t)dt. \tag{7.7.13}$$

The positive constants A and B are called the lower and upper bounds of the frame, respectively.

A sequence g_{mn} is said to be complete if the only element orthogonal to g_{mn} in $l^2(Z^2)$-space is a zero element. It is easy to verify that the frame is complete. Examining the inequality on the left side of Eq.(7.7.12) shows that when the frame $\psi_{mn}(t)$ is orthogonal to the function $f(t)$, i.e., $< f, \psi_{mn} >= 0$, there is

$$0 \le A||f||^2 \le \sum_{-\infty}^{\infty}\sum_{-\infty}^{\infty} 0 = 0 \Rightarrow f = 0, \tag{7.7.14}$$

that is, the frame is complete.

Definition 7.7.6. *(Snug frame and tight frame) Let $\{\psi_{mn}\}$ be a frame, if $B/A \approx 1$, then ψ_{mn} is said to be a snug frame. In particular, when $A = B$, then ψ_{mn} is said to be a tight frame.*

The snug frame is also known as an almost-tight frame.

Proposition 7.7.2. *If $\{g_k(t)\}$ is a tight frame with $A = B = 1$ and all frame elements have unit norm, then frame $\{g_k(t)\}$ is a standard orthogonal basis.*

Proof. Let g_l be a fixed element in the frame. Since $A = B = 1$,

$$||g_l||^2 = \sum_{k\in K} |\langle g_k, g_l\rangle|^2 = ||g_l||^4 + \sum_{k\neq l} |\langle g_k, g_l\rangle|^2 \tag{7.7.15}$$

is obtained from the definition of the frame. Since $||g_l||^4 = ||g_l||^2 = 1$, the above equation implies that $< g_k, g_l >= 0$ holds for all $k \neq l$. That is, the frame $\{g_k(t)\}$ is an orthogonal basis. Since the norm of each frame element is equal to 1, $\{g_k(t)\}$ is a standard orthogonal basis. □

Definition 7.7.7. *(Frame Operator) Let $\{g_k, k \in K\}$ be a known frame, if*

$$Tf = \sum_{k \in K} \langle f, g_k \rangle g_k \tag{7.7.16}$$

is an operator that maps function $f \in L^2(R)$ to $Tf \in L^2(R)$, then T is said to be a frame operator.

The following are the properties of the frame operator $\{g_k, k \in K\}$.
Property 1 The frame operator is bounded.
Property 2 The frame operator is self-adjoint, that is, $\langle f, Th \rangle = \langle f, h \rangle$ holds for all functions f and h.
Property 3 The frame operator is a positivity operator, that is., $\langle f, Th \rangle > 0$.
Property 4 The frame operator is reversible, that is, T^{-1} exists.

Definition 7.7.8. *(Exact frame) If the wavelet frame $\{\psi_{mn}\}$ is a set of independent sequences, it is said to be an exact frame.*

In the sense that it is no longer a frame after removing any element, the exact frame can be understood as "just the right frame". In wavelet analysis, the exact frame is often called the Riesz basis. Due to the importance of the Riesz basis, its strict definition is given here.

Definition 7.7.9. *(Riesz Basis) If the family of discrete wavelet basis functions $\{\psi_{j,k}(t) : j, k \in Z\}$ is linearly independent, and there exist two positive constants A and B ($0 < A \leq B < \infty$) such that*

$$A||\{c_{j,k}\}||_2^2 \leq \sum_{j=-\infty}^{\infty} \sum_{k=-\infty}^{\infty} |c_{j,k}\psi_{j,k}|^2 \leq B||\{c_{j,k}\}||_2^2 \tag{7.7.17}$$

holds for all square summable sequences $\{c_{j,k}\}$, where

$$||\{c_{j,k}\}||_2^2 = \sum_{j=-\infty}^{\infty} \sum_{k=-\infty}^{\infty} |c_{j,k}|^2 \leq \infty, \tag{7.7.18}$$

then the two-dimensional sequence $\{\psi_{j,k}(t) : j, k \in Z\}$ is said to be a Riesz basis within $L^2(R)$, and the constants A and B are called the Riesz lower and upper bounds, respectively.

Theorem 7.7.1. [57, 58] *Let $\psi(t) \in L^2(R)$, and $\psi_{j,k}(t)$ be a wavelet generated by $\psi(t)$, then the following three statements are equivalent:*

(1) $\{\psi_{j,k}\}$ *is the Riesz basis of* $L^2(R)$;
(2) $\{\psi_{j,k}\}$ *is the exact frame of* $L^2(R)$;
(3) $\{\psi_{j,k}\}$ *is a frame of* $L^2(R)$ *and is also a linearly independent family, i.e.,* $\sum_j \sum_k c_{j,k}$ $\psi_{j,k}(t) = 0$ *means* $c_{j,k} \equiv 0$, *and the Riesz bound is the same as the frame bound.*

So far, we have obtained three necessary conditions for a basic wavelet or a mother wavelet $\psi(t)$ to be used as a wavelet transform:
(1) The complete reconstruction condition Eq. (7.7.11) is equivalent to the admissible condition Eq. (7.7.3);
(2) The stability condition equation Eq. (7.7.13) for the basic wavelet $\psi(t)$;
(3) The linear independence condition of the wavelet family $\{\psi_{j,k}\}$, i.e., the Riesz basis or linear independence basis condition Eq. (7.7.18).

By Gram-Schmidt orthogonalization, a Riesz basis can be transformed into a standard orthogonal basis[72].

A wavelet $\psi(t) \in L^2(R)$ is called a Riesz wavelet if the family of discrete functions $\{\psi_{j,k}(t)\}$ generated by it according to Eq. (7.7.20) is a Riesz basis.

Definition 7.7.10. *(Orthogonal wavelet) A Riesz wavelet* $\{\psi(t)\}$ *is called an orthogonal wavelet if the generated family of discrete wavelets* $\{\psi_{j,k}(t) : j, k \in Z\}$ *satisfies the orthogonality condition*

$$\langle \psi_{j,k}, \psi_{m,n} \rangle = \delta(j-m)\delta(k-n), \quad \forall j,k,m,n \in Z. \tag{7.7.19}$$

Definition 7.7.11. *(Semiorthogonal wavelet) A Riesz wavelet* $\{\psi(t)\}$ *is called a semiorthognal wavelet if the generated family of discrete wavelets* $\{\psi_{j,k}(t)\}$ *satisfies the "cross-scale orthogonality"*

$$\langle \psi_{j,k}, \psi_{m,n} \rangle = 0, \quad \forall j,k,m,n \in Z \text{ and } j \neq m. \tag{7.7.20}$$

Since the semi-orthogonal wavelets can be transformed into orthogonal wavelets by standard orthogonalization, the semi-orthogonal wavelets will not be discussed later.

Definition 7.7.12. *(Non-orthogonal wavelet) If a Riesz wavelet* $\psi(t)$ *is not a semi-orthogonal wavelet, it is called a non-orthogonal wavelet.*

Definition 7.7.13. *(Biorthogonal wavelet) A Riesz wavelet* $\{\psi(t)\}$ *is called a biorthogonal wavelet if the wavelet families* $\{\psi_{j,k}(t)\}$ *and* $\{\tilde{\psi}_{j,k}(t)\}$ *generated by* $\psi(t)$ *and its dual* $\tilde{\psi}(t)$ *are biorthogonal Riesz bases*

$$\langle \psi_{j,k}, \tilde{\psi}_{m,n} \rangle = \delta(j-m)\delta(k-n), \quad \forall j,k,m,n \in Z. \tag{7.7.21}$$

The orthogonality defined above is the orthogonality of a single function itself, while biorthogonality refers to the orthogonality between two functions. Note that the biorthogonal wavelet does not involve the orthogonality of $\psi(t)$ and $\psi_{j,k}(t)$ itself. ObviouslyAnal wavelet must be a biorthogonal wavelet, but a biorthogonal wavelet is not necessarily an orthogonal wavelet. Therefore, an orthogonal wavelet is a special case of a biorthogonal wavelet.

Here are several typical wavelet functions.

(1) Gaussian wavelet

The wavelet function is a Gaussian function, i.e.,

$$\psi(t) = e^{-t^2/2}. \tag{7.7.22}$$

This wavelet is continuously differentiable. Its first-order derivative is

$$\psi'(t) = -te^{-t^2/2}. \tag{7.7.23}$$

(2) Mexican hat wavelet

The second-order derivative

$$\psi(t) = (t^2 - 1)e^{-t^2/2} \tag{7.7.24}$$

of a Gaussian wavelet is called a Mexican hat wavelet, because its waveform is similar to a Mexican hat.

Both Gaussian wavelet and Mexican hat wavelet do not satisfy the orthogonal condition, so they are non-orthogonal wavelets.

(3) Gabor wavelet

The Gabor function is defined as

$$G(t) = g(t - b)e^{j\omega t}. \tag{7.7.25}$$

It is the kernel function of the windowed Fourier transform introduced earlier, where $g(t)$ is a basis function, often taken as a Gaussian function. If the scale parameter a is taken as the scaled form of the Gabor function, that is, the Gabor wavelet

$$\psi(t) = \frac{1}{\sqrt{a}} g\left(\frac{t-b}{a}\right) e^{j\omega t} \tag{7.7.26}$$

is obtained.

(4) Morlet wavelet is defined as

$$\psi(t) = \frac{1}{\sqrt{a}} g\left(\frac{t-b}{a}\right) e^{j\omega t/a}. \tag{7.7.27}$$

It is very similar to the Gabor wavelet, but the frequency modulation term is different.

Gaussian wavelet and Mexican hat wavelet are real wavelet functions, while Gabor wavelet and Morlet wavelet are complex wavelet functions. The first three wavelets satisfy the admissible condition Eq. (7.7.3), while Morlet wavelets only approximately satisfy the admissible condition.

7.8 Multiresolution Analysis

It is no exaggeration to say that without the Fast Fourier Transform (FFT), Fourier analysis cannot be practically applied. Similarly, without the fast wavelet transform (FWT), wavelet analysis would only be a theoretical artifact in signal processing.

In 1989, Mallat proposed a fast algorithm for computing the orthogonal wavelet transform using quadrature mirror filters[147], which is now conventionally referred to as the fast wavelet transform. Later, this method has been extended to non-orthogonal wavelet basis functions. Since the design of the quadrature mirror filters is based on the multiresolution analysis of the signal, and wavelet analysis itself is multiresolution analysis, it is necessary to introduce the theory and methods of multiresolution analysis.

Consider the approximation of a strictly square integrable function $u(t) \in L^2(R)$ using multiple resolutions. If this function is a signal, then "approximating it with variable resolution 2^j" can also be equivalently described as "analyzing the signal with resolution 2^j". Therefore, multiresolution approximation and multiresolution analysis are equivalent.

Let $s(t)$ be a square integrable function, i.e., $s(t) \in L^2(R)$ means

$$\int_{-\infty}^{\infty} |s(t)|^2 dt < \infty. \tag{7.8.1}$$

Definition 7.8.1. *Multiresolution analysis within space $L^2(R)$ means constructing a subspace or chain $\{V_j : j \in Z\}$ within space $L^2(R)$ such that it has the following properties:*

(1) Inclusiveness

$$\cdots \subset V_{-2} \subset V_{-1} \subset V_0 \subset V_1 \subset V_2 \subset \cdots$$

or abbreviated as $V_j \subset V_{j+1}, \forall j \in Z$.

(2) Approximation (decreasing and increasing)

$$\begin{aligned} lim_{j\to+\infty} V_j &= L^2(R) \quad i.e., \cup_{j<N} V_j = L^2(R), \quad \forall N \quad (decreasing) \\ lim_{j\to+\infty} V_j &= 0 \quad i.e., \cap_{j<N} V_j = \{0\}, \quad \forall N \quad (increasing) \end{aligned}$$

(3) Translation invariance

$$s(t) \in V_j \Leftrightarrow s(t-k) \in V_j, \quad \forall k \in Z,$$

and scalable

$$s(t) \in V_j \Leftrightarrow s(2t) \in V_{j+1}.$$

(4) Existence of Riesz basis: There exists a function $\phi(t)$ whose translation $\{\phi(t-k), k \in Z\}$ forms the Riesz basis of the reference subspace V_0.

The physical interpretation of the above properties of the subspace column or subspace chain $\{V_j : j \in Z\}$ is as follows:

Inclusion: Lower resolution corresponds to coarser signal content, which corresponds to a larger subspace.

Approximation: The union of all multiresolution analysis subspaces represents the entire space of the square integrable function $\phi(t)$, i.e., the $L^2(R)$ space. In addition, from the inclusiveness, the intersection of all subspaces $V_j, j \in Z$ should be zero space.

Translation invariance and scalable: The translation of function $s(t)$ does not change its shape and its time resolution remains unchanged, so $s(t)$ and $s(t-k)$ belong to the same subspace. The increase of the time scale means that the function is expanded and its time resolution decreases, so the subspace V_j is required to have similar scalability, i.e., $s(t) \in V_j \Leftrightarrow s(2t) \in V_{j+1}$.

Existence of Riesz basis: The subspace V_0 is used as the reference space. With $\{\phi(t-k), k \in Z\}$ as the Riesz basis of the subspace V_0, this basis function can be used to expand the signal $f(t)$ to be approximated. The function $\phi(t)$ is called the generator of multiresolution analysis. Since multiresolution analysis is also called multiscale analysis, the generator $\phi(t)$ of multiresolution analysis is usually called scale function.

It should be pointed out that there are two kinds of symbols in multiresolution analysis:

(1) Daubechies symbol[72] defines the resolution of subspace V_j as 2^{-j}. Therefore, as j decreases, the value of 2^{-j} increases, i.e., the resolution of subspace V_j decreases. In this case, the inclusiveness is $V_j \in V_{j-1}$ and the scalability is $\phi(t) \in V_j \Leftrightarrow \phi(2t) \in V_{j-1}$. And $\lim_{j\to-\infty} V_j = L^2(R)$.

(2) Malllat symbol[147] defines the resolution of the subspace of V_j as $2^j V_j \to L^2(R), j \to +\infty$. Therefore, the smaller the value of j, the smaller the value of 2^{-j}, i.e., the higher the resolution of the subspace V_j. Thus, the inclusiveness is $V_j \in V_{j+1}$ and the scalability is $\phi(t) \in V_j \Leftrightarrow \phi(2t) \in V_{j+1}$. And $\lim_{j\to\infty} V_j = L^2(R)$. This is the notation used in this book.

When reading other documents, readers should pay attention to the difference between these two kinds of symbols.

From the scalability and inclusiveness, we know that $\phi(\frac{t}{2}) \in V_{-1} \subset V_0$ is $\phi(\frac{t}{2}) \in V_0$, so $\phi(\frac{t}{2})$ can be expanded by the Riesz basis function $\{\phi(t-k), k \in Z\}$ of the subspace V_0. Let the expansion formula be

$$\phi\left(\frac{t}{2}\right) = \sqrt{2}\sum_{k=-\infty}^{\infty} h(k)\phi(t-k) \tag{7.8.2}$$

or its equivalent be

$$\phi(t) = \sqrt{2}\sum_{k=-\infty}^{\infty} h(k)\phi(2t-k). \tag{7.8.3}$$

This equation is the two-scale difference equation of the scaling function, where $\{h(k)\}$ is a square summable sequence.

The spectrum of the scaling function is defined as

$$\Phi(\omega) = \frac{1}{2}\sum_{t=-\infty}^{\infty} \phi(t)e^{-j\omega t}. \tag{7.8.4}$$

Define the filter

$$H(\omega) = \sum_{k=-\infty}^{\infty} \frac{h(k)}{\sqrt{2}}e^{-j\omega k}. \tag{7.8.5}$$

Note that the filter $H(\omega)$ is equivalent to the discrete Fourier transform of $h(k)$ in the sense that it only differs by a constant factor $\frac{1}{\sqrt{2}}$. It is easy to verify that $H(\omega)$ is a periodic function with period 2π.

From Eq. (7.8.3) to Eq. (7.8.5), we can obtain

$$\begin{aligned}\Phi(\omega) &= \frac{1}{2}\sum_{t=-\infty}^{\infty}\left[\sqrt{2}\sum_{k=-\infty}^{\infty} h(k)\phi(2t-k)\right]e^{-j\omega t}\\ &= \sum_{k=-\infty}^{\infty}\frac{h(k)}{\sqrt{2}}\Phi\left(\frac{\omega}{2}\right)e^{-j\omega k/2} \quad \text{(make variable substitution } 2t-k=u)\\ &= H\left(\frac{\omega}{2}\right)\Phi\left(\frac{\omega}{2}\right).\end{aligned} \tag{7.8.6}$$

When $\omega = 0$, the above equation gives the result $\Phi(0) = H(0)\Phi(0)$. As long as $\Phi(0) \neq 0$, then there must be $H(0) = 1$. This shows that the filter $H(\omega)$ is a low-pass filter.

Similarly, the two-scale difference equation of the wavelet function is

$$\psi(t) = \sqrt{2}\sum_{k=-\infty}^{\infty} g(k)\phi(2t-k), \tag{7.8.7}$$

where $\{g(k)\}$ is a square summable sequence.

The spectrum of the wavelet function is defined as

$$\Psi(\omega) = \frac{1}{2}\sum_{t=-\infty}^{\infty} \psi(t)e^{-j\omega t}. \tag{7.8.8}$$

Similar to Eq. (7.8.5), a filter

$$G(\omega) = \sum_{k=-\infty}^{\infty} \frac{g(k)}{\sqrt{2}}e^{-j\omega k} \tag{7.8.9}$$

is defined, then there is

$$\begin{aligned}\Psi(\omega) &= \frac{1}{2}\sum_{t=-\infty}^{\infty}\left[\sqrt{2}\sum_{k=-\infty}^{\infty} g(k)\phi(2t-k)\right]e^{-j\omega t}\\ &= \sum_{k=-\infty}^{\infty}\frac{g(k)}{\sqrt{2}}\Phi\left(\frac{\omega}{2}\right)e^{-j\omega k/2} \quad \text{(make variable substitution } 2t-k=u)\\ &= G\left(\frac{\omega}{2}\right)\Phi\left(\frac{\omega}{2}\right).\end{aligned} \tag{7.8.10}$$

Making the variable substitution $\omega' = \omega/2$, Eq. (7.8.6) gives

$$\Phi\left(\frac{\omega}{2}\right) = H\left(\frac{\omega}{4}\right)\Phi\left(\frac{\omega}{4}\right). \tag{7.8.11}$$

And so on, finally, there is

$$\Phi(\omega) = \prod_{k=1}^{\infty} H\left(\frac{\omega}{2^k}\right)\Phi(0). \tag{7.8.12}$$

To make the spectrum $\Phi(\omega)$ of the scaling function only related to $H(\omega)$, let

$$\phi(0) = \int_{-\infty}^{\infty} \phi(t)dt = 1 \tag{7.8.13}$$

and call it the admissible condition of the scaling function. In this way, Eq. (7.8.12) is simplified to

$$\Phi(\omega) = \prod_{k=1}^{\infty} H\left(\frac{\omega}{2^k}\right). \tag{7.8.14}$$

This shows that the spectrum $\Phi(\omega)$ of the scaling function $\phi(t)$ is completely determined by the low-pass filter $H(\omega)$. In other words, if the low-pass filter $H(\omega)$ is given, the spectrum $\Phi(\omega)$ of the scaling function is uniquely determined, and its inverse Fourier transform, the scaling function $\phi(t)$, is also uniquely determined. Therefore, the generation of a suitable scaling function is attributed to the design of the low-pass filter $H(\omega)$, which is independent of the initial value of this function.

Substituting the admissible condition $\Psi(0) = 0$ of wavelet and the admissible condition $\Phi(0) = 1$ of scaling function into Eq. (7.8.10), we immediately have $G(0) = 0$. This shows that the filter $G(\omega)$ is a high-pass filter.

Let W_j be the complementary spaces of V_{j+1} within V_j, i.e., these subspaces satisfy the relation

$$V_{j+1} = V_j \oplus W_j, \tag{7.8.15}$$

where $\oplus$ is the direct sum of the subspaces. The so-called direct sum means that every element of subspace V_{j+1} can be written in a unique form as the sum of an element of subspace W_j and an element of subspace V_j.

Since the subspace V_j is used to approximate the original signal or function with resolution 2^j, the subspace V_j contains the rough "image information" for approximating the original signal or function with resolution 2^j, and the subspace W_j which contains the detail information needed from the approximation with resolution 2^j to the approximation with resolution 2^{j+1}.

If the translation set $\{\psi(t-k) : k \in Z\}$ of function $\psi(t)$ is the Riesz basis of the subspace W_0, then the function $\psi(t)$ is called a wavelet function, or wavelet for short. Then, the set $\{\phi_{j,k} : j, k \in Z\}$ of wavelet functions is the basis function of $L^2(R)$. Subspaces V_j and W_j are sometimes referred to as scaling subspace and wavelet subspace respectively, for which the scaling and wavelet functions are used as the basis functions, respectively.

The main purpose of the multiresolution analysis is to construct the desired wavelet using the scaling function. In order to make the set $\{\phi(t-k) : k \in Z\}$ to approximate even the simplest functions (such as constants), it is natural to assume that the scaling function and its integer time translations obey the so-called "unit decomposition", i.e.,

$$\sum_{k=-\infty}^{\infty} \phi(t-k) = 1, \quad \forall t \in R. \tag{7.8.16}$$

To summarize the above discussion, the scaling function in the multiresolution analysis should satisfy two basic constraints: (1) admissible condition Eq. (7.8.13); (2) unit decomposition Eq. (7.8.16).

7.9 Orthogonal Filter Banks

Once the scaling function $\phi(t)$ of multiresolution analysis has been determined, the wavelet function $\psi(t)$ can also be constructed. According to the constructed orthogonal scaling function or biorthogonal scaling function, the obtained wavelets are called orthogonal wavelet and biorthogonal wavelet respectively; the corresponding multiresolution analysis is orthogonal and biorthogonal multiresolution analysis.

This section mainly discusses how to construct orthogonal wavelets and biorthogonal wavelets.

7.9.1 Orthogonal wavelet

The two-scale difference equations of the scaling function $\phi(t)$ and the wavelet function $\psi(t)$ are

$$\phi(t) = \sqrt{2}\sum_{k=0}^{N-1} h(k)\phi(2t-k), \tag{7.9.1}$$

$$\psi(t) = \sqrt{2}\sum_{k=0}^{N-1} g(k)\phi(2t-k), \tag{7.9.2}$$

where $h(k)$ and $g(k)$ are the coefficients of the low-pass filter $H(\omega)$ and the high-pass filter $G(\omega)$, respectively. Therefore, the construction of the scaling function and wavelet function depends on the design of the filter banks.

In the following, several well-known orthogonal wavelets are described.

Haar wavelet: its scaling function is called the Haar scaling function, defined as

$$\phi(t) = \begin{cases} 1, & 0 \le t \le 1 \\ 0, & \text{Others} \end{cases}. \tag{7.9.3}$$

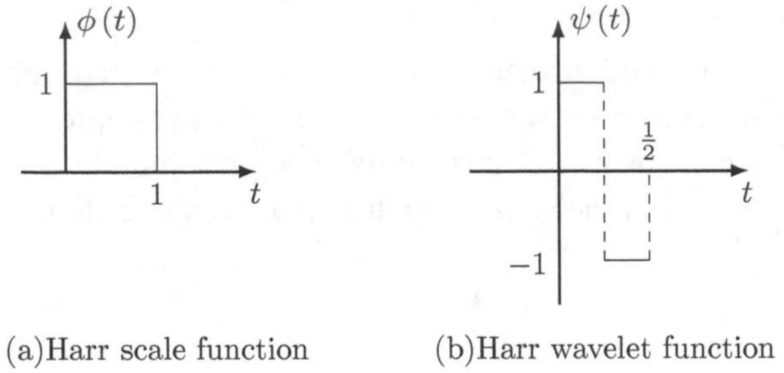

Fig. 7.9.1: Haar scaling function $\phi(t)$ and Haar wavelet $\psi(t)$

The expression of the Haar wavelet function is

$$\psi(t) = \mathcal{X}_{[0,1/2]}(x) - \mathcal{X}_{[1/2,1]}(x) = \begin{cases} 1, & 0 \le t \le 0.5 \\ -1, & 0.5 \le t \le 1 \\ 0, & \text{Others} \end{cases}, \tag{7.9.4}$$

where $\mathcal{X}_{[a,b]}(x)$ is a box function, defined as

$$\mathcal{X}_{[a,b]}(x) = \begin{cases} 1, & a \le t < b \\ 0, & \text{Others} \end{cases}. \tag{7.9.5}$$

Fig. 7.9.1 (a) and (b) show the waveforms of the Haar scaling function and Haar wavelet, respectively.

It is easy to see from Fig. 7.9.1 that the Haar scaling function $\phi(t)$ and the Haar wavelet $\psi(t)$ are standard orthogonal functions, respectively, and the scaling function and wavelet function are also orthogonal to each other.

Shannon wavelet is defined as

$$\psi_{\text{Shannon}}(t) = \frac{\sin(2\pi t) - \sin(\pi t)}{\pi t}. \tag{7.9.6}$$

Haar wavelet and Shannon wavelet are less used in practice, because Haar wavelet is not smooth, while Shannon wavelet is smooth but decays very slowly.

Daubechies wavelet: Daubechies proposes the following iterative method[111] to construct the scaling function.

Algorithm 7.9.1. *Iterative Construction Algorithm of scaling function*
Step 1 Let the initial value $\phi^{(0)}(t) = p_{[0,1)}(t)$, *where*

$$p_{[0,1)}(x) = \begin{cases} 1, & t \in [0,1) \\ 0, & \textit{Others} \end{cases} \tag{7.9.7}$$

is a rectangular window function defined in the interval $[0, 1)$.

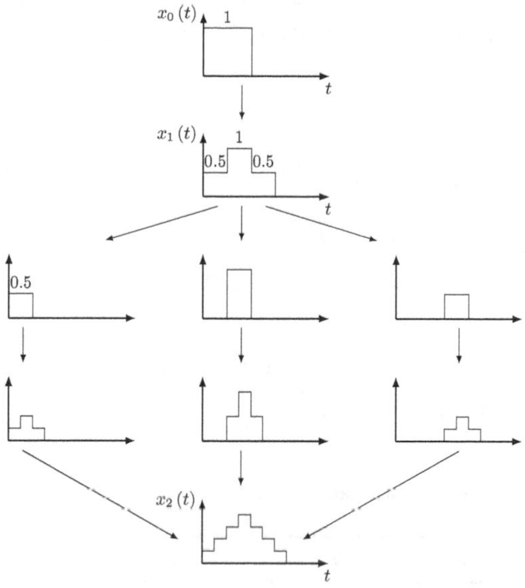

Fig. 7.9.2: The first two recursive steps of the iterative construction of a scaling function

Step 2 Compute

$$\phi^{(i+1)}(t) = \sqrt{2}\sum_{k=0}^{N-1} h(k)\phi^{(i)}(2t-k). \tag{7.9.8}$$

Step 3 Determine whether $\phi^{(i)}(t)$ converges or not. If it converges, stop the iteration; otherwise, let $i \leftarrow i+1$ and return to Step 2 to continue the iteration until the algorithm converges.

It has been proved that after iterations, $\phi^{(i)}(t)$ converges to $\phi(t)$, differing by at most a multiplier factor[111]. As an example, Fig. 7.9.2 illustrates the first two recursive steps of the iterative construction of the scaling function, where the filter coefficients $h(k) = \frac{\sqrt{2}}{2}\left\{\frac{1}{2}, 1, \frac{1}{2}\right\}$.

The construction algorithm of Daubechies orthogonal wavelet[111] is as follows.

Algorithm 7.9.2. *Daubechies standard orthonormal wavelet construction algorithm*
Step 1 Select the length N of the scaling filter $H(\omega)$.
Step 2 Let

$$P(z) = \sum_{k=0}^{N-1} \binom{N-1+k}{k} \left(\frac{2-z-z^{-1}}{4}\right)^k + \left(\frac{2-z-z^{-1}}{4}\right)^N R\left(\frac{z+z^{-1}}{4}\right), \tag{7.9.9}$$

where $R(z)$ is a polynomial of odd order such that $P(e^{j\omega})$ is nonnegative for all ω.

Step 3 Decompose $P(z)$ into $P(z) = Q(z)Q(z^{-1})$, where $Q(z)$ is a polynomial of z. The most common method is to take the roots of $P(z)$ in the unit circle to form the polynomial $Q(z)$.

Step 4 Construct the scaling function

$$H(\omega) = \left(\frac{1+e^{-j\omega}}{2}\right)^M Q(e^{-j\omega}) \tag{7.9.10}$$

and let $G(\omega) = e^{-j\omega}H^*(\omega+\pi)$ or $g(k) = (-1)^k h^*(1-k)$.

Step 5 Use Algorithm 7.9.1 to iteratively construct the scaling function $\phi(t)$ and use Eq. (7.9.2) to construct the wavelet function $\psi(t)$.

For the convenience of readers, Table 7.9.1 lists the 4th, 6th, and 8th order scaling filter coefficients used by the Daubechies orthogonal wavelet[178].

Tab. 7.9.1: The low-pass filter coefficients for Daubechies orthogonal wavelets

N	n	$h(n)$	N	n	$h(n)$	N	n	$h(n)$
4	0	0.482962913145	6	0	0.332670552950	8	0	0.230377813309
	1	0.836516303738		1	0.806891509311		1	0.714846570553
	2	0.224143868042		2	0.459877502118		2	0.630880767930
	3	-0.129409522551		3	-0.135011020010		3	-0.027983769417
				4	-0.085441273882		4	-0.187034811719
				5	0.035226291882		5	0.030841381836
							6	0.032883011667
							7	-0.010597401785

7.9.2 Fast Orthogonal Wavelet Transform

The frequency of a non-stationary signal varies with time, and this variation can be divided into two parts: slow-varying and fast-varying. The slow-varying part corresponds to the low-frequency part of the non-stationary signal and represents the main contour or rough image of the signal, while the fast-varying part corresponds to the high-frequency part of the signal and represents the details of the signal. Similarly, any image can be decomposed into two parts: contour edge (low frequency) and detail texture (high frequency). It is on this basis that a famous pyramidal algorithm for image decomposition and reconstruction was developed. Its basic idea is that the original image $f(x,y)$ is regarded as a discrete approximation $A_0 f$ with resolution $2^0 = 1$, and it can then be decomposed into the sum of a coarse approximation $A_J f$ with resolution 2^J and a number of successive detail approximations $D_j f$ with resolution $2^j (0 < j < J)$.

Inspired by the pyramidal algorithm mentioned above and combined with multiresolution analysis, Mallat proposed a pyramidal multiresolution decomposition and synthesis algorithm for signal[147], which is customarily called the fast orthogonal wavelet transform algorithm, often abbreviated as Mallat algorithm. The status of the orthogonal fast wavelet transform algorithm in wavelet analysis is quite similar to that of FFT in classical Fourier analysis.

The idea of the orthogonal fast wavelet transform algorithm is as follows: assuming that the discrete approximation $A_j f$ of a function or signal $f(t) \in L^2(R)$ at the resolution 2^j has been calculated, the discrete approximation $A_{j-1}f(t)$ of $f(t)$ at the resolution 2^{j-1} can be obtained by filtering $A_j f(t)$ with a discrete low-pass filter H.

Let $\phi(t)$ and $\psi(t)$ be the scaling function and wavelet function of function $f(t)$ at 2^j resolution approximation, respectively, then the discrete approximation $A_j f(t)$ and detail $D_j f(t)$ can be expressed as

$$A_j f(t) = \sum_{k=-\infty}^{\infty} c_{j,k}\phi_{j,k}(t) \text{ and } D_j f(t) = \sum_{k=-\infty}^{\infty} d_{j,k}\psi_{j,k}(t), \tag{7.9.11}$$

where $c_{j,k}$ and $d_{j,k}$ are the scaling (or rough image) and wavelet (or detail) coefficients at 2^j resolution, respectively.

If $A_j f(t)$ is decomposed into the sum of rough image $A_{j-1}f(t)$ and detail $D_{j-1}f(t)$

$$A_j f(t) = A_{j-1}f(t) + D_{j-1}f(t), \tag{7.9.12}$$

where

$$A_{j-1}f(t) = \sum_{m=-\infty}^{\infty} c_{j-1,m}\phi_{j-1,m}(t), \tag{7.9.13}$$

$$D_{j-1}f(t) = \sum_{m=-\infty}^{\infty} d_{j-1,m}\psi_{j-1,m}(t), \tag{7.9.14}$$

then we have

$$\sum_{m=-\infty}^{\infty} c_{j-1,m}\phi_{j-1,m}(t) + \sum_{m=-\infty}^{\infty} d_{j-1,m}\psi_{j-1,m}(t) = \sum_{m=-\infty}^{\infty} c_{j,m}\phi_{j,m}(t). \tag{7.9.15}$$

Next, we study the relationship between $c_{j-1,k}$ and $c_{j,m}$, as well as the relationship between $d_{j-1,k}$ and $d_{j,m}$. Note that the scaling functions $\phi(t)$ and $\psi(t)$ are (standard) orthogonal functions, respectively. First, from the two-scale difference equation of the scaling function, we get

$$\phi_{j-1,k}(t) = 2^{(j-1)/2}\phi(2^{j-1}t - k) = 2^{(j-1)/2} \cdot \sqrt{2}\sum_{i=-\infty}^{\infty} h(i)\phi(2^j t - 2k - i).$$

Making variable substitution $m' = 2k + i$, and the above equation becomes

$$\phi_{j-1,k}(t) = \sum_{m'=-\infty}^{\infty} h(m' - 2k)2^{j/2}\phi(2^j t - m') = \sum_{m'=-\infty}^{\infty} h(m' - 2k)\phi_{j,m'}(t). \tag{7.9.16}$$

Multiplying both sides of Eq. (7.9.16) by $\phi^*_{j,m}(t)$ and taking the integral about t, then using the orthogonality of $\phi_{j,k}(t)$ yields

$$\langle \phi_{j-1,k}, \phi_{j,m} \rangle = h(m-2k).$$

Taking the complex conjugate we obtain

$$\langle \phi_{j,m}, \phi_{j-1,k} \rangle = h^*(m-2k). \tag{7.9.17}$$

Similarly, using the two-scale difference equation of the wavelet function we have

$$\begin{aligned} \psi_{j-1,k}(t) &= 2^{(j-1)/2}\psi(2^{(j-1)}(t-k) \\ &= 2^{(j-1)/2}\sqrt{2}\sum_{i=-\infty}^{\infty} g(i)\phi(2^j t - 2k - i) \\ &= \sum_{m'=-\infty}^{\infty} g(m'-2k)\phi_{j,m'}(t). \end{aligned} \tag{7.9.18}$$

Multiplying both sides of Eq. (7.9.18) by $\phi^*_{j,m}(t)$ and taking the integral about t, we get

$$\langle \phi_{j,m}, \psi_{j-1,k} \rangle = g^*(m-2k). \tag{7.9.19}$$

Multiplying a suitable function on both sides of Eq. (7.9.15) and making the integral about t, then using the relevant orthogonality, the following three important results can be obtained.

(1) Multiplying $\phi^*_{j-1,k}(t)$ and using Eq. (7.9.17), we have

$$c_{j-1,k} = \sum_{m=-\infty}^{\infty} h^*(m-2k)c_{j,m}. \tag{7.9.20}$$

(2) Multiplying $\psi^*_{j-1,k}(t)$ and using Eq. (7.9.19), we have

$$d_{j-1,k} = \sum_{m=-\infty}^{\infty} g^*(m-2k)d_{j,m}. \tag{7.9.21}$$

(3) Multiplying $\phi^*_{j,k}(t)$, using Eqs. (7.9.17) and (7.9.19), we have

$$c_{j,k} = \sum_{m=-\infty}^{\infty} h(m-2k)c_{j-1,m} + \sum_{m=-\infty}^{\infty} g(m-2k)d_{j-1,m}. \tag{7.9.22}$$

Defining infinite dimensional vectors $\mathbf{c}_j = [c_{j,k}]_{k=-\infty}^{\infty}$, $\mathbf{d}_j = [d_{j,k}]_{k=-\infty}^{\infty}$ and matrices $\mathbf{H} = [H_{m,k}]_{m,k=-\infty}^{\infty}$, $\mathbf{G} = [G_{m,k}]_{m,k=-\infty}^{\infty}$, where $H_{m,k} = h^*(m-2k)$, and $G_{m,k} = g^*(m-2k)$, then Eq. (7.9.20) to Eq. (7.9.22) can be abbreviated as

$$\begin{cases} \mathbf{c}_{j-1} = \mathbf{H}\mathbf{c}_j \\ \mathbf{d}_{j-1} = \mathbf{G}\mathbf{d}_j \end{cases} \quad j = 0, -1, \cdots, -J+1 \tag{7.9.23}$$

and

$$\mathbf{c}_j = \mathbf{H}^* \mathbf{c}_{j-1} + \mathbf{G}^* \mathbf{d}_{j-1}, \quad j = -J+1, \cdots, -1, 0, \tag{7.9.24}$$

where $\mathbf{H}^*$ and $\mathbf{G}^*$ are the conjugate matrices of $\mathbf{H}$ and $\mathbf{G}$, respectively.

Eq. (7.9.23) is the fast orthogonal wavelet transform algorithm (or Mallat pyramidal decomposition algorithm), while Eq. (7.9.24) is the inverse orthogonal wavelet transform algorithm (or Mallat pyramidal reconstruction algorithm), which are respectively shown in Fig. 7.9.3 (a) and (b), and the pyramidal structure of these two algorithms is clear when drawn in vertical form.

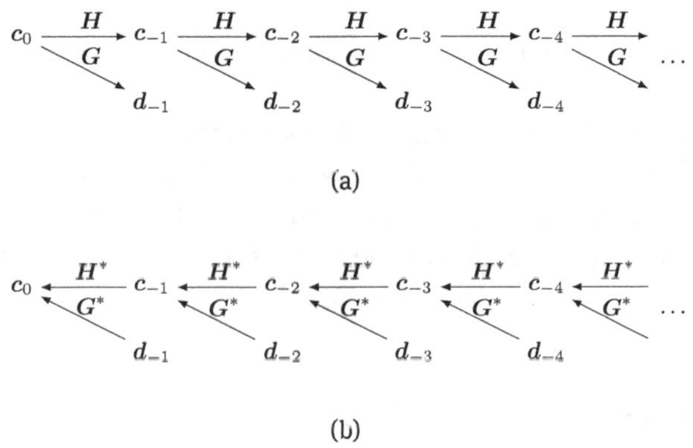

Fig. 7.9.3: (a) The fast orthogonal wavelet transform algorithm; (b) The inverse transform algorithm

The low-pass filter H and high-pass filter G form a filter bank, and the conjugate filter bank (H^*, G^*) decomposes the original signal and is called the analysis filter bank. Filter bank (H, G) is used to reconstruct the signal, that is, to obtain the signal reconstruction of orthogonal multiresolution analysis[147], called synthetic filter bank. The left half of Fig. 7.9.4 shows the signal analysis schematic diagram for orthogonal multiresolution analysis, and the right half shows the signal reconstruction schematic diagram.

In the figure, $\downarrow 2$ denotes downsampling (i.e., sampling of one out of every two samples), while $\uparrow 2$ is upsampling (i.e., inserting a zero between every two samples), or interpolation.

The finite impulse response filter (FIR) is easy to implement, and the linear phase is the prerequisite for maintaining the signal without distortion. Therefore, from the practical application, it is natural to expect that both the analysis filter bank and synthesis filter bank can be constructed with FIR filters with a linear phase.

The orthogonal multiresolution analysis includes four basic operations: filtering, downsampling, upsampling (i.e., interpolation), and reconstruction. The operation form composed of them is called the conjugate quadratic filter subband coding method

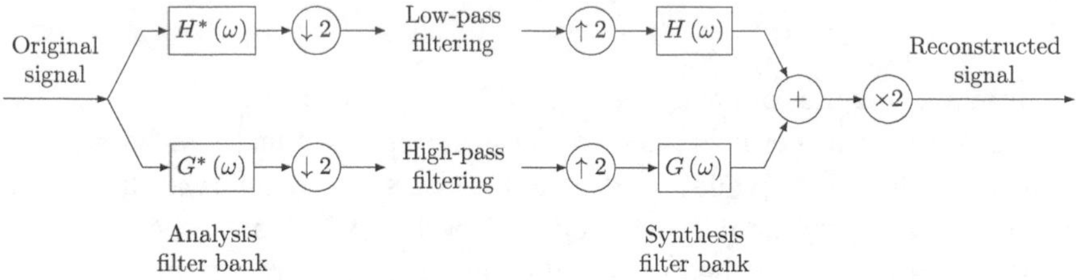

Fig. 7.9.4: The signal reconstruction schematic diagram of orthogonal multiresolution analysis

in signal processing, originally proposed by Smith and Barnwell in 1986 as an image processing method[196].

The advantage of using conjugate quadratic filters to realize the fast orthogonal wavelet transform is that only two filters $H(\omega)$ and $G(\omega)$ need to be designed. However, unless both $H(\omega)$ and $G(\omega)$ are taken as Haar filters, they cannot be both FIR and linear phase[58]. Unfortunately, the wavelets generated by the Haar filter are discontinuous and nonsmooth, so this wavelet has no practical application.

7.10 Biorthgonal Filter Bank

The low-pass filter H and high-pass filter G of the orthogonal filter bank can not be FIR and linear phase at the same time. An effective way to overcome this major drawback is not only to use the filter bank (H, G) but also to add another non-conjugate filter bank $(\tilde{H}, \tilde{G})$, to expand the freedom of filter design. This is the basic starting point of the biorthogonal filter bank.

7.10.1 Biorthogonal Multiresolution Analysis

Compared with the orthogonal wavelet transform, although the synthetic filter bank (H, G) is used, the analysis filter bank $(\tilde{H}^*, \tilde{G}^*)$ used in biorthogonal wavelet transform is no longer the conjugate form of the synthetic filter bank (H, G), but the conjugate form of another filter bank $(\tilde{H}, \tilde{G})$, as shown in Fig. 7.10.1.

Since two filters are added to the orthogonal wavelet, there is more freedom in filter design, thus it is possible to implement all the four filters $H(\omega)$, $G(\omega)$, $(\tilde{H}(\omega)$ and $\tilde{G}(\omega)$ with linear phase FIR filter.

Let the functions constructed by filters $H(\omega)$ and $\tilde{H}(\omega)$ be the scaling function $\phi(t)$ and the dual scaling function $\tilde{\phi}(t)$, respectively, and the function constructed by filters $G(\omega)$ and $\tilde{G}(\omega)$ be the wavelet function $\psi(t)$ and the dual wavelet function $\tilde{\psi}(t)$. Biorthogonal multiresolution analysis actually consists of two multiresolution

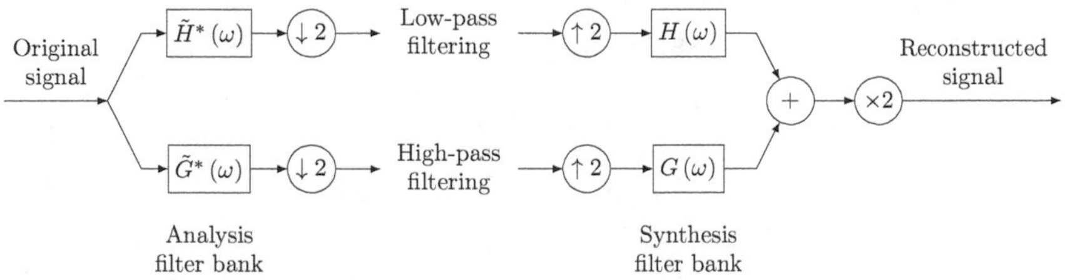

Fig. 7.10.1: The signal reconstruction schematic diagram of biorthogonal multiresolution analysis

analyses: one is the multiresolution analysis generated by the scaling function $\phi(t)$ and the wavelet function $\psi(t)$, i.e., $V_{j+1} = V_j \oplus W_j$; the other is the dual multiresolution analysis generated by the dual scaling function $\tilde{\phi}(t)$ and the dual wavelet function $\tilde{\psi}(t)$, i.e., $\tilde{V}_{j+1} = \tilde{V}_j \oplus \tilde{W}_j$.

It should be noted that the wavelet subspace W_j is no longer the orthogonal complement of the scaling subspace V_j, and the dual wavelet subspace $\tilde{W}_j$ is not the orthogonal complement of the dual scaling subspace $\tilde{V}_j$. However, the orthogonal complement relationships

$$\left.\begin{array}{l} W_j \perp \tilde{V}_j \\ \tilde{W}_j \perp V_j \end{array}\right\} \tag{7.10.1}$$

still exist between these four subspaces, that is, the wavelet subspace W_j is the orthogonal complement of dual scaling subspace $\tilde{V}_j$ in $\tilde{V}_{j+1}$, while dual wavelet subspace $\tilde{W}_j$ is the orthogonal complement of scaling subspace V_j in $Vj + 1$. From Eq.(7.10.1), it is easy to know

$$\tilde{W}_j \perp W_{j'} \quad \forall j \neq j'. \tag{7.10.2}$$

The multiresolution analysis subspaces corresponding to $\tilde{W}_j$ and W_j are V_j and $\tilde{V}_j$, respectively. Note that the scaling function $\phi(t)$ still satisfies the admissible condition Eq. (7.8.13) and the unit decomposition Eq. (7.8.16).

Now, the task is how to design the filter banks (H, G) and $\tilde{H}, \tilde{G})$ to construct the biorthogonal wavelet function. Therefore, it is necessary to analyze the constraints between these four filters from the complete reconstruction of the signal[58],

Let the Fourier transform of the discrete signal $s(n)$ be

$$S(\omega) = \sum_{n=-\infty}^{\infty} s(n)e^{-jn\omega}. \tag{7.10.3}$$

Using the filters $\tilde{H}$ and $\tilde{G}$, the discrete signal $s(n)$ is transformed into an approximation sequence $a(n)$ and a detail sequence $d(n)$, whose Fourier transforms are defined as

$$A(\omega) = \frac{1}{2}[\tilde{H}^*(\omega)S(\omega) + \tilde{H}^*(\omega+\pi)S(\omega+\pi)], \tag{7.10.4}$$

$$D(\omega) = \frac{1}{2}[\tilde{G}^*(\omega)S(\omega) + \tilde{G}^*(\omega+\pi)S(\omega+\pi)], \tag{7.10.5}$$

respectively. Thus, the frequency domain form of the reconstructed signal $r(n)$ can be written as

$$R(\omega) = \alpha(\omega)S(\omega) + \beta(\omega)S(\omega+\pi), \tag{7.10.6}$$

where

$$\alpha(\omega) = H(\omega)\tilde{H}^*(\omega) + G(\omega)\tilde{G}^*(\omega), \tag{7.10.7}$$

$$\beta(\omega) = H(\omega)\tilde{H}^*(\omega+\pi) + G(\omega)\tilde{G}^*(\omega+\pi). \tag{7.10.8}$$

When $\alpha = 1$ and $\beta = 0$ hold for all $\Omega \in [-\pi, \pi]$, Eq. (7.10.6) gives the expected complete reconstruction result $R(\omega) = S(\omega)$, which is expressed as $r(n) = s(n)$ in time domain, i.e., the complete reconstruction of the discrete signal $s(n)$ is realized. That is, the conditions for complete signal reconstruction are

$$H(\omega)\tilde{H}^*(\omega) + G(\omega)\tilde{G}^*(\omega) = 1, \tag{7.10.9}$$

$$H(\omega)\tilde{H}^*(\omega+\pi)S(\omega) + G(\omega)\tilde{G}^*(\omega+\pi) = 0. \tag{7.10.10}$$

Definition 7.10.1. *Consider the filter bank (A, B) and its dual filter bank $(\tilde{A}, \tilde{B})$, let*

$$\mathbf{M} = \begin{bmatrix} A(\omega) & A(\omega+\pi) \\ B(\omega) & B(\omega+\pi) \end{bmatrix} \quad \textit{and} \quad \tilde{\mathbf{M}} = \begin{bmatrix} \tilde{A}(\omega) & \tilde{A}(\omega+\pi) \\ \tilde{B}(\omega) & \tilde{B}(\omega+\pi) \end{bmatrix}, \tag{7.10.11}$$

if

$$\tilde{\mathbf{M}}^H\mathbf{M} = \mathbf{I}_2 \quad \textit{or} \quad \mathbf{M}^T\tilde{\mathbf{M}}^* = \mathbf{I}_2 \tag{7.10.12}$$

where $\mathbf{I}_2$ is a 2×2 identity matrix, then (A, B) and $(\tilde{A}, \tilde{B})$ are called biorthogonal filter banks.

It can be proved (left as an exercise) that the filter banks (H, G) and $(\tilde{H}, \tilde{G})$ satisfy the above definition, so they are biorthogonal filter banks.

Let $z = e^{j\omega}$, then Eq. (7.10.9) and Eq. (7.10.10) can be written as

$$H(z)\tilde{H}(z^{-1}) + G(z)\tilde{G}(z^{-1}) = 1,$$

$$H(z)\tilde{H}(-z^{-1}) + G(z)\tilde{G}(-z^{-1}) = 0,$$

respectively, and their solutions are

$$H(z) = \frac{\Delta_H}{\Delta} \quad \text{and} \quad G(z) = \frac{\Delta_G}{\Delta}, \tag{7.10.13}$$

where

$$\Delta = \begin{vmatrix} \tilde{H}(z^{-1}) & \tilde{G}(z^{-1}) \\ \tilde{H}(-z^{-1}) & \tilde{G}(-z^{-1}) \end{vmatrix}, \tag{7.10.14}$$

$$\Delta_H = \begin{vmatrix} 0 & \tilde{G}(z^{-1}) \\ 1 & \tilde{G}(-z^{-1}) \end{vmatrix} = \tilde{G}(-z^{-1}), \tag{7.10.15}$$

$$\Delta_G = \begin{vmatrix} \tilde{H}(z^{-1}) & 1 \\ \tilde{H}(-z^{-1}) & 0 \end{vmatrix} = -\tilde{H}(-z^{-1}). \tag{7.10.16}$$

Obviously, to make $H(z)$ and $G(z)$ avoid infinite impulse response solutions, the determinant $\triangle$ of the equation must be a monomial az. For simplicity, choose $\triangle = -z$, then Eq. (7.10.14) gives

$$[-z^{-1}\tilde{G}(-z^{-1})]\tilde{H}(z^{-1}) + [-z^{-1}\tilde{G}(z^{-1})]\tilde{H}(-z^{-1})] = 1. \tag{7.10.17}$$

In this case, from Eqs. (7.10.13) and (7.10.14) we can get

$$H(z) = \frac{\tilde{G}(-z^{-1})}{-z} = -z^{-1}\tilde{G}(-z^{-1}) \quad \text{or} \quad H(-z) = \tilde{G}(z^{-1}). \tag{7.10.18}$$

Substituting them into Eq. (7.10.17) yields

$$H(z)\tilde{H}(z^{-1}) + H(-z)\tilde{H}(-z^{-1}) = 1. \tag{7.10.19}$$

The filter $\tilde{H}(\omega)$ satisfying this condition is called the dual filter of $H(\omega)$.

On the other hand, from Eqs. (7.10.13) and (7.10.16), it is easily obtain $G(z) = z^{-1}\tilde{H}(-z^{-1})$. Combining it with the solution $H(-z) = z^{-1}\tilde{G}(z^{-1})$ obtained earlier, we have

$$G(z) = z^{-1}\tilde{H}(-z^{-1}) \quad \text{or} \quad G(\omega) = e^{-j\omega}\tilde{H}^*(\omega + \pi), \tag{7.10.20}$$

$$\tilde{G}(z) = z^{-1}H(-z^{-1}) \quad \text{or} \quad \tilde{G}(\omega) = e^{-j\omega}H^*(\omega + \pi). \tag{7.10.21}$$

Substituting the above two equations into Eq. (7.10.17), we obtain

$$G(z)\tilde{G}(z^{-1}) + G(-z)\tilde{G}(-z^{-1}) = 1. \tag{7.10.22}$$

The filter $\tilde{G}(\omega)$ satisfying this condition is called the dual filter of $G(\omega)$.

7.10.2 Design of Biorthogonal Filter Banks

Eq. (7.10.19) can be rewritten as

$$P(z) + P(-z) = 1, \tag{7.10.23}$$

where $P(z) = H(z)\tilde{H}(z^{-1})$.

Summarizing the above discussion, the following steps can be derived for the design of biorthogonal filter banks:

(1) Determine the filters $H(z)$ and $\tilde{H}(z)$ from the factorization $P(z) = H(z)\tilde{H}(z^{-1})$ of the solution $P(z)$ of Eq. (7.10.23);
(2) Designing filters $G(z)$ and $\tilde{G}(z)$ using Eq. (7.10.20) and Eq. (7.10.21), respectively.

For the FIR structure of filter bank (H, G) satisfying Eq. (7.10.23), Vetterli and Herley proved the following important result[216].

Proposition 7.10.1. *The linear phase real FIR filters $H(z)$ and $\tilde{H}(z)$ satisfying the complete reconstruction condition have one of the following forms:*
(1) Both the filter $H(z)$ and $\tilde{H}(z)$ are symmetric and odd length, and their lengths differ by an odd multiple of 2.
(2) One filter is symmetric and the other is antisymmetric. Both filters are of even length, and their lengths are equal or differ by an even multiple of 2.
(3) One filter is of odd length and the other filter is of even length, and the zeros of the two filters are all on the unit circle. The two filters are either symmetric, or one is symmetric and the other is antisymmetric.

Note that the filter of the form (3) has almost no practical significance and is a trivial solution.

The filter bank based on the complete reconstruction of the signal will lead to the biorthogonal scaling function and wavelet function. The proof is as follows.
(1) Eq. (7.10.19) means that the same odd power terms of z in $H(z)\tilde{H}(z^{-1})$ and $H(-z)\tilde{H}(-z^{-1})$ cancel each other, while all even power terms of z should be equal to zero, and the zero power term of $H(z)\tilde{H}(z^{-1})$ is equal to $\frac{1}{2}$. If let

$$H(z) = \sum_{k=-\infty}^{\infty} \frac{h(k)}{\sqrt{2}} z^{-k} \quad \text{and} \quad \tilde{H}(z) = \sum_{k=-\infty}^{\infty} \frac{\tilde{h}(k)}{\sqrt{2}} z^{-k}, \tag{7.10.24}$$

then the inverse z transform of $H(z)\tilde{H}(z)$ is

$$\sum_k h(k)\tilde{h}(k-2n) = \delta(n). \tag{7.10.25}$$

This shows that the low-pass filter coefficient $h(k)$ and its dual low-pass filter coefficient $\tilde{h}(k)$ are biorthogonal.
(2) Eq. (7.10.22) means that the same odd power terms of z in $G(z)\tilde{G}(z^{-1})$ and $G(-z)\tilde{G}(-z^{-1})$ cancel each other, while all even power terms of z should be equal to zero, and the zero power term of $G(z)\tilde{G}(z^{-1})$ is equal to $\frac{1}{2}$. If let

$$G(z) = \sum_{k=-\infty}^{\infty} \frac{g(k)}{\sqrt{2}} z^{-k} \quad \text{and} \quad \tilde{G}(z) = \sum_{k=-\infty}^{\infty} \frac{\tilde{g}(k)}{\sqrt{2}} z^{-k}, \tag{7.10.26}$$

then the inverse z transform of $G(z)\tilde{G}(z)$ is

$$\sum_k g(k)\tilde{g}(k-2n) = \delta(n). \tag{7.10.27}$$

This shows that the high-pass filter coefficient $g(k)$ and its dual high-pass filter coefficient $\tilde{g}(k)$ are biorthogonal.

(3) Using Eq. (7.10.20), we can obtain

$$G(z)\tilde{H}(z) = z^{-1}\tilde{H}(z^{-1})\tilde{H}(z^{-1}). \tag{7.10.28}$$

Notice that $\tilde{H}(z^{-1})\tilde{H}(z^{-1})$ has only zero and even power terms of z with coefficients not equal to zero, while the coefficients of all odd terms are zero, so $G(z)\tilde{H}(z) = z^{-1}\tilde{H}(z^{-1})\tilde{H}(z^{-1})$ has only odd power terms of z with coefficients not equal to zero, while the coefficients of zero and all even power terms are zero. This means that the inverse z transform of $G(z)\tilde{H}(z)$ is

$$\sum_k g(k)\tilde{h}(k-2n) = 0, \quad \forall n. \tag{7.10.29}$$

Similarly, it can be proved that

$$\sum_k h(k)\tilde{g}(k-2n) = 0, \quad \forall n. \tag{7.10.30}$$

To summarize the above discussion, an important conclusion can be drawn: the analytic filter bank $(\tilde{H}, \tilde{G})$ and the synthetic filter bank (G, H) satisfy the complete reconstruction condition generate biorthogonal scaling functions and wavelet functions.

7.10.3 Biorthogonal Wavelet and Fast Biorthogonal Transform

The above theory of filter bank can be used to design biorthogonal wavelet function.

First, consider the iterative construction of the scaling function and the dual scaling function.

(1) A unit DC signal $U(x) = 1$ (where $x \in [0, 1]$) is used as the initial value $\phi^{(0)}(x)$ of the scaling function iteration. If let the filter $H^{(0)} = h^{(0)}(0) = 1$, then the initial value can be written as

$$\phi^{(0)}(x) = U(x) = h^{(0)}(0) = 1, \quad 0 \le x \le 1. \tag{7.10.31}$$

(2) Taking the $2\downarrow$ sampling (downsampling) for $\phi^{(0)}(x)$, and then passing the sampling result through the filter $H(z)$ to obtain the first iteration $\phi^{(1)}(x)$ of the scaling function, as shown in the left half of Fig. 7.10.2(a). Since using 2 downsampling and then filtering is equivalent to filtering with $H(z^2)$ and then downsampling, the left half is equivalently drawn as the right half. If let the equivalent filter

$$H^{(1)}(z) = H(z^2) = \prod_{m=1}^{1} H(z^{2^m}), \tag{7.10.32}$$

then the scaling function generated by the first iteration is a piecewise invariant function and can be expressed as

$$\phi^{(1)}(x) = 2^{1/2}h^{(1)}(k), \quad 2^{-1}k \le x < 2^{-1}(k+1), \tag{7.10.33}$$

where $h^{(1)}(k)$ is the k-th coefficient of the filter $H^{(1)}(z)$. Note that the length of $H^{(1)}(z)$ is twice as long as $H(z)$.

(3) Then, taking the $2\downarrow$ sampling of $\phi^{(0)}(x)$ and filtering to obtain the second iteration $\phi^{(2)}(x)$ of the scaling function. And so on, the scaling function produced by the i-th iteration is the output of $U(x)$ through the i-th cascade of downsampling + filtering, as shown in Fig. 7.10.2(b). Obviously, this result can be equivalently drawn as Fig. 7.10.2.

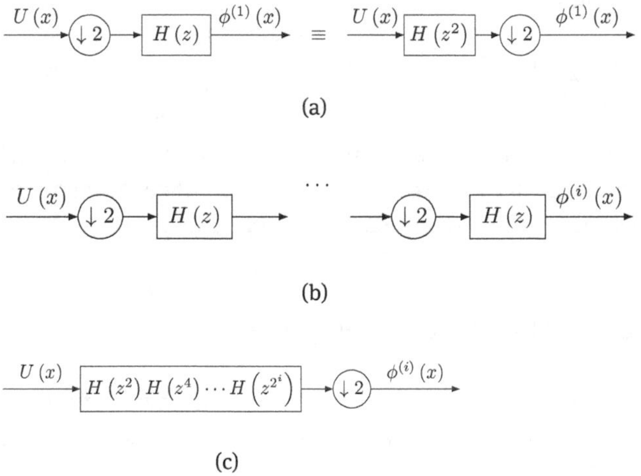

Fig. 7.10.2: Iterative generation of the scaling function

The z-transform of the equivalent filter is

$$H^{(i)}(z) = \prod_{m=1}^{i} H(z^{2^m}). \tag{7.10.34}$$

When $U(x)$ is used as input, the output of the equivalent filter is

$$\phi^{(i)}(x) = 2^{i/2}h^{(i)}(k), \quad 2^{-i}k \le x < 2^{-i}(k+1), \tag{7.10.35}$$

where $h^{(i)}(k)$ denotes the k-th coefficient of the filter $H^{(i)}(z)$.

If the filter $H(z)$ in Fig. 7.10.2 is replaced by $\tilde{H}(z)$, then the output of the equivalent filter is

$$\tilde{\phi}^{(i)}(x) = 2^{i/2}\tilde{h}^{(i)}(k), \quad 2^{-i}k \le x < 2^{-i}(k+1), \tag{7.10.36}$$

where $\tilde{h}^{(i)}(k)$ denotes the k-th coefficient of the filter $\tilde{H}^{(i)}(z)$.

In the process of iteratively constructing the scaling function and the dual scaling function, using the formula

$$\psi^{(i)}(x) = \sqrt{2}\sum_{k=0}^{L-1} g^{(i)}(k)\phi^{(i)}(2x-k), \quad 2^{-i}k \le x < 2^{-i}(k+1), \tag{7.10.37}$$

$$\tilde{\psi}^{(i)}(x) = \sqrt{2}\sum_{k=0}^{L-1} \tilde{g}^{(i)}(k)\tilde{\phi}^{(i)}(2x-k), \quad 2^{-i}k \le x < 2^{-i}(k+1), \tag{7.10.38}$$

the wavelet function $\psi^{(i)}(x)$ and the dual wavelet function $\tilde{\psi}^{(i)}(x)$ can be constructed, respectively.

Eq. (7.10.35) to Eq. (7.10.38) constitute the iterative construction algorithm of the scaling function $\phi(x)$, dual scaling function $\tilde{\phi}(x)$, wavelet function $\psi(x)$ and dual wavelet function $\tilde{\psi}(x)$.

If the filter bank satisfies the complete reconstruction condition and the low-pass filter $H(z)$ and $\tilde{H}(z)$ can guarantee $\phi^{(i)}(x)$ and $\tilde{\phi}^{(i)}(x)$ to converge to continuous functions, respectively, then the constructed scaling function and wavelet function satisfy the biorthogonality condition

$$\langle \phi(x-n), \tilde{\phi}(x-l)\rangle = \delta_{nl}, \tag{7.10.39}$$

$$\langle \psi(x-n), \tilde{\psi}(x-l)\rangle = \delta_{nl}, \tag{7.10.40}$$

$$\langle \phi(x-n), \tilde{\psi}(x-l)\rangle = 0, \quad \forall n, l, \tag{7.10.41}$$

$$\langle \psi(x-n), \tilde{\phi}(x-l)\rangle = 0, \quad \forall n, l, \tag{7.10.42}$$

i.e., the constructed wavelet function is biorthogonal.

In wavelet analysis, frame theory, multiresolution analysis, and filter bank theory are closely related to each other. A more detailed discussion of frame theory and filter bank can be found in the literature [201].

The fast wavelet transform formula Eq. (7.9.23) and inverse wavelet transform formula Eq. (7.9.24) can be easily generalized to the biorthogonal wavelet transform algorithm

$$\begin{cases} \mathbf{c}_{j-1} = \tilde{\mathbf{H}}\mathbf{c}_j \\ \mathbf{d}_{j-1} = \tilde{\mathbf{G}}\mathbf{c}_j \end{cases} \quad j = 0, -1, \cdots, -J+1 \tag{7.10.43}$$

and inverse biorthogonal wavelet transform algorithm

$$\mathbf{c}_j = \mathbf{H}\mathbf{c}_{j-1} + \mathbf{G}\mathbf{d}_{j-1} \quad j = -J+1, \cdots, -1, 0, \tag{7.10.44}$$

where $\mathbf{c}_j = [c_{j,k}]_{k=-\infty}^{\infty}$, $\mathbf{d}_j = [d_{j,k}]_{k=-\infty}^{\infty}$, $\tilde{\mathbf{H}} = [\tilde{h}(m-2k)]_{m,k=-\infty}^{\infty}$ and $\tilde{\mathbf{G}} = [\tilde{g}(m-2k)]_{m,k=-\infty}^{\infty}$.

Fig. 7.10.3 (a) and (b) show the fast biorthogonal wavelet transform algorithm and the inverse transform algorithm, respectively.

When using orthogonal signal transform and non-orthogonal signal transform, we need to pay attention to the following matters[28].

Precautions for orthogonal signal transform:

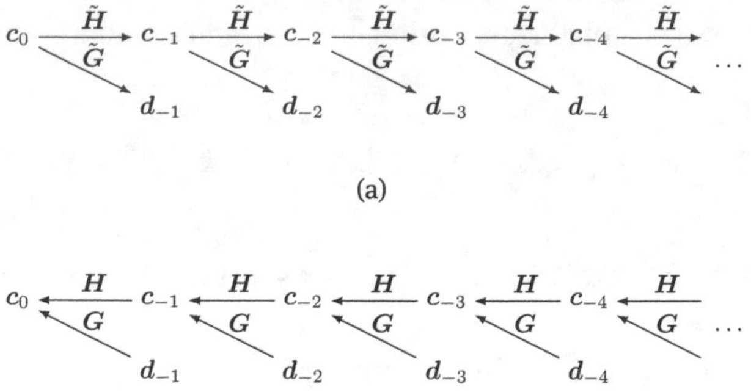

$$c_0 \xrightarrow{\tilde{H}} c_{-1} \xrightarrow{\tilde{H}} c_{-2} \xrightarrow{\tilde{H}} c_{-3} \xrightarrow{\tilde{H}} c_{-4} \xrightarrow{\tilde{H}} \cdots$$
$$\searrow_{\tilde{G}} \quad \searrow_{\tilde{G}} \quad \searrow_{\tilde{G}} \quad \searrow_{\tilde{G}} \quad \searrow_{\tilde{G}}$$
$$d_{-1} \qquad d_{-2} \qquad d_{-3} \qquad d_{-4}$$

(a)

$$c_0 \xleftarrow{H} c_{-1} \xleftarrow{H} c_{-2} \xleftarrow{H} c_{-3} \xleftarrow{H} c_{-4} \xleftarrow{H} \cdots$$
$$\nwarrow_{G} \quad \nwarrow_{G} \quad \nwarrow_{G} \quad \nwarrow_{G} \quad \nwarrow_{G}$$
$$d_{-1} \qquad d_{-2} \qquad d_{-3} \qquad d_{-4}$$

(b)

Fig. 7.10.3: (a) The fast orthogonal wavelet transform algorithm; (b) The inverse transform algorithm

(1) The orthogonal transform of a signal represents the approximation of the signal at a certain time period and these approximations cannot be used to extrapolate or predict the value of the signal outside this time period.

(2) When using orthogonal signal transform, it is difficult and often impossible to obtain a priori information about the input signal source.

(3) If the real components of the signal are not orthogonal to each other, the orthogonal transform does not decompose the signal into their real components.

(4) The orthogonal signal transform is not suitable for irregular sampling signal[3].

Precautions for non-orthogonal signal transform:

(1) If a signal transform uses a set of orthogonal continuous basis functions which are digitized using irregular sampling, then the digitized signal transform is non-orthogonal.

(2) When the observation time interval is shorter than the time period of the signal, the signal transform must use a non-orthogonal transform.

(3) If the real components of the signal are not orthogonal to each other, the signal transform must use a non-orthogonal transform.

---

3 Uniform sampling refers to equally spaced sampling, and nonuniform sampling refers to non-equally spaced sampling. However, nonuniform sampling can be either nonuniformly sampled according to some rules (nonuniform sampling) or completely irregularly sampled (irregular sampling).

Summary

This chapter first introduces two basic forms of signal local transform, the analytic signal, and the exclusion principle. Then, focusing on the linear transform of time-frequency signal analysis, the short-time Fourier transform, critical sampling, and oversampling Gabor transform are discussed; and the fractional-order Fourier transform is introduced.

Wavelet analysis is an important part of this chapter, which includes two basic problems: the design of the wavelet and the fast algorithm for the wavelet transform. We discuss the design method of wavelet and the realization of fast wavelet transform in detail from the frame theory, multiresolution analysis, and filter bank theory.

Exercises

7.1 Find the instantaneous frequency $\omega_i(t)$ and average frequency $\bar{\omega}$ of the following signals:

(1) Normalized Gaussian signal

$$s(t) = g(t) = \left(\frac{\alpha}{\pi}\right)^{1/4} \exp\left(-\frac{\alpha}{2}t^2\right), \quad \alpha > 0$$

(2) Linear FM signal with Gaussian envelope

$$s(t) = g(t)e^{jmt^2}$$

7.2 Let

$$Y(t, \omega) = \int_{-\infty}^{\infty} y(u)\gamma^*(u - t)e^{-j\omega u}du$$

be the short-time Fourier transform. Express $Y(t, \omega)$ in terms of the Fourier transform of $y(t)$ and $\gamma(t)$, and use this representation to show why $\gamma(t)$ is required to be a narrow-band function?

7.3 Prove the following properties of the short-time Fourier transform:

(1) Short-time Fourier transform is a linear time-frequency representation;

(2) Short-time Fourier transform is frequency shift invariant

$$\tilde{z}(t) = z(t)e^{j\omega_0 t} \rightarrow \mathrm{STFT}_{\tilde{z}}(t, \omega) = \mathrm{STFT}_z(t, \omega - \omega_0)$$

7.4 Let the window function be

$$g(t) = \left(\frac{\alpha}{\pi}\right)^{1/4} \exp\left(-\frac{\alpha}{2}t^2\right).$$

Find the short-time Fourier transform $\mathrm{STFT}(t, \omega)$ of the Gaussian signal

$$s(t) = \left(\frac{\beta}{\pi}\right)^{1/4} \exp\left(-\frac{\beta}{2}t^2\right)$$

7.5 Prove that the signal $z(t)$ can be recovered or reconstructed using an inverse short-time Fourier transform, i.e.,

$$z(t) = \frac{1}{g^*(0)} \int_{-\infty}^{\infty} \text{STFT}(t,f) e^{j2\pi ft} df$$

7.6 Let $\mathbf{e}_1 = [1,0]^T$ and $\mathbf{e}_2 = [0,1]^T$. Defining $\mathbf{g}_1 = \mathbf{e}_1, \mathbf{g}_2 = -0.5\mathbf{e}_1 + 0.5\sqrt{3}\mathbf{e}_2$ and $\mathbf{g}_3 = -0.5\mathbf{e}_1 - 0.5\sqrt{3}\mathbf{e}_2$. Does $\{\mathbf{g}_i, i = 1,2,3\}$ constitute a framework? If not, explain the reason; If so, what kind of framework is it?

7.7 Prove that any frame $\{\mathbf{g}_k, k \in K\}$ is a complete set of L_2 spaces.

7.8 Prove that the filter $G(\omega)$ that generates the standard orthogonal wavelet satisfies the condition

$$|G(\omega)|^2 + |G(\omega + \pi)|^2 = 1$$

7.9 Let P be a probability measure whose support region is $[-\epsilon, \epsilon] \subset \left[-\frac{\pi}{3}, \frac{\pi}{3}\right]$. The Fourier transform of a scaling function is known to be

$$\Phi(\omega) = \left[\int_{\omega-\pi}^{\omega+\pi} dP\right]^{1/2}$$

which is the nonnegative square root of an integral, prove that this scaling function is orthogonal.

7.10 Prove that the filter bank (H, G) and the dual filter bank $(\tilde{H}, \tilde{G})$ satisfying the signal complete reconstruction condition

$$\begin{aligned} H(\omega)\tilde{H}^*(\omega) + G(\omega)\tilde{G}^*(\omega) &= 1 \\ H(\omega)\tilde{H}^*(\omega + \pi) + G(\omega)\tilde{G}^*(\omega + \pi) &= 0 \end{aligned}$$

are biorthogonal filter banks.

7.11 Let $\tilde{f}(t)$ be the biorthogonal dual function of $f(t)$, i.e.,

$$\langle f(t-n), \tilde{f}(n-k) \rangle = \delta(n-k)$$

and $F(\omega)$ and $\tilde{F}(\omega)$ are the Fourier transform of $f(t)$ and $\tilde{f}(t)$, respectively. Prove

$$\sum_{k=-\infty}^{\infty} F(\omega + 2k\pi)\tilde{F}^*(\omega + 2k\pi) = 1, \quad \forall \omega$$

7.12 Prove that if the two-scale difference equation of the scaling function $\phi(t)$ is rewritten as the equation of $\phi_{jk}(t)$, then there is

$$\phi_{jk}(t) = \sum_{l} h(l - 2k)\phi_{j+1,l}(t)$$

7.13 Prove that the scaling function $\phi(2t)$ can be expanded into a sum of two series using the scaling function $\phi(t)$ and the wavelet function $\psi(t)$

$$\phi(2t-k)=\sum_{l}\tilde{h}(k-2l)\phi(t-l)+\sum_{l}\tilde{g}(k-2l)\psi(t-l)$$

7.14 Let the low-pass filter be

$$H(\omega)=\begin{cases}1, & |\omega|\le|\frac{\pi}{2}|\\ 0, & \text{Others}\end{cases}$$

and

$$G(\omega)=-e^{-j\omega}H^{*}(\omega+\pi)$$

Find the wavelet function $\psi(t)$ generated by $G(\omega)$.

7.15 Let $g(t)=\sqrt{2\lambda}e^{-\lambda t}, t\ge 0$. Prove that the biorthogonal function $\gamma(t)$ of $g(t)$ is given by

$$\gamma(t)=\begin{cases}-\frac{e^{\lambda t}}{\sqrt{2\lambda}}, & -1\le t<0\\ \frac{e^{\lambda t}}{\sqrt{2\lambda}}, & 0<t\le 1\\ 0, & \text{Others}\end{cases}$$

7.16 Let

$$H_{\text{Zak}}(t,f)=\sum_{k=-\infty}^{\infty}h(t-k)e^{-j2\pi kf}$$

be the Zak transform of function $\{h(t)\}$. Prove that for any integer n, there is always

$$H_z(t-n,f)=e^{j2\pi nf}H_z(t,f)$$

7.17 Let $H_{\text{Zak}}(t,f)$ and $G_{\text{Zak}}(t,f)$ be the Zak transform of function $h(t)$ and $g(t)$, respectively. Given

$$\langle h,g_{mn}\rangle^2=\int_0^1\int_0^1 H_{\text{Zak}}(t,f)G^{*}_{\text{Zak}}(t,f)e^{-j2\pi(mt-nf)}dtdf$$

try to deduce the identity

$$\sum_m\sum_n|\langle h,g_{mn}\rangle|^2=\int_0^1\int_0^1|H_{\text{Zak}}(t,f)|^2|G^{*}_{\text{Zak}}(t,f)|^2dtdf$$

from this relationship, and use the above identity to prove that $\{g_{mn}(t)\}$ is a frame if and only if

$$A\le|G_{\text{Zak}}(t,f)|^2\le B$$

holds almost everywhere in $(t,f)\in[0,1]\times[0,1]$. *Hint: Using the Poisson summation formula.*

7.18 Let

$$y(t) = \sum_{m=-\infty}^{\infty} \sum_{n=-\infty}^{\infty} a_{mn} g(t-m) d^{j2\pi nt}$$

is the Gabor expansion of function $y(t)$. Prove that the Gabor expansion coefficient a_{mn} can be determined by

$$a_{mn} = \int_0^1 \int_0^1 \frac{Y_{\text{Zak}}(t,f)}{G_{\text{Zak}}(t,f)} e^{-j2\pi(nt+mf)} dt df$$

where $Y_{\text{Zak}}(t,f)$ and $G_{\text{Zak}}(t,f)$ are Zak transform of $y(t)$ and $h(t)$, respectively.

8 Quadratic Time-frequency Distribution

In the previous chapter, three linear time-frequency representations of short-time Fourier transform, wavelet transform and Gabor transform are discussed, which use a joint function of time and frequency (taken in the form of a linear transform) to describe the variation of the signal spectrum with time. Similarly, the joint function of time and frequency can be used to describe the variation of the energy density of the signal with time. The energy expression of a non-stationary signal is called time-frequency distribution for short. Since the energy itself is the quadratic representation of the signal, the time-frequency distribution is a kind of nonlinear transformation of the non-stationary signal ("energized" quadratic transformation).

This chapter will start with the general theory of time-frequency distribution and introduce the forms, mathematical properties of various time-frequency distributions, and how to improve their time-frequency aggregation performance.

8.1 The General Theory of Time-frequency Distribution

Although the linear time-frequency representations such as short-time Fourier transform, Gabor transform, and wavelet transform can effectively describe the local performance of a non-stationary signal, the quadratic time-frequency representation is a more intuitive and reasonable signal representation when the time-frequency representation is used to describe the energy variation of non-stationary signal, because the energy itself is a quadratic representation.

Many quadratic time-frequency representations can be used to roughly represent energy. Two prominent examples are spectrogram and scalogram. The spectrogram is defined as the square of the modulus of the short-time Fourier transform

$$\mathrm{SPEC}(t,\omega) = |\mathrm{STFT}(t,\omega)|^2. \tag{8.1.1}$$

While the scalogram is defined as the square of the modulus of the wavelet transform

$$\mathrm{SCAL}(a,b) = |\mathrm{WT}(a,b)|^2. \tag{8.1.2}$$

However, the description of energy distribution by spectrogram and scalogram is very rough, because they do not satisfy the more stringent requirements for energy distribution.

In order to more accurately describe the energy distribution of non-stationary signals with time, it is necessary to investigate other "energized" quadratic time-frequency representations with better performance. Since these time-frequency representations can describe the energy density distribution of signals, they are often referred to as time-frequency distributions. In fact, the time-frequency distribution outperforms the spectrogram and scalogram in many properties. In order to better understand the

https://doi.org/10.1515/9783110475562-008

various time-frequency distributions, it is necessary to discuss their basic concepts and the basic properties required for them before studying specific time-frequency distributions.

8.1.1 Definition of the Time-frequency Distribution

We are no stranger to the quadratic (bilinear) transform $z(t)z^*(t)$ of complex signals since it is used in stationary signals to obtain the correlation function and power spectrum, that is,

$$R(\tau) = \int_{-\infty}^{\infty} z(t)z^*(t-\tau)dt, \tag{8.1.3}$$

$$P(\omega) = \int_{-\infty}^{\infty} R(\tau)e^{-j\omega\tau}d\tau. \tag{8.1.4}$$

In addition to the asymmetric form of Eq. (8.1.3), the autocorrelation function can also be defined in symmetric form

$$R(\tau) = \int_{-\infty}^{\infty} z(t+\frac{\tau}{2})z^*(t-\frac{\tau}{2})dt, \tag{8.1.5}$$

The above-mentioned definitions of the autocorrelation function and power spectrum of stationary signals can be easily generalized to non-stationary signals, and the time-varying autocorrelation function $R_z(t,\tau)$ in symmetric form is more useful than the asymmetric form in the analysis of non-stationary signals because the symmetric form of bilinear transformation $z(t+\frac{\tau}{2})z^*(t-\frac{\tau}{2})$ of the signal $z(t)$ can show some important characteristics of non-stationary signals. However, when a bilinear transform similar to Eq. (8.1.5) is used for a non-stationary signal, in order to reflect the local time-domain characteristics of the signal, a sliding window processing similar to that in the short-time Fourier transform should be made, while weighting along the τ axis to obtain the time-varying correlation function

$$R(t,\tau) = \int_{-\infty}^{\infty} \phi(u-t,\tau)z(u+\frac{\tau}{2})z^*(u-\frac{\tau}{2})du. \tag{8.1.6}$$

where $\phi(t,\tau)$ is the window function and $R(t,\tau)$ is the "local correlation function". The Fourier transform of the local correlation function gives the time-varying power spectrum, which is the time-frequency distribution of the signal energy

$$P(t,\omega) = \int_{-\infty}^{\infty} R(t,\tau)e^{-j\omega\tau}d\tau. \tag{8.1.7}$$

This shows that the time-frequency distribution $P(t, \omega)$ can also be defined by using the local correlation function $R(t, \tau)$. In fact, if different forms of local correlation functions are taken, different definitions of time-frequency distribution can be obtained. This will be discussed in detail in the subsequent sections.

8.1.2 Basic Properties of Time-frequency Distribution

Since it is the representation of the energy distribution of a non-stationary signal, the time-frequency distribution should have some basic properties.

Property 1 The time-frequency distribution must be real (and hopefully nonnegative).

Property 2 The integration of the time-frequency distribution with respect to time t and ω should give the total energy E of the signal, i.e.,

$$\frac{1}{2\pi}\int_{-\infty}^{\infty}\int_{-\infty}^{\infty} P(t, \omega)dtd\omega = E \quad \text{(Total energy of signal).} \tag{8.1.8}$$

Property 3 Edge characteristics

$$\int_{-\infty}^{\infty} P(t, \omega)dt = |Z(\omega)|^2 \quad \text{and} \quad \frac{1}{2\pi}\int_{-\infty}^{\infty} P(t, \omega)d\omega = |z(t)|^2, \tag{8.1.9}$$

i.e., the integration of the time-frequency distribution with respect to time t and frequency ω gives the spectral density of the signal at frequency ω and the instantaneous power of the signal at moment t, respectively.

Property 4 The first order moment of the time-frequency distribution gives the instantaneous frequency $\omega_i(t)$ and the group delay $\tau_g(\omega)$, i.e.,

$$\omega_i(t) = \frac{\int_{-\infty}^{\infty} \omega P(t, \omega)d\omega}{\int_{-\infty}^{\infty} P(t, \omega)d\omega} \quad \text{and} \quad \tau_g(\omega) = \frac{\int_{-\infty}^{\infty} tP(t, \omega)dt}{\int_{-\infty}^{\infty} P(t, \omega)dt}. \tag{8.1.10}$$

Property 5 Finite time support

$$z(t) = 0 \quad (|t| > t_0) \Rightarrow P(t, \omega) = 0 \quad (|t| > t_0), \tag{8.1.11}$$

and finite frequency support

$$Z(\omega) = 0 \quad (|\omega| > \omega_0) \Rightarrow P(t, \omega) = 0 \quad (|\omega| > \omega_0). \tag{8.1.12}$$

Finite support is a fundamental property proposed for the time-frequency distribution from the perspective of energy. In signal processing, as an engineering approximation, the signal is often required to have a finite time width and finite bandwidth. If a signal $z(t)$ only takes a non-zero value in a certain time interval, and the signal spectrum $Z(\omega)$ only takes a non-zero value in a certain frequency interval, then the signal $z(t)$ and its spectrum are said to be finite support. Similarly, if the time-frequency distribution of the signal is equal to zero outside the total support area of $z(t)$ and $Z(\omega)$, it is said that the time-frequency distribution is finite support. Cohen[63] proposed that an ideal time-frequency distribution should also have a finite support property, that is, where $z(t)$ and $Z(\omega)$ are equal to zero, the time-frequency distribution $P(t, \omega)$ should also be equal to zero.

It should be pointed out that as a representation of energy density, the time-frequency distribution should not only be real but also nonnegative. However, as we will be seen in the next, the actual time-frequency distribution is hardly guaranteed to take a positive value.

Like other linear functions, the linear time-frequency representation satisfies the linear superposition principle, which brings great convenience to the analysis and processing of multicomponent signals, because we can first analyze and process each single component signal separately, and then simply superpose the results. Unlike the linear time-frequency representation, the quadratic time-frequency distribution no longer obeys the linear superposition principle, which makes the time-frequency analysis of multicomponent signals no longer as simple as the processing of the linear time-frequency representation. For example, it is easy to see from the definition of the spectrogram that the spectrogram of the sum of two signals $z_1(t) + z_2(t)$ is not equal to the sum of the individual signal spectrogram, i.e.,

$$\mathrm{SPEC}_{z_1+z_2}(t, \omega) \neq |\mathrm{STFT}_{z_1+z_2}(t, \omega)|^2, \tag{8.1.13}$$

$$\mathrm{SPEC}_{z_1}(t, \omega) + \mathrm{SPEC}_{z_2}(t, \omega) \neq |\mathrm{STFT}_{z_1}(t, \omega)|^2 + |\mathrm{STFT}_{z_2}(t, \omega)|^2. \tag{8.1.14}$$

That is, the linear structure of the STFT is broken in the quadratic spectrogram.

Similar to the linear time-frequency representation obeying the "linear superposition principle", any quadratic time-frequency representation satisfies the so-called "quadratic superposition principle". Therefore, it is necessary to focus on it.

Let

$$z(t) = c_1 z_1(t) + c_2 z_2(t) \tag{8.1.15}$$

then any quadratic time-frequency distribution obeys the following quadratic superposition principle

$$P_z(t, \omega) = |c_1|^2 P_{z_1}(t, \omega) + |c_2|^2 P_{z_2}(t, \omega) + c_1 c_2^* P_{z_1,z_2}(t, \omega) + c_2 c_1^* P_{z_2,z_1}(t, \omega), \tag{8.1.16}$$

where $P_z(t, \omega) = P_{z,z}(t, \omega)$ represents the "self time-frequency distribution" of signal $z(t)$ (referred to as "signal term" or "self term"), which is a bilinear function of $z(t)$; $P_{z_1,z_2}(t, \omega)$ represents the "mutual time-frequency distribution" of signal components $z_1(t)$ and $z_2(t)$, (referred to as "cross term"), which is a bilinear function of $z_1(t)$ and $z_2(t)$. The cross term is usually equal to the interference.

Extending the principle of quadratic superposition to the p-component signal $z(t) = \sum_{k=1}^{p} c_k z_k(t)$, the following general rule can be obtained:

(1) Each signal component $c_k z_k(t)$ has a self (time-frequency distribution) component,i.e., a signal term $|c_k|^2 P_{z_k}(t, \omega)$;
(2) Each pair of signal components $c_k z_k(t)$ and $c_l z_l(t)$ (where $k \neq l$) has a corresponding mutual (time-frequency distribution) component, i.e., the cross term $c_k c_l^* P_{z_k, z_l}(t, \omega) + c_l c_k^* P_{z_l, z_k}(t, \omega)$.

Therefore, for a p-component signal $z(t)$, the time-frequency distribution $P_z(t, \omega)$ will contain p signal terms as well as $\binom{p}{2} = p(p-1)/2$ pairwise combine of cross-terms. Since the number of cross-terms increases quadratically with the increase of the number of signal components, the more signal components are, the more serious the cross-terms are.

In most practical applications, the main purpose of time-frequency signal analysis is to extract the signal components and suppress the cross-terms that exist as interference. Therefore, it is usually desired that a time-frequency distribution should have as strong a signal term as possible and as weak a cross term as possible. It can be said that cross-term suppression is both the focus and difficulty in the design of time-frequency distribution. This will be the main topic of discussion throughout the subsequent sections of this chapter.

8.2 The Wigner-Ville Distribution

The properties of the time-frequency distribution are divided into macroscopic properties (e.g., real-valued, total energy conservation) and local properties (e.g., edge characteristics, instantaneous frequency, etc.). In order to correctly describe the local energy distribution of the signal, it is hoped that wherever the signal has local energy, the time-frequency distribution also gathers in these places, which is the time-frequency local aggregation, and it is one of the important indicators to measure the time-frequency distribution. A time-frequency distribution is an impractical time-frequency distribution, even if it has ideal macroscopic properties if spurious signals appear locally (i.e., poor local aggregation). In other words, it is better that some macroscopic properties are not satisfied, the time-frequency distribution should also have good local characteristics.

How to get a time-frequency distribution with good local aggregation? Since the Wigner-Ville distribution is the first time-frequency distribution to be introduced, and all other time-frequency distributions can be regarded as the windowing form of the Wigner-Ville distribution, the Wigner-Ville distribution is regarded as the mother of all time-frequency distributions. In this section, we will discuss and analyze this time-frequency distribution first.

8.2.1 Mathematical Properties

It is pointed out in the previous section that taking different forms of the local correlation function yields different time-frequency distributions. Now consider a simple and effective form of local correlation function: using the time impulse function $\phi(u-t,\tau)=\delta(u-t)$ (without limitation for τ, but taking the instantaneous value in the time-domain) as the window function, i.e., the local correlation function is taken as

$$R_z(t,\tau)=\int_{-\infty}^{\infty}\delta(u-t)z(u+\frac{\tau}{2})z^*(u-\frac{\tau}{2})du=z(t+\frac{\tau}{2})z^*(t-\frac{\tau}{2}), \tag{8.2.1}$$

which is called the instantaneous correlation function or bilinear transformation of signal $z(t)$.

The Fourier transform of the local correlation function $R_z(t,\tau)$ with respect to the lag variable τ is

$$W_z(t,\omega)=\int_{-\infty}^{\infty}z(t+\frac{\tau}{2})z^*(t-\frac{\tau}{2})e^{-j\omega\tau}d\tau. \tag{8.2.2}$$

Since this distribution was first introduced in quantum mechanics by Wigner[228] in 1932, and Ville[112] proposed it as a signal analysis tool in 1948, it is now customarily called the Wigner-Ville distribution.

The Wigner-Ville distribution can also be defined as

$$W_Z(\omega,t)=\frac{1}{2\pi}\int_{-\infty}^{\infty}Z(\omega+\frac{\upsilon}{2})Z^*(\omega-\frac{\upsilon}{2})e^{j\upsilon\tau}d\upsilon \tag{8.2.3}$$

by the signal spectrum $Z(\omega)$, where υ is the frequency offset. Note that the Wigner-Ville distribution involves four parameters: time t, time delay τ, frequency ω and frequency offset υ.

The main mathematical properties of the Wigner-Ville distribution are discussed below.

(1) Real-valued: The Wigner-Ville distribution $W_z(t,\omega)$ is a real function of t and ω.
(2) Time-shift invariance: If $\tilde{z}(t)=z(t-t_0)$, then $W_{\tilde{z}}(t,\omega)=W_z(t-t_0,\omega)$.
(3) Frequency-shift invariance: If $\tilde{z}(t)=z(t)e^{j\omega_0 t}$, then $W_{\tilde{z}}(t,\omega)=W_z(t,\omega-\omega_0)$.
(4) Time edge characteristics: The Wigner-Ville distribution satisfies the time edge characteristic $\frac{1}{2\pi}\int_{-\infty}^{\infty}W_z(t,\omega)d\omega=|z(t)|^2$ (instantaneous power).

In addition to the above basic properties, the Wigner-Ville distribution also has some other basic properties, as shown in Table 8.2.1.

The time edge property P4 and the frequency edge property P5 show that the Wigner-Ville distribution is not guaranteed to be positive in the entire time-frequency plane. In other words, the Wigner-Ville distribution violates the principle that real

Tab. 8.2.1: The important mathematical properties of the Wigner-Ville distribution

P1(Real-valued)	$W_z^*(t,\omega) = W_z(t,\omega)$
P2(Time-shift invariance)	$\tilde{z}(t) = z(t-t_0) \Rightarrow W_{\tilde{z}}(t,\omega) = W_z(t-t_0,\omega)$
P3(Frequency-shift invariance)	$\tilde{z}(t) = z(t)e^{j\omega_0 t} \Rightarrow W_{\tilde{z}}(t,\omega) = W_z(t,\omega-\omega_0)$
P4(Time edge characteristics)	$\frac{1}{2\pi}\int_{-\infty}^{\infty} W_z(t,\omega)d\omega = \lvert z(t)\rvert^2$
P5(Frequency edge characteristics)	$\int_{-\infty}^{\infty} W_z(t,\omega)dt = \lvert Z(\omega)\rvert^2$
P6(Instantaneous frequency)	$\omega_i(t) = \frac{\int_{-\infty}^{\infty} tW_z(t,\omega)d\omega}{\int_{-\infty}^{\infty} W_z(t,\omega)d\omega}$
P7(Group delay)	$\tau_g(\omega) = \frac{\int_{-\infty}^{\infty} tW_z(t,\omega)dt}{\int_{-\infty}^{\infty} W_z(t,\omega)dt}$
P8(Finite time support)	$z(t) = 0(t \notin [t_1,t_2]) \Rightarrow W_z(t,\omega) = 0(t \notin [t_1,t_2])$
P9(Finite frequency support)	$Z(\omega) = 0(\omega \notin [\omega_1,\omega_2]) \Rightarrow W_z(t,\omega) = 0(\omega \notin [\omega_1,\omega_2])$
P10(Moyal formula)	$\frac{1}{2\pi}\int_{-\infty}^{\infty}\int_{-\infty}^{\infty} W_x(t,\omega)W_y(t,\omega)d\omega = \lvert\langle x,y\rangle\rvert^2$
P11(Convolution property)	$\tilde{z}(t) = \int_{-\infty}^{\infty} z(u)h(t-u)du \Rightarrow W_{\tilde{z}}(t,\omega) = \int_{-\infty}^{\infty} W_z(u,\omega)W_h(t-u,\omega)du$
P12(Product property)	$\tilde{z}(t) = z(t)h(t) \Rightarrow W_{\tilde{z}}(t,\omega) = \frac{1}{2\pi}\int_{-\infty}^{\infty} W_z(t,\upsilon)W_h(t,\omega-\upsilon)d\upsilon$
P13(Fourier transform property)	$W_z(\omega,t) = 2\pi W_z(t,-\omega)$

time-frequency energy distribution must not be negative. This sometimes leads to unexplained results.

Taking the complex conjugate of Eq. (8.2.3) and using the real-valued $W_z^*(t,\omega) = W_z(t,\omega)$, the definition Eq. (8.2.3) of the Wigner-Ville distribution can be equivalently written as

$$W_Z(\omega,t) = \int_{-\infty}^{\infty} Z^*(\omega+\frac{\upsilon}{2})Z(\omega-\frac{\upsilon}{2})e^{-j\upsilon\tau}d\upsilon. \tag{8.2.4}$$

Example 8.2.1 Wigner-Ville distribution of complex harmonic signals

When the signal $z(t) = e^{j\omega_0 t}$ is a single complex harmonic signal, its Wigner-Ville distribution is

$$\begin{aligned} W_z(t,\omega) &= \int_{-\infty}^{\infty} \exp\left[j\omega_0\left(t+\frac{\tau}{2}-t+\frac{\tau}{2}\right)\right]\exp(-j\omega\tau)d\tau \\ &= \int_{-\infty}^{\infty} \exp[-j(\omega-\omega_0)\tau]d\tau \\ &= 2\pi\delta(\omega-\omega_0). \end{aligned} \tag{8.2.5}$$

And when the signal $z(t) = e^{j\omega_1 t} + e^{j\omega_2 t}$ is a superposition of two complex harmonic signals, its Wigner-Ville distribution is

$$\begin{aligned} W_z(t,\omega) &= W_{\text{auto}}(t,\omega) + W_{\text{cross}}(t,\omega) \\ &= W_{z_1}(t,\omega) + W_{z_2}(t,\omega) + 2\text{Re}[W_{z_1,z_2}(t,\omega)]. \end{aligned} \tag{8.2.6}$$

In the equation, the signal term, i.e., the self term, is

$$W_z(t,\omega) = 2\pi[\delta(\omega-\omega_1)+\delta(\omega-\omega_2)] \tag{8.2.7}$$

and the cross-term is

$$
\begin{aligned}
W_{z_1,z_2}(t,\omega) &= \int_{-\infty}^{\infty} \exp\left[j\omega_1\left(t+\frac{\tau}{2}\right) - j\omega_2\left(t-\frac{\tau}{2}\right)\right]\exp(-j\omega\tau)d\tau \\
&= \exp[(\omega_1-\omega_2)t]\int_{-\infty}^{\infty} \exp\left[-j\left(\omega-\frac{\omega_1+\omega_2}{2}\right)\tau\right]d\tau \\
&= 2\pi\delta(\omega-\omega_m)\exp(\omega_d t),
\end{aligned}
\tag{8.2.8}
$$

where $\omega_m = \frac{1}{2}(\omega_1+\omega_2)$ is the average of the two frequencies and $\omega_d = \omega_1 - \omega_2$ is the difference between the two frequencies. Therefore, the Wigner-Ville distribution of the two complex harmonic signals can be expressed as

$$
W_z(t,\omega) = 2\pi[\delta(\omega-\omega_1)+\delta(\omega-\omega_2)] + 4\pi\delta(\omega-\omega_m)cos(\omega_d t). \tag{8.2.9}
$$

This shows that the signal term of the Wigner-Ville distribution $W_{\text{auto}}(t,\omega)$ is a band impulse function along the straight line of the two frequencies of the complex harmonic signals and the amplitude is 2π. The two frequencies of the complex harmonic signals can be correctly detected by the signal term. In addition to the signal term, there is a relatively large cross term at the average frequency ω_m of the two frequencies, whose envelope is $2\pi cos(\omega_d)$, which is related to the difference between the two frequencies.

This example can also be extended to the case of a superposition of p complex harmonic signals: the signal term of the Wigner-Ville distribution is expressed as a p-band impulse function along a straight line at each harmonic frequency. If the sample data of signal $z(t)$ is limited, the signal term of the Wigner-Ville distribution is distributed in the dorsal fin shape of a fish along a straight line at each harmonic frequency and is no longer the ideal band impulse function. It can also be seen from this example that the cross-terms of the Wigner-Ville distribution are serious.

8.2.2 Relationship to Evolutive Spectrum

Defining the time-varying autocorrelation function of the signal $z(t)$ as

$$
R_z(t,\tau) = E\left\{z\left(t+\frac{\tau}{2}\right)z^*\left(t-\frac{\tau}{2}\right)\right\}. \tag{8.2.10}
$$

Similar to the power spectrum of the stationary signal, the Fourier transform of the time-varying autocorrelation function is defined as the time-varying power spectrum

$$
S_z(t,\omega) = \int_{-\infty}^{\infty} R_z(t,\tau)e^{-j2\pi\tau f}d\tau \tag{8.2.11}
$$

$$
= \int_{-\infty}^{\infty} E\left\{z\left(t+\frac{\tau}{2}\right)z^*\left(t-\frac{\tau}{2}\right)\right\}e^{-j\omega\tau}d\tau \tag{8.2.12}
$$

of $z(t)$. The time-varying power spectrum is often called an evolutive spectrum. If the positions of the two operators of mathematical expectation and integral are exchanged, then Eq. (8.2.12) gives the result

$$S_z(t,\omega) = E\left\{\int_{-\infty}^{\infty} z\left(t+\frac{\tau}{2}\right) z^*\left(t-\frac{\tau}{2}\right) e^{-j\omega\tau} d\tau\right\}. \tag{8.2.13}$$

This shows that the evolutive spectrum of the signal $z(t)$ is equal to the mathematical expectation of the Wigner-Ville distribution $W_z(t,\omega)$ of the signal, that is, there is

$$S_z(t,\omega) = E\{W_z(t,\omega)\}. \tag{8.2.14}$$

Because of this relationship, the evolutive spectrum is sometimes called the Wigner-Ville spectrum.

It is well known that in the correlation analysis of stationary random signals, the coherence of two random signals $x(t)$ and $y(t)$ is defined as

$$\alpha_{xy}(\omega) = \frac{S_{xy}(\omega)}{\sqrt{S_x(\omega)S_y(\omega)}}, \tag{8.2.15}$$

where $S_{xy}(\omega)$ is the mutual power spectrum of $x(t)$ and $y(t)$, while $S_x(\omega)$ and $S_y(\omega)$ are the power spectra of $x(t)$ and $y(t)$, respectively.

Similarly, the coherence of two non-stationary random signals $x(t)$ and $y(t)$ can be defined. Since this coherence is a function of two variables, time and frequency, it is called the time-frequency coherence function, defined as

$$\alpha_{xy}(t,\omega) = \frac{S_{xy}(t,\omega)}{\sqrt{S_x(t,\omega)S_y(t,\omega)}}, \tag{8.2.16}$$

where $S_x(t,\omega)$ and $S_y(t,\omega)$ are the (self) evolutive spectra or Wigner-Ville spectra of signals $x(t)$ and $y(t)$, respectively, while $S_{xy}(t,\omega)$ is the mutual evolutive spectrum or mutual Wigner-Ville spectrum of $x(t)$ and $y(t)$, defined as

$$S_{xy}(t,\omega) = \int_{-\infty}^{\infty} E\left\{x\left(t+\frac{\tau}{2}\right) y^*\left(t-\frac{\tau}{2}\right)\right\} e^{-j\omega\tau} d\tau = E\{W_{xy}(t,\omega)\}. \tag{8.2.17}$$

From the definition of Eq. (8.2.16), it is easy to prove that the time-frequency coherence function has the following properties.

Property 1 $0 \le |\alpha_{xy}(t,\omega)| \le 1$.

Property 2 If x and y are uncorrelated at time t, i.e., $R_{xy}(t+\frac{\tau}{2}, t-\frac{\tau}{2}) = 0, \forall\tau$, then $S_{xy}(t,\omega) = 0$.

Property 3 If $x(t)$ and $y(t)$ are the outputs obtained by non-stationary signal $q(t)$ through linear shift-invariant filters H_1 and H_2, respectively, then

$$\alpha_{xy}(t,\omega) = \frac{S_q(t,\omega) ** W_{h_1,h_2}(t,\omega)}{\sqrt{S_q(t,\omega) ** W_{h_1}(t,\omega)}\sqrt{S_q(t,\omega) ** W_{h_2}(t,\omega)}}, \tag{8.2.18}$$

where ** denotes the two-dimensional convolution and

$$W_{h_1,h_2}(t,\omega) = \int_{-\infty}^{\infty} h_1(t+\frac{\tau}{2})h_2^*(t-\frac{\tau}{2})e^{-j\omega\tau}d\tau \tag{8.2.19}$$

is the mutual Wigner-Ville distribution of the impulse responses $h_1(t)$ and $h_2(t)$ of filters H_1 and H_2.

8.2.3 Signal Reconstruction based on Wigner-Ville Distribution

Now consider how the discrete signal $z(n)$ can be recovered or reconstructed from the Wigner-Ville distribution. Let $z(n)$ have length $2L + 1$ and the discrete Wigner-Ville distribution is defined as

$$W_z(n,k) = 2\sum_{m=-L}^{L} z(n+m)z^*(n-m)e^{-j4\pi km/N}. \tag{8.2.20}$$

Taking the inverse discrete Fourier transform of both sides of the above equation and making variable substitutions $n = \frac{n_1+n_2}{2}$ and $m = \frac{n_1-n_2}{2}$, we have

$$\frac{1}{2}\sum_{k=-L}^{L} W_z\left(\frac{n_1+n_2}{2},k\right)e^{j2\pi(n_1-n_2)k/N} = z(n_1)z^*(n_2). \tag{8.2.21}$$

If let $n_1 = 2n$ and $n_2 = 0$, then Eq. (8.2.21) gives the result

$$\frac{1}{2}\sum_{k=-L}^{L} W_z(n,k)e^{j4\pi nk/N} = z(2n)z^*(0). \tag{8.2.22}$$

This shows that the sampled signal $z(2n)$ of an even index can be reconstructed uniquely by the discrete Wigner-Ville distribution $W_z(n,k)$ differing by up to a complex multiplicative constant $z^*(0)$. Similarly, if let $n_1 = 2n - 1$ and $n_2 = 1$ in Eq. (8.2.21), then we have

$$\frac{1}{2}\sum_{k=-L}^{L} W_z(n,k)e^{j4\pi(n-1)k/N} = z(2n-1)z^*(1). \tag{8.2.23}$$

which means that the sampled signal $z(2n - 1)$ of an odd ordinal number can be recovered uniquely by $W_z(n,k)$ differing by up to a complex multiplicative constant $z^*(1)$.

Eqs. (8.2.22) and (8.2.23) show that the inverse problem of the discrete Wigner-Ville distribution can be decomposed into two smaller inverse problems: finding samples with even index and finding samples with odd index, i.e., if the discrete Wigner-Ville distribution of the discretely sampled signal $z(n)$ is known, the sampled signals with even index and odd index can be uniquely recovered within the following range of complex exponential constants, respectively.

$$\frac{z^*(0)}{|z(0)|} = e^{j\phi_e} \quad \text{and} \quad \frac{z^*(1)}{|z(1)|} = e^{j\phi_o}. \tag{8.2.24}$$

This result tells us that if two adjacent (one even and one odd) nonzero sample values of signal $z(n)$ are known, the accurate reconstruction of the discrete signal $z(n)$ can be realized.

So, how to realize the above signal reconstruction or synthesis? Find the instantaneous correlation function to obtain

$$k_z(n, m) = \frac{1}{2}\sum_{k=-L}^{L} W_z(n, k)e^{j4\pi km/N} = z(n + m)z^*(n - m). \tag{8.2.25}$$

As mentioned above, it is only necessary to solve samples with even index $z_e(n') = z(2n')$. In order to reconstruct $z(2n')$, we construct a $\frac{N}{2} \times \frac{N}{2}$-dimensional matrix $\mathbf{A}_e$ whose elements are

$$a_e(i, j) = k_z(i + j, i - j) = z_e(i)z_e^*(j). \tag{8.2.26}$$

The rank of $\mathbf{A}_e$ is equal to 1 and its Eigen decomposition is

$$\mathbf{A}_e = \sum_{i=1}^{N/2} \lambda_i \mathbf{e}_i \mathbf{e}_i^* = \lambda_1 \mathbf{e}_1 \mathbf{e}_1^*, \tag{8.2.27}$$

where λ_i and $\mathbf{e}_i$ are the i-th eigenvalue and the corresponding eigenvector of $\mathbf{A}_e$, respectively. Since the rank of $\mathbf{A}_e$ is equal to 1, only the first eigenvalue is nonzero, while the other $\frac{N}{2} - 1$ eigenvalues are all zero. Thus, the reconstruction with even samples of the signal is given by

$$\mathbf{z}_e = \sqrt{\lambda_1}\mathbf{e}_1 e^{j\phi_e}, \tag{8.2.28}$$

where

$$\mathbf{z}_e = [z(0), z(2), \cdots, z(N/2)]^T. \tag{8.2.29}$$

Again, it shows that the samples with an even index can be reconstructed accurately within the range of one complex exponential parameter. This means that phase loss will occur when using the Wigner-Ville distribution to reconstruct the signal, so this kind of signal reconstruction is not suitable for the situation where phase information is required.

In order to reduce the cross term interference, Boashash[30] suggests using the Wigner-Ville distribution $W_{z_a}(t, \omega)$ of the analytic signal $z_a(t)$. Since the analytic signal is a handband function with only positive frequency components, $W_{z_a}(t, \omega)$ can avoid the cross-terms caused by negative frequency components.

However, it is important to note that the analytic signal $z_a(t)$ is different from the original real signal $z(t)$ in many aspects. For example, the analytic signal $z_a(t)$ of the original time-limited signal $z(t) = 0, t \notin [t_1, t_2]$ is no longer time-limited because the analytic signal is bandlimited. Therefore, for a real signal, whether to use its Wigner-Ville distribution $W_z(t, \omega)$ directly or the Wigner-Ville distribution $W_{z_a}(t, \omega)$ of the analytic signal should be carefully selected. However, the Wigner-Ville distribution of analytic signal should be used as much as possible when the suppression of cross-terms is the main consideration.

8.3 Ambiguity Function

As mentioned earlier, the Wigner-Ville distribution is a Fourier transform of the time-delay parameter τ of the instantaneous correlation function $k_z(t, \tau) = z(t + \frac{\tau}{2})z^*(t - \frac{\tau}{2})$. If the Fourier transform is applied to the time variable t of the instantaneous correlation function, another important time-frequency distribution function

$$\begin{aligned} A_z(\tau, \upsilon) &= \int_{-\infty}^{\infty} z\left(t + \frac{\tau}{2}\right) z^*\left(t - \frac{\tau}{2}\right) e^{-j\upsilon t} dt \\ &= \int_{-\infty}^{\infty} k_z(t, \tau) e^{-j\upsilon t} dt \\ &= \mathcal{F}_{t \to \upsilon}[k_z(t, \tau)] \end{aligned} \tag{8.3.1}$$

is obtained. In radar signal processing, it is called the radar ambiguity function.

The radar ambiguity function is mainly used to analyze the resolution performance of the radar signal after matched filtering of the transmit signal. As the impulse response of the matched filter is proportional to the conjugate reciprocal of the signal, when the radar regards the general target as a "point" target, the waveform of the echo signal is the same as the transmitted signal, but with a different time delay τ and different frequency offset υ (i.e., Doppler angular frequency), which makes the ambiguity function become the two-dimensional response of the matched filter output of the radar signal to τ and υ.

In non-stationary signal processing, the ambiguity function is defined differently: the inverse Fourier transform of the instantaneous correlation function with respect to time t, rather than the Fourier transform, that is, the ambiguity function is defined as

$$A_z(\tau, \upsilon) = \mathcal{F}^{-1}_{t \to \upsilon}[k_z(t, \tau)] = \int_{-\infty}^{\infty} z\left(t + \frac{\tau}{2}\right) z^*\left(t - \frac{\tau}{2}\right) e^{j\upsilon t} dt. \tag{8.3.2}$$

The ambiguity function can also be defined as

$$A_Z(\upsilon, \tau) = \int_{-\infty}^{\infty} Z^*\left(\omega + \frac{\upsilon}{2}\right) Z\left(\omega - \frac{\upsilon}{2}\right) e^{j\omega\tau} d\omega \tag{8.3.3}$$

by the Fourier transform $Z(\omega)$ of the signal.

Comparing the ambiguity function and the Wigner-Ville distribution, we know that they are both bilinear transformations of the signal or some kind of linear transformation of the instantaneous correlation function $k_z(t,\tau)$. The latter transforms to the time-frequency plane, which represents the energy distribution and is the energy domain representation; the latter transforms to the time-delay frequency offset plane, which represents the correlation and is the correlation domain representation. Since the ambiguity function represents correlation and the Wigner-Ville distribution represents the energy distribution, then the Wigner-Ville distribution should be some kind of Fourier transform of the ambiguity function. From the definition Eq. (8.3.2) of the Wigner-Ville distribution, it is easy to prove that

$$\begin{aligned} W_z(t,\omega) &= \int_{-\infty}^{\infty} z\left(t+\frac{\tau}{2}\right) z^*\left(t-\frac{\tau}{2}\right) e^{-j\omega\tau} d\tau \\ &= \int_{-\infty}^{\infty}\int_{-\infty}^{\infty} z\left(u+\frac{\tau}{2}\right) z^*\left(u-\frac{\tau}{2}\right) e^{-j\omega\tau}\delta(u-t) du d\tau \\ &= \frac{1}{2\pi}\int_{-\infty}^{\infty}\int_{-\infty}^{\infty}\int_{-\infty}^{\infty} z\left(u+\frac{\tau}{2}\right) z^*\left(u-\frac{\tau}{2}\right) e^{-j\omega\tau} e^{j\upsilon(u-t)} du d\tau d\upsilon \\ &= \frac{1}{2\pi}\int_{-\infty}^{\infty}\int_{-\infty}^{\infty}\left[\int_{-\infty}^{\infty} z\left(u+\frac{\tau}{2}\right) z^*\left(u-\frac{\tau}{2}\right) e^{j\upsilon u} du\right] e^{-j(\upsilon t+\omega\tau)} d\upsilon d\tau . \end{aligned}$$

The well-known Fourier transform $\frac{1}{2\pi}\int_{-\infty}^{\infty} e^{j\upsilon(u-t)} d\upsilon = \delta(u-t)$ is used. Substituting the definition of the ambiguity function Eq. (8.3.2) into the above equation, we immediately have

$$W_z(t,\omega) = \frac{1}{2\pi}\int_{-\infty}^{\infty}\int_{-\infty}^{\infty} A_z(\tau,\upsilon) e^{-j(\upsilon t+\omega\tau)} d\upsilon d\tau . \tag{8.3.4}$$

That is, the Wigner-Ville distribution is equivalent to the two-dimensional Fourier transform of the ambiguity function, only differing by a constant factor $\frac{1}{2\pi}$.

According to the definition Eq. (8.3.2), it is not difficult to prove that the ambiguity function has the following properties.

P1 (Conjugate symmetry) The ambiguity function is conjugate symmetric, i.e., it has

$$A_z(\tau,\upsilon) = A_z^*(-\tau,-\upsilon). \tag{8.3.5}$$

P2 (Time-shift ambiguity) The mode of the ambiguity function is insensitive to the time shift, i.e., there is

$$\tilde{z}(t) = z(t-t_0) \Rightarrow A_{\tilde{z}}(\tau,\upsilon) = A_z(\tau,\upsilon) e^{j2\pi t_0\upsilon} . \tag{8.3.6}$$

P3 (Frequency-shift ambiguity) The mode of the ambiguity function is also insensitive to the frequency shift, i.e., there is

$$\tilde{z}(t) = z(t)e^{j\omega_0 t} \Rightarrow A_{\tilde{z}}(\tau, \upsilon) = A_z(\tau, \upsilon)e^{j2\pi f_0 \tau}. \tag{8.3.7}$$

P4 (Time-delay edge characteristics)

$$A_z(0, \upsilon) = \frac{1}{2\pi}\int_{-\infty}^{\infty} Z\left(\omega - \frac{1}{2}\upsilon\right) Z^*\left(\omega + \frac{1}{2}\upsilon\right) d\omega. \tag{8.3.8}$$

P5 (Frequency offset edge characteristics)

$$A_z(\tau, 0) = \int_{-\infty}^{\infty} z\left(t + \frac{\tau}{2}\right) z^*\left(t - \frac{\tau}{2}\right) dt. \tag{8.3.9}$$

P6 (Total energy retention)

$$A_z(0, 0) = \int_{-\infty}^{\infty} |z(t)|^2 dt = \frac{1}{2\pi}\int_{-\infty}^{\infty} |Z(\omega)|^2 d\omega = E(\text{Total energy of signal}). \tag{8.3.10}$$

P7 (Instantaneous frequency)

$$\omega_i(t) = \frac{\int_{-\infty}^{\infty} \left[\frac{\partial A_z(\tau,\upsilon)}{\partial \tau}\right]_{\tau=0} e^{j\upsilon t} d\upsilon}{\int_{-\infty}^{\infty} A_z(0, \upsilon)e^{j\upsilon t} d\upsilon}. \tag{8.3.11}$$

P8 (Group delay)

$$\tau_g(\upsilon) = \frac{\int_{-\infty}^{\infty} \left[\frac{\partial A_z(\tau,\upsilon)}{\partial \upsilon}\right]_{\upsilon=0} e^{j\upsilon \tau} d\tau}{\int_{-\infty}^{\infty} A_z(0, \upsilon)e^{j\upsilon \tau} d\tau}. \tag{8.3.12}$$

P9 (Finite delay support) If $z(t) = 0, t \notin [t_1, t_2]$, then $A_z(\tau, \upsilon) = 0, \tau > t_2 - t_1$.

P10 (Finite frequency offset support) If $Z(\omega) = 0, t \notin [\omega_1, \omega_2]$, then $A_z(\tau, \upsilon) = 0, \upsilon > \omega_2 - \omega_1$.

P11 (Moyal formula)

$$\frac{1}{2\pi}\int_{-\infty}^{\infty}\int_{-\infty}^{\infty} A_z(\tau, \upsilon)A_x^*(\tau, \upsilon)d\tau d\upsilon = \left|\int_{-\infty}^{\infty} z(t)x^*(t)dt\right|^2 = |\langle z, x\rangle|^2 \tag{8.3.13}$$

In particular, if $x(t) = z(t)$, then

$$\frac{1}{2\pi}\int_{-\infty}^{\infty}\int_{-\infty}^{\infty} |A_z(\tau, \upsilon)|^2 d\tau d\upsilon = \left[\int_{-\infty}^{\infty} |z(t)|^2 dt\right]^2 \tag{8.3.14}$$

is called the volume invariance of the ambiguity function.

P12 (Convolution property) If $z(t) = x(t) * h(t) = \int_{-\infty}^{\infty} x(u)h(t-u)du$, then

$$A_z(\tau, \upsilon) = \int_{-\infty}^{\infty} A_x(\tau, \upsilon')A_h(\tau - \tau', \upsilon)d\tau'. \tag{8.3.15}$$

P13 (Product property) If $z(t) = x(t)h(t)$ then

$$A_z(\tau, \upsilon) = \frac{1}{2\pi} \int_{-\infty}^{\infty} A_x(\tau, \theta)A_h(\tau, \upsilon - \theta)d\theta. \tag{8.3.16}$$

P14 (Fourier transform) The relationship between the ambiguity function $A_Z(\tau, \upsilon)$ of the Fourier transform $Z(\omega)$ of the signal $z(t)$ and the ambiguity function $A_z(\tau, \upsilon)$ of the original signal is

$$A_Z(\tau, \upsilon) = 2\pi A_z(\tau, -\upsilon). \tag{8.3.17}$$

For convenience, the above properties of the ambiguity function are summarized in Table 8.3.1.

Tab. 8.3.1: The important mathematical properties of the ambiguity function

P1(Conjugate symmetry)	$A_z(\tau, \upsilon) = A_z^*(-\tau, -\upsilon)$
P2(Time-shift ambiguity)	$\tilde{z}(t) = z(t - t_0) \Rightarrow A_{\tilde{z}}(\tau, \upsilon) = A_z(\tau, \upsilon)e^{j2\pi t_0 \upsilon}$
P3(Frequency-shift ambiguity)	$\tilde{z}(t) = z(t)e^{j\omega_0 t} \Rightarrow A_{\tilde{z}}(\tau, \upsilon) = A_z(\tau, \upsilon)e^{j2\pi f_0 \tau}$
P4(Time-delay edge characteristics)	$A_z(0, \upsilon) = \frac{1}{2\pi} \int_{-\infty}^{\infty} Z\left(\omega - \frac{1}{2}\upsilon\right) Z^*\left(\omega + \frac{1}{2}\upsilon\right) d\omega$
P5(Frequency offset edge characteristics)	$A_z(\tau, 0) = \int_{-\infty}^{\infty} z\left(t + \frac{\tau}{2}\right) z^*\left(t - \frac{\tau}{2}\right) dt$
P6(Total energy retention)	$A_z(0, 0) = \int_{-\infty}^{\infty} \lvert z(t)\rvert^2 dt = \frac{1}{2\pi} \int_{-\infty}^{\infty} \lvert Z(\omega)\rvert^2 d\omega = E$
P7(Instantaneous frequency)	$\omega_I(t) = \frac{\int_{-\infty}^{\infty} \left[\frac{\partial A_z(\tau,\upsilon)}{\partial \tau}\right]_{\tau=0} e^{j\upsilon t} d\upsilon}{\int_{-\infty}^{\infty} A_z(0,\upsilon)e^{j\upsilon t} d\upsilon}$
P8 (Group delay)	$\tau_g(\upsilon) = \frac{\int_{-\infty}^{\infty} \left[\frac{\partial A_z(\tau,\upsilon)}{\partial \upsilon}\right]_{\upsilon=0} e^{j\upsilon \tau} d\tau}{\int_{-\infty}^{\infty} A_z(0,\upsilon)e^{j\upsilon\tau} d\tau}$
P9(Finite delay support)	$z(t) = 0, t \notin [t_1, t_2] \Rightarrow A_z(\tau, \upsilon) = 0, \tau > t_2 - t_1$
P10(Finite frequency offset support)	$Z(\omega) = 0, t \notin [\omega_1, \omega_2] \Rightarrow A_z(\tau, \upsilon) = 0, \upsilon > \omega_2 - \omega_1$
P11(Moyal formula)	$\frac{1}{2\pi} \int_{-\infty}^{\infty} \int_{-\infty}^{\infty} \lvert A_z(\tau, \upsilon)\rvert^2 d\tau d\upsilon = \left[\int_{-\infty}^{\infty} \lvert z(t)\rvert^2 dt\right]^2 = \lvert\langle z, x\rangle\rvert^2$
P12(Convolution property)	$z(t) = x(t) * h(t) \Rightarrow A_z(\tau, \upsilon) = \int_{-\infty}^{\infty} A_x(\tau, \upsilon')A_h(\tau - \tau', \upsilon)d\tau'$
P13(Prodct property)	$z(t) = x(t)h(t) \Rightarrow A_z(\tau, \upsilon) = \frac{1}{2\pi} \int_{-\infty}^{\infty} A_x(\tau, \theta)A_h(\tau, \upsilon - \theta)d\theta$
P14(Fourier transform)	$A_Z(\tau, \upsilon) = 2\pi A_z(\tau, -\upsilon)$

Since the ambiguity function is also a time-frequency distribution of signals, it must also obey the principle of quadratic superposition, that is, there are cross-terms, so it is necessary to examine the mutual ambiguity function of two signals.

The mutual ambiguity function of signal $z(t)$ and $g(t)$ is defined as

$$A_{z,g}(\tau, \upsilon) = \int_{-\infty}^{\infty} z\left(t + \frac{\tau}{2}\right) g^*\left(t - \frac{\tau}{2}\right) e^{jt\upsilon} dt. \tag{8.3.18}$$

Note that in the general case, the mutual ambiguity function takes complex values and it has no conjugate symmetry, i.e.,

$$A_z(\tau, \nu) \neq A_z^{*}(\tau, \nu). \tag{8.3.19}$$

The Wigner-Ville distribution $W_z(t, \omega)$ is centered on the signal parameter (t_0, ω_0), while the ambiguity function $A_z(\tau, \theta)$ is centered on the origin (0, 0) and is an oscillating waveform whose phase $\omega_0\tau - t_0\theta$ is related to the time shift t_0 and frequency modulation ω_0 of the signal.

For multicomponent signals, the ambiguity functions of each signal component are centered at the origin (0, 0) and mixed. In this sense, this time-frequency function is ambiguous for each signal component. However, all the cross-terms of the ambiguity function are generally far from the origin. As the two-dimensional Fourier transform of the ambiguity function is the Wigner-Ville distribution, the energy domain representation (time-frequency distribution) and correlation domain representation (ambiguity function) of a non-stationary signal are equally important in the analysis and processing of the non-stationary signal, just as the time domain and frequency domain representation of the stationary signal.

8.4 Cohen's Class Time-frequency Distribution

Since the appearance of the Wigner-Ville distribution in 1948, it has been widely used in many fields. In practice, it has been found that the Wigner-Ville distribution should be improved for different application needs, resulting in a series of other forms of time-frequency distribution. In 1966, Cohen[62] found that many time-frequency distributions are just variations of the Wigner-Ville distribution, and they can be expressed in a uniform form. Generally speaking, in this unified representation, different time-frequency distributions only add different kernel functions to the Wigner-Ville distribution, and the requirements for various properties of the time-frequency distribution are reflected in the constraints on the kernel functions. This unified time-frequency distribution is now conventionally called Cohen's class time-frequency distribution.

8.4.1 Definition of Cohen's Class Time-frequency Distribution

In reference [62], Cohen defined the time-varying autocorrelation function as

$$R_z(t, \tau) = \frac{1}{2\pi} \int_{-\infty}^{\infty} A_z(\tau, \nu)\phi(\tau, \nu)e^{-j\nu t} d\nu, \tag{8.4.1}$$

where $A_z(\tau, \nu)$ is the ambiguity function of the signal $z(t)$, defined by Eq. (8.3.2); $\phi(\tau, \nu)$ is called the kernel function. Since the Fourier transform of the time-varying autocorrelation function gives the time-frequency distribution, there is

$$C_z(t,\omega) \stackrel{\text{def}}{=} \int_{-\infty}^{\infty} R_z(t,\tau)e^{-j\omega\tau}d\tau = \int_{-\infty}^{\infty}\int_{-\infty}^{\infty} A_z(\tau,\upsilon)\phi(\tau,\upsilon)e^{-j(\upsilon t+\omega\tau)}d\tau d\upsilon. \tag{8.4.2}$$

The time-frequency distributions with this form are conventionally referred to as Cohen's class time-frequency distribution. As will be seen later, most of the existing time-frequency distributions belong to Cohen's class distribution, but they take different kernel functions.

Substituting the convolution theorem

$$G(\omega) = \int_{-\infty}^{\infty} s_1(x)s_2(x)e^{-j\omega x}dx = \frac{1}{2\pi}S_1(\omega) \star S_2(\omega) \tag{8.4.3}$$

into Eq. (8.4.1), then the time-varying autocorrelation function can be rewritten as

$$R_z(t,\tau) = \upsilon \overset{\mathcal{F}}{\rightarrow} t[A_z(\tau,\upsilon)] \star \upsilon \overset{\mathcal{F}}{\rightarrow} t[\phi_z(\tau,\upsilon)] \tag{8.4.4}$$

$$= \left[z\left(t+\frac{\tau}{2}\right)z^*\left(t-\frac{\tau}{2}\right)\right] \star \psi(t,\tau) \tag{8.4.5}$$

$$= \int_{-\infty}^{\infty} z\left(u+\frac{\tau}{2}\right)z^*\left(u-\frac{\tau}{2}\right)\psi(t-u,\tau)du, \tag{8.4.6}$$

where

$$\psi(t,\tau) = \upsilon \overset{\mathcal{F}}{\rightarrow} t[\phi(\tau,\upsilon)] = \int_{-\infty}^{\infty} \phi(\tau,\upsilon)e^{-j\upsilon t}d\upsilon. \tag{8.4.7}$$

Subsitituting Eq. (8.4.6) into Eq. (8.4.2) yields an alternative definition of Cohen's class distribution

$$C_z(t,\omega) = \int_{-\infty}^{\infty}\int_{-\infty}^{\infty} z\left(u+\frac{\tau}{2}\right)z^*\left(u-\frac{\tau}{2}\right)\psi(t-u,\tau)e^{-j\omega\tau}dud\tau. \tag{8.4.8}$$

Eqs. (8.4.2) and (8.4.8) are commonly used defining formulas for Cohen's class distribution, and they are equivalent to each other.

Consider a special choice of the kernel function $\phi(\tau,\upsilon) \equiv 1$. At this time, the definition of Eq. (8.4.2) degenerates to the right side of Eq. (8.3.4), i.e., the Cohen's class distribution with window function $\phi(\tau,\upsilon) \equiv 1$ is the Wigner-Ville distribution. In addition, we know from Eq. (8.4.7), the kernel function corresponding to $\phi(\tau,\upsilon) \equiv 1$ is $\psi(t,\tau) = \delta(t)$, so we can also get $C_z(t,\omega) = W_z(t,\omega)$ from Eq. (8.4.8), which proves again that the Wigner-Ville distribution is Cohen's class distribution with $\phi(\tau,\upsilon) \equiv 1$.

The Wigner-Ville distribution and the ambiguity function have four variables t, ω, τ and υ. In principle, any two of them can form a two-dimensional distribution. For example, the instantaneous correlation function $k_z(t,\tau)$ itself is also a two-dimensional distribution, which takes time t and time-shift τ as variables. The fourth two-dimensional

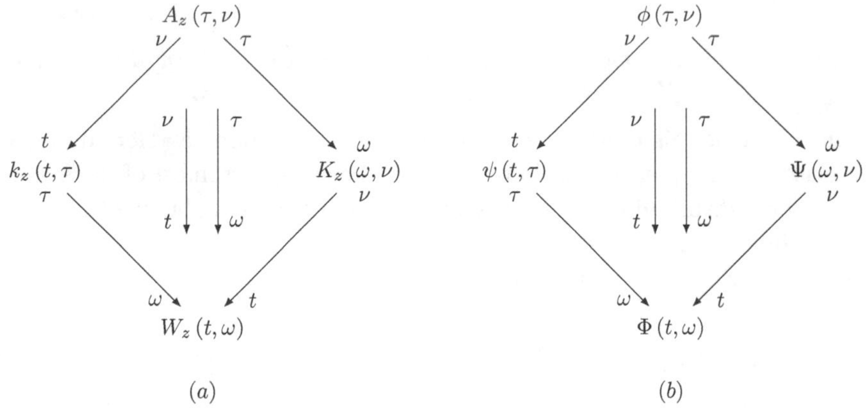

Fig. 8.4.1: (a) Relationship of the four distributions of Cohen's class; (b) Relationship between the four kernel functions

distribution takes frequency ω and frequency offset υ as variables, which is called the point spectrum correlation function, that is, $K_z(\omega, \upsilon)$. According to the previous definition, the Fourier transform of the point spectrum correlation function to υ should be the Wigner-Ville distribution; and if we follow the definition of the instantaneous correlation function, the point spectrum correlation function should be defined as $Z\left(\omega + \frac{\upsilon}{2}\right) Z^*\left(\omega - \frac{\upsilon}{2}\right)$. In fact, calculating the Fourier transform of $Z\left(\omega + \frac{\upsilon}{2}\right) Z^*\left(\omega - \frac{\upsilon}{2}\right)$, we can obtain

$$\int_{-\infty}^{\infty} Z\left(\omega + \frac{\upsilon}{2}\right) Z^*\left(\omega - \frac{\upsilon}{2}\right) e^{-j\upsilon t} d\upsilon = W_z(-t, \omega), \tag{8.4.9}$$

and it violates the Wigner-Ville distribution. In order to solve this contradiction, the point spectrum correlation function needs to be defined as

$$K_z(\omega, \upsilon) = Z^*\left(\omega + \frac{\upsilon}{2}\right) Z\left(\omega - \frac{\upsilon}{2}\right). \tag{8.4.10}$$

Wigner-Ville distribution, ambiguity function, instantaneous correlation function, and point spectrum correlation function are the four basic distributions of Cohen's class. In order to facilitate readers to understand intuitively and vividly, Fig. 8.4.1(a) shows the transformation relations among these four distributions, while Fig. 8.4.1(b) shows the transformation relations among the four kernel functions $\psi(t, \tau)$, $\Psi(\omega, \upsilon)$, $\Phi(t, \omega)$ and $\phi(\tau, \upsilon)$.

From Fig. 8.4.1(a), the relationship between the Wigner-Ville distribution, the ambiguity function, the instantaneous correlation function $z\left(t + \frac{\tau}{2}\right) z^*\left(t - \frac{\tau}{2}\right)$and the point spectrum correlation function $Z\left(\omega+\frac{\upsilon}{2}\right)Z^{*}\left(\omega-\frac{\upsilon}{2}\right)$ can be written as

$$W_z(t,\omega)=\int_{-\infty}^{\infty} z\left(t+\frac{\tau}{2}\right)z^{*}\left(t-\frac{\tau}{2}\right)e^{-j\omega\tau}d\tau, \tag{8.4.11}$$

$$W_Z(\omega,t)=\int_{-\infty}^{\infty} Z^{*}\left(\omega+\frac{\upsilon}{2}\right)Z\left(\omega-\frac{\upsilon}{2}\right)e^{-j\upsilon t}d\upsilon, \tag{8.4.12}$$

$$A_z(\tau,\upsilon)=\int_{-\infty}^{\infty} z\left(t+\frac{\tau}{2}\right)z^{*}\left(t-\frac{\tau}{2}\right)e^{j\upsilon t}dt, \tag{8.4.13}$$

$$A_Z(\upsilon,\tau)=\int_{-\infty}^{\infty} Z^{*}\left(\omega+\frac{\upsilon}{2}\right)Z\left(\omega-\frac{\upsilon}{2}\right)e^{j\omega\tau}d\omega, \tag{8.4.14}$$

$$W_z(t,\omega)=\frac{1}{2\pi}\int_{-\infty}^{\infty}\int_{-\infty}^{\infty} A_z(\tau,\upsilon)e^{-j(\upsilon t+\omega\tau)}d\upsilon d\tau. \tag{8.4.15}$$

These equations are consistent with the corresponding equations defined or deduced earlier.

In addition, from Fig. 8.4.1(b), it is possible to write the relationship between the four nuclear functions

$$\psi(t,\tau)=\int_{-\infty}^{\infty}\phi(\tau,\upsilon)e^{-j\upsilon t}d\upsilon, \tag{8.4.16}$$

$$\Psi(\omega,\upsilon)=\int_{-\infty}^{\infty}\phi(\tau,\upsilon)e^{-j\omega\tau}d\tau, \tag{8.4.17}$$

$$\Phi(t,\omega)=\int_{-\infty}^{\infty}\psi(t,\omega)e^{-j\omega\tau}d\tau, \tag{8.4.18}$$

$$\Phi(t,\omega)=\int_{-\infty}^{\infty}\Psi(\omega,\upsilon)e^{-j\upsilon t}d\upsilon, \tag{8.4.19}$$

$$\Phi(t,\omega)=\int_{-\infty}^{\infty}\int_{-\infty}^{\infty}\phi(\tau,\upsilon)e^{-j(\upsilon t+\omega\tau)}d\upsilon d\tau, \tag{8.4.20}$$

$$\Psi(\omega,\upsilon)=\int_{-\infty}^{\infty}\int_{-\infty}^{\infty}\psi(t,\tau)e^{j(\upsilon t-\omega\tau)}dtd\tau. \tag{8.4.21}$$

According to the relationship shown in Fig. 8.4.1, readers can also write other equations between two kernel functions.

It is necessary to point out that in some literature (for example, Reference [178] and [179]), the ambiguity function is directly defined using the definition of the radar ambiguity function Eq. (8.3.1). At this time, all Fourier transform about υ in Fig. 8.4.1, Eqs. (8.4.11)~(8.4.15), and Eqs. (8.4.16)~(8.4.21) need to be changed to the inverse Fourier transform, and all inverse Fourier transform about υ need to be changed to the Fourier transform as well. This point is expected to be noted by the reader when reading other literature. However, from Fig. 8.4.1, the ambiguity function is defined as the inverse Fourier transform of $z\left(t+\frac{\tau}{2}\right)z^*\left(t-\frac{\tau}{2}\right)$ with respect to υ, which brings a very easy rule to remember the Fourier transform relationships between the two time-frequency distributions and the Fourier transform relationships between the two kernel functions.

8.4.2 Requirements for Kernel Function

When $\phi(\tau,\upsilon)=1$, i.e., without kernel function, Eq. (8.4.2) is simplified to Eq. (8.3.4), that is, the Cohen's class time-frequency distribution gives the Wigner-Ville distribution. In other words, Cohen's class distribution is a filtered form of the Wigner-Ville distribution.

Since Cohen's class distribution is the filtering result of the Wigner-Ville distribution, the kernel function will naturally cause some changes in the properties of the original Wigner-Ville distribution. Therefore, if the changed time-frequency distribution is still required to satisfy some basic properties, the kernel function should be subject to certain restrictions.

1. Total Energy and Edge Characteristics

If Cohen's class distribution $C_z(t,\omega)$ is required to be a joint distribution of energy density, it is expected to satisfy two edge characteristics: the integral over the frequency variables is equal to the instantaneous power $|z(t)|^2$, while the integral over the time variables gives the energy density spectrum $|Z(\omega)|^2$.

The integral of Eq. (8.4.2) with respect to the frequency ω gives the result

$$\begin{aligned}\frac{1}{2\pi}\int_{-\infty}^{\infty}C_z(t,\omega)d\omega &= \int_{-\infty}^{\infty}\int_{-\infty}^{\infty}\int_{-\infty}^{\infty} z\left(u+\frac{\tau}{2}\right)z^*\left(u-\frac{\tau}{2}\right)\\ &\quad\times \delta(\tau)e^{j2\pi\upsilon(u-t)}\phi(\tau,\upsilon)d\tau d\upsilon du\\ &= \int_{-\infty}^{\infty}\int_{-\infty}^{\infty}|z(u)|^2 e^{j2\pi\upsilon(u-t)}\phi(0,\upsilon)d\upsilon du.\end{aligned}$$

Obviously, the only choice to make the above equation equal to $|z(t)|^2$ is $\int_{-\infty}^{\infty}\phi(0,\upsilon)$ $e^{j2\pi(u-t)}d\upsilon=\delta(t-u)$, which means

$$\phi(0,\upsilon)=1. \tag{8.4.22}$$

Similarly, if $\int_{-\infty}^{\infty}C_z(t,\omega)dt=|Z(\omega)|^2$ is desired, then the kernel function must satisfy

$$\phi(\tau,0)=1. \tag{8.4.23}$$

For the same reason, it is also generally desired that the total energy of the signal (normalized energy) remains constant, i.e.,

$$\frac{1}{2\pi}\int_{-\infty}^{\infty}\int_{-\infty}^{\infty} C_z(t,\omega)d\omega dt = 1 = \text{Total energy}. \tag{8.4.24}$$

Therefore, it is necessary to take

$$\phi(0,0) = 1. \tag{8.4.25}$$

This condition is called the normalization condition, which is weaker than the edge conditions Eqs. (8.4.22) and (8.4.23). That is, there may exist some time-frequency distribution whose total energy is the same as the total energy of the signal, but the edge characteristics may not be satisfied.

2. Real-valued

The bilinear distribution is generally not guaranteed to be positive, but as a measure of energy, it should at least be required to be real. Taking the complex conjugate of Eq. (8.4.2) and letting $C_z^*(t,\omega) = C_z(t,\omega)$, it is easy to prove that Cohen's class distribution is real-valued if and only if the kernel function satisfies the condition

$$\phi(\tau,v) = \phi^*(-\tau,-v). \tag{8.4.26}$$

Table 8.4.1 summarizes the constraints that the kernel function should satisfy to make Cohen's class distribution have some basic properties. The kernel function requirements $1 \sim 10$ in the table are proposed by Classen and Mecklenbrauker[61].

Tab. 8.4.1: Requirements of kernel function for the basic properties of Cohen's class time-frequency distribution

No	Basic Property	Requirement of the kernel function $\phi(\tau, \upsilon)$
1	Time-shift invariance	Independent of the time variable t
2	Frequency-shift invariance	Independent of the frequency variable ω
3	Real-valued	$\phi(\tau,\upsilon) = \phi^*(-\tau,-\upsilon)$
4	Time edge characteristics	$\phi(0,\upsilon) = 1$
5	Frequency edge characteristics	$\phi(\tau,0) = 1$
6	Instantaneous frequency characteristics	$\phi(0,\upsilon) = 1$ and $\frac{\partial}{\partial\tau}\phi(\tau,\upsilon)\|_{\tau=0} = 0$
7	Group delay characteristics	$\phi(\tau,0) = 1$ and $\frac{\partial}{\partial\upsilon}\phi(\tau,\upsilon)\|_{\upsilon=0} = 0$
8	Positivity	$\phi(\tau,\upsilon)$ is the ambiguity function of any window function $\gamma(t)$
9	Finite time support	$\psi(t,\tau) = \int_{-\infty}^{\infty}\phi(\tau,\upsilon)e^{-j\upsilon t}d\upsilon = 0$ (where $\|t\| > \frac{\|\tau\|}{2}$)
10	Finite frequency support	$\Psi(\omega,\upsilon) = \int_{-\infty}^{\infty}\phi(\tau,\upsilon)e^{-j\omega\tau}d\tau = 0$ (where $\|\omega\| > \frac{\|\upsilon\|}{2}$)
11	Moyal formula	$\|\phi(\tau,\upsilon)\| = 1$
12	Convolution property	$\phi(\tau_1+\tau_2,\upsilon) = \phi(\tau_1,\upsilon)\phi(\tau_2,\upsilon)$
13	Prodct property	$\phi(\tau,\upsilon_1+\upsilon_2) = \phi(\tau,\upsilon_1)\phi(\tau,\upsilon_2)$
14	Fourier transform	$\phi(\tau,\upsilon) = \phi(\upsilon,-\tau)$, $\forall\tau$ and υ

8.5 Performance Evaluation and Improvement of Time-frequency Distribution

Most applications of time-frequency signal analysis are related to multicomponent extraction of non-stationary signals. It is usually expected that time-frequency signal analysis has the following functions:
(1) It can determine the number of signal components in a signal;
(2) It can identify signal components and cross-terms;
(3) It can distinguish two or more signal components that are very close to each other on the time-frequency plane;
(4) It can estimate the instantaneous frequency of each component of the signal.

Determining whether a time-frequency distribution has these functions involves the performance evaluation of time-frequency distribution. In order to select an appropriate time-frequency distribution in the practical application of time-frequency signal analysis, it is necessary to understand the advantages and disadvantages of various time-frequency distributions. The advantages and disadvantages of a time-frequency distribution are mainly determined by its time-frequency aggregation and cross-terms. They are analyzed and discussed respectively.

8.5.1 Time-frequency Aggregation

Just as the typical stationary signal is a Gaussian signal, the non-stationary signal also has a typical signal, which is linear frequency modulation (LFM) signal[4]. As the name suggests, the LFM signal is a signal whose frequency varies according to a linear law over time. Now, it has been widely recognized that any kind of time-frequency analysis can not be used as a time-frequency analysis tool for non-stationary signals if it does not provide good time-frequency aggregation performance for LFM signals.

Since the time-frequency distribution is used to describe the time-varying or local time-frequency characteristics of non-stationary signals, it is natural to expect it to be well time-frequency localized, i.e., it is required to be highly aggregated in the time-frequency plane. This property is called the time-frequency aggregation of the time-frequency distribution.

Consider a single-component LFM signal

$$z(t) = e^{j(\omega_0 t + \frac{1}{2} m t^2)} \tag{8.5.1}$$

4 The LFM signal is also called chirp signal

with amplitude 1, which is widely used in detection systems such as radar, sonar, and seismic. The bilinear transformation of the single-component LFM signal is

$$\begin{aligned} z\left(t+\frac{\tau}{2}\right) z^*\left(t-\frac{\tau}{2}\right) &= \exp\left\{j\left[\omega_0\left(t+\frac{\tau}{2}\right)+\frac{1}{2}m\left(t+\frac{\tau}{2}\right)^2\right]\right\} \\ &\quad \times \exp\left\{-j\left[\omega_0\left(t-\frac{\tau}{2}\right)+\frac{1}{2}m\left(t-\frac{\tau}{2}\right)^2\right]\right\} \\ &= \exp[j(\omega_0+mt)\tau]. \end{aligned} \tag{8.5.2}$$

Thus the Wigner-Ville distribution of the LFM signal can be obtained as

$$\begin{aligned} W_{\mathrm{LFM}}(t,\omega) &= \int_{-\infty}^{\infty} z\left(t+\frac{\tau}{2}\right) z^*\left(t-\frac{\tau}{2}\right) e^{-j2\pi\tau f} d\tau \\ &= \int_{-\infty}^{\infty} \exp[j(\omega_0+mt)\tau]\exp(-j\omega\tau) d\tau \\ &= \delta[\omega-(\omega_0+mt)]. \end{aligned} \tag{8.5.3}$$

The integration result

$$\int_{-\infty}^{\infty} e^{-j[\omega-(\omega_0+mt)]\tau} d\tau = \delta[\omega-(\omega_0+mt)] \tag{8.5.4}$$

is used here.

It can be seen from Eq. (8.5.3) that the Wigner-Ville distribution of the single-component LFM signal is an impulse line spectrum distributed along the line $\omega = \omega_0 + mt$, that is, the amplitude of the time-frequency distribution concentrates on the straight line representing the variation law of the instantaneous frequency of the signal. Therefore, the Wigner-Ville distribution has ideal time-frequency aggregation in the sense of the best frequency modulation law of the LFM signal. Note that the conclusion that the Wigner-Ville distribution is an impulse line spectrum is only applicable to LFM signals of infinite length. In practice, the length of the signal is always finite, and the Wigner-Ville distribution is in the shape of the dorsal fin of a fish.

In fact, for a single-component LFM signal, no matter how to choose the window function $\phi(\tau, \upsilon)$ to get the Cohen distribution, it is impossible to give better time-frequency aggregation than the Wigner-Ville distribution with the window function $\phi(\tau, \upsilon) = 1$. This conclusion is not surprising, because according to the exclusion principle, the bandwidth of the infinite width window function $\phi(\tau, \upsilon) = 1$ is zero, so it has the highest frequency resolution. But the problem is, the LFM signal is a non-stationary signal, why can we take the infinite width window function? This is mainly due to the quadratic stationarity of the single component LFM signal. From the quadratic signal expression Eq. (8.5.2) of the LFM signal, it is easy to find that its time-varying autocorrelation function is

$$R(t,\tau') = E\{e^{j(\omega_0+mt)\tau} e^{-j(\omega_0+m(t-\tau'))\tau}\} = e^{jm\tau\tau'}, \tag{8.5.5}$$

which is independent of time t. That is to say, the quadratic signal $z\left(t+\frac{\tau}{2}\right)z^*\left(t-\frac{\tau}{2}\right)$ of the single component LFM signal is second-order stationary. This is the reason why the window function $\phi(\tau, \nu) = 1$ with infinite time width can be added to the quadratic signal of the single-component LFM signal (it is not to the signal itself). However, for slightly more complex signals, the situation is quite different. For example, if $z(t) = \sum_{i=1}^{2} e^{j(\omega_i t + \frac{1}{2}m_i t^2)}$ is the superposition of two LFM signals, the autocorrelation function of the quadratic signal $z\left(t+\frac{\tau}{2}\right)z^*\left(t-\frac{\tau}{2}\right)$ is a function of time, i.e., the quadratic signal is not second-order stationary. This means that the window function $\phi(\tau, \nu) = 1$ is no longer optimal, i.e., the Wigner-Ville distribution needs to be improved.

8.5.2 Cross-Term Suppression

For any multicomponent signal, there is a cross term in the quadratic time-frequency distribution, which comes from the cross interaction between different signal components in the multicomponent signal. The signal terms in the time-frequency distribution correspond to each component of the signal itself, and they are consistent with the physical properties of the signal for which the time-frequency distribution has finite support. That is, if a priori knowledge of the signal $z(t)$ and its spectrum is given, the signal terms appear in the time-frequency plane only in those places where we want them to appear. In contrast to the case of the signal terms, the cross-terms are the interference products in time-frequency distribution, and they exhibit results in the time and/or frequency domains that contradict the physical properties of the original signal. Therefore, one of the main problems of the time-frequency distribution is how to suppress its cross-terms.

There are two key filtering methods for cross-term suppression:

(1) Ambiguity domain filtering;
(2) Filtering with kernel functions.

Ambiguity domain filtering is a result of applying radar theory to the time-frequency distribution. In the previous discussion of the ambiguity function, we have highlighted an important fact: in the ambiguity domain, the cross-terms tend to move away from the origin, while the signal terms cluster near the origin. It is useful to remember this important fact because the Wigner-Ville distribution is a two-dimensional Fourier transform of the ambiguity function. Therefore, a natural way to reduce the cross-terms is to filter the ambiguity function in the ambiguity domain to filter out the cross-terms; then, the Wigner-Ville distribution is derived from the two-dimensional Fourier transform of the ambiguity function.

All Cohen's class distributions can be regarded as the filtering form of the Wigner-Ville distribution using kernel function $\phi(\tau, \nu) \neq 1$ or $\psi(t, \tau) \neq \delta(t)$, and the purpose of filtering is to suppress the cross-terms. Therefore, the kernel function of Cohen's class distribution has become a hot research in time-frequency signal analysis. Before discussing how to choose the kernel function, it is necessary to analyze the cross term of the Wigner-Ville distribution. When analyzing the effect of cross-terms, the tone signal and the LFM signal are often used. For simplicity, a stationary tone signal is used here as an example.

For a stationary tone signal $z(t) = e^{j\omega_0 t}$, from the equivalent defining equation (8.4.8) of the Cohen's class time-frequency distribution, we have

$$\begin{aligned} C(t,\omega) &= \int_{-\infty}^{\infty}\int_{-\infty}^{\infty} \psi(t-u,\tau) e^{j\omega_0(u+\frac{\tau}{2})} e^{-j\omega_0(u-\frac{\tau}{2})} e^{-j\omega\tau} du d\tau \\ &= \int_{-\infty}^{\infty}\int_{-\infty}^{\infty} \psi(t-u,\tau) e^{-j(\omega-\omega_0)\tau} du d\tau \\ &= \int_{-\infty}^{\infty}\int_{-\infty}^{\infty} \psi(t',\tau) e^{-j[0\cdot t'+\tau(\omega-\omega_0)]} dt' d\tau . \end{aligned}$$

Applying the relationship between $\Psi(\omega, \upsilon)$ and $\psi(t, \tau)$ (see Fig 8.4.1) to the last equation above yields

$$C(t,\omega) = \Psi(\omega-\omega_0, 0). \tag{8.5.6}$$

This shows that Cohen's class time-frequency distribution of tone signal is directly the kernel function value $\Psi(\omega-\omega_0, 0)$, but with an offset in the frequency of the kernel function (the offset is the input frequency ω_0).

Now consider the two-tone signal

$$z(t) = z_1(t) + z_2(t) = e^{j\omega_1 t} + e^{j\omega_2 t}, \quad \omega_1 < \omega_2 \tag{8.5.7}$$

whose Cohen's class time-frequency distribution consists of signal term and cross term

$$\begin{aligned} C(t,\omega) &= C_{\text{auto}}(t,\omega) + C_{\text{cross}}(t,\omega) \\ &= C_{\text{auto}}(t,\omega) + C_{z_1,z_2}(t,\omega) + C_{z_2,z_1}(t,\omega), \end{aligned} \tag{8.5.8}$$

where the signal term is

$$C_{\text{auto}}(t,\omega) = \Psi(\omega-\omega_1, 0) + \Psi(\omega-\omega_2, 0) \tag{8.5.9}$$

and the first cross term is

$$\begin{aligned} C_{z_1,z_2}(t,\omega) &= \int_{-\infty}^{\infty}\int_{-\infty}^{\infty} \psi(t-u,\tau) e^{j\omega_1(u+\frac{\tau}{2})} e^{-j\omega_2(u-\frac{\tau}{2})} e^{-j\omega\tau} du d\tau \\ &= \int_{-\infty}^{\infty}\int_{-\infty}^{\infty} \psi(t-u,\tau) e^{-j(\omega_2-\omega_1)u} e^{-j(\omega-\frac{\omega_1+\omega_2}{2})\tau} du d\tau \\ &= e^{-j(\omega_2-\omega_1)} \int_{-\infty}^{\infty}\int_{-\infty}^{\infty} \psi(u',\tau) e^{j[(\omega_2-\omega_1)u'-(\omega-\frac{\omega_1+\omega_2}{2})\tau]} du' d\tau . \end{aligned}$$

Using the relationship between $\psi(t,\tau)$ and $\Psi(\omega,\upsilon)$, we get

$$C_{z_1,z_2}(t,\omega) = \Psi\left(\omega - \frac{\omega_1+\omega_2}{2}, \omega_2-\omega_1\right) e^{j(\omega_1-\omega_2)t}. \tag{8.5.10}$$

Similarly, the second cross term is

$$C_{z_2,z_1}(t,\omega) = \Psi\left(\omega - \frac{\omega_1+\omega_2}{2}, \omega_2-\omega_1\right) e^{j(\omega_2-\omega_1)t}. \tag{8.5.11}$$

Therefore, the sum of the cross-terms of the two-tone signals is

$$\begin{aligned} C_{\text{cross}}(t,\omega) &= C_{z_1,z_2}(t,\omega) + C_{z_2,z_1}(t,\omega) \\ &= 2\text{Re}\left[\Psi\left(\omega - \frac{\omega_1+\omega_2}{2}, \omega_2-\omega_1\right) e^{j(\omega_2-\omega_1)t}\right]. \end{aligned} \tag{8.5.12}$$

Then can the cross-terms in the Wigner-Ville distribution be completely eliminated? Eq. (8.5.12) tells us that this is not possible unless a meaningless kernel function $\Psi(\omega,\upsilon) \equiv 0$ is chosen (in this case, all signal terms are also equal to zero).

Next, consider how to suppress the cross-terms represented by Eq. (8.5.12).

1. Weak Finite Support of the Cross-terms

When Cohen[63] expressed the time-frequency distribution in a uniform form, he proposed that an ideal time-frequency distribution should also have the finite support property, that is, wherever the signal $z(t)$ and its spectrum $Z(\omega)$ are equal to zero, the Cohen's class distribution $C(t,\omega)$ should also be equal to zero. This means that the kernel function should satisfy the conditions[60]

$$\psi(t,\tau) = \int_{-\infty}^{\infty} \phi(\tau,\upsilon) e^{j\upsilon t} d\upsilon = 0, \quad |t| > \frac{|\tau|}{2} \tag{8.5.13}$$

and

$$\Psi(\omega,\upsilon) = \int_{-\infty}^{\infty} \phi(\tau,\upsilon) e^{-j\omega\tau} d\tau = 0, \quad |\omega| > \frac{|\upsilon|}{2}. \tag{8.5.14}$$

However, these two conditions can only guarantee the "weak finite support" of the time-frequency distribution, which has a limited effect on cross-term suppression. To see this clearly, substituting Eq. (8.5.14) into the kernel function $\Psi(\omega,\upsilon)$ in Eq. (8.5.12) yields

$$\Psi\left(\omega - \frac{\omega_1+\omega_2}{2}, \omega_2-\omega_1\right) = 0, \quad \text{if } \left|\omega - \frac{\omega_1+\omega_2}{2}\right| > \frac{|\omega_2-\omega_1|}{2}. \tag{8.5.15}$$

From the above equation and Eq. (8.5.12), the cross term is equal to zero when $\omega < \omega_1$ or $\omega > \omega_2$. That is to say, only the cross-terms outside the region $[\omega_1,\omega_2]$ are suppressed.

2. Strong Finite Support of the Cross-terms

The question is, can the cross-terms within the region $[\omega_1,\omega_2]$ also be suppressed? Since the two-tone signal takes values at frequencies ω_1 and ω_2, and the cross-terms of any time-frequency distribution can not be completely suppressed, it is natural to ask:

can the cross-terms of the Winger-Ville distribution only appear at two signal frequencies, while the cross-terms of other frequencies can be suppressed? This requirement is called strong finite frequency support of the cross-terms. Similarly, if the cross term is equal to zero when there is no signal, it is called strong finite time support of the cross term.

The question is how to make the cross-terms have the desired strong finite time support and strong finite frequency support. For this purpose, consider

$$K(\omega;\Psi) = \Psi\left(\omega - \frac{\omega_1+\omega_2}{2}, \omega_2-\omega_1\right) \tag{8.5.16}$$

which represents the envelope of the cross-terms of the two-tone signal.

From Eq. (8.5.16), it is easy to see that the values of the cross-term envelope at signal frequency ω_1 and ω_2 are

$$K(\omega;\Psi)|_{\omega=\omega_1} = \Psi\left(\frac{\omega_1-\omega_2}{2}, \omega_2-\omega_1\right) = \Psi(\omega,\upsilon)|_{\upsilon=\omega_2-\omega_1,\omega=\upsilon/2} \tag{8.5.17}$$

$$K(\omega;\Psi)|_{\omega=\omega_2} = \Psi\left(\frac{\omega_2-\omega_1}{2}, \omega_2-\omega_1\right) = \Psi(\omega,\upsilon)|_{\upsilon=\omega_2-\omega_1,\omega=-\upsilon/2} \tag{8.5.18}$$

respectively. Thus, the nature of the cross-term of the two-tone signal is revealed. The above results show that in order to prevent cross-terms of the time-frequency distribution from appearing at nonsignal frequencies, it is sufficient to add the following constraint to the kernel function $\Psi(\omega,\upsilon)$:

$$\Psi(\omega,\upsilon) = 0, \quad \forall \quad |\omega| \neq \frac{|\upsilon|}{2}. \tag{8.5.19}$$

This is the frequency domain constraint that must be satisfied by the kernel function $\Psi(\omega,\upsilon)$ when the cross-terms have strong finite frequency support properties.

Similarly, the constraint

$$\psi(t,\tau) = 0, \quad \forall \quad |t| \neq \frac{|\tau|}{2} \tag{8.5.20}$$

ensures that the cross-terms will not appear in the time period when the signal $z(t)$ is equal to zero.

The cross term strong finite support constraints Eqs. (8.5.19) and (8.5.20) were proposed by Loughlin et al[143]. Although it is derived for the case of the tone signals, it can be shown that these two conditions hold for any signals. Readers interested in this proof can refer to [143].

It should be noted that the suppression of the cross term and the maintenance of the signal term are a pair of contradictions because the reduction of the cross term will inevitably have a flattening negative effect on the signal term. If the kernel function satisfies the strong finite support conditions of the cross term, Eqs. (8.5.19) and (8.5.20), although the cross term is completely suppressed in the time-frequency region where there is no signal term, it is always accompanied by the presence of the cross term wherever the signal term exists. This coexistence of the signal term and the cross term is obviously disadvantageous to signal recovery.

8.5.3 Other Typical Time-frequency Distributions

Around the cross-term suppression and the improvement of the Wigner-Ville distribution, many specific Cohen's class distributions have been proposed.

Table 8.5.1 lists some commonly used Cohen's class distributions and their corresponding kernel functions. The kernel functions in the table are all fixed functions, and their design is independent of the signal to be analyzed.

Several typical Cohen's class distributions are described below.

1. The Choi-Williams Distribution

The signal term of the ambiguity function, i.e. the self term, is centered at the origin (0, 0), while the mutual ambiguity function, i.e., the cross-terms, occurs far away from the origin. In order to suppress the ambiguity function far away from the origin, Choi and Williams[54] introduced the exponential kernel function

$$\phi(\tau,\upsilon)=\exp[-\alpha(\tau\upsilon)^2] \tag{8.5.21}$$

into Cohen's class distribution. The inverse Fourier transform of this exponential kernel function is

$$\psi(t,\tau)=\int_{-\infty}^{\infty}\phi(\tau,\upsilon)e^{j\upsilon t}d\upsilon=\frac{1}{\sqrt{4\pi\alpha\tau^2}}\exp\left(-\frac{1}{4\alpha\tau^2}t^2\right). \tag{8.5.22}$$

Substituting Eq. (8.5.22) into Cohen's class distribution definition Eq. (8.4.8), the expression of the Choi-Williams distribution

$$\mathrm{CWD}_z(t,\omega)=\int_{-\infty}^{\infty}\frac{1}{\sqrt{4\pi\alpha\tau^2}}\exp\left(-\frac{(t-u)^2}{4\alpha\tau^2}\right)z\left(t+\frac{\tau}{2}\right)z^*\left(t-\frac{\tau}{2}\right)e^{-j\omega\tau}dud\tau \tag{8.5.23}$$

is obtained.

In the following, the functions of the Choi-Williams distribution in cross items suppression are given.

(1) From Eq. (8.5.21), it is easy to verify $\phi(0,0)=1$, $\phi(0,\upsilon)=1$ and $\phi(\tau,0)=1$. This shows that the exponential kernel function does not affect the ambiguity functions on the origin (0, 0), the horizontal (τ axis), and vertical (υ axis) axes. Therefore, if the cross-terms of the ambiguity function appear on the horizontal and vertical axes, they will not be suppressed and thus the corresponding cross-terms in the time-frequency distribution will not be suppressed.

(2) Since $\phi(\tau,\upsilon)<1$, if $\tau\neq 0$ and $\upsilon\neq 0$, the cross-terms of the ambiguity function outside the coordinate axes can be suppressed to some extent so that the cross-terms of the time-frequency distribution corresponding to these ambiguity cross-terms can be reduced.

Tab. 8.5.1: Typical Cohen's class time-frequency distributions and the corresponding kernel functions

Cohen's class time-frequency distributions	$\psi(t,\tau)$	$\phi(\tau,\upsilon)$
B distribution	$\left(\frac{\lvert\tau\rvert}{\cosh^2(t)}\right)^{\sigma}$	$\int_{-\infty}^{\infty}\psi(t,\tau)e^{-j\upsilon t}dt$
Born-Jordan distribution (BJD)	$\begin{cases}\frac{1}{\lvert\tau\rvert}, & \lvert\tau\rvert\geq 2\lvert t\rvert\\ 0, & \text{others}\end{cases}$	$\frac{\sin(\tau\upsilon/2)}{\tau\upsilon/2}$
Butterworth distribution (BUD)	$\mathcal{F}^{-1}\phi(\tau,\upsilon)$	$\frac{1}{1+(\tau/\tau_0)^{2M}+(\upsilon/\upsilon_0)^{2N}}$
Choi-Williams distribution (CWD)	$\frac{1}{\sqrt{4\pi\alpha\tau^2}}\exp\left(-\frac{1}{4\alpha\tau^2}t^2\right)$	$\exp[-\alpha(\tau\upsilon)^2]$
Cone-shaped distribution (CSD)	$\begin{cases}g(\tau), & \lvert\tau\rvert\geq 2\lvert t\rvert\\ 0, & \text{others}\end{cases}$	$2g(\tau)\frac{\sin(\tau\upsilon/2)}{\upsilon/2}$
Generalized exponential distribution (GED)	$\mathcal{F}^{-1}\Psi(\tau,\upsilon)$	$\exp\left[-\left(\frac{\tau}{\tau_0}\right)^{2M}\left(\frac{\upsilon}{\upsilon_0}\right)^{2N}\right]$
Generalized Wigner distribution (GWD)	$\delta(t+\alpha\tau)$	$\exp(j\alpha\tau\upsilon)$
Levin distribution (LD)	$\delta\left(t+\frac{\lvert\tau\rvert}{2}\right)$	$\exp(j\pi\lvert\tau\rvert\upsilon)$
Page distribution (PD)	$\delta\left(t-\frac{\lvert\tau\rvert}{2}\right)$	$\exp(-j\pi\lvert\tau\rvert\upsilon)$
Pseudo Wigner distribution (PWD)	$\delta(t)\eta\left(\frac{\tau}{2}\right)\eta^*\left(-\frac{\tau}{2}\right)$	$\eta\left(\frac{\tau}{2}\right)\eta^*\left(-\frac{\tau}{2}\right)$
Real-valued generalized Wigner distribution (RGWD)	$\frac{1}{2}[\delta(t+\alpha\tau)+\delta(t-\alpha\tau)]$	$\cos(2\pi\alpha\tau\upsilon)$
Reduced interference distribution (RID)	$\frac{1}{\lvert\tau\rvert}s\left(\frac{t}{\tau}\right)^{1)}$	$S(\tau\upsilon)^{1)}$
Rihaczek distribution (RD)	$\delta\left(t-\frac{\tau}{2}\right)$	$\exp(j\pi\tau\upsilon)$
Smoothed pseudo Wigner distribution (SPWD)	$g(t)\eta\left(\frac{\tau}{2}\right)\eta^*\left(-\frac{\tau}{2}\right)$	$\eta\left(\frac{\tau}{2}\right)\eta^*\left(-\frac{\tau}{2}\right)G(\upsilon)$
Spectrogram (SPEC)	$\gamma\left(-t-\frac{\tau}{2}\right)\gamma^*\left(-t+\frac{\tau}{2}\right)$	$A_\gamma(-\tau,-\upsilon)$
Wigner-Ville distribution (WVD)	$\delta(t)$	1

The function $s(\alpha)\leftrightarrow S(\beta)$ satisfies the condition: $\alpha=0$ then $s(\alpha)=0$, $S(\beta)\in R$, $S(0)=1$, and $\frac{d}{d\beta}S(\beta)|_{\beta=0}=0$.

2. The Reduced Interference Distribution

The Choi-Williams distribution can suppress the cross-terms outside the horizontal and vertical axes in the ambiguity domain but still retains the cross-terms on the horizontal and vertical axes. Considering the general principle of cross-term suppression, since the signal term is generally located near the origin in the ambiguity plane, i.e. the (τ, υ) plane, while the cross-terms occur far away from the origin, it is natural to expect the kernel function $\phi(\tau, \upsilon)$ to be a two-dimensional low-pass filter, i.e.,

$$|\phi(\tau, \upsilon)| \ll 1, \quad \text{for} \quad |\tau\upsilon| \gg 0. \tag{8.5.24}$$

This specific condition was proposed by Williams and Jeong[114] based on the idea of reference [54]. The low-pass kernel function can be designed by the following two steps[114].

Step 1 Design a real-valued window function $h(t)$ that satisfies the following:

R1: $\int_{-\infty}^{\infty} h(t) = 1$;

R2: $h(t) = h(-t)$;

R3: $h(t) = 0$, where $|t| > 0.5$;

R4: The Fourier transform $H(\omega)$ of $h(t)$ is differentiable and has a low-pass characteristic, i.e., for large frequency ω, the amplitude response of the filter is much less than 1, or $|H(\omega)| \ll 1$.

Step 2 The kernel function is taken as

$$\phi(\tau, \upsilon) = H(\tau\upsilon). \tag{8.5.25}$$

Cohen's class distributions with such kernel functions are called reduced interference distribution (RID) by Jeong and Williams. Many window functions can be used as the real-valued window function $h(t)$. Examples of the window functions given by Jeong and Williams include triangular window, generalized Hamming window, and truncated Gaussian window. It can be easily seen that the RID has the following kernel functions in the domains:

$$\psi_{\text{RID}}(t, \tau) = \frac{1}{|\tau|} h\left(\frac{t}{\tau}\right), \tag{8.5.26}$$

$$\Psi_{\text{RID}}(\omega, \upsilon) = \frac{1}{|\upsilon|} h\left(\frac{\omega}{\upsilon}\right), \tag{8.5.27}$$

$$\phi_{\text{RID}}(\tau, \upsilon) = H(\tau\upsilon), \tag{8.5.28}$$

$$\Phi_{\text{RID}}(t, \omega) = \int_{-\infty}^{\infty} h\left(\frac{t}{\tau}\right) e^{-j2\pi\tau\omega} d\tau. \tag{8.5.29}$$

Note that Eqs. (8.5.26) and (8.5.27) show that the kernel function in the frequency-frequency offset domain, i.e., the (ω, υ) plane, has the same shape as the kernel function in the time- time delay domain, i.e., the (t, τ) plane.

The RID distribution can be expressed as the integral form of the basic function $h(t)$

$$\mathrm{RID}_z(t, \omega; h) = \int_{-\infty}^{\infty}\int_{-\infty}^{\infty} \frac{1}{|\tau|} h\left(\frac{u-t}{\tau}\right) z\left(u+\frac{\tau}{2}\right) z^*\left(u-\frac{\tau}{2}\right) e^{-j\omega\tau}\, du\, d\tau. \tag{8.5.30}$$

To calculate the RID distribution, the generalized correlation function at time instant t can be defined as

$$R_z'(t, \tau; h) = \int_{-\infty}^{\infty} \frac{1}{|\tau|} h\left(\frac{u-t}{\tau}\right) z\left(u+\frac{\tau}{2}\right) z^*\left(u-\frac{\tau}{2}\right) du. \tag{8.5.31}$$

Thus, the calculation of the RID distribution is converted into the calculation of the Fourier transform of the generalized correlation function

$$\mathrm{RID}_z(t, \omega; h) = \int_{-\infty}^{\infty} R_z'(t, \tau; h) e^{-j\omega\tau}\, d\tau. \tag{8.5.32}$$

For the convenience of the reader, the requirements of the cross-term suppression for the kernel function are summarized as follows.

(1) Weak finite time support of the cross-terms

$$\psi(t, \tau) = \int_{-\infty}^{\infty} \phi(\tau, \upsilon) e^{j\upsilon t}\, d\upsilon = 0, \quad |t| > \frac{|\tau|}{2}. \tag{8.5.33}$$

(2) Weak finite frequency support of the cross-terms

$$\Psi(\omega, \upsilon) = \int_{-\infty}^{\infty} \phi(\tau, \upsilon) e^{-j\omega\tau}\, d\tau = 0, \quad |\omega| > \frac{|\upsilon|}{2}. \tag{8.5.34}$$

(3) Strong finite time support of the cross-terms

$$\psi(t, \tau) = 0, \quad |t| \neq \frac{|\tau|}{2}. \tag{8.5.35}$$

(4) Strong finite frequency support of the cross-terms

$$\Psi(\omega, \upsilon) = 0, \quad |\omega| \neq \frac{|\upsilon|}{2}. \tag{8.5.36}$$

(5) The Reduced cross-terms

$$|\phi(\tau, \upsilon)| \ll 1, \quad \text{for} \quad |\tau\upsilon| \gg 0. \tag{8.5.37}$$

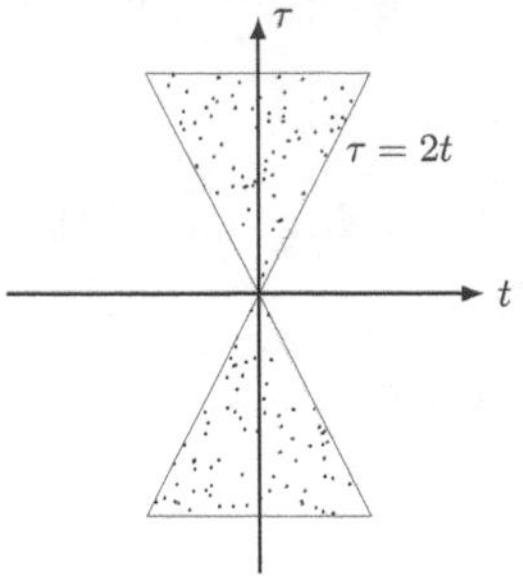

Fig. 8.5.1: Cone-shaped kernel function

3. The Zhao-Atlas-Marks Distribution (Cone-shaped Distribution)

Another well-known time-frequency distribution is called the cone-shaped distribution, which is named after the conical shape of its kernel function. This distribution was proposed by Zhao et al.[255] and its kernel function is taken as

$$\psi(t,\tau) = \begin{cases} g(\tau), & |\tau| \geq 2|t| \\ 0, & \text{Others} \end{cases} . \tag{8.5.38}$$

Fig. 8.5.1 shows this kernel function, which is shaped like a cone.

The representation of this cone-shaped kernel function in the ambiguity domain (τ, υ) is

$$\psi(\tau,\upsilon) = \int_{-\infty}^{\infty} \psi(t,\tau) e^{-j\upsilon t} dt = g(\tau) \int_{-\tau/2}^{\tau/2} e^{-j\upsilon t} dt = 2g(\tau) \frac{\sin(\tau\upsilon/2)}{\upsilon} . \tag{8.5.39}$$

In particular, if let

$$g(\tau) = \frac{1}{\tau} e^{-\alpha\tau^2} , \tag{8.5.40}$$

then the cone-shaped kernel function is

$$\phi(\tau,\upsilon) = \frac{\sin(\tau\upsilon/2)}{\tau\upsilon/2} e^{-\alpha\tau^2} , \quad \alpha > 0. \tag{8.5.41}$$

Different from the exponential kernel function Eq. (8.5.22) which can not suppress the cross-terms on the coordinate axis, the cone-shaped kernel function shown in Eq. (8.5.41) can suppress the cross-terms on the τ axis.

4. Variants of the Wigner-Ville Distribution

The following are several modified distributions of the Winger-Ville distribution.

(1) The pseudo Wigner-Ville distribution (PWD)

Eq. (8.5.26) shows that the RID distribution achieves cross-terms suppression by adding the window function $\psi(t,\tau) = h(\frac{t}{\tau})$ to the variables t and τ. Since the Wigner-Ville distribution corresponds to a Cohen's class distribution with a window function

$\psi(t,\tau) = \delta(t)$, the simplest way to modify this distribution is to add a window function $h(\tau)$ to the variable τ to reduce the cross term. The modified Wigner-Ville distribution is conventionally called the pseudo Wigner-Ville distribution (PWD) and is defined as

$$\mathrm{PWD}_z(t,\omega) \overset{def}{=} \int_{-\infty}^{\infty} z\left(t+\frac{\tau}{2}\right) z^*\left(t-\frac{\tau}{2}\right) h(\tau)e^{-j\omega\tau}d\tau = W_z(t,\omega) \overset{\omega}{*} H(\omega), \qquad (8.5.42)$$

where $\overset{\omega}{*}$ represents the convolution about the frequency variable ω. Note that the window function $h(\tau)$ should satisfy the requirements R1to R4, i.e., $H(\omega)$ should essentially be a low-pass filter.

(2) The smoothed Wigner-Ville distribution (SWD)

Another approach to reducing the cross-terms of the Wigner-Ville distribution is to directly smooth the Wigner-Ville distribution $W_z(t,\omega)$ to obtain the so-called smoothed Wigner-Ville distribution

$$\mathrm{SWD}_z(t,\omega) = W_z(t,\omega) \overset{t}{*}\overset{\omega}{*} G(t,\omega), \qquad (8.5.43)$$

where $\overset{t\omega}{**}$ represents the two-dimensional convolution of time and frequency, and $G(t,\omega)$ is a smoothing filter.

Interestingly, the spectrogram can be seen as a special case of the smoothed Wigner-Ville distribution. From the definition of the spectrum and short-time Fourier transform, it is easy to know

$$\begin{aligned}\mathrm{SPEC}(t,\omega) &= |\mathrm{STFT}(t,\omega)|^2 \\ &= \int_{-\infty}^{\infty} z(u)\gamma^*(u-t)e^{-j\omega u}du \int_{-\infty}^{\infty} z^*(s)\gamma(s-t)e^{j\omega s}ds \\ &= W_z(t,\omega) \overset{t}{*}\overset{\omega}{*} W_\gamma(-t,\omega),\end{aligned} \qquad (8.5.44)$$

where $W_\gamma(-t,\omega)$ is the time-reversal form of the Wigner-Ville distribution $W_\gamma(t,\omega)$ of the window function $\gamma(t)$ of the short-time Fourier transform. The above formula shows that the spectrogram is a two-dimensional convolution of the signal and the Wigner Ville distribution of the window function. In particular, if the smoothing filter

$$G(t,\omega) = W_\gamma(-t,\omega) \qquad (8.5.45)$$

is selected in the smoothed Wigner-Ville distribution Eq. (8.5.43), the smoothed Wigner-Ville distribution Eq. (8.5.43) degenerates to spectrogram Eq. (8.5.44).

(3) The smoothed pseudo Wigner-Ville distribution (SPWD)

Eq. (8.5.26) enlightens us that the RID distribution obtains ideal cross-term suppression effect by adding combined window function $h\left(\frac{t}{\tau}\right)$ to t and τ. In fact, the idea of adding window functions to t and τ is also applicable to the Wigner-Ville distribution, but the added window function is different from the window function $h\left(\frac{t}{\tau}\right)$ of the RID distribution. Specifically, a window function $g(t)h(\tau)$ is used to smooth the t and τ by adding $g(t)$ and $h(\tau)$. The Wigner-Ville distribution obtained by this modification is called the smoothed pseudo Wigner-Ville distribution, defined as

$$\mathrm{SPWD}_z(t,\omega)=\int_{-\infty}^{\infty}\int_{-\infty}^{\infty} g(u)h(\tau)z\left(t-u+\frac{\tau}{2}\right)z^*\left(t-u-\frac{\tau}{2}\right)e^{-j\omega\tau}du d\tau, \qquad (8.5.46)$$

where $g(t)$ and $h(\tau)$ are two real even window functions with $h(0)=g(0)=1$.

(4) The modified smoothed pseudo Wigner-Ville distribution (MSPWD)

Auger and Flandrin[15] found that the performance of the smoothed pseudo-Wigner-Ville distribution will be further improved after a proper "reassignment" (modification), and the smoothed pseudo Wigner-Ville distribution after reassignment is called modified smoothed pseudo Wigner-Ville distribution (MSPWD), i.e.,

$$\mathrm{MSPWD}_z(t,\omega)=\int_{-\infty}^{\infty}\int_{-\infty}^{\infty} \mathrm{SPWD}_z(t',\omega')\delta[t-\hat{t}(t',\omega')]\delta[\omega-\hat{\omega}(t',\omega')]dt'd\omega' \qquad (8.5.47)$$

where $\hat{t}(t',\omega')$ and $\hat{\omega}(t',\omega')$ represent the time and frequency points after reassignment, respectively.

It should be noted that the Wigner-Ville distribution and its three generalized forms are bilinear time-frequency representations, but the modified smoothed pseudo Wigner-Ville distribution is not. Moreover, what it loses is only this bilinear, which can retain other properties of the Wigner-Ville distribution.

5. B Distribution

B distribution is a time-frequency distribution proposed by Barkat and Boashash[19]. In the B distribution, the kernel function is taken as

$$\psi(t,\tau)=\left(\frac{|\tau|}{\cosh^2(t)}\right)^{\sigma}, \qquad (8.5.48)$$

where σ is a constant, and its choice is application dependent, but its range is between 0 and 1, i.e., $0<\sigma\le 1$.

B distribution is defined as

$$B(t,\omega)\stackrel{def}{=}\int_{-\infty}^{\infty}\int_{-\infty}^{\infty} z\left(u+\frac{\tau}{2}\right)z^*\left(u-\frac{\tau}{2}\right)\frac{|\tau|^{\sigma}}{\cosh^{2\sigma}(u-t)}e^{-j\omega\tau}du d\tau. \qquad (8.5.49)$$

It is easy to show that the B distribution satisfies most of the properties of the time-frequency distribution. In particular, the B distribution has the following important properties.

Property 1 The B distribution is real since $\phi(\tau,\nu)=\phi^*(-\tau,0\nu)$;

Property 2 The B distribution is time-shift invariant since the kernel function $\phi(\tau,\nu)$ is not a function of time;

Property 3 The B distribution is frequency-shift invariant since the kernel function $\phi(\tau, \upsilon)$ is not a function of frequency;

Property 4 The first moment of the B distribution yields the instantaneous frequency

$$f_i(t) = \frac{\int_{-\infty}^{\infty} f B_z(t,f)df}{\int_{-\infty}^{\infty} B_z(t,f)df} \tag{8.5.50}$$

since the kernel function satisfies

$$\left.\frac{\partial\phi(\tau,\upsilon)}{\partial\tau}\right|_{\upsilon=0} = 0 \quad \text{and} \quad \phi(0,\upsilon) = \text{constant.} \tag{8.5.51}$$

The above repeatedly emphasizes the harmful side of the cross-term and introduces various kernel function design methods to reduce the cross-term in detail. It should be noted that the reader should not be under the illusion that the cross-term is just a black sheep, which has hundreds of harm but no benefit. In fact, in some important signal processing applications, the cross-term can be a useful asset. For example, when using coherent radar to detect icebergs floating on the surface of the ocean, the cross-term is a reflection of the existence of the detection target (iceberg) and is therefore useful. Readers interested in this application can refer to reference [101].

Summary

This chapter first focuses on the mother of time-frequency distributions, the Wigner-Ville distribution, followed by a discussion of the relationship between it and the ambiguity functions, and the unified form of the time-frequency distributions, the Cohen's class time-frequency distribution, which is the windowed form of the Wigner-Ville distribution.

The evaluation of the performance of time-frequency distribution is mainly determined by the time-frequency aggregation and the cross-term suppression, and the latter becomes the focus and difficulty of time-frequency distribution research and application. Various improved time-frequency distributions can be obtained by selecting different window functions. Being familiar with various time-frequency distributions will help to select an appropriate time-frequency distribution for the application.

This chapter concludes with a presentation of the time-frequency analysis of polynomial FM signals. In particular, as a product of the combination of time-frequency signal analysis and higher-order statistical analysis, the trispectrum of the Wigner-Ville distribution is a powerful analytical tool for polynomial FM signals.

Exercises

8.1 Compute the Wigner-Ville distributions of time domain δ function $z(t) = \delta(t - t_0)$ and frequency domain δ function $Z(\omega) = \delta(\omega - \omega_0)$.
8.2 $z(t) = e^{j\frac{1}{2}mt^2}$ is a linear frequency modulation (LFM) signal, where m is the frequency modulation slope. Find its Wigner-Ville distribution.
8.3 Find the Wigner-ville distribution of the Gaussian signal $z(t) = \frac{1}{\sqrt{\sigma}} e^{-\pi t^2/\sigma^2}$.
8.4 Signal

$$z(t) = \left(\frac{\alpha}{\pi}\right)^{1/4} \exp\left(\frac{\alpha}{2} t^2\right)$$

is a normalized Gaussian signal with unit energy. Find its Wigner-Ville distribution.
8.5 A non-stationary signal is composed of two Gaussian functions

$$z(t) = \left(\frac{\alpha}{\pi}\right)^{1/4} \left[\exp\left(-\frac{\alpha}{2}(t - t_1)^2 + j\omega_1 t\right) + \exp\left(-\frac{\alpha}{2}(t - t_2)^2 + j\omega_2 t\right)\right],$$

where $t_1 > t_2$ and $\omega_1 > \omega_2$. Prove that the signal term (self term)

$$W_{\text{auto}}(t, \omega) = 2\sum_{i=1}^{2} \exp\left(-\alpha(t - t_i)^2 - \frac{1}{\alpha}(\omega - \omega_i)\right)$$

and the cross-term

$$W_{\text{cross}}(t, \omega) = 4\exp\left[-\alpha(t - t_m)^2 - \frac{1}{\alpha}(\omega - \omega_m)^2\right] \cos[(\omega - \omega_m)t_d + \omega_d t]$$

of the Wigner-Ville distribution of signal $z(t)$, where

$$t_m = \frac{1}{2}(t_1 + t_2) \quad \text{and} \quad \omega_m = \frac{1}{2}(\omega_1 + \omega_2)$$

are the average of the time delay and the average of the frequency of the two harmonic signals respectively, while

$$t_d = t_1 - t_2 \quad \text{and} \quad \omega_d = \omega_1 - \omega_2$$

are the difference in time delay and frequency of the two harmonic signals, respectively.
8.6 The frequency domain definition of the mutual Wigner-Ville distribution of signal $x(t)$ and $g(t)$ is

$$W_{X,G}(t, \omega) = \frac{1}{2\pi} \int_{-\infty}^{\infty} X\left(\omega + \frac{\upsilon}{2}\right) G^*\left(\omega - \frac{\upsilon}{2}\right) e^{j\upsilon t} d\upsilon.$$

Prove

$$\int_{-\infty}^{\infty} W_{X,G}(t, \omega) dt = \int_{-\infty}^{\infty} W_x(t, \omega) dt.$$

8.7 Prove that the definition formula of the instantaneous frequency

$$\omega_i(t) = \frac{\int_{-\infty}^{\infty} \omega W_x(t, \omega) d\omega}{\int_{-\infty}^{\infty} W_x(t, \omega) d\omega}.$$

Hint: Using the integral formula

$$\int_{-\infty}^{\infty} \omega e^{-j\omega\tau} d\omega = \frac{2\pi}{j} \frac{\partial \delta(\tau)}{\partial \tau}.$$

8.8 Let $x_a(t) = x(t) + j\hat{x}(t)$ be the analytic signal of the real signal $x(t)$, where $\hat{x}(t)$ is the Hilbert transform of $x(t)$. Try to find the relationship between the Wigner-Ville transform of the analytic signal $x_a(t)$ and the Wigner-Ville transform of the real signal $x(t)$.

8.9 Compute the ambiguity function of time domain δ signal $z(t) = \delta(t - t_0)$ and frequency domain δ signal $Z(\omega) = \delta(\omega - \omega_0)$.

8.10 Find the ambiguity function of the LFM signal $z(t) = e^{\frac{1}{2}mt^2}$.

8.11 Find the ambiguity function of the Gaussian signal $z(t) = \frac{1}{\sqrt{\sigma}} e^{-\pi t^2/\sigma^2}$.

8.12 Find the Wigner-Ville distribution and ambiguity function of a single Gaussian signal

$$z(t) = \left(\frac{\alpha}{\pi}\right)^{1/4} \exp\left[-\frac{\alpha}{2}(t - t_0)^2 + j\omega_0 t\right].$$

8.13 Find the self-term and cross-term of the ambiguity function of the signal superimposed by two Gaussian functions

$$z(t) = \left(\frac{\alpha}{\pi}\right)^{1/4} \left[\exp\left(-\frac{\alpha}{2}(t - t_1)^2 + j\omega_1 t\right) + \exp\left(-\frac{\alpha}{2}(t - t_2)^2 + j\omega_2 t\right)\right],$$

where $t_1 > t_2$ and $\omega_1 > \omega_2$.

8.14 Prove that when the kernel function $\phi(\tau, \upsilon)$ takes the ambiguity function of any window function $\gamma(t)$, the Cohen's class time-frequency distribution is equivalent to the spectrogram, i.e., it has non-negativity.

8.15 Prove the time shift-invariance and frequency shift-invariance of Cohen's class time-frequency distribution.

8.16 Prove the unitary form of Cohen's class time-frequency distribution or Moyal formula

$$\frac{1}{2\pi}\langle C_z, C_x\rangle = |\langle z, x\rangle|^2.$$

8.17 Let

$$\mathrm{SWD}_z(t, \omega) = \frac{1}{2\pi} \int_{-\infty}^{\infty} \int_{-\infty}^{\infty} \Phi(\theta, \upsilon) W_z(t - \theta, \omega - \upsilon) d\upsilon d\theta$$

be the modified Wigner-Ville distribution of signal $z(t)$. If

$$\Phi(t, \omega) = (0.5\pi t_0 \omega_0)^{-1/2} e^{-(t/t_0)^2 - (\omega/\omega_0)^2}.$$

Prove that $\mathrm{SWD}_z(t, \omega)$ is non-negative in the special case $t_0\omega_0 = 1$.

9 Blind Signal Separation

The study of blind signal separation (BBS) originated from a paper published by Jutten and Herault in 1991 [115]. Later in 1994, another BSS technique named as independent component analysis (ICA) method is proposed by Common[64]. It is their pioneering works that have greatly promoted the research on blind signal separation, making it a research hotspot in signal processing, machine learning, and neural computation over the last three decades. With the extensive application as the background, the theory and methods of BSS have gained rapid development. At the same time, it also promoted and enriched the development of the theory and methods of signal processing, machine learning, and neural computation significantly, and gained wide application in many fields such as data communication, multimedia communication, image processing, speech processing, biomedical signal processing, radar and wireless communications and so on.

In this chapter, we will introduce the main theory, methods, and some typical applications of blind signal separation.

9.1 Basic Theory of Blind Signal Processing

Before introducing the theory and method related to BSS, it is necessary to begin with a broader vision from the problem of blind signal processing.

9.1.1 A Brief Introduction to Blind Signal Processing

Blind signal processing can be divided into two categories:
(1) Fully-blind signal processing: techniques using only the output (or observed) data of the system.
(2) Semi-blind signal processing: techniques using not only the output data of the system but also the input or some statistical characteristic of the system.

Fully-blind signal processing is relatively difficult since only a little information is available, while semi-blind signal processing is easier since more information is available compared to the fully-blind case.

Take the problem of system identification shown in Fig.9.1.1 as an example, system identification techniques can be classified into the following three catcategories
(1) White box technique: structure of the system is known (be transparent inside);
(2) Gray box technique: partial structure of the system is known (can be obtained through physical observation);
(3) Black box technique: structure of the system is unknown.

https://doi.org/10.1515/9783110475562-009

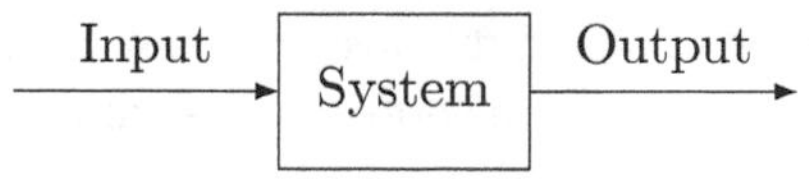

Fig. 9.1.1: System identification

The following are some typical examples of blind signal processing.

(1) Radar target detection: the target (such as a fighter and missile) is a non-cooperative object.
(2) Seismic exploration (to get the reflection coefficient of the earth layer): the earth layer is a non-cooperative object.
(3) Mobile communication: the user is a cooperative object and thus the transmitted signal can be encoded with some structural feature through design.

The main branches of blind signal processing include blind signal separation, blind beamforming, blind channel estimation, blind system identification and target recognition, blind equalization, and so on.

Blind signal processing is an important branch of blind signal processing and is closely related to other important branches of blind signal processing (blind beamforming, etc.).

9.1.2 Model and Basic Problem of BSS

Blind signal separation (BSS), is also known as blind source separation. Its mathematical model can be formulated uniformly as

$$\boldsymbol{x}(t) = \boldsymbol{A}\boldsymbol{s}(t) . \tag{9.1.1}$$

There are two different interpretations of this model.

Model of the array signal processing: A physical model where $\boldsymbol{A}$ is the response matrix of a sensor array, representing the channel through which the signal is transmitted, with a clear physical meaning. After the signal is transmitted through the channel, it is received or observed by the sensor array and becomes the array response $\boldsymbol{x}(t) = \boldsymbol{A}\boldsymbol{s}(t)$.

Model of the blind signal separation: A physically independent mathematical model where $\boldsymbol{A}$ is a mixing matrix whose elements represent the linear mixing coefficient corresponding to each source. Thus there is no explicitly physical parameter and $\boldsymbol{A}\boldsymbol{s}(t)$ essentially represents the linear mixing result of multiple signals.

The basic problem of BSS is to identify the mixing matrix $\boldsymbol{A}$ or recover all source signals $\boldsymbol{s}(t)$ using only the observation vector $\boldsymbol{x}(t)$ without any a priori knowledge about $\boldsymbol{A}$.

A typical engineering application of BSS is the cocktail party problem: there are n guests talking at the party and m sensors are used to obtain the observation data. It is desired to separate the mixed conversation signals to obtain certain conversations of interest.

The term "blind" has two meanings: (1) the source can not be observed; (2) the mixing matrix is unknown, that is, how the signals are mixed in unknown. Therefore, blind signal separation is a natural choice when it is difficult to establish the mathematical model of the transmission process from the source signal to the sensor, or prior knowledge about the transmission is not available.

Consider using m sensors to observe n source signals $\boldsymbol{s}_1(t), \cdots, \boldsymbol{s}_n(t)$, and then use the mixed signal $\boldsymbol{As}(t)$ for blind signal separation. Under this case, the observed signal is a $m \times 1$ vector $\boldsymbol{x}(t) = [x_1(t), \cdots, x_m(t)]^{\mathrm{T}} = \boldsymbol{As}(t)$, the mixing matrix A is a $m \times n$ matrix, and the source signal is $\boldsymbol{s}(t) = [s_1(t), \cdots, x_n(t)]^{\mathrm{T}}$. It is usually required that the number of sensors must not be less than the number of unknown source signals, i.e., $m \geqslant n$.

The ideal result of signal separation can be expressed as

$$\boldsymbol{s}(t) = \boldsymbol{A}^{\dagger}\boldsymbol{x}(t) = \left(\boldsymbol{A}^{\mathrm{T}}\boldsymbol{A}\right)^{-1}\boldsymbol{A}^{\mathrm{T}}\boldsymbol{x}(t). \tag{9.1.2}$$

However,since the mixing matrix $\boldsymbol{A}$ is unknown, it is impossible to find the Moore-Penrose inverse matrix $\left(\boldsymbol{A}^{\mathrm{T}}\boldsymbol{A}\right)^{-1}\boldsymbol{A}^{\mathrm{T}}$, which makes $\boldsymbol{s}(t) = \boldsymbol{A}^{\dagger}\boldsymbol{x}(t)$ impossible to realize.

The mixing matrix $bdsymA$ is unknown, resulting in two uncertainties or ambiguities in BSS. Denote the mixing matrix as $\boldsymbol{A} = [\boldsymbol{a}_1, \cdots, \boldsymbol{a}_n]$, then we have

$$\boldsymbol{x}(t) = \boldsymbol{As}(t) = \sum_{i=1}^{n} \boldsymbol{a}_i \boldsymbol{s}_i(t) = \sum_{i=1}^{n} \frac{\boldsymbol{a}_i}{\alpha_i} \alpha_i \boldsymbol{s}_i(t). \tag{9.1.3}$$

In other words, the following two types of uncertainties exist in BSS.

(1) *Uncertainty of the source signal ordering.* If permuting the column vector $\boldsymbol{a}_i$ and $\boldsymbol{a}_j$ of the mixing matrix as well as permuting the order of the source signal s_i and s_j, the mixed signal $\boldsymbol{x}(t) = \sum_{i=1}^{n} \boldsymbol{a}_i \boldsymbol{s}_i(t)$ remains unchanged. That is, the mixed signal $\boldsymbol{x}(t)$ does not contain any information about the order of the individual source signals.

(2) *Uncertainty of the source signal amplitude.* If dividing the column vector $\boldsymbol{a}_i$ by a nonzero complex-valued constant α_i and multiplying the corresponding source signal by the same α_i, the mixed signal $\boldsymbol{x}(t) = \boldsymbol{As}(t) = \sum_{i=1}^{n} \frac{\boldsymbol{a}_i}{\alpha_i} \alpha_i \boldsymbol{s}_i(t) = \sum_{i=1}^{n} \boldsymbol{a}_i \boldsymbol{s}_i(t)$ remains unchanged. That is, the true amplitude and phase of any source signal can not be identified uniquely by the mixed signal $\boldsymbol{x}(t)$.

The basic principle of BSS (see Fig.9.1.2) is to design an $n \times m$ separation matrix (or demixing matrix) $\boldsymbol{B}$ for an $m \times n$ unknown mixing matrix such that its output

$$\boldsymbol{y}(t) = \boldsymbol{Bx}(t) = \boldsymbol{BAs}(t) = \hat{\boldsymbol{s}}(t) \tag{9.1.4}$$

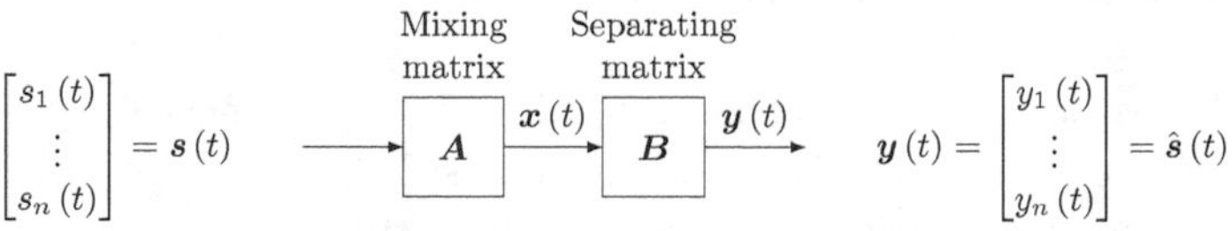

Fig. 9.1.2: The basic principle of the BSS

is an estimate of the source signal vector $\boldsymbol{s}(t)$.
Obviously, if the separation matrix is designed to satisfy

$$\boldsymbol{B} = \boldsymbol{G}\boldsymbol{A}^{\dagger} \tag{9.1.5}$$

where $\boldsymbol{G}$ is an $n \times n$ generalized permutation matrix, which refers to a matrix that contains exactly one nonzero element in each row and each column. Then the separation output can be expressed as

$$\boldsymbol{y}(t) = \left(\boldsymbol{G}\boldsymbol{A}^{\dagger}\right)\boldsymbol{A}\boldsymbol{s}(t) = \boldsymbol{G}\left(\boldsymbol{A}^{\mathrm{T}}\boldsymbol{A}\right)^{-1}\boldsymbol{A}^{\mathrm{T}}\boldsymbol{A}\boldsymbol{s}(t) = \boldsymbol{G}\boldsymbol{s}(t) = \hat{\boldsymbol{s}}(t), \tag{9.1.6}$$

which shows that since the mixing matrix $\boldsymbol{A}$ is unknown, there exist two types of uncertainty in identifying the Moore-Penrose inverse matrix $\boldsymbol{A}^{\dagger}$, that is, uncertainty in the ordering of its column vectors and uncertainty in the element magnitudes, which leads to two types of uncertainty in the results of the BSS as

(1) *Uncertainty in the ordering of the separation signal* $\boldsymbol{G}\boldsymbol{s}(t) = \hat{\boldsymbol{s}}(t)$*;*
(2) *Uncertainty in the waveform (i.e., amplitude and phase) of the separated signals* $\boldsymbol{G}\boldsymbol{s}(t) = \hat{\boldsymbol{s}}(t)$.

These two types of uncertainty are consistent with that of the mixed signal. It is that the mixed signal can not distinguish the ordering of the source signals and the true amplitude of each signal, so the BSS $\boldsymbol{y}(t) = \boldsymbol{B}\boldsymbol{x}(t) = \boldsymbol{G}\boldsymbol{s}(t) = \hat{\boldsymbol{s}}(t)$ using only the mixed signal will lead to the uncertainty of the ordering and waveform of the separated signal .

It is worth pointing out that these two types of uncertainties are allowed by signal separation, since the separated signal $\boldsymbol{y}(t) = \boldsymbol{G}\boldsymbol{s}(t)$ is a copy of the source signal $\boldsymbol{s}(t)$, which ensures that the two essential requirements of signal separation are met.

(1) The ordering of the original signals is not the main object of our interest when performing signal separation.
(2) The separated signal should be "hi-fidelity" to the original signal. The fixed initial phase difference between the separated and the original signal may be corrected by proper phase compensation, while a fixed amplitude difference only stands for that a signal is amplified or attenuated by a fixed scale, and would not have any influence on the fidelity of signal waveform.

9.1.3 Basic Assumption and Performance Requirement of BSS

In order to realize the two basic purposes of blind signal separation, namely, separating the mixed signal and keeping a high fidelity, it is necessary the make some basic assumptions and propose some basic performance requirements.

1. Basic Assumption of BSS

In order to make up for the insufficiency of the given information in the BSS problem, it is usually necessary to make some basic assumptions as follows to the BSS:

Assumption 1 Each component of the signal vector $\mathbf{s}(t)$ is independent at any time t.

Assumption 2 There is at most one Gaussian signal among all the components $s_i(t)$ of $\mathbf{s}(t)$.

Assumption 3 The $m \times n$ mixing matrix to be of full column rank, and $m \leqslant n$.

Assumption 4 The components of $\mathbf{s}(t)$ have unit variance.

These assumptions are reasonable for most applications for the following reasons.

(1) Assumption 1 is a crucial assumption that a signal can be blind separated. Although this assumption is a strong statistical hypothesis but a physically very plausible one since it is expected to be verified whenever the source signals arise from physically separated systems.
(2) Since the linear mixture of two Gaussian signals is still a Gaussian signal, Assumption 2 is a very natural one.
(3) Assumption 3 requires that the number m of the sensors be not less than that n of the independent source signals, that is, it requires that the model of BSS be well-determined ($m = n$) or over-determined $m > n$. In practice, most applications belong to well-determined or over-determined BSS problems. In the under-determined ($m < n$) BSS, the number of the sensors is less than that of the independent source signals, which will be discussed in the last topic of this chapter.
(4) Assumption 4 is related to the ambiguity of signal separation. The incomplete identification of the mixing matrix $\boldsymbol{A}$ is called the uncertainty of $\boldsymbol{A}$. Since $\boldsymbol{A}$ has uncertainty, without loss of generality, it is natural to assume that the source signal has unit variance, that is, the dynamic variations of the amplitude and phase of each source signal can be merged into the corresponding column vectors of $\boldsymbol{A}$.

Assumption 2 allows there is at most one source signal to be Gaussian, all the other sources must be non-Gaussian ones. As mentioned in Chapter 6, the non-Gaussian signal can be classified as sub-Gaussian signal and supper-Gaussian signal. It is worth noting that the methods suitable for separating sub-Gaussian and super-Gaussian are somewhat different.

2. Basic Performance Requirement of the BSS

The basic performance requirement of the BSS is that the separated signal must have equivariance.

Definition 9.1.1. *A BSS method is called to have equivariance, if its separating output $\mathbf{y}(t) = \mathbf{Bx}(t)$ obtained using the observed signal $\mathbf{x}(t) = \mathbf{As}(t)$ does not change with the variations of the mixing matrix $\mathbf{A}$ when the source signal $\mathbf{s}(t)$ is given.*

BSS methods may be divided into online methods and offline methods. Offline BSS methods are also known as batch-processing or fixed BSS methods, while online BSS methods are usually called as adaptive BSS methods.

Equivariance is a basic performance requirement for batch-processing methods. The batch-processing separators are named as equivariance (batch-processing) signal separators if their outputs satisfy the equivariance requirement, and they would exhibit a uniform BSS performance consequently.

BSS that uses equivariance separators is called equivariance BSS. In signal separation, the performance of any equivariant signal separation algorithm is completely independent of the mixing matrix (i.e., the channel of signal transmission), which is the meaning of the so-called "uniform performance". Otherwise, a blind signal separator suitable for some linear mixed signals may not be suitable for other linear mixed forms. Obviously, a blind signal separator without uniform performance cannot be generalized and lacks the value of practical applications.

Definition 9.1.2. *Let $\mathbf{W}(k)$ be the separation matrix or ing matrix of a BSS algorithm, and construct a mixing-demixing system $\mathbf{C}(k) = \mathbf{W}(k)\mathbf{A}$. A BSS algorithm is called to have equivariance if $\mathbf{C}(k)$ satifies*

$$\mathbf{C}(k+1) = \mathbf{C}(k) - \eta(k)\mathbf{H}(\mathbf{C}(k)\mathbf{s}(k))\mathbf{C}(k) \tag{9.1.7}$$

and the matrix function $\mathbf{H}(\mathbf{C}(k)\mathbf{s}(k))$ of matrix $\mathbf{C}(k)\mathbf{s}(k)$ is independent of the mixing matrix $\mathbf{A}$.

"The performance of a batch-processing signal separation algorithm is independent of how the source signals are mixed" is a basic performance desired for any BSS algorithm to have. It is worth pointing out that if a batch-processing BSS algorithm does not have equivarance, its corresponding adaptive algorithm must not have equivarance.

9.2 Adaptive Blind Signal Separation

If a batch-processing BSS algorithm with equivariance is modified to an adaptive one, the "uniform performance" of the original batch algorithm would be inherited by its corresponding adaptive algorithm.

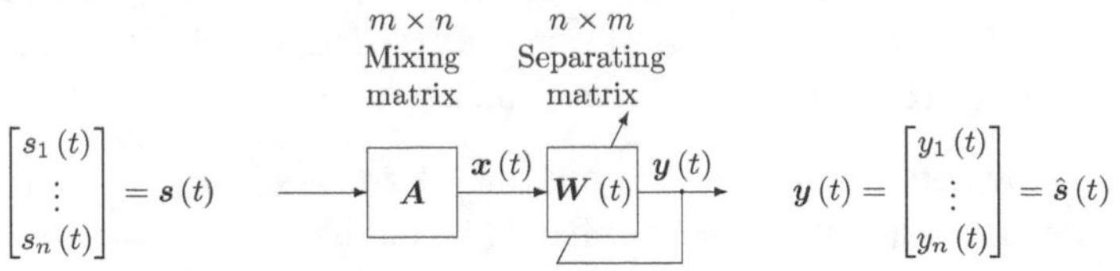

Fig. 9.2.1: The block diagram of the adaptive BSS

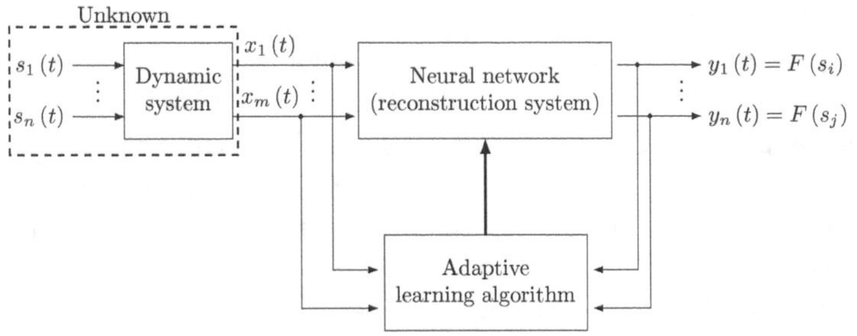

Fig. 9.2.2: Functional block diagram illustrating a general BSS

9.2.1 Neural Network Implementation of Adaptive Blind Signal Separation

Fig.9.2.1 shows the block diagram of the adaptive BSS, where the separating matrix $\boldsymbol{B}$ of the batch-processing algorithm is replaced by the weighting matrix $\boldsymbol{W}(k)$ which is adjusted adaptively by the (machine) learning algorithms.

Adaptive BSS is usually realized using a neural network, the principle of which is shown in Fig.9.2.2.

More specifically, the structure of the feed-forward neural network of adaptive BSS is shown in Fig.9.2.3, in which Fig.9.2.3(a) is a block diagram and Fig.9.2.3(b) is the detailed structure of the mixing model and a basic feed-forward neural network of the adaptive BSS.

As shown in Fig.9.2.3(b), let the weight coefficient of the feed-forward neural network be w_{ij}, $i = 1, \cdots, n$; $j = 1, \cdots, m$ and the $n \times m$ weight matrix at time t be $\boldsymbol{W}(t) = \left[w_{ij}(t)\right]$, then we have

$$\boldsymbol{y}(t) = \boldsymbol{W}(t)\boldsymbol{x}(t), \tag{9.2.1}$$

where $\boldsymbol{W}(t)$ is known to be an unmixing matrix or demixing matrix for BSS while synaptic weight matrix for the neural network.

As emphasized in the previous section, the goal of BSS is to find a separation matrix $\boldsymbol{B}$ such that $\boldsymbol{BA} = \boldsymbol{G}$, where $\boldsymbol{G}$ is a generalized permutation matrix. Therefore, it is desired to use the adaptive learning algorithm so that the converged synaptic weight

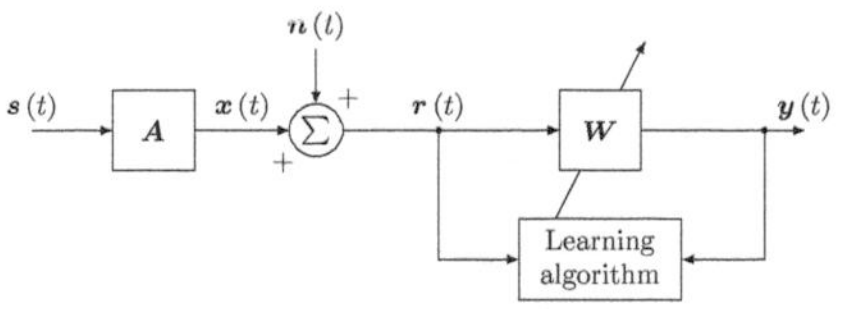

(a) General block diagram

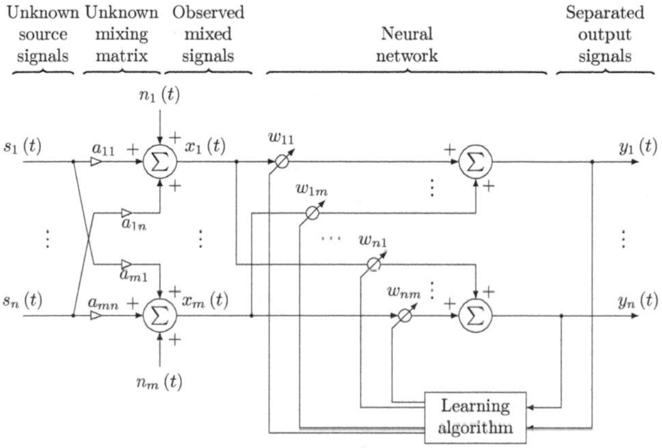

(b) The detailed architecture of mixing and separating models

Fig. 9.2.3: Illustration of the mixing model and a basic feed-forward neural network of adaptive BSS[10]

matrix $\boldsymbol{W}_\infty$ satisfies the following relation

$$\boldsymbol{W}_\infty \boldsymbol{A} = \boldsymbol{G}. \tag{9.2.2}$$

That is to say, the separation performance of a neural network at time t is evaluated by the composite matrix $\boldsymbol{T}(t) \overset{\text{def}}{=} \boldsymbol{W}(t)\boldsymbol{A}$, and this matrix describes the "separation precision" of all the separated independent components of the signal in the mixing-separating model $\boldsymbol{y}(t) = \boldsymbol{T}(t)\boldsymbol{s}(t)$.

Although what is illustrated in Fig.9.2.3 is a feed-forward neural network, a fully connected feedback (recurrent) neural network is also applicable. Let the weight matrix of a simple recurrent neural network at time t to be $\hat{\boldsymbol{W}}(t)$, then

$$y_i(t) = x_i(t) - \sum_{j=1}^{n} \hat{w}_{ij}(t) y_j(t)\,. \tag{9.2.3}$$

Schematic illustrations of a simple recurrent neural network ($m = n$), a feed-forward-feedback cascaded model, and a hybrid model are shown in Fig.9.2.4(a) (c).

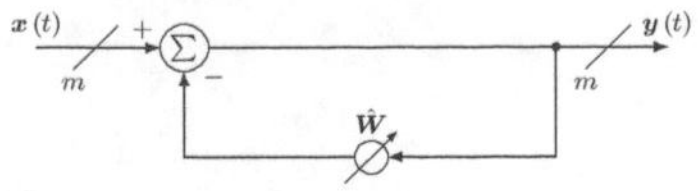

(a) The recurrent neural network model

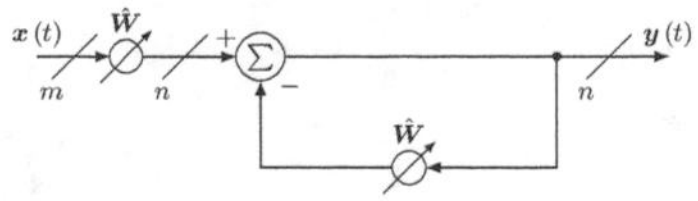

(b) The feed-forward-feedback cascaded model

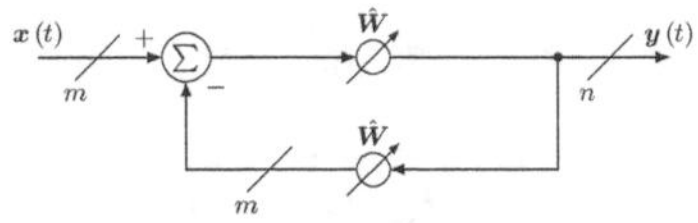

(c) The hybrid model

Fig. 9.2.4: The neural network structure of adaptive BSS[10]

It should be noted that, for the ideal memoryless case, the basic model of the recurrent and feed-forward neural network are equivalent under the condition that

$$\boldsymbol{W}(t) = \left[\boldsymbol{I} + \hat{\boldsymbol{W}}(t)\right]^{-1} . \tag{9.2.4}$$

Let the mathematical model of the recurrent neural network be

$$\boldsymbol{y}(t) = \left[\boldsymbol{I} + \hat{\boldsymbol{W}}(t)\right]^{-1} \boldsymbol{x}(t) , \tag{9.2.5}$$

if a linear adaptive neural network is employed to perform the BSS, its output would be

$$\boldsymbol{y}(t) = NN(\boldsymbol{W}, \boldsymbol{x}(t)) = \boldsymbol{W}\boldsymbol{x}(t) , \tag{9.2.6}$$

where $\boldsymbol{W}$ is the neural network weight matrix and $NN(\cdot)$ represents a linear neural network function.

The task of the neural network is to adjust the weight matrix (or demixing weight) $\boldsymbol{W}$ adaptively by sample training to

$$\boldsymbol{W} = \boldsymbol{\Lambda}\boldsymbol{P}\boldsymbol{A}^{\dagger} = \boldsymbol{G}\boldsymbol{A}^{\dagger} , \tag{9.2.7}$$

where $\boldsymbol{\Lambda}$ is a nonsingular diagonal matrix, $\boldsymbol{P}$ is a permutation matrix, and $\boldsymbol{G}$ is a generalized permutation matrix.

In most cases, the core problem of adaptive BSS is the learning algorithm of separating (or demixing) matrix, which is an unsupervised machine learning problem.

The basic idea of unsupervised machine learning is to extract statistically independent features as input representations without losing information.

When the mixing model is nonlinear, the source signals usually can not be recovered directly from the observed mixtures, unless some prior knowledge or information about the signal and (or) the mixing model is available. In the following sections of this chapter, only BSS in the case of the linear mixing model is discussed.

9.2.2 Quasiidentity Matrix and Contrast Function

The normalization of the variance of each source signal only resolves the uncertainty of the amplitude of each element of the mixing matrix $\boldsymbol{A}$, while the ordering of each column of $\boldsymbol{A}$ and the initial phase of each source remain uncertain. In order to describe and resolve these two kinds of the uncertainty of mixing matrix $\boldsymbol{A}$, Cardoso and Laheld[44] introduce the concept of "quasiidentity" between two matrices into the problem of BSS.

Definition 9.2.1. *(Quasiidentity Matrix): $m \times n$ matrix $\boldsymbol{A}$ and matrix $\boldsymbol{U}$ is called to be quasiidentity matrix, denoted as $\boldsymbol{A} \doteq \boldsymbol{U}$, if $\boldsymbol{A} = \boldsymbol{U}\boldsymbol{G}$ where $\boldsymbol{G}$ is a $n \times n$ generalized permutaion matrix.*

The BSS problem can also be described as follows: identify the demixing matrix (or separating matrix) $\boldsymbol{W} = \boldsymbol{A}\boldsymbol{G}$ which is a quasiidentity matrix to the mixing matrix $\boldsymbol{A}$ and/or obtain a copy or estimate $\hat{\boldsymbol{x}}(n)$ of the source signal based only on the sensor output $\boldsymbol{x}(t)$.

Just as any optimization algorithm needs one or more objective functions, an adaptive BSS algorithm also needs an objective function. In the area of adaptive BSS, "contrast function" is often used to call the objective function.

Definition 9.2.2. *(Contrast Function)[64] A contrast function of a $n \times 1$ vector $\boldsymbol{y}$, denoted as $C(\boldsymbol{y})$, is a mapping from n-dimensional complex-valued space $\mathbb{C}^n$ to a positive real-valued function $\mathbb{R}^+$ to satisfying the following three requirements.*

(1) The contrast function $C(\boldsymbol{y})$ does not change is the elements y_i of $\boldsymbol{y}$ permuted. In other words, $C(\boldsymbol{P}\boldsymbol{y}) = C(\boldsymbol{y})$ always holds for any permutation matrix $\boldsymbol{P}$.

(2) The contrast function $C(\boldsymbol{y})$ is invariant by "scale" of elements y_i of $\boldsymbol{y}$. In other words, $C(\boldsymbol{D}\boldsymbol{y}) = C(\boldsymbol{y})$ always holds for any diagonal invertible matrix $\boldsymbol{D}$.

(3) If $\boldsymbol{y}$ has independent components, then

$$C(\boldsymbol{B}\boldsymbol{y}) \leqslant C(\boldsymbol{y}) \tag{9.2.8}$$

where $\boldsymbol{B}$ is an arbitrary invertible matrix.

Requirement (1) and (2) can be expressed as $C(\boldsymbol{G}\boldsymbol{y}) = C(\boldsymbol{y})$ by the generalized permutation matrix G. From this result and $\boldsymbol{y} = \boldsymbol{W}\boldsymbol{x}$ it is easy to know

$$C(\boldsymbol{y}) = C(\boldsymbol{W}\boldsymbol{x}) = C(\boldsymbol{G}\boldsymbol{W}\boldsymbol{x})\,. \tag{9.2.9}$$

If $\boldsymbol{W}$ is a separating matrix, and its output is $\boldsymbol{y} = \boldsymbol{W}\boldsymbol{x}$, the separating matrix can be given by the solution that maximizes the contrast function

$$\hat{\boldsymbol{W}} = \arg\max C(\boldsymbol{y}) = \arg \max_{\mathrm{E}\{\|\boldsymbol{W}\boldsymbol{x}\|_F\}=1} C(\boldsymbol{G}\boldsymbol{W}\boldsymbol{x}) \tag{9.2.10}$$

or minimizes the contrast function

$$\hat{\boldsymbol{W}} = \arg\min C(\boldsymbol{y}) = \arg \min_{\mathrm{E}\{\|\boldsymbol{W}\boldsymbol{x}\|_F\}=1} C(\boldsymbol{G}\boldsymbol{W}\boldsymbol{x}), \tag{9.2.11}$$

according to the specific choice of different contrast functions. These discussion suggests that the separating matrix $\boldsymbol{G}\boldsymbol{A}^{\dagger}$ obtained by means of maximizing (or minimizing) the contrast function is an estimate of $\boldsymbol{G}\boldsymbol{W}$, which is just a quasiidentity matrix $\boldsymbol{G}\boldsymbol{A}^{\dagger}$ to the unknown ideal separating matrix $\boldsymbol{W} = \boldsymbol{A}^{\dagger}$. Obviously, the function that does not satisfy Eq. (9.2.9) cannot be used as the contrast function for BSS, because this would result in the solution of the optimization algorithm not being quasiidentity to the Moore-Penrose inverse $\boldsymbol{A}^{\dagger}$ of the mixing matrix, thus making it impossible for the separated output vector to be a copy of the source signal.

The core problem of BSS is the design of the contrast function and optimization algorithm, and with the contrast function differs, the corresponding BSS algorithm will differ accordingly. In the following sections of this chapter, we will focus on several representative BSS algorithms according to the different choices of the contrast function.

Four basic problems to be solved by the neural network approach to BSS can be summarized as:

(1) the existence and identifiability of the inverse system;
(2) the stability of the inverse system model;
(3) the convergency, converge rate, and how to avoid falling into local extreme points of a machine learning algorithm;
(4) the reconstruction precision of the source signals.

9.3 Indenpent Component Analysis

Independent component analysis (ICA) is a pioneering method proposed by Common for BSS in 1994. As the term suggests, the basic purpose of ICA is to determine the linear transformation matrix or separating matrix $\boldsymbol{W}$ such that each component $y_i(t)$ of its output vector $\boldsymbol{y}(t) = \boldsymbol{W}\boldsymbol{x}(t)$ is as statistically independent as possible. such that each component y of its output vector Q is as statistically independent as possible.

9.3.1 Mutual Information and Negentropy

In order to realize independent component analysis, it is necessary to have a measure to measure the statistical independence between the components of the signal vector.

Definition 9.3.1. *The mutual information between each component of a signal vector* $\boldsymbol{y} = [y_1, \cdots, y_n]^T$, *ususally denoted as* $I(\boldsymbol{y})$, *is defined as*

$$I(\boldsymbol{y}) = \int p_y(y_1, \cdots, y_n) \log \frac{p_y(y_1, \cdots, y_n)}{\prod_{i=1}^n p_i(y_i)} dy_1 \cdots dy_n. \tag{9.3.1}$$

Mutual information is a nonnegative quantity, that is

$$I(\boldsymbol{y}) \geq 0. \tag{9.3.2}$$

always hold.

Definition 9.3.2. *Let* $\boldsymbol{y}(t) = \boldsymbol{Wx}(t)$ *be the output vector of a neural network whose probability density function and its decomposition form are* $p_y(\boldsymbol{y}, \boldsymbol{W})$ *and* $\tilde{p}_y(\boldsymbol{y}, \boldsymbol{W})$ *respectively, then the Kullaback-Leibler (K-L) divergency between* $p_y(\boldsymbol{y}, \boldsymbol{W})$ *and* $\tilde{p}_y(\boldsymbol{y}, \boldsymbol{W},)$

$$D(\boldsymbol{y}) = KL\left[p_y(\boldsymbol{y}, \boldsymbol{W}) \parallel \tilde{p}_y(\boldsymbol{y}, \boldsymbol{W})\right] \stackrel{\text{def}}{=} \int p_y(\boldsymbol{y}, \boldsymbol{W}) \log \frac{p_y(\boldsymbol{y}, \boldsymbol{W})}{\tilde{p}_y(\boldsymbol{y}, \boldsymbol{W})} d\boldsymbol{y} \tag{9.3.3}$$

is called the dependence between each component of the output signal vector $\boldsymbol{y}$. *In which, the decomposition form of the density function* $\tilde{p}_y(\boldsymbol{y}, \boldsymbol{W})$ *is the product of the marginal probability density function of each component of* $\boldsymbol{y}$, *that is,* $\tilde{p}_y(\boldsymbol{y}, \boldsymbol{W}) = \prod_{i=1}^n p_i(y_i, \boldsymbol{W})$.

By comparing Definition 9.3.1 with Definition 9.3.2, it is easy to find that the mutual information and dependence between each component of the output signal vector $\boldsymbol{y}$ are equivalent, that is

$$I(\boldsymbol{y}) = D(\boldsymbol{y}) \geq 0. \tag{9.3.4}$$

Obviously, if each component $y_i(t)$ of the output $\boldsymbol{y}(t) = \left[y_1(t), \cdots, y_n(t)\right]^{\mathrm{T}}$ is statistically independent to each other, then $I(\boldsymbol{y}) = D(\boldsymbol{y}) = 0$ because of $p_y(\boldsymbol{y}, \boldsymbol{W}) = \tilde{p}_y(\boldsymbol{y}, \boldsymbol{W})$. Conversely, if $I(\boldsymbol{y}) = D(\boldsymbol{y}) = 0$ then each component of the neural network outputs is statistically independent to each other. In other words, there is an equivalent relation

$$I(\boldsymbol{y}) = D(\boldsymbol{y}) = 0 \iff y_i(t), \quad i = 1, \cdots, n \quad \text{being statistically indenpendent.} \tag{9.3.5}$$

And minimize the mutual information (or dependence) of the separated output vector inspires the idea of the ICA technique.

The equivalent relation described in Eq. (9.3.5) may also be expressed as: mutual information is the contrast function of ICA, that is[64]

$$I(\boldsymbol{y}) = 0 \quad \text{iff} \quad \boldsymbol{W} = \boldsymbol{\Lambda P A}^{\dagger} = \boldsymbol{G A}^{\dagger}. \tag{9.3.6}$$

On the other hand, entropy is the basic concept of information theory. The entropy of a random variable can be interpreted as the degree of information that the observation of the random variable gives. The more “random”, i.e. unpredictable and unstructured the variable is, the larger its entropy[109]. The mutual information can be expressed in terms of entropy as

$$I(\boldsymbol{y}) = D(\boldsymbol{y}) = -\boldsymbol{H}(\boldsymbol{y}, \boldsymbol{W}) + \sum_{i=1}^{n} H(y_i, \boldsymbol{W}). \tag{9.3.7}$$

According to Eq. (9.3.7), the condition for the the mutual information to be minimum is that the entropy $H(\boldsymbol{y}, \boldsymbol{W})$ takes the maximum value $\sum_{i-1}^{n} H(\boldsymbol{y}_i, \boldsymbol{W})$. Therefore, minimizing the mutual information is equivalent to maximizing the entropy, i.e., the independent component analysis method is equivalent to the maximum entropy method.

Let $\boldsymbol{y} = [y_1, \cdots, y_n]^{\mathrm{T}}$, then its differential entropy can be expressed as

$$H(\boldsymbol{y}, \boldsymbol{W}) = -\int f(\boldsymbol{y}) \log f(\boldsymbol{y})\, \mathrm{d}\boldsymbol{y} \tag{9.3.8}$$

$$= -\int \cdots \int f(y_1, \cdots, y_n) \log f(y_1, \cdots, y_n)\, \mathrm{d}y_1 \cdots \mathrm{d}y_n . \tag{9.3.9}$$

A fundamental result of information theory is that a Gaussian variable has the largest entropy among all random variables of equal variance. This means that entropy could be used as a measure of non-Gaussianity. Therefore, using the entropy of the Gaussian random variable, differential entropy can be normalized to the negentropy [108]

$$J(\boldsymbol{y}) = \boldsymbol{H}(\boldsymbol{y}_{Gauss}, \boldsymbol{W}) - \boldsymbol{H}(\boldsymbol{y}, \boldsymbol{W}), \tag{9.3.10}$$

where $\boldsymbol{y}_{Gauss}$ is a Guassian random vector output of the separating matrix $\boldsymbol{W}$, and has the same variance matrix $\boldsymbol{V} = \left[V_{ij}\right]$ as the non-Gaussian random vector output $\boldsymbol{y}$.

Negentropy has the following outstanding properties [64].

(1) Negentropy is always non-negative, i.e., $J(\boldsymbol{y}) \geq 0$.

(2) Negentropy is invariant for invertible linear transformations.

(3) Negentropy may be interpreted as a measure of non-Gaussianity, and it is zero if and only if $\boldsymbol{y}$ has a Gaussian distribution.

It is widely known that the entropy of a Gaussian random vector is

$$\boldsymbol{H}(\boldsymbol{y}_{Gauss}, \boldsymbol{W}) = \frac{1}{2}[n + n\log(2\pi) + \log\det \boldsymbol{V}], \tag{9.3.11}$$

and the entropy of the component $y_{i,Gauss}$ of a Gaussian random vector is

$$\boldsymbol{H}\left(y_{i,Gauss}, \boldsymbol{W}\right) = \frac{1}{2}[1 + \log(2\pi) + \log\det V_{ii}] . \tag{9.3.12}$$

Substituting Eq. (9.3.10) into Eq. (9.3.7) and then using Eqs. (9.3.11) and (9.3.12), the mutual information can be rewritten as

$$\begin{aligned} I(\boldsymbol{y}) &= J(\boldsymbol{y}) + \sum_{i=1}^{n} H(y_i, \boldsymbol{W}) - \boldsymbol{H}(\boldsymbol{y}_{Gauss}, \boldsymbol{W}) \\ &= J(\boldsymbol{y}) - \sum_{i=1}^{n} J(y_i) + \sum_{i=1}^{n} H(y_i, \boldsymbol{W}) - \boldsymbol{H}(\boldsymbol{y}_{Gauss}, \boldsymbol{W}) \\ &= J(\boldsymbol{y}) - \sum_{i=1}^{n} J(y_i) + \frac{1}{2}\left(\sum_{i=1}^{n} \log V_{ii} - \log\det \boldsymbol{V}\right), \end{aligned}$$

that is[64]

$$I(\boldsymbol{y}) = J(\boldsymbol{y}) - \sum_{i=1}^{n} J(y_i) + \frac{1}{2}\left(\log \frac{\prod_{i=1}^{n} V_{ii}}{\det \boldsymbol{V}}\right). \quad (9.3.13)$$

Particularly, if each component of $\boldsymbol{y}$ is uncorrelated, its covariance matrix $\boldsymbol{V} = \operatorname{diag}(V_{11}, \cdots, V_{nn})$ will be a diagonal matrix. In this case, we can learn from Eq. (9.3.13) that the negentropy of a random vector with all components to be uncorrelated can be expressed as

$$I(\boldsymbol{y}) = J(\boldsymbol{y}) - \sum_{i=1}^{n} J(y_i). \quad (9.3.14)$$

Eq. (9.3.1) reveals that the essence of the ICA method for finding the separation matrix $\boldsymbol{W}$ is that minimizing the mutual information $\boldsymbol{I}(y)$ of the output vector $\boldsymbol{y} = \boldsymbol{W}\boldsymbol{x}$ is equivalent to maximizing the negetropy $\sum_{i=1}^{n} J(y_i)$, thus achieving the goal that the components of output $\boldsymbol{y}$ are as independent of each other as possible.

9.3.2 Natural Gradient Algorithm

In order to measure the independence between the outputs, in addition to using the second-order correlation function between them, we must also use the high-order statistics between them. Therefore, the same nonlinear transformation must be applied to each component of the output vector $\boldsymbol{y}(t)$. Let the nonlinear transform be $g(\cdot)$, then

$$\boldsymbol{z}(t) = g(\boldsymbol{y}(t)) = [g(y_1(t)), \cdots, g(y_n(t))]^{\mathrm{T}}. \quad (9.3.15)$$

It has been proved[12] that

$$\max H(\boldsymbol{y}, \boldsymbol{W}) = \max H(\boldsymbol{z}, \boldsymbol{W}) \quad (9.3.16)$$

and

$$\frac{\partial H(\boldsymbol{z}, \boldsymbol{W})}{\partial \boldsymbol{W}} = \boldsymbol{W}^{-\mathrm{T}} - \mathrm{E}\left\{\phi(\boldsymbol{y})\boldsymbol{y}^{\mathrm{T}}\right\}, \quad (9.3.17)$$

where

$$\phi(\boldsymbol{y}) = \left[-\frac{g''(y_1)}{g'(y_1)}, \cdots, -\frac{g''(y_n)}{g'(y_n)}\right]^{\mathrm{T}}, \quad (9.3.18)$$

and $g'(y_i)$ and $g''(y_i)$ are the first and second derivatives of the nonlinear transform function $g(y_i)$, respectively.

The standard gradient (or else, absolute gradient) algorithm to update the weight matrix $\boldsymbol{W}$ can be described as

$$\frac{\mathrm{d}\boldsymbol{W}}{\mathrm{d}t} = \eta \frac{\partial H(\boldsymbol{z}, \boldsymbol{W})}{\partial \boldsymbol{W}} = \eta\left(\boldsymbol{W}^{-\mathrm{T}} - \mathrm{E}\left\{\phi(\boldsymbol{y})\boldsymbol{y}^{\mathrm{T}}\right\}\right). \quad (9.3.19)$$

If the expectation term $\mathrm{E}\left\{\phi(\boldsymbol{y})\boldsymbol{y}^{\mathrm{T}}\right\}$ in the gradient algorithm is replaced by the instantaneous value $\phi(\boldsymbol{y})\boldsymbol{y}^{\mathrm{T}}$, the stochastic gradient algorithm

$$\frac{\mathrm{d}\boldsymbol{W}}{\mathrm{d}t} = \eta\left(\boldsymbol{W}^{-\mathrm{T}} - \phi(\boldsymbol{y})\boldsymbol{y}^{\mathrm{T}}\right). \quad (9.3.20)$$

is obtained, which was proposed by Bell and Sejnowski in 1995[23].

The main drawback of the stochastic gradient algorithm for a parametric system is that it converges very slowly. Therefore, it is desirable to find an optimization algorithm that can not only maintain the simplicity and numerical stability of the stochastic gradient algorithm but also gain a good asymptotic convergency. In addition, it is also desirable that the performance of the optimization algorithm is independent of the mixing matrix $\boldsymbol{A}$, so that the algorithm can work well even when $\boldsymbol{A}$ is nearly singular. Fortunately, it is completely achievable, since it has been proved that (see, e.g., reference [233] and [12] for more details) that if an invertible matrix $\boldsymbol{G}^{-1}$ is applied to the matrix

$$\frac{\mathrm{d}\boldsymbol{W}}{\mathrm{d}t} = \eta \boldsymbol{G}^{-1} \frac{\partial H(\boldsymbol{z}, \boldsymbol{W})}{\partial \boldsymbol{W}} \tag{9.3.21}$$

in the stochastic gradient algorithm, then the convergence performance and numerical stability of the algorithm will be significantly improved.

Starting from the Riemann structure of the parametric space of the matrix $\boldsymbol{W}$, Yang and Amari[233] proved in 1997 that the natural selection of $\boldsymbol{G}$ is

$$\boldsymbol{G}^{-1} \frac{\partial H(\boldsymbol{z}, \boldsymbol{W})}{\partial \boldsymbol{W}} = \frac{\partial H(\boldsymbol{z}, \boldsymbol{W})}{\partial \boldsymbol{W}} \boldsymbol{W}^{\mathrm{T}} \boldsymbol{W}. \tag{9.3.22}$$

Thus, Eq. (9.3.19) of the standard algorithm can be improved to the standard (or real) natural gradient algorithm

$$\frac{\mathrm{d}\boldsymbol{W}}{\mathrm{d}t} = \eta \left(\boldsymbol{W}^{-\mathrm{T}} - \mathrm{E}\left\{ \boldsymbol{\phi}(\boldsymbol{y}) \boldsymbol{y}^{\mathrm{T}} \right\} \right) \boldsymbol{W}^{\mathrm{T}} \boldsymbol{W} = \eta \left(\boldsymbol{I} - \mathrm{E}\left\{ \boldsymbol{\phi}(\boldsymbol{y}) \boldsymbol{y}^{\mathrm{T}} \right\} \right) \boldsymbol{W}. \tag{9.3.23}$$

And Eq. (9.3.20) of the stochastic algorithm can also be modified to the stochastic natural gradient algorithm

$$\frac{\mathrm{d}\boldsymbol{W}}{\mathrm{d}t} = \eta \left(\boldsymbol{W}^{-\mathrm{T}} - \boldsymbol{\phi}(\boldsymbol{y}) \boldsymbol{y}^{\mathrm{T}} \right) \boldsymbol{W}^{\mathrm{T}} \boldsymbol{W} = \eta \left(\boldsymbol{I} - \boldsymbol{\phi}(\boldsymbol{y}) \boldsymbol{y}^{\mathrm{T}} \right) \boldsymbol{W}. \tag{9.3.24}$$

The standard gradient algorithm formula Eq. (9.3.19), the stochastic gradient algorithm formula Eq. (9.3.20), the standard natural gradient algorithm formula Eq. (9.3.23), and the natural stochastic gradient algorithm formula Eq. (9.3.19) using respectively the following gradients

(1) *Real or "absolute" gradient* : $\nabla f(\boldsymbol{W}) = \frac{\partial H(\boldsymbol{W})}{\partial \boldsymbol{W}}$;
(2) *Stochastic gradient* : the instantaneous value of $\hat{\nabla} f(\boldsymbol{W}) = \nabla f(\boldsymbol{W})$;
(3) *Real natural gradient* : $\frac{\partial H(\boldsymbol{W})}{\partial \boldsymbol{W}} \boldsymbol{W}^{\mathrm{T}} \boldsymbol{W} = \nabla f(\boldsymbol{W}) \boldsymbol{W}^{\mathrm{T}} \boldsymbol{W}$;
(4) *Stochastic natural gradient* : $\hat{\nabla} f(\boldsymbol{W}) \boldsymbol{W}^{\mathrm{T}} \boldsymbol{W}$;

Stochastic natural gradient and stochastic natural gradient algorithms are usually called as natural gradient and natural gradient algorithm for short, respectively.

The discrete form of the updation formula of the natural gradient algorithm is

$$\boldsymbol{W}_{k+1} = \boldsymbol{W}_k + \eta_k \left(\boldsymbol{I} - \boldsymbol{\phi}(\boldsymbol{y}_k) \boldsymbol{y}_k^{\mathrm{T}} \right) \boldsymbol{W}_k. \tag{9.3.25}$$

According to Eq. (9.3.23) of the real natural gradient algorithm, it is easy to know that the condition for the neural network to converge to the equilibrium point is

$$\mathrm{E}\left\{\boldsymbol{\phi}(\boldsymbol{y})\boldsymbol{y}^{\mathrm{T}}\right\} = \boldsymbol{I} \text{ or } \mathrm{E}\left\{\phi(y_i(t))\, y_j(t)\right\} = \delta_{ij}. \tag{9.3.26}$$

In addition to the natural gradient, there is also relative gradient[44]

$$\frac{\partial H(\boldsymbol{z}, \boldsymbol{W})}{\partial \boldsymbol{W}} \boldsymbol{W}^{\mathrm{T}}. \tag{9.3.27}$$

and the blind signal separation algorithm that uses relative gradient is called the Equivariant Adaptive Separating via Independence (EASI), which can be expressed as

$$\boldsymbol{W}_{k+1} = \boldsymbol{W}_k - \eta_k \left[\boldsymbol{y}_k \boldsymbol{y}_k^{\mathrm{T}} - \boldsymbol{I} + \boldsymbol{\phi}(\boldsymbol{y}_k)\boldsymbol{y}_k^{\mathrm{T}} - \boldsymbol{y}_k \boldsymbol{\phi}^{\mathrm{T}}(\boldsymbol{y}_k)\right] \boldsymbol{W}_k, \tag{9.3.28}$$

and was proposed by Cardoso and Leheld in 1996[44].

The normalized form of the EASI algorithm is known as the normalized EASI algorithm with the update formula[44]

$$\boldsymbol{W}_{k+1} = \boldsymbol{W}_k - \eta_k \left[\frac{\boldsymbol{y}_k \boldsymbol{y}_k^{\mathrm{T}} - \boldsymbol{I}}{1 + \eta_k \boldsymbol{y}_k \boldsymbol{y}_k^{\mathrm{T}}} + \frac{\boldsymbol{\phi}(\boldsymbol{y}_k)\boldsymbol{y}_k^{\mathrm{T}} - \boldsymbol{y}_k \boldsymbol{\phi}^{\mathrm{T}}(\boldsymbol{y}_k)}{1 + \eta_k |\boldsymbol{y}_k \boldsymbol{\phi}^{\mathrm{T}}(\boldsymbol{y}_k)|}\right] \boldsymbol{W}_k. \tag{9.3.29}$$

As a simplification and improvement of the EASI algorithm, Cruces et al. proposed the iterative inversion algorithm[9] in 2000

$$\boldsymbol{W}_{k+1} = \boldsymbol{W}_k - \eta_k \left[\boldsymbol{\phi}(\boldsymbol{y}_k)\boldsymbol{f}^{\mathrm{T}}(\boldsymbol{y}_k) - \boldsymbol{I}\right] \boldsymbol{W}_k, \tag{9.3.30}$$

and the corresponding normalized iterative inversion algorithm[9] is

$$\boldsymbol{W}_{k+1} = \boldsymbol{W}_k - \eta_k \left[\frac{\boldsymbol{\phi}(\boldsymbol{y}_k)\boldsymbol{f}^{\mathrm{T}}(\boldsymbol{y}_k) - \boldsymbol{I}}{1 + \eta_k |\boldsymbol{f}^{\mathrm{T}}(\boldsymbol{y}_k)\boldsymbol{\phi}(\boldsymbol{y}_k)|}\right] \boldsymbol{W}_k. \tag{9.3.31}$$

The iterative inversion algorithm incorporates the natural gradient algorithm and the ESAI algorithm as simplified algorithms:

(1) If $f(\boldsymbol{y}_k) = \boldsymbol{y}_k$, then Eq. (9.3.30) of the iterative inversion algorithm is simplified to Eq. (9.3.25) of the natural gradient algorithm, and Eq. (9.3.31) of the normal iterative inversion algorithm is simplified to the formula of the normalized natural gradient algorithm

$$\boldsymbol{W}_{k+1} = \boldsymbol{W}_k - \eta_k \left[\frac{\boldsymbol{\phi}(\boldsymbol{y}_k)\boldsymbol{y}_k^{\mathrm{T}} - \boldsymbol{I}}{1 + \eta_k |\boldsymbol{y}_k^{\mathrm{T}} \boldsymbol{\phi}(\boldsymbol{y}_k)|}\right] \boldsymbol{W}_k. \tag{9.3.32}$$

(2) If

$$\boldsymbol{\phi}(\boldsymbol{y}_k)\boldsymbol{f}^{\mathrm{T}}(\boldsymbol{y}_k) = \boldsymbol{y}_k \boldsymbol{y}_k^{\mathrm{T}} + \boldsymbol{\phi}(\boldsymbol{y}_k)\boldsymbol{y}_k^{\mathrm{T}} - \boldsymbol{y}_k \boldsymbol{\phi}^{\mathrm{T}}(\boldsymbol{y}_k), \tag{9.3.33}$$

then Eq. (9.3.30) of the iterative inversion algorithm becomes Eq. (9.3.28) of the EASI algorithm, and Eq. (9.3.31) of the normalized iterative inversion algorithm becomes Eq. (9.3.2) of the normalized EASI algorithm.

The EASI algorithm and the iterative inversion algorithm can be regarded as variations of the natural gradient algorithm.

These three gradient algorithms discussed previously are only suitable for the instances in which only sub-Gaussian signal or super-Gaussian signal is involved in the mixed signals. For the cases that both sub-Gaussian and super-Gaussian signals are mixed in the signal to be separated simultaneously, adaptive algorithms such as the generalized ICA algorithm[135] or the flexible ICA algorithm[55]. In general, these algorithms are very complicated.

9.3.3 Implementation of the Natural Gradient Algorithm

Let $\phi(\mathbf{y}(t)) = [\phi_1(y_1), \cdots, \phi_m(y_m)]^{\mathrm{T}}$ denote the nonlinear transform vector. When implementing the natural gradient algorithm, two practical problems must be solved, namely, the selection of activation function $\phi_i(y_i)$ and the selection of adaptive step η_k.

1. Selection of Activation Function

The equilibrium point of the ICA algorithm must be stable. A sufficient and necessary condition for the activation function $\phi_i(y_i)$ to satisfy the stability of equilibrium point is[13, 12]:

$$\mathrm{E}\left\{y_i^2\phi_i'(y_i)\right\} + 1 > 0, \tag{9.3.34}$$

$$\mathrm{E}\left\{\phi_i'(y_i)\right\} > 0, \tag{9.3.35}$$

$$\mathrm{E}\left\{y_i^2\right\}\mathrm{E}\left\{y_j^2\right\}\mathrm{E}\left\{\phi_i'(y_i)\right\}\mathrm{E}\left\{\phi_j'(y_j)\right\} > 0, \tag{9.3.36}$$

where $y_i = y_i(t)$ is the source signal extracted from the i-th output, and $\phi_i'(y_i) = \frac{\mathrm{d}\phi_i(y_i)}{\mathrm{d}y_i}$ is the first derivative of the activation function $\phi_i(y_i)$.

A commonly used activation function $\phi_i(y_i)$ whose equilibrium point satisfies the stationary condition is[10]:

$$\phi_i(y_i) = \begin{cases} \alpha y_i + \tanh(\gamma y_i), \ (\alpha > 0, \gamma > 2) & \text{for super-Gaussian signal,} \\ \alpha y_i + y_i^3, \ (\alpha > 0) & \text{for sub-Gaussian signal.} \end{cases} \tag{9.3.37}$$

And here we give several alternative of the activation function $\phi_i(y_i)$ as follows:

(1) Odd activation function[13]

$$\phi_i(y_i) = |y_i|^p \operatorname{sgn}(y_i), \ p = 1, 2, 3, \cdots, \tag{9.3.38}$$

where $\operatorname{sgn}(u)$ is the sign function.

(2) Symmetrical sigmoidal odd funciton[13]

$$\phi_i(y_i) = \tanh(\gamma y_i), \ \gamma > 0. \tag{9.3.39}$$

(3) Odd quadratic function[119]

$$\phi_i(y_i) = \begin{cases} y_i^2 + y_i, & y_i \geqslant 0, \\ -y_i^2 + y_i, & y_i < 0. \end{cases} \tag{9.3.40}$$

The following is the updation formula of the natural gradient algorithm proposed by Yang[233]

$$\kappa_3^i(t) = \kappa_3^i(t-1) - \mu(t)\left[\kappa_3^i(t-1) - y_i^3(t)\right], \tag{9.3.41}$$

$$\kappa_4^i(t) = \kappa_4^i(t-1) - \mu(t)\left[\kappa_4^i(t-1) - y_i^4(t) + 3\right], \tag{9.3.42}$$

$$\alpha_i\left(\kappa_3^i(t), \kappa_4^i(t)\right) = -\frac{1}{2}\kappa_3^i(t) + \frac{9}{4}\kappa_3^i(t)\kappa_4^i(t), \tag{9.3.43}$$

$$\beta_i\left(\kappa_3^i(t), \kappa_4^i(t)\right) = -\frac{1}{6}\kappa_4^i(t) + \frac{3}{2}\left(\kappa_3^i(t)\right)^2 + \frac{3}{4}\left(\kappa_4^i(t)\right)^2, \tag{9.3.44}$$

$$\phi_i(y_i(t)) = \alpha_i\left(\kappa_3^i(t), \kappa_4^i(t)\right) y_i^2(t) + \beta_i\left(\kappa_3^i(t), \kappa_4^i(t)\right) y_i^3(t), \tag{9.3.45}$$

$$\boldsymbol{W}(t+1) = \boldsymbol{W}(t) + \eta(t)\left(\boldsymbol{I} - \phi(\boldsymbol{y}(t))\boldsymbol{y}^{\mathrm{T}}(t)\right)\boldsymbol{W}(t). \tag{9.3.46}$$

2. Selection of the Adaptive Step

In the adaptive ICA algorithms, the selection of the learning rate η_k plays a key role in the convergence of the algorithm. The simplest way is to use a fixed learning rate. Similar to the general gradient algorithm, its disadvantage is that if the learning rate is large, the algorithm converges quickly, but the separation precision of the signal, (i.e., steady-state performance) is poor. On the contrary, if the learning is small, the steady-state performance is good, but the algorithm converges slowly. A better approach is to use a time-varying learning rate, namely, a variable step. Furthermore, variable step methods can be divided into non-adaptive variable step methods and adaptive variable step methods. For example, the learning rate based on annealing rules[10] and the learning rate of exponential decay[232] are two typical non-adaptive steps. The adaptive step, also known as the learning of learning rate, is a technique proposed by Amari as early as 1967[11]. In particular, here we mainly introduce several adaptive step methods for the ICA algorithm, such as the algorithm in which every weight coefficient is updated in an individual step respectively[59] and the variable step algorithm that is based on auxiliary variables[159].

Here we introduce a fuzzy inference system for adaptive learning step determination of blind signal separation[142]. Its basis is that the relationships among the second- and higher order correlation measures which reflect the separation state of source signals at each time instant. Based on this, the output signal with a larger correlation measure should use a larger learning rate to speed up its capturing, and the output signal with a smaller correlation measure should use a smaller learning rate to track finely.

Although the mutual information is a perfect measure of the dependence between the components of a blind signal separation neural network, it cannot be used to evaluate the dependence between output components at separation stages due to the use of the unknown probability density distribution of each source signal. Therefore, it is necessary to introduce another measure available for practical applications. These practical measures can be classified into second-order correlation measure r_{ij} and higher-order correlation measure hr_{ij} which is defined as

$$r_{ij} = \frac{C_{ij}}{\sqrt{C_{ii}C_{jj}}} = \frac{\operatorname{cov}\left[y_i(t), y_j(t)\right]}{\operatorname{cov}[y_i(t)]\operatorname{cov}\left[y_j(t)\right]}, \quad i,j = 1, \cdots, l;\ i \neq j, \tag{9.3.47}$$

$$hr_{ij} = \frac{HC_{ij}}{\sqrt{H_{ii}C_{jj}}} = \frac{\operatorname{cov}\left[\phi(y_i(t)), y_j(t)\right]}{\operatorname{cov}[\phi(y_i(t))]\operatorname{cov}\left[y_j(t)\right]}, \quad i,j = 1, \cdots, l;\ i \neq j, \tag{9.3.48}$$

where

$$\bar{m}_x = \frac{1}{N}\sum_{t=1}^{N} x(t), \tag{9.3.49}$$

$$\operatorname{cov}[x(t)] = \frac{1}{N}\sum_{t=1}^{N} |x(t) - \bar{m}_x|^2, \tag{9.3.50}$$

$$\operatorname{cov}[x(t), y(t)] = \frac{1}{N}\sum_{t=1}^{N} [x(t) - \bar{m}_x]\,[y(t) - \bar{m}_y]^* . \tag{9.3.51}$$

The adaptive updation formula of the second-order correlation and higher-order correlation can be summarized, respectively, as[142]

$$\bar{m}_{y_i}(t) = \lambda\frac{t-1}{t}\bar{m}_{y_i}(t-1) + \frac{1}{t}y_i(t), \tag{9.3.52}$$

$$\Delta_{y_i}(t) = \bar{m}_{y_i}(t) - \bar{m}_{y_i}(t-1), \tag{9.3.53}$$

$$\begin{aligned} C_{ij}(t) = \lambda\frac{t-1}{t}&\left[C_{ij}(t-1) + \Delta_{y_i}(t)\Delta_{y_j}(t)\right] \\ &+ \frac{1}{t}\left[y_i(t) - \bar{m}_{y_i}(t)\right]\left[y_j(t) - \bar{m}_{y_j}(t)\right]^*, \end{aligned} \tag{9.3.54}$$

and

$$\bar{m}_{\phi_i}(t) = \lambda\frac{t-1}{t}\bar{m}_{\phi_i}(t-1) + \frac{1}{t}\phi(y_i(t)), \tag{9.3.55}$$

$$\Delta_{\phi_i}(t) = \bar{m}_{\phi_i}(t) - \bar{m}_{\phi_i}(t-1), \tag{9.3.56}$$

$$\begin{aligned} HC_{ij}(t) = \lambda\frac{t-1}{t}&\left[HC_{ij}(t-1) + \Delta_{\phi_i}(t)\Delta_{\phi_j}(t)\right] \\ &+ \frac{1}{t}\left[\phi(y_i(t)) - \bar{m}_{\phi_i}(t)\right]\left[y_j(t) - \bar{m}_{y_j}(t)\right]^*, \end{aligned} \tag{9.3.57}$$

$$H_{ii}(t) = \lambda\frac{t-1}{t}\left[H_{ii}(t-1) + \Delta^2_{\phi_i}(t)\right] + \frac{1}{t}\left[\phi(y_i(t)) - \bar{m}_{\phi_i}(t)\right]^2, \tag{9.3.58}$$

where $i, j = 1, \cdots, l$, and λ is the forgetting factor.

Let

$$D_i(t) \overset{\text{def}}{=} D(y_i(t)) = \sqrt{\frac{1}{l-1} \sum_{j=1, j\neq i}^{l} r_{ij}^2(t)}, \ i = 1, \cdots, l, \tag{9.3.59}$$

$$HD_i(t) \overset{\text{def}}{=} HD(y_i(t)) = \sqrt{\frac{1}{l-1} \sum_{j=1, j\neq i}^{l} hr_{ij}^2(t)}, \ i = 1, \cdots, l \tag{9.3.60}$$

represent the second-order correlation measure and higher-order correlation measure of $y_i(t)$ relative to all the other output components $y_j(t)$, $j \neq i$.

According to the second-order correlation measure and higher-order correlation measure, the fuzzy rule for blind signal separation can be obtained as follows.

(1) If both $D_i(t)$ and $HD_i(t)$ are sufficiently small, we may say that the output component $y_i(t)$ are almost independent of all the other output components, i.e., the separation state of $y_i(t)$ from the other components is good.
(2) If either $D_i(t)$ or $HD_i(t)$ is not small, then the output component $y_i(t)$ is at least correlated with another output component, i.e., the separation state of $y_i(t)$ is not good.
(3) If both $D_i(t)$ and $HD_i(t)$ is large, we may consider $y_i(t)$ to be correlated strongly with the other components, i.e., the separation state of $y_i(t)$ is worse.

Specifically, the implementation rules of the fuzzy rules for blind signal separation may be described as follows[142]

(1) If $D_i(t)$ and $HD_i(t)$ are less than 0.1, then the separation state is quite good.
(2) If $D_i(t)$ and $HD_i(t)$ are located at the interval $[0.1, 0.2]$, then the moderate separation state is reached.
(3) If $D_i(t)$ and $HD_i(t)$ are greater than 0.2, then the separation state is poor.

If the second-order correlation measure r_{ij} of two separated output components satisfies $r_{ij} \approx 1$, then these two components are copies of each other, and thus the j-th row of the demixing matrix should be deleted from the matrix in order to remove the redundant output component $y_j(t)$. Hence, the $m \times n$ demixing matrix $\boldsymbol{W}$ reduced to a $l \times n$ matrix, where $l = m - 1$. With this procedure continues, the rows of the $l \times n$ demixing matrix will be reduced to $l = m - 1, \cdots, n$ successively until a number n satisfies that there are exactly n independent output components separated from the mixed signals.

3. Performance Evaluation

In the research and implementation of adaptive blind signal separation, it is often desired to compare the performance of several different algorithms. A commonly used evaluation metric is the root-mean-square error (RMSE) of the cross-talking error[234]

$$E = \sum_{i=1}^{l} \left(\sum_{j=1}^{n} \frac{|b_{ij}|}{\max_k |b_{ik}|} - 1 \right) + \sum_{j=1}^{n} \left(\sum_{i=1}^{l} \frac{|b_{ij}|}{\max_k |b_{kj}|} - 1 \right), \tag{9.3.61}$$

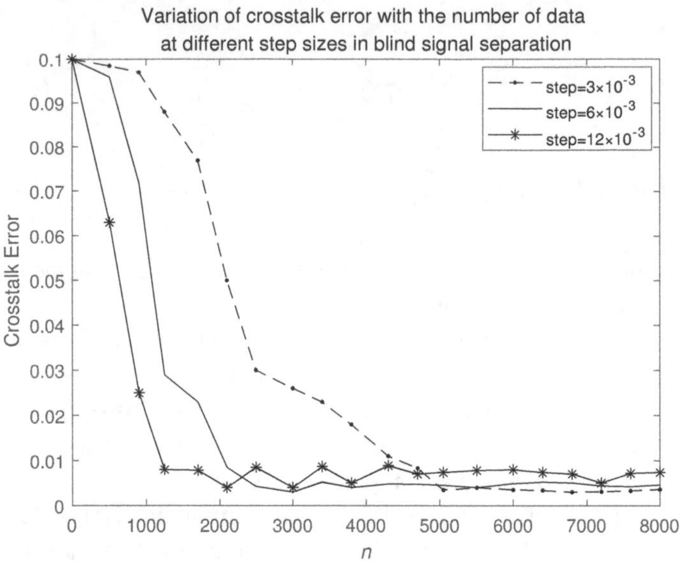

Fig. 9.3.1: Cross-talking error Vs. length of sample N

where $\boldsymbol{B} = [b_{ij}] = \boldsymbol{WA}$ is a matrix obtaind by right mulitplying the $l \times m$ separating matrix $\boldsymbol{W}$ with a $m \times m$ mixing matrix $\boldsymbol{A}$, and $\max_k |b_{ik}| = \{|b_{i1}|, \cdots, |b_{in}|\}$ and $\max_k |b_{kj}| = \{|b_{1j}|, \cdots, |b_{lj}|\}$, respectively.

As illustrated by Fig.9.3.1, by plotting the cross-talking error curve with the length of sample N, we may intuitively compare the convergence and tracking performance of the same blind separation algorithm with different steps or different blind separation algorithms.

9.3.4 Fixed-Point Algorithm

For cases where adaptive separation is not required, the fixed-point algorithm proposed by Hyvarinen[108] is a fast and numerically stable ICA algorithm. The contrast function of this algorithm is defined as

$$J_G(\boldsymbol{w}) = \left[\mathrm{E}\left\{G\left(\boldsymbol{w}^{\mathrm{T}}\boldsymbol{x}\right)\right\} - \mathrm{E}\{G(\boldsymbol{v})\}\right]^2, \tag{9.3.62}$$

where $G(\cdot)$ is practically any nonquadratic function, and $\boldsymbol{v}$ is a Gaussian random vector of zero mean and unit variance, and $\boldsymbol{w}$ is a weight vector satisfies $\mathrm{E}\left\{\left(\boldsymbol{w}^{\mathrm{T}}\boldsymbol{x}\right)^2\right\} = 1$. Based on the principle of separated one-by-one, the ICA of n source signals becomes the solution of n constrained optimization subproblems formulated as

$$w_i = \arg\min \sum_{i=1}^{n} J_G(w_i), \; i = 1, \cdots, n \tag{9.3.63}$$

with constraint $\mathrm{E}\left\{\left(\boldsymbol{w}_k^{\mathrm{T}}\boldsymbol{x}\right)\left(\boldsymbol{w}_j^{\mathrm{T}}\boldsymbol{x}\right)\right\} = \delta_{jk}$.

The solution to the above optimization problem can be obtained using the fixed-point ICA algorithm

$$\boldsymbol{w}_{p+1} \leftarrow \boldsymbol{w}_{p+1} - \sum_{j=1}^{p} \boldsymbol{w}_{p+1}^{\mathrm{T}} \boldsymbol{C} \boldsymbol{w}_j \boldsymbol{w}_j, \tag{9.3.64}$$

$$\boldsymbol{w}_{p+1} \leftarrow \frac{\boldsymbol{w}_{p+1}}{\sqrt{\boldsymbol{w}_{p+1}^{\mathrm{T}} \boldsymbol{C} \boldsymbol{w}_{p+1}}}, \tag{9.3.65}$$

where $\boldsymbol{C} = \mathrm{E}\left\{\boldsymbol{x}\boldsymbol{x}^{\mathrm{T}}\right\}$ is the covariance matrix of the observation data. Eqs. (9.3.64) and (9.3.65) can be computed iteratively until $\boldsymbol{w}_{p+1}$ converges. In addition, it should be noted that data $\boldsymbol{x}$ need to be prewhitened in the fixed-point ICA algorithm.

The following are three choices of the contrast function $G(\cdot)$ used in the fixed-point ICA algorithm

$$G_1(u) = \frac{1}{a_1} \log\cosh(a_1 u), \tag{9.3.66}$$

$$g_1(u) = \tanh(a_1 u), \tag{9.3.67}$$

$$G_2(u) = -\frac{1}{a_2} \exp\left(-a_2 u^2/2\right), \tag{9.3.68}$$

$$g_2(u) = u \exp\left(-a_2 u^2/2\right), \tag{9.3.69}$$

$$G_3(u) = \frac{1}{4} u^4, \tag{9.3.70}$$

$$g_3(u) = u^3, \tag{9.3.71}$$

where $g_i(u)$ is the first derivative of the corresponding contrast function $G_i(u)$.

Here are also some considerations for selecting the contrast function $G(u)$:

(1) $G_1(u)$ is suitable for the cases where both the sub-Gaussian and super-Gaussian signals coexist.

(2) When the independent source signals are super-Gaussian signals with very large kurtosis or when robustness is very important, $G_2(u)$ may be better.

(3) Using $G_3(u)$ for estimating sub-Gaussian signals.

The core of the fixed-point ICA algorithm is the compression mapping performed in Eq. (9.3.64). The idea of using compression mapping techniques to separate signals one-by-one in ICA was first proposed in the literature [73]. Although this literature introduces the adaptive ICA algorithm based on compression mapping, the effectiveness of the algorithm still needs to be proved. The fixed-point ICA algorithm is also known as the fast ICA algorithm.

9.4 Nonlinear Principal Component Analysis

The main idea of independent component analysis is to make each component of the output vector $\boldsymbol{y}(t) = \boldsymbol{W}\boldsymbol{x}(t)$ as statistically independent as possible by minimizing mutual information or maximizing entropy.

Strict statistical independence means that the statistics of each order are uncorrelated. If the cross-correlation function between two stochastic processes is statistically uncorrelated and a higher-order cross-correlation function (third-order or fourth-order cross-correlation) between them is also statistically uncorrelated, the two stochastic processes can be approximated to be statistically independent.

That the cross-correlation function of each component of the output vector $\boldsymbol{z}(t) = \boldsymbol{B}\boldsymbol{x}(t)$ to be statistically uncorrelated may be realized easily by means of pre-whitening. If each component of the output vector is subjected to appropriate nonlinear transformation, the statistical uncorrelation of the third-order or fourth-order cross-correlation function of each component can be realized. To sum up, pre-whitening and nonlinear transformation is the basic idea of the nonlinear principal component analysis method for blind signal separation.

9.4.1 Pre-whitening

Assume that the mean value of the $m \times 1$ observation data vector $\boldsymbol{x}(t)$ is already zeroing. Pre-whitening is a commonly used technique in the field of signal processing field, the procedure of which may be described as

(1) Compute the $m \times m$ autocorrelation matrix $\hat{\boldsymbol{R}} = \frac{1}{N}\sum_{t=1}^{N} \boldsymbol{x}(t)\boldsymbol{x}^{\mathrm{H}}(t)$ according to the $m \times 1$ observation vector $\boldsymbol{x}(t)$.
(2) Compute the eigenvalue decomposition of $\hat{\boldsymbol{R}}$ and denote the greatest n eigenvalues as $\lambda_1, \cdots, \lambda_n$ and the corresponding eigenvector as $\boldsymbol{h}_1, \cdots, \boldsymbol{h}_n$.
(3) Take the mean of the $m - n$ smallest eigenvalue of $\hat{\boldsymbol{R}}$ as the variance estimation $\hat{\sigma}^2$ of the additive white noise.
(4) Compute the $n \times n$ pre-whitened data vector $\boldsymbol{z}(t) = [z_1(t), \cdots, z_n(t)]^{\mathrm{T}} = \boldsymbol{B}\boldsymbol{x}(t)$, where $\boldsymbol{B} = \left[(\lambda_1 - \hat{\sigma}^2)\boldsymbol{h}_1, \cdots, (\lambda_n - \hat{\sigma}^2)\boldsymbol{h}_n\right]^{\mathrm{H}}$ is a $m \times n$ pre-whiten matrix.

Now, the components of the $n \times n$ pre-whitened data vector $\boldsymbol{z}(t) = \boldsymbol{B}\boldsymbol{x}(t)$ are already second-order statistically uncorrelated. The next step of blind signal separation is how to make each component of the separated output $\boldsymbol{y} = \boldsymbol{W}\boldsymbol{z}(t)$ be higher-order statistically uncorrelated with each other, so as to realize the approximate statistical independence between each component of the separated output.

9.4.2 Linear Principal Component Analysis

The purpose of blind signal separation is to separate or extract n (where $n \leqslant m$) source signals from m observation signals. From the perspective of information theory, there is information redundancy in the m observation signals, which span the m-dimensional observation space, while the separated n signals form a lower dimensional space (n-dimensional signal space) without information redundancy. The linear transformation from a high-dimensional space with information redundancy to a low-dimensional space without information redundancy is called dimension reduction. A commonly used method of dimension reduction is principal component analysis (PCA).

By performing transformation using the orthogonal matrix $\boldsymbol{Q}$, the observation signals $\boldsymbol{x}(t)$ with additive white noise, which are not statistically independent can be transformed into m components $\boldsymbol{y}(t) = \boldsymbol{Q}[\boldsymbol{x}(t) + \boldsymbol{e}(t)]$ which are orthogonal to each other. Among these resulted m new components, n signal components that have higher power may be regarded as the n principal components of the m new components, which are also named principal components for short. Correspondingly, data or signal analysis using only n principal components in the m-dimensional data vector is called principal component analysis.

Definition 9.4.1. *Let $\boldsymbol{R}_x$ be the autocorrelation matrix of a m-dimensional data vector $\boldsymbol{x}(t)$ and assume $\boldsymbol{R}_x$ has n principal eigenvalues. The n eigenvector corresponding to these principal eigenvalues are called the principal components of the data vector $\boldsymbol{x}(t)$.*

The main steps and ideas of PCA are as follows.

(1) *Dimension Reduction* Synthesize n principal components from m random variables using

$$\tilde{x}_j(t) = \sum_{i=1}^{m} h_{ij}^{*} x_i(t) = \boldsymbol{h}_j^{\mathrm{H}} \boldsymbol{x}(t), \quad j = 1, \cdots, n, \tag{9.4.1}$$

where $\boldsymbol{h}_j = \left[h_{1j}, \cdots, h_{mj}\right]^{\mathrm{T}}$ and $\boldsymbol{x}(t) = [x_1(t), \cdots, x_m(t)]^{\mathrm{T}}$, respectively.

(2) *Orthgonalization* To make the principal components orthogonal to each other and each with unit variance (normalization), i.e.,

$$\langle x_i(t), x_j(t)\rangle = \delta_{ij} = \begin{cases} 1, & i = j \\ 0, & \text{otherwise} \end{cases} \tag{9.4.2}$$

then it is known from

$$\langle \tilde{x}_i(t), \tilde{x}_j(t)\rangle = \boldsymbol{x}(t)^{\mathrm{H}} \boldsymbol{h}_i^{\mathrm{H}} \boldsymbol{h}_j \boldsymbol{x}(t) = \begin{cases} 1, & i = j \\ 0, & i \neq j \end{cases} \tag{9.4.3}$$

that the coefficient vector $\boldsymbol{h}_i$ must be selected to satisfy the normal orthogonal condition $\boldsymbol{h}_i^{\mathrm{H}} \boldsymbol{h}_j = \delta_{ij}$

(3) *Power maximization* If we choose $\boldsymbol{h}_i = \boldsymbol{u}_i,\ i = 1, \cdots, n,$, where $\boldsymbol{u}_i,\ (i = 1, \cdots, n)$ is the eigenvectors corresponding to the n large eigenvalues $\lambda_1 \geqslant \cdots \geqslant \lambda_n$ of the

autocorrelation matrix $\boldsymbol{R}_x = \mathrm{E}\{\boldsymbol{x}(t)\boldsymbol{x}^{\mathrm{H}}(t)\}$, it is easy to compute the energy of each non-redundant component as

$$\begin{aligned} E_{\tilde{x}_i} &= \mathrm{E}\{|\tilde{x}_i(t)|^2\} = \mathrm{E}\left\{\boldsymbol{h}_i^{\mathrm{H}}\boldsymbol{x}(t)\left[\boldsymbol{h}_i^{\mathrm{H}}\boldsymbol{x}(t)\right]^*\right\} \\ &= \boldsymbol{u}_i^{\mathrm{H}}\mathrm{E}\{\boldsymbol{x}(t)\boldsymbol{x}^{\mathrm{H}}(t)\}\boldsymbol{u}_i = \boldsymbol{u}_i^{\mathrm{H}}\boldsymbol{R}_x\boldsymbol{u}_i \\ &= \boldsymbol{u}_i^{\mathrm{H}}[\boldsymbol{u}_1, \boldsymbol{u}_2, \cdots, \boldsymbol{u}_m]\begin{bmatrix} \lambda_1 & & & 0 \\ & \lambda_2 & & \\ & & \ddots & \\ 0 & & & \lambda_m \end{bmatrix}\begin{bmatrix} \boldsymbol{u}_1^{\mathrm{H}} \\ \boldsymbol{u}_2^{\mathrm{H}} \\ \vdots \\ \boldsymbol{u}_m^{\mathrm{H}} \end{bmatrix}\boldsymbol{u}_i \\ &= \lambda_i . \end{aligned} \tag{9.4.4}$$

Since the eigenvalues are arranged in nondegenerate order so

$$E_{\tilde{x}_1} \geqslant E_{\tilde{x}_2} \geqslant \cdots \geqslant E_{\tilde{x}_n} . \tag{9.4.5}$$

Therefore, according to the magnitude of energy, $\tilde{x}_1(t)$ is often called the first principal component, $\tilde{x}_2(t)$ the second principal component and so on.
Note the $m \times m$ autocorrelation matrix

$$\boldsymbol{R}_x = \mathrm{E}\left\{\boldsymbol{x}(t)\boldsymbol{x}^{\mathrm{H}}(t)\right\} = \begin{bmatrix} \mathrm{E}\{|x_1(t)|^2\} & \mathrm{E}\{x_1(t)x_2^*(t)\} & \cdots & \mathrm{E}\{x_1(t)x_m^*(t)\} \\ \mathrm{E}\{x_2(t)x_1^*(t)\} & \mathrm{E}\{|x_2(t)|^2\} & \cdots & \mathrm{E}\{x_2(t)x_m^*(t)\} \\ \vdots & \vdots & \ddots & \vdots \\ \mathrm{E}\{x_m(t)x_1^*(t)\} & \mathrm{E}\{x_m(t)x_2^*(t)\} & \cdots & \mathrm{E}\{|x_m(t)|^2\} \end{bmatrix} , \tag{9.4.6}$$

and use the definition and property of matrix trace to know

$$\mathrm{tr}(\boldsymbol{R}_x) = \mathrm{E}\left\{|x_1(t)|^2\right\} + \mathrm{E}\left\{|x_2(t)|^2\right\} + \cdots + \mathrm{E}\left\{|x_m(t)|^2\right\} = \lambda_1 + \lambda_2 + \cdots + \lambda_m . \tag{9.4.7}$$

Furthermore, if the autocorrelation matrix $\boldsymbol{R}_x$ has only n large eigenvalues, then there is

$$\mathrm{E}\left\{|x_1(t)|^2\right\} + \mathrm{E}\left\{|x_2(t)|^2\right\} + \cdots + \mathrm{E}\left\{|x_m(t)|^2\right\} \approx \lambda_1 + \lambda_2 + \cdots + \lambda_n . \tag{9.4.8}$$

In summary, it can be concluded that the basic idea of principal component analysis is to transform the original m statistically correlated random data into n mutually orthogonal principal components through three steps including dimension reduction, orthogonalization, and power maximization, and the sum of the energy of these principal components should be approximately equal to the sum of the energy of the original m random data.

Definition 9.4.2. [229] *Let $\boldsymbol{R}_x$ be the autocorrelation matrix of a m-dimensional data vector $\boldsymbol{x}$ and assume it has m pricipal eigenvalues and $m - n$ minor (i.e., smaller) eigenvalues.*

Those $m - n$ eigenvectors that corresponding to the $m - n$ minor eigenvalues is called the minor component of the data vector $\mathbf{x}(t)$.

Data analysis or signal analysis using only $m - n$ minor components of the data vector is called minor component analysis (MCA). The principal component analysis can give the contour and main information of the analyzed signal and image. In contrast, the minor component analysis may provide signal details and image texture. The minor component analysis is widely used in many fields. For example, minor component analysis has been used for frequency estimation[148, 149], blind beamforming[95], moving target indication[124], clutter cancellation[18], etc. In pattern recognition, when principal component analysis could not recognize two object signals, the minor component analysis should be further conducted to compare the details of the information they contain.

By constructing the cost function of principal component analysis

$$J(\boldsymbol{W}) = \mathrm{E}\left\{\| \boldsymbol{x}(t) - \boldsymbol{W}\boldsymbol{W}^{\mathrm{H}}\boldsymbol{x}(t) \|^2\right\}, \tag{9.4.9}$$

reference [231] proves that the solution of the optimization problem $\min J(\boldsymbol{W})$ will hold several important properties as follows:

(1) The global minimum of the objective function $J(\boldsymbol{W})$ is given by $\boldsymbol{W} = \boldsymbol{U}_r\boldsymbol{Q}$, where $\boldsymbol{U}_r$ is composed of r pricipal eigenvectors of the autocorrelation matrix $\boldsymbol{R} = \mathrm{E}\left\{\boldsymbol{x}\boldsymbol{x}^{\mathrm{H}}\right\}$, and $\boldsymbol{Q}$ is an arbitrary unitary matrix.
(2) The resulted solution matrix $\boldsymbol{W}$ must be a quasiunitary (or semiorthogonal) matrix, i.e., $\boldsymbol{W}^{\mathrm{H}}\boldsymbol{W} = \boldsymbol{I}$.

The above important properties show that, if PCA is used for blind signal processing, it will bring the following advantages or benefits:

(1) The stationary point when the blind signal separation algorithm converges must be the global minimum point of the optimization algorithm it uses.
(2) Since the solution of the blind signal separation algorithm is $\boldsymbol{W} = \boldsymbol{U}_r\boldsymbol{Q}$, the separated signal can be obtained directly by

$$\hat{\boldsymbol{s}}(t) = \boldsymbol{W}^{\dagger}\boldsymbol{x}(t) = \left(\boldsymbol{W}^{\mathrm{H}}\boldsymbol{W}\right)^{-1}\boldsymbol{W}\boldsymbol{x}(t) = \boldsymbol{W}^{\mathrm{H}}\boldsymbol{x}(t), \tag{9.4.10}$$

and it is unnecessary to compute the Moore-Penrose inverse matrix $\boldsymbol{W}^{\dagger}$, which simplifies the calculation.

Define the exponentially weighted objective function

$$\begin{aligned} J_1(\boldsymbol{W}(t)) &= \sum_{i=1}^{t} \beta^{t-i} \| \boldsymbol{x}(i) - \boldsymbol{W}(t)\boldsymbol{W}^{\mathrm{H}}(t)\boldsymbol{x}(i) \|^2 \\ &= \sum_{i=1}^{t} \beta^{t-i} \| \boldsymbol{x}(i) - \boldsymbol{W}(t)\boldsymbol{y}(i) \|^2, \end{aligned} \tag{9.4.11}$$

where $0 \leqslant \beta \leqslant 1$ is a forgetting factor, and $\boldsymbol{y}(i) = \boldsymbol{W}^H(t)\boldsymbol{x}(i)$.

Based on the exponentially weighted objective function $J_1(\boldsymbol{W}(t))$, Yang proposed a Projection Approximation Subspace Tracking (PAST) algorithm for linear principal component analysis in 1995[231].

Algorithm 9.4.1. *PAST algorithm for linear principal component analysis*
Step 1 Choose $\boldsymbol{P}(0)$ *and* $\boldsymbol{W}(0)$ *suitably*
Step 2 For $t = 1, 2, \cdots$, *compute*

$$\begin{aligned}
\boldsymbol{y}(t) &= \boldsymbol{W}^H(t-1)\boldsymbol{x}(t) \\
\boldsymbol{h}(t) &= \boldsymbol{P}(t-1)\boldsymbol{y}(t) \\
\boldsymbol{g}(t) &= \frac{\boldsymbol{h}(t)}{\left[\beta + \boldsymbol{y}^H(t)\boldsymbol{h}(t)\right]} \\
\boldsymbol{P}(t) &= \frac{1}{\beta}\mathrm{Tri}\left[\boldsymbol{P}(t-1) - \boldsymbol{g}(t)\boldsymbol{h}^H(t)\right]
\end{aligned}$$

$$\begin{aligned}
\boldsymbol{e}(t) &= \boldsymbol{x}(t) - \boldsymbol{W}(t-1)\boldsymbol{y}(t) \\
\boldsymbol{W}(t) &= \boldsymbol{W}(t-1) + \boldsymbol{e}(t)\boldsymbol{g}^H(t)
\end{aligned}$$

where the operator $\mathrm{Tri}[\boldsymbol{A}]$ *indicates that only the upper (or lower) triangular part of matrix* $\boldsymbol{A}$ *is calculated and its Hermitian transposed version is copied to another lower (or upper) triangular part.*

9.4.3 Nonlinear Principal Component Analysis

Linear Principal component analysis can not be applied directly to the blind signal separation, because the principal components of the autocorrelation matrix $\boldsymbol{R}_x$ are only the principal components of the second-order statistics, which is not sufficient for the blind separation of non-Gaussian signals.

In order to solve the problem of blind separation of non-Gaussian signals, a nonlinear transformation

$$\boldsymbol{y}(t) = g\left(\boldsymbol{W}^{\mathrm{H}}\boldsymbol{x}(t)\right) \tag{9.4.12}$$

must be introduced before applying the principal component analysis method, where $g(\boldsymbol{u}(t)) = [g(u_1(t)), \cdots, g(u_m(t))]^{\mathrm{T}}$ is the nonlinear transformation vector.

The purpose of performing the nonlinear transformation is to introduce the higher-order statistics of a stochastic signal into the analysis.

The vector $\boldsymbol{y}(t)$ of the linear principal component analysis objective function Eq. (9.4.11) is replaced by the nonlinear transformation $\boldsymbol{y}(t) = g\left(\boldsymbol{W}^{\mathrm{H}}\boldsymbol{x}(t)\right)$, i.e., the objective function of the exponentially weighted nonlinear principal component analysis is obtained[119]. Therefore, the PAST algorithm for linear principal component analysis algorithm is generalized to the PAST algorithm for nonlinear principal component analysis[119].

Algorithm 9.4.2. *PAST algorithm for nonlinear principal component analysis of blind signal separation*
For $t = 1, 2, \cdots$, *compute*

$$\begin{aligned}
\boldsymbol{y}(t) &= g\left(\boldsymbol{W}^H(t-1)\boldsymbol{x}(t)\right)\\
\boldsymbol{h}(t) &= \boldsymbol{P}(t-1)\boldsymbol{y}(t)\\
\boldsymbol{g}(t) &= \frac{\boldsymbol{h}(t)}{\left[\beta + \boldsymbol{y}^H(t)\boldsymbol{h}(t)\right]}\\
\boldsymbol{P}(t) &= \frac{1}{\beta}\mathrm{Tri}\left[\boldsymbol{P}(t-1) - \boldsymbol{g}(t)\boldsymbol{h}^H(t)\right]\\
\boldsymbol{e}(t) &= \boldsymbol{x}(t) - \boldsymbol{W}^H(t-1)\boldsymbol{y}(t)\\
\boldsymbol{W}(t) &= \boldsymbol{W}(t-1) + \boldsymbol{e}(t)\boldsymbol{g}^H(t)
\end{aligned}$$

where the nonlinear transform can be an odd quadratic function[132]*, i.e.,*

$$g(u) = -\frac{E\left\{|u|^2\right\}p_u'(u)}{p_u(u)}, \tag{9.4.13}$$

where $p_u(u)$ *and* $p_u'(u)$ *are the probability density function of random variable* u *and its first derivative.*

In a simple case, the nonlinear transformation function can be[119]*:*

$$g(u) = \begin{cases} u^2 + u, & u \geqslant 0 \\ -u^2 + u, & u < 0 \end{cases}. \tag{9.4.14}$$

Here, we may also make a brief comparison between the PAST algorithm of nonlinear principal component analysis and the natural gradient algorithm for blind signal separation as follows:

(1) The natural gradient algorithm belongs to the LMS algorithm, while the PAST algorithm belongs to the RLS algorithm. Generally speaking, LMS algorithms is a point update without memory. Since only the data at the current time is used, the utilization rate of information is low, so the convergence is slow. The RLS algorithm is a block update with memory, which uses the data (blocks) of the current and several previous moments, and the utilization rate of information is high. Therefore, the RLS algorithm converges faster than the LMS algorithm.
(2) The natural gradient algorithm and nonlinear principal component analysis both use the nonlinear transformation of signal to introduce higher-order statistics.

9.5 Joint Diagonalization of Matrices

In addition to the independent component analysis (ICA) method and the nonlinear principal component analysis (NPCA) method, the blind signal separation problem can also be solved by the joint diagonalization of matrices.

9.5.1 Blind Signal Separation and Joint Diagonalization of Matrices

Considering the array received signal model

$$\boldsymbol{x}(n) = \boldsymbol{A}\boldsymbol{s}(n) + \boldsymbol{v}(n), \quad n = 1, 2, \cdots \tag{9.5.1}$$

in the presence of additive noise, the main purpose of blind signal separation is to find the demixing matrix $\boldsymbol{W} \doteq \boldsymbol{A} = \boldsymbol{AG}$ that is essentially equal to the mixing matrix $\boldsymbol{A}$, and then realize blind signal separation by $\hat{s}(n) = \boldsymbol{W}^{\dagger}\boldsymbol{x}(n)$.

Two mainstream algorithms, ICA and NPCA, for finding the separating matrix $\boldsymbol{W}$ which is essentially equal to the mixing matrix $\boldsymbol{A}$ were introduced in the previous section, respectively. Next, another mainstream method, the joint diagonalization of matrices, is discussed below.

Unlike the ICA and NPCA methods, the assumptions on additive noise are more relaxed here.

(1) The additive noise is a temporally white, spatially colored Gaussian noise, i.e., the autocorrelation matrix is

$$\boldsymbol{R}_v(k) = \mathrm{E}\left\{\boldsymbol{v}(t)\boldsymbol{v}^{\mathrm{H}}(t-k)\right\} = \delta(k)\boldsymbol{R}_v = \begin{cases} \boldsymbol{R}_v, & k = 0 \quad \text{(Spatially colored)} \\ \boldsymbol{O}, & k \neq 0 \quad \text{(Temporally white)} \end{cases}.$$

The temporally white means that the additive noise on each sensor is Gaussian white noise, while the spatially colored means that the additive Gaussian white noise of different sensors may be correlated.

(2) The n source signals are statistically independent, i.e., $\mathrm{E}\left\{\boldsymbol{s}(t)\boldsymbol{s}^{\mathrm{H}}(t-k)\right\} = \boldsymbol{D}_k$ (Diagonal matrix).

(3) The source signals independent of the additive noise, i.e., $\mathrm{E}\left\{\boldsymbol{s}(t)\boldsymbol{v}^{\mathrm{H}}(t-k)\right\} = \boldsymbol{O}$ (Zero matrix).

Under the above assumptions, the autocorrelation matrix of the array output vector is

$$\begin{aligned} \boldsymbol{R}_x(k) &= E\left\{\boldsymbol{x}(t)\boldsymbol{x}^{\mathrm{H}}(t-k)\right\} \\ &= \mathrm{E}\left\{\left[\boldsymbol{A}\boldsymbol{s}(t) + \boldsymbol{v}(t)\right]\left[\boldsymbol{A}\boldsymbol{s}(t-k) + \boldsymbol{v}(t-k)\right]^{\mathrm{H}}\right\} \\ &= \boldsymbol{A}E\left\{\boldsymbol{s}(t)\boldsymbol{s}^{\mathrm{H}}(t-k)\right\}\boldsymbol{A}^{\mathrm{H}} + \mathrm{E}\left\{\boldsymbol{v}(t)\boldsymbol{v}^{\mathrm{H}}(t-k)\right\} \\ &= \begin{cases} \boldsymbol{A}\boldsymbol{D}_0\boldsymbol{A}^{\mathrm{H}} + \boldsymbol{R}_v, & k = 0 \\ \boldsymbol{A}\boldsymbol{D}_k\boldsymbol{A}^{\mathrm{H}}, & k \neq 0 \end{cases}. \end{aligned} \tag{9.5.2}$$

This result shows that the effect of the temporally white, spatially colored Gaussian noise $\boldsymbol{v}(n)$ can be completely suppressed if the K autocorrelation matrices $\boldsymbol{R}_x(k)$, $k = 1, \cdots, K$ without noise influence (lag $k \neq 0$) are used.

Given K autocorrelation matrices $\boldsymbol{R}_x = \boldsymbol{R}_x(k)$, $k = 1, \cdots, K$, the joint diagonalization is

$$\boldsymbol{R}_k = \boldsymbol{W}\boldsymbol{\Sigma}_k\boldsymbol{W}^{\mathrm{H}}, \quad k = 1, \cdots, K, \tag{9.5.3}$$

where $\boldsymbol{W}$ is called the joint diagonalizer of the K autocorrelation matrices $\boldsymbol{R}_1, \cdots, \boldsymbol{R}_k$.

Comparing Eqs. (9.5.2) and (9.5.3), it is easy to know that the joint diagonalizer $\boldsymbol{W}$ is not necessarily a mixing matrix, but must be a matrix that is essentially equivalent to the mixing matrix A,

$$\boldsymbol{W} \doteq \boldsymbol{A} = \boldsymbol{AG}, \tag{9.5.4}$$

where $\boldsymbol{G}$ is a generalized permutation matrix.

The above analysis shows that the demixing matrix (or separation matrix) in blind signal separation can also be obtained by the joint diagonalization of matrices. A prominent advantage of joint diagonalization is that it can theoretically completely suppress the effect of the temporally white, spatially colored additive Gaussian noise.

Once the mixing matrix has been solved by the joint diagonalization, blind signal separation $\hat{s}(n) = \boldsymbol{W}^{\dagger}\boldsymbol{x}(n)$ can be performed.

The joint diagonalization of multiple matrices was first proposed by Flury in 1984 when considering the common principal component analysis of K covariance matrices[160]. Later, Cardoso and Souloumiac[46] in 1996 and Belochrani et al.[25] in 1997 proposed the approximate joint diagonalization of multiple cumulant matrices and covariance matrices from the perspective of blind signal separation, respectively. Since then, joint diagonalization has been widely studied and applied in the field of blind signal separation.

The mathematical problem of joint diagonalization is: given K $m \times m$ symmetric matrix $\boldsymbol{A}_1, \cdots, \boldsymbol{A}_K$, the joint diagonalization seeks a $m \times n$ full column rank matrix $\boldsymbol{W}$ making the K matrices diagonalize at the same time (joint diagonalization)

$$\boldsymbol{A}_k - \boldsymbol{W}\boldsymbol{\Lambda}_k\boldsymbol{W}^{\mathrm{H}}, \quad k - 1, \cdots, K, \tag{9.5.5}$$

where $\boldsymbol{W} \in \mathbb{C}^{m\times n}$ is called as joint diagonalizer, and $\boldsymbol{\Lambda}_k \in \mathbb{R}^{n\times n}$, $k = 1, \cdots, K$ are diagonal matrices.

Joint diagonalization $\boldsymbol{A}_k = \boldsymbol{W}\boldsymbol{\Lambda}_k\boldsymbol{W}^{\mathrm{H}}$ is exact joint diagonalization. However, the actual joint diagonalization is an approximate joint diagonalization. Given the set of matrices $\mathcal{A} = \{\boldsymbol{A}_1, \cdots, \boldsymbol{A}_K\}$, the approximate joint diagonalization problem seeks a joint diagonalizer $\boldsymbol{W} \in \mathbb{C}^{m\times n}$ and K associated $n \times n$ diagonal matrices $\boldsymbol{\Lambda}_1, \cdots, \boldsymbol{\Lambda}_K$ to minimize the objective function[45, 46]

$$\min J_1(\boldsymbol{W}, \boldsymbol{\Lambda}_1, \cdots, \boldsymbol{\Lambda}_K) = \min \sum_{k=1}^{K} \alpha_k \parallel \boldsymbol{W}^{\mathrm{H}}\boldsymbol{A}_k\boldsymbol{W} - \boldsymbol{\Lambda}_k \parallel_F^2 \tag{9.5.6}$$

or[219, 235]

$$\min J_2(\boldsymbol{W}, \boldsymbol{\Lambda}_1, \cdots, \boldsymbol{\Lambda}_K) = \min \sum_{k=1}^{K} \alpha_k \parallel \boldsymbol{A}_k - \boldsymbol{W}\boldsymbol{\Lambda}_k\boldsymbol{W}^{\mathrm{H}} \parallel_F^2, \tag{9.5.7}$$

where $\alpha_1, \cdots, \alpha_K$ are the positive weights. For simplicity, assume $\alpha_1 = \cdots = \alpha_K = 1$.

9.5.2 Orthogonal Approximate Joint Diagonalization

The so-called orthogonal approximate joint diagonalization requires that the $m \times n$ ($m \geq n$) dimensional joint diagonalizer must be a semi-orthogonal matrix $\boldsymbol{W}^{\mathrm{H}}\boldsymbol{W} = \boldsymbol{I}_n$. Therefore, the orthogonal approximate joint diagonalization problem is a constrained optimization problem

$$\min J_1(\boldsymbol{W}, \boldsymbol{\Lambda}_1, \cdots, \boldsymbol{\Lambda}_K) = \min \sum_{k=1}^{K} \| \boldsymbol{W}^{\mathrm{H}}\boldsymbol{A}_k\boldsymbol{W} - \boldsymbol{\Lambda}_k \|_F^2, \tag{9.5.8}$$

$$\text{subject to} \quad \boldsymbol{W}^{\mathrm{H}}\boldsymbol{W} = \boldsymbol{I}, \tag{9.5.9}$$

or

$$\min J_2(\boldsymbol{W}, \boldsymbol{\Lambda}_1, \cdots, \boldsymbol{\Lambda}_K) = \min \sum_{k=1}^{K} \| \boldsymbol{A}_k - \boldsymbol{W}\boldsymbol{\Lambda}_k\boldsymbol{W}^{\mathrm{H}} \|_F^2, \tag{9.5.10}$$

$$\text{subject to} \quad \boldsymbol{W}^{\mathrm{H}}\boldsymbol{W} = \boldsymbol{I}. \tag{9.5.11}$$

In many engineering applications, only the joint diagonalization matrix $\boldsymbol{W}$ is used instead of the diagonal matrices $\boldsymbol{\Lambda}_1, \cdots, \boldsymbol{\Lambda}_K$. Therefore, it is a practical problem to convert the objective function of the approximate joint diagonalization problem into a function containing only the joint diagonalization matrix $\boldsymbol{W}$.

1. Off-diagonal Function Minimization Method

A $m \times m$ matrix $\boldsymbol{M}$ is said to be normal matrix if $\boldsymbol{M}\boldsymbol{M}^{\mathrm{H}} = \boldsymbol{M}^{\mathrm{H}}\boldsymbol{M}$.

Spectral theorem[107]: A normal matrix $\boldsymbol{M}$ is unitarily diagonalizable, if there exists a unitary matrix $\boldsymbol{U}$ and a diagonal matrix $\boldsymbol{D}$ such that $\boldsymbol{M} = \boldsymbol{U}\boldsymbol{D}\boldsymbol{U}^{\mathrm{H}}$.

In numerical analysis, the "off" of a $m \times m$ matrix $\boldsymbol{M} = [M_{ij}]$, written as off($\boldsymbol{M}$), is defined as the sum of squares of the absolute values of all non-diagonal elements, ie.,

$$\mathrm{off}(\boldsymbol{M}) \stackrel{\mathrm{def}}{=} \sum_{i=1, i \neq j}^{m} \sum_{j=1}^{n} |M_{ij}|^2. \tag{9.5.12}$$

From the spectrum theorem, if $\boldsymbol{M} = \boldsymbol{U}\boldsymbol{D}\boldsymbol{U}^{\mathrm{H}}$, where $\boldsymbol{U}$ is unitary matrix and $\boldsymbol{D}$ is diagonal with distinct diagonal elements, then matrix $\boldsymbol{M}$ may be unitarily diagonalized only by unitary matrix $\boldsymbol{V} \doteq \boldsymbol{U}$ that are essentially equal to $\boldsymbol{U}$. That is, if off($\boldsymbol{V}^{\mathrm{H}}\boldsymbol{M}\boldsymbol{V}$) = 0, then $\boldsymbol{V} \doteq \boldsymbol{U}$.

If all non-principal diagonal elements of the square matrix $\boldsymbol{M}$ are extracted to form a matrix

$$[\boldsymbol{M}_{\mathrm{off}}]_{ij} = \begin{cases} 0, & i = j \\ M_{ij}, & i \neq j \end{cases} \tag{9.5.13}$$

called off matrix, then the off function is the square of the Frobenius norm of the off matrix

$$\mathrm{off}(\boldsymbol{M}) = \| \boldsymbol{M}_{\mathrm{off}} \|_F^2 . \tag{9.5.14}$$

Using the off function, the orthogonal approximate joint diagonalization problem can be expressed as[45, 46]

$$\min J_{1a}(\boldsymbol{W}) = \sum_{k=1}^{K} \text{off}(\boldsymbol{W}^{\text{H}}\boldsymbol{A}_k\boldsymbol{W}) = \sum_{k=1}^{K}\sum_{i=1,i\neq j}^{n}\sum_{j=1}^{n} |(\boldsymbol{W}^{\text{H}}\boldsymbol{A}_k\boldsymbol{W})_{ij}|^2. \tag{9.5.15}$$

The orthogonal joint diagonal of the matrices $\boldsymbol{A}_1, \cdots, \boldsymbol{A}_K$ can be achieved by implementing a series of Given rotations on the off-diagonal elements of these matrices. The product of all Given rotation matrices gives the orthogonal joint diagonalizer $\boldsymbol{W}$. This is the Jacobi algorithm for orthogonal approximate joint diagonalization proposed by Cardoso et al[45, 46].

2. Diagonal Function Maxmization Method

The diagonal function of a square matrix can be a scalar function, a vector function, or a matrix function.

(1) Diagonal function. $\text{diag}(\boldsymbol{B}) \in \mathbb{R}$ is a diagonal function of the $m \times m$ matrix $\boldsymbol{B}$, defined as

$$\text{diag}(\boldsymbol{B}) \stackrel{\text{def}}{=} \sum_{i=1}^{m} |B_{ii}|^2. \tag{9.5.16}$$

(2) Diagonal vector function. The diagonal vector function of $m \times m$ matrix $\boldsymbol{B}$, denoted as $\boldsymbol{diag}(\boldsymbol{B}) \in \mathbb{C}^m$, is a vector that aligns the diagonal elements of matrix $\boldsymbol{B}$, i.e.,

$$\boldsymbol{diag}(\boldsymbol{B}) \stackrel{\text{def}}{=} [B_{11}, \cdots, B_{mm}]^{\text{T}}. \tag{9.5.17}$$

(3) Diagonal matrix funcion. The diagonal matrix function of $m \times m$ matrix $\boldsymbol{B}$, denoted as $\boldsymbol{Diag}(\boldsymbol{B}) \in \mathbb{C}^{m\times m}$, is a diagonal matrix consisting of the diagonal elements of matrix $\boldsymbol{B}$, i.e., there is

$$\boldsymbol{Diag}(\boldsymbol{B}) \stackrel{\text{def}}{=} \begin{bmatrix} B_{11} & & 0 \\ & \ddots & \\ 0 & & B_{mm} \end{bmatrix}. \tag{9.5.18}$$

The minimization of $\text{off}(\boldsymbol{B})$ can be equivalent to the maximization of diagonal function $\text{diag}(\boldsymbol{B})$, that is, there is

$$\min \text{off}(\boldsymbol{B}) = \max \text{diag}(\boldsymbol{B}). \tag{9.5.19}$$

Therefore, Eq. (9.5.15) can be rewritten as[219]

$$\max J_{1b}(\boldsymbol{W}) = \sum_{k=1}^{K} \text{diag}(\boldsymbol{W}^{\text{H}}\boldsymbol{A}_k\boldsymbol{W}) = \sum_{k=1}^{K}\sum_{i=1}^{n} |(\boldsymbol{W}^{\text{H}}\boldsymbol{A}_k\boldsymbol{W})_{ii}|^2. \tag{9.5.20}$$

The following is the orthogonal approximate joint diagonalization algorithm for blind signal separation[24].

Algorithm 9.5.1. *Orthogonal approximate joint diagonalization algorithm for blind signal separation*

Step 1 *Estimate the* $m \times m$ *autocorrelation matrix* $\hat{\mathbf{R}} = \frac{1}{N}\sum_{t=1}^{N}\mathbf{x}(t)\mathbf{x}^H(t)$ *from the* $m \times 1$ *observed data vector* $\mathbf{x}(t)$ *with zero mean. Compute the eigenvalue decomposition of* $\hat{\mathbf{R}}$ *and denote by* $\lambda_1, \cdots, \lambda_n$ *the* n *largest eigenvalues and* $\boldsymbol{h}_1, \cdots, \boldsymbol{h}_n$ *the corresponding eigenvectors of* $\hat{\mathbf{R}}$.

Step 2 *Pre-whitening: The estimation* $\hat{\sigma}^2$ *of the noise variance is the average of the* $m - n$ *smallest eigenvalues of* $\hat{\mathbf{R}}$. *The whitened* $n \times n$ *signal vector is* $\boldsymbol{z}(t) = [z_1(t), \cdots, z_n(t)]^T = \boldsymbol{W}\boldsymbol{x}(t)$, *where* $\boldsymbol{W} = [(\lambda_1 - \hat{\sigma}^2)^{-\frac{1}{2}}\boldsymbol{h}_1, \cdots, (\lambda_n - \hat{\sigma}^2)^{-\frac{1}{2}}\boldsymbol{h}_n]^H$ *is the* $m \times n$ *pre-whitening matrix.*

Step 3 *Compute the* K *autocorrelation matrix of the pre-whitening data vectors,* $\hat{\mathbf{R}}_z(k) = \frac{1}{N}\sum_{t=1}^{N} z(t)z^*(t-k)$, $k = 1, \cdots, K$.

Step 4 *Joint diagonalization: Take the orthogonal approximate joint diagonalization for the* K *autocorrelation matrices* $\hat{\mathbf{R}}_z(k)$

$$\hat{\mathbf{R}}_z(k) = \boldsymbol{U}\boldsymbol{\Sigma}_k\boldsymbol{U}^H, \quad k = 1, \cdots, K, \tag{9.5.21}$$

the unitary matrix $\boldsymbol{U}$ *as a joint diagonalizer is obtained.*

Step 5 *Blind signal separation: The source signals are estimated as* $\hat{s}(t) = \boldsymbol{U}^H\boldsymbol{z}(t)$, *and/or the mixing matrix* $\hat{\mathbf{A}} = \boldsymbol{W}^{\dagger}\boldsymbol{U}$.

9.5.3 Nonorthogonal Approximate Joint Diagonalization

The advantage of orthogonal joint diagonalization is that there will be no trivial solution, i.e., zero solution ($\boldsymbol{W} = \mathbf{0}$) and degenerate solution (i.e., singular solution), the disadvantage of orthogonal joint diagonalization is that the observation data vector must be pre-whitened first.

Pre-whitening has two main disadvantages.

(1) The pre-whitening phase seriously affects the performance of the signal separation because the errors of whitening are not corrected in the later signal separation, which easily causes the propagation and spread of errors.

(2) The pre-whitening phase can practically distort the weighted least squares criterion. It would attain exact diagonalization of the selected matrix at the possible cost of poor diagonalization of the others.

The nonorthogonal joint diagonalization is the joint diagonalization without the constraint $\boldsymbol{W}^{\mathrm{H}}\boldsymbol{W} = \boldsymbol{I}$, which becomes the mainstream joint diagonalization method in blind signal separation. The advantage of nonorthogonal joint diagonalization is that it does not have the two disadvantages of whitening, while the disadvantage is that there may be trivial solutions and degenerate solutions.

Typical algorithms for nonorthogonal joint diagonalization include Pham's iterative algorithm that minimizes an information theoretic criterion[174], Vander Veen's Newton iterative subspace fitting algorithm[214], Yeredor's AC-DC algorithm[235], etc.

The AC-DC algorithm separates the coupled optimization problem

$$J_{\text{WLS2}}(\boldsymbol{W},\boldsymbol{\Lambda}_1,\cdots,\boldsymbol{\Lambda}_K)=\sum_{k=1}^{K}\alpha_k\parallel \boldsymbol{A}_k-\boldsymbol{W}\boldsymbol{\Lambda}_k\boldsymbol{W}^{\text{H}}\parallel_F^2$$
$$=\sum_{k=1}^{K}\alpha_k\parallel \boldsymbol{A}_k-\sum_{n=1}^{N}\lambda_n^{[k]}\boldsymbol{w}_k\boldsymbol{w}_k^{\text{H}}\parallel_F^2$$

into two decoupled optimization problems. This algorithm consists of two stages.

(1) The alternating columns (AC) phase minimizes $J_{\text{WLS2}}(\boldsymbol{W})$ with respect to a selected column of $\boldsymbol{W}$ while keeping its other columns,as well as $\boldsymbol{\Lambda}_1,\cdots,\boldsymbol{\Lambda}_K$ fixed.
(2) The diagonal centers (DC) phase minimizes $J_{\text{WLS2}}(\boldsymbol{W},\boldsymbol{\Lambda}_1,\cdots,\boldsymbol{\Lambda}_K)$ with respect to all the diagonal matrices $\boldsymbol{\Lambda}_1,\cdots,\boldsymbol{\Lambda}_K$ while keeping $\boldsymbol{W}$ fixed.

A simple way to avoid the trivial solution is to add the constraint $\boldsymbol{Diag}(\boldsymbol{B})=\boldsymbol{I}$. However, the main drawback of nonorthogonal joint diagonalization is that the joint diagonalizer $\boldsymbol{W}$ may be singular or has a large condition number. A solution that is singular or has a large condition number is called a degenerate solution. The degenerate solution of the nonorthogonal joint diagonalization problem was proposed and solved in Reference [137].

In order to avoid the trivial solution and any degenerate solution simultaneously, Li and Zhang proposed the following objective function minimization[137]

$$\min f(\boldsymbol{W})=\sum_{k=1}^{K}\alpha_k\sum_{i=1}^{N}\sum_{j=1,j\neq i}^{N}\left|\left[\boldsymbol{W}^{\text{H}}\boldsymbol{A}_k\boldsymbol{W}\right]_{ij}\right|^2-\beta\ln\left|\det(\boldsymbol{W})\right|, \tag{9.5.22}$$

where α_k $(1\le k\le K)$ are the positive weights, β is a positive number, and ln is the natural logarithm.

The above cost function can be divided into the sum of the squared off-diagonal error term

$$f_1(\boldsymbol{W})=\sum_{k=1}^{K}\alpha_k\sum_{i=1}^{N}\sum_{j=1,j\neq i}^{N}\left|\left[\boldsymbol{W}^{\text{H}}\boldsymbol{A}_k\boldsymbol{W}\right]_{ij}\right|^2 \tag{9.5.23}$$

and the minus log determinant term

$$f_2(\boldsymbol{W})=-\ln\left|\det(\boldsymbol{W})\right|. \tag{9.5.24}$$

An obvious advantage of the cost function Eq. (9.5.22) is that $\boldsymbol{W}=\boldsymbol{O}$ or singular, $f_2(\boldsymbol{W})\to+\infty$. Therefore, the minimization of the cost function $f(\boldsymbol{W})$ can avoid the trivial solution and any degenerate solutions simultaneously.

In addition, the following important results were proved in Reference [137].

(1) $f_1(\boldsymbol{W})$ is lower unbounded if and only if there exists a nonsingular matrix $\boldsymbol{W}$ that diagonalizes all the given matrices $\boldsymbol{A}_k$, $k=1,\cdots,K$ exactly. In other words, in the approximate joint diagonalization, $f(\boldsymbol{W})$ is lower bounded.

(2) The minimization of the cost function $f(\boldsymbol{W})$ is independent of the penalty factor β. This means that β can be chosen to any infinite value, and simply choose $\beta = 1$, thus avoiding that the performance of the penalty function method depends on the selection of the penalty parameters.

The joint diagonalization has been widely used in common problems, such as blind signal separation[158, 235], blind beamforming[45], time-delay estimation[236], frequency estimation[155], array signal processing[221], blind MIMO equalization[65], and blind MIMO system identification[52], and so on.

9.6 Blind Signal Extraction

Blind signal separation usually separates all source signals simultaneously. In practice, the number of sensors (i.e., the number of mixed signals) may be large, but the number of source signals of interest is relatively small. If blind signal separation is still used to separate all source signals, it will increase the unnecessary computational complexity and cause waste of computational resources. In this case, it is necessary to separate or extract only a few source signals of interest, while keeping the uninteresting source signals in the mixed signal. This is the blind signal extraction (BSE) problem.

9.6.1 Orthognal Blind Signal Extraction

Consider the same mixed signal model

$$\boldsymbol{x}(n) = \boldsymbol{AS}(n) \tag{9.6.1}$$

as blind signal separation, where $\boldsymbol{x} \in \mathbb{C}^{m\times 1}$, $\boldsymbol{A} \in \mathbb{C}^{m\times n}$, and $\boldsymbol{s}(n) \in \mathbb{C}^{n\times 1}$, and $m \gg n$.

Consider the pre-whitening of the observed signal vector. Let $m \times m$ matrix $\boldsymbol{M} = \boldsymbol{R}_x^{-1/2}$, where $\boldsymbol{R}_x = \mathrm{E}\{\boldsymbol{x}(n)\boldsymbol{x}^{\mathrm{H}}(n)\}$ is the autocorrelation matrix of the observed data vector $\boldsymbol{x}(n)$. It is easy to verify that

$$\boldsymbol{z}(n) = \boldsymbol{Mx}(n) \tag{9.6.2}$$

satisfies $\mathrm{E}\{\boldsymbol{z}(n)\boldsymbol{z}^{\mathrm{T}}(n)\} = \boldsymbol{I}$. This shows that the new observed signal vector $\boldsymbol{z}(n)$ is the pre-whitening result of the original observed signal vector $\boldsymbol{x}(n)$, i.e., the matrix $\boldsymbol{M}$ is the pre-whitening matrix of $\boldsymbol{x}(n)$.

Now design a $m \times 1$ demixing vector $\boldsymbol{w}_1$ to extract a (not necessarily the original first) source signal in the source signal vector, i.e.,

$$s_k(n) = \boldsymbol{w}^{\mathrm{H}}\boldsymbol{z}(n), \quad k \in \{1, \cdots, n\} \tag{9.6.3}$$

is a source signal $s_k(n)$ in the source signal vector.

Design a $(m-1)\times m$ matrix

$$\boldsymbol{B}=\begin{bmatrix}\boldsymbol{b}_1^{\mathrm{T}}\\ \vdots\\ \boldsymbol{b}_{m-1}^{\mathrm{T}}\end{bmatrix} \tag{9.6.4}$$

with each of its row vectors orthogonal to the demixing vector $\boldsymbol{w}_1$, i.e., $\boldsymbol{b}_i^{\mathrm{H}}\boldsymbol{w}_1=0$, $i=1,\cdots,m-1$.

From $s_k(n)=\boldsymbol{w}^{\mathrm{H}}\boldsymbol{z}(n)$ and $\boldsymbol{b}_i^{\mathrm{H}}\boldsymbol{w}_1=0$, it is easy to know that the $(m-1)\times 1$ vector

$$\boldsymbol{z}_1(n)=\boldsymbol{B}\boldsymbol{z}(n)=\begin{bmatrix}\boldsymbol{b}_1^{\mathrm{T}}\\ \vdots\\ \boldsymbol{b}_{m-1}^{\mathrm{T}}\end{bmatrix}\boldsymbol{z}(n) \tag{9.6.5}$$

will not contain any component of the extracted resource signal $s_k(n)$, $k\in\{1,\cdots,n\}$. The $(m-1)\times m$ matrix $\boldsymbol{B}$ is called the block matrix since it blocks some source signal, and $\boldsymbol{B}\boldsymbol{z}(n)$ is essentially a deflation of the pre-whitened observed signal vector.

For the $(m-1)\times 1$ new observed signal vector $\boldsymbol{z}_1$ that blocks $s_k(n)$, a new $(m-1)\times 1$ demixing matrix $\boldsymbol{w}_2$ can be designed to extract another source signal in the source signal vector. Then, a new $(m-2)\times(m-1)$ block matrix $\boldsymbol{B}$ is designed to block the newly extracted source signal. This continues until all source signals of interest are extracted.

Since the above method uses the orthogonality of vectors in the signal extraction process, it is called orthogonal blind signal extraction.

9.6.2 Nonorthogonal Blind Signal Extraction

Orthogonal blind signal extraction requires pre-whitening of the observed signal vector, which introduces whitening errors. At the same time, it is also not easy to implement in real-time. Therefore, it is necessary to use non-orthogonal blind signal extraction.

The instantaneous linear mixed source model can be rewritten as

$$\boldsymbol{x}(t)=\boldsymbol{A}\boldsymbol{s}(t)=\sum_{j=1}^{n}\boldsymbol{a}_j s_j(t)=\boldsymbol{a}_1 s_1(t)+\sum_{j=2}^{n}\boldsymbol{a}_j s_j(t). \tag{9.6.6}$$

If the source signal $s\,(t)$ is the desired extracted signal, then the first term of the above equation is the desired term, and the second term (summation term) is the interference term.

Let $\boldsymbol{u}$ and $\boldsymbol{a}_1$ have the same column space, i.e.,

$$U=\operatorname{span}(\boldsymbol{u})=\operatorname{span}(\boldsymbol{a}_1). \tag{9.6.7}$$

It represents the signal subspace of the desired signal $s_1(t)$.

If let the column space formed by $\boldsymbol{n}$ and $\boldsymbol{a}_2, \cdots, \boldsymbol{a}_n$ is the same as

$$N = \text{span}(\boldsymbol{n}) = \text{span}(\boldsymbol{a}_2, \cdots, \boldsymbol{a}_n), \tag{9.6.8}$$

then N represents the subspace of interference signal $s_2(t), \cdots, s_n(t)$, and

$$N^{\perp} = \text{span}(\boldsymbol{n}^{\perp}) \tag{9.6.9}$$

denotes the orthogonal complement space of the interference signal $s_2(t), \cdots, s_n(t)$.

If making the oblique projection $\boldsymbol{E}_{U|N^{\perp}}$ of the observed signal vector $\boldsymbol{x}(t)$ onto the signal subspace U of the desired signal along the orthogonal complement subspace $N^{\perp}$ of the interference signals, then the desired signal $s_1(t)$ will be extracted and all the interference signals $s_2(t), \cdots, s_n(t)$ will be suppressed. In other words, multiplying both sides of Eq. (9.6.6) left by the oblique projection matrix $\boldsymbol{E}_{U|N^{\perp}}$, we have

$$\boldsymbol{E}_{U|N^{\perp}}\boldsymbol{x}(t) = \boldsymbol{a}_1 s_1(t) = \boldsymbol{u}s_1(t), \tag{9.6.10}$$

where the oblique projection $\boldsymbol{E}_{U|N^{\perp}}$ is

$$\boldsymbol{E}_{U|N^{\perp}} = \boldsymbol{u}(\boldsymbol{u}^{\mathrm{H}}\boldsymbol{P}_{N^{\perp}}^{\perp}\boldsymbol{u})^{-1}\boldsymbol{u}^{\mathrm{H}}\boldsymbol{P}_{N^{\perp}}^{\perp}. \tag{9.6.11}$$

Note that the orthogonal projection $\boldsymbol{P}_{N^{\perp}}^{\perp}\boldsymbol{y}$ of any vector to the orthogonal complement space $N^{\perp}$ is on the vector space N, that is,

$$\boldsymbol{P}_{N^{\perp}}^{\perp}\boldsymbol{y} = \alpha\boldsymbol{n}, \quad \forall \boldsymbol{y} \neq \boldsymbol{0}. \tag{9.6.12}$$

Therefore, the oblique projection matrix

$$\boldsymbol{E}_{U|N^{\perp}} = \boldsymbol{u}(\boldsymbol{u}^{\mathrm{H}}\boldsymbol{n})^{-1}\boldsymbol{n}^{\mathrm{H}} = \frac{\boldsymbol{u}\boldsymbol{n}^{\mathrm{H}}}{\boldsymbol{u}^{\mathrm{H}}\boldsymbol{n}}. \tag{9.6.13}$$

Substituting Eq. (9.6.13) into Eq. (9.6.10),

$$\frac{\boldsymbol{u}\boldsymbol{n}^{\mathrm{H}}}{\boldsymbol{u}^{\mathrm{H}}\boldsymbol{n}}\boldsymbol{x}(t) = \boldsymbol{u}s_1(t) \tag{9.6.14}$$

is obtained immediately. Since the above equation holds for all vectors $\boldsymbol{u}$ satisfying $\text{span}(\boldsymbol{u}) = \text{span}(\boldsymbol{a}_1)$, we have

$$s_1(t) = \frac{\boldsymbol{n}^{\mathrm{H}}\boldsymbol{x}(t)}{\boldsymbol{u}^{\mathrm{H}}\boldsymbol{n}}. \tag{9.6.15}$$

This is the extraction formula for the source signal $s_1(t)$. The problem is how to find the vectors $\boldsymbol{u}$ and $\boldsymbol{n}$.

On the other hand, consider the joint diagonalization of the autocorrelation matrices $\boldsymbol{R}_k = \frac{1}{T}\sum_{t=1}^{T}\boldsymbol{x}(t)\boldsymbol{x}(t-k)$, $k = 1, \cdots, K$

$$\boldsymbol{R}_k\boldsymbol{u} = \boldsymbol{U}\boldsymbol{\Sigma}_k\boldsymbol{U} = \sum_{i=1}^{n}\lambda_k(i)\boldsymbol{u}_i\boldsymbol{u}_i^{\mathrm{H}}, \quad k = 1, \cdots, K. \tag{9.6.16}$$

Let $\boldsymbol{u} = \boldsymbol{u}_1$, then we have

$$\boldsymbol{R}_k\boldsymbol{u} = \boldsymbol{U}\boldsymbol{\Sigma}_k\boldsymbol{U}^{\mathrm{H}}\boldsymbol{u} = d_k\boldsymbol{n}, \quad k = 1, \cdots, K. \tag{9.6.17}$$

Let $\boldsymbol{d} = [d_1, \cdots, d_n]^{\mathrm{T}}$, from Eq. (9.6.17), it is easy to get the optimization problem

$$\min J(\boldsymbol{u}, \boldsymbol{n}, \boldsymbol{d}) = \min \frac{1}{2} \sum_{k=1}^{K} \| \boldsymbol{R}_k\boldsymbol{u} - d_k\boldsymbol{n} \|_2^2, \tag{9.6.18}$$

$$\text{subject to } \| \boldsymbol{n} \| = \| \boldsymbol{d} \| = 1. \tag{9.6.19}$$

which is an optimization problem with three variable vectors $\boldsymbol{u}$, $\boldsymbol{n}$, and $\boldsymbol{d}$. It can be decoupled into the following three sub-optimization problems.

(1) The optimization of vector $\boldsymbol{u}$

Fix $\boldsymbol{n}$ and $\boldsymbol{d}$. This is an unconstrained optimization problem

$$J_u(\boldsymbol{u}) = J(\boldsymbol{u}, \boldsymbol{n}, \boldsymbol{d}). \tag{9.6.20}$$

Using the conjugate gradient

$$\frac{\partial J_u(\boldsymbol{u})}{\partial \boldsymbol{u}^*} = \sum_{k-1}^{K} \boldsymbol{R}_k^{\mathrm{H}}(\boldsymbol{R}_k\boldsymbol{u} - d_k\boldsymbol{n}) = \boldsymbol{0}, \tag{9.6.21}$$

the closed solution

$$\boldsymbol{u} = \left(\sum_{k=1}^{K} \boldsymbol{R}_k^{\mathrm{H}}\boldsymbol{R}_k\right)^{-1} \left(\sum_{k=1}^{K} d_k\boldsymbol{R}_k^{\mathrm{H}}\boldsymbol{n}\right) \tag{9.6.22}$$

can bc obtaincd.

(2) The optimization of vector $\boldsymbol{d}$

Fix $\boldsymbol{u}$ and $\boldsymbol{n}$. Since the constraint $\| \boldsymbol{d} \| - 1$ exists, the optimization of $\boldsymbol{d}$ is a constrained optimization problem. Its Lagrange objective function is

$$J_d(\boldsymbol{d}) = J(\boldsymbol{u}, \boldsymbol{n}, \boldsymbol{d}) + \lambda_d(\boldsymbol{d}^{\mathrm{H}}\boldsymbol{d} - 1). \tag{9.6.23}$$

Using the conjugate gradient

$$\frac{\partial J_d(\boldsymbol{d})}{\partial \boldsymbol{d}^*} = -\begin{bmatrix} \boldsymbol{n}^{\mathrm{H}}\boldsymbol{R}_1\boldsymbol{u} \\ \vdots \\ \boldsymbol{n}^{\mathrm{H}}\boldsymbol{R}_k\boldsymbol{u} \end{bmatrix} + (1 + \lambda_d)\boldsymbol{d} = \boldsymbol{0} \tag{9.6.24}$$

and the constraint $\| \boldsymbol{d} \| = 1$, the closed solution

$$\boldsymbol{d} = \frac{1}{\sum_{k=1}^{K} \| \boldsymbol{n}^{\mathrm{H}}\boldsymbol{R}_k\boldsymbol{u} \|^2} \begin{bmatrix} \boldsymbol{n}^{\mathrm{H}}\boldsymbol{R}_1\boldsymbol{u} \\ \vdots \\ \boldsymbol{n}^{\mathrm{H}}\boldsymbol{R}_k\boldsymbol{u} \end{bmatrix} \tag{9.6.25}$$

can be obtained.

(3) The optimization of vector $\boldsymbol{n}$

Fix $\boldsymbol{u}$ and $\boldsymbol{d}$. Since the constraint $\| \boldsymbol{n} \| = 1$ exists, the optimization of $\boldsymbol{n}$ is a constrained optimization problem. Its Lagrange objective function is

$$J_n(\boldsymbol{n}) = J(\boldsymbol{u}, \boldsymbol{n}, \boldsymbol{d}) + \lambda_n(\boldsymbol{n}^{\mathrm{H}}\boldsymbol{n} - 1). \tag{9.6.26}$$

Using the conjugate gradient

$$\frac{\partial J_n(\boldsymbol{n})}{\partial \boldsymbol{n}^*} = -\sum_{k=1}^{K} d_k^*(\boldsymbol{R}_k\boldsymbol{u} - d_k\boldsymbol{n}) + \lambda_n\boldsymbol{n} = \boldsymbol{0} \tag{9.6.27}$$

and the constraint $\| \boldsymbol{n} \| = 1$, the closed solution

$$\boldsymbol{n} = \frac{1}{\sum_{k=1}^{K} \| d_k^*\boldsymbol{R}_k\boldsymbol{u} \|^2} \sum_{k=1}^{K} d_k^*\boldsymbol{R}_k\boldsymbol{u} \tag{9.6.28}$$

can be obtained.

The above sequential blind extraction algorithm via approximate joint diagonalization was proposed in Reference [138] and is summarized in the following.

Algorithm 9.6.1. *Sequential blind extraction algorithm via approximate joint diagonalization*

Initialization: vector $\boldsymbol{d}, \boldsymbol{n}$.

Step 1 Compute the autocorrelation matrix $\boldsymbol{R}_k = \frac{1}{T}\sum_{t=1}^{T} \boldsymbol{x}(t+k)\boldsymbol{x}^*(t), k = 1, \cdots, K$.

Step 2 Compute vector $\boldsymbol{u}, \boldsymbol{d}$, *and* $\boldsymbol{n}$ *using Eqs.* (9.6.22), (9.6.25), *and* (9.6.28). *Repeat this step until convergence.*

Step 3 Extract some source signal using Eq. (9.6.15)*and the converged* $\boldsymbol{u}$ *and* $\boldsymbol{n}$.

Step 4 Delete the extracted signal from the observed data using deflation. Then the above steps are repeated for the deflation observed data to extract the next source signal. This blind "signal extraction + deflation" is repeated until all the source signals of interest are extracted.

9.7 Blind Signal Separation of Convolutively Mixed Sources

The linear mixing methods of signals are usually divided into instantaneous linear mixing and convolutively linear mixing. As the name suggests, instantaneous linear mixing is the linear mixing of multiple sources at a certain time, excluding the source at any other time. Instantaneous linear mixing, also known as memoryless mixing, has a memoryless channel: the source signal is transmitted through the memoryless channel without any time delay between the observed data and the source signal. Convolutively linear mixing is memorized mixing, and its transmission channel is a memorized channel; after the source signal is transmitted through the memorized channel, the mixed data with time delay is generated, and the channel is equivalent to having the function of memorizing the data of previous moments. The previous sections of this chapter focus on the blind signal separation of instantaneous linear mixed sources, and this section discusses the blind signal separation of convolutively mixed.

9.7.1 Convolutively Mixed Sources

Assume that the mixing channel is a linear time-invariant multiple input-multiple output (MIMO) system, which is excited by a $n \times l$ dimensional source signal vector $\boldsymbol{s}(t) = [s_1(t), \cdots, s_n(t)]^{\mathrm{T}}$, and the source signal vector is unobservable. The mixing MIMO system or channel has a $m \times n$ dimensional transfer function matrix $\boldsymbol{A}(z)$, whose elements $a_{ij}(z)$ are polynomials of a time delay operator z. The output end uses m sensors to observe the convoluted mixed signal, and the observed data vector is $\boldsymbol{x}(t) = [x_1(t), \cdots, x_m(t)]^{\mathrm{T}}$.

There are two commonly used representation models for convolutively mixed sources.

(1) Z-transform domain product model

$$\tilde{\boldsymbol{x}}(z) = \boldsymbol{A}(z)\tilde{\boldsymbol{s}}(z) + \tilde{\boldsymbol{e}}(z), \tag{9.7.1}$$

where $\tilde{\boldsymbol{s}}(z)$, $\tilde{\boldsymbol{e}}(z)$, and $\tilde{\boldsymbol{x}}(z)$ are Z transforms of the source signal vector $\boldsymbol{s}(t)$, the additive noise $\boldsymbol{e}(t)$ and the observation data vector $\boldsymbol{x}(t)$, respectively. Since $\boldsymbol{s}(t)$, $\boldsymbol{e}(t)$ and $\boldsymbol{x}(t)$ are not polynomial forms of z and are independent of z, $\tilde{\boldsymbol{s}}(z) = \boldsymbol{s}(t)$, $\tilde{\boldsymbol{e}}(z) = \boldsymbol{e}(t)$, and $\tilde{\boldsymbol{x}}(z) = \boldsymbol{x}(t)$. Therefore, the convolutional mixing formula Eq. (9.7.1) in the Z transform domain can be written as

$$\boldsymbol{x}(t) = \boldsymbol{A}(z)\boldsymbol{s}(t) + \boldsymbol{e}(t). \tag{9.7.2}$$

(2) Time domain convolutively model

The transfer function matrix $\boldsymbol{A}(z)$ of the MIMO system is usually assumed to be the finite impulse response polynomial matrix in the z domain

$$\boldsymbol{A} = \begin{bmatrix} A_{11}(z) & A_{12}(z) & \cdots & A_{1n}(z) \\ A_{21}(z) & A_{12}(z) & \cdots & A_{2n}(z) \\ \vdots & \vdots & \ddots & \vdots \\ A_{m1}(z) & A_{12}(z) & \cdots & A_{mn}(z) \end{bmatrix}, \tag{9.7.3}$$

where $A_{ij}(z)$ represents the mixing FIR filter between the jth input (source signal) and the ith output (observation signal). Mixing FIR filter can describe the acoustic reverberation phenomenon in the actual indoor environment and the multipath problems in wireless communications.

Matrix $\boldsymbol{A}(z)$ is called the mixing FIR filter matrix.

Let the maximum order of FIR filter $\boldsymbol{A}(z)$ be L and the filter coefficients be $a_{ij}(k)$, $k = 0, 1, \cdots, L$, there is

$$A_{ij}(z) = \sum_{k=0}^{L} a_{ij}(k)z^{-k}, \quad i = 1, \cdots, m; \quad j = 1, \cdots, n. \tag{9.7.4}$$

Therefore, formula Eq. (9.7.2) of the product model in the Z transform domain can be expressed as a time domain convolution model

$$x_i(t) = \sum_{j=1}^{n}\sum_{k=0}^{L} a_{ij}(k)s_j(t-k) + e_i(t), \quad i = 1, \cdots, m \tag{9.7.5}$$

in the form of elements.

The problem of blind signal separation of convolutively mixed sources is to separate the source signal $s_i(t)$, $i = 1, \cdots, n$ only by using the observation data vector $\boldsymbol{x}(t)$. Therefore, it is necessary to design a demixing (or deconvoluted) FIR polynomial matrix $\boldsymbol{W}(z) \in \mathbb{C}^{n\times m}$ so that the output

$$\boldsymbol{y}(t) = \boldsymbol{W}(z)\boldsymbol{x}(t) \tag{9.7.6}$$

of the demixing FIR polynomial matrix is a copy of the source signal vector $\boldsymbol{s}(k)$.

In order to perform blind signal separation of convolutively mixed sources, the following assumptions need to be made.

Assumption 1 The source signal $s_i(t)$, $i = 1, \cdots, n$ is a non-Gaussian zero mean independently identically distributed random signal, and each component is statistically independent.

Assumption 2 The additive observation noise vector $\boldsymbol{e}(t)$ is negligible.

Assumption 3 The number of source signals, n, is less than or equal to the number of sensors, m, that is, $n \le m$.

Assumption 4 $\boldsymbol{A}(z)$ is a $m\times n$ dimensional FIR polynomial matrix, the rank of $\boldsymbol{A}(z)$ is full column rank for every non-zero time delay z, i.e., $\text{rank}(\boldsymbol{A}(z)) = n$, $\forall z = 1, \cdots, \infty$.

Under the conditions of Assumption 2, substituting Eq. (9.7.6) into Eq. (9.7.2) yields

$$\boldsymbol{y}(t) = \boldsymbol{W}(z)\boldsymbol{A}(z)\boldsymbol{s}(t). \tag{9.7.7}$$

Therefore, in the case of ideal blind signal separation, the $n \times m$ demixing FIR polynomial matrix $\boldsymbol{W}(z)$ and the $m \times n$ mixing FIR polynomial matrix $\boldsymbol{A}(z)$ should satisfy the following relationship

$$\boldsymbol{W}(z)\boldsymbol{A}(z) = \boldsymbol{I} \quad \text{or} \quad \boldsymbol{W}(z) = \boldsymbol{A}^{\dagger}(z), \quad \forall z = 1, \cdots, \infty, \tag{9.7.8}$$

where $\boldsymbol{I}$ is a $n \times n$ identity matrix.

However, since $\boldsymbol{A}(z)$ is unknown, it is impossible to design the demixing FIR polynomial matrix $\boldsymbol{W}(z) = \boldsymbol{A}^{\dagger}(z)$. Therefore, the design objective should be relaxed to

$$\boldsymbol{W}(z) = \boldsymbol{G}(z)\boldsymbol{A}^{\dagger}(z), \quad \forall z = 1, \cdots, \infty, \tag{9.7.9}$$

where $\boldsymbol{G}(z) = [\boldsymbol{g}_1(z), \cdots, \boldsymbol{g}_n(z)]^{\mathrm{T}}$ is the generalized exchange FIR polynomial vector. Each generalized exchange FIR polynomial vector $\boldsymbol{g}_i(z)$ has exactly one nonzero element (FIR polynomial) $d_{i'}z^{-\tau_{i'}}$, which is the i'-th element of $\boldsymbol{g}_i(z)$. However, i' is not necessarily the same as i, and the nonzero elements of any two vectors $\boldsymbol{g}_{i_1}(z)$ and $\boldsymbol{g}_{i_2}(z)$, $i_1 \neq i_2$, cannot appear in the same position.

Like the blind separation of instantaneous linear mixed sources, the blind separation of convolutively linear mixed sources also has two uncertainties: (1) the uncertainty of the ordering of the separated sources and (2) the uncertainty of the amplitude of each separated signal.

Blind separation of convolutively linear mixed sources can be divided into the following three types.

(1) Time domain blind signal separation methods: the convolutively mixed source is transformed into an equivalent instantaneous mixed source, and the design and adaptive updating of the demixing matrix are performed in the time domain.
(2) Frequency domain blind signal separation method: the time domain observation data are transformed into the frequency domain, and the design and adaptive update of the demixing matrix are performed in the frequency domain.
(3) Time-frequency domain blind signal separation method: using the quadratic time-frequency distribution of non-stationary observation data, the design and adaptive updating of the demixing matrix is performed in the time-frequency domain.

The following three sections will introduce the above three methods respectively.

9.7.2 Time Domain Blind Signal Separation of Convolutively Mixed Sources

Since $\boldsymbol{W}(z)\boldsymbol{A}(z) = \boldsymbol{I}$ cannot be realized, the design of blind signal separation needs to consider how to relize $\boldsymbol{W}(z) = \boldsymbol{G}(z)\boldsymbol{A}^{\dagger}(z)$ or $\boldsymbol{W}(z)\boldsymbol{A}(z) = \boldsymbol{G}(z)$.

Denote the demixing FIR filter matrix as

$$\boldsymbol{W}(z) = \begin{bmatrix} \boldsymbol{w}_1^{\mathrm{T}}(z) \\ \vdots \\ \boldsymbol{w}_n^{\mathrm{T}}(z) \end{bmatrix}, \tag{9.7.10}$$

where $\boldsymbol{w}_i^{\mathrm{T}} = [w_{i1}(z), \cdots, w_{im}(z)]$ denotes the i-th row vector of $\boldsymbol{W}(z)$. Thus, $\boldsymbol{W}(z)\boldsymbol{A}(z) = \boldsymbol{G}(z)$ can be equivalently written as

$$\boldsymbol{w}_i^{\mathrm{T}}(z)\boldsymbol{A}(z) = \boldsymbol{g}_i^{\mathrm{T}}(z), \quad i = 1, \cdots, n. \tag{9.7.11}$$

Note that the row vector $\mathbf{g}_i^{\mathrm{T}}(z)$ on the right side of Eq. (9.7.11) has only one non-zero term $d_{i'}z^{-\tau_{i'}}$, which is the i-th element of the row vector $\mathbf{g}_i^{\mathrm{T}}(z)$, but i' is not necessarily the same as i.

Therefore, the i-th component of the output vector $\mathbf{y}(t) = \boldsymbol{W}(z)\boldsymbol{A}(z)\mathbf{s}(t)$ of the demixing FIR polynomial matrix $\boldsymbol{W}(z)$ can be expressed as

$$y_i(t) = \mathbf{w}_i^{\mathrm{T}}(z)\boldsymbol{A}(z)\mathbf{s}(t) = \mathbf{g}_i^{\mathrm{T}}(z)\mathbf{s}(t) = d_{i'}s_{i'}(t-\tau_{i'}), \quad i = 1, \cdots, n. \tag{9.7.12}$$

In other words, the i-th component $y_i(t)$ of the output vector $\mathbf{y}(t)$ is a copy of the i'-th source signal $s_{i'}(t)$ and may differ from $s_{i'}(t)$ by a scale factor $d_{i'}$ and a time delay $\tau_{i'}$.

Since the mixing FIR polynomial matrix $\boldsymbol{A}(z)$ is unobservable and the source signal $s_i(t)$ is unkonwn, the following assumption is added to $s_i(t)$.
Assumption 5 Each source signal $s_i(t)$ has a unit variance.

Under the condition of Assumption 5, the variance of the output signal $y_i(t)$ is constrained to 1, that is, Eq. (9.7.12) is simplified to

$$y_i(t) = \mathbf{w}_i^{\mathrm{T}}(z)\boldsymbol{A}(z)\mathbf{s}(t) = s_{i'}(t-\tau_{i'}) = s_{i'}(t)z^{-\tau_{i'}}, \quad i = 1, \cdots, n. \tag{9.7.13}$$

This shows that one of the n source signals can be extracted from the convolutively mixed signal.

The expression of FIR filter $\mathbf{w}_i^{\mathrm{T}}(z) = [w_{i1}(z), \cdots, w_{im}(z)]$ satisfying the blind signal separation Eq. (9.7.13) of the convolutively mixed source is derived below.

Let the maximum order of the demixing FIR filter $\mathbf{w}_i^{\mathrm{T}}(z)$ be K, i.e.,

$$w_{ij}(z) = \sum_{k=0}^{K} w_{ij}(k)z^{-k}, \quad i = 1, \cdots, n; j = 1, \cdots, m. \tag{9.7.14}$$

The order K of the demixing FIR filter should be greater than or equal to the order L of the mixing FIR filter, i.e. $K \geq L$.

The specific expression of the separated signal of the demixing FIR filter shown in Eq. (9.7.13) is

$$\begin{aligned} y_i(t) = \mathbf{w}_i^{\mathrm{T}}(z)\mathbf{x}(t) &= [w_{i1}(z), \cdots, w_{im}(z)] \begin{bmatrix} x_1(t) \\ \vdots \\ x_m(t) \end{bmatrix} \\ &= \sum_{k=0}^{K} w_{i1}(k)x_1(t-k) + \cdots + \sum_{k=0}^{K} w_{im}(k)x_m(t-k). \end{aligned} \tag{9.7.15}$$

In the matrix representation of the FIR filter, the underlined vectors and matrices are commonly used to represent the vectors and matrices related to the FIR filter, which are referred to as FIR vectors and FIR matrices for short.

The multiplication of two FIR matrices $\underline{\boldsymbol{A}}$ and $\underline{\boldsymbol{B}}$ is defined as [132]

$$\underline{\boldsymbol{A}} \cdot \underline{\boldsymbol{B}} = \begin{bmatrix} a_{11} & \cdots & a_{1n} \\ \vdots & \ddots & \vdots \\ a_{m1} & \cdots & a_{mn} \end{bmatrix} \begin{bmatrix} b_{11} & \cdots & b_{1p} \\ \vdots & \ddots & \vdots \\ b_{n1} & \cdots & b_{np} \end{bmatrix} = \begin{bmatrix} \sum_{j=1}^{n} a_{1j} * b_{j1} & \cdots & \sum_{j=1}^{n} a_{1j} * b_{jk} \\ \vdots & \ddots & \vdots \\ \sum_{j=1}^{n} a_{mj} * b_{j1} & \cdots & \sum_{j=1}^{n} a_{mj} * b_{jk} \end{bmatrix}, \tag{9.7.16}$$

where $*$ represents convolution. For example, if $\underline{\boldsymbol{b}} = \boldsymbol{s}(t) = [s_1(t), \cdots, s_n(t)]^{\mathrm{T}}$, then

$$\underline{\boldsymbol{A}} \cdot \boldsymbol{s}(t) = \begin{bmatrix} \sum_{j=1}^{n} \sum_{k=0}^{K} a_{1j}(k) s_j(t-k) & \cdots & \sum_{j=1}^{n} \sum_{k=0}^{K} a_{1j}(k) s_j(t-k) \\ \vdots & \ddots & \vdots \\ \sum_{j=1}^{n} \sum_{k=0}^{K} a_{mj}(k) s_j(t-k) & \cdots & \sum_{j=1}^{n} \sum_{k=0}^{K} a_{mj}(k) s_j(t-k) \end{bmatrix}. \tag{9.7.17}$$

Therefore, the convolutively mixing model can be expressed as

$$\boldsymbol{x}(t) = \underline{\boldsymbol{A}} \cdot \boldsymbol{s}(t). \tag{9.7.18}$$

Using the FIR matrix, Eq. (9.7.15) can be simplified as

$$y_i(t) = \underline{\boldsymbol{w}}_i^{\mathrm{T}}(t) \underline{\boldsymbol{x}}(t), \quad i = 1, \cdots, n, \tag{9.7.19}$$

where

$$\underline{\boldsymbol{w}}_i^{\mathrm{T}} = [w_{i1}(0), w_{i1}(1), \cdots, w_{i1}(K), \cdots, w_{im}(0), w_{im}(1), \cdots, w_{im}(K)], \tag{9.7.20}$$

$$\underline{\boldsymbol{x}}(t) = [x_1(t), x_1(t-1), \cdots, x_1(t-K), \cdots, x_m(t), x_m(t-1), \cdots, x_m(t-K)]^{\mathrm{T}}. \tag{9.7.21}$$

The FIR vector $\underline{\boldsymbol{w}}_i$ is the demixing vector, also known as the extraction vector.

Let $i = 1, \cdots, n$, then Eq. (9.7.19) can be expressed in terms of FIR vectors and FIR matrices as

$$\boldsymbol{y}(t) = \underline{\boldsymbol{W}\boldsymbol{x}}(t). \tag{9.7.22}$$

Compared with the instantaneous linear mixed source, the blind signal separation of the convolutively linear mixed source has the following differences.

(1) The separation or demixing model for instantaneous linear mixing is $\boldsymbol{y}(t) = \boldsymbol{W}\boldsymbol{x}(t)$, while the separation model for convolutively linear mixing is $\boldsymbol{y}(t) = \underline{\boldsymbol{W}\boldsymbol{x}}(t)$.

(2) The dimension of the demixing matrix $\boldsymbol{W}$ of the instantaneous linear mixed source is $n \times m$, while the dimension of the demixing FIR matrix $\underline{\boldsymbol{W}}$ of the convolutively linear mixed source is $n \times (K+1)m$.

(3) The observation vector $\boldsymbol{x}(t)$ of the instantaneous linear mixed model is an $m \times 1$ vector whose elements include only $x_1(t), \cdots, x_n(t)$, while the observation vector

$\underline{\boldsymbol{x}}(t)$ of the convolutively linear mixed model is a $(K+1)m \times 1$ dimensional FIR vector whose elements include $x_1(t), \cdots, x_m(t)$ and their delays $x_i(t-1), \cdots, x_i(t-K)$, $i = 1, \cdots, m$. In other words, the elements of the observation vector of the instantaneous linear mixed model are all memoryless observations, while the elements of the observation vector of the convolutively linear mixed model are the observation data themselves and their delays (memory length is K).

The above comparison reveals the following relationships between instantaneous linear mixing and convolutively linear mixing.

(1) As long as the $m \times 1$ observation data vector $\boldsymbol{x}(t)$ and the $n \times m$ demixing matrix $\boldsymbol{W}$ of the instantaneous linear mixing model are replaced by the FIR data vector $\underline{\boldsymbol{x}}(t)$ and the $n \times (K+1)m$ demixing FIR matrix $\underline{\boldsymbol{W}}$ of the convolutively linear mixing model, then the blind signal separation algorithm of the instantaneous linear mixed source can be used for the blind signal separation of the convolutively linear mixed source in principle.

(2) The instantaneous linear mixed source is a special case of the convolutively linear mixed source in the case of the single-tap FIR filter.

Since delayed observations are used, the convolutively linear mixed source is also called a delayed linear mixed source.

For example, the natural gradient algorithm for the blind signal separation of convolutively mixed sources is

$$\underline{\boldsymbol{W}}_{k+1} = \underline{\boldsymbol{W}}_k + \eta_k[\boldsymbol{I} - \phi(\boldsymbol{y}_k)\boldsymbol{y}_k^{\mathrm{H}}]\underline{\boldsymbol{W}}_k. \tag{9.7.23}$$

Similarly, it is easy to generalize other algorithms for blind signal separation of instantaneous mixed sources (e.g., EASI algorithm and iterative inverse algorithm) to the corresponding time domain algorithms for blind signal separation of convolutively mixed sources.

9.7.3 Frequency Domain Blind Signal Separation of Convolutively Mixed Sources

Consider the noise-free case of the convolutively mixture model Eq. (9.7.5)

$$x_i(t) = \sum_{j=1}^{n}\sum_{k=0}^{L} a_{ij}(k)s_j(t-k) = \boldsymbol{a}_i^{\mathrm{T}}\boldsymbol{s}(t), \quad i = 1, \cdots, m, \tag{9.7.24}$$

where $\boldsymbol{a}_i^{\mathrm{T}}$ is the i-th row vector of the mixing matrix $\boldsymbol{A}$.

The sampling frequency f_s is used to discretely sample the time domain observation signal $x_i(t)$, and the N-point discrete short-time Fourier transform

$$X_i(\tau, f) = \sum_{p=-N/2}^{N/2} x_i(\tau+p)h(p)e^{-j2\pi fp} \tag{9.7.25}$$

is performed on $x_i(t)$, where $f \in \{0, \frac{1}{N}f_s, \cdots, \frac{N-1}{N}f_s\}$ is the frequency point, $h(p)$ is the smoothing window function with a time width of τ.

On the other hand, the discrete short-time Fourier transform is performed on both sides of the time-domain expression Eq. (9.7.24) of the convolutively mixing model, we can obtain the frequency domain expression

$$X_i(\tau, f) = \sum_{j=1}^{n} a_{ij}(f)S_j(\tau, f) \tag{9.7.26}$$

of the linear mixing model immediately, where $S_j(\tau, f)$ is the short-time Fourier transform of the j-th source signal $s_j(t)$ and $a_{ij}(f)$ is the frequency response of the j-th source signal $s_j(t)$ to the i-th sensor.

Take $i = 1, \cdots, m$ and let $\hat{\mathbf{x}}(\tau, f) = [X_1(\tau, f), \cdots, X_m(\tau, f)]^{\mathrm{T}}$ represents the short-time Fourier transform vector of the observed data. Thus, Eq. (9.7.26) can be expressed as

$$\hat{\mathbf{x}}(\tau, f) = \sum_{j=1}^{n} \mathbf{a}_j(f)S_j(\tau, f), \tag{9.7.27}$$

where $\mathbf{a}_j^{\mathrm{T}}(f) = [a_{1j}(f), \cdots, a_{mj}(f)]$ represents the frequency response of the j-th source signal to all sensors.

The objective of blind signal separation for convolutively linear mixed sources is to design a $n \times m$ frequency domain demixing matrix $\mathbf{W}(f)$ such that its output

$$\hat{\mathbf{y}}(\tau, f) = [Y_1(\tau, f), \cdots, Y_n(\tau, f)]^{\mathrm{T}} = \mathbf{W}(f)\hat{\mathbf{x}}(\tau, f) \tag{9.7.28}$$

is a frequency domain vector with independent components.

Compared with instantaneous linear mixed sources, the frequency domain blind signal separation of convolutively linear mixed sources has the following similarities and differences.

(1) The separation model of instantaneous linear mixing is $\mathbf{y}(t) = \mathbf{W}\mathbf{x}(t)$, while the frequency domain separation model of convolutively linear mixing is $\hat{\mathbf{y}}(\tau, f) = \mathbf{W}(f)\hat{\mathbf{x}}(\tau, f)$.
(2) The time domain separation matrix $\mathbf{W}$ of the instantaneous linear mixed source and the frequency domain separation matrix of the convolutively mixed source have the same dimension, both are $n \times m$.
(3) The observation vector $\mathbf{x}(t)$ of instantaneous linear mixed source is $m \times 1$ vector, and its element is $x_1(t), \cdots, x_n(t)$; while the equivalent observation vector $\hat{\mathbf{x}}(\tau, f)$ of convolutively linear mixed source is the short-time Fourier transform of observation data.

The above comparison reveals the relationship between the time domain blind signal separation of instantaneous linear mixing and the frequency domain of convolutively linear mixing: as long as the $m \times 1$ observation data vector $\mathbf{x}(t)$ and $n \times m$ separation matrix $\boldsymbol{W}$ in the instantaneous linear mixing model are replaced by the frequency domain (short-time Fourier transform) vector $\hat{\boldsymbol{x}}(\tau, f)$ and $n \times m$ frequency domain separation matrix $\boldsymbol{W}(f)$ of the convolutional linear mixing model, respectively, both ICA and nonlinear PCA algorithms for instantaneous linear mixed source separation can be used or frequency domain blind signal separation of convolutional linear mixed sources.

Algorithm 9.7.1. *Frequency domain blind signal separation of convolutively linear mixed sources*

Step 1 Preprocessing: Make the observed data vector with zero-mean.

Step 2 Estimation of the number of sources: Compute the eigenvalue decomposition of the autocorrelation matrix $\boldsymbol{R}_x = \frac{1}{T}\sum_{t=1}^{T} \boldsymbol{x}(t)\boldsymbol{x}^H(t)$ *and the number of the large eigenvalues gives an estimation of the number of sources n.*

Step 3 Update of the frequency domain separation matrix: Calculate the short-time Fourier transform $X_i(\tau, f), i = 1, \cdots, m$ *of the observed data, and adaptively update the* $n \times m$ *frequency domain separation matrix* $\boldsymbol{W}(f)$ *using the ICA or nonlinear PCA algorithm for blind signal separation in the frequency domain.*

Step 4 Signal separation: Calculate the short-time Fourier transform $Y_j(\tau, f), j = 1, \cdots, n$ *of the separated output using Eq. (9.7.28), and then perform the inverse short-time Fourier transform to recover or reconstruct the source signal.*

9.7.4 Time-Frequency Domain Blind Signal Separation of Convolutively Mixed Sources

Considering the non-stationary signal $z(t)$, the commonly used time-frequency signal representation is the following quadratic time-frequency distributions.

(1) Cohen's class of time-frequency distribution

$$\rho_{zz}^{\text{cohen}}(t, f) = \int_{-\infty}^{\infty}\int_{-\infty}^{\infty} \phi(v, \tau) z(t + v + \frac{\tau}{2}) z^*(t + v - \frac{\tau}{2}) e^{-j2\pi f\tau} dv d\tau. \tag{9.7.29}$$

(2) Wigner-Ville distribution (WVD)

$$\rho_{zz}^{\text{wvd}}(t, f) = \int_{-\infty}^{\infty} z(t + \frac{\tau}{2}) z^*(t - \frac{\tau}{2}) e^{-j2\pi f\tau} d\tau. \tag{9.7.30}$$

The advantage of the WVD is the high time-frequency resolution, and the disadvantage is the computational complexity and the existence of the cross-terms.

(3) Short-time Fourier transform (STFT)

$$\text{STFT}_z(t, f) = \int_{-\infty}^{\infty} z(t) h(\tau - t) e^{-j2\pi f\tau} d\tau, \tag{9.7.31}$$

where $h(t)$ is a window function. STFT is a linear time-frequency representation, and its quadratic time-frequency distribution is the spectrogram

$$\rho_{zz}^{\text{spec}}(t,f) = |\text{STFT}_z(t,f)|^2. \tag{9.7.32}$$

The advantages of the STFT and the spectrogram are the computational simplicity and the free of cross-terms. The disadvantage is that the time-frequency resolution is much lower than that of the Wigner- Ville distribution.

(4) Masked Wigner-Ville distribution (MWVD)

$$\rho_{zz}^{\text{mwvd}}(t,f) = \rho_{zz}^{\text{wvd}}(t,f) \cdot \rho_{zz}^{\text{spec}}(t,f). \tag{9.7.33}$$

The masked Wigner-Ville distribution combines the advantages of the Wigner-Ville distribution (high resolution) with those of the spectrogram (free of cross-terms). It is called the masked Wigner-Ville distribution because $\rho_{zz}^{\text{mwvd}}(t,f)$ acts as a mask for the cross-terms of the Wigner-Ville distribution.

The short-time Fourier transform and spectrogram are particularly suitable for speech or audio signals, while the Wigner-Ville distribution is more suitable for frequency modulation (FM) signals.

For signal vector $\boldsymbol{z}(t) = [z_1(t), \cdots, z_m(t)]^{\text{T}}$, its time-frequency representation involves the quadratic cross time-frequency distribution between the signal components $z_1(t)$ and $z_2(t)$.

(1) Cohen's class of cross time-frequency distribution

$$\rho_{z_1 z_2}^{\text{cohen}}(t,f) = \int_{-\infty}^{\infty}\int_{-\infty}^{\infty} \phi(\nu,\tau) z_1(t+\nu+\frac{\tau}{2}) z_2^*(t+\nu-\frac{\tau}{2}) e^{-j2\pi f\tau} d\nu d\tau. \tag{9.7.34}$$

(2) Cross Wigner-Ville distribution

$$\rho_{z_1 z_2}^{\text{wvd}}(t,f) = \int_{-\infty}^{\infty} z_1(t+\frac{\tau}{2}) z_2^*(t-\frac{\tau}{2}) e^{-j2\pi f\tau} d\tau. \tag{9.7.35}$$

(3) Cross spectrogram

$$\rho_{z_1 z_2}^{\text{spec}}(t,f) = \text{STFT}_{z_1}(t,f)\text{STFT}_{z_2}^*(t,f). \tag{9.7.36}$$

(4) Masked cross Wigner-Ville distribution

$$\rho_{z_1 z_2}^{\text{mwvd}}(t,f) = \rho_{z_1 z_2}^{\text{wvd}}(t,f) \cdot \rho_{z_1 z_2}^{\text{spec}}(t,f). \tag{9.7.37}$$

Wigner-Ville distribution has not only high temporal resolution, but also high frequency resolution. However, it introduces the cross-terms when the signal is a multi-component signal. These cross-terms come from the interaction between different signal components.

In order to perform a time-frequency analysis of non-stationary signals, it is necessary to use a "clear" time-frequency distribution of the signal, which reveals the characteristics of the signal and is free from any "ghost" components. In other words, the time-frequency distribution is expected to be free of cross-terms while maintaining a high time-frequency resolution.

As a high resolution quadratic time-frequency distribution, the B distribution[19] is defined as

$$\rho_{zz}^{\mathrm{B}}(t,f) = \int_{-\infty}^{\infty}\int_{-\infty}^{\infty} \left(\frac{|\tau|}{\cosh(u)}\right)^{\sigma} z(t-u+\frac{\tau}{2})z^{*}(t-u-\frac{\tau}{2})e^{-j2\pi f\tau}\,du\,d\tau, \tag{9.7.38}$$

where $0 \le \sigma \le 1$ is a real parameter.

A noise thresholding procedure

$$T_{\mathrm{th}}(t,f) = \begin{cases} T(t,f), & \text{if} \quad T(t,f) > \varepsilon \\ 0, & \quad\text{Others} \end{cases} \tag{9.7.39}$$

is used to remove those points that have low energy time-frequency distribution in the time-frequency domain that is smaller than the threshold value ε.

The threshold usually selected as $\varepsilon = 0.01 \max T(t,f), (t,f) \in \Omega$ or $\varepsilon = 0.05$.

For a noiseless or cross-terms-free time-frequency distribution, the number of signal components at a moment t_0 can be estimated by the number of peaks in the slices $T(t_0,f)$ of the TFD. By searching and calculating the peak of each TFD slice, a histogram of the number of peaks at different moments can be obtained. The maximum number of peaks in this histogram gives an estimation of the signal number. Algorithm 9.7.2 gives the steps for estimation of the number of signals.

Algorithm 9.7.2. *Estimation of the number of signals*
Step 1 Compute the TFD slices $T(t_0,f)$ for each moment $t_0 = 1,\cdots,t_{\max}$.
Step 2 Search and calculate the number of peaks of each TFD slice.
Step 3 Obtain the histogram of the number of peaks at different moments $t_0 = 1,\cdots,t_{\max}$.
Step 4 The maximum peak number of the histogram is taken as the estimation of the signal components number.

Assuming that the vector $\boldsymbol{z}(t) = [z_1(t),\cdots,z_m(t)]^{\mathrm{T}}$ is composed of m source signals emitted in different spaces or observed signals from m sensors at different locations, the cross-TFD of different elements of vector $\boldsymbol{z}(t)$ is called the spatial time-frequency distribution (STFD). The matrix is composed of the spatial time-frequency distribution

$$\boldsymbol{D}_{zz}(t,f) = \begin{bmatrix} \rho_{z_1z_1}(t,f) & \cdots & \rho_{z_1z_m}(t,f) \\ \vdots & \ddots & \vdots \\ \rho_{z_mz_1}(t,f) & \cdots & \rho_{z_mz_m}(t,f) \end{bmatrix} \tag{9.7.40}$$

is called the STFD matrix[24]. Here, $\rho_{z_iz_j}(t,f)$ represent the cross-TFD between two signals $z_i(t)$ and $z_j(t)$. According to different non-stationary signals, the cross-TFD is usually $\rho_{z_iz_j}^{\text{cohen}}(t,f)$, $\rho_{z_iz_j}^{\text{wvd}}(t,f)$, $\rho_{z_iz_j}^{\text{spec}}(t,f)$, or $\rho_{z_iz_j}^{\text{mwvd}}(t,f)$.

For the unmixed source signal vector $\boldsymbol{s}(t) = [s_1(t), \cdots, s_n(t)]^{\mathrm{T}}$, we call an auto-source TF point (t_a, f_a) a point at which there is a true energy $\rho_{s_is_j}(t,f)$, $i = 1, \cdots, n$ contribution/concentration of source or sources in the TF domain, and a cross-source TF point (t,f) a point at which there is a "false" energy $\rho_{s_is_j}(t,f)$, $i \neq j$ contribution (due to the cross-term effect of quadratic TFDs)[5]. The area composed of auto-source TF points is called the auto-source TF area, and the area composed of cross-source TF points is called the cross-source TF area. Note that, at other points with no energy contribution, the TFD value is ideally equal to zero.

Let Ω_1 and Ω_2 be the TF supports (i.e., the definition domain of time-frequency distribution) of two sources $s_1(t)$ and $s_2(t)$. If $\Omega_1 \cap \Omega_2 = \emptyset$, then $s_1(t)$ and $s_2(t)$ is called disjoint in TF domain. On the contrary, the two non-stationary signals are said to be nondisjoint in the TF domain.

Disjoint in the TF domain is a rather strict restriction for non-stationary signals. In fact, the sources are nondisjoint in some local TF domains.

The following assumptions are usually made when using TFD for underdetermined blind signal separation[5].

Assumption 1 The column vector of the mixing matrix $\boldsymbol{A} = [\boldsymbol{a}_1, \cdots, \boldsymbol{a}_n]$ are pairwise linearly independent. That is, for any $i \neq j$, we have $\boldsymbol{a}_i$ and $\boldsymbol{a}_j$ linearly independent.

Assumption 2 The number of sources p that contribute their energy at any TF point (t,f) (i.e., with nonzero TFD) is strictly less than the number of sensors m, that is, $p < m$.

Assumption 3 For each source, there exists a region in the TF domain, where this source exists alone.

Consider the blind signal separation model

$$\boldsymbol{x}(t) = \boldsymbol{A}\boldsymbol{s}(t) \quad \text{or} \quad x_i(t) = \boldsymbol{a}_i^{\mathrm{T}}\boldsymbol{s}(t), \tag{9.7.41}$$

where $\boldsymbol{a}_i^{\mathrm{T}} \in \mathbb{C}^{1\times n}$, $i = 1, \cdots, m$ is the ith row vector of the mixing matrix, we have

$$\boldsymbol{A} = \begin{bmatrix} \boldsymbol{a}_1^{\mathrm{T}} \\ \vdots \\ \boldsymbol{a}_m^{\mathrm{T}} \end{bmatrix} \quad \text{and} \quad \boldsymbol{A}^{\mathrm{H}} = [\boldsymbol{a}_1^*, \cdots, \boldsymbol{a}_m^*]. \tag{9.7.42}$$

Next, taking Cohen's class TFD as an example, the expression of the STFD matrix $\boldsymbol{D}_{\boldsymbol{xx}}(t,f)$ of the observed data is derived.

The Cohen's class time-frequency distribution between the elements $x_i(t)$ and $x_k(t)$ of the observed data vector $\boldsymbol{x}(t)$ is

$$\begin{aligned}\rho_{x_i x_k}(t,f) &= \int_{-\infty}^{\infty}\int_{-\infty}^{\infty} \phi(\nu,f) x_i(t+\nu+\frac{\tau}{2}) x_k^*(t+\nu-\frac{\tau}{2}) e^{-j2\pi f\tau} d\nu d\tau \\ &= \int_{-\infty}^{\infty}\int_{-\infty}^{\infty} \phi(\nu,f)\left[\boldsymbol{a}_i^{\mathrm{T}}\boldsymbol{s}\left(t+\nu+\frac{\tau}{2}\right)\right]\left[\boldsymbol{a}_k^{\mathrm{T}}\boldsymbol{s}\left(t+\nu-\frac{\tau}{2}\right)\right]^* e^{-j2\pi f\tau} d\nu d\tau \\ &= \boldsymbol{a}_i^{\mathrm{T}} \boldsymbol{D}_{ss}(t,f)\boldsymbol{a}_k^*, \end{aligned} \tag{9.7.43}$$

where $\boldsymbol{D}_{ss}(t,f) = [\rho_{s_i s_j}(t,f)]_{ij}$ is the STFD matrix of the source signal. Here, $\boldsymbol{a}_k^{\mathrm{T}}\boldsymbol{s}(t') = \boldsymbol{s}^{\mathrm{T}}(t')\boldsymbol{a}_k$ and $[\boldsymbol{a}_k^{\mathrm{T}}\boldsymbol{s}(t')]^* = \boldsymbol{s}^{\mathrm{H}}(t')\boldsymbol{a}_k'$ are used to obtain the final result of the above formula.

Eq. (9.7.43) can be written as the STFD matrix of the observation data

$$\begin{aligned}\boldsymbol{D}_{xx}(t,f) &= \begin{bmatrix} \rho_{x_1x_1}(t,f) & \cdots & \rho_{x_1x_m}(t,f) \\ \vdots & \ddots & \vdots \\ \rho_{x_mx_1}(t,f) & \cdots & \rho_{x_mx_m}(t,f) \end{bmatrix} \\ &= \begin{bmatrix} \boldsymbol{a}_1^{\mathrm{T}}\boldsymbol{D}_{ss}(t,f)\boldsymbol{a}_1^* & \cdots & \boldsymbol{a}_1^{\mathrm{T}}\boldsymbol{D}_{ss}(t,f)\boldsymbol{a}_m^* \\ \vdots & \ddots & \vdots \\ \boldsymbol{a}_m^{\mathrm{T}}\boldsymbol{D}_{ss}(t,f)\boldsymbol{a}_1^* & \cdots & \boldsymbol{a}_m^{\mathrm{T}}\boldsymbol{D}_{ss}(t,f)\boldsymbol{a}_m^* \end{bmatrix} \\ &= \begin{bmatrix} \boldsymbol{a}_1^{\mathrm{T}} \\ \vdots \\ \boldsymbol{a}_m^{\mathrm{T}} \end{bmatrix} \boldsymbol{D}_{ss}(t,f)[\boldsymbol{a}_1^*, \cdots, \boldsymbol{a}_m^*]. \end{aligned} \tag{9.7.44}$$

Substituting Eq. (9.7.42) into Eq. (9.7.44), the concise expression for the STFD matrix

$$\boldsymbol{D}_{xx}(t,f) = \boldsymbol{A}\boldsymbol{D}_{ss}(t,f)\boldsymbol{A}^{\mathrm{H}} \tag{9.7.45}$$

of the observed data can be obtained.

Next, two time-frequency blind signal separation algorithms for solving the time-frequency representation matrix Eq. (9.7.45) of blind signal separation are introduced.

1. Joint Diagonalization Algorithm

Since the off-diagonal elements of the source STFD matrix $\boldsymbol{D}_{ss}(t,f)$ represent the mutual TFD between different source signals, the mutual TFD is equal to zero when any two source signals are mutually uncorrelated. Therefore, $\boldsymbol{D}_{ss}(t,f)$ is a diagonal matrix.

Under the condition that $\boldsymbol{D}_{ss}(t,f)$ is a diagonal matrix, Eq. (9.7.45) shows that, given K fixed time-frequency points (t_k,f_k), $k=1,\cdots,K$, the joint diagonalization of the STFD matrix $\boldsymbol{D}_{xx}(t_k,f_k)$, $k=1,\cdots,K$ of the observed data

$$\boldsymbol{D}_{xx}(t_k,f_k) = \boldsymbol{W}\boldsymbol{\Sigma}_k\boldsymbol{W}^{\mathrm{H}} \tag{9.7.46}$$

can be used to obtain the matrix $W \doteq A$ essentially equal to the mixed matrix A can be obtained. Then, the blind separation result of the source signal vector $\mathbf{s}(t)$ can be obtained by using $\hat{\mathbf{s}}(t) = \mathbf{W}^{\dagger}\mathbf{x}(t)$.

The following is the orthogonal approximate joint diagonalization algorithm for blind signal separation using time-frequency distribution, which was proposed by Belouchrani and Amin in 1998[24].

Algorithm 9.7.3. *Orthogonal joint diagonalization time-frequency blind signal separation algorithm*

Step 1 Eigenvalue decomposition: Estimate the $m \times m$ *autocorrelation matrix* $\hat{\mathbf{R}} = \frac{1}{N}\sum_{t=1}^{N}\mathbf{x}(t)\mathbf{x}^H(t)$ *from the zero-mean* $m \times 1$ *observed data vector* $\mathbf{x}(t)$*. Then compute the eigenvalue decomposition of* $\hat{\mathbf{R}}$ *and denote the n largest eigenvalues by* $\lambda_1, \cdots, \lambda_n$ *and* $\mathbf{h}_1, \cdots, \mathbf{h}_n$ *the corresponding eigenvectors.*

Step 2 Prewhitening: The average of the $m - n$ *smallest eigenvalues of* $\hat{\mathbf{R}}$ *is taken as the estimate* $\hat{\sigma}^2$ *of the noise variance. Then the prewhitened* $n \times n$ *data vector is* $\mathbf{z}(t) = [z_1(t), \cdots, z_n(t)]^T = \mathbf{W}\mathbf{x}(t)$*, where* $\mathbf{W} = [(\lambda_1 - \hat{\sigma}^2)^{-1/2}\mathbf{h}_1, \cdots, (\lambda_n - \hat{\sigma}^2)^{-1/2}\mathbf{h}_n]^H$ *is the* $m \times n$ *prewhitening matrix.*

Step 3 Estimate the STFD matrix: Compute the K STFD matrices of the prewhitened data vector $\mathbf{D}_{xx} = \mathbf{D}_{zz}(t_k, f_k)$*,* $k = 1, \cdots, K$.

Step 4 Joint diagonalization: Take orthogonal approximate joint diagonalization for the K STFD matrices $\mathbf{D}_k$

$$\mathbf{D}_k = \mathbf{U}\boldsymbol{\Sigma}_k\mathbf{U}^H, \quad k = 1, \cdots, K, \tag{9.7.47}$$

obtain the unitary matrix $\mathbf{U}$ *as the joint diagonalizer.*

Step 5 Blind signal separation: The source signals are estimated as $\hat{\mathbf{s}}(t) = \mathbf{U}^H\mathbf{z}(t)$*, and/or the mixing matrix* $\mathbf{A}$ *is estimated as* $\hat{\mathbf{A}} = \mathbf{W}^{\dagger}\mathbf{U}$.

2. Cluster-based Quadratic Time-Frequency Blind Signal Separation Algorithm

In Algorithm 9.7.2, the number of source signals is estimated by searching the peak number of the time-frequency distribution slices. The following algorithm uses vector clustering to estimate the number of source signals[5].

Algorithm 9.7.4. *Cluster-based quadratic time-frequency blind signal separation algorithm*

Step 1 Sample STFD estimation: Compute

$$\left[\mathbf{D}_{\mathbf{xx}}^{wvd}(t,f)\right]_{ij} = \rho_{x_ix_j}^{wvd}(t,f), \tag{9.7.48}$$

$$\left[\mathbf{D}_{\mathbf{xx}}^{stft}(t,f)\right]_{ij} = \rho_{x_ix_j}^{stft}(t,f), \tag{9.7.49}$$

$$\left[\mathbf{D}_{\mathbf{xx}}^{mwvd}(t,f)\right]_{ij} = \rho_{x_ix_j}^{wvd}(t,f) \cdot \rho_{x_ix_j}^{stft}(t,f). \tag{9.7.50}$$

Step 2 Selection of effective time-frequency point of signal: If

$$\frac{\|\mathbf{D}_{\mathbf{xx}}^{mwvd}(t_p, f_q)\|}{\max_f\left\{\|\mathbf{D}_{\mathbf{xx}}^{mwvd}(t_p, f)\|\right\}} > \varepsilon, \tag{9.7.51}$$

then time-frequency point (t_p, f_q) is selected and kept as an effective point. Here, ε is a small threshold and typically $\varepsilon = 0.05$.

Step 3 Vector clustering and effective TFD estimation: Compute the spatial direction for each effective time-frequency point $(t_p, f_q) \in \Omega$

$$\boldsymbol{a}(t_a, f_a) = \frac{\boldsymbol{diag}\{\boldsymbol{D}_{\boldsymbol{xx}}^{stft}(t_a, f_a)\}}{\|\boldsymbol{diag}\{\boldsymbol{D}_{\boldsymbol{xx}}^{stft}(t_a, f_a)\|}, \tag{9.7.52}$$

where $\boldsymbol{diag}\{\boldsymbol{B}\}$ denotes a vector with diagonal elements of matrix $\boldsymbol{B}$. Without loss of generality, let the first entry of the spatial direction vector be real and positive. After obtaining $\{\boldsymbol{a}(t_a, f_a) | (t_a, f_a) \in \Omega\}$, they are clustered into n classes using the unsupervised clustering algorithm.

Step 4 Source TFD estimation

$$\hat{\rho}_{s_i}^{wvd}(t, f) = \begin{cases} tr\left(\boldsymbol{D}_{\boldsymbol{xx}}^{wvd}(t, f)\right), & (t, f) \in \Omega \\ 0, & Others \end{cases}. \tag{9.7.53}$$

Step 5 Signal separation: Reconstruct signal $s_i(t)$ using the TFD $\hat{\rho}_{s_i}^{wvd}(t, f)$ of the signal.

Table 9.7.1 compares three blind signal separation methods of convolutively linear mixed source signals in the time domain, frequency domain, and time-frequency domain.

Tab. 9.7.1: Comparison of time domain, frequency domain, and time-frequency domain blind signal separation

Method	Mixing Model	Separation Algorithm	Separation Signal
Time domain	$\boldsymbol{x}(t) = \underline{\boldsymbol{A}}\boldsymbol{s}(t)$	$\boldsymbol{y}_k = \boldsymbol{W}_k\boldsymbol{x}(t),\ \boldsymbol{W}_1 = \boldsymbol{I}$ $\underline{\boldsymbol{W}}_{k+1} = \underline{\boldsymbol{W}}_k + \eta_k[\boldsymbol{I} - \phi(\boldsymbol{y}_k)\boldsymbol{y}_k^{\mathrm{H}}]\underline{\boldsymbol{W}}_k$	$\hat{\boldsymbol{s}}(t) = \boldsymbol{y}(t)$
Frequency domain	$\hat{\boldsymbol{x}}(\tau, f) = \sum_{j=1}^{n} \boldsymbol{a}_j(f) S_j(\tau, f)$	$\hat{\boldsymbol{y}}_k = \boldsymbol{W}_k^f\hat{\boldsymbol{x}}(\tau, f),\ \boldsymbol{W}_1^f = \boldsymbol{I}$ $\boldsymbol{W}_{k+1}^f = \boldsymbol{W}_k^f + \eta_k[\boldsymbol{I} - \phi(\hat{\boldsymbol{y}}_k)\hat{\boldsymbol{y}}_k^{\mathrm{H}}]\boldsymbol{W}_k^f$	$\hat{\boldsymbol{y}}(\tau, f) \xrightarrow{\mathrm{ISTFT}} \hat{\boldsymbol{s}}(t)$
time-frequency domain	$\boldsymbol{D}_{xx}(t, f) = \boldsymbol{A}\boldsymbol{D}_{xx}(t, f)\boldsymbol{A}^{\mathrm{H}}$	$\boldsymbol{D}_{xx}(t, f) = \boldsymbol{U}\boldsymbol{\Sigma}_k\boldsymbol{U}^{\mathrm{H}}$	$\hat{\boldsymbol{s}}(t) = \boldsymbol{U}^{\mathrm{H}}\boldsymbol{x}(t)$

Blind signal separation has been widely used in speech signal processing, image processing and imaging, communication signal processing, medical signal processing, radar signal processing, and so on.

When designing and applying blind signal separation algorithms, the following matters need to be noted[193].

(1) The closer the mixing matrix is to a singular matrix the harder the separation task is for algorithms that do not exhibit the equivariant behavior. In the presence of noise, the task becomes harder also for equivariant algorithms.
(2) Whether it is instantaneous mixing or convolutively mixing, the probability density functions of the signals have an effect. Usually the closer the signals are to Gaussian distributions, the harder the separation gets.

(3) The performance of the blind signal separation algorithm may vary greatly for the narrowband and wideband signals. Therefore, tests of the blind signal separation algorithm should include both of these two signals. A good blind signal separation algorithm should have good performance for both narrowband and wideband signals.
(4) A good blind signal separation algorithm should have good performance for both slow time-varying signals and fast time-varying signals.

The following is the Internet address of some international scholars, research groups, and some program codes that conduct blind signal separation research abroad.

http://www.bsp.brain.riken.go.jp/ICALAB
http://www.islab.brain.riken.go.jp/shiro
http://www.cnl.salk.edu/tewon/ICA/Code/
http://www .cnl.salk.edu/ tony/ica.html
http://www.lis.inpg.fr/demos/sep-sourc/ICAdemo
http://www.cis.hut.fi/projects/ica/
http://www.cis.hut.fi/projects/ica/fastica/

Summary

This chapter first introduces the basic theory of blind signal separation and then introduces three mainstream methods of blind signal separation for linear mixed sources: independent component analysis, nonlinear principal component analysis method, and joint diagonalization method. Then, the blind signal extraction methods for linear mixed sources are discussed. The following are the basic principles and comparisons of these four methods.

Independent component analysis method: Through nonlinear transform, the mutual information between each separated output signal component is minimized, i.e., statistically independent.

Nonlinear principal component analysis method: Through nonlinear transform, the main signal components in the mixed observation data are extracted, and these principal components are made to be approximately statistically independent from each other.

Joint diagonalization of matrices: The blind signal separation is achieved directly by using the joint diagonalizer (which is a matrix essentially equal to the unknown mixed matrix) of a set of autocorrelation matrices of the observed data vectors (it is the matrix that is essentially equal to the unknown mixture matrix).

Blind signal extraction method: Only the signal component with the largest energy is extracted, and the next signal component with the largest energy is extracted after deflation, so as to extract all the signal components of interest.

In addition, this chapter also introduces the time domain blind signal separation method, frequency domain blind signal separation method, and time-frequency domain blind signal separation method for convolutiivelyl mixed sources. The basic principles and comparison of these three methods are as follows.

Time domain blind signal separation method: Using the FIR matrix, the convolutively mixed source model is transformed into the time domain representation model of the linear mixed source, and then the blind signal separation can be achieved by using the independent component analysis or nonlinear principal component analysis.

Frequency domain blind signal separation method: For the short-time Fourier transform of the observed data, the independent component analysis or nonlinear principal component analysis method of frequency domain blind signal separation is used to adaptively update the frequency domain demixing matrix to obtain the short-time Fourier transform of the separated signal, and then the time-domain reconstruction of the separated source is realized through the inverse short-time Fourier transform.

Time-frequency domain blind signal separation method: The joint diagonalizer of the spatial time-frequency distribution (STFD) matrix of the observed data gives the estimation of the mixing matrix directly, thus realizing the blind separation of the non-stationary signals.

Exercises

9.1 Let $\mathbf{x}(t) = \mathbf{A}\mathbf{s}(t)$ be $n \times 1$ sensor observed vector, where $\mathbf{A}$ is the $m \times n$ mixing matrix and $\mathbf{s}(t)$ is the $n \times 1$ source vector. Now we want to design a $m \times m$ pre-whitening matrix $\mathbf{W}$ so that $\mathbf{y}(t) = \mathbf{W}\mathbf{x}(t)$ is a standard white noise. Try to find the relationship between the pre-whitening matrix $\mathbf{W}$ and the mixing matrix $\mathbf{A}$.

9.2 Consider the mixed signal model $\mathbf{x}(k) = \mathbf{A}\mathbf{s}(k)$, where the elements a_{ij} of the mixing matrix $\mathbf{A} = [a_{ij}]_{i=1,j=1}^{m,n}$ are arbitrarily selected from the random variables uniformly distributed over $[-1, +1]$, and the number of sources is $n = 3$. If the source $s_1(k) = \sin(200k)$ be a sine wave signal, $s_2(k) = \text{sgn}(50k + 6\cos(45k))$ be a sign signal, and $s_3(k)$ be a uniformly distributed signal over $[-1, +1]$. Select $k = 1, 2, \cdots, 512$.

(1) Under-determined mixing: take $m = 2$ and draw the waveforms of $\mathbf{x}_1(k)$ and $\mathbf{x}_2(k)$, respectively.

(2) Well-dtermined mixing: take $m = 3$ and draw the waveform of $s_i(k)$, $i = 1, 2, 3$, respectively.

(3) Over-determined mixing: take $m = 5$ and draw the waveform of $s_i(k)$, $i = 1, \cdots, 5$, respectively.

9.3 Consider the following mixed model

(1) Unit linear transformation

$$\begin{bmatrix} x_1(t) \\ x_2(t) \end{bmatrix} = \begin{bmatrix} 2 & 1 \\ 1 & 1 \end{bmatrix} \begin{bmatrix} s_1(t) \\ s_2(t) \end{bmatrix}$$

(2) The unit linear transformation is the same as (1), but the two sources are exchanged.

$$\begin{bmatrix} x_1(t) \\ x_2(t) \end{bmatrix} = \begin{bmatrix} 2 & 1 \\ 1 & 1 \end{bmatrix} \begin{bmatrix} s_2(t) \\ s_1(t) \end{bmatrix}$$

(3) The unit linear transformation is the same as (1), but the polarities of the two sources are opposite.

$$\begin{bmatrix} x_1(t) \\ x_2(t) \end{bmatrix} = \begin{bmatrix} 2 & 1 \\ 1 & 1 \end{bmatrix} \begin{bmatrix} -s_1(t) \\ -s_2(t) \end{bmatrix}$$

(4) Rotation linear transformation and the source is the same as (1)

$$\begin{bmatrix} x_1(t) \\ x_2(t) \end{bmatrix} = \begin{bmatrix} \cos\alpha & -\sin\alpha \\ \sin\alpha & \cos\alpha \end{bmatrix} \begin{bmatrix} s_1(t) \\ s_2(t) \end{bmatrix}$$

(5) Ordinary linear transformation, the source is the same as (1)

$$\begin{bmatrix} x_1(t) \\ x_2(t) \end{bmatrix} = \begin{bmatrix} a_{11} & a_{12} \\ a_{21} & a_{22} \end{bmatrix} \begin{bmatrix} s_1(t) \\ s_2(t) \end{bmatrix}$$

For example, the elements a_{ij} of the mixing matrix $\boldsymbol{A}$ are arbitrarily selected random variables uniformly distributed over $[-1, +1]$.

The sources are divided into three cases.

(a) $s_1(t)$ is a Gaussian signal $\mathcal{N}(0, 1)$ and $s_2(t)$ is a signal uniformly distributed over $[-1, +1]$;

(b) $s_1(t)$ is a signal uniformly distributed over $[-1, +1]$ and $s_2(t)$ is a signal uniformly distributed over $[-0.5, +0.5]$;

(c) $s_1(t)$ is a Gaussian signal $\mathcal{N}(0, 1)$ and $s_2(t)$ is also a Gaussian signal, but the variance is different, such as $s_2(t) \backsim \mathcal{N}(0, 1)$.

Try to draw the sample distribution diagram (x_1, x_2) of the mixed signals for each of the five different mixing models in three rows: the first row corresponds to source distribution (a), the second row corresponds to source distribution (b), and the third line corresponds to source distribution (c).

Try to answer the following questions based on the distribution diagram of the observed samples.

(1) Can the mixed signals be separated in the case of different mixing matrices? Why?

(2) In the case that the sources obey which distribution, the mixed signal cannot be separated, can be separated one or two sources?

9.4 Consider the stochastic gradient algorithm for blind signal separation

$$\boldsymbol{W}(k+1) = \boldsymbol{W}(k) + \mu[\boldsymbol{I} - \phi(\boldsymbol{y}(k))\boldsymbol{y}^{\mathrm{H}}(k)],$$

its continuous time stochastic gradient algorithm is

$$\dot{\boldsymbol{W}}(t) = \mu(t)[\boldsymbol{I} - \phi(\boldsymbol{y}(t))\boldsymbol{y}^{\mathrm{H}}(t)].$$

It has been proved[13] that the solution of the separation matrix of the learning algorithm is a stable equilibrium point if and only if the following three conditions are satisfied

$$m_i + 1 > 0, \quad k_i > 0, \quad \sigma_i^2\sigma_j^2 k_i k_j > 1, \quad \forall i, j(i \neq j),$$

where

$$\begin{aligned} m_i &= \mathrm{E}\left\{y_i^2\dot{\phi}_i(y_i)\right\} = p\mathrm{E}\left\{|y_i|^{p+1}\right\} > 0, \\ k_i &= \mathrm{E}\left\{\dot{\phi}_i(y_i)\right\} = p\mathrm{E}\left\{|y_i|^{p-1}\right\} > 0, \\ \sigma_i^2 &= E\left\{y_i^2\right\}, \end{aligned}$$

where p is an integer, $\phi(y_i)$ is the activation function of the stochastic gradient algorithm, and $\dot{\phi} = \frac{d\phi(t)}{dt}$.

If the odd activation function $\phi_i(y_i) = |y_i|^p \mathrm{sgn}(y_i)$, $p = 1, 2, 3, \cdots$, is selected, try to verify that the three conditions of the above equilibrium point are met.

9.5 The same as the above problem, but the activation function takes the symmetric S-shaped odd function $\phi_i(y_i) = \tanh(\gamma y_i)$, where γ is any positive real number. Prove that the three conditions for verifying the equilibrium point are also met.

9.6 (linear mixed static signals) The source signal vector is known to consist of five non-Gaussian signals

$$\boldsymbol{s}(t) = [\mathrm{sgn}(\cos(2\pi 155t)), \sin(2\pi 800t), \sin(2\pi 300t{+}6\cos(2\pi 600t)), \sin(2\pi 90t), r(t)],$$

where $r(t)$ is a uniformly distributed signal over $[-1, +1]$. The mixed signals are generated by the $m \times p$ mixing matrix $\boldsymbol{A}$ through $\boldsymbol{x}(t) = \boldsymbol{A}\boldsymbol{s}(t)$. The sampling period is $T_s = 0.0001$s, and the data length is taken as $N = 5000$. For the following cases, the blind signal separation using the natural gradient algorithm, the nonlinear principal component analysis algorithm, and the fixed-point ICA algorithm is performed independently for 100 computer simulation experiments. Draw the waveform of each separation output in any operation, and plot the curve of the statistical average of crosstalk error of various algorithms with the number of iterative samples.

(1) Well-determined blind signal separation: $m = p = 5$, no additive noise.
(2) Overdetermined blind signal separation: $m = 8$, $p = 5$, divided into no additive noise and additive noise $n(t) \sim \mathcal{N}(0, 1)$.

9.7 (Linear mixed image signals) Consider the linear mixed images: sample by frame and then connect in sequence to form the discrete sampling value of the image. The elements of the mixing matrix are still generated by random variables that are uniformly distributed over $[-1, +1]$.

(1) Multiple face images are selected as the sources and blind separation of image signals is performed using the natural gradient algorithm, nonlinear principal component analysis method, and fixed-point ICA method, and the separation results are recovered into images, which are compared with the original images.

(2) Different images such as human faces, natural scenery, and traffic are selected for the linear mixing. Blind separation of image signals is performed using the same algorithm as above, and the separation results are recovered into images and compared with the original images.

9.8 (Linear mixed dynamic signals) Consider the separation of mixed speech: the speech signal is first converted into an analog electrical signal using a microphone, and then the discrete sample values are obtained by A/D conversion. The source signals are two male and two female signals, and each of the four people reads the same sentences for 3 seconds. Each speech signal was first sampled and A/D converted at 22050 Hz/8 bit to obtain the respective discrete sample values. The samples are then mixed by a linear mixing model $\boldsymbol{x}(k) = \boldsymbol{A}\boldsymbol{s}(k)$, and the number of samples is taken as $N = 2048$, where the elements $a_{ij}(i = 1, \cdots, 6; j = 1, \cdots, 4)$ of the mixing matrix $\boldsymbol{A}$ are chosen as uniformly distributed random variables over $[-1, +1]$.

(1) Draw the continuous-time waveforms of the four speech signals and the linear mixed signals.

(2) Use the natural gradient algorithm, the nonlinear principal component analysis method, the fixed-point ICA algorithm, and the nonorthogonal joint diagonalization method to perform blind signal separation and compare the results of the separated signals.

(3) Conduct the computer experiments 100 times and plot the statistical mean and deviation of the crosstalk error obtained by various blind signal separation methods.

(4) Select the separation results of several independent experiments randomly and play them to text the separation effect of each speech signal.

9.9 (Convolutively mixed dynamic signals) The four speech signals are the same as the previous problem, but they are convolutional linear mixed

$$x_i(t) = \sum_{j=1}^{4} \sum_{k=0}^{3} a_{ij}(k) s_j(t-k) = \boldsymbol{a}_i^{\mathrm{T}} \boldsymbol{s}(t), \quad i = 1, \cdots, 6; t = 1, \cdots, 2048,$$

where the 6×4 mixing matrix $\boldsymbol{A}$ is the same as above.

(1) Draw the waveforms of four speech signals and convolution linear mixed signals respectively.

(2) Use the time domain blind signal separation, frequency domain blind signal separation (algorithm 9.7.1), and time-frequency domain blind signal separation of the

convolutively linear mixed source to carry out computer separation experiments and compare the separation results.

(3) Conduct the computer experiments 100 times independently and plot the statistical mean and deviation curves of the crosstalk errors of various blind signal separation methods.

(4) Select the separation results of several independent experiments randomly and play to test the separation effect of each speech signal.

10 Array Signal Processing

When multiple signal sources exist in space, it is often necessary to locate these spatial signals to track or detect those of interest and suppress those considered to be interference. Therefore, it is necessary to use an array antenna to receive multiple spatial signals. The analysis and processing of spatial signals received by an array antenna are collectively referred to as array signal processing.

Array signal processing has been widely used, with typical applications including[215]:
Radar: phased-array radar, air traffic control radar, synthetic aperture radar, etc;
Sonar: source localization and classification;
Communications: directional transmission and reception, sector broadcast in satellite communications, mobile communications, etc;
Imaging: ultrasonic imaging, optical imaging, tomographic imaging, etc;
Geophysical exploration: earth crust mapping, oil exploration, etc;
Astrophysical exploration: high-resolution imaging of the universe;
Biomedicine: fetal heart monitoring, tissue hyperthermia, hearing aids, etc.

The main problems of array signal processing are[102, 126]

(1) Beamforming - make the mainlobe of the array pattern point to the desired direction;
(2) Nulling forming - align the zero point of the antenna with all interference directions;
(3) Direction of arrival estimation (DOA) - super resolution estimation of the DOA of spatial signals.

The first special issue on array signal processing was published in 1964 in the IEEE Transactions on Antenna Propagation Transactions[1]. Since then, IEEE journals have published several special issues on array signal processing[2, 3].

This chapter will introduce the basic theory, main methods, and typical applications for array signal processing.

10.1 Coordinate Representation of Array

A beamformer is a signal processor used in conjunction with a sensor array to provide a versatile form of spatial filtering[215]. The sensor array collects spatial samples of propagating wave field, which are processed by the beamformer. The objective is to estimate the signal arriving from a desired direction in the presence of noise and interfering signals. A beamformer performs spatial filtering to separate multiple spatial signals that have overlapping frequency content but originate from different spatial locations. The spatial signal from the desired direction is called the desired signal, and all other spatial signals in the undesired directions are collectively referred to as interfering signals.

https://doi.org/10.1515/9783110475562-010

10.1.1 Array and Noise

An array is the arrangement of multiple sensors or antennas. The sensor or antenna element of the array is referred to as the array element. The layout forms of the array elements can be generally divided into categorises as uniform, nonuniform, and random distribution.

According to the arrangement or layout rules of the array elements, the array is divided into the following three categories.

(1) Linear array: the sensors aligned with a line, divided into uniform linear array, nonuniform linear array (sparse array and random linear array).
(2) Planar array: composed of multiple uniform linear array or uniform circular array, nonuniform circular array.
(3) Cubic array: multiple planar arrays aligned with equal intervals or cylindrical arrays (multiple uniform circular arrays concentric arrangement).

The transmitted sources in space or the receiving signals of the sensors are collectively referred to as spatial signals. The main characteristics of spatial signals are

Time-domain characteristics: modulated (such as linear frequency modulation signal, BPSK signal, etc.) and non-modulated signal.

Frequency-domain characteristics: narrowband signals and wideband signals.

Spatial characteristics: near-field signals (spherical wave signals) and far-field signals (planar wave signals).

Noise (or interference) is classified as follows.

$$\text{Noise}\begin{cases}\text{By interference mode}\begin{cases}\text{Additive noise}\\ \text{Multiplicative noise}\end{cases}\\ \text{By statistical characteristics}\begin{cases}\text{White noise}\begin{cases}\text{Time domain white noise}\\ \text{Spatial white noise}\end{cases}\\ \text{Colored noise}\begin{cases}\text{Time domain colored noise}\\ \text{Spatial colored noise}\end{cases}\end{cases}\\ \text{By probability density}\begin{cases}\text{Gaussian noise}\\ \text{Non-Gaussian noise}\end{cases}\end{cases}$$

where

Additive noise: signal added by noise;
Multiplicative noise: signal multiplied by noise;
Time-domain white noise: statistically uncorrelated noise at different moments;
Time-domain colored noise: statistically correlated noise at different moments;
Spatial white noise: the noise of different array element statistically uncorrelated;
Spatial colored noise: the noise of different array element statistically correlated.

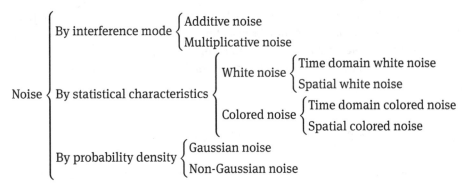

10.1.2 Coordinate System of Array

Consider p signal sources propagating in space, all of which are narrowband signals. Now, a sensor array with M omnidirectional elements is used to receive these signals.

The coordinate system representations of a planar array and linear array are discussed below.

1. Planar Array Coordinate System

Assuming that the array is arranged on the $x-y$ plane, take the first element as the origin O of the planner array coordinate system (i.e., the time reference point of the planar array), as shown in Figure 10.1.1.

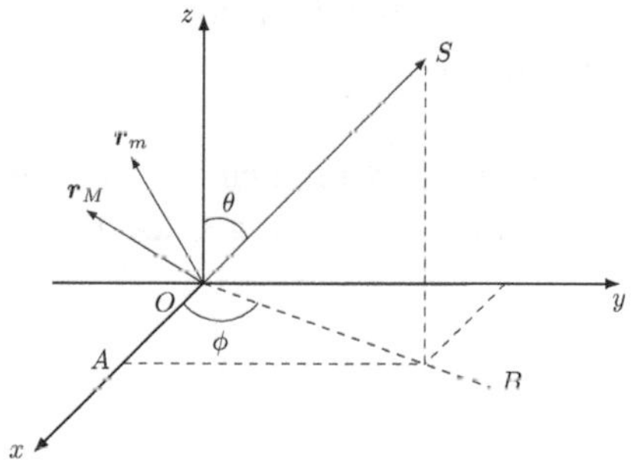

Fig. 10.1.1: Planar array coordinate system

The core problem of array signal processing is to locate the spatial source using omnidirectional sensor or antenna array. As is widely known, the location of a source is determined by three-dimensional parameters of the source: the distance OS from the origin to the source (source spatial distance), the angle θ between the source and the normal of the $x-y$ plane (z-axis), i.e., the source elevation angle, and the source azimuth angle ϕ. In array signal processing, only the elevation angle θ and the azimuth angle ϕ need to be considered.

For the direction cosine u_i of the i-th source with respective to the x-axis , the directional cosine v_j with respective to the y-axis, and the elevation angle θ and azimuth angle ϕ, there exist a relationship between them as

$$u_i + jv_i = \sin\theta_i e^{j\phi_i}. \tag{10.1.1}$$

Thus, if the cosine u_i in the x-axis direction and the cosine v_i in the y-axis direction of the i-th source are obtained, the elevation angle θ and azimuth angle ϕ of the source can be determined as

$$\phi_i = \arctan(u_i/v_i), \quad \theta_i = \arctan(u_i/\sin\phi_i). \tag{10.1.2}$$

Therefore, main parameters need to be estimated by a two-dimensional array signal processing algorithm is x-axis cosine u and y-axis cosine v of each source.
The vertical distance SB from the source S to the $x-y$ plane is called the target down-range distance, and the vertical distance AB from S to the normal plane $x-z$ is called the target cross-range distance. The resolution of source localization is divided into target down-range resolution and cross-range resolution.

The time taken by a plane wave departing from the i-th source in direction (ϕ_i, θ_i) and received by the m-th element is given by

$$\tau_m(\phi_i, \theta_i) = \frac{\langle \mathbf{r}_m, \hat{\mathbf{v}}(\phi_i, \theta_i) \rangle}{c}, \tag{10.1.3}$$

where $\mathbf{r}_m$ is the position vector of the m-th element to the first element, i.e., the reference point (the origin of the coordinate system), $\hat{\mathbf{v}}(\phi_i, \theta_i)$ is the unit vector in direction (ϕ_i, θ_i), c is the propagation speed of the plane wave front.

2. Uniform Linear Array Coordinate System

Assuming that a linear array of equispaced elements with element spacing d aligned with the x-axis such that the first element is situated at the origin (i.e., the time reference point), as shown in Figure 10.1.2.

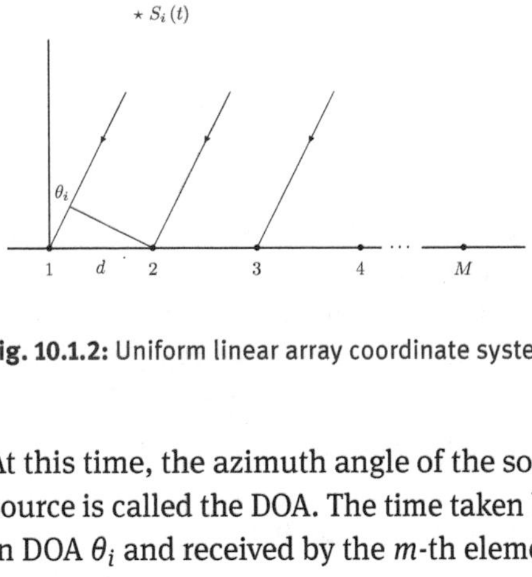

Fig. 10.1.2: Uniform linear array coordinate system

At this time, the azimuth angle of the source $\phi = 0$, and the elevation angle θ of the source is called the DOA. The time taken by a plane wave arriving from the i-th source in DOA θ_i and received by the m-th element is given by

$$\tau_m(\theta_i) = \frac{d}{c}(m-1)\cos\theta_i, \quad m = 1, 2, \cdots, M. \tag{10.1.4}$$

The signal induced on the reference element due to the i-th source is norrmally expressed in complex-valued notation as

$$x_1(t) = m_i(t)e^{j2\pi f_0 t} = m_i(t)e^{j\omega_0 t}, \tag{10.1.5}$$

where $m_i(t)$ is the complex-valued modulation function, $\omega_0 = 2\pi f_0$ is the frequency of the source. The structure of the complex-valued modulation function is determined by the particular modulation used in the array signal processing system. For example, in mobile communication systems, the typical modulation functions are concentrated as follows[91].

(1) For a frequency division multiple access (FDMA) communication system, it is a frequency-modulated signal given by

$$m_i(t) = A_i e^{j\xi_i(t)}, \tag{10.1.6}$$

where A_i is the amplitude of the i-th user and $\xi_i(t)$ is the message of the i-th user at moment t.

(2) For a time division multiple access (TDMA) communication system, it is a time-modulated signal given by

$$m_i(t) = \sum_n d_i(n)p(t - n\Delta), \tag{10.1.7}$$

where $p(t)$ is the sampling pulse, the amplitude $d_i(n)$ is the message symbol of the i-th user, and Δ is the sampling imterval.

(3) For a code division multiple access (CDMA) communication system, the modulation function is

$$m_i(t) = d_i(n)g(t), \tag{10.1.8}$$

where $d_i(n)$ denotes the message sequence of the i-th user and $g(t)$ is a peseudo-random binary sequence having the values +1 or −1.

Assuming that the wavefront arrives at the i-th element $\tau_m(\phi_i, \theta_i)$ seconds earlier before it arrives at the reference element. Under this assumption, it is readily to have $\tau_1(\phi_i, \theta_i) = 0$, and the signal induced on the m-th element due to the i-th source can be then formulated as

$$x_m(t) = m_i(t)e^{j\omega_0[t+\tau_m(\phi_i,\theta_i)]}, \quad m = 1, 2, \cdots, M. \tag{10.1.9}$$

For the first array element to be the reference element, its observation signal is $x_1(t) = m_i(t)e^{j\omega_0 t}$, since $\tau_1(\phi_i, \theta_i) = 0$.

Suppose the m-th element has additive observation noise $e_m(t)$, we will find that

$$x_m(t) = m_i(t)e^{j\omega_0[t+\tau_m(\phi_i,\theta_i)]} + e_m(t), \quad m = 1, 2, \cdots, M. \tag{10.1.10}$$

The additive noise $e_m(t)$ usually includes background noise of the m-th element, observation error, and electronic noise generated in the channel from the i-th source to the m-th element.

10.2 Beamforming and Spatial Filtering

A beamformer is essentially a spatial filter that extracts the desired signal. The desired signal is identified by time-delay steering of the sensor output: any signal incident on the array from the direction of interest will appear at the output of the steering filter as an exact copy of the signal itself; any other signal that does not have this property is considered as noise or interference. The purpose of a beamformer is to minimize the influence of noise or interference at the output of the array while maintaining the frequency response of the desired signal.

Each element in the array is usually omnidirectional (i.e., omnidirectional). Such a sensor array is called an omnidirectional array, which can not only greatly reduce the cost of the array, but also facilitate the detection of sources from any direction of 360°.

Using a set of finite impulse response (FIR) filters, the observation signals from a set of omnidirectional elements (or sensors) can be properly weighted and summed to form a beam (major lobe) that is directly aligned with a source of interest (desired source) and that minimizes the effect of other undesired signals (collectively referred to as noise or interference) leaked from the sidelobe. Since this filter is neither a time-domain filter nor a frequency-domain filter, but only uses several taps as the coefficients of the filter, it is called a (spatial) beamformer and is a spatial filter.

The relationship between the FIR filter and the beamformer is analyzed below.

10.2.1 Spatial FIR Filter

Since the envelope of the narrowband signal changes slowly, the envelope of the same signal received by each element of the uniform linear array is the same. If the spatial signal $s_i(n)$ is far enough away from the element such that the wave front of its electronic wave reaching each element is a plane, such a spatial signal is called a far-field signal. On the contrary, if the spatial signal $s_i(n)$ is close to the element and the wave front of the electronic wave arriving at each element is a spherical wave, then it is called a near-field signal.

The directional angle of the far-field signal $s_i(n)$ arriving at each element is the same, denoted by θ_i, and is called the DOA, which is defined as the angle between the direct line of the signal $s_i(n)$ arriving at the element and the normal direction of the array. Taking the first element as the reference point (referred to as the reference element), that is, the received signal $s_i(n)$ of spatial signal $s_i(n)$ on the reference element is equal to $s_i(n)$. There is a delay (or advanced) in the arrival of this signal to the other elements with respective to the reference element. Let the phase difference caused by the delay of signal $s_i(n)$ propagation at the second element be ω_i, then there is a relationship between the DOA θ_i and the phase difference ω_i

$$\omega_i = 2\pi \frac{d}{\lambda} \sin(\theta_i), \tag{10.2.1}$$

where d is the distance between two adjacent elements and λ is the signal wavelength. The element distance d should satisfy the half-wavelength condition, i.e., $d \leq \lambda/2$, otherwise the phase difference ω_i may be greater than π, resulting in the so-called direction ambiguity, that is, θ_i and $\theta_i + \pi$ can all be the DOA of signal $s_i(t)$. Obviously, since it is an uniform linear array, the phase difference between the waves of signal $s_i(n)$ arriving at m-th element and those arriving at the reference element is $(m-1)\omega_i = 2\pi\frac{d}{\lambda}(m-1)\sin(\theta_i)$. Therefore, the received signal $s_i(n)$ at the m-th element is $s_i(n)e^{j(m-1)\omega_i}$.

If the array consists of M elements, then the vector

$$\boldsymbol{a}(\theta_i) \stackrel{\text{def}}{=} [1, e^{j\omega_i}, \cdots, e^{-j(M-1)\omega_i}]^T = [a_1(\theta_1), \cdots, a_M(\theta_i)]^T, \tag{10.2.2}$$

consisting of the phase differences of the signal $s_i(n)$ arriving at each element is called the response vector, direction vector, or streering vector of the signal $s_i(n) = s_i e^{j\omega_i n}$[215]. The geometric interpretation of the steering vector $\boldsymbol{a}(\theta_i)$ is a vector pointing to the i-th source, i.e., the beam.

The m-th element $e^{j(m-1)\omega_i}$ of the steering vector $\boldsymbol{a}(\theta_i)$ describes the phase difference of the wave propagation of the i-th source at the m-th array element, with respective to the reference element.

If there are p signals in the far-field (where $p \le M$), then the observed or received signal $x_m(n)$ at the m-th element is

$$x_m(n) = \sum_{i=1}^{p} a_m(\omega_i)s_i(n) + e_m(n), \quad m = 1, \cdots, M, \tag{10.2.3}$$

where $e_m(n)$ denotes the additive observation noise on the m-th element. The observations of all the M elements form an $M \times 1$ observation data vector as

$$\boldsymbol{x}(n) = [x_1(n), \cdots, x_M(n)]^T. \tag{10.2.4}$$

Similarly, an $M \times 1$ observation noise vector

$$\boldsymbol{e}(n) = [e_1(n), \cdots, e_M(n)]^T \tag{10.2.5}$$

can be defined. In this way, Eq. (10.2.3) can be expressed in vector form as

$$\boldsymbol{x}(n) = \sum_{i=1}^{p} \boldsymbol{a}(\omega_i)s_i(n) + \boldsymbol{e}(n) = \boldsymbol{A}(\omega)\boldsymbol{s}(n) + \boldsymbol{e}(n), \tag{10.2.6}$$

where

$$\boldsymbol{A}(\omega) = [\boldsymbol{a}(\omega_1), \cdots, \boldsymbol{a}(\omega_p)] \tag{10.2.7}$$

$$= \begin{bmatrix} 1 & 1 & \cdots & 1 \\ e^{-j\omega_1} & e^{-j\omega_2} & \cdots & e^{-j\omega_p} \\ \vdots & \vdots & \vdots & \vdots \\ e^{-j(M-1)\omega_1} & e^{-j(M-1)\omega_2} & \cdots & e^{-j(M-1)\omega_p} \end{bmatrix}, \tag{10.2.8}$$

$$\boldsymbol{s}(n) = [s_1(n), \cdots, s_p(n)]^T \tag{10.2.9}$$

are the $M \times p$ array response matrix (or direction matrix, transmission matrix) and the $p \times 1$ signal vector, respectively. A matrix with the structure shown in Eq. (10.2.8) is said to be an Vandermonde matrix. The characteristic of Vandermonde matrix is that if $\omega_1, \cdots, \omega_p$ are different from each other, then its columns are independent of each other, that is, the Vandermonde matrix has full column rank.

In array signal processing, one sample is called one snapshot. Assuming that the received signal $x_m(1), \cdots, x_m(N)$ of N snapshots is observed on each element, where $m = 1, \cdots, M$. The beamforming problem can be formulated as follows: only using these observations, how to find the DOA θ_d of a desired signal $s_d(n)$ so as to achieve the localization of the desired spatial source.

Fig.10.2.1 shows the schematic diagram of the spatial FIR filter for beamforming.

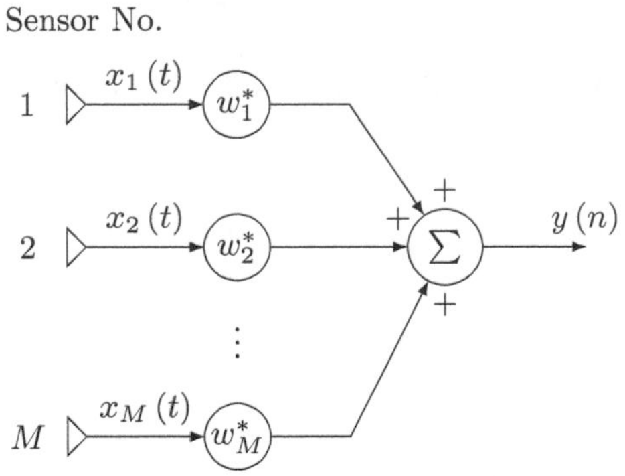

Fig. 10.2.1: Spatial FIR filter

In the figure, the signal $x_1(n), \cdots, x_M(n)$ from each element are mutiplied by the weight vector $\boldsymbol{w} = [w_1, \cdots, w_M]^{\mathrm{T}}$ of the spatial FIR filter and summed to form the array output

$$y(n) = \sum_{m=1}^{M} w_m^* x_m(n). \tag{10.2.10}$$

Since each sensor generally uses a quadrature receiver to generate biphasic (in-phase and quadrature) data, the observation data and the tap coefficients FIR filter take complex values.

Taking the first sensor as reference, if the signal propagation delay between two adjacent sensors is $\triangle$, then the frequency response of the FIR filter $\boldsymbol{w} = [w_1, \cdots, w_M]^{\mathrm{T}}$ can be expressed as

$$r(\omega) = \sum_{m=1}^{M} w_m^* \mathrm{e}^{-\mathrm{j}(m-1)\omega\triangle} = \boldsymbol{w}^{\mathrm{H}}\boldsymbol{d}(\omega), \tag{10.2.11}$$

where

$$\boldsymbol{d}(\omega) = [1, \mathrm{e}^{-\mathrm{j}\omega\triangle}, \cdots, \mathrm{e}^{-\mathrm{j}(M-1)\omega\triangle}]^{\mathrm{T}} \tag{10.2.12}$$

represents the response of the FIR filter to a complex exponential signal of frequency ω and $d_m(\omega) = \mathrm{e}^{-\mathrm{j}(m-1)\omega\triangle}$, $m = 1, \cdots, M$ is the phase of the complex exponential at each tap of the FIR filter relative to the tap associated with w_1.

The signal received by a sensor is usually assumed to be zero-mean, and the variance or expected power of the FIR filter output is thereby

$$|E\{y(n)\}|^2 = \boldsymbol{w}^{\mathrm{H}} E\{\boldsymbol{x}(n)\boldsymbol{x}^{\mathrm{H}}(n)\}\boldsymbol{w} = \boldsymbol{w}\boldsymbol{R}_{xx}\boldsymbol{w}. \tag{10.2.13}$$

If the received data $\boldsymbol{x}(n)$ of the sensor is wide sense stationary, then $\boldsymbol{R}_{xx} = E\{\boldsymbol{x}(n)\boldsymbol{x}^{\mathrm{H}}(n)\}$, the data covariance matrix, is independent of time. Although we often encounter non-stationary data, the wide sense stationary assumption is used to develop statistically optimal FIR filters and evaluate steady-state performance.

Using the frequency response vector $\boldsymbol{d}(\omega)$ of the FIR filter, the frequency response of the sensor observation data vector can be expressed as $\hat{\boldsymbol{x}}(\omega) = S(\omega)\boldsymbol{d}(\omega)$, where $S(\omega) = \mathcal{F}\{x(n)\}$ is the spectrum of the desired spatial source $s(n)$. Let $P_s(\omega) = |S(\omega)|^2 = S(\omega)S^*(\omega)$ represent the power spectrum of $s(n)$, and suppose the spectrum $S(\omega)$ of the spatial source $s(n)$ has no energy outside of the spectral band $[\omega_a, \omega_b]$. then the data covariance matrix can be formulated as

$$\boldsymbol{R}_{xx} = E\{\boldsymbol{x}(n)\boldsymbol{x}^{\mathrm{H}}(n)\} = E\{\hat{\boldsymbol{x}}(\omega)\hat{\boldsymbol{x}}^{\mathrm{H}}(\omega)\} = \frac{1}{2\pi}\int_{\omega_a}^{\omega_b} P_s(\omega)\boldsymbol{d}(\omega)\boldsymbol{d}^{\mathrm{H}}(\omega)d\omega, \tag{10.2.14}$$

where the frequency band $[\omega_a, \omega_b]$ is narrow enough and corresponds to a narrowband signal.

It should be noted that $\boldsymbol{d}(\omega)$ is only the frequency response vector of the FIR filter, instead of the steering vector because ω is the frequency, not the DOA θ. Therefore, an interesting and important question is, what is the relationship between the DOA and the frequency ω? Can the FIR filter and beamformer be analogized or interchanged? This is exactly the problem to be discussed below.

10.2.2 Broadband Beamformer

The broadband beamformer samples the signal waveform propagating in the spatial field and time field, which is suitable for the signal with obvious frequency bandwidth (broadband signal).

As shown in Fig.10.2.2, the output $y(n)$ at moment n of the broadband beamformer can be expressed as

$$y(n) = \sum_{m=1}^{M}\sum_{p=0}^{K-1} w_{m,p}^{*} x_m(n-p), \tag{10.2.15}$$

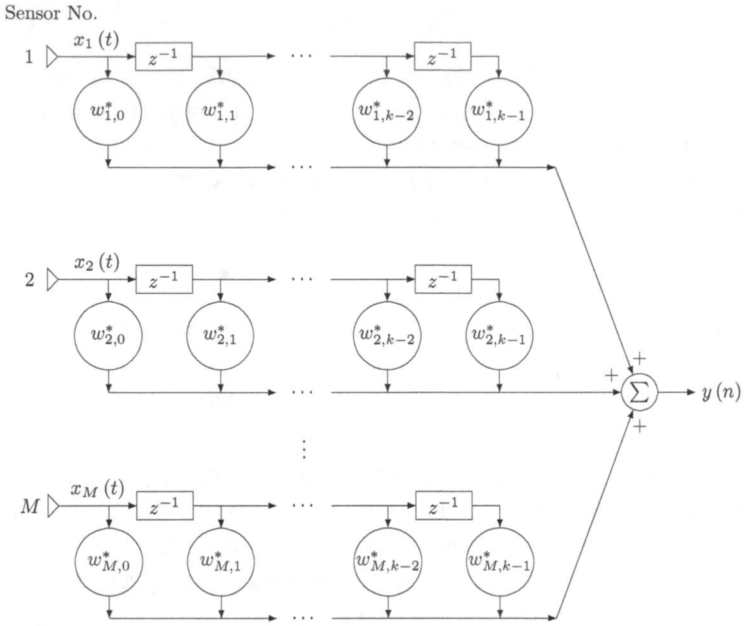

Fig. 10.2.2: Broadband beamformer

where $K - 1$ is the number of delays in each of the M sensor channels.

The output of the FIR filter and the broadband beamformer can be expressed uniformly in vector form as

$$y(n) = \mathbf{w}^{\mathrm{H}}\mathbf{x}(n), \tag{10.2.16}$$

where

$$\text{FIR filter:}\begin{cases} \mathbf{w} &= [w_1, \cdots, w_M]^{\mathrm{T}}, \\ \mathbf{x}(n) &= [x_1(n), \cdots, x_M(n)]^{\mathrm{T}}, \end{cases}$$

$$\text{Beamforming:}\begin{cases} \mathbf{w} &= [w_{1,0}, \cdots, w_{1,K-1}, \cdots, w_{M,0}, \cdots, w_{M,K-1}]^{\mathrm{T}}, \\ \mathbf{x}(n) &= [x_1(n), \cdots, x_1(n-K+1), \cdots, x_M(n), \cdots, x_M(n-K+1)]^{\mathrm{T}}. \end{cases}$$

The response of broadband beamformer is defined as the amplitude and phase presented to a complex-valued plane wave as a function of the location and frequency of the source. The spatial location of a source is generally a three-dimensional quantity, but often we are only concerned with one-dimensional or two-dimensional DOA, not the distance between the source and the sensor.

When a sensor array samples a spatially propagating far-field signal, it is usually assumed that the far-field signal is a complex-valued plane wave with DOA θ and frequency ω. For convenience let the first sensor be the reference point and the phase be zero. This implies $x_1(n) = \mathrm{e}^{\mathrm{j}\omega n}$ and $x_m(n) = \mathrm{e}^{\mathrm{j}\omega[n-\triangle_m(\theta)]}$, $m = 2, \cdots, M$. $\triangle_m(\theta)$

represents the time delay due to propagation from the first to the m-th sensor. Especially, the time delay from the first to itself is $\triangle_1(\theta) = 0$.

Substituting the observation signal $x_m(n) = \mathrm{e}^{\mathrm{j}\omega[n-\triangle_m(\theta)]}$ of the m-th sensor into the expression of the broadband beamformer output, we can obtain

$$y(n) = \mathrm{e}^{\mathrm{j}\omega k} \sum_{m=1}^{M} \sum_{p=0}^{K-1} w_{m,p}^{*} \mathrm{e}^{-\mathrm{j}\omega[\triangle_m(\theta)+p]} = \mathrm{e}^{\mathrm{j}\omega k} r(\theta, \omega), \tag{10.2.17}$$

where $r(\theta, \omega)$ is the frequency response of the broadband beamformer and can be expressed in vector form as

$$r(\theta, \omega) = \sum_{m=1}^{M} \sum_{p=0}^{K-1} w_{m,p}^{*} \mathrm{e}^{-\mathrm{j}\omega[\triangle_m(\theta)+p]} = \boldsymbol{w}^{\mathrm{H}} \boldsymbol{d}(\theta, \omega) \tag{10.2.18}$$

where the elements $w_{m,p}$ of

$$\boldsymbol{w} = [\boldsymbol{w}_1^{\mathrm{T}}, \cdots, \boldsymbol{w}_M^{\mathrm{T}}]^{\mathrm{T}} = [w_{1,0}, \cdots, w_{1,K-1}, \cdots, w_{M,0}, \cdots, w_{M,K-1}]^{\mathrm{T}} \tag{10.2.19}$$

represent the weight coefficients of the m-th FIR filter $\boldsymbol{w} = [w_{m,0}, w_{m,1}, \cdots, w_{m,K-1}]^{\mathrm{T}}$ to the sensor observation data $x_m(n-p)$. The elements of $\boldsymbol{d}(\theta, \omega)$ correspond to the complex exponentials $\mathrm{e}^{-\mathrm{j}\omega[\triangle_m(\theta)+p]}$. In general, it can be expressed as

$$\begin{aligned} \boldsymbol{d}(\theta, \omega) &= [d_1(\theta, \omega), d_2(\theta, \omega), \cdots, d_{MK}(\theta, \omega)]^{\mathrm{T}} \\ &= [1, \mathrm{e}^{\mathrm{j}\omega\tau_2(\theta,\omega)}, \mathrm{e}^{\mathrm{j}\omega\tau_3(\theta,\omega)}, \cdots, \mathrm{e}^{\mathrm{j}\omega\tau_{MK}(\theta,\omega)}]^{\mathrm{T}}, \end{aligned} \tag{10.2.20}$$

where the subscripts of $d_i(\theta, \omega)$, $i = 1, 2, \cdots, MK$ are

$$i = (m-1)K + p + 1, \quad m = 1, \cdots, M; \quad p = 0, 1, \cdots, K-1. \tag{10.2.21}$$

Let the first sensor be taken as the reference and its time delay be zero, that is, $\tau_1(\theta, \omega) = 0$, then the time delay $\tau_i(\theta, \omega) = -[\triangle_m(\theta) + p]$ includes two parts:

(1) the time delay $-\triangle_m(\theta, \omega)$ due to propagation from the first to the M-th sensor;
(2) the time delay $-p$ of the m-th FIR filter $\boldsymbol{w}_m$ due to the tap delay from the zero phase reference $w_{m,0}$ to the $w_{m,p}$.

Using the steering vector $\boldsymbol{d}(\theta, \omega)$ of the broadband beamformer, the frequency response of the sensor observation data vector can be expressed as $\hat{\boldsymbol{x}}(\omega) = \mathcal{F}\{\boldsymbol{x}(n)\} = S(\omega)\boldsymbol{d}(\theta, \omega)$. Suppose the spatial source $s(n)$ has no energy outside of the spectral band $[\omega_a, \omega_b]$. At this time, the data covariance matrix is

$$\boldsymbol{R}_{xx} = E\{\boldsymbol{x}(n)\boldsymbol{x}^{\mathrm{H}}(n)\} = E\{\hat{\boldsymbol{x}}(\omega)\hat{\boldsymbol{x}}^{\mathrm{H}}(\omega)\} = \frac{1}{2\pi} \int_{\omega_a}^{\omega_b} P_s(\omega) \boldsymbol{d}(\theta, \omega) \boldsymbol{d}^{\mathrm{H}}(\theta, \omega) d\omega, \tag{10.2.22}$$

where $P_s(\omega) = |S(\omega)|^2$ represents the power spectrum of the source with its frequency band $[\omega_a, \omega_b]$ being relatively wide, corresponding to a broadband signal.

10.2.3 Analogy and Interchange between Spatial FIR Filter and Beamformer

The vector notation in the frequency response $r(\theta, \omega) = \mathbf{w}^{\mathrm{H}}\boldsymbol{d}(\theta, \omega)$ of the broadband beamformer suggests a vector space interpretation of beamforming. This point of view is useful both in beamformer design and analysis. Since the weight vector $\boldsymbol{w}$ and the array response vectors $\boldsymbol{d}(\theta, \omega)$ are vectors in a MK-dimensional vector space, the cosine of the angle between $\boldsymbol{w}$ and $\boldsymbol{d}(\theta, \omega)$, namely,

$$\cos \alpha = \frac{\langle \boldsymbol{w}, \boldsymbol{d}(\theta, \omega)\rangle}{\| \boldsymbol{w} \|_2 \cdot \| \boldsymbol{d}(\theta) \|_2} = \frac{\mathbf{w}^{\mathrm{H}}\boldsymbol{d}(\theta, \omega)}{\| \boldsymbol{w} \|_2 \cdot \| \boldsymbol{d}(\theta) \|_2}, \tag{10.2.23}$$

determine the response $r(\theta, \omega)$ of the beamformer. Specifically, for some DOA and frequency combination (θ_0, ω_0), if the angle between $\boldsymbol{w}$ and $\boldsymbol{d}(\theta_0, \omega_0)$ is 90°, i.e., they are orthogonal to each other, then the response $r(\theta_0, \omega_0) = \mathbf{w}^{\mathrm{H}}\boldsymbol{d}(\theta_0, \omega_0)$ of the beamformer is zero. This shows that the signal on DOA θ_1 is completely cancelled, thus constituting the nulling froming.On the contrary, if $\boldsymbol{w} = \boldsymbol{d}(\theta_1, \omega_1)$, the output f of the beamformer $r(\theta_1, \omega_1) = \mathbf{w}^{\mathrm{H}}\boldsymbol{d}(\theta_1, \omega_1) = \| \boldsymbol{d}(\theta_1, \omega_1) \|_2^2$ reaches a maximum, i.e., the spatial signal on DOA θ_1 is detected with the maximum intensity.

Consider two sources at different locations and/or frequencies, say (θ_1, ω_1) and (θ_2, ω_2), the abiltiy to descriminate between them is determinded by the cosine

$$\cos \alpha = \frac{\langle \boldsymbol{d}(\theta_1, \omega_1), \boldsymbol{d}(\theta_2, \omega_2)\rangle}{\| \boldsymbol{d}(\theta_1, \omega_1) \|_2 \cdot \| \boldsymbol{d}(\theta_2, \omega_2) \|_2} \tag{10.2.24}$$

of the angle between their array response vectors[67] $\boldsymbol{d}(\theta_1, \omega_1)$ and $\boldsymbol{d}(\theta_2, \omega_2)$. Obviously, if $\boldsymbol{w} = \boldsymbol{d}(\theta_1, \omega_1)$ and $\boldsymbol{d}(\theta_1, \omega_1) \perp \boldsymbol{d}(\theta_2, \omega_2)$, then the two sources can be fully identified. Otherwise, even if $\boldsymbol{w} = \boldsymbol{d}(\theta_1, \omega_1)$ is selected, but $\boldsymbol{d}(\theta_1, \omega_1)$ is not orthogonal to $\boldsymbol{d}(\theta_2, \omega_2)$, then although $r(\theta_1, \omega_1) = M$ is maximum, $r(\theta_2, \omega_2) \neq 0$, so the two spatial sources will be detected at the same time, only the intensity of detection is different, thus forming the mainlobe and sidelobe.

The above analysis suggests that if $\boldsymbol{w} = \boldsymbol{d}(\theta_d, \omega)$ is selected and appropriate constraints are imposed, it is possible to extract only the spatial sources with DOA θ and suppress all sources with the other DOA. The source is called a narrowband signal with center frequency ω, if the covariance matrix $\boldsymbol{R}_{xx}$ can be expressed as an outer product in the form

$$\boldsymbol{R}_{xx} = \sigma_s^2 \boldsymbol{d}(\theta, \omega_0)\boldsymbol{d}^{\mathrm{H}}(\theta_1, \omega_0) \tag{10.2.25}$$

of rank 1[215], where

$$\sigma_s^2 = \frac{1}{a\pi} \int_{\omega_a}^{\omega_b} P_s(\omega) d(\omega) \tag{10.2.26}$$

is the variance or average power of the narrowband source.

The observation time bandwidth product (TBWP) is a basic parameter that determines whether a source can be viewed as narrowband[36, 66]. As the observation time interval is increased, the bandwidth must decrease. On the contrary, as the observation time interval is decreased, the bandwidth must increase.

The analogy between an omnidirectional narrowband uniform linear array and a single-channel FIR filer is illustrated in Fig.10.2.3[215].

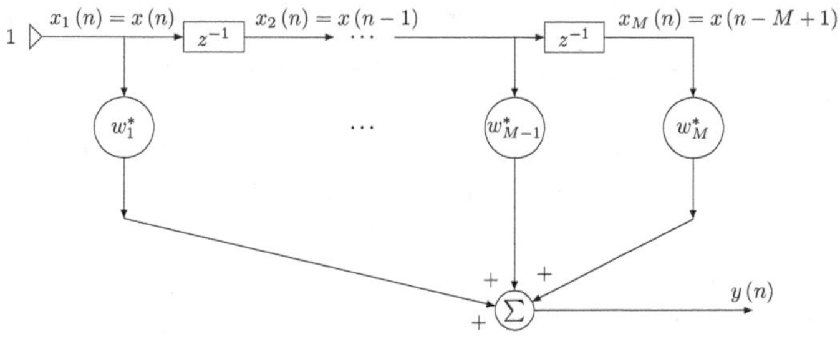

(a) A single-channel FIR filter

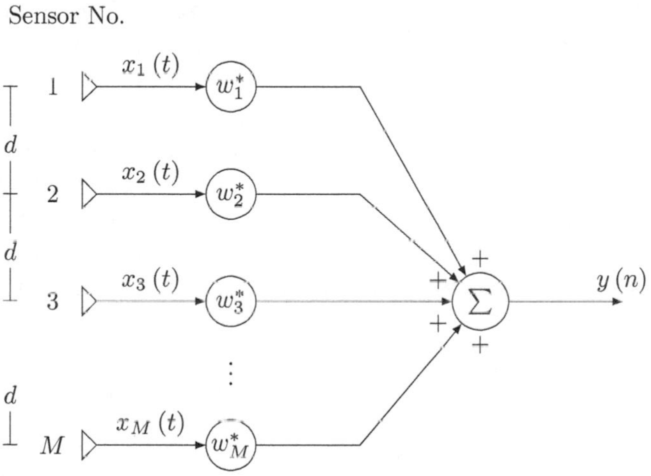

(b) A omnidirectional narrowband ULA

Fig. 10.2.3: The analogy between an omnidirectional narrowband ULA and a single-channel FIR filter.

The correspondence between FIR filtering and beamforming is closest when the beamformer operates at a single temporal frequency ω_0 and the array geometry is linear and uniform. Letting the sensor spaceing be d, propagation velocity be c, and θ represent DOA relative to broadside (perpendicular to the array), then we have[215]

$$\tau_i(\theta) = (i-1)\frac{d}{c}\sin\theta. \tag{10.2.27}$$

In this case, the relationship between temporal frequency ωin the frequency response $\boldsymbol{d}(\omega)$ of the spatial FIR filter and the steering vector $\boldsymbol{d}(\theta, \omega_0)$ of the beamformer is

$$\omega = \frac{\omega_0 d}{c} \sin \theta. \tag{10.2.28}$$

Thus, temporal frequency in a spatial FIR filter corresponds to the sine of DOA in a narrowband, uniform linear beamformer. Therefore, complete interchange of FIR filter and beamformer is possible for this special case provided the mapping between frequency and DOA is accounted for.

The general effects of spatial sampling are similar to temporal sampling. Spatial sampling may also produce spatial aliasing, which corresponds to ambiguity or uncertainty in source locations[215]. This implies that sources at different locations may have the same array response vector.

(1) For narrowband sources, $\boldsymbol{d}(\theta_1, \omega_0) = \boldsymbol{d}(\theta_2, \omega_0)$. This can occur if the sensors are spaced too far apart. If the sensors are too close together, spatial discrimination suffers as a result of the smaller than necessary aperture.
(2) For broadband sources, another type of ambiguity occurs when a source at one location and frequency cannot be distinguished from a source at a different location and frequency, i.e., the steeing vetcor $\boldsymbol{d}(\theta_1, \omega_1) = \boldsymbol{d}(\theta_2, \omega_2)$. For example, this occurs in an uniform linear array whenever $\omega_1 \sin \theta_1 = \omega_2 \sin \theta_2$.

The array aperture determines the beamwidth of the mainlobe generated by the array beamformer. The way to prevent the narrowband source identification from ambiguity is to properly configure the sensor spacing of the uniform array to meet the required aperture requirements. In addition, the introduction of temporal sample technique with a single spatial sample may also prevent the broadband source identification from ambiguity.

Consider an FIR filter with M taps to separate a complex-valuedfrequency component ω_0 from other frequency components. At this point, the desired FIR frequency response is

$$r(\omega) = \boldsymbol{w}^{\mathrm{H}} \boldsymbol{d}(\omega) = \begin{cases} 1, & \omega = \omega_0, \\ 0, & \omega \neq \omega_0. \end{cases} \tag{10.2.29}$$

A common solution to this desired response problem is to choose the FIR filter response with frequency ω_0, that is

$$\boldsymbol{w} = \boldsymbol{d}(\omega_0). \tag{10.2.30}$$

The FIR filter with the desired frequency response shown in Eq. (10.2.29) is equivalent to a spectral line enhancer with frequency ω_0, which only outputs the source with frequency ω_1 and suppresses all other frequency sources.

The actual response $r(\omega)$ is not a binary function of $\{1, 0\}$ but a function characterized by a mainlobe and many sidelobes. The amplitude squared $|r(\omega)|^2$ of the frequency response $r(\omega)$ is called the beampattern of the FIR filter.

Since $\mathbf{w} = \mathbf{d}(\omega_0)$, each element $w_m = d_m(\omega) = \mathrm{e}^{-\mathrm{j}(m-1)\omega\triangle}$ of $\mathbf{w}$ has unit magnitude. The mainlobe or beamwidth is in contradiction with the size of the sidelobe. Therefore, tapping or windowing the amplitudes of the elements of $\mathbf{w}$ permits trading of mainlobe or beamwidth against sidelobe levels to form the response into a desired shape.

By extension, the desired frequency response of the broadband beamformer, i.e., the desired steering vector

$$r_d(\theta,\omega) = \mathbf{w}^{\mathrm{H}}\mathbf{d}(\theta,\omega) = \begin{cases} 1, & \omega = \omega_0, \\ 0, & \omega \neq \omega_0. \end{cases} \tag{10.2.31}$$

The magnitude square of the actual steering vector is referred to as the beampattern of the broadband beamformer.

Consider source $s(n) = m_s(n)\mathrm{e}^{\mathrm{j}\omega_0 k}$, where $m_s(n)$ and ω_0 are the modulation function and frequency of the source respectively. Assume that the power of the source $s(n)$ is $P_s = E\{|s(n)|^2\} = E\{|m_s(n)|^2\}$, and the direction angle of the source observed from the array is θ_0.

The selected beamformer vector is equal to the steering vector $\mathbf{d}(\theta_0,\omega_0)$ of the source $s(n)$, i.e.,

$$\mathbf{w} = \mathbf{d}(\theta_0,\omega_0) = [1, \mathrm{e}^{\mathrm{j}\omega_0\tau_2(\theta_0,\omega_0)}, \cdots, \mathrm{e}^{\mathrm{j}\omega_0\tau_M(\theta_0,\omega_0)}]^{\mathrm{T}}. \tag{10.2.32}$$

The observation data of the source $s(n)$ by the m-th sensor is

$$x_{s,m}(n) = m_s(n)\mathrm{e}^{\mathrm{j}\omega_0[k+\tau_m(\theta_0,\omega_0)]}. \tag{10.2.33}$$

Therefore, the array observation vector generated by the source $s(n)$ in the observation direction is

$$\mathbf{x}_s(n) = [x_{s,1}(n), \cdots, x_{s,M}(n)]^{\mathrm{T}} = m_s(n)\mathrm{e}^{\mathrm{j}\omega_0}\mathbf{s}_0(n), \tag{10.2.34}$$

where

$$\mathbf{s}_0(n) = [1, \mathrm{e}^{\mathrm{j}\omega_0\tau_2(\theta_0,\omega_0)}, \cdots, \mathrm{e}^{\mathrm{j}\omega_0\tau_M(\theta_0,\omega_0)}]^{\mathrm{T}}. \tag{10.2.35}$$

When $\mathbf{w} = \mathbf{d}(\theta_0,\omega_0)$, the output of the beamformer is

$$y(n) = \mathbf{w}^{\mathrm{H}}\mathbf{x}_s(n) = m_s(n)\mathrm{e}^{\mathrm{j}\omega_0 k}, \tag{10.2.36}$$

and its average power is

$$P_y = E\{|y(n)|^2\} = E\{|m_s(n)|^2\} = P_s. \tag{10.2.37}$$

This indicates that mean output power P_y of the beamformer steered in the observation direction is equal to the power P_s of the source $s(n)$ in the observation direction θ_0. This process is similar to steering the array mechanically in the observation direction except that is done electronically.

The main differences between electronic steering and mechanical steering are as follows.

(1) Mechanical steering is only suitable for directional antennas, while electronic steering is suitable for omnidirectional antennas or sensors.
(2) Mechanical steering rotates the antenna mechanically to the desired direction, while for electronic steering, it is done electronically by adjusting the phase of the beamformer to achieve the pointing of the array.
(3) The aperture of an electronically steered array is different from that of a mechanically steered array[91].

The design objective of the broadband beamformer is to choose $\boldsymbol{w}$ so the actual response $r(\theta, \omega) = \boldsymbol{w}^{\mathrm{H}}\boldsymbol{d}(\theta, \omega)$ approximates a desired response $r_d(\theta, \omega)$.

10.3 Linearly-Constrained Adaptive Beamformer

Beamformers can be classified as data independent and statistically optimal[215]. The weights in a data independent beamformer do not depend on the array data and are chosen to present a specified response for all signal and interference scenarios. The weights in a statistically optimal beamformer are chosen based on the statistics of the array data to optimize the array response. In general, the statistically optimal beamformer places nulls in the directions of interfering sources in an attempt to maximize the signal-to-noise ratio at the beamformer output.

The desired signal and noise (or interference) are spatial position changing and waveform changing with time. In order to track the spatial-varying and time-varying desired sources, the statistically optimal beamformer should also be time-varying, with taps that can be adjusted at different times. Therefore, the beamformer is essentially a spatial-time two-dimensional filter or signal processor.

The optimal implementation of a time-varying beamformer is an adaptive beamformer. In order to generate only one mainlobe while minimizing the other sidelobes, it is necessary to impose constraints to the adaptive beamformer. The simplest and most effective constraint is the linear constraint. An adaptive beamformer that obeys the linear constraint is referred to as a linearly constrained adaptive beamformer.

Linearly constrained adaptive beamformer, also known as linearly constrained adaptive array processing, has two main implementations: the direct form and the generalized sidelobe canceling form. These two linearly constrained adaptive beamformers are introduced below.

10.3.1 Classical Beamforming

The design task of the broadband beamformer is to choose $\boldsymbol{w}$ so that the actual response $r(\theta, \omega) = \boldsymbol{w}^{\mathrm{H}}\boldsymbol{d}(\theta, \omega)$ approximates a desired response $r_d(\theta, \omega)$.

Consider the p DOA and frequencies at point $(\theta_1, \omega_1), \cdots, (\theta_p, \omega_p)$ so that the actual response of the beamformer approximates the desired response. Let the steering vector matrix and response vector of these points be

$$\boldsymbol{A} = [\boldsymbol{a}(\theta_1, \omega_1), \cdots, \boldsymbol{a}(\theta_p, \omega_p)], \tag{10.3.1}$$

$$\boldsymbol{r}_d = [r(\theta_1, \omega_1), \cdots, r(\theta_p, \omega_p)]^{\mathrm{T}}. \tag{10.3.2}$$

The general optimization criterion for solving the approximation problems is to minimize the l_q norm of the error vector between the actual and the desired response vector

$$\boldsymbol{w}_{\text{opt}} = \arg\min_{\boldsymbol{w}} E\{\| \boldsymbol{A}^{\mathrm{H}}\boldsymbol{w} - \boldsymbol{r}_d \|_q\}, \tag{10.3.3}$$

$$\text{subject to } f(\boldsymbol{w}) = 0. \tag{10.3.4}$$

where $f(\boldsymbol{w})$ represents the linear constrain and $\| \boldsymbol{x} \|_q$ represents the l_q norm of vector $\boldsymbol{x} = [x_1, \cdots, x_p]^{\mathrm{T}}$, that is,

$$\| \boldsymbol{x} \|_q = (x_1^q + \cdots + x_p^q)^{1/q}. \tag{10.3.5}$$

The most commonly used vector norm is l_2 norm, i.e., Frobenius norm, and the corresponding optimization criterion is

$$\boldsymbol{w}_{\text{opt}} = \arg\min_{\boldsymbol{w}} E\{\| \boldsymbol{A}^{\mathrm{H}}\boldsymbol{w} - \boldsymbol{r}_d \|_2^2\}, \tag{10.3.6}$$

$$\text{subject to } f(\boldsymbol{w}) = 0. \tag{10.3.7}$$

Assuming that the steering vector $\boldsymbol{A}$ has full row rank, then the optimal beamformer is the least squares solution

$$\boldsymbol{w}_{\text{opt}} = (\boldsymbol{A}\boldsymbol{A}^{\mathrm{H}})^{-1}\boldsymbol{A}\boldsymbol{r}_d. \tag{10.3.8}$$

of the optimization problem. Choosing different optimization criteria and/or constraints $f(\boldsymbol{w})$, different types of optimal beamformers will be obtained. Several classical types of optimal beamformers are introduced below.

1. Multiple Sidelobe Canceller

The multiple sidelobe canceller (MSC) is perhaps the earliest statistically optimal beamformer. It was proposed by Applebaum in 1966 in the form of a technical report and formally published as a paper in the IEEE Transactions on Antenna Propagation in 1976[14]. An MSC consists of a "main channel" and one or more "auxiliary channels". The main channel can be either a single high gain antenna or a data independent beamformer. It has a highly directional response. When the main channel is pointed in the desired signal direction, the interfering signals enter through the main channel sidelobes. There is no desired signal in the auxiliary channels, while only interference signals are received. The goal of the auxiliary channels is to cancel the main channel interference component.

2. Beamformer Using Reference Signal
If the desired signal $y_d(n)$ were known, then the weights could be chosen to minimize the error between the beamformer output $y(n) = \mathbf{w}^{\mathrm{H}}\mathbf{x}(n)$ and the reference signal as depicted in Fig.10.3.1.

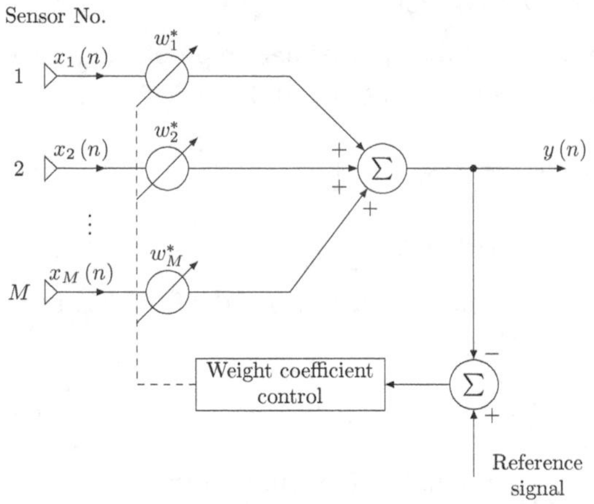

Fig. 10.3.1: Beamformer using reference signal

In practical applications, enough may be known about the desired signal to generate a signal that closely represents it. This signal is called a reference signal. The beamformer using reference signal was proposed by Widrow on 1967[225].

3. Maximization of Signal-to-Noise Ratio

The criterion of the maximization of signal-to-noise ratio beamformer is to maximize the SNR $\boldsymbol{w}^{\mathrm{H}}\boldsymbol{R}_s\boldsymbol{w}/(\boldsymbol{w}^{\mathrm{H}}\boldsymbol{R}_n\boldsymbol{w})$. This SNR maximization problem is actually a generalized Rayleigh quotient maximization problem, which is also equivalent to solving the generalized eigenvalue problem $\boldsymbol{R}_s\boldsymbol{w} = \lambda_{\max}\boldsymbol{R}_n\boldsymbol{w}$.

The maximization of signal-to-noise ratio beamformer was proposed by Monzingo and Miller in 1980[157].

4. Linearly Constrained Minimum Variance Beamformer

In many applications, the desired signal may be of unknown strength, resulting in the reference signal cannot being generated and preventing utilization of the reference signal approach. Or because the covariance matrices $\boldsymbol{R}_s$ and $\boldsymbol{R}_n$ of signal and noise cannot be estimated, the maximum SNR beamformer cannot be realized. In addition, the desired signal may always be present in the auxiliary channels which makes the MSC doesn't work. These limitations can be overcome by applying linear constraints to the beamformer vector of the linearly constrained minimum variance beamformer.

The linearly constrained minimum variance beamformer was proposed by Frost on 1972[81], which minimizes the variance of the output under the linear constraints and can be formulated as

$$\min_{w} w^{\mathrm{H}} R_{xx} w, \tag{10.3.9}$$

$$\text{subject to } C^{\mathrm{H}} w = g, \tag{10.3.10}$$

where the $m \times p$ constrain matrix C represents the constraints of the m-element array on the p spatial signals, and the $p \times 1$ vector g is reference vector.

Using the Lagrange multiplier method, it is easy to obtain the solution of the above linear constrained minimum variance optimization problem as

$$w_{\mathrm{opt}} = R_{xx}^{-1} C (C^{\mathrm{H}} R_{xx}^{-1} C^{-1} g), \tag{10.3.11}$$

which is called the linear constrained minimum variance beamformer.

Table 10.3.1 lists and compares the parameter definitions, output representations, optimization criterion, closed-form solutions, advantages and disadvantages of the above four classical optimal beamformers.

Tab. 10.3.1: Classical optimal beamformers

Type	MSC	Reference Signal	Max SNR	LCMV
Definitions	x_a - auxiliary data y_m - primary data $r_{ma} = E\{y_m^* x_a\}$ $R_a = E\{x_a x_a^{\mathrm{H}}\}$	x-array data y_d - desired signal $r_{xd} = E\{x y_d^*\}$ $R_{xx} = E\{x x^{\mathrm{H}}\}$	$x = s + n$- array data s - signal component n - noise component $R_s = E\{s s^{\mathrm{H}}\}$ $R_n = E\{n n^{\mathrm{H}}\}$	x-array data C - constraint matrix g - reference vector $R_{xx} = E\{x x^{\mathrm{H}}\}$
Output	$y = y_m - w_a^{\mathrm{H}} x_a$	$y = w^{\mathrm{H}} x$	$y = w^{\mathrm{H}} x$	$y = w^{\mathrm{H}} x$
Criterion	$\min_{w_a} E\{\lvert y_m - w_a^{\mathrm{H}} x_a \rvert^2\}$	$\min E\{\lvert y - y_d \rvert^2\}$	$\min_{w} \frac{w^{\mathrm{H}} R_s w}{w^{\mathrm{H}} R_n w}$	$\min_{w} E\{w^{\mathrm{H}} R_{xx} w\}$ s.t. $c^{\mathrm{H}} w = g$
Optimal Weights	$w_a = R_a^{-1} r_{ma}$	$w = R_{xx}^{-1} r_{xd}$	$R_n^{-1} R_s w = \lambda_{\max} w$	$w = R_{xx}^{-1} C (C^{\mathrm{H}} R_{xx}^{-1} C)^{-1} g$
Advantage	Simple	Direction of desired signal can be unkonwn	Maximization of SNR	Flexible and general constraints
Disadvantages	Requires absence of desired signal from auxiliary channels for weight determination	Must generate reference signal	Must know R_s and R_n, and perform generalized eigenvalue decomposition for weights	Computation of constrained weight vector

According to the choice of the linear constraints, the minimum variance beamformer has different applications. Several typical application examples are listed below.

Example 10.3.1 In multi-user detection for wireless communication, user 1 with spreading code vector s_1 is constrained as $w^{\mathrm{H}} s_1 = 1$, at this point, $C = s_1$ and $g = 1$, so the optimal detector for user 1 is

$$w_{\mathrm{opt}} = \frac{R_{xx}^{-1} s_1}{s_1^{\mathrm{H}} R_{xx}^{-1} s_1}. \tag{10.3.12}$$

Example 10.3.2 When the constraint matrix

$$C = [a(\theta_i - \triangle\theta_i), a(\theta_i), a(\theta_i + \triangle\theta_i)], \quad g = [1, 1, 1]^T, \tag{10.3.13}$$

for the i-th spatial signal is chosen, the minimum variance beamformer can improve the robustness of the estimation of the DOA θ_i and reduce the fluctuation of the DOA estimation result caused by the slow moving of the i-th spatial signal.

Example 10.3.3 When a spatial signal is known to be in θ_0 direction and possibly near θ_0 direction, if

$$C = [a(\theta_0), a'(\theta_0), \cdots, a^{(k)}(\theta_0)], \quad g = [1, 0, \cdots, 0]^T, \tag{10.3.14}$$

is chosen, then the minimum variance beamformer can spread the mainlobe aligned with DOA θ_0, where $a^{(k)}$ denotes the k-th order derivative of the steering vector $a(\theta)$ with respect to the DOA θ.

In particular, when $C = a(\theta)$ and $g = 1$ are selected, the linearly constrained minimum variance beamformer becomes

$$w_{opt} = \frac{R_{xx}^{-1}a(\theta)}{a^H(\theta)R_{xx}^{-1}a(\theta)}, \tag{10.3.15}$$

commonly called the minimum variance distortionless response (MVDR) beamformer, which was proposed by Capon in 1969[42].

The linearly constrained minimum variance beamformer is a generalization of the MVDR beamformer from single linear constraint condition to multiple linear constraint conditions.

In practical applications, beamformers are often required to be able to adjust in real-time adaptively. For this reason, the adaptive implementation of beamformer is discussed below.

10.3.2 Direct Implementation of Adaptive Beamforming

Assume that there are M omnidirectional sensors and $x_m(n)$ is the sampled output of the m-th time-delayed sensor

$$x_m(n) = s(n) + n_m(n), \tag{10.3.16}$$

where $s(n)$ is the desired signal and $n_m(n)$ represents the totality of noise and interference (including all other undesired signals) observed by the m-th sensor. A beamformed output signal $y(n)$ can be formulated as the sum of the delayed and weighted outputs of the M sensors, i.e.,

$$y(n) = \sum_{m=1}^{M}\sum_{i=-K}^{K} a_{m,i}x_m(n - \tau_i), \tag{10.3.17}$$

where $a_{m,i}$ represents the weight coefficients used for the m-th sensor signal $x_m(n-\tau_i)$ at delay τ_i. Each channel uses a filter (whose length is $2K+1$) to weight and adjust their $2K+1$ delays and that the zero time reference is at the filter midpoint.

Let

$$\boldsymbol{a}_i = [a_{1,i}, a_{2,i}, \cdots, a_{M,i}]^{\mathrm{T}}, \tag{10.3.18}$$

$$\boldsymbol{x}(n) = [x_1(n), x_2(n), \cdots, x_M(n)]^{\mathrm{T}} \tag{10.3.19}$$

represent the i-th weight coefficient vector and sensor output vector of the M spatial filters, respectively, then the output of the beamformer can be expressed in vector form as

$$y(n) = \sum_{i=-K}^{K} \boldsymbol{a}_i^{\mathrm{H}} \boldsymbol{x}(n-\tau_i), \tag{10.3.20}$$

where

$$\boldsymbol{x}(n-\tau_i) = s(n-\tau_i)\mathbf{1} + \boldsymbol{n}(n-\tau_i). \tag{10.3.21}$$

Note that $\mathbf{1}$ is an M-dimensional vector with all elements equal to 1, which is called the summing vector. $\boldsymbol{n}(n) = [n_1(n), n_2(n), \cdots, n_M(n)]^{\mathrm{T}}$ represents the additive noise (or interference) vector on the M sensors.

Prescribed gain and phase response for the desired signal is ensured by constraining the sum of channel weights at each delay point to be specific values. Let $f(i)$ be the sum for the set of spatial filter weight coefficients at delay i, that is,

$$\boldsymbol{a}_i^{\mathrm{H}}\mathbf{1} = f(i). \tag{10.3.22}$$

Substituting Eq. (10.3.21) into Eq. (10.3.20) and using Eq. (10.3.22), the output of the spatial filter corresponding to the desired signal is

$$y_s(n) = \sum_{i=-K}^{K} f(i)s(n-\tau_i) + \sum_{i=-K}^{K} \boldsymbol{a}_i^{\mathrm{H}} \boldsymbol{n}(n-\tau_i). \tag{10.3.23}$$

If $\boldsymbol{a}_i$ is orthogonal to the noise or interference subspace span$\{\boldsymbol{n}(n)\}$, the noise or interference can be completely suppressed, that is, the second term of Eq. (10.3.23) is zero. At this time, Eq. (10.3.23) is simplified as

$$y_s(n) = \sum_{i=-K}^{K} f(i)s(n-\tau_i). \tag{10.3.24}$$

Eq. (10.3.24) shows that $f(i)$ represent the impulse response of a FIR filter have length $2K+1$. The problem now is : how to constrain the response $f(i)$ of the FIR beamformer?

One commonly used constant is that of zero distortion in which

$$f(i) = \delta(i) = \begin{cases} 1, & i = 0, \\ 0, & \text{otherwise.} \end{cases} \tag{10.3.25}$$

Under this constraint, we have $y_s(n) = s(n)$, that is, the FIR filter is like an ideal beamformer, which produces only one beam (directly pointed to the desired signal). Given this, such an FIR filter is called FIR beamforming.

The zero distortion constraint Eq. (10.3.25) can be normalized as

$$\boldsymbol{f}^{\mathrm{T}}\mathbf{1} = 1, \tag{10.3.26}$$

where

$$\boldsymbol{f}[f(-K), \cdots, f(0), \cdots, f(K)]^{\mathrm{T}} \tag{10.3.27}$$

is the impulse response vector of the FIR beamformer.

The typical adaptive implementation of the FIR beamformer is[96]

$$\boldsymbol{a}_i(k+1) = \boldsymbol{a}_i(n) + \triangle_i(n), \tag{10.3.28}$$

where $i = -K, \cdots, 0, \cdots, K$ is the time delay and the correction term used is the Frost linearly-constrained error correction method[96], i.e.,

$$\triangle_i(n) = \mu_k y(n)[q_x(n-i)\mathbf{1} - \boldsymbol{x}(n-i)] - q_{a,i}(n)\mathbf{1} + \frac{1}{M}f(i)\mathbf{1}, \tag{10.3.29}$$

where

$$q_x(n-i) = \frac{1}{M}\boldsymbol{x}^{\mathrm{T}}(n-i)\mathbf{1}, \tag{10.3.30}$$

$$q_{a,i}(n) = \frac{1}{M}\boldsymbol{a}_i^{\mathrm{T}}(n)\mathbf{1}. \tag{10.3.31}$$

The adaptive step size μ is a scalar which controls both the convergence rate and tracking performance of the adaptive beamformer and it can be choosen as

$$\mu_k = \frac{\alpha}{P(n)}, \tag{10.3.32}$$

where $P(n)$ is the total power of the sampled signal

$$P(n) = \sum_{m=1}^{M}\sum_{i=-K}^{K} x_m^2(n-i). \tag{10.3.33}$$

Convergence of the algorithm is ensured if $0 < \alpha < 1$.

Fig.10.3.2 illustrates the direct form implementation of the linearly constrained adaptive beamforming[96]. In this figure, time-delay steering elements $\tau_1, \cdots, \tau_M$ are used to point the array in the direction of interest. Each coefficient (the tapped delay in the figure) in the beamformer is updated by the adaptive algorithm.

10.3.3 Generlized Sidelobe Canceling Form of Adaptive Beamforming

The direct form implementation of linearly constrained adaptive beamforming directly generates a beam pointing to the desired signal. The linearly constrained adaptive

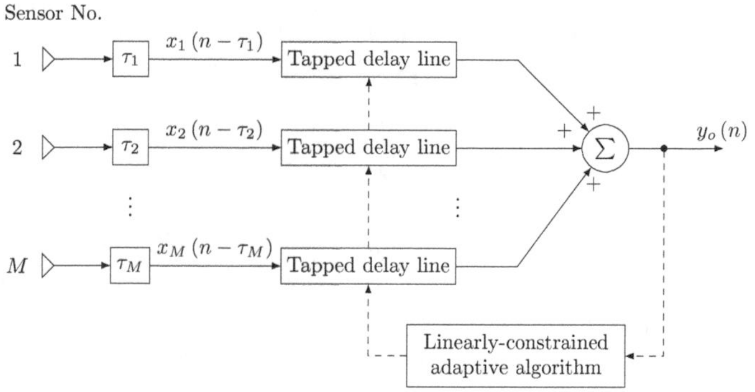

Fig. 10.3.2: Direct implementation of linearly-constrained adaptive beamforming

beamforming can also be implemented in an indirect form, which is called the generalized sidelobe canceling form, which was proposed by Griffiths and Jim in 1982[96].

Fig.10.3.3 shows the schematic diagram of the generlized sidelobe canceling form of adaptive beamforming.

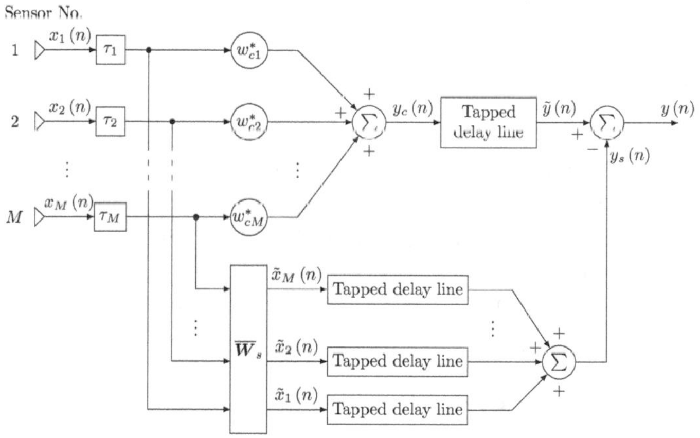

Fig. 10.3.3: Generlized sidelobe canceling form of linearly-constrained adaptive beamforming

The adaptive beamforming based on the generalized sidelobe canceling consists of two distinct substructures which are shown as the upper and lower processing paths. The upper is the fixed beamformer path and the lower is the sidelobe canceling path.

(1) Fixed beamformer path

For the fixed beamformer, each element of the weight coefficient vector $\boldsymbol{w}_c = [w_{c1}, w_{c2}, \cdots, w_{cM}]^{\mathrm{T}}$ is a fixed constant which produce non-adaptive beamformed signal

$$y_c(n) = \boldsymbol{w}_c^{\mathrm{T}}\boldsymbol{x}(n) = \sum_{m=1}^{M} w_{cm}x_m(n - \tau_m). \tag{10.3.34}$$

The weights $\boldsymbol{w}_c$ of the fixed beamformer can be chosen so as to trade off the relationship between array beamwidth and average sidelobe level[76]. One widely used method employs Chebyshev polynomials to design the $\boldsymbol{w}_c$.

Without a priori knowledge of the desired signal, the weight coefficient w_{cm} of the fixed beamformer can be designed in the form of RAKE receiver.

$$w_{cm} = \frac{|x_m(n)|^2}{|x_1(n)|^2 + \cdots + |x_M(n)|^2}, \quad m = 1, \cdots, M. \tag{10.3.35}$$

Therefore, the non-adaptive beamforming signal is composed of a desired signal with large energy and an interference signal.

In some applications, some a priori knowledge of the desired signal may be known. For example, in a CDMA system of wireless communication, the characteristic waveform matrix $\boldsymbol{C} = [\boldsymbol{c}_1, \cdots, \boldsymbol{c}_M]$ of each user is known to the base station. At this time, the weight coefficient vector of the fixed beamformer (called multi-user detection in wireless communication systems) is determined by

$$\boldsymbol{w}_c = (\boldsymbol{C}^{\mathrm{H}}\boldsymbol{C})^{-1}\boldsymbol{C}^{\mathrm{H}}\boldsymbol{g}, \tag{10.3.36}$$

where $\boldsymbol{g}$ is the gain vector, which has only one nonzero element 1. The position number of the nonzero element 1 represents that the user of this number is the desired user. A fixed beamformer can not only produce the main beam (mainlobe) pointing to the desired signal, but also it will inevitably produce several sidelobes pointing to the main interference source at the same time.

All weight coefficient w_{cm}, $m = 1, \cdots, M$ are assumed to be nonzero and are normalized to have a sum of unity, i.e.,

$$\boldsymbol{w}_c^{\mathrm{T}}\boldsymbol{1} = 1. \tag{10.3.37}$$

The output of the filter

$$\tilde{y}_c(n) = \sum_{i=-K}^{K} f(i)y_c(n - i) \tag{10.3.38}$$

is obtained by filtering the output of the fixed beamformer using the FIR filter with coefficient $f(i)$, where the coeffcent $f(i)$ of the filter $\boldsymbol{f} = [f(-K), \cdots, f(0), \cdots, f(K)]^{\mathrm{T}}$ obeys the constraint condition $\boldsymbol{f}^{\mathrm{T}}\boldsymbol{1} = 1$.

(2) Sidelobe canceling path

The sidelobe canceling path consists of a $(M - 1) \times M$ matrix preprocessor $\bar{\boldsymbol{W}}_s$ followed by a set of tapped-delay lines. Each tapped-delay line contains $2K + 1$ weights.

The purpose of $\bar{\boldsymbol{W}}_S$ is to block the desired signal $s(n)$ from the lower path. Since $s(n)$ is common to each of the steered sensor outputs, the blocking is ensured if the rows of $\bar{\boldsymbol{W}}_S$ sum up to zero. To see this clearly, the delayed observation vector is denoted as

$$\boldsymbol{x}(n-\tau) = [x_1(n-\tau_1), \cdots, x_M(n-\tau_M)]^{\mathrm{T}}. \tag{10.3.39}$$

Using the matrix preprocessor $\bar{\boldsymbol{W}}_S$ to preprocess $\boldsymbol{x}(n-\tau)$, the output

$$\tilde{x}(n) = \bar{\boldsymbol{W}}_S \boldsymbol{x}(n-\tau) = \begin{bmatrix} \boldsymbol{b}_1^{\mathrm{T}} \\ \vdots \\ \boldsymbol{b}_{M-1}^{\mathrm{T}} \end{bmatrix} \boldsymbol{x}(n-\tau) \tag{10.3.40}$$

contains $M-1$ components. Here, the sum of each row of the $(M-1)\times M$ matrix preprocessor $\bar{\boldsymbol{W}}_S$ is zero, i.e.,

$$\boldsymbol{b}_m^{\mathrm{T}}\mathbf{1} = 0, \quad \forall \quad m = 1, \cdots, M-1. \tag{10.3.41}$$

Since $s(n)$ is common to each of the steered sensor outputs, i.e., $\boldsymbol{x}(n-\tau) = s(n)\mathbf{1} + \boldsymbol{n}(n-\tau)$, from Eqs. (10.3.40) and (10.3.41), we have

$$\tilde{x}(n) = \bar{\boldsymbol{W}}_S \boldsymbol{x}(n-\tau) = \begin{bmatrix} \boldsymbol{b}_1^{\mathrm{T}} \\ \vdots \\ \boldsymbol{b}_{M-1}^{\mathrm{T}} \end{bmatrix} [s(n)\mathbf{1} + +\boldsymbol{n}(n-\tau)] = \begin{bmatrix} \boldsymbol{b}_1^{\mathrm{T}} \\ \vdots \\ \boldsymbol{b}_{M-1}^{\mathrm{T}} \end{bmatrix} \boldsymbol{n}(n-\tau).$$

That is, the desired signal is blocked, leaving only the linearly mixed sidelobe signal.

The lower path of the sidelobe canceler $\tilde{\boldsymbol{a}}_i$ is a $(M-1)\times M$ vector

$$\tilde{\boldsymbol{a}}_i = [\tilde{a}_{1,i}, \cdots, \tilde{a}_{M-1,i}]^{\mathrm{T}}, \quad i = -K, \cdots, K. \tag{10.3.42}$$

This sidelobe canceler generates a scalar output as the sum of delayed and weighted elements of $\tilde{\boldsymbol{x}}(n)$

$$y_A(n) = \sum_{i=-K}^{K} \tilde{\boldsymbol{a}}_i^{\mathrm{T}} \tilde{\boldsymbol{x}}(n-i). \tag{10.3.43}$$

Note that, $y_A(n)$ only contains the interferences, i.e., sidelobes. In contrast, $\tilde{y}_c(n)$ contains the desired signal (mainlobe) and interferences (sidelobes), and the difference

$$y(n) = \tilde{y}_c(n) - \tilde{y}_A(n) \tag{10.3.44}$$

is used as the output of the beamformer based on the generalized sidelobe canceling structure. In other words, the sidelobes are canceled, and the generalized sidelobe canceler only outputs the mainlobe signal, which is equivalent to beamforming for the desired signal.

The adaptive update algorithm of the sidelobe canceler is

$$\tilde{\boldsymbol{a}}_i(k+1) = \tilde{\boldsymbol{a}}_i(n) + \mu y(n)\tilde{\boldsymbol{x}}(n-i). \tag{10.3.45}$$

Eqs. (10.3.35), (10.3.37) and (10.3.41) constitute the adaptive algorithm of the generalized sidelobe canceller in Reference [96].

The beamformer uses a weighted sum of the observation data from the sensor array to form a scalar output signal. The weight coefficients determine the spatial filtering characteristics of the beamformer. If the source signals are at different spatial locations, the beamformer can separate these spatial source signals even if they have overlapping frequency components. The statistically optimal beamformers optimize its response by selecting the weight coefficients based on the statistics of the observation data. Now that the data statistics are usually unknown, so statistically optimal beamformer is usually implemented using adaptive FIR spatial filters.

10.4 Multiple Signal Classification (MUSIC)

As mentioned previously, array signal processing can be classified into two major techniques: beamforming and DOA estimation. The representative method of DOA estimation is high-resolution spatial spectrum estimation.

Power spectral density describes the distribution of signal power with frequency and is a frequency domain representation of the signal. Since the main task of array signal processing is to estimate signal spatial parameters (location parameters of the source), it is of great importance to extend and promote the concept of power spectral density to the spatial domain. This generalized power spectrum is often referred to as the spatial spectrum. The spatial spectrum describes the distribution of the spatial parameters of the signal.

10.4.1 Spatial Spectrum

Consider minimizing the average output energy of N snapshots, that is

$$\min_{\boldsymbol{w}} \frac{1}{N}\sum_{n=1}^{N}|y(n)|^2 = \min_{\boldsymbol{w}} \frac{1}{N}\sum_{n=1}^{N}|\boldsymbol{w}^{\mathrm{H}}\boldsymbol{x}(n)|^2. \tag{10.4.1}$$

This criterion for designing the weight vector $\boldsymbol{w}$ is called the minimum output energy (MOE) criterion. Let

$$\hat{\boldsymbol{R}}_X = \frac{1}{N}\sum_{t=1}^{N}\boldsymbol{x}(n)\boldsymbol{x}^{\mathrm{H}}(n) \tag{10.4.2}$$

be the sample autocovariance matrix of the signal vector $x(n)$, then the MOE criterion can be rewritten as

$$\min_{w} \frac{1}{N}\sum_{n=1}^{N}|y(n)|^2 = \min_{w} \boldsymbol{w}^{\mathrm{H}}\left(\frac{1}{N}\sum_{n=1}^{N}\boldsymbol{x}(n)\boldsymbol{x}^{\mathrm{H}}(n)\right)\boldsymbol{w} = \min_{w} \boldsymbol{w}^{\mathrm{H}}\hat{\boldsymbol{R}}_X\boldsymbol{w}.$$

When $N \to \infty$, the above equation becomes

$$\mathrm{E}\left\{|y(n)|^2\right\} = \lim_{N\to\infty} \frac{1}{N} \sum_{n=1}^{N} |y(n)|^2 = \boldsymbol{w}^{\mathrm{H}} \boldsymbol{R}_{xx} \boldsymbol{w}. \tag{10.4.3}$$

Note that the signal vector observed by the array can be formulated as

$$\boldsymbol{x}(n) = \boldsymbol{a}(\omega_k) s_k(n) + \sum_{i=1, i\neq k}^{p} \boldsymbol{a}(\omega_i) s_i(n) + \boldsymbol{e}(n), \tag{10.4.4}$$

where the first term on the right side of the equal sign is the desired signal, while the second term is the sum of other signals that is wished to be rejected (collectively referred to as interference signals), and the third term is the additive noise term. Substituting Eq. (10.4.4) into Eq. (10.4.3), it follows that

$$\mathrm{E}\left\{|y(n)|^2\right\} = \mathrm{E}\left\{|s_k(n)|^2\right\} \left|\boldsymbol{w}^{\mathrm{H}} \boldsymbol{a}(\omega_k)\right|^2 + \sum_{i=1, i\neq k}^{p} \mathrm{E}\left\{|s_i(n)|^2\right\} \left|\boldsymbol{w}^{\mathrm{H}} \boldsymbol{a}(\omega_i)\right|^2 + \sigma^2 |\boldsymbol{w}|^2 \tag{10.4.5}$$

by assuming the additive noise $e_1(n), \cdots, e_m(n)$ to have identical variance.

From Eq. (10.4.5), it is easy to see that if the weight vector $\boldsymbol{w}$ satisfies the constraint condition

$$\boldsymbol{w}^{\mathrm{H}} \boldsymbol{a}(\omega_k) = \boldsymbol{a}^{\mathrm{H}}(\omega_k) \boldsymbol{w} = 1, \text{ (Beamforming condition)}, \tag{10.4.6}$$

and

$$\boldsymbol{w}^{\mathrm{H}} \boldsymbol{a}(\omega_i) = 0, \quad \omega_i \neq \omega_k, \text{ (Zero-point formation conditions)}, \tag{10.4.7}$$

simultaneously, then the weight vector would only extract the desired signal, while rejecting all other interference signals. For this case, Eq. (10.4.5) is simplified to

$$\mathrm{E}\left\{|y(n)|^2\right\} = \mathrm{E}\left\{|s_k(n)|^2\right\} + \sigma^2 |\boldsymbol{w}|^2. \tag{10.4.8}$$

It is worth pointing out that the average output energy of the beamformer is still the same as the above formula by only minimizing the output energy $\mathrm{E}\left\{|y(n)|^2\right\}$ under the constraints of beamforming conditions, that is, the zero-point forming conditions can be established automatically. Therefore, the optimal beamformer design becomes one that minimizes the output energy $\mathrm{E}\left\{|y(n)|^2\right\}$ under constraint in Eq. (10.4.6).

Next, the Lagrange multiplier method is used to solve this optimization problem. For this reason, the objective function

$$J(\boldsymbol{w}) = \boldsymbol{w}^{\mathrm{H}} \boldsymbol{R}_{xx} \boldsymbol{w} + \lambda \left[1 - \boldsymbol{w}^{\mathrm{H}} \boldsymbol{a}(\omega_k)\right] \tag{10.4.9}$$

is constructed according to Eqs. (10.4.3) and (10.4.6). From $\frac{\partial J(\boldsymbol{w})}{\partial \boldsymbol{w}^{\mathrm{H}}} = 0$, we have $\boldsymbol{R}_{xx}\boldsymbol{w} - \lambda \boldsymbol{a}(\omega_k) = 0$. Therefore, the optimal beamformer that minimizes the output energy is obtained

$$\boldsymbol{w}_{\mathrm{opt}} = \lambda \boldsymbol{R}_{xx}^{-1} \boldsymbol{a}(\omega_k). \tag{10.4.10}$$

Substituting this beamformer into the constraint condition in Eq. (10.4.6), it can be seen that

$$\lambda = \frac{1}{a^{H}(\omega_k) R_{xx}^{-1} a(\omega_k)}, \tag{10.4.11}$$

because the Lagrange multiplier λ is a real number.

Substituting Eq. (10.4.11) into Eq. (10.4.10), we immediately get the optimal beamformer that minimize the output energy

$$w_{opt} = \frac{R_{xx}^{-1} a(\omega_k)}{a^{H}(\omega_k) R_{xx}^{-1} a(\omega_k)}, \tag{10.4.12}$$

which is just the minimum variance distortionless response (MVDR) beamformer proposed by Capon in 1969[42]. The basic principle of MVDR is to minimize the power contributed by any interference from the undesired DOA, while keeping the signal power constant in the observation direction. Therefore, it may be regarded as a sharp bandpass filter.

Eq. (10.4.12) shows that the design of the optimal beamformer for the k-th signal source is determined by the estimation of the frequency ω_k of that signal. In order to determine the frequency of p signals $\omega_1, \cdots, \omega_p$, Capon[42] defines "spatial spectrum" as

$$P_{Capon}(\omega) = \frac{1}{a^{H}(\omega) R_{xx}^{-1} a(\omega)}, \tag{10.4.13}$$

and $\omega_1, \cdots, \omega_p$ corresponding to the peaks are set as the frequency of p signals.

The spatial spectrum defined by Eq. (10.4.13) is customarily called the Capon spatial spectrum. Since the optimal filter used by Capon's spatial spectrum is similar to the form of maximum likelihood estimation for estimating the amplitude of a sinusoidal wave with known frequency in Gaussian random noise, Eq. (10.4.13) is often mistakenly called "maximum likelihood spectrum estimation". Now, in many documents, this popular name is still used.

Once the frequencies $\omega_1, \cdots, \omega_p$ of p sources are estimated using the spatial spectrum, Eq. (10.2.1), namely, $\omega_i = 2\pi \frac{d}{\lambda} \sin\theta_i$ can be used to find the DOA $\theta_i, i = 1, \cdots, p$ of each source in the case of an uniform linear array. In other words, the estimation of the DOA is actually equivalent to the estimation of the spatial spectrum.

10.4.2 Signal Subspace and Noise Subspace

In order to estimate the spatial spectrum, consider the array observation model

$$x(n) = A(\omega)s(n) + e(n) = \sum_{i=1}^{p} a(\omega_i) s_i(n) + e(n), \tag{10.4.14}$$

where $A(\omega) = [a(\omega_1), \cdots, a(\omega_p)]$. In the case of uniform linear arrays, the steering vector is

$$a(\omega) = \left[1, e^{j\omega}, \cdots, e^{j\omega(M-1)}\right]^{T}. \tag{10.4.15}$$

For the array observation model, the following assumptions are usually made.

Assumption 1 For different values of ω_i, the vectors $\boldsymbol{a}(\omega_i)$ are linearly independent of each other;

Assumption 2 Each element of the additive noise vector $\boldsymbol{e}(t)$ is a complex-valued white noise with zero mean and variance σ^2, and is independent of each other;

Assumption 3 The matrix $\boldsymbol{P} = \mathrm{E}\left\{s(n)s^{\mathrm{H}}(n)\right\}$ is nonsingular, that is, $\mathrm{rank}(\boldsymbol{P}) = p$.

For uniform linear arrays, Assumption 1 is automatically satisfied. Assumption 2 means that the additive white noise vector $e(n)$ satisfies the following conditions

$$\mathrm{E}\{\boldsymbol{e}(n)\} = \mathbf{0}, \quad \mathrm{E}\left\{\boldsymbol{e}(n)\boldsymbol{e}^{\mathrm{H}}(n)\right\} = \sigma^2\boldsymbol{I}, \quad \mathrm{E}\left\{\boldsymbol{e}(n)\boldsymbol{e}^{\mathrm{T}}(n)\right\} = \boldsymbol{O}, \tag{10.4.16}$$

where $\mathbf{0}$ and $\boldsymbol{O}$ represent zero vector and zero matrix respectively. If each signal is transmitted independently, Assumption 3 is automatically satisfied. Therefore, the above three assumptions are just general assumptions, which are easily met in practice.

Under the assumption of $1 \sim 3$, it is easy to get from Eq. (10.4.14) that

$$\begin{aligned}\boldsymbol{R}_{xx} &\stackrel{\mathrm{def}}{=} \mathrm{E}\left\{\boldsymbol{x}(n)\boldsymbol{x}^{\mathrm{H}}(n)\right\}\\ &= \boldsymbol{A}(\omega)\mathrm{E}\left\{\boldsymbol{s}(n)\boldsymbol{s}^{\mathrm{H}}(n)\right\}\boldsymbol{A}^{\mathrm{H}}(\omega) + \sigma^2\boldsymbol{I}\\ &= \boldsymbol{APA}^{\mathrm{H}} + \sigma^2\boldsymbol{I},\end{aligned} \tag{10.4.17}$$

where $\boldsymbol{A} = \boldsymbol{A}(\omega)$. It can be seen that $\boldsymbol{R}_{xx}$ is an Hermitian symmetric matrix. Let its eigenvalue decomposition be

$$\boldsymbol{R}_{xx} = \boldsymbol{U\Sigma U}^{\mathrm{H}}, \tag{10.4.18}$$

where $\boldsymbol{\Sigma} = \mathrm{diag}\left(\sigma_1^2, \cdots, \sigma_M^2\right)$.

Since $\boldsymbol{A}$ has full rank, so $\mathrm{rank}\left(\boldsymbol{APA}^{\mathrm{H}}\right) = \mathrm{rank}(\boldsymbol{P}) = p$. Assuming that the number of signal sources p is less than that of the sensors M, that is, $p < M$. Then, according to $\boldsymbol{U}^{\mathrm{H}}\boldsymbol{R}_{xx}\boldsymbol{U} = \boldsymbol{\Sigma}$, it can be derived that

$$\begin{aligned}\boldsymbol{U}^{\mathrm{H}}\boldsymbol{R}_{xx}\boldsymbol{U} &= \boldsymbol{U}^{\mathrm{H}}\boldsymbol{APA}^{\mathrm{H}}\boldsymbol{U} + \sigma^2\boldsymbol{U}^{\mathrm{H}}\boldsymbol{U}\\ &= \mathrm{diag}\left(\alpha_1^2, \cdots, \alpha_p^2, 0, \cdots, 0\right) + \sigma^2\boldsymbol{I} = \boldsymbol{\Sigma},\end{aligned} \tag{10.4.19}$$

where $\alpha_1^2, \cdots, \alpha_p^2$ is the eigenvalue of the autocovariance matrix $\boldsymbol{APA}^{\mathrm{H}}$ of the observation signal $\boldsymbol{Ax}(n)$ without additive noise.

Eq. (10.4.19) shows that the eigenvalue of the autocovariance matrix $\boldsymbol{R}_{xx}$ is

$$\lambda_i = \sigma_i^2 = \begin{cases}\alpha_i^2 + \sigma^2, & i = 1, \cdots, p,\\ \sigma^2, & i = p+1, \cdots, M.\end{cases} \tag{10.4.20}$$

In other words, when there is additive observation white noise, the eigenvalues of the autocovariance matrix of the observation data vector $x(n)$ consist of two parts: the first

p eigenvalues are equal to the sum of α_i^2 and the additive white noise variance σ^2, and the following $m-p$ eigenvalues are all equal to the variance of additive white noise.

Obviously, when the signal-to-noise ratio is high enough to make α_i^2 significantly larger than the additive white noise variance σ^2, it is easy to differentiate the first p large eigenvalues $\alpha_i^2+\sigma^2$ and the following $m-p$ small eigenvalues σ^2 of the matrix $\boldsymbol{R}_{xx}$. These p principal eigenvalues are called signal eigenvalues, and the remaining $m-p$ secondary eigenvalues are called noise eigenvalues. According to the signal eigenvalues and noise eigenvalues, the column vector of the eigenvalue matrix $\boldsymbol{U}$ can be divided into two parts, namely

$$\boldsymbol{U} = [\boldsymbol{S}, \boldsymbol{G}]\,, \tag{10.4.21}$$

where

$$\boldsymbol{S} = [\boldsymbol{s}_1, \cdots, \boldsymbol{s}_p] = [\boldsymbol{u}_1, \cdots, \boldsymbol{u}_p]\,, \tag{10.4.22}$$

$$\boldsymbol{G} = \left[\boldsymbol{g}_1, \cdots, \boldsymbol{g}_{m-p}\right] = \left[\boldsymbol{u}_{p+1}, \cdots, \boldsymbol{u}_m\right] \tag{10.4.23}$$

are composed of signal eigenvectors and noise eigenvectors.

Note that $\langle \boldsymbol{S}, \boldsymbol{S}\rangle = \boldsymbol{S}^{\mathrm{H}}\boldsymbol{S} = \boldsymbol{I}$, so the projection matrix

$$\boldsymbol{P}_S \stackrel{\text{def}}{=} \boldsymbol{S}\langle \boldsymbol{S}, \boldsymbol{S}\rangle^{-1}\boldsymbol{S}^{\mathrm{H}} = \boldsymbol{S}\boldsymbol{S}^{\mathrm{H}}\,, \tag{10.4.24}$$

$$\boldsymbol{P}_n \stackrel{\text{def}}{=} \boldsymbol{G}\langle \boldsymbol{G}, \boldsymbol{G}\rangle^{-1}\boldsymbol{G}^{\mathrm{H}} = \boldsymbol{G}\boldsymbol{G}^{\mathrm{H}} \tag{10.4.25}$$

respectively represent the projection operator of the signal subspace and the noise subspace, and there are

$$\boldsymbol{P}_n = \boldsymbol{G}\boldsymbol{G}^{\mathrm{H}} = \boldsymbol{I} - \boldsymbol{S}\boldsymbol{S}^{\mathrm{H}} = \boldsymbol{I} - \boldsymbol{P}_s\,. \tag{10.4.26}$$

10.4.3 MUSIC Algorithm

Multiple signals can be classified using the concept of subspace.

Since $\boldsymbol{R}_{xx} = \boldsymbol{A}\boldsymbol{P}\boldsymbol{A}^{\mathrm{H}} + \sigma^2\boldsymbol{I}$, and $\boldsymbol{R}_{xx}\boldsymbol{G} = \boldsymbol{A}\boldsymbol{P}\boldsymbol{A}^{\mathrm{H}}\boldsymbol{G} + \sigma^2\boldsymbol{G}$, there is

$$\boldsymbol{R}_{xx}\boldsymbol{G} = \left[\boldsymbol{S}, \boldsymbol{G}\right]\boldsymbol{\Sigma}\begin{bmatrix}\boldsymbol{S}^{\mathrm{H}}\\ \boldsymbol{G}^{\mathrm{H}}\end{bmatrix}\boldsymbol{G} = [\boldsymbol{S}; \boldsymbol{G}]\boldsymbol{\Sigma}\begin{bmatrix}\boldsymbol{O}\\ \boldsymbol{I}\end{bmatrix} = \sigma^2\boldsymbol{G}\,. \tag{10.4.27}$$

Using the result of Eq. (10.4.27), we immediately get

$$\boldsymbol{A}\boldsymbol{P}\boldsymbol{A}^{\mathrm{H}}\boldsymbol{G} = \boldsymbol{O}\,. \tag{10.4.28}$$

Then there is

$$\boldsymbol{G}^{\mathrm{H}}\boldsymbol{A}\boldsymbol{P}\boldsymbol{A}^{\mathrm{H}}\boldsymbol{G} = \boldsymbol{O}\,. \tag{10.4.29}$$

As is known, $t^H Q t = 0$ holds if and only if $t = \mathbf{0}$, so the sufficient and necessary condition for Eq. (10.4.29) to hold is

$$G^H A = O. \tag{10.4.30}$$

Substituting $A = [a(\omega_1), \cdots, a(\omega_p)]$ into Eq. (10.4.30), it yields that

$$G^H a(\omega_i) = \mathbf{0}, \quad i = 1, \cdots, p, \tag{10.4.31}$$

or in scalar form that

$$|G^H a(\omega_i)|_2^2 = a^H(\omega_i) G G^H a(\omega_i) = 0, \quad i = 1, \cdots, p, \tag{10.4.32}$$

which is named as null spectrum. Obviously, when $\omega \neq \omega_1, \cdots, \omega_p$, the obtained spectrum is nonzero spectrum due to $a^H(\omega) G G^H a(\omega) \neq 0$. In other words, the spatial parameters $\omega_1, \cdots, \omega_p$ satisfying the null spectrum are the spatial frequency estimations of the p sources.

In practical applications, the null spectrum definition in Eq. (10.4.32) is often rewritten as a function similar to the power spectrum

$$P_{\text{MUSIC}}(\omega) = \frac{a^H(\omega) a(\omega)}{|a^H(\omega) G|_2^2} = \frac{a^H(\omega) a(\omega)}{a^H(\omega) G G^H a(\omega)}. \tag{10.4.33}$$

This is formally similar to the Capon spatial spectrum defined in Eq. (10.4.13), with the difference that the covariance matrix R_{xx} of the Capon spatial spectrum is replaced by the noise subspace GG^H.

Eq. (10.4.33) take the ω values $\omega_1, \cdots, \omega_p$ of the peaks to give the frequencies of the p signal sources, so that the DOA $\theta_1, \cdots, \theta_p$ can be obtained from Eq. (10.2.1).

Since the spatial spectrum defined by Eq. (10.4.33) can distinguish (i.e., classify) multiple spatial signals, it is called multiple signal classification (MUSIC) spatial spectrum, first proposed by Schmidt[192], Bienvenu and Kopp[27] independently at an academic conferences in 1979. Later, Schmidt republished his paper in the IEEE Transactions on Antenna Propagation in 1986[191]. It is worth pointing out that Eq. (10.4.31) is the basic formula of the MUSIC spatial spectrum.

The separation of mixed multiple signals using MUSIC spatial spectrum is called multiple signal classification method, or MUSIC method for short. As will be seen later, various extensions of the MUSIC method are developed based on Eq. (10.4.31).

The MUSIC spatial spectrum estimation method has become a representative topic of signal processing and has been used extensively. Since the spatial spectrum is defined by the noise subspace GG^H, the MUSIC method is a noise subspace method.

In practical applications, ω is usually divided into hundreds of equally spaced bins to get

$$\omega_i = 2\pi i \Delta f. \tag{10.4.34}$$

For example, take $\Delta f = \frac{0.5}{500} = 0.001$, and then substitute each value of ω_i into the MUSIC spatial spectrum definition in Eq. (10.4.33) to find all the value of ω corresponding to the peak value. As a consequence, the MUSIC algorithm needs to perform a global search on the frequency axis to get those p peaks, and thus the amount of calculation is relatively large.

In order to improve the performance of the MUSIC algorithm, several variants have been proposed. Readers interested in them may refer to the literature [194]. In the following, one of them will be introduced, and its basis is the maximum likelihood method. Specifically, this modified MUSIC algorithm maximizes the likelihood value of the variable

$$\epsilon_i = \boldsymbol{a}^{\mathrm{H}}(\omega)\boldsymbol{g}_i, \quad i = 1, \cdots, m-p. \tag{10.4.35}$$

Note that the basic MUSIC algorithm is to minimize $\sum_{i=1}^{m-p} |\epsilon_i|^2$. Sharman and Durrani[194] proved that the asymptotic (large sample N) estimator of Eq. (10.4.35) with the largest likelihood is maximized by the following function

$$P_{\mathrm{MUSIC}}(\omega) = \frac{\boldsymbol{a}^{\mathrm{H}}(\omega)\hat{\boldsymbol{U}}\boldsymbol{a}(\omega)}{\boldsymbol{a}^{\mathrm{H}}(\omega)\boldsymbol{G}\boldsymbol{G}^{\mathrm{H}}\boldsymbol{a}(\omega)}, \tag{10.4.36}$$

where

$$\hat{\boldsymbol{U}} = \sigma^2 \sum_{k=1}^{p} \frac{\lambda_k}{\left(\sigma^2 - \lambda_k\right)^2} \boldsymbol{u}_k \boldsymbol{u}_k^{\mathrm{H}}. \tag{10.4.37}$$

Algorithm 10.4.1. *Improved MUSIC algorithm*[194]

Step 1 Compute the eigenvalue decomposition of the sample covariance matrix $\boldsymbol{R}_{xx}$, and get its principal eigenvalues $\lambda_1, \cdots, \lambda_p$ and secondary eigenvalues σ^2, and store the principal eigenvector $\boldsymbol{u}_1, \cdots, \boldsymbol{u}_p$.

Step 2 Use Eq. (10.4.36) to calculate the MUSIC spectrum $P_{MUSIC}(\omega_i)$, where $\omega_i = (i-1)\Delta\omega$, and the step size $\Delta\omega$ can be taken as $2\pi \cdot 0.001$, for instance.

Step 3 Search for the p peaks of $P_{MUSIC}(\omega)$, to give the estimation of the MUSIC spatial parameter $\omega_1, \cdots, \omega_p$, and then the estimation of the DOA $\theta_1, \cdots, \theta_p$ from $\omega_i = 2\pi\frac{d}{\lambda}\sin\theta_i$.

Stoica and Nehorai[198] analyzed the estimation performance of the MUSIC algorithm and proved the following conclusions:

(1) The estimation error $\left(\hat{\boldsymbol{u}}_i - \boldsymbol{u}_i\right)$ of the eigenvector $\boldsymbol{u}_i$ follows an asymptotic (for large samples N) joint Gaussian distribution with its mean to be a vector of all zero;
(2) The estimation error $\left(\hat{\boldsymbol{\omega}}_i - \boldsymbol{\omega}_i\right)$ also follows an asymptotic joint Gaussian distribution with its mean to be a vector of all zero;
(3) Assume that the function $r(\omega) = \boldsymbol{a}(\omega)\hat{\boldsymbol{U}}\boldsymbol{a}(\omega)$ satisfies the constrained condition $r(\omega_i) \neq 0, i = 1, \cdots, p$, then the maximization of the MUSIC spectrum $\boldsymbol{P}_{\mathrm{MUSIC}}(\omega)$ defined by Eq. (10.4.36) and Eq. (10.4.33) $\hat{\omega}_1, \cdots, \hat{\omega}_p$ follows the same asymptotic distribution.

10.5 Extensions of MUSIC Algorithm

The previous section focuses on the basic MUSIC method of spatial spectrum estimation. In some practical applications, it is necessary to make important extensions from the basic MUSIC method to improve the computational effectiveness or estimation performance of the MUSIC method. These applications include:

(1) Some two sources may be coherent;
(2) Real-time applications require to avoid searching for spatial spectrum peaks;
(3) The resolution of the MUSIC method needs to be further improved.

This section will focus on these important applications and introduce several extensions of the MUSIC method: decoherent MUSIC method, root-MUSIC method, beamspace MUSIC (BS-MUSIC) method, and so on.

10.5.1 Decoherent MUSIC Algorithm

Due to the influence of multipath transmission or artificial interference, the signals received by the array from different directions are sometimes coherent, which will lead to the source covariance matrix $\boldsymbol{P}$ to be rank-dificient.

As a simplest example, consider using two array elements to receive two spatial signals

$$\boldsymbol{x}(t) = \boldsymbol{A}\boldsymbol{s}(t) + \boldsymbol{n}(t) = s_1(t)\boldsymbol{a}(\omega_1) + s_2(t)\boldsymbol{a}(\omega_2) + \boldsymbol{n}(t), \tag{10.5.1}$$

where $\boldsymbol{A} = [\boldsymbol{a}(\theta_1), \boldsymbol{a}(\theta_2)]$. Consider two narrowband spatial signals $s_1(t) = s_1 e^{j\omega_1 t}$ and $s_2(t) = s_2 e^{j\omega_2 t}$. Therefore, the array correlation matrix $\boldsymbol{R}_x = \mathrm{E}\left\{\boldsymbol{x}(t)\boldsymbol{x}^{\mathrm{H}}(t)\right\}$ is

$$\boldsymbol{R}_x = \boldsymbol{A}\boldsymbol{E}\begin{bmatrix} s_1(t)s_1^*(t) & s_1(t)s_2^*(t) \\ s_2(t)s_1^*(t) & s_2(t)s_2^*(t). \end{bmatrix}\boldsymbol{A}^{\mathrm{H}} = \boldsymbol{A}\begin{bmatrix} \sigma_1^2 & \rho_{12}\sigma_1\sigma_2^* \\ \rho_{12}^*\sigma_2\sigma_1^* & \sigma_2^2 \end{bmatrix}\boldsymbol{A}^{\mathrm{H}}. \tag{10.5.2}$$

If the two spatial signals are coherent, then $|\rho_{12}| = 1$, so that the determinant

$$\begin{vmatrix} \sigma_1^2 & \rho_{12}\sigma_1\sigma_2^* \\ \rho_{12}^*\sigma_2\sigma_1^* & \sigma_2^2 \end{vmatrix} = 0.$$

That is, the rank of the 2×2 array correlation matrix $\boldsymbol{R}_x$ is 1, which suggests its rank dificiency. At this time, the array correlation matrix $\boldsymbol{R}_x$ has only one eigenvalue, which is neither related to the DOA θ_1 of signal $s_1(t)$ or the DOA θ_2 of signal $s_2(t)$. In order to see this fact more clearly, let the coherent signal $s_2(t) = C \cdot s_1(t)$, where C is a nonzero complex constant. Then, the array signal vector would be reduced to

$$\boldsymbol{x}(t) = s_1(t)\left[\boldsymbol{a}(\theta_1) + C\boldsymbol{a}(\theta_2)\right] = s_1(t)\boldsymbol{b}(\theta),$$

where $\boldsymbol{b}(\theta) = \boldsymbol{a}(\theta_1) + C\boldsymbol{a}(\theta_2)$ is the equivalent steering vector of $s_1(t)$ in the coherent case. Obviously, the DOA θ would be neither θ_1 nor θ_2. In other words, if only one uniform linear array is used to observe two coherent spatial signals, it would be impossible to estimate the DOA of either spatial signal.

If two of the p spatial signals are coherent, then the rank of the $p \times p$ array correlation matrix $\boldsymbol{R}_x$ is equal to $p - 1$, which is rank-deficient, resulting in that the dimension of the signal subspace is also $p - 1$. The MUSIC method will not be able to estimate the DOA of either of the two coherent spatial signals anymore.

In order to solve the rank deficiency of the array correlation matrix in the presence of coherent signals, it is necessary to introduce additionally a nonlinear correlation array vector. For uniform linear arrays, taking the "reverse array vector" as such an vector is an effective option.

Let $\boldsymbol{J}$ be the $L \times L$ permutation matrix, that is, all the elements on the anti-diagonal of $\boldsymbol{J}$ are 1, while all the other elements are equal to 0, then for an uniform linear array, there is $\boldsymbol{J}\boldsymbol{a}^{*}(\theta) = \mathrm{e}^{-\mathrm{j}(L-1)\phi}\boldsymbol{a}(\theta)$. Then, the corresponding array covariance matrix can be derived as

$$\boldsymbol{R}_{\mathrm{B}} = \boldsymbol{J}\boldsymbol{R}_x^{*}\boldsymbol{J} = \boldsymbol{A}\boldsymbol{\Phi}^{-(L-1)}\boldsymbol{P}\boldsymbol{\Phi}^{-(L-1)}\boldsymbol{A}^{\mathrm{H}} + \sigma^2\boldsymbol{I}, \tag{10.5.3}$$

where $\boldsymbol{\Phi}$ is a diagonal matrix, and the diagonal elements are $\mathrm{e}^{\mathrm{j}m\phi}(m = 1, \cdots, M)$. Find the average of the (forward) array covariance matrix $\boldsymbol{R}_{xx}$ and the reverse array covariance matrix $\boldsymbol{R}_{\mathrm{B}}$ and we can get the forward and reverse array covariance matrix

$$\boldsymbol{R}_{\mathrm{FB}} = \frac{1}{2}(\boldsymbol{R}_{xx} + \boldsymbol{R}_{\mathrm{B}}) = \frac{1}{2}\left(\boldsymbol{R}_{xx} + \boldsymbol{J}\boldsymbol{R}_x^{*}\boldsymbol{J}\right) = \boldsymbol{A}\tilde{\boldsymbol{P}}\boldsymbol{A}^{\mathrm{H}} + \sigma^2\boldsymbol{I}, \tag{10.5.4}$$

where the new source covariance matrix $\tilde{P} = \frac{1}{2}\left(\boldsymbol{P} + \Phi^{-(L-1)}\boldsymbol{P}\Phi^{-(L-1)}\right)$ usually has full rank.

Any algorithm based on the covariance matrix can be modified into the forward and reverse form like the algorithm mentioned above by simply replacing $\hat{\boldsymbol{R}}_x$ with $\hat{\boldsymbol{R}}_{\mathrm{FB}}$. The transformation $\hat{\boldsymbol{R}}_x \rightarrow \hat{\boldsymbol{R}}_{\mathrm{FB}}$ is also used to improve the estimated variance in noncoherent situations.

Spatial smoothing technique is another effective method to deal with coherent or highly correlated signals. The basic idea is to divide the N element uniform linear array into $L = N - M + 1$ overlapping subarrays, where each subarray is composed of M array elements. Figure 10.5.1 shows an example of a 7-element array divided into three subarrays.

In the case of an uniform linear array, since each subarray with M array elements have the same array manifold, the steering vector of each subarray can be denoted as

$$a_M(\theta) = \left[1, \mathrm{e}^{\mathrm{j}\pi\theta}, \cdots, \mathrm{e}^{\mathrm{j}\pi(M-1)\theta}\right]^{\mathrm{T}}. \tag{10.5.5}$$

Therefore, the observation signal vector of the first subarray is

$$\boldsymbol{x}_1(t) = A_M s(t) + n_1(t), \tag{10.5.6}$$

and the observation signal vector of the second subarray is

$$\boldsymbol{x}_2(t) = \boldsymbol{A}_M\boldsymbol{D}\boldsymbol{s}(t) + \boldsymbol{n}_2(t), \tag{10.5.7}$$

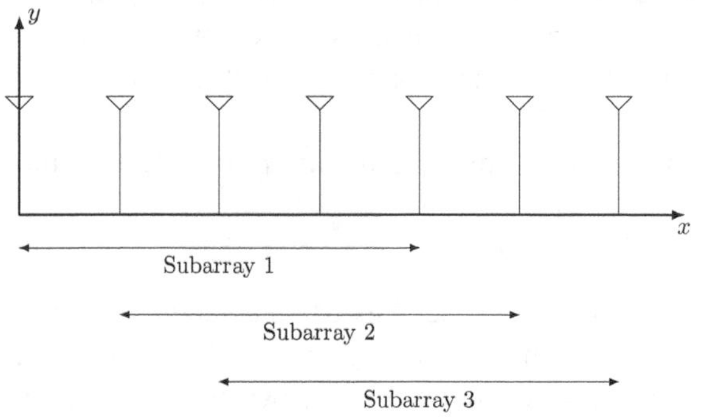

Fig. 10.5.1: Uniform linear array divided into three subarrays

where $\boldsymbol{D} = \text{diag}\left(e^{j\pi\sin\theta_1}, \cdots, e^{j\pi\theta_p}\right)$ is a diagonal matrix of $p \times p$, and $n_2(t)$ is the additive white noise of the second subarray. Similarly, the observation signal vector of the L-th subarray is

$$\boldsymbol{x}_L(t) = \boldsymbol{A}_M \boldsymbol{D}^{L-1}\boldsymbol{s}(t) + \boldsymbol{n}_L(t). \tag{10.5.8}$$

If the average of the signal vectors of the L subarrays is taken as the spatial smoothing array signal vector

$$\bar{\boldsymbol{x}}(t) = \frac{1}{\sqrt{L}}\sum_{i=1}^{L}\boldsymbol{x}_i(t), \tag{10.5.9}$$

then the spatial smoothing array correlation matrix can be derived as

$$\bar{\boldsymbol{R}}_x = \text{E}\left\{\bar{\boldsymbol{x}}(t)\bar{\boldsymbol{x}}^{\text{H}}(t)\right\} = \boldsymbol{A}_M\left[\frac{1}{L}\sum_{i=1}^{L}\boldsymbol{D}^{(i-1)}\boldsymbol{R}_s\boldsymbol{D}^{-(i-1)}\right]\boldsymbol{A}_M^{\text{H}} + \sigma_n^2, \tag{10.5.10}$$

or simply as

$$\bar{\boldsymbol{R}}_x = \boldsymbol{A}_M\bar{\boldsymbol{R}}_s\boldsymbol{A}_M^{\text{H}}, \tag{10.5.11}$$

where

$$\bar{\boldsymbol{R}}_s = \frac{1}{L}\sum_{i=1}^{L}\boldsymbol{D}^{(i-1)}\boldsymbol{R}_s\boldsymbol{D}^{-(i-1)} \tag{10.5.12}$$

represents the result of spatial smoothing of the source correlation matrix $\boldsymbol{R}_s = \text{E}\left\{\boldsymbol{s}(t)\boldsymbol{s}^{\text{H}}(t)\right\}$ through L subarrays, which is also named as the spatial smoothing source correlation matrix.

In the case that the two sources in p spatial information sources are coherent, although the rank of the $p \times p$ source correlation matrix $\boldsymbol{R}_s$ is equal to $p - 1$, namely the rank is deficient, the rank of the $p \times p$ spatial smoothing source correlation matrix $\bar{\boldsymbol{R}}_s$ is equal to p, which is a full-rank matrix.

Since the rank of the spatial smoothing source correlation matrix $\overline{\boldsymbol{R}}_S$ is equal to p, as long as the DOA of the p spatial signals is different, then the rank of the $M \times M$ spatial smoothing array correlation matrix $\overline{\boldsymbol{R}}_X$ is the same as the rank of $\overline{\boldsymbol{R}}_S$, which is also equal to p.

Let the eigenvalue decomposition of the correlation matrix of the $M \times M$ spatial smoothing array be

$$\overline{\boldsymbol{R}}_X = \boldsymbol{U}\boldsymbol{\Sigma}\boldsymbol{U}^{\mathrm{H}}, \tag{10.5.13}$$

where the first p of the M eigenvalues of the eigenvalue matrix $\boldsymbol{\Sigma}$ are large eigenvalues, corresponding to p spatial signals; the remaining $M - p$ small eigenvalues are the corresponding variance of the array observation noise. The eigenvectors corresponding to these small eigenvalues form the spatial smoothing noise subspace $\bar{\boldsymbol{G}}\bar{\boldsymbol{G}}^{\mathrm{H}}$, where

$$\overline{\boldsymbol{G}} = \left[\boldsymbol{u}_{p+1}, \cdots, \boldsymbol{u}_M\right]. \tag{10.5.14}$$

Therefore, we only need to replace the original N elements steering vector $\boldsymbol{a}(\theta) = \left[1, \mathrm{e}^{\mathrm{j}\pi\theta}, \cdots, \mathrm{e}^{\mathrm{j}\pi(N-1)\theta}\right]^{\mathrm{T}}$ in the MUSIC spatial spectrum with the M element subarray steering vector $\boldsymbol{a}_M(\theta) = \left[1, \mathrm{e}^{\mathrm{j}\pi\theta}, \cdots, \mathrm{e}^{\mathrm{j}\pi(M-1)\theta}\right]^{\mathrm{T}}$, and replace the noise subspace with $\bar{G}\bar{G}^{\mathrm{H}}$, then the basic MUSIC method can be extended to the decoherent MUSIC method, and its spatial spectrum is

$$P_{\mathrm{DECMUSIC}}(\theta) = \frac{\boldsymbol{a}_M^{\mathrm{H}}(\theta)\boldsymbol{a}_M(\theta)}{\boldsymbol{a}_M^{\mathrm{H}}(\theta)\overline{\boldsymbol{G}\boldsymbol{G}}^{\mathrm{H}}\boldsymbol{a}_M(\theta)}. \tag{10.5.15}$$

The disadvantage of spatial smoothing is that the effective aperture of the array is reduced because the subarrays are smaller than the original array. The more subarrays there are, the smaller the effective aperture of the array. Generally, it is divided into two subarrays. However, despite this aperture loss, the spatial smoothing transform alleviates the limitations of all subspace estimation techniques and is able to preserve the computational validity of the one-dimensional spectral search.

10.5.2 Root-MUSIC Algorithm

The root-MUSIC method is a polynomial root finding form of the MUSIC method, proposed by Barabell[17].

The peaks of the basic MUSIC spatial spectrum is equivalent to $\boldsymbol{a}^{\mathrm{H}}(\omega)\boldsymbol{U}_n = \boldsymbol{0}^{\mathrm{T}}$ or $\boldsymbol{a}^{\mathrm{H}}(\omega)\boldsymbol{u}_j = 0, j = p+1, \cdots, M$, where $\boldsymbol{u}_{p+1}, \cdots, \boldsymbol{u}_M$ are the secondary eigenvectors of the array sample covariance matrix $\hat{R}_X$.

If let $\boldsymbol{p}(z) = \boldsymbol{a}(\omega)\big|_{z=\mathrm{e}^{\mathrm{j}\omega}}$, then we have

$$\boldsymbol{p}(z) = \left[1, z, \cdots, z^{M-1}\right]^{\mathrm{T}}. \tag{10.5.16}$$

The vector inner product $u_i^{\mathrm{H}}p(z)$ gives a polynomial representation

$$p(z) = \boldsymbol{u}_i^{\mathrm{H}}\boldsymbol{p}(z) = u_{1,j}^{*} + u_{2,j}^{*}z + \cdots + u_{M,j}^{*}z^{M-1}, \tag{10.5.17}$$

where $u_{i,j}$ is the (i, j) element of the $M \times (M - p)$ eigenvector matrix $\boldsymbol{U}_n$. Therefore, the equation $\boldsymbol{a}^{\mathrm{H}}(\omega)\boldsymbol{u}_j = 0$ or $\boldsymbol{u}_j^{\mathrm{H}}\boldsymbol{a}(\omega) = 0, j = p + 1, \cdots, M$ for the basic MUSIC spatial spectrum can be equivalently expressed as

$$p_i(z) = \boldsymbol{u}_j^{\mathrm{H}}\boldsymbol{p}(z) = 0, \quad j = p + 1, \cdots, M. \tag{10.5.18}$$

The above equation can be combined together as $\boldsymbol{U}_n^{\mathrm{H}}\boldsymbol{p}(z) = 0$ or $\left|\boldsymbol{U}_n^{\mathrm{H}}\boldsymbol{p}(z)\right|_2^2 = 0$, and that is,

$$\boldsymbol{p}^{\mathrm{H}}(z)\boldsymbol{U}_n\boldsymbol{U}_n^{\mathrm{H}}\boldsymbol{p}(z) = 0, \quad z = \mathrm{e}^{\mathrm{j}\omega_1}, \cdots, \mathrm{e}^{\mathrm{j}\omega_p}. \tag{10.5.19}$$

In other words, as long as the roots z_i of the polynomial $p^{\mathrm{H}}(z)\boldsymbol{U}_n\boldsymbol{U}_n^{\mathrm{H}}\boldsymbol{p}(z)$, which should be on the unit circle, were found, we can get the spatial parameters $\omega_1, \cdots, \omega_p$ immediately. This is the basic idea of the root-MUSIC.

However, the Eq. (10.5.19) is not a polynomial of z, because it also contains the power term of z^*. Since we are only interested in the value of z on the unit circle, we can use $\boldsymbol{p}^{\mathrm{T}}\left(z^{-1}\right)$ instead of $\boldsymbol{p}^{\mathrm{H}}(z)$ which gives the root-MUSIC polynomial

$$p(z) = z^{M-1}\boldsymbol{p}^{\mathrm{T}}\left(z^{-1}\right)\hat{\boldsymbol{U}}_n\hat{\boldsymbol{U}}_n^{\mathrm{H}}p(z). \tag{10.5.20}$$

Now, $p(z)$ is a polynomial of order $2(M - 1)$, and its roots are several mirror-image pairs with respect to the unit circle. Among them, the phase of the p roots $\hat{z}_1, \hat{z}_2, \cdots, \hat{z}_p$ with the largest amplitude provides the DOA estimation, and there is

$$\hat{\theta}_i = \arccos\left[\frac{1}{kd}\arg\left(\hat{z}_m\right)\right], \quad i = 1, \cdots, p. \tag{10.5.21}$$

It has been proved[198] that MUSIC and root-MUSIC have the same asymptotic performance, but for small samples the performance of the root-MUSIC method is significantly better than that of MUSIC.

10.5.3 Minimum Norm Algorithm

The minimum norm method is suitable for uniform linear arrays, and its basic idea is to search for the peak of the spatial spectrum

$$P_{\mathrm{MN}}(\omega) = \frac{\boldsymbol{a}^{\mathrm{H}}(\omega)\boldsymbol{a}(\omega)}{\left|\boldsymbol{a}^{\mathrm{H}}(\omega)\boldsymbol{w}\right|^2} \tag{10.5.22}$$

to estimate the DOA[78]. In this formula, $\boldsymbol{w}$ is the array weight vector, which belongs to the noise subspace and has the smallest norm $\min |\boldsymbol{w}|_2$.

In addition to Eq. (10.4.31), i.e., $\boldsymbol{a}^{\mathrm{H}}(\omega)\boldsymbol{G} = 0^{\mathrm{T}}$, the basic formula of MUSIC method can also be expressed by the projection formula.

With regard to the projection of the steering vector on the subspace spanned by the array element observation vector, there are the following basic facts:

(1) Only when the spatial frequency is $\omega \in \{\omega_1, \cdots, \omega_p\}$, the projection of the steering vector $\boldsymbol{a}(\omega)$ on the signal subspace $\boldsymbol{S}\boldsymbol{S}^{\mathrm{H}}$ is the steering vector itself, while the projection on the noise space $\boldsymbol{G}\boldsymbol{G}^{\mathrm{H}} = \boldsymbol{I} - \boldsymbol{S}\boldsymbol{S}^{\mathrm{H}}$ should be equal to zero vector.

(2) When the spatial frequency is not equal to the spatial frequency of any signal source, the projection of the steering vector $\boldsymbol{a}(\omega)$ on the signal subspace $\boldsymbol{S}\boldsymbol{S}^{\mathrm{H}}$ should be equal to zero vector, and the projection on the noise space $\boldsymbol{G}\boldsymbol{G}^{\mathrm{H}}$ is equal to the steering vector itself.

Obviously, these basic facts described above can be expressed as

$$\boldsymbol{G}\boldsymbol{G}^{\mathrm{H}}\boldsymbol{a}(\omega) = \begin{cases} \boldsymbol{0}, & \omega = \omega_1, \cdots, \omega_p \\ \boldsymbol{a}(\omega), & \text{else}, \end{cases}, \tag{10.5.23}$$

or

$$\boldsymbol{a}^{\mathrm{H}}(\omega)\boldsymbol{G}\boldsymbol{G}^{\mathrm{H}} = \begin{cases} \boldsymbol{0}^{\mathrm{T}}, & \omega = \omega_1, \cdots, \omega_p \\ \boldsymbol{a}^{\mathrm{H}}(\omega), & \text{else}. \end{cases}. \tag{10.5.24}$$

Note that for an uniform linear array with M array elements, the steering vector is $\boldsymbol{a}(\omega) = \left[1, \mathrm{e}^{-\mathrm{j}\omega}, \cdots, \mathrm{e}^{-\mathrm{j}(M-1)\omega}\right]^{\mathrm{T}}$, so $\boldsymbol{a}^{\mathrm{H}}(\omega)\boldsymbol{e}_1 = 1$, where $\boldsymbol{e}_1$ is a basis vector with the first element to be 1 and all other elements to be zero.

Therefore, Eq. (10.5.24) can be rewritten in a scalar form as

$$a^{\mathrm{H}}(\omega)GG^{\mathrm{H}}e_1 = \begin{cases} 0, & \omega = \omega_1, \cdots, \omega_p, \\ 1, & \text{else}. \end{cases} \tag{10.5.25}$$

The corresponding spatial spectrum formulation is

$$P_{\mathrm{MN}}(\omega) = \frac{\boldsymbol{a}^{\mathrm{H}}(\omega)\boldsymbol{a}(\omega)}{\left|\boldsymbol{a}^{\mathrm{H}}(\omega)\boldsymbol{G}\boldsymbol{G}^{\mathrm{H}}\boldsymbol{e}_1\right|^2}, \tag{10.5.26}$$

which is equivalent to taking

$$w = \boldsymbol{G}\boldsymbol{G}^{\mathrm{H}}\boldsymbol{e}_1 \tag{10.5.27}$$

as the array weight vector in Eq. (10.5.22). The spatial spectrum $P_{\mathrm{MN}}(\omega)$ defined by Eq. (10.5.26) is called the minimum norm spatial spectrum.

Just as the basic MUSIC method requires one-dimensional search, the minimum norm method also needs to search for spectral peaks. Similarly, there is also a root-finding minimum norm method.

The minimum norm method has the following properties[122, 91]:

(1) Deviation: smaller than the basic MUSIC method;

(2) Resolution: higher than the basic MUSIC method.

10.5.4 First Principal Vector MUSIC Algorithm

The basic MUSIC method, the decoherent MUSIC method, as well as the root-MUSIC method introduced above all estimate the spatial spectrum based on the observation data of M elements directly, and is collectively referred to as the element space (ES) MUSIC method, abbreviated as ES-MUSIC method.

There is another important extension of the ES-MUSIC method, which can also improve the spatial spectrum estimation resolution of the MUSIC method. The method, which was proposed by Buckley and Xu [37] in 1990, is called the first principal vector (FIrst priNcipal vEctor, FINE) method.

The core idea of the FINE method is to use the signal covariance matrix. For the array model $\boldsymbol{x}(t) = \boldsymbol{A}\boldsymbol{s}(t) + \boldsymbol{n}(t)$, the array covariance matrix is

$$\boldsymbol{R}_{xx} = \mathrm{E}\left\{\boldsymbol{x}(t)\boldsymbol{x}^{\mathrm{H}}(t)\right\} = \boldsymbol{R}_s + \sigma_n^2\boldsymbol{I} = \boldsymbol{A}\boldsymbol{P}_s\boldsymbol{A}^{\mathrm{H}}, \tag{10.5.28}$$

where $\boldsymbol{R}_s$ is the signal covariance matrix, which represents the covariance matrix of the signal arriving at the array through channel transmission, and σ_n^2 represents the variance of the additive noise on each array element. Note that the signal covariance matrix $\boldsymbol{R}_s$ and the source covariance matrix $\boldsymbol{P}_s = \mathrm{E}\left\{\boldsymbol{s}(t)\boldsymbol{s}^{\mathrm{H}}(t)\right\}$ are different concepts.

Let the eigenvalue decomposition of the sample covariance matrix

$$\hat{\boldsymbol{R}}_x = \frac{1}{N}\sum_{t=1}^{N}\boldsymbol{x}(t)\boldsymbol{x}^{\mathrm{H}}(t),$$

be

$$\hat{\boldsymbol{R}}_x = \sum_{i=1}^{M}\lambda_i\boldsymbol{u}_i\boldsymbol{u}_i^{\mathrm{H}}, \tag{10.5.29}$$

and the $M \times (Mp)$ matrix consisting of the eigenvectors corresponding to the smaller eigenvalues $\lambda_{p+1} \approx \cdots \approx \lambda_M = \hat{\sigma}_n^2$ be

$$G = \left[\boldsymbol{u}_{p+1}, \cdots, \boldsymbol{u}_M\right], \tag{10.5.30}$$

then the basic MUSIC spatial spectrum is

$$P_{\mathrm{MUSIC}}(\omega) = \frac{\boldsymbol{a}^{\mathrm{H}}(\omega)\boldsymbol{a}(\omega)}{\left|\boldsymbol{a}^{\mathrm{H}}(\omega)\boldsymbol{G}\right|_2^2}, \tag{10.5.31}$$

and the eigenvalue decomposition of the sample signal covariance matrix is

$$\hat{\boldsymbol{R}}_s = \hat{\boldsymbol{R}}_x - \hat{\sigma}_n^2\boldsymbol{I} = \sum_{i=1}^{M}\eta_i\boldsymbol{v}_i\boldsymbol{v}_i^{\mathrm{H}}. \tag{10.5.32}$$

Assume that the K principal eigenvectors of the sample signal covariance matrix form a $M \times K$ matrix

$$\boldsymbol{V} = [\boldsymbol{v}_1, \cdots, \boldsymbol{v}_K]. \tag{10.5.33}$$

Using the eigenvectors of the sample covariance matrix $\boldsymbol{R}_x$ to form $M\times(M-p)$ secondary eigenvector matrix $\boldsymbol{G}$, if the sample signal covariance matrix $\boldsymbol{R}_s$ is known, then its principal eigenvectors can be used to form the $M \times K$ source principal eigenvector matrix $\boldsymbol{V}$.

Furthermore, from the singular value decomposition of the $K \times (M-p)$ matrix product of $\boldsymbol{V}^{\mathrm{H}}\boldsymbol{G}$

$$\boldsymbol{V}^{\mathrm{H}}\boldsymbol{G} = \boldsymbol{Y}\boldsymbol{\Sigma}\boldsymbol{Z}^{\mathrm{H}}, \tag{10.5.34}$$

we can get the $(M-p)\times(M-p)$ right singular vector matrix $\boldsymbol{Z}$ and then get the $M\times(M-p)$ matrix $\boldsymbol{T}$ as

$$\boldsymbol{T} = \boldsymbol{G}\boldsymbol{Z} = \left[\boldsymbol{t}_{\mathrm{FINE}}, \boldsymbol{t}_2, \cdots, \boldsymbol{t}_{M-p}\right], \tag{10.5.35}$$

in which the first column vector t_{FINE} is named as the first principal vector of matrix T, and the first several column vectors are called the primary principal vectors of T, denoted as $\boldsymbol{T}_{\mathrm{FINES}}$.

The $M \times (Mp)$ matrix $\boldsymbol{T} = \boldsymbol{G}\boldsymbol{Z}$ has an important property: the matrix is located in the noise space $\boldsymbol{G}\boldsymbol{G}^{\mathrm{H}}$ of the array observation vector $\boldsymbol{x}(t)$, because the projection of the matrix T to the noise space is equal to the matrix itself, that is

$$\boldsymbol{G}\boldsymbol{G}^{\mathrm{H}}\boldsymbol{T} = \boldsymbol{G}\boldsymbol{G}^{\mathrm{H}}\boldsymbol{G}\boldsymbol{Z} = \boldsymbol{G}\boldsymbol{Z} = \boldsymbol{T}. \tag{10.5.36}$$

Furthermore, multiply both sides of $\boldsymbol{a}^{\mathrm{H}}(\omega)\boldsymbol{G} = 0^{\mathrm{T}}$ of the basic MUSIC method with the nonsingular matrix $\boldsymbol{Z}$, it can be derived that

$$\boldsymbol{a}^{\mathrm{H}}(\omega)\boldsymbol{G}\boldsymbol{Z} = \boldsymbol{a}^{\mathrm{H}}(\omega)\boldsymbol{T} = \boldsymbol{0}^{\mathrm{T}}, \quad \omega = \omega_1, \cdots, \omega_p. \tag{10.5.37}$$

Substituting Eq. (10.5.36) into Eq. (10.5.37) yields

$$\boldsymbol{a}^{\mathrm{H}}(\omega)\boldsymbol{T} = \boldsymbol{a}^{\mathrm{H}}(\omega)\boldsymbol{G}\boldsymbol{G}^{\mathrm{H}}\boldsymbol{T} = \boldsymbol{0}^{\mathrm{T}}, \quad \omega = \omega_1, \cdots, \omega_p. \tag{10.5.38}$$

This leads to an important null spectrum formula

$$\left|\boldsymbol{a}^{\mathrm{H}}(\omega)\boldsymbol{G}\boldsymbol{G}^{\mathrm{H}}\boldsymbol{t}_{\mathrm{FINE}}\right|^2 = 0, \quad \omega = \omega_1, \cdots, \omega_p, \tag{10.5.39}$$

or

$$\left|\boldsymbol{a}^{\mathrm{H}}(\omega)\boldsymbol{G}\boldsymbol{G}^{\mathrm{H}}\boldsymbol{T}_{\mathrm{FINES}}\right|_2^2 = 0, \quad \omega = \omega_1, \cdots, \omega_p. \tag{10.5.40}$$

The corresponding first principal vector (FINE) spatial spectrum[37] is

$$P_{\mathrm{FINE}}(\omega) = \frac{\boldsymbol{a}^{\mathrm{H}}(\omega)\boldsymbol{a}(\omega)}{\left|\boldsymbol{a}^{\mathrm{H}}(\omega)\boldsymbol{G}\boldsymbol{G}^{\mathrm{H}}\boldsymbol{T}_{\mathrm{FINE}}\right|^2}, \tag{10.5.41}$$

and the multi-principal vector spatial spectrum is

$$P_{\mathrm{Fines}}(\omega) = \frac{\boldsymbol{a}^{\mathrm{H}}(\omega)\boldsymbol{a}(\omega)}{\left|\boldsymbol{a}^{\mathrm{H}}(\omega)\boldsymbol{G}\boldsymbol{G}^{\mathrm{H}}\boldsymbol{T}_{\mathrm{FINES}}\right|_2^2}. \tag{10.5.42}$$

Compared with the minimum norm spatial spectrum $P_{\mathrm{MN}}(\omega)$ defined by Eq. (10.5.26), the FINE spatial spectrum $P_{\mathrm{FINE}}(\omega)$ is just to replace the basic vector e_1 in Eq. (10.5.26) with the first principal vector t_{FINE}.

Literature [37] proved that the spatial spectral estimation resolution of the MUSIC method can be enhanced by using the first principal eigenvector of the covariance matrix.

The FINE method has the following properties[91, 230].

Deviation: less than the deviation of the MUSIC method;

Variance: smaller than the minimum norm method;

Resolution: better than the resolution of MUSIC and minimum norm method;

Advantages: good performance when the signal-to-noise ratio is low.

The key of the FINE spatial spectrum method is the estimation of the signal covariance matrix $\boldsymbol{R}_s$ to obtain the source principal eigenvector matrix $\boldsymbol{V}$.

10.6 Beamspace MUSIC Algorithm

Different from the ES-MUSIC method, the MUSIC method not only uses the array element observation data, but also uses the spatial beam output, which can be named as the beamspace MUSIC method, and abbreviated as the BS-MUSIC method.

10.6.1 BS-MUSIC Algorithm

Consider the array observation model

$$\boldsymbol{x}(n) = \boldsymbol{A}\boldsymbol{s}(n) + \boldsymbol{e}(n) = \sum_{i=1}^{p} \boldsymbol{a}(\omega_i)\, s_i(n) + \boldsymbol{e}(n), \tag{10.6.1}$$

where the array observation vector is $\boldsymbol{x}(n) = \left[x_1(n), \cdots, x_M(n)\right]^{\mathrm{T}}$, array response matrix is $\boldsymbol{A} = [\boldsymbol{a}(\omega_1), \cdots, \boldsymbol{a}(\omega_p)]$, and the steering vector is $\boldsymbol{a}(\omega) = \left[1, \mathrm{e}^{\mathrm{j}\omega}, \cdots, \mathrm{e}^{\mathrm{j}\omega(M-1)}\right]^{\mathrm{T}}$, and the additive white Gaussian noise $e_1(n), \cdots, e_M(n)$ has the same variance σ_e^2.

The covariance matrix of the array observation vector is

$$\boldsymbol{R}_{xx} = \mathrm{E}\left\{\boldsymbol{x}(n)\boldsymbol{x}^{\mathrm{H}}(n)\right\} = \boldsymbol{A}\boldsymbol{P}\boldsymbol{A}^{\mathrm{H}} + \sigma_e^2\boldsymbol{I}. \tag{10.6.2}$$

where $\boldsymbol{P} = \mathrm{E}\left\{\boldsymbol{s}(n)\boldsymbol{s}^{\mathrm{H}}(n)\right\}$ is nonsingular.

The discrete space Fourier transform (DSFT) of M array element observation data $x_m(n), m = 1, \cdots, M$ is defined as[197]

$$X(u; n) = \sum_{m=0}^{M-1} x_m(n)\mathrm{e}^{-\mathrm{j}m\pi u}. \tag{10.6.3}$$

The relationship between the space parameter u and the wave arrival direction angle θ is $u = \sin\theta$. In the above formula, $-1 \leqslant u \leqslant 1$ corresponds to the angle interval $-90^\circ \leqslant \theta \leqslant 90^\circ$, and this area is called the visible region of the array.

Discrete space Fourier transform can be calculated by discrete Fourier transform. It should be noted that the $M \times 1$ data vector of the n snapshot is $x(n) = \left[x_1(n), \cdots, x_M(n)\right]^{\mathrm{T}}$, and M point discrete space Fourier transform will give the M equally spaced samples in discrete space Fourier transform with the space parameter interval $0 \leqslant u \leqslant 2$.

Define $M \times 1$ discrete Fourier transform beamforming weight vector be

$$\boldsymbol{v}_M(u) = \left[1, \mathrm{e}^{\mathrm{j}\pi u}, \cdots, \mathrm{e}^{\mathrm{j}\,(M-1)\pi u}\right]^{\mathrm{T}}. \tag{10.6.4}$$

It has a Vandermonde structure. Thus, the linear transformation

$$X(u; n) = v_M^{\mathrm{H}}(u)\boldsymbol{x}(n) \tag{10.6.5}$$

gives the discrete space Fourier transform of the n snapshot in the space parameter u.

Split the space parameter $0 \leqslant u \leqslant 2$ into M equal parts by utilizing equal intervals $\Delta u = 2/M$, and define the $M \times B$ beam forming matrix as

$$\boldsymbol{W} = \frac{1}{\sqrt{M}}\left[v_M(0), \boldsymbol{v}_M\left(\frac{2}{M}\right), \cdots, \boldsymbol{v}_M\left((B-1)\frac{2}{M}\right)\right] \tag{10.6.6}$$

$$= \frac{1}{\sqrt{M}}\begin{bmatrix} 1 & 1 & \cdots & 1 \\ 1 & \mathrm{e}^{\mathrm{j}2\pi/M} & \cdots & \mathrm{e}^{\mathrm{j}2\pi(B-1)/M} \\ \vdots & \vdots & \vdots & \vdots \\ 1 & \mathrm{e}^{\mathrm{j}2\pi(M-1)/M} & \cdots & \mathrm{e}^{\mathrm{j}2\pi(M-1)(B-1)/M} \end{bmatrix} \tag{10.6.7}$$

$$= \frac{1}{\sqrt{M}}\begin{bmatrix} 1 & 1 & \cdots & 1 \\ 1 & w & \cdots & w^{B-1} \\ \vdots & \vdots & \vdots & \vdots \\ 1 & w^{M-1} & \cdots & w^{(M-1)(B-1)} \end{bmatrix}_{w=\mathrm{e}^{2\pi/M}}, \tag{10.6.8}$$

which has the structure of the Fourier matrix, and is called the beamspace Fourier matrix. According to the geometric progression summation formula $a_1 + a_1 q + \cdots + a_1 q^n = \frac{a_1(1-q^n)}{1-q}$, $1 + w + \cdots + w^{M-1} = 0$. With this result, it is easy to verify

$$\boldsymbol{W}^{\mathrm{H}}\boldsymbol{W} = \boldsymbol{I}. \tag{10.6.9}$$

Then, using the beamforming matrix $\boldsymbol{W}$ to perform linear transformation on the $M \times 1$ observation data vector $\boldsymbol{x}(n)$, the $B \times 1$ beamspace snapshot vector can be derived as

$$\tilde{\boldsymbol{x}}(n) = \begin{bmatrix} X(0; n) \\ X\left(\frac{2}{M}; n\right) \\ \vdots \\ X\left((B-1)\frac{2}{M}\right) \end{bmatrix} = \boldsymbol{W}^{\mathrm{H}}\boldsymbol{x}(n) = \boldsymbol{W}^{\mathrm{H}}\boldsymbol{A}\boldsymbol{s}(n) + \boldsymbol{W}^{\mathrm{H}}\boldsymbol{e}(n)$$

$$= \boldsymbol{B}\boldsymbol{s}(n) + \boldsymbol{W}^{\mathrm{H}}\boldsymbol{e}(n), \tag{10.6.10}$$

where $\boldsymbol{B} = \boldsymbol{W}^{\mathrm{H}}\boldsymbol{A} = [\boldsymbol{b}(\omega_1), \cdots, \boldsymbol{b}(\omega_B)]$ is the beamspace steering vector matrix, and $\boldsymbol{b}(\omega) = \boldsymbol{W}^{\mathrm{H}}\boldsymbol{a}(\omega)$. In other words, the beamspace snapshot vector $\tilde{\boldsymbol{x}}(n)$ is the B point discrete space Fourier transform of the array element space snapshot vector $\boldsymbol{x}(n)$.

Consider the covariance matrix of the $B \times 1$ beamspace snapshot vector (referred as the beam space covariance matrix)

$$\boldsymbol{R}_{\tilde{x}\tilde{x}} = \mathrm{E}\left\{\boldsymbol{W}^{\mathrm{H}}\boldsymbol{x}(n)\boldsymbol{x}^{\mathrm{H}}(n)\boldsymbol{W}\right\} = \boldsymbol{W}^{\mathrm{H}}\boldsymbol{APA}^{\mathrm{H}}\boldsymbol{W} + \sigma_e^2\boldsymbol{I}. \tag{10.6.11}$$

Let the eigenvalue decomposition of the beamspace sample covariance matrix be

$$\hat{\boldsymbol{R}}_{\tilde{x}\tilde{x}} = \sum_{i=1}^{i} \boldsymbol{u}_i\boldsymbol{u}_i^{\mathrm{H}}, \tag{10.6.12}$$

and the estimated number of large eigenvalues be K, then the number of small eigenvalues is $B - K$. Then, the signal eigenvectors $u_1, \cdots, u_K$ span the signal subspace $\boldsymbol{U}_s\boldsymbol{U}_s^{\mathrm{H}}$ with $\boldsymbol{U}_s = [\boldsymbol{u}_1, \cdots, \boldsymbol{u}_K]$, while the eigenvectors associated to the small eigenvalues span the noise subspace $\boldsymbol{U}_n\boldsymbol{U}_n^{\mathrm{H}}$ with $\boldsymbol{U}_n = [\boldsymbol{u}_{K+1}, \cdots, \boldsymbol{u}_B]$.

From Eq. (10.6.11), we have

$$\begin{aligned}
\boldsymbol{U}_n^{\mathrm{H}}\hat{\boldsymbol{R}}_{\tilde{x}\tilde{x}}\boldsymbol{U}_n &= \boldsymbol{U}_n^{\mathrm{H}}\boldsymbol{W}^{\mathrm{H}}\boldsymbol{APA}^{\mathrm{H}}\boldsymbol{WG} + \hat{\sigma}_e^2\boldsymbol{I} \\
&= \boldsymbol{U}_n^{\mathrm{H}}[\boldsymbol{U}_s, \boldsymbol{U}_n]\begin{bmatrix}\boldsymbol{\Sigma} + \hat{\sigma}_e^2\boldsymbol{I} & \boldsymbol{O} \\ \boldsymbol{O} & \hat{\sigma}_e^2\boldsymbol{I}\end{bmatrix}\begin{bmatrix}\boldsymbol{U}_s^{\mathrm{H}} \\ \boldsymbol{U}_n^{\mathrm{H}}\end{bmatrix}\boldsymbol{U}_n \\
&= \hat{\sigma}_e^2\boldsymbol{I}.
\end{aligned}$$

Comparing line 1 and line 3 of the above equation, it is easy to see that $\boldsymbol{U}_n^{\mathrm{H}}\boldsymbol{W}^{\mathrm{H}}\boldsymbol{APA}^{\mathrm{H}}\boldsymbol{WU}_n$ is equal to zero matrix. Since the matrix $\boldsymbol{P}$ is nonsingular, there is $\boldsymbol{A}^{\mathrm{H}}\boldsymbol{WU}_n = \boldsymbol{O}$, or

$$\boldsymbol{a}^{\mathrm{H}}(\omega)\boldsymbol{WU}_n = \boldsymbol{0}^{\mathrm{T}}, \quad \omega = \omega_1, \cdots, \omega_p. \tag{10.6.13}$$

Therefore, the beamspace null spectrum can be derived as

$$\left|\boldsymbol{a}^{\mathrm{H}}\boldsymbol{WU}_n\right|_2^2 = \boldsymbol{a}^{\mathrm{H}}(\omega)\boldsymbol{WU}_n\boldsymbol{U}_n^{\mathrm{H}}\boldsymbol{W}^{\mathrm{H}}\boldsymbol{a}(\omega) = 0, \quad \omega = \omega_1, \cdots, \omega_p, \tag{10.6.14}$$

while the beamspace MUSIC spatial spectrum[194] can be derived as

$$P_{\text{BS-MUSIC}}(\omega) = \frac{\boldsymbol{a}^{\mathrm{H}}(\omega)\boldsymbol{a}(\omega)}{\boldsymbol{a}^{\mathrm{H}}(\omega)\boldsymbol{WU}_n\boldsymbol{U}_n^{\mathrm{H}}\boldsymbol{W}^{\mathrm{H}}\boldsymbol{a}(\omega)}. \tag{10.6.15}$$

Algorithm 10.6.1. *Beamspace MUSIC algorithm*

Step 1 Use B point discrete space Fourier transform to calculate the beamspace observation data vector $\tilde{\boldsymbol{x}}(n) = \boldsymbol{W}^{H}\boldsymbol{x}(n)$, where $n = 1, \cdots, N$.

Step 2 Calculate the eigenvalue decomposition of the sample covariance matrix $\boldsymbol{R}_{\tilde{x}\tilde{x}}$ from Eq. (10.6.11), *and obtain its principal eigenvalues $\lambda_1, \cdots, \lambda_K$ and secondary eigenvalues $\lambda_{K+1}, \cdots, \lambda_B$, and construct the secondary eigenvector matrix $\boldsymbol{U}_n = [\boldsymbol{u}_{K+1}, \cdots, \boldsymbol{u}_B]$.*

Step 3 Use Eq. (10.6.15) *to calculate the beamspace MUSIC spectrum* $P_{BS\text{-}MUSIC}(\omega_i)$, *where* $\omega_i = (i-1)\Delta\omega$, *and the grid* $\Delta\omega$ *can be taken as* $2\pi \cdot 0.001$ *etc.*

Step 4 Search to determine the p peaks of $P_{BS\text{-}MUSIC}(\omega)$, *and give the estimated value of the MUSIC spatial parameter* $\omega_1, \cdots, \omega_p$. *Then, estimate the DOA from* $\omega_i = 2\pi\frac{d}{\lambda}\sin\theta_i$ $\theta_1, \cdots, \theta_p$.

10.6.2 Comparison of BS-MUSIC and ES-MUSIC

Compared with the array element space MUSIC methods, beamspace MUSIC methods have the following advantages[237].

(1) In the case of low signal-to-noise ratio, beamspace MUSIC is better than sensor array element space MUSIC, since the beamforming of beamspace MUSIC can provide processing gain. When there is only a single narrowband signal, the beamforming gain is equal to the number of sensors. Note that the beamforming gain plays an important role in overcoming the obvious signal power attenuation.

(2) The basic assumption of array element space MUSIC is that the source is a point target. The extended target violates this assumption so as to it cannot be handled correctly by the array element space MUSIC method. Since there are no assumptions about the target characteristics (type and size), the beamspace MUSIC is more attractive. It can be used to perform point target imaging as well as extended target imaging. This ability is essential for indoor imaging (such as through-wall radar imaging): due to the small separation distance, limited bandwidth and aperture, many targets behind the wall can be classified as spatially extended targets.

(3) In applications such as imaging, the array element space MUSIC requires two-dimensional interpolation to obtain sampled data along the rectangular grid. While in the beamspace MUSIC, the sampled data along the rectangular grid is obtained by beamforming instead of interpolation. Avoiding interpolation is a popular step because even an outlier in the data set can cause large interpolation errors.

It is assumed that all antennas of the imaging system have the same characteristics. The beamforming position can be anywhere in the antenna radiation pattern. Therefore, the number of beams can be arbitrarily selected. However, since DFT implementation requires beams to be evenly spaced apart, the number of beams needs to be determined according to the beam spacing. The imaging array antenna can provide a good foundation for the formation of different beam directions through the corresponding delay and the resolution of the beamformer. In order to ensure that all targets in the scene can be detected and imaged, the beam spacing should be smaller than the beamwidth. The beam spacing also determines the target resolution.

Comparing the array element space MUSIC and the beamspace MUSIC algorithm shown in the table 10.6.1, an important conclusion can be drawn: as long as the steering vector in the null spectrum $\boldsymbol{a}(\omega)$ is replaced by $\boldsymbol{W}^{\mathrm{H}}\boldsymbol{a}(\omega)$, and the noise subspace

Tab. 10.6.1: Comparision of the relation and difference between array element space MUSIC (ES-MUSIC) and beamspace MUSIC (BS-MUSIC) algorithm.

Method	ES-MUSIC	BS-MUSIC
Model	$\boldsymbol{x}(n) = \boldsymbol{A}\boldsymbol{s}(n) + \boldsymbol{e}(n)$	$\tilde{\boldsymbol{x}}(n) = \boldsymbol{B}\boldsymbol{s}(n) + \boldsymbol{W}^{\mathrm{H}}\boldsymbol{e}(n)$
Data	The original observation data $x_m(n)$	Discrete space Fourier transform $\tilde{x}(n) = \boldsymbol{v}_M^{\mathrm{H}}\boldsymbol{x}(n)$
Steering vector	$\boldsymbol{a}(\omega) = \left[1, \mathrm{e}^{\mathrm{j}\omega}, \cdots, \mathrm{e}^{\mathrm{j}(M-1)\omega}\right]^{\mathrm{T}}$	$\boldsymbol{b}(\omega) = \boldsymbol{W}^{\mathrm{H}}\boldsymbol{a}(\omega)$
Beamforming matrix	$\boldsymbol{I}$	$\boldsymbol{W} = \frac{1}{\sqrt{M}}\left[\boldsymbol{v}_M(0), \boldsymbol{v}_M\left(\frac{2}{M}\right), \cdots, \boldsymbol{v}_M\left((B-1)\frac{2}{M}\right)\right]$
Data vector	$\boldsymbol{x}(n) = \left[x_1(n), \cdots, x_M(n)\right]^{\mathrm{T}}$	$\tilde{\boldsymbol{x}}(n) = \boldsymbol{W}^{\mathrm{H}}\boldsymbol{x}(n) \quad (B \times 1)$
Covariance matrix	$\hat{\boldsymbol{R}}_{xx} = \frac{1}{N}\sum_{n=1}^{N}\boldsymbol{x}(n)\boldsymbol{x}^{\mathrm{H}}(n) \quad (M \times M)$	$\hat{\boldsymbol{R}}_{\tilde{x}\tilde{x}} = \frac{1}{N}\sum_{n=1}^{N}\tilde{\boldsymbol{x}}(n)\tilde{\boldsymbol{x}}^{\mathrm{H}}(n) \quad (B \times B)$
Eigenvalue Decomposition	$\hat{\boldsymbol{R}}_{xx} = \sum_{i=1}^{M}\lambda_i\boldsymbol{u}_i\boldsymbol{u}_i^{\mathrm{H}}$	$\hat{\boldsymbol{R}}_{\tilde{x}\tilde{x}} = \sum_{i=1}^{B}\lambda_i\boldsymbol{u}_i\boldsymbol{u}_i^{\mathrm{H}}$
Secondary eigenvector matrix	$\boldsymbol{G} = \left[\boldsymbol{u}_{p+1}, \cdots, \boldsymbol{u}_M\right]$	$\boldsymbol{U}_n = [\boldsymbol{u}_{K+1}, \cdots, \boldsymbol{u}_B]$
Noise subspace	$\boldsymbol{G}\boldsymbol{G}^{\mathrm{H}} = \sum_{i=p+1}^{M}\boldsymbol{u}_i\boldsymbol{u}_i^{\mathrm{H}}$	$\boldsymbol{U}_n\boldsymbol{U}_n^{\mathrm{H}} = \sum_{i=K+1}^{B}\boldsymbol{u}_i\boldsymbol{u}_i^{\mathrm{H}}$
Spatial spectrum	$P_{\text{ES-MUSIC}}(\omega) = \frac{\boldsymbol{a}^{\mathrm{H}}(\omega)\boldsymbol{a}(\omega)}{\boldsymbol{a}^{\mathrm{H}}(\omega)\boldsymbol{G}\boldsymbol{G}^{\mathrm{H}}\boldsymbol{a}(\omega)}$	$P_{\text{BS-MUSIC}}(\omega) = \frac{\boldsymbol{a}}{\boldsymbol{a}^{\mathrm{H}}(\omega)\boldsymbol{W}\boldsymbol{U}_n\boldsymbol{U}_n^{\mathrm{H}}\boldsymbol{W}^{\mathrm{H}}\boldsymbol{a}(\omega)}$
DOA θ	$\omega = 2\pi\frac{d}{\lambda}\sin\theta$	$\sin\theta$

of the array element space $\boldsymbol{G}\boldsymbol{G}^{\mathrm{H}}$ is replaced by the noise subspace of the beamspace $\boldsymbol{U}_n\boldsymbol{U}_n^{\mathrm{H}}$, the array element space MUSIC spatial spectrum is generalized to the beamspace MUSIC spatial spectrum. Given this, $\boldsymbol{a}(\omega)$ and $\boldsymbol{b}(\omega) = \boldsymbol{W}^{\mathrm{H}}\boldsymbol{a}(\omega)$ are respectively called the array element space steering vector and the beamspace steering vector.

Using these two correspondences between the array element space and the beamspace, it is easy to obtain the following spatial spectrum of the beamspace:

(1) Beamspace minimum norm spatial spectrum

$$P_{\text{BS-MN}}(\omega) = \frac{\boldsymbol{a}^{\mathrm{H}}(\omega)\boldsymbol{a}(\omega)}{\left|\boldsymbol{a}^{\mathrm{H}}(\omega)\boldsymbol{W}\boldsymbol{U}_n\boldsymbol{U}_n^{\mathrm{H}}\boldsymbol{e}_1\right|^2}. \tag{10.6.16}$$

(2) Beamspace first principal vector spatial spectrum

$$P_{\text{BS-FINE}}(\omega) = \frac{\boldsymbol{a}^{\mathrm{H}}(\omega)\boldsymbol{a}(\omega)}{\left|\boldsymbol{a}^{\mathrm{H}}(\omega)\boldsymbol{W}\boldsymbol{U}_n\boldsymbol{U}_n^{\mathrm{H}}\boldsymbol{t}_{\text{FINE}}\right|^2}. \tag{10.6.17}$$

(3) Beamspace multi-principal vector spatial spectrum

$$P_{\text{BS-FINES}}(\omega) = \frac{\boldsymbol{a}^{\mathrm{H}}(\omega)\boldsymbol{a}(\omega)}{\left|\boldsymbol{a}^{\mathrm{H}}(\omega)\boldsymbol{W}\boldsymbol{U}_n\boldsymbol{U}_n^{\mathrm{H}}\boldsymbol{T}_{\text{FINES}}\right|^2}. \tag{10.6.18}$$

Let

$$\boldsymbol{p}(z) = \boldsymbol{a}(\omega)\big|_{z=\mathrm{e}^{\mathrm{j}\omega}}, \tag{10.6.19}$$

then the beamspace MUSIC null spectrum formula $\left|\boldsymbol{a}^{\mathrm{H}}(\omega)\boldsymbol{W}\boldsymbol{G}\right|_2^2 = 0$ can be rewritten in the form of the polynomial of z as

$$p^{\mathrm{H}}(z)\boldsymbol{W}\boldsymbol{U}_n\boldsymbol{U}_n^{\mathrm{H}}\boldsymbol{W}^{\mathrm{H}}\boldsymbol{p}(z) = 0. \tag{10.6.20}$$

Premultiplying the above formula with z^{M-1}, then we have

$$p_{\text{BS-ROOT-MUSIC}}(z) = z^{M-1}\boldsymbol{p}^{\mathrm{T}}\left(z^{-1}\right)\boldsymbol{W}\boldsymbol{U}_n\boldsymbol{U}_n^{\mathrm{H}}\boldsymbol{W}^{\mathrm{H}}\boldsymbol{p}(z) = 0. \tag{10.6.21}$$

This is the beamspace root-MUSIC polynomial.

Comparing the array element space root-MUSIC polynomial and the beamspace root-MUSIC polynomial, we can get the following two substitution relations: when $\boldsymbol{p}(z)$ is replaced by $\boldsymbol{W}^{\mathrm{H}}\boldsymbol{p}(z)$, and the noise subspace of the array element space $\boldsymbol{G}\boldsymbol{G}^{\mathrm{H}}$ is replaced by the noise subspace of the beam space $\boldsymbol{U}_n\boldsymbol{U}_n^{\mathrm{H}}$, the root-MUSIC polynomial in the array element space is generalized to the root-MUSIC polynomial in the beam space. Using these two substitution relations, the following extensions of the MUSIC polynomial for finding roots in beam space are immediately obtained:

(1) Beamspace rooting-minimum norm polynomial

$$p_{\text{BS-ROOT-MN}}(z) = z^{M-1}\boldsymbol{p}^{\mathrm{T}}\left(z^{-1}\right)\boldsymbol{W}\boldsymbol{U}_n\boldsymbol{U}_n^{\mathrm{H}}\boldsymbol{e}_1\boldsymbol{e}_1^{\mathrm{T}}\boldsymbol{U}_n\boldsymbol{U}_n^{\mathrm{H}}\boldsymbol{W}^{\mathrm{H}}\boldsymbol{p}(z) = 0. \tag{10.6.22}$$

(2) Beamspace rooting-first principal vector polynomial

$$p_{\text{BS-ROOT-FINE}}(z) = z^{M-1}\boldsymbol{p}^{\mathrm{T}}\left(z^{-1}\right)\boldsymbol{W}\boldsymbol{U}_n\boldsymbol{U}_n^{\mathrm{H}}\boldsymbol{t}_{\text{FINE}}\boldsymbol{t}_{\text{FINE}}^{\mathrm{H}}\boldsymbol{U}_n\boldsymbol{U}_n^{\mathrm{H}}\boldsymbol{W}^{\mathrm{H}}\boldsymbol{p}(z) = 0. \tag{10.6.23}$$

(3) Beamspace rooting-Multi-principal vector polynomial

$$p_{\text{BS-ROOT-FINES}}(z) = z^{M-1}\boldsymbol{p}^{\mathrm{T}}\left(z^{-1}\right)\boldsymbol{W}\boldsymbol{U}_n\boldsymbol{U}_n^{\mathrm{H}}\boldsymbol{T}_{\text{FINES}}\boldsymbol{T}_{\text{FINES}}^{\mathrm{H}}\boldsymbol{U}_n\boldsymbol{U}_n^{\mathrm{H}}\boldsymbol{W}^{\mathrm{H}}\boldsymbol{p}(z) = 0. \tag{10.6.24}$$

Summarizing the above discussion and analysis, the following main points of the beamspace MUSIC method and its various extensions can be extracted:

(1) Since the beam transformation matrix $\boldsymbol{W}$ is an $M \times B$ Fourier matrix, linear transformation of $\tilde{\boldsymbol{x}}(n) = \boldsymbol{W}^{\mathrm{H}}\boldsymbol{x}(n)$ can be effectively performed using methods such as FFT.
(2) Replace the null spectrum steering vector $\boldsymbol{a}(\omega)$ (or polynomial vector $\boldsymbol{p}(z)$) with $\boldsymbol{W}^{\mathrm{H}}\boldsymbol{a}(\omega)$ (or $\boldsymbol{W}\boldsymbol{p}(z)$), and replace the noise subspace of the array element observation vector $\boldsymbol{G}\boldsymbol{G}^{\mathrm{H}}$ with the noise subspace of the beam observation vector $\boldsymbol{U}_n\boldsymbol{U}_n^{\mathrm{H}}$, various MUSIC method in the array element space (including various root-finding MUSIC methods) would become the corresponding MUSIC methods in the beam space directly.
(3) If the spatial Fourier transform domain is regarded as the spatial frequency domain of the spatial signal, the beamspace MUSIC method based on the discrete space Fourier transform and its extension are a space-time-frequency three-dimensional signal processing, while the array element space MUSIC method and its extension are a space-time two-dimensional signal processing.
(4) Because more information in the spatial frequency domain is used, various beamspace MUSIC methods can be expected to have higher DOA estimation resolution than the corresponding array element space MUSIC methods.

Table 10.6.2 summarizes the basic MUSIC and its various extensions.

MUSIC method uses the spatial spectrum to estimate the DOA of the source, which is a subspace method. The DOA can also be estimated with the help of eigenvalues or

Tab. 10.6.2: MUSIC method and its various extensions.

Method	Spatial spectrum or polynomial
ES-MUSIC	$P_{\text{ES-MUSIC}}(\omega)=\frac{\boldsymbol{a}^{\text{H}}(\omega)\boldsymbol{a}(\omega)}{\boldsymbol{a}^{\text{H}}(\omega)\boldsymbol{G}\boldsymbol{G}^{\text{H}}\boldsymbol{a}(\omega)}$
BS-MUSIC	$P_{\text{BS-MUSIC}}(\omega)=\frac{\boldsymbol{a}^{\text{H}}(\omega)\boldsymbol{a}(\omega)}{\boldsymbol{a}^{\text{H}}(\omega)\boldsymbol{W}\boldsymbol{U}_n\boldsymbol{U}_n^{\text{H}}\boldsymbol{W}^{\text{H}}\boldsymbol{a}(\omega)}$
ES-MN	$P_{\text{ES-MN}}(\omega)=\frac{\boldsymbol{a}^{\text{H}}(\omega)\boldsymbol{a}(\omega)}{\lvert\boldsymbol{a}^{\text{H}}(\omega)\boldsymbol{G}\boldsymbol{G}^{\text{H}}\boldsymbol{e}_1\rvert^2}$
BS-MN	$P_{\text{BS-MN}}(\omega)=\frac{\boldsymbol{a}^{\text{H}}(\omega)\boldsymbol{a}(\omega)}{\lvert\boldsymbol{a}^{\text{H}}(\omega)\boldsymbol{W}\boldsymbol{U}_n\boldsymbol{U}_n^{\text{H}}\boldsymbol{e}_1\rvert^2}$
ES-FINE	$P_{\text{ES-MN}}(\omega)=\frac{\boldsymbol{a}^{\text{H}}(\omega)\boldsymbol{a}(\omega)}{\lvert\boldsymbol{a}^{\text{H}}(\omega)\boldsymbol{G}\boldsymbol{G}^{\text{H}}\boldsymbol{t}_{\text{FINE}}\rvert^2}$
BS-FINE	$P_{\text{BS-MN}}(\omega)=\frac{\boldsymbol{a}^{\text{H}}(\omega)\boldsymbol{a}(\omega)}{\lvert\boldsymbol{a}^{\text{H}}(\omega)\boldsymbol{W}\boldsymbol{U}_n\boldsymbol{U}_n^{\text{H}}\boldsymbol{t}_{\text{FINE}}\rvert^2}$
ES-ROOT-MUSIC	$P_{\text{ES-MN}}(\omega)=\frac{\boldsymbol{a}^{\text{H}}(\omega)\boldsymbol{a}(\omega)}{\lvert\boldsymbol{a}^{\text{H}}(\omega)\boldsymbol{G}\boldsymbol{G}^{\text{H}}\boldsymbol{T}_{\text{FINES}}\rvert_2^2}$
BS-ROOT-MUSIC	$P_{\text{BS-MN}}(\omega)=\frac{\boldsymbol{a}^{\text{H}}(\omega)\boldsymbol{a}(\omega)}{\lvert\boldsymbol{a}^{\text{H}}(\omega)\boldsymbol{W}\boldsymbol{U}_n\boldsymbol{U}_n^{\text{H}}\boldsymbol{T}_{\text{FINES}}\rvert_2^2}$
BS-FINES	$(z)=z^{M-1}\boldsymbol{p}^{\text{T}}\left(z^{-1}\right)\boldsymbol{G}\boldsymbol{G}^{\text{H}}\boldsymbol{p}(z)=0$
ES-ROOT-MUSIC	$p_{\text{ES-ROOT-MUSIC}}(z)=z^{M-1}\boldsymbol{p}^{\text{T}}\left(z^{-1}\right)\boldsymbol{G}\boldsymbol{G}^{\text{H}}\boldsymbol{p}(z)=0$
BS-ROOT-MUSIC	$p_{\text{BS-ROOT-MUSIC}}(z)=z^{M-1}\boldsymbol{p}^{\text{T}}\left(z^{-1}\right)\boldsymbol{W}\boldsymbol{U}_n\boldsymbol{U}_n^{\text{H}}\boldsymbol{W}^{\text{H}}\boldsymbol{p}(z)=0$
ES-ROOT-MN	$p_{\text{ES-ROOT-MN}}(z)=z^{M-1}\boldsymbol{p}^{\text{T}}\left(z^{-1}\right)\boldsymbol{G}\boldsymbol{G}^{\text{H}}\boldsymbol{e}_1\boldsymbol{e}_1^{\text{T}}\boldsymbol{G}\boldsymbol{G}^{\text{H}}\boldsymbol{p}(z)=0$
BS-ROOT-MN	$p_{\text{BS-ROOT-MN}}(z)=z^{M-1}\boldsymbol{p}^{\text{T}}\left(z^{-1}\right)\boldsymbol{W}\boldsymbol{U}_n\boldsymbol{U}_n^{\text{H}}\boldsymbol{e}_1\boldsymbol{e}_1^{\text{T}}\boldsymbol{U}_n\boldsymbol{U}_n^{\text{H}}\boldsymbol{W}^{\text{H}}\boldsymbol{p}(z)=0$
ES-ROOT-FINE	$p_{\text{ES-ROOT-FINE}}(z)=z^{M-1}\boldsymbol{p}^{\text{T}}\left(z^{-1}\right)\boldsymbol{G}\boldsymbol{G}^{\text{H}}\boldsymbol{t}_{\text{FINE}}\boldsymbol{t}_{\text{FINE}}^{\text{H}}\boldsymbol{G}\boldsymbol{G}^{\text{H}}\boldsymbol{p}(z)=0$
BS-ROOT-FINE	$p_{\text{BS-ROOT-FINE}}(z)=z^{M-1}\boldsymbol{p}^{\text{T}}\left(z^{-1}\right)\boldsymbol{W}\boldsymbol{U}_n\boldsymbol{U}_n^{\text{H}}\boldsymbol{t}_{\text{FINE}}\boldsymbol{t}_{\text{FINE}}^{\text{H}}\boldsymbol{U}_n\boldsymbol{U}_n^{\text{H}}\boldsymbol{W}^{\text{H}}\boldsymbol{p}(z)=0$
ES-ROOT-FINES	$p_{\text{ES-ROOT-FINE}}(z)=z^{M-1}\boldsymbol{p}^{\text{T}}\left(z^{-1}\right)\boldsymbol{G}\boldsymbol{G}^{\text{H}}\boldsymbol{T}_{\text{FINES}}\boldsymbol{T}_{\text{FINES}}^{\text{H}}\boldsymbol{G}\boldsymbol{G}^{\text{H}}\boldsymbol{p}(z)=0$
BS-ROOT-FINES	$p_{\text{BS-ROOT-FINES}}(z)=z^{M-1}\boldsymbol{p}^{\text{T}}\left(z^{-1}\right)\boldsymbol{W}\boldsymbol{U}_n\boldsymbol{U}_n^{\text{H}}\boldsymbol{T}_{\text{FINES}}\boldsymbol{T}_{\text{FINES}}^{\text{H}}\boldsymbol{U}_n\boldsymbol{U}_n^{\text{H}}\boldsymbol{W}^{\text{H}}\boldsymbol{p}(z)=0$

generalized eigenvalues. This type of method is the Estimating Signal Parameters via Rotational Invariance Techniques (ESPRIT) method that will be introduced in the next section.

10.7 Estimating Signal Parameters via Rotational Invariance Techniques

Estimating Signal Parameters via Rotational Invariance Techniques (ESPRIT) algorithm, first proposed by Roy, et.al. in 1986, is an abbreviation for the method of Estimating Signal Parameters via Rotational Invariance Techniques[186]. ESPRIT becomes a typical method and is widely used in modern signal processing field up to now.

Similar to MUSIC, ESPRIT algorithms can also be divided into two categories: i.e., element space (ES) algorithms and beam space (BS) algorithms. Furthermore, there is unitary ESPRIT method especially designed for complex-valued signals.

10.7.1 Basic ESPRIT Algorithm

Consider p complex harmonic signals in white noise,

$$x(n) = \sum_{i=1}^{p} s_i e^{jn\omega_i} + w(n), \tag{10.7.1}$$

where s_i and $\omega_i \in (-\pi, \pi)$ denotes the amplitude and phase of the i-th signal,respectively. Assuming $w(n)$ is a white complex gaussian noise process with zero mean and variance of σ^2, namely

$$\begin{aligned} \mathrm{E}\left\{w(n)w^*(l)\right\} &= \sigma^2\delta(n-l), \\ \mathrm{E}\left\{w(n)w(l)\right\} &= 0, \quad \forall k, l. \end{aligned}$$

Define a new process $y(n) \overset{\text{def}}{=} x(n+1)$. Choose $M > p$, and introduce the following $M \times 1$ vectors

$$\boldsymbol{x}(n) \overset{\text{def}}{=} [x(n), x(n+1), \cdots, x(n+M-1)]^{\mathrm{T}}, \tag{10.7.2}$$

$$\boldsymbol{w}(n) \overset{\text{def}}{=} [w(n), w(n+1), \cdots, w(n+M-1)]^{\mathrm{T}}, \tag{10.7.3}$$

$$\begin{aligned} \boldsymbol{y}(n) &\overset{\text{def}}{=} [y(n), y(n+1), \cdots, y(n+M-1)]^{\mathrm{T}}, \\ &= [x(n+1), x(n+2), \cdots, x(n+M)]^{\mathrm{T}}, \end{aligned} \tag{10.7.4}$$

$$\boldsymbol{a}(\omega_i) \overset{\text{def}}{=} [1, e^{j\omega_i}, \cdots, e^{j(M-1)\omega_i}]^{\mathrm{T}}, \tag{10.7.5}$$

then $x(n)$ in Eq. (10.7.1) can be expressed in vector form as

$$\boldsymbol{x}(n) = \boldsymbol{A}\boldsymbol{s}(n) + \boldsymbol{w}(n), \tag{10.7.6}$$

and $y(n) = x(n+1)$ can also be formulated in vector form as

$$\boldsymbol{y}(n) = \boldsymbol{x}(n+1) = \boldsymbol{A\Phi s}(n) + \boldsymbol{w}(n+1), \tag{10.7.7}$$

where

$$\boldsymbol{A} \overset{\text{def}}{=} [\boldsymbol{a}(\omega_1), \boldsymbol{a}(\omega_2), \cdots, \boldsymbol{a}(\omega_p)], \tag{10.7.8}$$

$$\boldsymbol{s}(n) \overset{\text{def}}{=} [s_1 e^{j\omega_1 n}, s_2 e^{j\omega_2 n}, \cdots, s_p e^{j\omega_p n}]^{\mathrm{T}}, \tag{10.7.9}$$

$$\boldsymbol{\Phi} \overset{\text{def}}{=} \mathrm{diag}(e^{j\omega_1}, e^{j\omega_2}, \cdots, e^{j\omega_p}). \tag{10.7.10}$$

Note that $\boldsymbol{\Phi}$ in Eq. (10.7.7) is an unitary matrix such that $\boldsymbol{\Phi}^{\mathrm{H}}\boldsymbol{\Phi} = \boldsymbol{\Phi}\boldsymbol{\Phi}^{\mathrm{H}} = \boldsymbol{I}$, which collects the two vectors $\boldsymbol{x}(n)$ and $\boldsymbol{y}(n)$. $\boldsymbol{A}$ is an $M \times p$ Vandermonde matrix. Since $\boldsymbol{y}(n) = \boldsymbol{x}(n+1)$, $\boldsymbol{y}(n)$ can be regarded as a translation version of $\boldsymbol{x}(n)$. In view of this, $\boldsymbol{\Phi}$ is named rotation operator, since translation is the simplest rotation.

Eqs. (10.7.6) $\sim$ (10.7.10) establish the signal model of ESPRIT algorithm. It can be proved that this signal model is also applicable to the uniform linear array with M array elements. Take the first element as reference, similar to Eq. (10.7.1), the observation signal of this array can be expressed as

$$x_1(n) = \sum_{i=1}^{p} s_i e^{j\omega_i n} + w_1(n) = \sum_{i=1}^{p} s_i(n) + w_1(n), \tag{10.7.11}$$

in which $s_i(n) = s_i e^{j\omega_i n}$ is the i-th source signal.

According to the structure of ULA, there exists a propagation phase difference $e^{j(m-1)\omega_i}$ when the source signal reaches the m-th element. Therefore, the signal observed by the m-th element is

$$x_m(n) = \sum_{i=1}^{p} s_i(n) e^{j(m-1)\omega_i} + w_m(n), \tag{10.7.12}$$

and Eq. (10.7.12) can be rewritten in matrix-vector form as

$$\boldsymbol{x}(n) = [x_1(n), x_2(n), \cdots, x_M(n)]^{\mathrm{T}} = \boldsymbol{A}\boldsymbol{s}(n) + \boldsymbol{w}(n), \tag{10.7.13}$$

where $\boldsymbol{A} = [\boldsymbol{a}(\omega_1), \boldsymbol{a}(\omega_2), \cdots, \boldsymbol{a}(\omega_p)]$,$\boldsymbol{s}(n) = [s_1(n), s_2(n), \cdots, s_p(n)]^{\mathrm{T}}$ and $\boldsymbol{a}(\omega_i) = [1, e^{j\omega_i}, \cdots, e^{j(M-1)\omega_i}]^{\mathrm{T}}$. Then, the translation vector of $\boldsymbol{x}(n)$ can be expressed as

$$\boldsymbol{y}(n) = \boldsymbol{x}(n+1) = \boldsymbol{A}\boldsymbol{s}(n+1) + \boldsymbol{w}(n+1) = \boldsymbol{A}\boldsymbol{\Phi}\boldsymbol{s}(n) + \boldsymbol{w}(n+1). \tag{10.7.14}$$

Since

$$\boldsymbol{s}(n+1) = \begin{bmatrix} s_1 e^{j\omega_1(n+1)} \\ \vdots \\ s_p e^{j\omega_p(n+1)} \end{bmatrix} = \begin{bmatrix} e^{j\omega_1} & & 0 \\ & \ddots & \\ 0 & & e^{j\omega_p} \end{bmatrix} \begin{bmatrix} s_1 e^{j\omega_1 n} \\ \vdots \\ s_p e^{j\omega_p n} \end{bmatrix} = \boldsymbol{\Phi}\boldsymbol{s}(n). \tag{10.7.15}$$

By comparing Eq. (10.7.13) with Eq. (10.7.6), it is easy to find that signal model in Eq. (10.7.14) for ULA is the same as the signal model in Eq. (10.7.7) of ESPRIT algorithm. Therefore, the key of ESPRIT algorithm is how to construct the observation vector $\boldsymbol{x}(n)$ and its translation vector $\boldsymbol{y}(n) = \boldsymbol{x}(n+1)$.

The autocovariance matrix of observation vector $\boldsymbol{x}(n)$ can be expressed as

$$\boldsymbol{R}_{xx} = \mathrm{E}\left\{\boldsymbol{x}(n)\boldsymbol{x}^{\mathrm{H}}(n)\right\} = \boldsymbol{A}\boldsymbol{P}\boldsymbol{A}^{\mathrm{H}} + \sigma^2\boldsymbol{I}, \tag{10.7.16}$$

where

$$\boldsymbol{P} = \mathrm{E}\left\{\boldsymbol{s}(n)\boldsymbol{s}^{\mathrm{H}}(n)\right\}, \tag{10.7.17}$$

is the covariance matrix of the signal vector. If the signals are not coherent, $\boldsymbol{P} = \mathrm{diag}\left(\mathrm{E}\left\{|s_1|^2\right\}, \cdots, \mathrm{E}\left\{|s_p|^2\right\}\right)$ is a nonnegative diagnal matrix of size $p \times p$, with each diagnoal element being the power of the corresponding source. In the ESPRIT algorithm, only nonsingularity of $\boldsymbol{P}$ is demanded, and diagnoality is not mandatory requirements.

The crosscovariance matrix between the observation signal vector $\boldsymbol{x}(n)$ and its translation $\boldsymbol{y}(n)$ can be written as

$$\boldsymbol{R}_{xy} = \mathrm{E}\left\{\boldsymbol{x}(n)\boldsymbol{y}^{\mathrm{H}}(n)\right\} = \boldsymbol{AP\Phi}^{\mathrm{H}}\boldsymbol{A}^{\mathrm{H}} + \sigma^2\boldsymbol{Z}, \tag{10.7.18}$$

in which $\sigma^2\boldsymbol{Z} = \mathrm{E}\left\{\boldsymbol{w}(n)\boldsymbol{w}^{\mathrm{H}}(n+1)\right\}$. It is easy to verify that $\boldsymbol{Z}$ is a spectial matrix with a size of $M \times M$

$$\boldsymbol{Z} = \begin{bmatrix} 0 & & & 0 \\ 1 & 0 & & \\ & \ddots & \ddots & \\ 0 & 0 & 1 & 0 \end{bmatrix}. \tag{10.7.19}$$

That is, all elements on the diagonal below the main diagonal are 1 and all others are 0.

Since the element of autocovariance matrix $[\boldsymbol{R}_{xx}]_{ij} = \mathrm{E}\left\{x(i)x^*(j)\right\} = R_{xx}(i-j) = R^*_{xx}(j-i)$, it follows that

$$\boldsymbol{R}_{xx} = \begin{bmatrix} R_{xx}(0) & R^*_{xx}(1) & \cdots & R^*_{xx}(M-1) \\ R_{xx}(1) & R_{xx}(0) & \cdots & R^*_{xx}(M-2) \\ \vdots & \vdots & \ddots & \vdots \\ R_{xx}(M-1) & R_{xx}(M-2) & \cdots & R_{xx}(0) \end{bmatrix}. \tag{10.7.20}$$

Similarly, the element of crosscovariance matrix

$$[\boldsymbol{R}_{xy}]_{ij} = \mathrm{E}\left\{x(i)y^*(j)\right\} = \mathrm{E}\left\{x(i)x^*(j+1)\right\} = R_{xx}(i-j-1) = R^*_{xx}(j-i+1)$$

satisfies

$$\boldsymbol{R}_{xy} = \begin{bmatrix} R^*_{xx}(1) & R^*_{xx}(2) & \cdots & R^*_{xx}(M) \\ R_{xx}(0) & R^*_{xx}(1) & \cdots & R^*_{xx}(M-1) \\ \vdots & \vdots & \ddots & \vdots \\ R_{xx}(M-2) & R_{xx}(M-3) & \cdots & R^*_{xx}(1) \end{bmatrix}. \tag{10.7.21}$$

Note that $R_{xx}(0) = R^*_{xx}(0)$.

Now the problem is: given autocovariance function $R_{xx}(0), R_{xx}(1), \cdots, R_{xx}(M)$, how to estimae the number p of the harmonic sources, as well as the frequency ω_i and power $|s_i|^2, i = 1, \cdots, p$ of each source.

After translation, the vector $\boldsymbol{x}(n)$ becomes $\boldsymbol{y}(n) = \boldsymbol{x}(n+1)$. However, since $\boldsymbol{R}_{xx} \stackrel{\text{def}}{=} \mathrm{E}\left\{\boldsymbol{x}(n)\boldsymbol{x}^{\mathrm{H}}(n)\right\} = \mathrm{E}\left\{\boldsymbol{x}(n+1)\boldsymbol{x}^{\mathrm{H}}(n+1)\right\} \stackrel{\text{def}}{=} \boldsymbol{R}_{yy}$, that is, $\boldsymbol{R}_{xx}$ is equal to $\boldsymbol{R}_{yy}$ exactly. This indicates that the translation maintains the invariance of the subspace corresponding to $\boldsymbol{x}(n)$ and $\boldsymbol{y}(n)$ respectively.

Performing eigenvalue decomposition on $\boldsymbol{R}_{xx}$, we can obtain its minimum eigenvalue $\lambda_{min} = \sigma^2$. Then, a new pair of matrices can be constructed as follows

$$\boldsymbol{C}_{xx} = \boldsymbol{R}_{xx} - \lambda_{min}\boldsymbol{I} = \boldsymbol{R}_{xx} - \sigma^2\boldsymbol{I} = \boldsymbol{APA}^{\mathrm{H}}, \tag{10.7.22}$$

$$\boldsymbol{C}_{xy} = \boldsymbol{R}_{xy} - \lambda_{min}\boldsymbol{Z} = \boldsymbol{R}_{xy} - \sigma^2\boldsymbol{Z} = \boldsymbol{AP\Phi}^{\mathrm{H}}\boldsymbol{A}^{\mathrm{H}}, \tag{10.7.23}$$

where $\{C_{xx}, C_{xy}\}$ is called a matrix pencil or a matrix pair.

The generalized eigenvalue decomposition of matrix pencil $\{C_{xx}, C_{xy}\}$ is defined as

$$C_{xx}u = \gamma C_{xy}u, \tag{10.7.24}$$

where γ and u are called the generalized eigenvalue and generalized eigenvector, respectively. Then, the tuple (γ, u) is called generalized eigenpair. If only the eigenvalues is of interest, the matrix pencil is usually written as $C_{xx} - \gamma C_{xy}$. In case that γ is not a generalized eigenvalue, then the matrix pencil $C_{xx} - \gamma C_{xy}$ has full rank, while γ that causes rank dificiency of matrix pencil $C_{xx} - \gamma C_{xy}$ is defined to be its generalized eigenvalue.

Invistigate the matrix pencil

$$C_{xx} - \gamma C_{xy} = AP(I - \gamma\Phi^{H})A^{H}. \tag{10.7.25}$$

Since A has full column rank and P is nonsigular, in terms of matrix rank, from Eq. (10.7.25) we can obtain

$$\text{rank}(C_{xx} - \gamma C_{xy}) = \text{rank}(I - \gamma\Phi^{H}). \tag{10.7.26}$$

If $\gamma \neq \omega_i, i = 1, \cdots, p$, the matrix $I - \gamma\Phi^{H}$ is a nonsigular, while if γ equals to $e^{j\omega_i}$, $I - \gamma\Phi^{H}$ is a singular matrix, namely, rank deficient matrix, due to $\gamma e^{-j\omega_i} = 1$. The derivation shows that $e^{j\omega_i}, i = 1, \cdots, p$ are all the generalized eigenvalues of matrix pencil $\{C_{xx}, C_{xy},\}$. This result can be summarized as the following theorem.

Theorem 10.7.1. *Let Γ be the generalized eigenvalue matrix of matrix pencil $\{C_{xx}, C_{xy}\}$, in which $C_{xx} = R_{xx} - \lambda_{min}I$ and $C_{xy} = R_{xy} - \lambda_{min}Z$, and $\lambda_{\min}$ is the smallest eigenvalue of autocovariance matrix R_{xx}. If matrix P is nonsigular, then matrix Γ and rotation operator matrix Φ satisfy*

$$\Gamma = \begin{bmatrix} \Phi & 0 \\ 0 & 0 \end{bmatrix}, \tag{10.7.27}$$

namely, nonzero elements of matrix Γ is an arrangement of the elements of the rotation operator matraix Φ.

By summarying the above analyses, we can obtain the basic ESPRIT algorithm as follows.

Algorithm 10.7.1. *Basic ESPRIT algorithm*

Step 1 Estimate the autocovariance function $R_{xx}(i), i = 0, \cdots, M$, using given observation data $x(1), \cdots, x(N)$;

Step 2 Construct the $M \times M$ autocovariance matrix R_{xx} and $M \times M$ crosscovariance matrix R_{xy} using $R_{xx}(i), i = 0, \cdots, M$;

Step 3 Perform eigenvalue decomposition of R_{xx}. For $M > p$. Take the mean value of the smallest eigenvalues as the estimation of noise power σ^2;

Step 4 Compute C_{xx} and C_{xy} using σ as well as $C_{xx} = R_{xx} - \lambda_{min}I$ and $C_{xy} = R_{xy} - \lambda_{min}Z$;

Step 5 Compute the generalized eigenvalue decomposition of matrix pencil $\boldsymbol{C}_{xx}$, $\boldsymbol{C}_{xy}$ and find p eigenvalues $\mathrm{e}^{\mathrm{j}\omega_i}$, $i = 1, \cdots, p$ that locate on the unit circle , which can be mapped to the frequency estimation of complex exponential signal directly, and then estimate the DOA $\theta_1, \cdots, \theta_p$ according to $\omega_i = 2\pi\frac{d}{\lambda}\sin\theta_i$.

Let the generalized eigenvector that corresponding to the generalized eigenvalue γ_i be $\boldsymbol{e}_i$. From its definition, $\boldsymbol{e}_i$ satifies

$$\boldsymbol{APA}^{\mathrm{H}}\boldsymbol{e}_i = \gamma_i \boldsymbol{AP\Phi}^{\mathrm{H}}\boldsymbol{A}^{\mathrm{H}}\boldsymbol{e}_i, \tag{10.7.28}$$

which is equivalent to

$$\boldsymbol{e}_i^{\mathrm{H}}\boldsymbol{AP}\left(\boldsymbol{I} - \gamma_i\boldsymbol{\Phi}^{\mathrm{H}}\right)\boldsymbol{A}^{\mathrm{H}}\boldsymbol{e}_i = 0, \tag{10.7.29}$$

from which it is obviously that the i-th diagonal elements of the diagonal matrix are zeros while all of its other diagnal elements are nonzero, which are denoted as × to indicate that they are not of interest for us. Namely,

$$\boldsymbol{P}\left(\boldsymbol{I} - \gamma_i\boldsymbol{\Phi}^{\mathrm{H}}\right) = \mathrm{diag}(\times, \cdots, \times, 0, \times, \cdots, \times). \tag{10.7.30}$$

Consequently, to make Eq. (10.7.29) holds, $\boldsymbol{e}_i^{\mathrm{H}}\boldsymbol{A}$ and $\boldsymbol{A}^{\mathrm{H}}\boldsymbol{e}_i$ must have the following form

$$\boldsymbol{e}_i^{\mathrm{H}}\boldsymbol{A} = [0, \cdots, 0, e_i^{*}\boldsymbol{a}(\omega_i), 0, \cdots, 0], \tag{10.7.31}$$

$$\boldsymbol{A}^{\mathrm{H}}\boldsymbol{e}_i = [0, \cdots, 0, \boldsymbol{a}^{\mathrm{H}}(\omega_i)e_i, 0, \cdots, 0)^{\mathrm{H}}. \tag{10.7.32}$$

That is to say, the generalized eigenvector $\boldsymbol{e}_i$ corresponding to the generalized eigenvalue γ_i is orthgonal to all other steering vectors $\boldsymbol{a}(\omega_j), j \neq i$ except for $\boldsymbol{a}(\omega_i)$. On the other hand, the element on (i, i) of the diagonal matrix $\gamma_i\boldsymbol{\Phi}$ is 1, that is

$$\gamma_i\boldsymbol{\Phi} = \mathrm{diag}\left(\mathrm{e}^{-\mathrm{j}\omega_1}, \cdots, \mathrm{e}^{-\mathrm{j}\omega_{i-1}}, 1, \mathrm{e}^{-\mathrm{j}\omega_{i+1}}, \cdots, \mathrm{e}^{-\mathrm{j}\omega_p}\right). \tag{10.7.33}$$

Substituting $\boldsymbol{C}_{xx} = \boldsymbol{APA}^{\mathrm{H}}$ into Eq. (10.7.29) yields

$$\boldsymbol{e}_i^{\mathrm{H}}\boldsymbol{AP}\gamma_i\boldsymbol{\Phi}^{\mathrm{H}}\boldsymbol{A}^{\mathrm{H}}\boldsymbol{e}_i = \boldsymbol{e}_i^{\mathrm{H}}\boldsymbol{C}_{xx}\boldsymbol{e}_i. \tag{10.7.34}$$

By further substituting Eqs. (10.7.32) and (10.7.33) into Eq. (10.7.34) and noticing that $\boldsymbol{P}$ is a diagonal matrix, we can derive that

$$\mathrm{E}\left\{|s_i(n)|^2\right\}|\boldsymbol{e}_i^{\mathrm{H}}\boldsymbol{a}(\omega_i)|^2 = \boldsymbol{e}_i^{\mathrm{H}}\boldsymbol{C}_{xx}\boldsymbol{e}_i, \tag{10.7.35}$$

in other words,

$$\mathrm{E}\left\{|s_i(n)|^2\right\} = \frac{\boldsymbol{e}_i^{\mathrm{H}}\boldsymbol{C}_{xx}\boldsymbol{e}_i}{|\boldsymbol{e}_i^{\mathrm{H}}\boldsymbol{a}(\omega_i)|^2}, \tag{10.7.36}$$

which is the signal power estimation formula in case all the sources are independent of each other.

10.7.2 Element Space ESPRIT

It is necessary for the basic ESPRIT algorithm to perform the generalized eigenvalue decomposition on matrix pencil $\{\boldsymbol{C}_{xx}, \boldsymbol{C}_{xy}\}$, which will cause the basic ESPRIT algorithm to have two drawbacks:
(1) Unable to utilize the singular value decomposition of the observation data matrix $\boldsymbol{X} = [\boldsymbol{x}(1), \cdots, \boldsymbol{x}(N)]$, which exihibits a better numerical stability than the generalized eigenvalue decomposition procedure necessary to be performed on the covariance matrix pencil;
(2) More difficult in adaptively updating the covariance matrix.

Using the concept of subarray, element space ESPRIT algorithm for ULA can overcome the above two drawbacks.

Consider an M-element ULA shown in Figure 10.7.1. The array is divided into two subarrays, in which one subarray is composed of the first to $M-1$-th array elements and the other subarray is composed of 2-nd to M-th array elements.

Denote the $M \times N$ observation data matrix of the original array as

$$\boldsymbol{X} = [\boldsymbol{x}(1), \cdots, \boldsymbol{x}(N)], \tag{10.7.37}$$

in which $\boldsymbol{x}(n) = [x_1(n), \cdots, x_m(n)]^{\mathrm{T}}$ is the observation data vector cosists of the observation signal at time n of all these m elements, and N is data length, i.e.,$n = 1, \cdots, N$. If denote signal matrix as

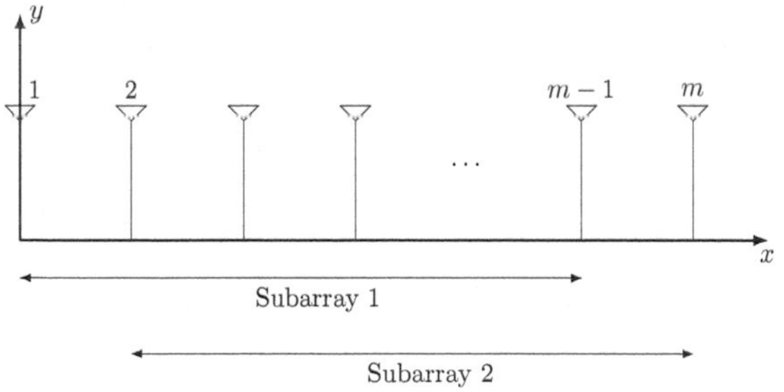

Fig. 10.7.1: Illustration of ULA divided into two subarrays

$$\boldsymbol{S} = [\boldsymbol{s}(1), \cdots, \boldsymbol{s}(N)], \tag{10.7.38}$$

in which

$$\boldsymbol{s}(n) = [s_1(n), \cdots, s_p(n)]^{\mathrm{T}}, \tag{10.7.39}$$

is signal vector. Then, For the data composed of N snapshots, Eq. (10.7.37) can be formulated as the matrix-vector equation

$$\boldsymbol{X} = [\boldsymbol{x}(1), \cdots, \boldsymbol{x}(N)] = \boldsymbol{A}\boldsymbol{s}, \tag{10.7.40}$$

where $\boldsymbol{A}$ is a $m \times p$ steering vector matrix.

Let $\boldsymbol{J}_1$ and $\boldsymbol{J}_2$ be two $(m-1) \times m$ selection matrix defined as

$$\boldsymbol{J}_1 = [\boldsymbol{I}_{m-1} \vdots \boldsymbol{0}_{m-1}], \tag{10.7.41}$$

$$\boldsymbol{J}_2 = [\boldsymbol{0}_{m-1} \vdots \boldsymbol{I}_{m-1}], \tag{10.7.42}$$

in which $\boldsymbol{I}_{m-1}$ denotes $(m-1) \times (m-1)$ identity matrix and $\boldsymbol{0}_{m-1}$denotes zero vector of size $(m-1) \times 1$.

Premultiply the observation data matrix $\boldsymbol{X}$ with the selection matrix $\boldsymbol{J}_1$ and $\boldsymbol{J}_2$ yields

$$\boldsymbol{X}_1 = \boldsymbol{J}_1\boldsymbol{X} = [\boldsymbol{x}_1(1), \cdots, \boldsymbol{x}_1(N)], \tag{10.7.43}$$

$$\boldsymbol{X}_2 = \boldsymbol{J}_2\boldsymbol{X} = [\boldsymbol{x}_2(1), \cdots, \boldsymbol{x}_2(N)], \tag{10.7.44}$$

respectively, where

$$\boldsymbol{x}_1(n) = [x_1(n), \cdots, x_{m-1}(n)]^{\mathrm{T}} \quad n = 1, \cdots, N, \tag{10.7.45}$$

$$\boldsymbol{x}_2(n) = [x_2(n), \cdots, x_m(n)]^{\mathrm{T}} \quad n = 1, \cdots, N. \tag{10.7.46}$$

In other words, the observation data submatrix $\boldsymbol{X}_1$ cosists of the first $m-1$ rows of the observation data matrix $\boldsymbol{X}$, which is equivalent to the observation data matrix of subarray 1; similar to $\boldsymbol{X}_1$, observation data submatrix $\boldsymbol{X}_2$ cosists of the last $m-1$ rows of the observation data matrix $\boldsymbol{X}$, which is equivalent to the observation data matrix of subarray 2.

Rewrite $\boldsymbol{A}$ as

$$\boldsymbol{A} = \begin{bmatrix} \boldsymbol{A}_1 \\ \text{last row} \end{bmatrix} = \begin{bmatrix} \text{first row} \\ \boldsymbol{A}_2 \end{bmatrix}, \tag{10.7.47}$$

and from the structure of array response matrix A, the relation bewteen submatrix $\boldsymbol{A}_1$ and $\boldsymbol{A}_2$ can be formulted as

$$\boldsymbol{A}_2 = \boldsymbol{A}_1\boldsymbol{\Phi}, \tag{10.7.48}$$

and it is also easy to verify that

$$\boldsymbol{X}_1 = \boldsymbol{A}_1\boldsymbol{S}, \tag{10.7.49}$$

$$\boldsymbol{X}_2 = \boldsymbol{A}_2\boldsymbol{S} = \boldsymbol{A}_1\boldsymbol{\Phi}\boldsymbol{S}. \tag{10.7.50}$$

Since $\boldsymbol{\Phi}$ is an unitary matrix, so $\boldsymbol{X}_1$ and $\boldsymbol{X}_2$ span the same signal subspace and noise subspace, that is to say, subarray 1 and subarray 2 have the same observation space (signal space and noise space). This is the physical interpretation for the translation invariance property of ULA.

The autocovariance matrix R_{xx} of the observation vector X can be written as

$$R_{xx} = APA^{\mathrm{H}} + \sigma^2 I = \begin{bmatrix} U_s & U_n \end{bmatrix} \begin{bmatrix} \Sigma_s & O \\ O & \sigma^2 I \end{bmatrix} \begin{bmatrix} U_s^{\mathrm{H}} \\ U_n^{\mathrm{H}} \end{bmatrix}$$

$$= \begin{bmatrix} U_s\Sigma_s & \sigma^2 U_n \end{bmatrix} \begin{bmatrix} U_s^{\mathrm{H}} \\ U_n^{\mathrm{H}} \end{bmatrix} = U_s\Sigma_s U_s^{\mathrm{H}} + \sigma^2 U_n U_n^{\mathrm{H}}. \qquad (10.7.51)$$

From $I - U_n U_n^{\mathrm{H}} = U_s U_s^{\mathrm{H}}$, Eq. (10.7.51) can be rearranged as

$$APA^{\mathrm{H}} + \sigma^2 U_s U_s^{\mathrm{H}} = U_s \Sigma_s U_s^{\mathrm{H}}. \qquad (10.7.52)$$

Postmultiplied at both side with U_s and noted that $U_s^{\mathrm{H}} U_s = I$, Eq. (10.7.52) can be rearranged as

$$U_s = AT, \qquad (10.7.53)$$

where

$$T = PA^{\mathrm{H}} U_s \left(\Sigma_s - \sigma^2 I\right)^{-1}, \qquad (10.7.54)$$

is a nonsingular matrix.

Although T is unknown, it will not hamper the following analysis at all as T is only a dummy argument and only its nonsingularity is used during the derivation below. Postmultiplied by T, Eq. (10.7.47) becomes

$$AT = \begin{bmatrix} A_1 T \\ \text{last row} \end{bmatrix} = \begin{bmatrix} \text{first row} \\ A_2 T \end{bmatrix}. \qquad (10.7.55)$$

Using the same block form, U_s can be divided into

$$U_s = \begin{bmatrix} U_1 \\ \text{last row} \end{bmatrix} = \begin{bmatrix} \text{first row} \\ U_2 \end{bmatrix}. \qquad (10.7.56)$$

Since $AT = U_s$, by comparing Eq. (10.7.55) with Eq. (10.7.56) we can get immediately

$$U_1 = A_1 T \quad \text{and} \quad U_2 = A_2 T. \qquad (10.7.57)$$

Further substituting $A_2 = A_1 \Phi$ into Eq. (10.7.57), we obtain

$$U_2 = A_1 \Phi T, \qquad (10.7.58)$$

and by combining $U_1 = A_1 T$ with Eq. (10.7.58), we can find

$$U_1 T^{-1} \Phi T = A_1 T T^{-1} \Phi T = A_1 \Phi T = U_2. \qquad (10.7.59)$$

Define

$$\Psi = T^{-1} \Phi T, \qquad (10.7.60)$$

in which Ψ is known to be a similarity transformation of matrix Φ. Since Ψ has identically the same eigenvalues as matrix Φ, namely, the eigenvalues of Ψ are also

$e^{j\phi_m}, m = 1, \cdots, M$. Substituting Eq. (10.7.60) into Eq. (10.7.59) yields an important relationship, that is

$$\boldsymbol{U}_2 = \boldsymbol{U}_1 \boldsymbol{\Psi}. \tag{10.7.61}$$

The results above can be concluded as the element space ESPRIT algorithm (ES-ESPRIT).

Algorithm 10.7.2. *ES-ESPRIT algorithm*

Step 1 Compute the singular value decomposition $\boldsymbol{X} = \boldsymbol{U\Sigma V}^H$ *for* $M \times N$ *observation data matrix* $\boldsymbol{X}$ *and take the left singular vectors corresponding to the p principal singular values to construct the matrix* $\boldsymbol{U}_s = [\boldsymbol{u}_1, \cdots, \boldsymbol{u}_p]$;

Step 2 Take the upper and lower $M - 1$ *rows of* $\boldsymbol{U}_s$ *to form* $\boldsymbol{U}_1$ *and* $\boldsymbol{U}_2$. *Compute the eigenvalue decompostion of the matrix* $\boldsymbol{\Psi} = \left(\boldsymbol{U}_1^H \boldsymbol{U}_1\right)^{-1} \boldsymbol{U}_1^H \boldsymbol{U}_2$ *to get its eigenvalues* $e^{j\omega_i}$ $(i = 1, \cdots, p)$, *which is identically the same as that of rotation operator* $\boldsymbol{\Phi}$, *and further to get the estimation of* ω_i $(i = 1, \cdots, p)$;

Step 3 Estimate DOA $\theta_1, \cdots, \theta_p$ *according to* $\hat{\omega}_i = 2\pi \frac{d}{\lambda} \sin \theta_i$.

By comparing algorithm 10.7.2 with algorithm 10.7.1, we can find two main differences between the basic ESPRIT and ES-ESPRIT algorithm as follows:

(1) It is necessary for the basic ESPRIT algorithm to compute the eigenvalue decomposition of the $M \times M$ autocovariance matrix $\boldsymbol{R}_{xx}$ and the generalized eigenvalue decomposition of the matrix pencil $\{\boldsymbol{C}_{xx}, \boldsymbol{C}_{xy}\}$; while it is necessary for the ES-ESPRIT algorithm to compute the singular value decomposition of the $M \times N$ observation data matrix $\boldsymbol{X}$ and the eigenvalue decomposition for the $M \times M$ matrix $\boldsymbol{\Psi}$.

(2) ES-ESPRIT algorithm can be easily generalized to the beam space ESPRIT algorithm, unitary ESPRIT algorithm, and beam space unitary ESPRIT algorithm, as detailed in the following subsections.

10.7.3 TLS-ESPRIT

The basic ESPRIT algorithm introduced above can be viewed as the least squares (LS) operator, which plays a role in mapping the original M-dimensional observation space to the constrained p-dimensional subspace with p being the spatial source number. For this reason, the basic ESPRIT algorithm is usually called as LS-ESPRIT algorithm. Roy and Kailath have pointed out in literature [185] that, LS-operator will usually encounter some potential numerical difficulties when performing generalized eigenvalue decomposition. It is widely known that SVD and totally least squares (TLS) can be used to convert an ill-conditioned generalized eigenvalue decomposition of larger $M \times M$ matrix pencil into a well-conditioned generalized eigenvalue problem of smaller $p \times p$ matrix pencil.

There are several TLS-ESPRIT algorithms and they need to perform different number of times of SVD during implementation. The algorithm proposed by Zhang and Liang[244] needs to perform SVD only once and thus is the TLS-ESPRIT algorithm with the highest computational efficiency.

Consider the generalized eigenvalue decomposition of matrix pencil $\{\boldsymbol{R}_1, \boldsymbol{R}_2\}$. Suppose that the SVD of matrix $\boldsymbol{R}_1$ to be

$$\boldsymbol{R}_1 = \boldsymbol{U}\boldsymbol{\Sigma}\boldsymbol{V}^{\mathrm{H}} = \begin{bmatrix} \boldsymbol{U}_1 & \boldsymbol{U}_2 \end{bmatrix} \begin{bmatrix} \boldsymbol{\Sigma}_1 & \boldsymbol{O} \\ \boldsymbol{O} & \boldsymbol{\Sigma}_2 \end{bmatrix} \begin{bmatrix} \boldsymbol{V}_1^{\mathrm{H}} \\ \boldsymbol{V}_2^{\mathrm{H}} \end{bmatrix}, \tag{10.7.62}$$

in which Σ_1 consists of p principal sigular value. Premultiplied and postmultiplied by $\boldsymbol{U}_1^{\mathrm{H}}$ and $\boldsymbol{V}_1$ respectively, $\boldsymbol{R}_1 - \gamma\boldsymbol{R}_2$ becomes

$$\boldsymbol{\Sigma}_1 - \gamma\boldsymbol{U}_1^{\mathrm{H}}\boldsymbol{R}_2\boldsymbol{V}_1, \tag{10.7.63}$$

with its generalized eigenvalues unchanged. Now, the generalized eigenvalue decomposition of higher order matrix pencil $\{\boldsymbol{R}_1, \boldsymbol{R}_2\}$ turns out to be the generalized eigenvalue decomposition of a lower order $p \times p$ matrix pencil $\{\boldsymbol{\Sigma}_1, \boldsymbol{U}_1^{\mathrm{H}}\boldsymbol{R}_2\boldsymbol{V}_1\}$.

Algorithm 10.7.3. *TLS-ESPRIT algorithm*

Step 1 Compute the eigenvalue decomposition of matrix $\boldsymbol{R}$*;*

Step 2 Compute matrix $\boldsymbol{C}_{xx} = \boldsymbol{R}_{xx} - \sigma^2\boldsymbol{I}$ *and* $\boldsymbol{C}_{xy} = \boldsymbol{R}_{xy} - \sigma^2\boldsymbol{Z}$ *using the smallest eigenvalue* σ^2 *of the matrix* $\boldsymbol{R}$ *obtained in Step 1;*

Step 3 Compute the sigular value decomposition of matrix $\boldsymbol{C}_{xx}$ *and determine its effective rank* p*, and construct* $\boldsymbol{\Sigma}_1, \boldsymbol{U}_1, \boldsymbol{V}_1$ *that correspodin͏g to the* p *principal sigular values;*

Step 4 Compute $\boldsymbol{U}_1^H\boldsymbol{C}_{xy}\boldsymbol{V}_1$*;*

Step 5 Compute the generalized eigenvalue decomposition for the obtained matrix pencil $\left\{\boldsymbol{\Sigma}_1, \boldsymbol{U}_1^H\boldsymbol{C}_{xy}\boldsymbol{V}_1\right\}$ *to get the generalized eigenvalues on the unit circle. Get the circular frequency of each complex exponential signal directly or estimate DOA* $\theta_1, \cdots, \theta_p$ *of each spatial sources according to* $\hat{\omega}_i = 2\pi\frac{d}{\lambda}\sin\theta_i$.

It is proved that the LS-ESPRIT and TLS-ESPRIT algorithms gives the same asymptotic estimation accuracy for large samples, and the TLS-ESPRIT always outperforms LS-ESPRIT for smaller samples. Furthermore, different from LS-ESPRIT, the TLS-ESPRIT algorithm takes both the effect of the noise of $\boldsymbol{C}_{xx}$ and that of $\boldsymbol{C}_{xy}$, therefore, TLS-ESPRIT is more plausible.

10.7.4 Beamspace ESPRIT Algorithm

When the number of the array element is too large, element space algorithms (such as ES-MUSIC and ES-ESPRIT) need to perform $O(M^3)$ times of eigenvalue decomposition operation in real-time implementation. One effective approach to overcome this drawback is to decrease the space dimension, namely, convert the original data space into lower B-dimensional ($B < M$) beam space using some kind of transformation (such as FFT).

The complexity of the beam space algorithm can be reduced from $O(M^3)$ of the element space algorithm to $O(B^3)$. Since the FLOPs of beam space transformation for each data vector is $O(MB)$, the sample covariance matrix can be real-time updated.

Using the discrete space Fourier matrix $\boldsymbol{W}$ defined in Eq. (10.6.5) to perform discrete time Fourier transforme of the element observation data vector $\boldsymbol{x}(n)$ and its time translation vector $\boldsymbol{y}(n) = \boldsymbol{x}(n+1)$ yields

$$\tilde{\boldsymbol{x}}(n) = \boldsymbol{W}^{\mathrm{H}}\boldsymbol{x}(n), \tag{10.7.64}$$

$$\tilde{\boldsymbol{y}}(n) = \boldsymbol{W}^{\mathrm{H}}\boldsymbol{x}(n+1). \tag{10.7.65}$$

By using the property of discrete space Fourier transform that $\boldsymbol{W}^{\mathrm{H}}\boldsymbol{W} = \boldsymbol{I}$, the autocovariance and crosscovariance matrix in beam space can be derived from Eqs. (10.7.16) and (10.7.18) respectively as

$$\begin{aligned}\boldsymbol{R}_{\tilde{x}\tilde{x}} &= \mathrm{E}\left\{\tilde{\boldsymbol{x}}(n)\tilde{\boldsymbol{x}}^{\mathrm{H}}(n)\right\} = \boldsymbol{W}^{\mathrm{H}}\boldsymbol{R}_{xx}\boldsymbol{W} = \boldsymbol{W}^{\mathrm{H}}(\boldsymbol{APA}^{\mathrm{H}} + \sigma^2\boldsymbol{I})\boldsymbol{W} \\ &= \boldsymbol{W}^{\mathrm{H}}\boldsymbol{APA}^{\mathrm{H}}\boldsymbol{W} + \sigma^2\boldsymbol{I},\end{aligned} \tag{10.7.66}$$

$$\begin{aligned}\boldsymbol{R}_{\tilde{x}\tilde{y}} &= \mathrm{E}\left\{\tilde{\boldsymbol{x}}(n)\tilde{\boldsymbol{x}}^{\mathrm{H}}(n+1)\right\} = \boldsymbol{W}^{\mathrm{H}}\boldsymbol{R}_{xy}\boldsymbol{W} = \boldsymbol{W}^{\mathrm{H}}(\boldsymbol{AP\Phi}^{\mathrm{H}}\boldsymbol{A}^{\mathrm{H}} + \sigma^2\boldsymbol{Z})\boldsymbol{W} \\ &= \boldsymbol{W}^{\mathrm{H}}\boldsymbol{AP\Phi}^{\mathrm{H}}\boldsymbol{A}^{\mathrm{H}}\boldsymbol{W} + \sigma^2\boldsymbol{Z}.\end{aligned} \tag{10.7.67}$$

Subsequently, after eleminating the noise, the autocovariance and crosscovariance matrix in beam space become respectively to

$$\boldsymbol{C}_{\tilde{x}\tilde{x}} = \boldsymbol{R}_{\tilde{x}\tilde{x}} - \sigma^2\boldsymbol{I} = \boldsymbol{W}^{\mathrm{H}}\boldsymbol{APA}^{\mathrm{H}}\boldsymbol{W}, \tag{10.7.68}$$

$$\boldsymbol{C}_{\tilde{x}\tilde{y}} = \boldsymbol{R}_{\tilde{x}\tilde{y}} - \sigma^2\boldsymbol{Z} = \boldsymbol{W}^{\mathrm{H}}\boldsymbol{AP\Phi}^{\mathrm{H}}\boldsymbol{A}^{\mathrm{H}}\boldsymbol{W}. \tag{10.7.69}$$

Study the beam space matrix pencil

$$\boldsymbol{C}_{\tilde{x}\tilde{x}} - \gamma\boldsymbol{C}_{\tilde{x}\tilde{y}} = \boldsymbol{W}^{\mathrm{H}}\boldsymbol{AP}(\boldsymbol{I} - \boldsymbol{\Phi}^{\mathrm{H}})\boldsymbol{A}^{\mathrm{H}}\boldsymbol{W}. \tag{10.7.70}$$

Since both the $M \times B$ Fourier matrix $\boldsymbol{W}$ and the $M \times p$ steering vector matrix $\boldsymbol{A}$ is full column rank matix, and $p \times p$ autocovariance matrix $\boldsymbol{P}$ of the signal sources is nonsigular, it is easy to find that the rank of beam space matrix pencil satisfies

$$\mathrm{rank}(\boldsymbol{C}_{\tilde{x}\tilde{x}} - \gamma\boldsymbol{C}_{\tilde{x}\tilde{y}}) = \mathrm{rank}(\boldsymbol{I} - \gamma\boldsymbol{\Phi}^{\mathrm{H}}), \tag{10.7.71}$$

which indicates that the generalized eigenvalues $\gamma_i = \mathrm{e}^{\mathrm{j}\omega_i, i=1,\cdots,p}$ of the beam space matrix pencil $\left\{\boldsymbol{C}_{\tilde{x}\tilde{x}}, \boldsymbol{C}_{\tilde{x}\tilde{y}}\right\}$ can be used directly to obtain the estimation result of spatial parameters $\omega_i, \cdots, \omega_p$ and thereby the estimation of DOA.

The above algorithm that estimates the DOA of spatial signal source by means of the generalized eigenvalue decomposition of the beam space matrix pencil $\left\{\boldsymbol{C}_{\tilde{x}\tilde{x}}, \boldsymbol{C}_{\tilde{x}\tilde{y}}\right\}$ is named as Beamspace ESPRIT or in short BS-ESPRIT algorithm, and was first proposed in 1994 by Xu et.al.[230].

Algorithm 10.7.4. *BS-ESPRIT algorithm 1*

Step 1 Compute the discrete space Fourier transform $\tilde{\boldsymbol{x}}(n) = \boldsymbol{W}^H\boldsymbol{x}(n)$ *and the sample autocovariance function* $R_{\tilde{x}\tilde{x}}(k) = \frac{1}{N}\sum_{n=0}^{N}\tilde{x}(n)\tilde{x}^*(n-k), k = 0, 1, \cdots, M$*;*

Step 2 Construct the B × B sample autocovariance and sample crosscovariance matrix respectively as

$$\boldsymbol{R}_{xy} = \begin{bmatrix} R^*_{\tilde{x}\tilde{x}}(1) & R^*_{\tilde{x}\tilde{x}}(2) & \cdots & R^*_{\tilde{x}\tilde{x}}(M) \\ R_{\tilde{x}\tilde{x}}(0) & R^*_{\tilde{x}\tilde{x}}(1) & \cdots & R^*_{\tilde{x}\tilde{x}}(M-1) \\ \vdots & \vdots & \ddots & \vdots \\ R_{\tilde{x}\tilde{x}}(M-2) & R_{\tilde{x}\tilde{x}}(M-3) & \cdots & R^*_{\tilde{x}\tilde{x}}(1) \end{bmatrix}, \tag{10.7.72}$$

$$\boldsymbol{R}_{\tilde{x}\tilde{y}} = \begin{bmatrix} R^*_{\tilde{x}\tilde{x}}(1) & R^*_{\tilde{x}\tilde{x}}(2) & \cdots & R^*_{\tilde{x}\tilde{x}}(M) \\ R_{\tilde{x}\tilde{x}}(0) & R^*_{\tilde{x}\tilde{x}}(1) & \cdots & R^*_{\tilde{x}\tilde{x}}(M-1) \\ \vdots & \vdots & \ddots & \vdots \\ R_{\tilde{x}\tilde{x}}(M-2) & R_{xx}(M-3) & \cdots & R^*_{\tilde{x}\tilde{x}}(1) \end{bmatrix}; \tag{10.7.73}$$

Step 3 Compute the eigenvalue decomposition $\boldsymbol{R}_{\tilde{x}\tilde{x}} = \sum_{i=1}^{B}\lambda_i\boldsymbol{u}_i\boldsymbol{u}_i^H$ *of the B×B beamspace sample autocovariance matrix* $\boldsymbol{R}_{\tilde{x}\tilde{x}}$ *and estimate the number of its larger eigenvalues as well as noise variance* $\sigma^2 = \frac{1}{B-p}\sum_{i=P+1}^{B}\lambda_i$*;*

Step 4 Compute $\boldsymbol{C}_{\tilde{x}\tilde{x}} = \boldsymbol{R}_{\tilde{x}\tilde{x}} - \sigma^2\boldsymbol{I}$ *and* $\boldsymbol{C}_{\tilde{x}\tilde{y}} = \boldsymbol{R}_{\tilde{x}\tilde{y}} - \sigma^2\boldsymbol{Z}$ *and the generalized eigenvalue decomposition of the beamspace matrix pencil* $\{\boldsymbol{C}_{\tilde{x}\tilde{x}}, \boldsymbol{C}_{\tilde{x}\tilde{y}}\}$ *to obtain p generalized eigenvalues* $\gamma_i = \mathrm{e}^{\mathrm{j}\omega_i}$ *on the unit circle. Then* $\gamma_i = \mathrm{e}^{\mathrm{j}\omega_i, i=1,\cdots,p}$ *can be used directly to obtain the estimation result of spatial parameters* $\omega_i, \cdots, \omega_p$ *and thereby the estimation of DOA* $\theta_1, \cdots, \theta_p$ *of each spatial sources according to* $\hat{\omega}_i = 2\pi\frac{d}{\lambda}\sin\theta_i$*.*

It is necessary for the implementation of BS-ESPRIT algorithm presented above to perform eigenvalue decomposition of beamspace sample autocovariance matrix and generalized eigenvalue decomposition of beamspace matrix pencil. By combining BS-ESPRIT with ES-ESPRIT, a BS-ESPRIT that needs no generalized eigenvalue decomposition of beamspace matrix pencil can be obtained.

Algorithm 10.7.5. *BS-ESPRIT algorithm 2*

Step 1 Construct M × N beamspace observation data matrix $\tilde{\boldsymbol{X}} = [\tilde{\boldsymbol{x}}(1), \cdots, \tilde{\boldsymbol{x}}(N)]$ *using the discrete space Fourier transform* $\tilde{\boldsymbol{x}}(n) = \boldsymbol{W}^H\boldsymbol{x}(n)$*;*

Step 2 Compute the singular value decomposition $\boldsymbol{X} = \tilde{U}\Sigma V^H$ *of the M × N observation data matrix* $\tilde{\boldsymbol{X}}$ *and take the left singular vectors corresponding to the p principal singular values to construct the matrix* $\boldsymbol{U}_s = [\boldsymbol{u}_1, \cdots, \boldsymbol{u}_p]$*;*

Step 3 Take the upper and lower M − 1 rows of $\boldsymbol{U}_s$ *to form* $\boldsymbol{U}_1$ *and* $\boldsymbol{U}_2$*, and compute the eigenvalue decompostion of the matrix* $\boldsymbol{\Psi} = \left(\boldsymbol{U}_1^H\boldsymbol{U}_1\right)^{-1}\boldsymbol{U}_1^H\boldsymbol{U}_2$ *to get its eigenvalues* $\mathrm{e}^{\mathrm{j}\omega_i}$ $(i = 1, \cdots, p)$*, which is identically the same as that of rotation operator* Φ*, and further to get the estimation* ω_i $(i = 1, \cdots, p)$*;*

Step 4 Estimate DOA $\theta_1, \cdots, \theta_p$ *according to* $\hat{\omega}_i = 2\pi\frac{d}{\lambda}\sin\theta_i$*.*

It should be pointed out that, although the derivation process of the ESPRIT algorithm makes them also suitable for complex-valued observation signal, the basic ESPRIT algorithm and its variations such as TLS-ESPRIT algorithm, ES-ESPRIT algorithm, BS-ESPRIT algorithm, and BS-ESPRIT algorithm 2 are more suitable for real-valued observation signal. For the case of complex-valued observation signals, better methods are the unitary ESPRIT algorithm and beamspace unitary ESPRIT algorithm, as will be discussed in the following section.

10.8 Unitary ESPRIT and Its Extensions

For one-dimensional ULA, the MUSIC and root-MUSIC algorithm determine the DOA of spatial sources through the location of spatial spectral peak or the root of the implicit polynomial, respectively; while the ESPRIT algorithms find them through the generalized eigenvalue of the matrix pencil composed by autocovariance and crosscovariance matrix. It is easy to verify that there exists a one-to-one correspondence between the spectral peaks, the roots on the unit circle of the corresponding polynomial or the generalized eigenvalues, and the DOA of spatial sources.

For a two-dimensional (planer) array, even if for uniform rectangular array (URA), direction of each spatial source involves two spatial parameters, namely azimuth ϕ and elevation θ. Consequently, it is necessary/naturally for two-dimensional MUSIC algorithms to perform a two-dimensional search to find every (ϕ_i, θ_i) pair, while for two-dimensional ESPRIT algorithm to perform nonlinear optimization.

One approach to avoid the two-dimensional search for MUSIC algorithm is to decouple the two-dimensional MUSIC problem into two decoupled one-dimensional MUSIC problems and perform azimuth search and elevation search, respectively. However, it will be somewhat a problem for this scheme to match each azimuth ϕ_i with its elevation counterpart θ_i. While an effective alternative of the two-dimensional ESPRIT algorithms to bypass the nonlinear optimization problem is the unitary ESPRIT algorithm.

10.8.1 Unitary ESPRIT Algorithm

Consider the $M \times N$ data matrix

$$\boldsymbol{X} = \left[\boldsymbol{x}(1), \cdots, \boldsymbol{x}(N)\right], \tag{10.8.1}$$

observed by a sensor array in which $x(t) = \left[x_1(t), \cdots, x_M(t)\right]^{\mathrm{T}}$ is the data vector observed by M array elements. Note that the uniform rectangular array can always be arranged into a linear one and $\boldsymbol{X}$ is usually a complex-valued matrix.

In principle, the basic ESPRIT algorithm, the TLS-ESPRIT algorithm, and the BS-ESPRIT algorithm may be used to estimate the desired DOA parameters. However, all these approaches have the same drawback, namely, only the data $x_1(t), \cdots, x_M(t)$

observed directly by the array is used instead of both the observation data and its conjugation $x_1^*(t), \cdots, x_M^*(t)$. Due to the well-known fact that a complex-valued data and its conjugation are complementary to each other as they contained different information. Therefore, if both $x_1(t), \cdots, x_M(t)$ and $x_1^*(t), \cdots, x_M^*(t)$ can be utilized, then the effective data length becomes equivalently doubled. Obviously, although using both the complex-valued observation data and its conjugation would somewhat increase the amount of computation, the estimation accuracy for the signal parameter would be effectively improved over the conventional ESPRIT algorithm without the demand for more array elements, as is just the problem that the unitary ESPRIT algorithm solves.

Specifically speaking, an unitary ESPRIT approach utilizing both the observation complex-valued data matrix $\boldsymbol{X} \in \mathbb{C}^{M\times N}$ and its conjugation matrix $\boldsymbol{X}^* \in \mathbb{C}^{M\times N}$, both of which together consititute a new $M \times 2N$ extended data matrix to estimate the spatial signal parameters $\omega_1, \cdots, \omega_p$. One simple kind of the extended data matrix may be

$$\boldsymbol{Z} = \left[\boldsymbol{X}, \boldsymbol{\Pi}_M \boldsymbol{X}^*\right], \tag{10.8.2}$$

where $\boldsymbol{\Pi}_M$ is an $M \times M$ real-valued permutation matrix with 1 on its main anti-diagonal and 0 elsewhere, namely

$$\boldsymbol{\Pi}_M = \begin{bmatrix} 0 & & & 1 \\ & & 1 & \\ & \cdot^{\cdot^{\cdot}} & & \\ 1 & & & 0 \end{bmatrix} \in \mathbb{R}^{M\times M}, \tag{10.8.3}$$

According to the structure of $\boldsymbol{\Pi}_M$, It is easy to verify that

$$\boldsymbol{\Pi}_M \boldsymbol{\Pi}_M^{\mathrm{T}} = \boldsymbol{I}_M. \tag{10.8.4}$$

The ESPRIT algorithm based on the extended data matrix $\boldsymbol{Z}$ is named as unitary ESPRIT algorithm, which outperforms the conventional ESPRIT algorithm in improving the estimation accuracy of DOA. However, since the column number of $\boldsymbol{Z}$ doubles to be $2N$ compared with that of the original observation data matrix and the data length N are usually large, how to reduce the computation amount required by the singular value decomposition of the extended data matrix naturally turns out to be a crucial problem. It is closely related to the structure of the extended data matrix in the unitary ESPRIT algorithms. An effective approach to deal with this problem is to construct a data matrix with central Hermitian symmetry.

Definition 10.8.1. *A complex-valued matrix $\boldsymbol{B} \in \mathbb{C}^{p\times q}$ is said to be a central Hermittian symmetric matrix, if*

$$\boldsymbol{\Pi}_p \boldsymbol{B}^* \boldsymbol{\Pi}_q = \boldsymbol{B}. \tag{10.8.5}$$

Haardt and Nossek[98] proposed to construct an $M \times 2N$ matrix

$$\boldsymbol{B} = \left[\boldsymbol{X}, \boldsymbol{\Pi}_M \boldsymbol{X}^* \boldsymbol{\Pi}_N\right] \in \mathbb{C}^{M\times 2N}, \tag{10.8.6}$$

as the extended data matrix. Easy to prove that this extended data matrix can achieves the purpose of increasing the data length as well as to be central Hermittian symmetric.

Denote the bijection transformation of a central Hermittian symmetric matrix as $\mathcal{T}(\boldsymbol{B})$, which is a real-valued matrix function defined as[134]

$$\mathcal{T}(\boldsymbol{B}) \stackrel{\text{def}}{=} \boldsymbol{Q}_M^{\mathrm{H}} \boldsymbol{B} \boldsymbol{Q}_{2N} = \boldsymbol{Q}_M^{\mathrm{H}} \left[\boldsymbol{X}, \boldsymbol{\Pi}_M \boldsymbol{X}^* \boldsymbol{\Pi}_N\right] \boldsymbol{Q}_{2N}, \tag{10.8.7}$$

where $\boldsymbol{Q}_M$ and $\boldsymbol{Q}_{2N}$ are called the left and right bijection transform matrix, respectively.

Haardt and Nossek[98] introduce an effective method to select the bijection transform:

(1) Divide the original observation data matrix into blocks as

$$\boldsymbol{X} = \begin{bmatrix} \boldsymbol{X}_1 \\ \boldsymbol{g}^{\mathrm{T}} \\ \boldsymbol{X}_2 \end{bmatrix}, \tag{10.8.8}$$

in which $\boldsymbol{X}_1$ and $\boldsymbol{X}_2$ have the same size as each other. Obviously, if the number m of rows of the observation data matrix $\boldsymbol{X} \in \mathbb{M} \times \mathbb{N}$ is even, the block matrix in Eq. (10.8.8) would not contain row vector $\boldsymbol{g}^{\mathrm{T}}$ any more.

(2) Select the left and right bijection transform matrix to be

$$\boldsymbol{Q}_M = \begin{cases} \dfrac{1}{\sqrt{2}} \begin{bmatrix} \boldsymbol{I}_{(M-1)/2} & \boldsymbol{0} & \mathrm{j}\boldsymbol{I}_{(M-1)/2} \\ \boldsymbol{0}^{\mathrm{T}} & \sqrt{2} & \boldsymbol{0}^{\mathrm{T}} \\ \boldsymbol{\Pi}_{(M-1)/2} & \boldsymbol{0} & -\mathrm{j}\boldsymbol{\Pi}_{(M-1)/2} \end{bmatrix}, & \text{if } M \text{ is odd,} \\ \dfrac{1}{\sqrt{2}} \begin{bmatrix} \boldsymbol{I}_{M/2} & \mathrm{j}\boldsymbol{I}_{M/2} \\ \boldsymbol{\Pi}_{M/2} & -\mathrm{j}\boldsymbol{\Pi}_{M/2} \end{bmatrix}, & \text{if } M \text{ is even,} \end{cases} \tag{10.8.9}$$

$$\boldsymbol{Q}_{2N} = \frac{1}{\sqrt{2}} \begin{bmatrix} \boldsymbol{I}_N & \mathrm{j}\boldsymbol{I}_N \\ \boldsymbol{\Pi}_N & -\mathrm{j}\boldsymbol{\Pi}_N \end{bmatrix} \in \mathbb{C}^{2N\times 2N}, \tag{10.8.10}$$

respectively.

This bijection described above exhibits two important properties:

(1) The bijection transformation of a central Hermittian symmetric matrix $\boldsymbol{B}$ can be derived as

$$\mathcal{T}(\boldsymbol{B}) = \begin{bmatrix} \mathrm{Re}(\boldsymbol{X}_1 + \boldsymbol{\Pi}\boldsymbol{X}_2^*) & \text{-}\mathrm{Im}(\boldsymbol{X}_1 - \boldsymbol{\Pi}\boldsymbol{X}_2^*) \\ \sqrt{2}\mathrm{Re}(\boldsymbol{g}^{\mathrm{T}}) & -\sqrt{2}\mathrm{Im}(\boldsymbol{g}^{\mathrm{T}}) \\ \mathrm{Im}(\boldsymbol{X}_1 + \boldsymbol{\Pi}\boldsymbol{X}_2^*) & \mathrm{Re}(\boldsymbol{X}_1 - \boldsymbol{\Pi}\boldsymbol{X}_2^*) \end{bmatrix} \in \mathbb{R}^{M\times 2N}, \tag{10.8.11}$$

(2) The central Hermittian symmetric matrix $\boldsymbol{B}$ itself satisfies

$$\boldsymbol{Q}_M \mathcal{T}(\boldsymbol{B}) \boldsymbol{Q}_{2N}^{\mathrm{H}} = \boldsymbol{Q}_M \boldsymbol{Q}_M^{\mathrm{H}} \boldsymbol{B} \boldsymbol{Q}_{2N} \boldsymbol{Q}_{2N}^{\mathrm{H}} = \boldsymbol{B}, \tag{10.8.12}$$

due to $\boldsymbol{Q}_M \boldsymbol{Q}_M^{\mathrm{H}} = \boldsymbol{I}_M$ and $\boldsymbol{Q}_{2N} \boldsymbol{Q}_{2N}^{\mathrm{H}} = \boldsymbol{I}_{2N}$.

Assume the sigular value decompositon of the resulted real-valued matrix $\mathcal{T}(\boldsymbol{B})$ after bijection transform to be

$$\mathcal{T}(\boldsymbol{B}) = \boldsymbol{U\Sigma V}^{\mathrm{H}} \quad (\boldsymbol{U} \in \mathbb{C}^{M\times M}, \ \ \boldsymbol{V} \in \mathbb{C}^{2N\times 2N}), \tag{10.8.13}$$

then combining Eq. (10.8.12) with Eq. (10.8.13) yields

$$\boldsymbol{B} = \boldsymbol{Q}_M \mathcal{T}(\boldsymbol{B}) \boldsymbol{Q}_{2N}^{\mathrm{H}} = (\boldsymbol{Q}_M \boldsymbol{U}) \boldsymbol{\Sigma} (\boldsymbol{V}^{\mathrm{H}} \boldsymbol{Q}_{2N}^{\mathrm{H}} s), \tag{10.8.14}$$

which means that the sigular value decomposition of the central Hermittian symmetric matrix $\boldsymbol{B}$ can be obtained directly to be $\boldsymbol{B} = (\boldsymbol{Q}_M\boldsymbol{U})\boldsymbol{\Sigma}(\boldsymbol{V}^{\mathrm{H}}\boldsymbol{Q}_{2N}^{\mathrm{H}})$ once the sigular value decomposition $\mathcal{T}(\boldsymbol{B}) = \boldsymbol{U\Sigma V}^{\mathrm{H}}$ of the resulted real-valued matrix $\mathcal{T}(\boldsymbol{B})$ is known due to that $\boldsymbol{Q}_M\boldsymbol{U}$ and $\boldsymbol{V}^{\mathrm{H}}\boldsymbol{Q}_{2N}^{\mathrm{H}}$ are $M \times M$ and $2N \times 2N$ unitary matrix, respectively.

The method introduced above can not only utilize the extended data with a doubled length of $\boldsymbol{X}$ but also bypass the problem caused by the singular value decomposition of the matrix with large number of columns by performing the sigular value decomposition of the corresponding central Hermittian symmetric matrix. Consequently, it becomes an efficient approach to get the singular value decomposition with a better accuracy than that of the direct singular value decomposition of the observation data matrix $\boldsymbol{X}$. This efficient approach is proposed by Hardt and Nossek in 1995[98].

In practice, it is unnecessary to compute the singular value decomposition of the central Hermitian symmetric matrix $\boldsymbol{B}$ since the estimation of DOA can be obtained immediately just by applying ES-ESPRIT algorithm to the matrix $\boldsymbol{U}_S$, which is composed of the first p columns of the left singular vector $\boldsymbol{U}_M = \boldsymbol{Q}_M\boldsymbol{U}$. To sum up, the unitary ESPRIT algorithm [98] can be described as below.

Algorithm 10.8.1. *Unitary ESPRIT Algorithm*

Given the observation data $x_1(t), \cdots, x_M(t)$, $t = 1, \cdots, N$, of a sensor array with M elements.

Step 1 Construct the observation data matrix $\boldsymbol{X} = \left[\boldsymbol{x}(1), \cdots, \boldsymbol{x}(n)\right]$;

Step 2 Divide $\boldsymbol{X}$ into blocks as Eq. (10.8.8) *and then compute the $M \times 2N$ real-valued matrix $\mathcal{T}(\boldsymbol{B})$ according to Eq.* (10.8.11) *with $\boldsymbol{Q}_M$ and $\boldsymbol{Q}_{2N}$ constructed as Eqs.* (10.8.9) *and* (10.8.10)*;*

Step 3 Compute the singular value decomposition $\mathcal{T}(\boldsymbol{B}) = \boldsymbol{U\Sigma V}^{H}$ and find the number p of its principal singular values, namely, effective rank of $\mathcal{T}(\boldsymbol{B})$;

Step 4 Compute the left singular vector matrix $\boldsymbol{U}_M = \boldsymbol{Q}_M\boldsymbol{U}$ of the extended central Hermittian symmetric matrix $\boldsymbol{B}$ and take the first p columns from $\boldsymbol{U}_M$ to construct $\boldsymbol{U}_S$, column vectors of which span the signal subspace of $\boldsymbol{B}$;

Step 5 Apply Step 2 and Step 3 of the ES-ESPRIT algorithm 10.7.2 to $\boldsymbol{U}_s$ to estimate the DOA.

To conclude, when applying the ESPRIT algorithm to complex-valued observation data, utilizing in addition the conjugate observation data is equivalent to doubling the data length, which is helpful to improve the estimation accuracy of the signal parameters.

By constructing an observation data matrix ("virtual", not necessary to be computed actually) with central Hermitian symmetry, the singular value decomposition of a complex-valued matrix is reduced to that of a real-valued matrix. Consequently, unitary ESPRIT is a computationally efficient ESPRIT algorithm under the complex-valued data case.

10.8.2 Beamspace Unitary ESPRIT Algorithm

By replacing the element space observation data matrix in the unitary ESPRIT algorithm discussed previously with the beam space observation data matrix, the beamspace unitary ESPRIT algorithm may immediately be developed from its element space counterpart.

Algorithm 10.8.2. *Beamspace Unitary ESPRIT Algorithm*

Given the observation data $x_1(t), \cdots, x_M(t),\ t = 1, \cdots, N$, *of a sensor array with* M *elements.*

Step 1 Compute the $M \times N$ *beamspace observation data martix* $\tilde{\mathbf{X}} = \mathbf{W}_M^H \mathbf{X}$ *by means of discrete Fourier transform;*

Step 2 Divide $\tilde{\mathbf{X}}$ *into blocks as Eq.* (10.8.8) *and then compute the* $M \times 2N$ *real-valued matrix* $\mathcal{T}(\mathbf{B})$ *according to Eq.* (10.8.11) *with* $\mathbf{Q}_M$ *and* $\mathbf{Q}_{2N}$ *constructed as Eqs.* (10.8.9) *and* (10.8.10);

Step 3 Compute the singular value decomposition $\mathcal{T}(\mathbf{B}) = \mathbf{U}\mathbf{\Sigma}\mathbf{V}^H$ *and find the number* p *of its principal singular values, namely, effective rank of* $\mathcal{T}(\mathbf{B})$;

Step 4 Compute the left singular vector matrix $\mathbf{U}_M = \mathbf{Q}_M \mathbf{U}$ *of the extended central Hermittian symmetric matrix* $\mathbf{B}$ *and take the first* p *columns from* $\mathbf{U}_M$ *to construct* $\mathbf{U}_s$, *column vectors of which span the signal subspace of* $\mathbf{B}$;

Step 5 Apply Step 2 and Step 3 of the ES-ESPRIT algorithm 10.7.2 to $\mathbf{U}_s$ *to estimate the DOA.*

The beamspace unitary ESPRIT algorithm is first proposed by Zoltowsk, et.al. in 1996, and is named as two-dimensional discrete Fourier transform beam space ESPRIT algorithm, or 2D-DFT-BS-ESPRIT algorithm for short[257].

Algorithm 10.8.3. *2D-DFT-BS-ESPRIT Algorithm*

Given the observation data $x_1(t), \cdots, x_M(t),\ t = 1, \cdots, N$, *of a sensor array with* M *elements.*

Step 1 Compute the $M \times N$ *beamspace observation data martix* $\mathbf{Y} = \mathbf{W}_M^H \mathbf{X}$ *by means of discrete Fourier transform;*

Step 2 Divide $\tilde{\mathbf{X}}$ *into blocks as Eq.* (10.8.8) *and then compute the* $M \times 2N$ *real-valued matrix* $\mathcal{T}(\mathbf{B})$ *according to Eq.* (10.8.11) *with* $\mathbf{Q}_M$ *and* $\mathbf{Q}_{2N}$ *constructed as Eqs.* (10.8.9) *and* (10.8.10);

Step 3a Find the solution $\boldsymbol{\Psi}_\mu$ *of the* $(M-1)MN \times d$ *matrix equation* $\boldsymbol{\Gamma}_{\mu 1}\boldsymbol{U}_S\boldsymbol{\Psi}_\mu = \boldsymbol{\Gamma}_{\mu 2}\boldsymbol{U}_S$. *Among which, both* $\boldsymbol{\Gamma}_{\mu 1} = \boldsymbol{I}_N \otimes \boldsymbol{\Gamma}_1$ *and* $\boldsymbol{\Gamma}_{\mu 2} = \boldsymbol{I}_N \otimes \boldsymbol{\Gamma}_2$ *are* $(M-1)N \times MN$ *real-valued matrix, where* $\boldsymbol{A} \otimes \boldsymbol{B} = [a_{ij}\boldsymbol{B}]$ *respresents the Kronecker product of two matrix* $\boldsymbol{A}$ *and* $\boldsymbol{B}$*;*

$$\boldsymbol{\Gamma}_1 = \begin{bmatrix} 1 & \cos\left(\frac{\pi}{M}\right) & 0 & 0 & \cdots & \cdots & 0 & 0 \\ 0 & \cos\left(\frac{\pi}{M}\right) & \cos\left(\frac{2\pi}{M}\right) & 0 & \cdots & \cdots & 0 & 0 \\ 0 & 0 & \cos\left(\frac{2\pi}{M}\right) & \cos\left(\frac{3\pi}{M}\right) & \cdots & \cdots & 0 & 0 \\ \vdots & \vdots & \vdots & \ddots & \ddots & \vdots & \vdots & \vdots \\ \vdots & \vdots & \vdots & \vdots & \ddots & \ddots & \vdots & \vdots \\ 0 & 0 & 0 & 0 & \cdots & \cos\left(\frac{(M-3)\pi}{M}\right) & \cos\left(\frac{(M-2)\pi}{M}\right) & 0 \\ 0 & 0 & 0 & 0 & \cdots & \cdots & \cos\left(\frac{(M-2)\pi}{M}\right) & \cos\left(\frac{(M-1)\pi}{M}\right) \end{bmatrix}$$

$$\boldsymbol{\Gamma}_2 = \begin{bmatrix} 1 & \sin\left(\frac{\pi}{M}\right) & 0 & 0 & \cdots & \cdots & 0 & 0 \\ 0 & \sin\left(\frac{\pi}{M}\right) & \sin\left(\frac{2\pi}{M}\right) & 0 & \cdots & \cdots & 0 & 0 \\ 0 & 0 & \sin\left(\frac{2\pi}{M}\right) & \sin\left(\frac{3\pi}{M}\right) & \cdots & \cdots & 0 & 0 \\ \vdots & \vdots & \vdots & \ddots & \ddots & \vdots & \vdots & \vdots \\ \vdots & \vdots & \vdots & \vdots & \ddots & \ddots & \vdots & \vdots \\ 0 & 0 & 0 & 0 & \cdots & \sin\left(\frac{(M-3)\pi}{M}\right) & \sin\left(\frac{(M-2)\pi}{M}\right) & 0 \\ 0 & 0 & 0 & 0 & \cdots & \cdots & \sin\left(\frac{(M-2)\pi}{M}\right) & \sin\left(\frac{(M-1)\pi}{M}\right) \end{bmatrix}$$

Step 3b Find the solution $\boldsymbol{\Psi}_\nu$ *of the* $(N-1)NM \times d$ *matrix equation* $\boldsymbol{\Gamma}_{\nu 1}\boldsymbol{U}_S\boldsymbol{\Psi}_\nu - \boldsymbol{\Gamma}_{\nu 2}\boldsymbol{U}_S$. *In which, both* $\boldsymbol{\Gamma}_{\nu 1} = \boldsymbol{\Gamma}_3 \otimes \boldsymbol{I}_N$ *and* $\boldsymbol{\Gamma}_{\nu 2} = \boldsymbol{\Gamma}_4 \otimes \boldsymbol{I}_N$ *are* $(N-1)M \times NM$ *real-valued matrices, and both* $\boldsymbol{\Gamma}_3$ *and* $\boldsymbol{\Gamma}_4$ *are* $(N-1) \times N$ *matrix. Moreover, the structure of matrix* $\boldsymbol{\Gamma}_3$ *and* $\boldsymbol{\Gamma}_4$ *are similar to that of matrix* $\boldsymbol{\Gamma}_1$ *and* $\boldsymbol{\Gamma}_2$*, respectively, while the difference lies in that their sizes are related to N instead of M;*

Step 4 Estimate the eigenvalue λ_i, $i = 1, \cdots, d$ *of the* $d \times d$ *matrix* $\boldsymbol{\Psi}_\mu + \mathrm{j}\boldsymbol{\Psi}_\nu$*;*

Step 5 Compute the direction cosine of the i-th source $i = 1, \cdots, d$*, namely,* $\mu_i = 2\arctan\left(Re(\lambda_i)\right)$ *with respective to x-axis and* $\nu_n = 2\arctan\left(Im(\lambda_i)\right)$ *with respective to y-axis, respectively;*

Step 6 Compute the azimuth $\phi_i = \arctan\left(\mu_i/\nu_i\right)$ *and the elevation* $\theta_i = \arctan\left(\mu_i/\sin\phi_i\right)$ *of the i-th source* $i = 1, \cdots, d$*, respectively.*

By comparison, the difference between the ESPRIT and MUSIC algorithms may be concluded as follows:

(1) MUSIC algorithms estimate the DOA using spatial spectral estimation; while ESPRIT algorithms usually estimate the DOA through the eigenvalue or the generalized eigenvalue.
(2) MUSIC algorithm estimate the spatial spectrum using the eigenvector corresponding to the noise variance, which belongs to the noise subspace approach; ESPRIT

algorithm estimate the spatial parameters using the (generalized) principal eigenvalue, and thus is regarded as signal subspace technique.

Summary

There are mainly two kinds of techniques in the array signal processing field, that is beamforming and DOA estimation. This chapter first focuses on the comparison and analysis between the beamformer and the spatial FIR filter, which is followed by the linearly constrained adaptive beamformer. As to the DOA estimation problem, the well-known MUSIC and ESPRIT algorithms are mainly discussed.

MUSIC is a subspace technique to estimate the spatial parameters of the signal source, which includes root-MUSIC, FINE-MUSIC as well as their variants.

ESPRIT is a rotation invariant technique to estimate the spatial parameters of the signal. Although none concept of the spectrum is involved, the frequency of complex harmonic sources may be obtained with a high accuracy via its idea of generalized eigenvalue decomposition. As for the two-dimensional planer array, several unitary ESPRIT methods are introduced.

In addition, the element space and beam space algorithms of both the MUSIC and ESPRIT algorithm are illustrated and compared in particular.

Exercises

10.1 Consider a two-dimensional planer arrray consists of m elements and the i-th element position to be $\boldsymbol{r}_i = [x_i, y_i]^{\mathrm{T}}$, $i = 1, \cdots, m$. A narrowband planer wave $s(t)$ with central frequency to be ω_0 incident upon this array and the transimssion direction vector to be $\boldsymbol{\alpha} = \frac{1}{c}[\cos\theta, \sin\theta]^{\mathrm{T}}$, in which c is the light velocity. Assume the signal received by the i-th element to be

$$z_i(t) = s\left(t - \boldsymbol{\alpha}^{\mathrm{T}}\boldsymbol{r}_i\right) \mathrm{e}^{\mathrm{j}\omega_0(t-\boldsymbol{\alpha}^{\mathrm{T}}\boldsymbol{r}_i)}, \quad i = 1, \cdots, m,$$

(1) Prove that the observation signal vector of the array to be

$$\boldsymbol{z}(t) = \begin{bmatrix} z_1(t) \\ \vdots \\ z_1(t) \end{bmatrix} = s(t)\,\mathrm{e}^{\mathrm{j}\omega_0 t} \begin{bmatrix} \mathrm{e}^{\mathrm{j}\frac{2\pi}{\lambda}(x_1\cos\theta + y_1\sin\theta)} \\ \vdots \\ \mathrm{e}^{\mathrm{j}\frac{2\pi}{\lambda}(x_m\cos\theta + y_m\sin\theta)} \end{bmatrix} = s(t)\mathrm{e}^{\mathrm{j}\omega_0 t}\boldsymbol{a}(\theta),$$

where $\boldsymbol{a}(\theta) = \left[\mathrm{e}^{\mathrm{j}\frac{2\pi}{\lambda}(x_1\cos\theta + y_1\sin\theta)}, \cdots, \mathrm{e}^{\mathrm{j}\frac{2\pi}{\lambda}(x_m\cos\theta + y_m\sin\theta)}\right]^{\mathrm{T}}$ is the steering vector of the narrowband signal $s(t)$.

(2) Try to give the expression of the observation signal vector of the sensor array, if the number of narrowband signal source is p, and is transmitted independently.

10.2 Beampattern of a beamformer $\boldsymbol{w}$ is defined as

$$P_{\boldsymbol{w}}(\theta) = \boldsymbol{w}^{\mathrm{T}}\boldsymbol{a}(\theta),$$

and the square of its modulus is named as the power pattern of an antenna. Consider that a spatial signal incidents on an uniform linear array consists of M elements with a DOA of θ_0.

(1) Prove that the power pattern of the antenna is

$$|P_{\boldsymbol{w}}(\theta)|^2 = \left|\frac{\sin\left(\frac{m(\phi-\phi_0)}{2}\right)}{\sin\left(\frac{(\phi-\phi_0)}{2}\right)}\right|,$$

in which $\phi = \sin\theta$ and $\phi_0 = \sin\theta_0$;

(2) Let $\theta_0 = \pi/4$, plot the power pattern and find its mainlobe beamwidth.

10.3 Let $\boldsymbol{R}_{xx} = \mathrm{E}\left\{\boldsymbol{x}(t)\boldsymbol{x}^{\mathrm{H}}(t)\right\}$ be the autocovariance matrix of the observation vector $\boldsymbol{x}(t)$ of a sensor array. The optimal beamformer is usually determined by the solution of a constrained optimization problem expressed as

$$\min_{\boldsymbol{w}} = \boldsymbol{w}^{\mathrm{H}}\boldsymbol{R}_{xx}\boldsymbol{w},$$
$$\text{subject to } f(\boldsymbol{w}) = 0,$$

where $f(\boldsymbol{w}) = 0$ is some kinds of constraint.

If $f(\boldsymbol{w})$ is chosen to be $f(\boldsymbol{w}) = \boldsymbol{w}^{\mathrm{H}}\boldsymbol{a}(\omega_0) - 1 = 0$, this optimation problem is known to be the miminum square error (MSE) criterion of optimal beamformer design. Prove that the optimal beamformer designed according to the MSE criterion is

$$\boldsymbol{w}_{\mathrm{opt}} = \frac{\boldsymbol{R}_{xx}^{-1}\boldsymbol{a}(\omega_0)}{\boldsymbol{a}(\omega_0)^{\mathrm{H}}\boldsymbol{R}_{xx}^{-1}\boldsymbol{a}(\omega_0)}.$$

10.4 Assume the signal vector of an sensor array to be $\boldsymbol{x} = \boldsymbol{x}_s + \boldsymbol{x}_n$, in which, the signal component $\boldsymbol{x}_s$ and the noise component $\boldsymbol{x}_n$ are statistically incorrelated. Prove that the optimal beamformer designed under the maximum signal-to-noise ratio criterion satisfies

$$\boldsymbol{R}_s\boldsymbol{w}_{opt} = \lambda_{\max}\boldsymbol{R}_s\boldsymbol{w}_{\mathrm{opt}}$$

where $\boldsymbol{R}_s = \mathrm{E}\left\{\boldsymbol{x}_s\boldsymbol{x}_s^{\mathrm{H}}\right\}$ and $\boldsymbol{R}_s = \mathrm{E}\left\{\boldsymbol{x}_n\boldsymbol{x}_n^{\mathrm{H}}\right\}$. In other words, the optimal beamformer vector $\boldsymbol{w}_{\mathrm{opt}}$ is the generalized eigenvector corresponding to the largest generalized eigenvalue of matrix pencil $(\boldsymbol{R}_s, \boldsymbol{R}_n)$.

10.5 Consider the received signal vector of a ULA

$$\boldsymbol{x}(t) = s_1(t)\boldsymbol{a}(\theta_1) + s_2(t)\boldsymbol{a}(\theta_2) + \boldsymbol{n}(t)$$

where $s_1(t)$ and $s_2(t)$ are coherent with each other, namely, $s_1(t) = c \cdot s_2(t)$ with c to be a complex-value constant. Moreover, $\boldsymbol{n}(t) = \left[n_1(t), n_2(t)\right]^{\mathrm{T}}$ denotes the additive white noise vector of the ULA. Prove that neither θ_1 nor θ_2 can be estimated from the signal vector $\boldsymbol{x}(t)$ of the ULA.

10.6 Consider three coherent spatial narrowband signals $s_i(t) = A_i \cos(2\pi f_0 t)$, $i = 1, 2, 3$, where $f_0 = 1000\text{Hz}$,and their amplitudes are $A_1 = 1$, $A_2 = 2$ and $A_3 = 3$, respecitvely. In addition, their DOAs are $\theta_1 = 5°$, $\theta_2 = 15°$ and $\theta_3 = 30°$. All these signals are received by a ULA cosists of 10 elements and the signal received by the k-th array element is

$$x_k(t) = \sum_{i=1}^{3} s_i(t) \exp\left(-\mathrm{j}\frac{2\pi}{\lambda}(k-1)\,d \sin\theta_i\right) + n_k(t), \quad k = 1, \cdots, 10.$$

Let $d = \lambda/5$ and the additive white noise follow $\mathcal{N}(0, \sigma^2)$, and the sample time is $t = 1, 2, \cdots, 1024$. Adjust the variance σ^2 of the white noise to get the SNR of 5 , 10 , 15 , 20 , 25 , 30 dB, respectively. Try to perform the DOA estimation for 20 runs independently using each algorithm including the basic MUSIC, the MUSIC base on two subarray spatial-sliding technique with each consists of 9 elements as well as the ESPRIT algorithm. And furthermore, plot the mean and mean square root error of their DOA estimation versus SNR, respectively.

10.7 Assume the simulated observation data be generated according to

$$x(n) = \sqrt{20}\sin(2\pi 0.2n) + \sqrt{2}\sin(2\pi 0.213n) + w(n),$$

where $w(n)$ is a Gaussian white noise with zero mean and unit variance, and choose $n = 1, \cdots, 128$.

(1) Code the program of MUSIC algorithm to conduct the simulated harmonic recovery experiments for 50 runs, and give the statistical result of the estimation.
(2) Code the program of root-MUSIC algorithm to conduct the simulated harmonic recovery experiments for 50 runs, and plot the statistical mean and deviation error curve of the frequency estimation result.

10.8 Use the same simulated data as the previous exercise. Code the program of the basic ESPRIT, SVD-TLS-based ESPRIT algorithm, respectively, and conduct the simulated harmonic recovery experiments each for 50 runs. plot the statistical mean and deviation error curve of the frequency estimation result.

Index

https://doi.org/10.1515/9783110475562-011

Bibliography

[1] Special issue on active and adaptive antennas. *IEEE Trans.Antennas Propagat.*, AP-12, 1964.

[2] Special issue on adaptive antennas. *IEEE Trans.Antennas Propagat.*, AP-24, 1976.

[3] Special issue on nderwater acoustic signal processing. *IEEE J. Oceanic Eng*, OE-12, 1987.

[4] H. Abdi. Signal detection theory (sdt). *In: Encyclopedia of Measurement and Statistics*, page 1–9, 2007.

[5] A. Aïssa-El-Bey, N. Linh-Trung, K. Abed-Meraim, A. Belouchrani, and Y. Grenier. Underdetermined blind separation of nondisjoint sources in the time-frequency domain. *IEEE Trans. Signal Process.*, 55(3):897–907, 2007.

[6] H. Akaike. Power spectrum estimation through autoregressive model fitting. *Annals of the Institute of Statistical Mathematics*, 21(1):407–419, 1969.

[7] H. Akaike. A new look at the statistical model identification. *IEEE Transactions on Automatic Control*, 19(6):716–723, 1974.

[8] L. Almeida. The fractional fourier transform and time-frequency representations. *IEEE Transactions on Signal Processing*, 42(11):3084–3091, 1994.

[9] S. A. C. Alvarez, A. Cichocki, and L. Castedo. An iterative inversion approach to blind source separation. *IEEE Trans. Neural Networks Learn. Syst.*, 11(6):1423–1437, 2000.

[10] S. Amari and A. Cichocki. Adaptive blind signal processing-neural network approaches. *Proceedings of the IEEE*, 86(10):2026–2048, 1998.

[11] S. I. Amari. A theory of adaptive pattern classifiers. *IEEE Trans. Electron. Comput.*, 16(3):299–307, 1967.

[12] S. I. Amari. Natural gradient works efficiently in learning. *Neural Comput.*, 10(2):251–276, 1998.

[13] S. I. Amari, T. Chen, and A. Cichocki. Stability analysis of learning algorithms for blind source separation. *Neural Networks*, 10(8):1345–1351, 1997.

[14] S. Applebaum. Adaptive arrays. *IEEE Transactions on Antennas and Propagation*, 24(5), 1976.

[15] F. Auger and P. Flandrin. Improving the readability of time-frequency and time-scale representations by the reassignment method. *IEEE Trans. Signal Process.*, 43(5):1068–1089, 1995.

[16] V. V. B. *Minimum variance beamforming. In: Adaptive Radar Detection and Estimation ed.Haykin H and Steinhardt A.* Wiley-Interscience, 1992.

[17] A. J. Barabell. Improving the resolution performance of eigenstructure-based direction-finding algorithms. In *IEEE International Conference on Acoustics, Speech, and Signal Processing, ICASSP '83, Boston, Massachusetts, USA*, pages 336–339. IEEE, 1983.

[18] S. Barbarossa, E. D"Addio, and G. Galati. Comparison of optimum and linear prediction technique for clutter cancellation. *IEE Proceedings. Part F-Communications, Radar and Signal Processing*, 134(3):277–282, 1987.

[19] B. Barkat and B. Boashash. A high-resolution quadratic time-frequency distribution for multicomponent signals analysis. *IEEE Trans. Signal Process.*, 49(10):2232–2239, 2001.

[20] M. S. Bartlett. *An Introduction to Stochastic Processes*. UK: Cambridge University Press, 1955.

[21] M. J. Bastiaans. A sampling theorem for the complex spectrogram, and gabor's expansion of a signal in gaussian elementary signals. *International Society for Optics and Photonics*, 20:594–597, 1981.

[22] F. Beaufays. Transform-domain adaptive filters: an analytical approach. *IEEE Trans. Signal Process.*, 43(2):422–431, 1995.

https://doi.org/10.1515/9783110475562-012

[23] A. J. Bell and T. J. Sejnowski. An information-maximization approach to blind separation and blind deconvolution. *Neural Comput.*, 7(6):1129–1159, 1995.

[24] A. Belouchrani and M. G. Amin. Blind source separation based on time-frequency signal representations. *IEEE Trans. Signal Process.*, 46(11):2888–2897, 1998.

[25] A. Belouchrani, K. A. Meraim, J. F. Cardoso, and E. Moulines. A blind source separation technique using second-order statistics. *IEEE Trans. Signal Process.*, 45(2):434–444, 1997.

[26] Y. Benjamini and Y. Hochberg. Controlling the false discovery rate: A practical and powerful approach to multiple testing. *Journal of the Royal Statistical Society. Series B: Methodological*, 57(1):289–300, 1995.

[27] G. Biemvenu and L. Kopp. Principè de la goniomgraveetre passive adaptive. *Pro. 7'eme Colloque Gresit, Nice Frace*, pages 106/1–106/10, 1979.

[28] I. Bilinskis and A. Mikelsons. *Randomized signal processing*. New York: Prentice-Hall, Inc., 1992.

[29] R. B. Blackmanand and J. W. Tukey. *The Measurement of Power Spectra*. New York: Dover Publications, Inc., 1959.

[30] B. Boashash. Note on the use of the wigner distribution for time-frequency signal analysis. *IEEE Trans. Acoust. Speech Signal Process.*, 36(9):1518–1521, 1988.

[31] C. Bonferroni. *Elementi di Statistica General*. Libereria Seber, 1930.

[32] G. E. P. Box and G. M. Jenkins. *Time Series Analysis-Forecasting and Control*. San Francisco, CA: Holden-Day, 1970.

[33] R. B. Brem, J. D. Storey, J. Whittle, and L. Kruglyak. Genetic interactions between polymorphisms that affect gene expression in yeast. *Nature*, 436:701–703, 2005.

[34] D. R. Brillinger. Asymptotic theory of estimate of k-th order spectra. *Spectral Analysis of Time Signals*, pages 153–188, 1967.

[35] D. R. Brillinger and M. Rosenblatt. Computation and interpretation of-th order spectra. *Spectral Analysis of Time*, pages 189–232, 1967.

[36] K. M. Buckley. Spatial/spectral filtering with linearly constrained minimum variance beamformers. *IEEE Trans. Acoust. Speech Signal Process.*, 35(3):249–266, 1987.

[37] K. M. Buckley and X. L. Xu. Spatial-spectrum estimation in a location sector. *IEEE Trans. Acoust. Speech Signal Process.*, 38(11):1842–1852, 1990.

[38] A. Bultheel and H. E. M. Sulbaran. Computation of the fractional fourier transform. *Applied and Computational Harmonic Analysis*, 16(3):182–202, 2004.

[39] J. P. Burg. Maximum entropy spectral analysis. *37th Ann. Int. Soc. Explar Geophysics meeting*, 1967.

[40] J. Cadzow. High performance spectral estimation–a new arma method. *IEEE Transactions on Acoustics, Speech, and Signal Processing*, 28(5):524–529, 1980.

[41] J. Cadzow. Spectral estimation: An overdetermined rational model equation approach. *Proceedings of the IEEE*, 70(9):907–939, 1982.

[42] J. Capon. High-resolution frequency-wavenumber spectrum analysis. *Proceedings of the IEEE*, 57(8):1408–1418, 1969.

[43] G. Carayannis, N. Kalouptsidis, and D. Manolakis. Fast recursive algorithms for a class of linear equations. *IEEE Trans. Acoustics, Speech, and Signal Processing*, 30(2):227–239, 1982.

[44] J. F. Cardoso and B. H. Laheld. Equivariant adaptive source separation. *IEEE Trans. Signal Process.*, 44(12):3017–3030, 1996.

[45] J. F. Cardoso and A. Souloumiac. Blind beamforming for non-gaussian signals. *Proceedings of the IEE, F*, 40(6):362–370, 1993.

[46] J. F. Cardoso and A. Souloumiac. Jacobi angles for simultaneous diagonalization. *SIAM J. Matrix Anal. Appl.*, 17(1):161–164, 1996.

[47] J. R. Carson and T. C. Fry. Variable frequency electric circuit theory with application to the theory of frequency-modulation. *Bell Labs Technical Journal*, 16(4):513–540, 1937.

[48] Y. Chan and J. Wood. A new order determination technique for ARMA processes. *IEEE Transactions on Acoustics, Speech, and Signal Processing*, 32(3):517–521, 1984.

[49] Y. T. Chan, R. V. Hattin, and J. B. Plant. The least squares estimation of time delay and its use in signal detection. In *IEEE International Conference on Acoustics, Speech, and Signal Processing, ICASSP '78, Tulsa, Oklahoma, USA, April 10-12, 1978*, pages 665–669. IEEE, 1978.

[50] Y. T. Chan, R. V. Hattin, and J. B. Plant. A parametric estimation approach to time-delay estimation and signal detection. *IEEE Transactions on Acoustics Speech and Signal Processing*, 28:8–16, 1980.

[51] V. Chandran and S. L. Elgar. Pattern recognition using invariants defined from higher order spectra- one dimensional inputs. *IEEE Trans. Signal Process.*, 41(1):205–212, 1993.

[52] B. Chen and A. P. Petropulu. Frequency domain blind MIMO system identification based on second and higher order statistics. *IEEE Trans. Signal Process.*, 49(8):1677–1688, 2001.

[53] N. I. Cho, C. H. Choi, and S. U. Lee. Adaptive line enhancement by using an IIR lattice notch filter. *IEEE Trans. Acoust. Speech Signal Process.*, 37(4):585–589, 1989.

[54] H. Choi and W. J. Williams. Improved time-frequency representation of multicomponent signals using exponential kernels. *IEEE Trans. Acoust. Speech Signal Process.*, 37(6):862–871, 1989.

[55] S. Choi, A. Cichocki, and S. Amari. Flexible independent component analysis. *J. VLSI Signal Process.*, 26(1-2):25–38, 2000.

[56] J. A. Chow. On estimating the orders of an autoregressive moving-average process with uncertain observations. *IEEE Transactions on Automatic Control*, 17(5):707–709, 1972.

[57] C. K. Chui. *An Introduction to Wavelets*. New York: Academic Press Inc., 1992.

[58] C. K. Chui. *Wavelets: A Tutorial in Theory and Applications*. NeNew York: Academic Press Inc., 1992.

[59] A. Cichocki, S. I. Amari, M. Adachi, and W. Kasprzak. Self-adaptive neural networks for blind separation of sources. *Proc 1996 International Symp.on Circuits and Systems*, 2(1):157 – 160, 1996.

[60] Classen, T., Mecklenbrauker, and W. The aliasing problem in discrete-time wigner distributions. *IEEE Transactions on Acoustics, Speech and Signal Processing*, 34:442 – 451, 1983.

[61] T. A. C. M. Classen and W. F. G. Mecklenbraeuker. The wigner distribution - a tool for time-frequency signal analysis. *Philips Journal of Research*, 35(4):217 – 250, 276 – 300, 1067 – 1072, 1980.

[62] Cohen and Leon. Generalized phase-space distribution functions. *Journal of Mathematical Physics*, 7(5):781–786, 1966.

[63] L. Cohen. *A Primer On Time-Frequency Analysis*. In: New Methods in Time-Frequency Anadios(Ed.by Boashash B), Sydney Australia:Longman Cheshire, 1992.

[64] P. Comon. Independent component analysis, A new concept? *Signal Process.*, 36(3):287–314, 1994.

[65] P. Comon and E. Moreau. Blind MIMO equalization and joint-diagonalization criteria. In *IEEE International Conference on Acoustics, Speech, and Signal Processing, ICASSP 2001, 7-11 May*, pages 2749–2752. IEEE.

[66] R. T. Compton. *Adaptive Antennas, Concepts and Performance*. Ne, New Jersey:Prentice- Hall, Englewood Cliffs, 1988.

[67] H. Cox. Resolving power and sensitivity to mismatch of optimum array processors. *The Journal of the Acoustical Society of America*, 54(3):771–785, 1973.

[68] S. S. D. *Statistical Inference*. Baltimore: Penguin Books, 1970.

[69] Z. X. Da. *Time series analysis: higher order statistics method*. Beijing: Tsinghua University Press, 1996.

[70] C. Darken and J. E. Moody. Towards faster stochastic gradient search. In *Advances in Neural Information Processing Systems 4, [NIPS Conference, Denver, Colorado, USA]*, pages 1009–1016, 1991.

[71] I. Daubechies. The wavelet transform, time-frequency localization and signal analysis. *IEEE Trans. Inf. Theory*, 36(5):961–1005, 1990.

[72] I. Daubechies. *Ten Lectures on Wavelets*. SIAM, 1992.

[73] N. Delfosse and P. Loubaton. Adaptive blind separation of independent sources: A deflation approach. *Signal Process.*, 45(1):59–83, 1995.

[74] A. N. Delopoulos and G. B. Giannakis. Strongly consistent identification algorithms and noise insensitive MSE criteria. *IEEE Trans. Signal Process.*, 40(8):1955–1970, 1992.

[75] J. Doherty and R. Porayath. A robust echo canceler for acoustic environments. *IEEE Trans. on Circuits and Systems II: Analog and Digital Signal Processing*, 44(5):389–396, 1997.

[76] C. L. Dolph. A current distribution for broadside arrays which optimizes the relationship between beam width and side-lobe level. *Proc Ire*, 35(6):335–348, 1946.

[77] R. Duda and P. E. Hart. *Pattern Classification and Scene Analysis*. New York: Wiley, 1973.

[78] V. T. Ermolaev and A. B. Gershman. Fast algorithm for minimum-norm direction-of-arrival estimation. *IEEE Trans. Signal Process.*, 42(9):2389–2394, 1994.

[79] J. R. Fonollosa and C. L. Nikias. Wigner-ville higher-order moment spectra: Definition, properties, computation and application to transient signal analysis. *IEEE Trans. Signal Process.*, 41(1):245–266, 1993.

[80] B. R. Frieden. Restoring with maximum likelihood and maximum entropy. *Journal of the Optical Society of America*, 62(4):511–518, 1972.

[81] O. Frost. An algorithm for linearly constrained adaptive array processing. *Proceedings of the IEEE*, 60(8):926–935, 1972.

[82] D. Gabor. Theory of communications. *J. Inst. Elec. Eng*, 93:429–457, 1946.

[83] W. S. Gan. Fuzzy step-size adjustment for the LMS algorithm. *Signal Process.*, 49(2):145–149, 1996.

[84] W. Gersch. Estimation of the autoregressive parameters of a mixed autoregressive moving-average time series. *IEEE Transactions on Automatic Control*, 15(5):583–588, 1970.

[85] G. Giannakis. Cumulants: A powerful tool in signal processing. *Proceedings of the IEEE*, 75(9):1333–1334, 1987.

[86] G. B. Giannakis. On the identifiability of non-gaussian arma models using cumulants. *IEEE Transactions on Automatic Control*, 35(1):18–26, 1990.

[87] G. B. Giannakis and J. M. Mendel. Identification of nonminimum phase systems using higher order statistics. *IEEE Trans. Acoust. Speech Signal Process.*, 37(3):360–377, 1989.

[88] G. B. Giannakis and J. M. Mendel. Cumulant-based order determination of non-gaussian ARMA models. *IEEE Trans. Acoust. Speech Signal Process.*, 38(8):1411–1423, 1990.

[89] G. B. Giannakis and A. Swami. On estimating noncausal nonminimum phase ARMA models of non-gaussian processes. *IEEE Trans. Acoust. Speech Signal Process.*, 38(3):478–495, 1990.

[90] G. O. Glentis, K. Berberidis, and S. Theodoridis. Efficient least squares adaptive algorithms for FIR transversal filtering. *IEEE Signal Process. Mag.*, 16(4):13–41, 1999.

[91] L. Godara. Application of antenna arrays to mobile communications. ii. beam-forming and direction-of-arrival considerations. *Proceedings of the IEEE*, 85(8):1195–1245, 1997.

[92] J. I. Gold and T. Watanabe. Perceptual learning. *Current Biology*, 20(2):R46–R48, 2010.

[93] G. H. Golub and C. F. L. Van. *Matrix Computations,2nd ed*. The John Hopiks Univ.Press, 1989.

[94] R. M. Gray. *Entropy and Information Theory*. New York: Springer-Verlag, 1990.

[95] Griffiths and W. R. J. Adaptive array processing. a tutorial. *IEE Proceedings H Microwaves, Optics and Antennas*, 130(1):137–142, 1983.

[96] L. Griffiths and C. Jim. An alternative approach to linearly constrained adaptive beamforming. *IEEE Trans. Antennas and Propagation*, 30(1):27–34, 1982.

[97] M. M. C. Group. *Cepstrum analysis and parametric methods for multispectral estimation*. Beijing: Higher Education Press, 1979.

[98] M. Haardt and J. A. Nossek. Unitary ESPRIT: how to obtain increased estimation accuracy with a reduced computational burden. *IEEE Trans. Signal Process.*, 43(5):1232–1242, 1995.

[99] R. W. Harris, D. M. Chabries, and F. A. Bishop. A variable step (VS) adaptive filter algorithm. *IEEE Trans. Acoust. Speech Signal Process.*, 34(2):309–316, 1986.

[100] S. Haykin. *daptive Filter Theory, 3rd Edit.* A.Prentice Hall, 1996.

[101] S. Haykin. Neural networks expand sp's horizons. *IEEE Signal Process. Mag.*, 13(2):24–49, 1996.

[102] S. Haykin, P. Reilly, J., V. Kezys, and E. Vertatschitsch. Some aspects of array signal processing. *Radar & Signal Processing IEE Proceedings F*, 139:1–26, 1992.

[103] F. B. Hildebrand. *Introduction to Numerical Analysis*. New York: MoGraw-Hill, 1956.

[104] Y. Hochberg. A sharper bonferroni procedure for multiple tests of significance. *Biometrika*, 75:800–802, 1988.

[105] S. Holm. A simple sequentially rejective multiple test procedure. *Scandinavian Journal of Statistics*, 6(2):65–70, 1979.

[106] M. L. Honig, U. Madhow, and S. Verdú. Blind adaptive multiuser detection. *IEEE Trans. Inf. Theory*, 41(4):944–960, 1995.

[107] R. A. Horn and C. R. Johnson. *Matrix Analysis*. London: Cambridge Univ.Press, 1956.

[108] A. Hyvärinen. Fast and robust fixed-point algorithms for independent component analysis. *IEEE Trans. Neural Networks*, 10(3):626–634, 1999.

[109] A. Hyvärinen and E. Oja. Independent component analysis: algorithms and applications. *Neural Networks*, 13(4-5):411–430, 2000.

[110] S. Ihara. Maximum entropy spectral analysis and ARMA processes. *IEEE Trans. Inf. Theory*, 30(2):377–380, 1984.

[111] Ingrid and Daubechies. Orthonormal bases of compactly supported wavelets. *Communications on Pure and Applied Mathematics*, pages 909–996, 1988.

[112] V. J. Theorie et applications de la notion de signal analytique. *Cables Et Transmissions*, 2A(9):61–74, 1948.

[113] B. D. Jawerth and W. Sweldens. An overview of wavelet based multiresolution analyses. *SIAM Rev.*, 36(3):377–412, 1994.

[114] J. Jeong and W. J. Williams. Kernel design for reduced interference distributions. *IEEE Trans. Signal Process.*, 40(2):402–412, 1992.

[115] C. Jutten and J. Hérault. Blind separation of sources, part I: an adaptive algorithm based on neuromimetic architecture. *Signal Process.*, 24(1):1–10, 1991.

[116] T. Kailath. An innovations approach to least-squares estimation–part I: Linear filtering in additive white noise. *IEEE Transactions on Automatic Control*, 13(6):646–655, 1968.

[117] T. Kailath. The innovations approach to detection and estimation theory. *Proceedings of the IEEE*, 58(5):680–695, 1970.

[118] S. Kapoor, S. Gollamudi, S. Nagaraj, and Y.-F. Huang. Adaptive multiuser detection and beamforming for interference suppression in cdma mobile radio systems. *IEEE Trans. Vehicular Technology*, 48(5):1341–1355, 1999.

[119] J. Karhunen, P. Pajunen, and E. Oja. The nonlinear PCA criterion in blind source separation: Relations with other approaches. *Neurocomputing*, 22(1-3):5–20, 1998.

[120] S. Karlin and H. T. Taylor. *A First Course in Stochastic Process*. New York: Academic Press, 1975.

[121] R. Kashyap. Inconsistency of the aic rule for estimating the order of autoregressive models. *IEEE Transactions on Automatic Control*, 25(5):996–998, 1980.

[122] M. Kaveh and A. J. Barabell. The statistical performance of the MUSIC and the minimum-norm algorithms in resolving plane waves in noise. *IEEE Trans. Acoust. Speech Signal Process.*, 34(2):331–341, 1986.

[123] S. M. Kay and S. L. Marple. Spectrum analysis—a modern perspective. *Proceedings of the IEEE*, 69(11):1380–1419, 1981.

[124] Klemm and R. Adaptive airborne mti: an auxiliary channel approach. *Communications, Radar and Signal Processing, IEE Proceedings F*, 134:269–276, 1987.

[125] C. Knapp and G. Carter. The generalized correlation method for estimation of time delay. *IEEE Transactions on Acoustics, Speech, and Signal Processing*, 24(4):320–327, 1976.

[126] H. Krim and M. Viberg. Two decades of array signal processing research: the parametric approach. *IEEE Signal Process. Mag.*, 13(4):67–94, 1996.

[127] R. Kumaresan. *Estimating the paramaters of exponentially damped or undamped sinusoidal signals in noise*. Ph.D.dissertation,University of Rhode Island, RI, 1982.

[128] M. A. Kutay. fracf:fast computation of the fractional fourier transform. http://www.ee.bilkent.edu.tr/haldun/fracF.m, 1996.

[129] R. H. Kwong and E. W. Johnston. A variable step size LMS algorithm. *IEEE Trans. Signal Process.*, 40(7):1633–1642, 1992.

[130] H. M. Lagunas, P. M. Santamaria, and A. Figueiras, Vidal. ARMA model maximum entropy power spectral estimation. *IEEE Transactions on Acoustics, Speech, and Signal Processing*, 32(5):984–990, 1984.

[131] M. A. Lagunas. Cepstrum constraints in ME spectral estimation. pages 1442–1445. Proc. IEEE ICASSP '83, 1983.

[132] R. Lambert. *Multichannel Blind Deconvolution: FIR Matrix Algebra and Separation of Multipath Mixtures*. Ph.D.Dissertation, Univ.of Southern California, Dept. of Electrical Eng., 1996.

[133] R. Lawrence and H. Kaufman. The kalman filter for the equalization of a digital communications channel. *IEEE Trans. Communication Technology*, 19(6):1137–1141, 1971.

[134] A. Lee. Centrohermitian and skew-centrohermitian matrices. *Linear Algebra and its Applications*, 29:205–210, 1980.

[135] T. W. Lee, M. A. Girolami, and T. J. Sejnowski. Independent component analysis using an extended infomax algorithm for mixed sub-gaussian and super-gaussian sources. *Neural Comput.*, 11(2):417–441, 1999.

[136] F. L. Lewis. *Optimal Estimation*. New York: John Wiley & Sons, Inc., 1986.

[137] X. L. Li and X. D. Zhang. Nonorthogonal joint diagonalization free of degenerate solution. *IEEE Trans. Signal Process.*, 55(5-1):1803–1814, 2007.

[138] X. L. Li and X. D. Zhang. Sequential blind extraction adopting second-order statistics. *IEEE Signal Process. Lett.*, 14(1):58–61, 2007.

[139] X. Liao and Z. Bao. Circularly integrated bispectra: Novel shift invariant features for high-resolution radar target recognition. *Electronics Letters*, 34(19):1879–1880, 1998.

[140] L. P. Liu, L. Zhang, and Z. G. Cai. A multiple hypothesis test and its application in econometrics (in chinese). *Statistical Research*, 24(4):26–30, 2007.

[141] L. Ljung, M. Morf, and D. Falconer. Fast calculations of gain matrices for recursive estimation schemes. *Int.Journal of Control*, 27:1–19, 1978.

[142] S. T. Lou and X. D. Zhang. Fuzzy-based learning rate determination for blind source separation. *IEEE Trans. Fuzzy Syst.*, 11(3):375–383, 2003.

[143] P. J. Loughlin, J. W. Pitton, and L. E. Atlas. Bilinear time-frequency representations: new insights and properties. *IEEE Trans. Signal Process.*, 41(2):750–767, 1993.

[144] O. Macchi. Optimization of adaptive identification for time-varying filters. *IEEE Trans. Automatic Control*, 31(3):283–287, 1986.

[145] O. Macchi. *Adaptive Processing: The LMS Approach with Applicatins in Transmission*. New York: John Wiley & Sons,Inc., 1995.

[146] J. Makhoul and L. Cosell. Adaptive lattice analysis of speech. *IEEE Trans. Acoustics, Speech, and Signal Processing*, 29(3):654–659, 1981.

[147] S. Mallat. A theory for multiresolution signal decomposition: The wavelet representation. *IEEE Trans. Pattern Anal. Mach. Intell.*, 11(7):674–693, 1989.

[148] G. Mathew and V. U. Reddy. Development and analysis of a neural network approach to pisarenko's harmonic retrieval method. *IEEE Trans. Signal Process.*, 42(3):663–667, 1994.

[149] G. Mathew and V. U. Reddy. Orthogonal eigensubspace estimation using neural networks. *IEEE Trans. Signal Process.*, 42(7):1803–1811, 1994.

[150] J. E. Mazo. On the independence theory of equalizer convergence. *Bell Syst. Tech. J.*, 38:963–993, 1979.

[151] M. Mboup, M. Bonnet, and N. J. Bershad. LMS coupled adaptive prediction and system identification: a statistical model and transient mean analysis. *IEEE Trans. Signal Process.*, 42(10):2607–2615, 1994.

[152] A. C. Mcbride and F. H. Kerr. On namias's fractional fourier transforms. *IMA Journal of Applied Mathematics*, (2):159–175, 1987.

[153] R. McDonough and W. Huggins. Best least-squares representation of signals by exponentials. *IEEE Transactions on Automatic Control*, 13(4):408–412, 1968.

[154] J. M. Mendel. Tutorial on higher-order statistics (spectra) in signal processing and system theory: theoretical results and some applications. *Proceedings of the IEEE*, 79(3):278–305, 1991.

[155] O. J. Micka and A. J. Weiss. Estimating frequencies of exponentials in noise using joint diagonalization. *IEEE Trans. Signal Process.*, 47(2):341–348, 1999.

[156] C. J. Miller, R. C. Nichol, and D. J. Batuski. Acoustic oscillations in the early universe and today. *Science*, 292(5525):2302–2303, 2001.

[157] R. Monzingo and T. Miller. *Introduction to Adaptive Arrays*. New York: Wiley & Sons, 1986.

[158] E. Moreau. A generalization of joint-diagonalization criteria for source separation. *IEEE Trans. Signal Process.*, 49(3):530–541, 2001.

[159] N. Murata, K. R. Müller, A. Ziehe, and S. I. Amari. Adaptive on-line learning in changing environments. In M. Mozer, M. I. Jordan, and T. Petsche, editors, *Advances in Neural Information Processing Systems 9, NIPS, Denver, CO, USA,*, pages 599–605. MIT Press, 1996.

[160] F. B. N. Common principal components in k groups. *Amer. Statist. Assoc.*, 79:892–897, 1984.

[161] C. Nadeu. Spectral estimation with rational modeling of the log spectrum. *Signal Processing*, 10(1):7–18, 1986.

[162] A. Nehorai. A minimal parameter adaptive notch filter with constrained poles and zeros. *IEEE Trans. Acoust. Speech Signal Process.*, 33(4):983–996, 1985.

[163] J. W. V. Ness. Asymptotic normality of bispectral estimation. *Ann. Math. Statistics*, 37:1257–1272, 1986.

[164] C. L. Nikias and F. Liu. Bicepstrum computation based on second- and third-order statistics with applications. In *1990 International Conference on Acoustics, Speech, and Signal Processing, ICASSP '90, Albuquerque, New Mexico, USA, April 3-6, 1990*, pages 2381–2385. IEEE, 1990.

[165] C. L. Nikias and R. Pan. Time delay estimation in unknown gaussian spatially correlated noise. *IEEE Trans. Acoust. Speech Signal Process.*, 36(11):1706–1714, 1988.

[166] C. L. Nikias and A. P. Petropulu. *Higher-Order Spectra Analysis*. New York: Prentice Hall, 1993.

[167] J. O'Neill. Discretetfds: A collection of matlab files for time-frequency analysis. ftp.mathworks.com/pub/contrib/v5/signal/DiscreteTFDs/, 1999.

[168] H. M. Özaktas, O. Arikan, M. A. Kutay, and G. Bozdagi. Digital computation of the fractional fourier transform. *IEEE Trans. Signal Process.*, 44(9):2141–2150, 1996.

[169] H. M. Ozaktas, Z. Zalevsky, and M. A. Kutay. *The Fractional Fourier Transform*. Chichester: Wiley, 2001.

[170] A. Papoulis. *Signal Analysis*. New York: McGraw-Hill, 1977.

[171] A. Papoulis. *Random Variables and Stochastic Processes*. New York: McGraw-Hill, 1984.

[172] E. Parzen. Some recent advances in time series modeling. *IEEE Transactions on Automatic Control*, 19(6):723–730, 1974.

[173] L. A. Pflug, G. E. Ioup, J. W. Ioup, and R. Field. Properties of higher-order correlations and spectra for bandlimited, deterministic transients. *The Journal of the Acoustical Society of America*, 91(2):975–988, 1992.

[174] D. T. Pham. Joint approximate diagonalization of positive definite hermitian matrices. *SIAM J. Matrix Anal. Appl.*, 22(4):1136–1152, 2001.

[175] H. V. Poor and X. Wang. Code-aided interference suppression for DS/CDMA communications. II. parallel blind adaptive implementations. *IEEE Trans. Commun.*, 45(9):1112–1122, 1997.

[176] B. Porat. *Digital Processing of Random Signals-Theory and Methods*. New Jersey: Prentice-Hall Inc., 1994.

[177] R. K. Potter, G. A. Kopp, and H. C. Green. *Visible Speech*. New York: Van Nostrand, 1947.

[178] A. D. Poularikas. *The Handbook of Formulas and Tables for Signal Processing*. CRC Press, Springer,IEEE Press, 1998.

[179] S. Qian and D. pang Chen. *Joint Time-Frequency Analysis: Methods and Applications*. New York: Prentice Hall, 1996.

[180] S. C. Qian. *Cepstrum analysis and parametric methods for multispectral estimation (in Chinese)*. Report on the annual meeting of China Signal Processing Society, 1996.

[181] K. R, D. H. K, and L. L. Y. The sound spectrograph. *J. Acoust. Soc. Amer.*, 18:19–49, 1948.

[182] D. Rao and S.-Y. Kung. Adaptive notch filtering for the retrieval of sinusoids in noise. *IEEE Trans. Acoustics, Speech, and Signal Processing*, 32(4), 1984.

[183] J. Rissanen. A universal prior for integers and estimation by minimum description length. *The Annals of Statistics*, 11(2):416–431, 1983.

[184] H. Robbins and S. Monro. A stochastic approximation method. *Annals of Mathematical Statistics*, 22(3):400–407, 1951.

[185] R. H. Roy and T. Kailath. Esprit-estimation of signal parameters via rotational invariance techniques. *IEEE Trans. Acoust. Speech Signal Process.*, 37(7):984–995, 1989.

[186] R. H. Roy, A. Paulraj, and T. Kailath. ESPRIT-A subspace rotation approach to estimation of parameters of cisoids in noise. *IEEE Trans. Acoust. Speech Signal Process.*, 34(5):1340–1342, 1986.

[187] V. S. *Multiuser Detection*. Cambridge:U.K.:Cambridge Univ. Press, 1998.

[188] K. Sasaki, T. Sato, and Y. Nakamura. Holographic passive sonar. *IEEE Transactions on Sonics and Ultrasonics*, 24(3):193–200, 1977.

[189] K. Sasaki, T. Sato, and Y. Yamashita. Minimum bias windows for bispectral estimation. *Journal of Sound and Vibration*, 40(1):139–148, 1975.

[190] A. H. Sayed and T. Kailath. A state-space approach to adaptive RLS filtering. *IEEE Signal Process. Mag.*, 11(3):18–60, 1994.

[191] R. Schmidt. Multiple emitter location and signal parameter estimation. *IEEE Trans. Antennas and Propagation*, 34(3):276–280, 1986.

[192] R. O. Schmidt. Multiple emitter location and signal parameters estimation. In *Proc Radc Spectrum Estimation Workshop*, pages 243–258, 1979.

[193] D. W. E. Schobben, K. Torkkola, and P. Smaragdis. Evaluation of blind signal separation methods. *Proc of Ica & Bss*, pages 261–266, 1999.

[194] K. C. Sharman and T. S. Durrani. A comparative study of modern eigenstructure methods for bearing estimation-a new high performance approach. In *1986 25th IEEE Conference on Decision and Control*, pages 1737–1742, 1986.

[195] R. J. Simes. An improved bonferroni procedure for multiple tests of significance. *Biometrika*, 73:751–754, 1986.

[196] M. J. T. Smith and T. P. B. III. Exact reconstruction techniques for tree-structured subband coders. *IEEE Trans. Acoust. Speech Signal Process.*, 34(3):434–441, 1986.

[197] B. D. Steinberg. *Principles of Aperture and Arruy System Design*. New York: Wiley, 1976.

[198] P. Stoica and A. Nehorai. Music, maximum likelihood, and cramer-rao bound. *IEEE Trans. Acoust. Speech Signal Process.*, 37(5):720–741, 1989.

[199] J. D. Storey. A direct approach to false discovery rates. *Journal of the Royal Statistical Society: Series B (Statistical Methodology)*, 64(3):479–498, 2002.

[200] J. D. Storey. The positive false discovery rate: A bayesian interpretation and the q-value. *Ann. Statist.*, 31:2013–2035, 2003.

[201] G. Strang and T. Nguyen. *Wavelets and Filter Banks*. Wellesley-Cambridge, 1996.

[202] A. Swami and J. M. Mendel. Closed-form recursive estimation of MA coefficients using autocorrelations and third-order cumulants. *IEEE Trans. Acoust. Speech Signal Process.*, 37(11):1794–1795, 1989.

[203] A. Swami and J. M. Mendel. ARMA parameter estimation using only output cumulants. *IEEE Trans. Acoust. Speech Signal Process.*, 38(7):1257–1265, 1990.

[204] A. Swami and J. M. Mendel. Cumulant-based approach to harmonic retrieval and related problems. *IEEE Trans. Signal Process.*, 39(5):1099–1109, 1991.

[205] A. Swami and J. M. Mendel. Identifiability of the AR parameters of an ARMA process using cumulants. *IEEE Transactions on Automatic Control*, 37(2):268–273, 1992.

[206] S. Takuso. Bispectral holography. *Journal of the Acoustical Society of America*, 62(2):404–408, 1977.

[207] M. K. Tsatsanis and Z. Y. Xu. Performance analysis of minimum variance CDMA receivers. *IEEE Trans. Signal Process.*, 46(11):3014–3022, 1998.

[208] J. K. Tugnait. Approaches of FIR system identification with noisy data using higher order statistics. *IEEE Trans. Acoust. Speech Signal Process.*, 38(7):1307–1317, 1990.

[209] J. K. Tugnait. New results on FIR system identification using higher order statistics. *IEEE Trans. Signal Process.*, 39(10):2216–2221, 1991.

[210] J. K. Tugnait. On time delay estimation with unknown spatially correlated gaussian noise using fourth-order cumulants and cross cumulants. *IEEE Trans. Signal Process.*, 39(6):1258–1267, 1991.

[211] J. K. Tugnait. Detection of non-gaussian signals using integrated polyspectrum. *IEEE Trans. Signal Process.*, 42(11):3137–3149, 1994.

[212] T. J. Ulrych and R. W. Clayton. Time series modelling and maximum entropy. *Physics of the Earth & Planetary Interiors*, 12(2-3):188–200, 1976.

[213] N. V. The fractional order fourier transform and its application to quantum mechanics. *Geoderma*, 25(3):241–265, 1980.

[214] A. J. van der Veen. Joint diagonalization via subspace fitting techniques. In *IEEE International Conference on Acoustics, Speech, and Signal Processing, ICASSP*, pages 2773–2776. IEEE, 2001.

[215] B. Van Veen and K. Buckley. Beamforming: a versatile approach to spatial filtering. *IEEE ASSP Magazine*, 5(2):4–24, 1988.
[216] M. Vetterli and C. Herley. Wavelets and filter banks: theory and design. *IEEE Trans. Signal Process.*, 40(9):2207–2232, 1992.
[217] S. W. Wavelets: What next? *Proceedings of the IEEE*, 84(4):680–685, 1996.
[218] H. Y. Wang. *Random digital signal processing*. Beijing: Science Press, 1988.
[219] M. Wax and J. Sheinvald. A least-squares approach to joint diagonalization. *IEEE Signal Process. Lett.*, 4(2):52–53, 1997.
[220] E. J. Wegman, J. B. Thomas, and S. C. Schwartz. *Topics in Non-Gaussian Signal Processing*. New York: Springer-Verlag, 1989.
[221] A. J. Weiss and B. Friedlander. Array processing using joint diagonalization. *Signal Process.*, 50(3):205–222, 1996.
[222] J. Wexler and S. Raz. Discrete gabor expansions. *Signal Process.*, 21(3):207–220, 1990.
[223] M. V. Wickerhauser. *Adapted Wavelet Analysis from Theory to Software*. Wellesley, MA: A. K. Peters, 1994.
[224] B. Widrow, J. Glover, J. McCool, J. Kaunitz, C. Williams, R. Hearn, J. Zeidler, J. Eugene Dong, and R. Goodlin. Adaptive noise cancelling: Principles and applications. *Proceedings of the IEEE*, 63(12):1692–1716, 1975.
[225] B. Widrow, P. Mantey, L. Griffiths, and B. Goode. Adaptive antenna systems. *Proceedings of the IEEE*, 55(12):2143–2159, 1967.
[226] B. Widrow, M. McCool, J. M.and Larimore, and C. Johnson. Stationary and nonstationary learning characteristics of the LMS adaptive filter. *Proceedings of the IEEE*, 64(8):1151–1162, 1976.
[227] B. Widrow and S. D. Stearns. *Adaptive Signal Processing*. New York: Prentice-Hall, 1985.
[228] E. P. Wigner. On the quantum correction for thermodynamic equilibrium. *Phy Rev*, 40(40):749–759, 1932.
[229] L. Xu, E. Oja, and C. Y. Suen. Modified hebbian learning for curve and surface fitting. *Neural Networks*, 5(3):441–457, 1992.
[230] X. L. Xu and K. M. Buckley. Bias analysis of the MUSIC location estimator. *IEEE Trans. Signal Process.*, 40(10):2559–2569, 1992.
[231] B. Yang. Projection approximation subspace tracking. *IEEE Trans. Signal Process.*, 43(1):95–107, 1995.
[232] H. H. Yang. Serial updating rule for blind separation derived from the method of scoring. *IEEE Trans. Signal Process.*, 47(8):2279–2285, 1999.
[233] H. H. Yang and S. I. Amari. Adaptive on-line learning algorithms for blind separation-maximization entropy and minimum mutual information. *Neural Comput.*, 9(7):1457–1482, 1997.
[234] H. H. Yang, S. I. Amari, and A. Cichocki. Information-theoretic approach to blind separation of sources in non-linear mixture. *Signal Process.*, 64(3):291–300, 1998.
[235] A. Yeredor. Non-orthogonal joint diagonalization in the least-squares sense with application in blind source separation. *IEEE Trans. Signal Process.*, 50(7):1545–1553, 2002.
[236] A. Yeredor. Time-delay estimation in mixtures. In *2003 IEEE International Conference on Acoustics, Speech, and Signal Processing, ICASSP '03, Hong Kong, April 6-10*, pages 237–240. IEEE, 2003.
[237] Y. S. Yoon and M. G. Amin. High-resolution through-the-wall radar imaging using beamspace MUSIC. *IEEE Transactions on Antennas and Propagation*, 56(6):1763–1774, 2008.
[238] J. Zeidler. Performance analysis of LMS adaptive prediction filters. *Proceedings of the IEEE*, 78(12):1781–1806, 1990.
[239] R. T. Zhang and K. T. Fang. *Introduction to multivariate analysis*. Beijing: Science Press, 1982.

[240] X. D. Zhang. *Matrix Analysis and Application*. Beijing: Tsinghua University Press, 1996.

[241] X. D. Zhang and Z. Bao. *Communication signal processing*. Beijing: National Defense Industry Press, 1998.

[242] X. D. Zhang and Z. Bao. *Nonstationary signal analysis and processing*. Beijing: National Defense Industry Press, 1998.

[243] X. D. Zhang and Y. Li. Harmonic retrieval in mixed gaussian and non-gaussian ARMA noises. *IEEE Trans. Signal Process.*, 42(12):3539–3543, 1994.

[244] X. D. Zhang and Y. C. Liang. Prefiltering-based ESPRIT for estimating sinusoidal parameters in non-gaussian ARMA noise. *IEEE Trans. Signal Process.*, 43(1):349–353, 1995.

[245] X. D. Zhang, Y. C. Liang, and Y. Li. A hybrid approach to harmonic retrieval in non-gaussian ARMA noise. *IEEE Trans. Inf. Theory*, 40(4):1220–1226, 1994.

[246] X. D. Zhang, Y. Shi, and Z. Bao. A new feature vector using selected bispectra for signal classification with application in radar target recognition. *IEEE Trans. Signal Process.*, 49(9):1875–1885, 2001.

[247] X. D. Zhang, Y. Song, and Y. Li. Adaptive identification of nonminimum phase ARMA models using higher order cumulants alone. *IEEE Trans. Signal Process.*, 44(5):1285–1288, 1996.

[248] X. D. Zhang and H. Takeda. An order recursive generalized least squares algorithm for system identification. *IEEE Transactions on Automatic Control*, 30(12):1224–1227, 1985.

[249] X. D. Zhang and H. Takeda. An approach to time series analysis and ARMA spectral estimation. *IEEE Trans. Acoust. Speech Signal Process.*, 35(9):1303–1313, 1987.

[250] X. D. Zhang and W. Wei. Blind adaptive multiuser detection based on kalman filtering. *IEEE Trans. Signal Process.*, 50(1):87–95, 2002.

[251] X. D. Zhang and Y. S. Zhang. Determination of the MA order of an ARMA process using sample correlations. *IEEE Trans. Signal Process.*, 41(6):2277–2280, 1993.

[252] X. D. Zhang and Y. S. Zhang. Singular value decomposition-based MA order determination of non-gaussian ARMA models. *IEEE Trans. Signal Process.*, 41(8):2657–2664, 1993.

[253] X. D. Zhang and Y. S. Zhang. FIR system identification using higher order statistics alone. *IEEE Trans. Signal Process.*, 42(10):2854–2858, 1994.

[254] X. D. Zhang and Y. Zhou. A novel recursive approach to estimating MA parameters of causal ARMA models from cumulants. *IEEE Trans. Signal Process.*, 40(11):2870–2873, 1992.

[255] Y. Zhao, L. E. Atlas, and R. J. M. II. The use of cone-shaped kernels for generalized time-frequency representations of nonstationary signals. *IEEE Trans. Acoust. Speech Signal Process.*, 38(7):1084–1091, 1990.

[256] D. Y. Zhu, C. O. Wu, and W. L. Qin. *Multivariate statistical analysis and software SAS*. Nanjing: Southeast University Press, 1999.

[257] M. D. Zoltowski, M. Haardt, and C. P. Mathews. Closed-form 2-d angle estimation with rectangular arrays in element space or beamspace via unitary ESPRIT. *IEEE Trans. Signal Process.*, 44(2):316–328, 1996.

Printed in the USA
CPSIA information can be obtained
at www.ICGtesting.com
JSHW051340180624
65043JS00005B/218

9 783110 475555